D0223426

Automotive Electricity and Electronics

Third Edition

Online Services

Delmar Online
To access a wide variety of Delmar products and services on the World Wide Web, point your browser to:

http://www.delmar.com/delmar.html
or email: info@delmar.com

thomson.com
To access International Thomson Publishing's home site for information on more than 34 publishers and 20,000 products, point your browser to:

http://www.thomson.com
or email: findit@kiosk.thomson.com

A service of I(T)P®

Automotive Electricity and Electronics

Third Edition

Al Santini

College of DuPage
Glen Ellyn, IL

Delmar Publishers

I(T)P® *an International Thomson Publishing company*

Albany • Bonn • Boston • Cincinnati • Detroit • London • Madrid
Melbourne • Mexico City • New York • Pacific Grove • Paris • San Francisco
Singapore • Tokyo • Toronto • Washington

NOTICE TO THE READER

Publisher does not warrant or guarantee any of the products described herein or perform any independent analysis in connection with any of the product information contained herein. Publisher does not assume, and expressly disclaims, any obligation to obtain and include information other than that provided to it by the manufacturer.

The reader is expressly warned to consider and adopt all safety precautions that might be indicated by the activities described herein and to avoid all potential hazards. By following the instructions contained herein, the reader willingly assumes all risks in connection with such instructions.

The publisher makes no representations or warranties of any kind, including but not limited to, the warranties of fitness for particular purpose or merchantability, nor are any such representations implied with respect to the material set forth herein, and the publisher takes no responsibility with respect to such material. The publisher shall not be liable for any special, consequential or exemplary damages resulting, in whole or in part, from the readers' use of, or reliance upon, this material.

Cover Design: Betty Kodela

Delmar Staff
Publisher: Robert D. Lynch
Acquisition Editor: Vernon Anthony
Project Editor: Thomas Smith
Production Coordinator: Karen Smith
Art and Design Coordinator: Cheri Plasse

COPYRIGHT © 1997
By Delmar Publishers
an International Thomson Publishing Company

The ITP logo is a trademark under license.

Printed in the United States of America

For more information, contact:

Delmar Publishers
3 Columbia Circle, Box 15015
Albany, NY 12212-5015

International Thomson Publishing Europe
Berkshire House
168-173 High Holborn
WC1V7AA London,
England

Thomas Nelson Australia
102 Dodds Street
South Melbourne, 3205
Victoria, Australia

Nelson Canada
1120 Birchmount Road
Scarborough, Ontario
Canada, M1K 5G4

International Thomson Editores
Campos Eliseos 385, Piso 7
Col Polanco
11560 Mexico D F Mexico

International Thomson Publishing GmbH
Königswinterer Strasse. 418
53227 Bonn
Germany

International Thomson Publishing Asia
221 Henderson Road
#05-10 Henderson Building
Singapore 0315

International Thomson Publishing—Japan
Kyowa Building, 3F
2-2-1 Hirakawacho
Chiyoda-ku, Tokyo 102
Japan

All rights reserved. Part of this work © 1988. No part of this work covered by the copyright hereon may be reproduced or used in any form, or by any means—graphic, electronic, or mechanical, including photocopying, recording, taping, or information storage and retrieval systems—without written permission of the publisher.

4 5 6 7 8 9 10 XXX 02 01 00 99 98

Library of Congress Cataloging-in-Publication Data

Santini, Al.
Automotive electricity and electronics / Al Santini. — 3rd ed.
p. cm.
Includes index.
ISBN 0-8273-6743-0 (perfect)
1. Automobiles—Electric equipment. 2. Automobiles—Electronic equipment. I. Title.
TL272.S346 1996
629.25'4—dc20 96-31187
CIP

Contents

Preface

Automotive Electricity and Electronics, Third Edition, has been written to fill the void in post secondary textbooks. Although there are a great number of textbooks and materials available, each one seems to cover a particular topic in depth, while only touching on others. In addition, many textbooks try to reproduce only factory circuits, which tends to date the material and the book. For this reason, generally I have avoided specific specialized circuits, and instead, have concentrated on component testing from a generic standpoint. We frequently find the same component on a variety of vehicles, so the testing procedure can be generic. This greatly simplifies the student's ability to comprehend the material. Generally, this text is divided up into categories or groups which span most, if not all, of the manufacturers. The real strength of this text is in its ability to teach, in a logical sequence, convertible skills. Convertible in that the knowledge that will be needed to repair tomorrow's vehicles will be found in the basics. This text is first and foremost a text of the basics. Circuit design, various applicable laws, wiring principles, and a hands-on practical approach in the use of test equipment has been incorporated into this text. It is hoped that the automotive electricity student using this approach will encounter sufficient positive experiences to serve as a foundation for the remainder of the text, which covers batteries, starting, charging, ignition (both distributed and distributerless), and accessories. Two additional chapters have been added covering computer control of fuel and ignition and the use of a digital storage scope to diagnose typical computer inputs and outputs.

Each system has two chapters devoted to it. The first is generally theory and information, while the second is devoted to testing and repair techniques common to the industry. On-vehicle testing is stressed using the most common equipment found in the field. Emphasis has been placed on digital multimeters, ignition oscilloscopes, scanners, and digital storage oscilloscopes. Two new chapters have been added to this third edition covering computer control and its testing. The generic approaches found in all the repair chapters should carry most technicians through most of the vehicles of today.

Finally, an emphasis has been placed on the readability of this text. Most students will find it easy to comprehend without insulting their intelligence. The relaxed style of writing should improve the enjoyment of a basic electricity class. As an instructor, please feel free to add or subtract material as needed. The text has been written with instructors in mind. The basic chapters should help make introductory material easier for the student to comprehend. Much of the information, after the basics, is of the stand-alone variety. This will allow you to pick and choose what will fit your syllabus and do your students the most good. ASE style questions appear at the end of each chapter, and OBD-II terminology has been incorporated. The material within the text is sufficient to allow for NATEF certification of Electrical and part of Engine Performance. I believe, if the students' first exposure to our industry is positive, they will become the technicians of tomorrow that our industry needs. This text strives to help make that first experience a positive one.

Many individuals have helped with this third edition. To mention them all would fill pages. I would like to mention a few, however. Al Engeldahl, the Coordinator of the Automotive Service Technology Progam and an instructor at the College of DuPage, has read virtually every word of the text, and his constructive suggestions are found throughout. His help and friendship are appreciated daily. I am currently in my 16th year at the College of DuPage and am thankful that it is the kind of community college where students not only can learn, but where instructors are encouraged to accomplish positive student oriented activities, such as this text. Tom Roesing, my Associate Dean, continues to be supportive of my writing and teaching activities. Fluke Corporation and especially Clem Clemenson have been very supportive of my efforts. Not only do they manufacture some of the best automotive test equipment available, but they are willing to help instructors learn how to improve their teaching. I need to mention all the great people I have come into contact with at Delmar Publishers, from my editor, Vern Anthony, to my salesman, Tom Mankowski, everyone is committed to producing the best that is available. In addition, my family continues to be fantastic. Without their support, this project would not have been completed. My daughter Amy, now in her senior year of college and my son, Keith, who is preparing to go off to college next year have endured hours of my staring at the computer screen. My hope for them is continued success in whatever they try to accomplish. I am proud of them on a daily basis. My wife, Carol and I celebrated our 25th wedding anniversary this past year and it is obvious to me that she is more than a lifelong partner and best friend. She continues to be supportive of my writing efforts and is frequently the gentle push of encouragement that I need to get back to the job at hand. I thank her and love her.

Acknowledgments (Reviewers)

Special thanks to Matt Lee of Ford Motor Company; Del Wright, Wright and Associates; James E. Quaintaince, Terra Technical College; Jim Armitage, Waubonsee Community College; George W. Behrens, Monroe Community College.

Introduction—Why Study Electricity/Electronics?

Students frequently have difficulty in automotive electricity courses. This statement probably does not come as a complete shock to you, for perhaps you are one of these "anti-electrical" students. The time has never been more appropriate than it is today to point out just why your thinking is wrong! Within this introduction we will discuss the vehicles of yesterday, today, and tomorrow. We will also point out what the future appears to be for the automotive diagnostician. This is done so that you will realize that times today require that you have a full and complete electrical background if you wish to be successful as a professional technician. Notice I have used the word "require"—not "nice to know" or "helpful," but required. Electrical control of the automobile is here to stay. Most industry sources feel we have just begun to see electrical systems take over the entire running of the vehicle. The future will no doubt be very bright for those who are electrically prepared and very dismal and dark for those who are prepared at a 1960s level! What level you will be at is entirely up to you.

Let us look at the vehicle of the past. When I speak of the past, I am not going to go back to the Model T. You are probably aware that its electrical system, or lack of a system, really cannot be compared to anything we have on the road today. With no accessories, diagnosis was a snap. Instead, I would like to talk about the vehicle of the sixties—the vehicles that are around 30+ years old. Some of them might still be seen on your local streets. These vehicles were really the beginning or foundation of what we see today. The manufacturers were beginning to see the consumer requesting accessories and conveniences that they felt would increase their driving comfort. Some of these options were electrical. It was a fun age for the mechanic. He, finally, was seeing vehicles that many "backyard" mechanics were having difficulty repairing, and he found his services and skills requested more often. The sixties found changes in charging systems, starting systems, and found vehicle air conditioning becoming more common. These vehicles, by today's standards, could still be referred to as electrically archaic. It was as different from the Model T as today's vehicle is different from it.

Electrical gingerbread was just appearing. Corvette decided that it would turn the headlights over with electrical motors. Radio antennas were being raised with electrical motors. Transmissions were being downshifted with electrical solenoids and FM radios featured search or scan capability. Alternators and power windows and seats were commonplace. Mechanics were scratching their heads and saying Where will it all end? The end was nowhere in sight—and it still is not in sight today!

Let us look at the wiring of the sixties vehicle. With relatively few accessories or engine controls, the main harnesses carrying wires under the dash and in the engine compartment were about 1/2 inch in diameter with connections or breaks at convenient locations. This made testing relatively simple. As a matter of fact, wiring diagrams were about a page in length. Aftermarket (not made by automobile manufacturers) components were readily available, and easily located on the vehicle. Relays (electromechanical switches) were in use in certain heavy current circuits. Electricity was in use in many of the accessory circuits.

However, when manufacturers were faced with an engine problem, they frequently cured it with mechanical components. Vacuum was used to move or control heating, doors, locks, and, parking brakes, among others. The vacuum system within the engine was becoming complicated. We found many vacuum lines going to vacuum actuators or motors. Fuel and spark delivery was controlled by vacuum. Transmissions were shifted by vacuum or a mechanical connection to the throttle.

If the above sounds complicated, put yourself in the shoes of technicians, who for years were just concerned with how well the engine ran and the vehicle braked and steered. Suddenly, faced with accessories and engine controls that were for the most part vacuum controlled, technicians realized that to stay current they were going to have to go back to school. They were going to have to subscribe to some type of service magazine to help them stay current, and attend night seminars designed to update their knowledge. Good current technicians found that their day did not end when they left the shop. They attended every seminar available, while automotive magazines became their monthly reading assignment. The industry responded with classes and articles that each year would fill in the gaps between what the technician knew and what was on new vehicles. The basic mechanical skills that the industry had were sufficient to carry us through the changes. We often heard mechanics saying "give me something new and I'll take it apart to figure out how it works." Our industry technicians were extremely mechanically inclined and could figure out how just about anything worked. The hourly wage for figuring out "how something worked" was less than $10.00 for the experienced professional technician.

Now let us look at the vehicle of today. Within the last thirty years, some of the new options or original equipment changes that are electrical include: rear window defoggers with timer, engine cooling fans, cycling air-conditioning compressors, keyless entry systems, alarm systems, intermittent wipers, automatic headlights, level systems, digital dashboards, computers that talk to the driver, and cathode ray tubes in the passenger compartment are commonplace today. In addition, think about the engine controls that are also electrical: electronic ignition, electronic voltage regulation, automatic idle speed, torque converter clutches, fuel injection, fuel pumps, and turbocharging. The list is endless and grows each year.

Now we also have vehicles that monitor the weather and change the fuel and ignition system to match the altitude. Computers as powerful as those that filled rooms in the sixties are installed on virtually all makes and models.

Where does this put the technicians today? First of all, notice the name change. Technician has become the preferred term, replacing the mechanic notation of the past. And what about the job? Obviously, as the vehicle became more electrical in nature, the task of repairing it also changed. Remember the quotation, "give me something new, I'll take it apart and figure out how it works?" Try taking apart a computer or a sealed ignition module and figuring out how it works. The days of being only mechanically inclined are gone for most technicians. Oh, it is true that some jobs still exist with the mechanical repair area that on-board electronics has not touched. Some individuals do make a living without getting involved with electronics. But to be realistic, the bread and butter of the automotive industry involves some degree of electricity. General technicians or diagnosticians must be able to understand electricity. They must be able to predict what will happen based on the information in a wiring diagram and their own electrical foundation. Once you, as the technician, understand how electricity behaves, the vehicles of today are not nearly as complicated as you might think.

We anticipate that the future will see changes in how and what on-board electronics control. There probably is no area of the vehicle that will not see electronic changes in the future. Multiple computers talking back and forth to one another, while they run the vehicle will become commonplace. Self-diagnostic capability, or the ability to tell the technician what circuit is at fault, is already on most systems. Economy, power, and driveability are all linked directly to the computer.

The role of the technician will be ultra important. In the few years that we have seen major electrical changes within the automotive industry, we have been able to conclude that today's vehicle has no match in terms of economy, power per cylinder, and driveability. The vehicles of the past have been surpassed two and three times over. Consumers generally like the way the vehicle of today performs, whether it is American or foreign made. No doubt about it, though, throw in an electrical malfunction and you can have one very sick running vehicle. Not only do the modern systems contribute to the overall good running of the vehicle, but they can and do contribute to the overall lack of performance if they have malfunctioned. Enter the diagnostician, armed with an electrical background, a wiring diagram, a diagnostic chart, and a tester. After carefully following test and repair procedures, we again have a satisfied customer and a vehicle that performs as it was designed.

The technicians of today and tomorrow will also be known for their ability to adapt or change with the times. Our industry lost some technicians in the last few years because of the inability to convert or adapt their skills over to a more modern approach. The vehicle of today or tomorrow will not be repaired with skills from the sixties. As the vehicle changes, so must you! We have many repair facilities today where some of the technicians specialize in automotive electronics. This way of running a service center may increase. The age of specialization among automotive technicians will only help the industry. Quicker diagnosis, more accurate repairs, and less come backs are the characteristics of specialists. Remember the wiring system of the sixties? Well, look at it now: harnesses 2 inches in diameter, wiring diagrams 15 pages long, and an impossible task of memorizing where components are even located. Component locators, special diagnostic testers, and hourly repair rates that are four or five times that of the sixties, these all characterize the diagnostician of today.

Well, where do you fit? Do you belong in the sixties or do you belong in today's automobile? The future is bright. The road is long. But the way is clear! The way to this specialty is partially in this text. We have compiled a foundation of information for you to read, think about, and do. If this text is being used in an organized automotive electricity class, the selected lab type activities will help you understand the principles presented. Do the lab activities with your classmates and discuss with your instructor not just what happened, but why it happened. The age of electronics is sometimes described as the age when the "what if" statement was born. Do not be afraid to ask "what if." Your instructor, with the help of this book, is in a position to answer those questions and help you become prepared for a career, not just a job!

CHAPTER 1

Voltage, Amperage, and Resistance

OBJECTIVES

After completing this chapter, you should be able to

- define electrical terms that are common to auto mechanics.
- use Ohm's law to figure circuit voltage, amperage, or resistance.
- understand the use of relays on vehicles.
- demonstrate electrical changes in series circuits, parallel circuits, and series-parallel circuits.

INTRODUCTION

The ability to diagnose many modern systems found on the vehicles of today lies in the technician's ability to understand what electricity has done or is doing in a particular circuit. Frequently, diagnosis is made easier by the fact that the technician can "mentally" figure out what is happening before the tools even leave the toolbox. To understand how electricity behaves within a circuit and to be able to analyze the circuit on the vehicle, one must have a good background of the fundamentals. This chapter contains the building blocks of the rest of the text. Learn these principles now and the remainder of the material will be easier. Keep reading the information until it becomes second nature to you. The chapter objectives, summaries, and the chapter checkups will help test your ability to comprehend the material.

Within this chapter, we will attempt to build your vocabulary and your comprehension of the necessary ideas and components you will use when wiring a circuit. We will discuss the sections of a circuit and build some simple, yet practical examples of working circuits. New words will be set in bold type and defined for you.

ELECTRICITY

Electricity is difficult to define unless we split it into its elements. To be able to do this we will take a brief look at the atom. All matter is composed of atoms. Everything, including yourself, contains atoms. They are the building blocks of everything you see. Refer to Figure 1-1 and notice that the center of the atom, called the nucleus, contains both positive-charged (+) particles and negative-charged (-) particles surrounded by moving electrons in orbits, like the planets around the sun. These orbiting electrons are what we will be concerned with as we look at electricity. If we could examine a length of copper wire very closely, we would notice that the outer electron orbit of the copper atom has only one electron in it. Notice that in Figure 1-2 there are a total of 29 electrons in four separate orbits around the nucleus. The first three orbits do not really concern us. The outermost orbit, or **valence orbit**, as it is called, is where the action is. With a little "push", we will be able to move this electron out of its orbit. The fewer the electrons in the valence orbit, the easier they move or travel between atoms. These electrons are sometimes referred to as being "free." They normally move between atoms that are very close together. This characteristic makes copper (or another atom with one or two electrons in the valence orbit) a good electrical wire. These free electrons can be bumped or pushed into adjacent or nearby atoms easily. When this bumping occurs, the second atom, which now has two electrons in its outermost orbit, will then give up one electron to a third atom, and so on. By pushing a series of electrons along a wire from one atom to the next, we have a pretty good simple definition of electricity: moving electrons. The more electrons that move, the more electricity. Copper or any other atom with one or two electrons in its valence orbit is called a **conductor**. Simply defined, a conductor is a material in which electrons move easily. We typically use copper, aluminum, or steel for our conductors in automobiles.

The opposite of a conductor is an **insulator**. Insulators are composed of materials whose atoms contain nearly full valence orbits. Full valence orbits tend to stabilize the atom and eliminate the movement of outer electrons between atoms. Insulators are used in electrical systems to ensure that electricity

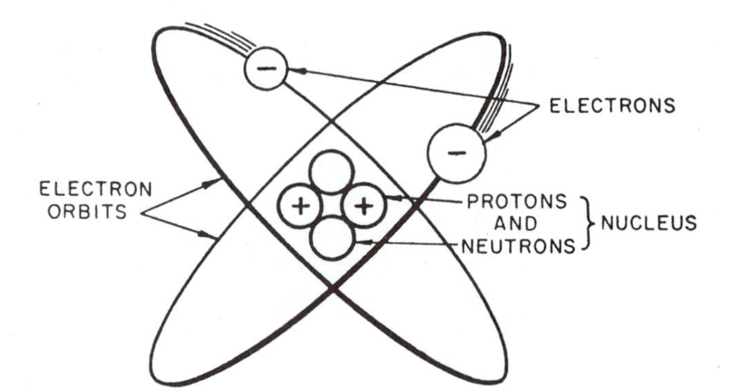

FIGURE 1-1 Electrons revolve around the nucleus.

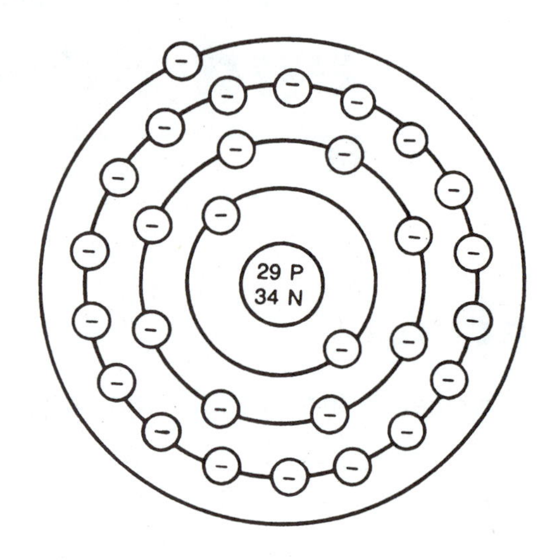

FIGURE 1-2 A copper atom.

arrives where we want it to go without being sidetracked. Paper, plastic, rubber, and cloth are frequently used as insulators in automobiles.

VOLTAGE

Up to this point, we have been talking about moving electrons and have defined them as electricity. Did you wonder why the electrons were moving in the first place? Voltage is the reason. Voltage is electrical pressure.

Electrons move when we apply a pressure against them. This pressure can be measured and, again, is called voltage. Defining voltage as pressure allows us to compare it to the pressure that forces water out of a faucet when you open it. Water pressure measured in pounds per square inch, or psi, is the push behind the water. Without some psi behind the water, opening the faucet would not accomplish anything. The water would remain in the pipes and not flow. Electricity, defined as moving electrons, would react in the same manner. Without pressure, or voltage, behind the electrons, they would not move and would remain in the wire, or conductor.

Voltage should be thought of as a pressure difference that will cause electron movement. Look at Figure 1-3. In the top example, water flows because there is a pressure of 12 psi from the left. But notice that water will also flow in the bottom example where a pressure of 12 psi is pushing against 4 psi and still pushes or moves the water to the right. The pressure difference will move the water from a higher to a lower pressure. Electricity is much the same. If a pressure of 12 volts is applied against a pressure of 4 volts, the electrons will move from the higher 12 volts toward the lower 4 volts because of the 8-volt difference.

We should also realize that voltage can be positive or negative. This is called **polarity** and is simply the type of charge or pressure on the electrons. Let's try some examples. Refer to Figure 1-4. Notice that the pressure difference determines the direction of electron movement. If there is a pressure differential, electrons will move. If, however, we apply equal pressure of the same polarity (+12 against +12), the end result is the same as if we applied no pressure (0 volts): no electron movement. Later you will see that the major source of voltage or pressure will be the battery or the charging system.

Keep in mind that the amount of pressure behind water will determine just how much or how fast the water flows. We can increase the amount of water flowing by increasing the pressure. Electricity is the same. Increasing the pressure behind the electrons will cause more of them to flow. Reducing the pressure will cause less of them to flow. Remember, it

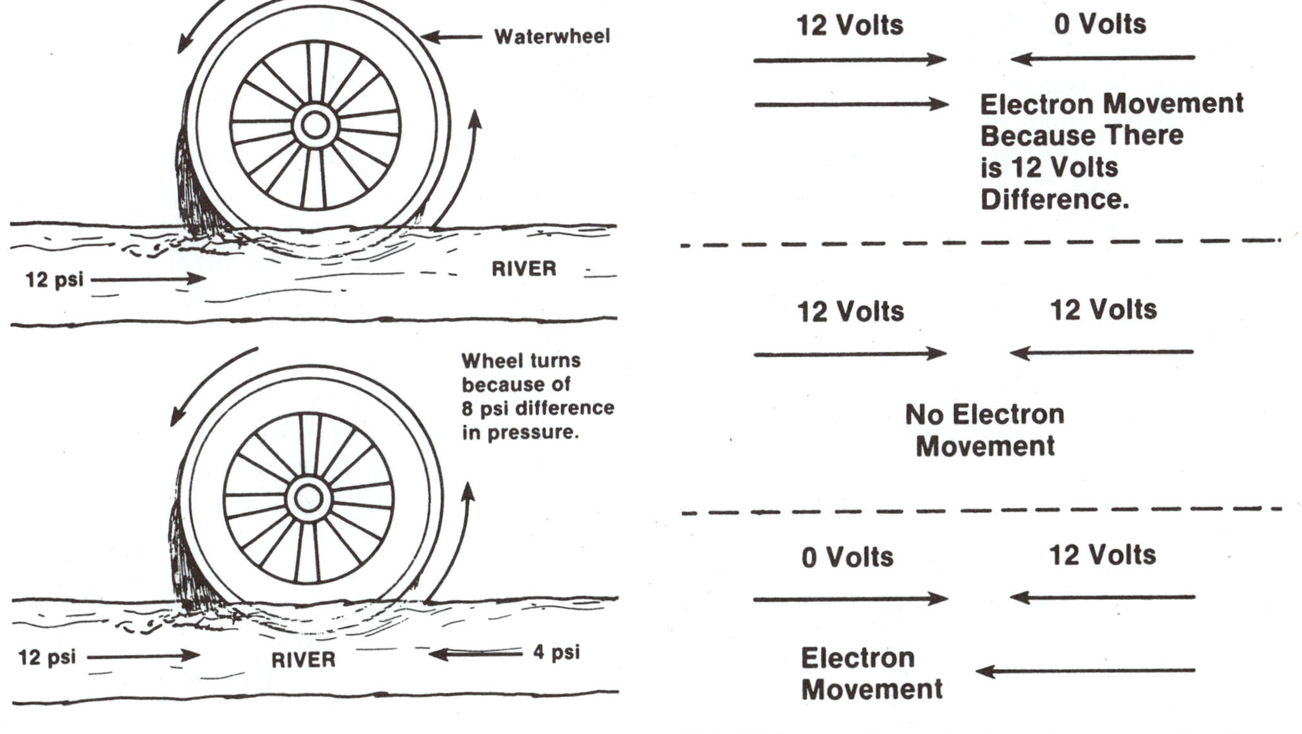

FIGURE 1-3 Waterwheels turn because of water pressure.

FIGURE 1-4 Current moves because of voltage differences.

is the pressure difference that actually causes the movement, not just the pressure.

CURRENT

Up to this point, we have been referring to moving electrons without trying to count or define them. This movement of electrons is called a **current**. Much the same as a current in a river, it flows through the wires. In reality, we frequently want to know the amount or quantity of electrons flowing, because it is this current that will do the work for us. It would be very impractical to actually count electrons because the number needed to do a reasonable amount of work is extremely large. A small automotive light bulb, for example, would require 6 billion electrons a second to light! Not a very easy number to work with. It would be similar to being paid each week in pennies! Scientists have helped us by defining the 6 billion electrons as an **ampere**, or simply an **amp**. Remember that the quantity of electrons moving is what will do the work we want done, and that this mass of electrons is called current and is measured in amperage, or amps.

Let's go back and pick up our water coming out of the faucet because of the pressure behind it example, Figure 1-5. What is the electrical equal to the water? If you said electrons or amperage, you are correct. Remember that another term for this electron mass is current. You will sometimes see "amps of current" printed. This is just another way of telling you how many electrons are actually working for you. Another way to remember current is to think of the current in a river. It is the moving water. The electrical current then becomes the moving electrons. Why are they moving? The pressure or voltage behind them is pushing them from one atom to the next atom.

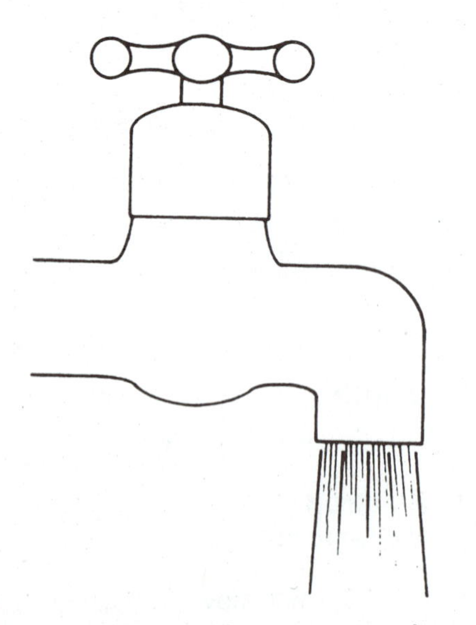

FIGURE 1-5 Water and current react the same.

It then becomes simple to realize that it is this quantity of pressurized electrons that will do work for us. Let's define work. Electricity working will give us light, heat, or magnetic field, Figure 1-6. It is important to realize that any time work is being done, we will be producing one or more of these results. Most of the time, this work is designed or engineered into the circuit, but sometimes it is not. A bad connection, for example, might get warm as electrons flow through it. The warmth is an indication of work being done even though we did not design the bad connection into the vehicle.

You might be thinking about all of the devices on the car that do work for us. All of them have some current or amperage flowing through them, and are doing work to produce light, heat, a magnetic field, or a combination of two or more. Certain devices, however, also give us motion. Starter motors, heater blower motors, windshield wipers, and so on, all give us motion, but they do so by setting up magnetic fields, as we will see in future chapters.

Work cannot be done by just voltage or just amperage. It takes both. If we need a waterwheel turned, it will take both water and pressure to get the work accomplished. Pressure only or water only will not turn the wheel! Electricity is the same. Voltage without current or current without voltage accomplishes nothing! When you leave your headlights on overnight and run the battery down, you are removing the pressure or voltage. When you try to start the car the next morning, you have plenty of current (electrons) because the whole vehicle is made up of

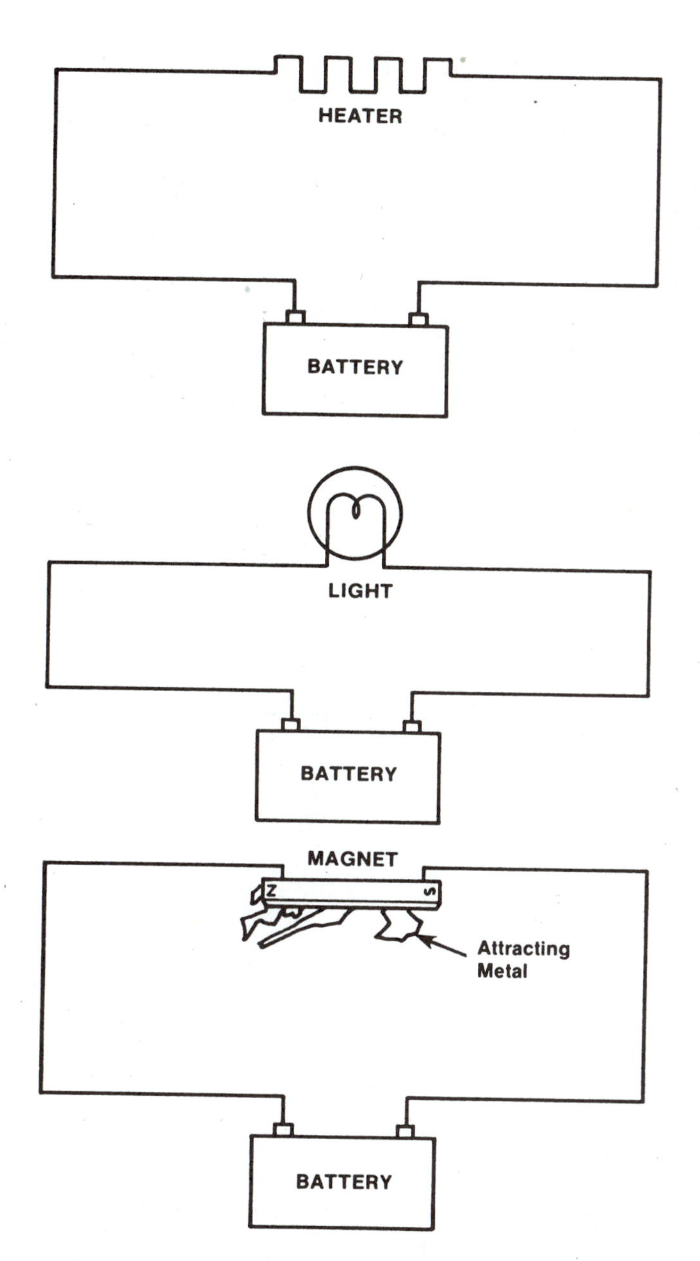

FIGURE 1-6 Loads develop heat, light, or magnetic fields.

them, but without voltage or pressure behind the current, nothing happens when you turn the key. No pressure with lots of current does not work!

WATTAGE

Another way of looking at work done besides amperage is to consider wattage. Wattage is the electrical means of monitoring just how much work is being done. Consider the light bulb in your home. It is rated in watts. A 60-watt bulb delivers one half the actual light of a 120-watt bulb. Wattage is the result of multiplying the voltage applied by the current flowing. Because it takes both pressure and current to actually do work, it makes sense that an indication of the amount of work accomplished would use both in its calculations. Your 60-watt bulb at home actually draws a 1/2 ampere of current at 120 volts, for example, 120 V × 0.5 A = 60 watts, Figure 1-7.

We usually do not use the wattage formula when we repair vehicles, but knowing that work accomplished can be measured in wattage is useful especially when dealing with some of the experiments we will try in Chapter 3. We also use wattage when trying to understand starting systems. If a certain amount of work must be accomplished, such as cranking an engine over, the amperage used will be based on the voltage applied. Mathematically, it figures out like this. If 2,000 watts of work will be needed to crank over an engine, 200 amps will be needed if the voltage can be kept up to 10 (200 × 10 = 2,000). However, if the voltage drops down to 5 volts, the amperage will have to go up to 400 to crank the engine over at the same speed. Our discussion of conductors will show why it will be harder to deliver 400 amps than 200 amps.

If you look at vehicles today, you will see mostly 12-volt systems, and yet years ago many systems were 6 volt. Getting the same amount of light, heat, and magnetic field out of a 6-volt system requires twice the amperage. Delivering and generating twice the amperage was not only expensive, but difficult to control. Another example of wattage at work can be seen when looking at large diesel engines. Most crank over at 24 volts rather than 12 volts because the starting system will need to draw half the current at 24 volts than it will at 12 volts to accomplish the same amount of work. Do not forget:

$$\text{Voltage} \times \text{Amperage} = \text{Wattage}$$

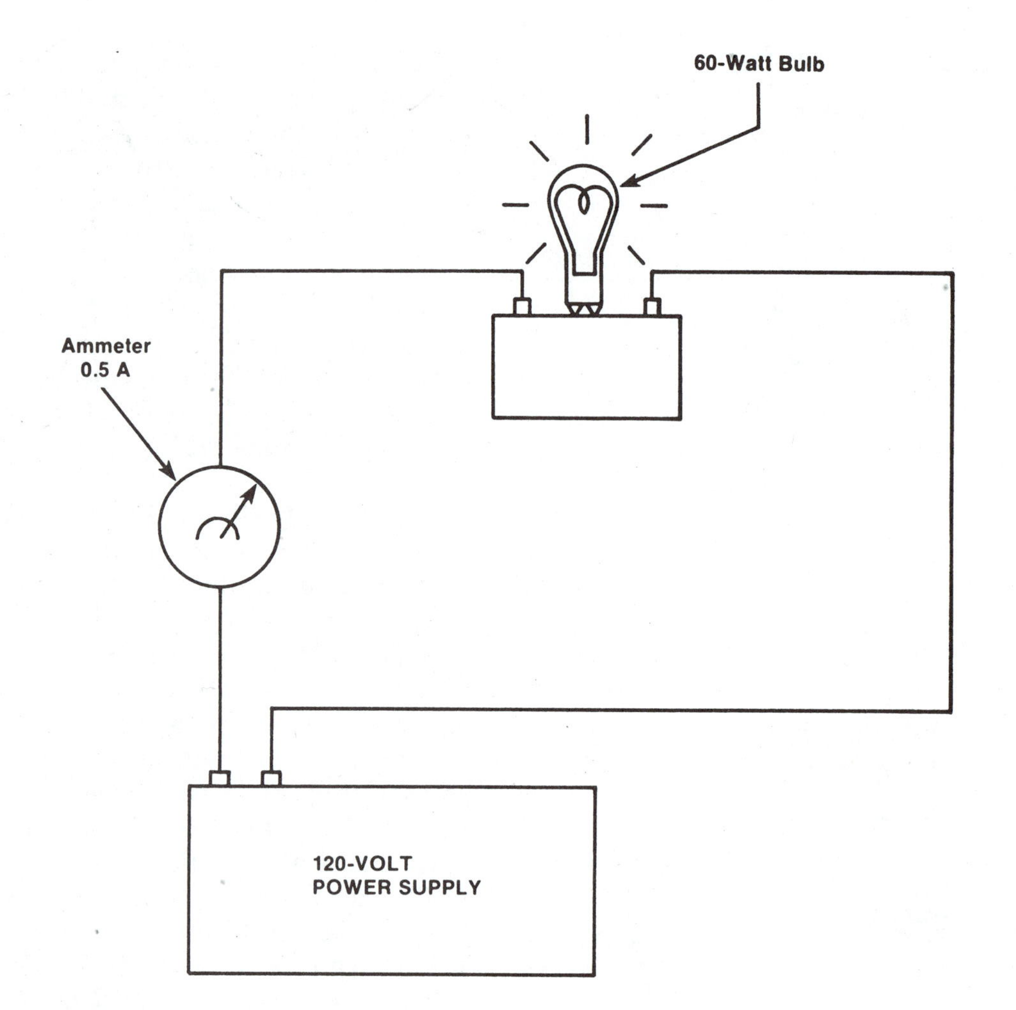

FIGURE 1-7 A 60-watt bulb draws 0.5 amp with 120 volts applied.

RESISTANCE

The last item necessary (voltage and amperage being the other two) is resistance. Resistance, measured in ohms, is the force that opposes the flow of electrons. This back pressure to electrons, along with the voltage will dictate the number of electrons able to flow through a circuit. When current is moving through electrical components, such as bulbs and motors, the amount of current flowing will be dependent on two other conditions: the voltage or pressure on the electrons and the resistance that the circuit puts up against the flow of electrons. In this way, resistance is the opposition to the flow of current. If we decrease the resistance, it will be easier for electrons to flow. If we decrease the resistance, more electrons will flow through the circuit and more work will be accomplished. Let's go back to our water example. Pressure against the water moved a current and turned the waterwheel. That waterwheel will take some effort to turn. It will offer some resistance to the water trying to push it. In essence, this water is equal to the electrical resistance that a light bulb offers to the flow of current. Resistance is measured in ohms and has the Greek symbol of the omega, Ω. As you look at resistance and measure it, always keep in mind that it is the opposition or force that will make it harder for current to flow.

Consider now the interrelationship between voltage, amperage, and resistance. This relationship is called **Ohm's law** and, simply stated, it tells us what the third or unknown value will be in a circuit if we know the remaining two. If you know the voltage and the resistance, you can determine the current flowing. Or, if you know the current and the voltage, you will be able to find the resistance. This relationship must be very clear in your mind because diagnosis of most circuits involves our looking for changes that occur in voltage, amperage, or resistance. To be practical, we very seldom sit alongside the automobile and use Ohm's law mathematically to repair a problem. We do, however, use the principles of Ohm's law in virtually every electrical repair. Let's look at this interrelationship closely.

If we understand that resistance is opposition, increasing or decreasing this opposition will have an inverse (or opposite) effect on current flow (assuming the voltage remained the same). In other words, if we increase the resistance in a circuit whose voltage has remained the same, we will see that more opposition will allow less current to flow. It will now be harder for current to flow through the circuit and therefore less will be able to get through the resistance. The opposite is also true. If we decrease the resistance within a circuit, the current will find itself in an easier path and more will flow. This inverse, or opposite, relationship assumes that the voltage remains the same. Remember our discussion of work. We discussed the amount of work done by the amount of current. The more current flowing, the more work accomplished. With this in mind, answer the following question: Which component has the greatest resistance, a taillight bulb or a starter motor? If you said taillight bulb, you are correct. Let's think through this example together. Which component is doing the least amount of work? The taillight is. This then means that it must be drawing the least current, and Ohm's law says that if the current is down, the resistance must be up. Were you able to follow that? Let's try the same example, but look at it from the starter side. The starter is turning over the engine. This represents a lot of work accomplished. Lots of work equals lots of current. Go back to Ohm's Law. Lots of current can only flow if the resistance is very low. The starter and a taillight bulb have been used as examples to emphasize that physical size has nothing to do with electrical resistance, Figure 1-8.

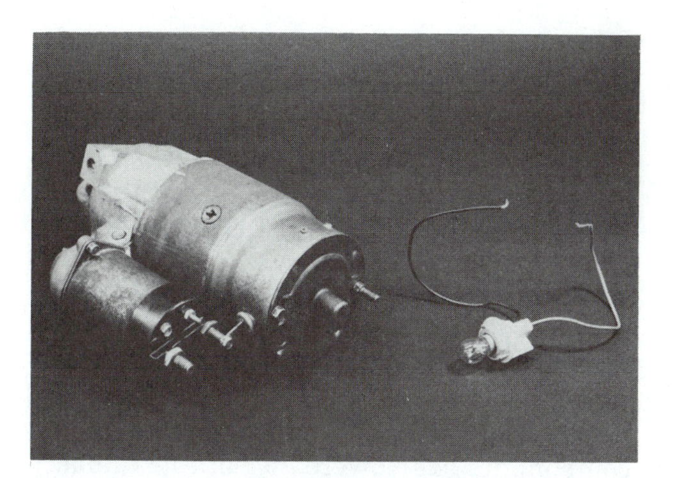

FIGURE 1-8 Load size and resistance are not equal.

To be specific, Ohm's law is best understood by looking at Figure 1-9. Putting your finger over the unknown value will leave behind the formula that you use to determine what the unknown is. If the voltage and the amperage are known, the resistance can be easily figured. Blocking over resistance, the unknown, leaves behind voltage over amperage. For example, if a 12-volt circuit has 2 amps flowing, a simple fraction with 12 on top and 2 on the bottom is the result of placing the known voltage and amperage into the formula. Dividing 12 by 2 shows that the circuit must have 6 ohms of resistance. How many watts of work are being accomplished?

FIGURE 1-9 Ohm's law.

Remember watts? Voltage times amperage! That is, 12 volts times 2 amps equal 24 watts of work. It all fits together; doesn't it?

Let's try another example. If 12 volts is applied to 12 ohms of resistance, how much current should be flowing? Cover the unknown value of amperage. This leaves voltage over resistance, or 12 over 12. Dividing 12 volts by 12 ohms will equal 1 amp flowing, Figure 1-10. Most of the actual examples on the modern vehicle are with approximately 12 volts applied because most components are powered off of the vehicle's 12-volt battery. But circuits with more or less voltage can just as easily be figured using Ohm's law. We will use Ohm's law more when we cover types of circuits toward the end of this chapter.

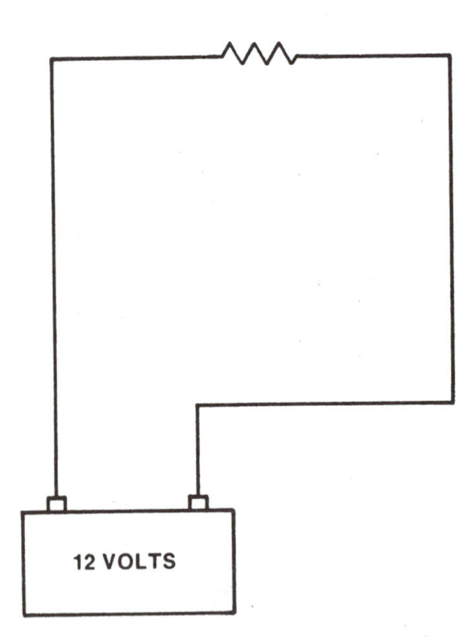

$$\frac{12 \text{ Volts}}{12 \text{ Ohms}} = 1 \text{ Amp}$$

FIGURE 1-10 Ohm's law at work.

INSULATORS AND CONDUCTORS

We will now begin to put the various principles that we have discussed into workable circuits. We should realize that any discussion of amps, ohms, watts, or whatever would not make much sense unless we apply it to actual circuits on the vehicle. To be able to analyze what happens with Ohm's law also requires a working knowledge of circuits. With this in mind, what is a circuit? Simply stated, a **circuit** is an electrical path from a source to a component that will do work and go back to the source. A circuit does work! Voltage pushing current through resistance results in light, heat, or magnetic field, which is defined as work.

Let's zero in on a circuit and analyze its components or parts. At this point, let's use a battery as a source of power, Figure 1-11. We will spend much more time on batteries later, discussing construction and service. For now, just remember it as a source of electrical pressure when charged. If the work to be accomplished will be to light a bulb, the next thing we need are conductors, or wires. Remember our discussion of atoms and valence electrons. A conductor was defined as a substance that has one or two electrons in its valence, or outermost orbit. This substance accepts and gives up electrons easily, which is another way of saying it conducts electricity.

The opposite (electrically) of a conductor is an insulator. It is defined as a substance

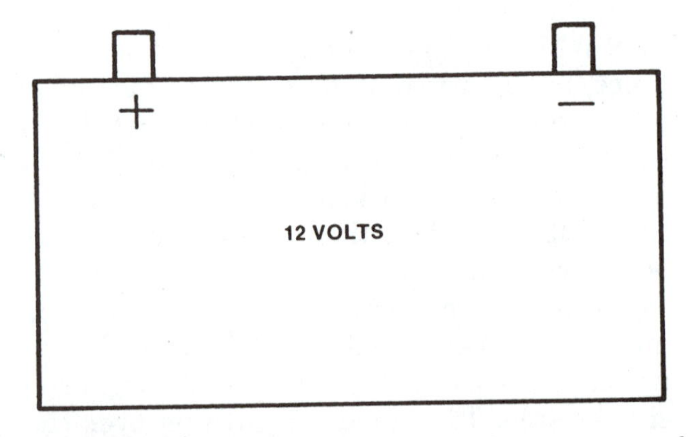

FIGURE 1-11 The battery is the primary source of power.

whose atoms do not "easily" accept and give up electrons. It is important to note that the difference between conductors and insulators is sometimes a fine line. Take for instance the human body. If you put your hands across a positive (+) and negative (–) battery terminal, the high resistance of your body prevents the electrons from flowing through you. Yet, when we touch the end of a spark plug wire, we certainly can feel those electrons flowing right through us! What is the difference? The pressure. We are insulators at low pressure; yet, if the pressure or voltage is raised up high enough, we become conductors. The 12.6 volts of the battery is low enough so that the resistance of our body could not be overcome. However, raise this voltage up to 30,000 volts and our high resistance will not prevent the flow of current. The ignition system currently in use today has in excess of 30,000 volts available and can easily pass small amounts of current through our body. Another example can be seen during any lightning storm. The extremely high voltage of lightning takes the normal insulating air and passes current right through the center of it, Figure 1-12. Given a high enough voltage, electrons can be pushed through almost anything!

This dictates that the insulation in use must match the voltage applied in any circuit. Low voltage requires less insulation than higher voltage. If you were to cut a spark plug wire in half, you would observe that the greatest percentage of the wire is insulation. Sometimes as high as 90% of the wire is insulation. Cutting a taillight wire in half would reveal the opposite, Figure 1-13. The greatest percentage of this wire would be the conductor, or center, which is the part that will be carrying the current. The size of the conductor will be determined by the amount of current flowing and the length of the circuit as seen in Table 1-1. The larger the number of the wire, the thinner it is. Notice that a larger wire (smaller number) is needed to carry the same amount of current longer distances. The greater length of the wire adds more resistance. The additional resistance must be compensated for with a larger conductor. Greater current or additional length will require larger conductors. Do not forget this when you repair a circuit. Always use an adequate size wire for the current and the length of the circuit with adequate insulation for the voltage. Remember, it is the job of the insulation to make sure that the current actually gets to where it is supposed to go. Without in-

FIGURE 1-12 High voltage at work.

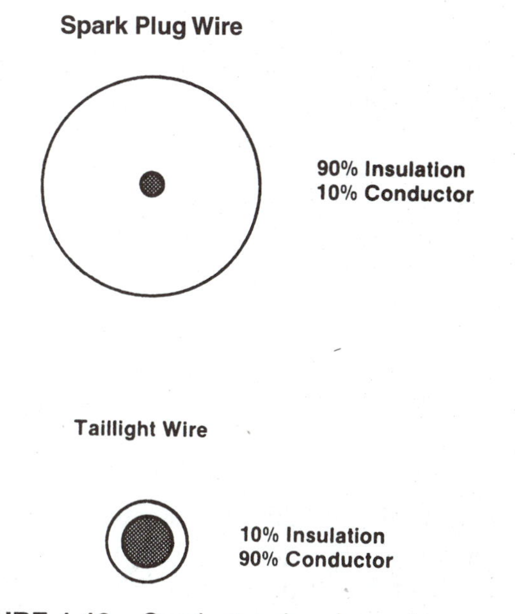

FIGURE 1-13 Conductor size determines current capacity, insulator size determines voltage capacity.

TABLE 1-1

Total Approximate Circuit Amperes	Wire Gauge (for Length in Feet)											
12 V	3′	5′	7′	10′	15′	20′	25′	30′	40′	50′	75′	100′
1.0	18	18	18	18	18	18	18	18	18	18	18	18
1.5	18	18	18	18	18	18	18	18	18	18	18	18
2	18	18	18	18	18	18	18	18	18	18	16	16
3	18	18	18	18	18	18	18	18	18	18	14	14
4	18	18	18	18	18	18	18	18	16	16	12	12
5	18	18	18	18	18	18	18	18	16	14	12	12
6	18	18	18	18	18	18	16	16	16	14	12	10
7	18	18	18	18	18	18	16	16	14	14	10	10
8	18	18	18	18	18	16	16	16	14	12	10	10
10	18	18	18	18	16	16	16	14	12	12	10	10
11	18	18	18	18	16	16	14	14	12	12	10	8
12	18	18	18	18	16	16	14	14	12	12	10	8
15	18	18	18	18	14	14	12	12	12	10	8	8
18	18	18	16	16	14	14	12	12	10	10	8	8
20	18	18	16	16	14	12	10	10	10	10	8	6
22	18	18	16	16	12	12	10	10	10	8	6	6
24	18	18	16	16	12	12	10	10	10	8	6	6
30	18	16	16	14	10	10	10	10	10	6	4	4
40	18	16	14	12	10	10	8	8	6	6	4	2
50	16	14	12	12	10	10	8	8	6	6	2	2
100	12	12	10	10	6	6	4	4	4	2	1	1/0
150	10	10	8	8	4	4	2	2	2	1	2/0	2/0
200	10	8	8	6	4	4	2	2	1	1/0	4/0	4/0

sulation, current would take the path of least resistance back to the battery, bypassing, or shorting, around the bulb or motor. This is called a **short circuit**. Because all of the metal used to make our vehicles is conductive, we would have one large short circuit on our hands without insulation.

LOADS

Following this current, which we now are assured will go where we want it to go, we eventually come to the component that will do the work the circuit was designed for: the load. The load in a circuit is the component whose resistance will produce light, heat, or magnetic field when current is pushed through it, Figure 1-14. A load does work! As we have mentioned previously, voltage pushing current through resistance forces the resistance to do work. It is the resistance of the circuit that controls how much current will actually flow and how much work will be done. We know this from Ohm's law. It is this resistance that we are concerned with. All circuits require a load if they are to do productive work for us. However, we should realize that sometimes unproductive work is done. The corroded connection ahead of the headlight is an example of electrical resistance. Not only will it have an effect on the brightness of the light, causing it to be dimmer, but, because it is resistance, when voltage pushes current through it, it will do some work—probably a slight amount of heat. This unwanted heat was produced because of the

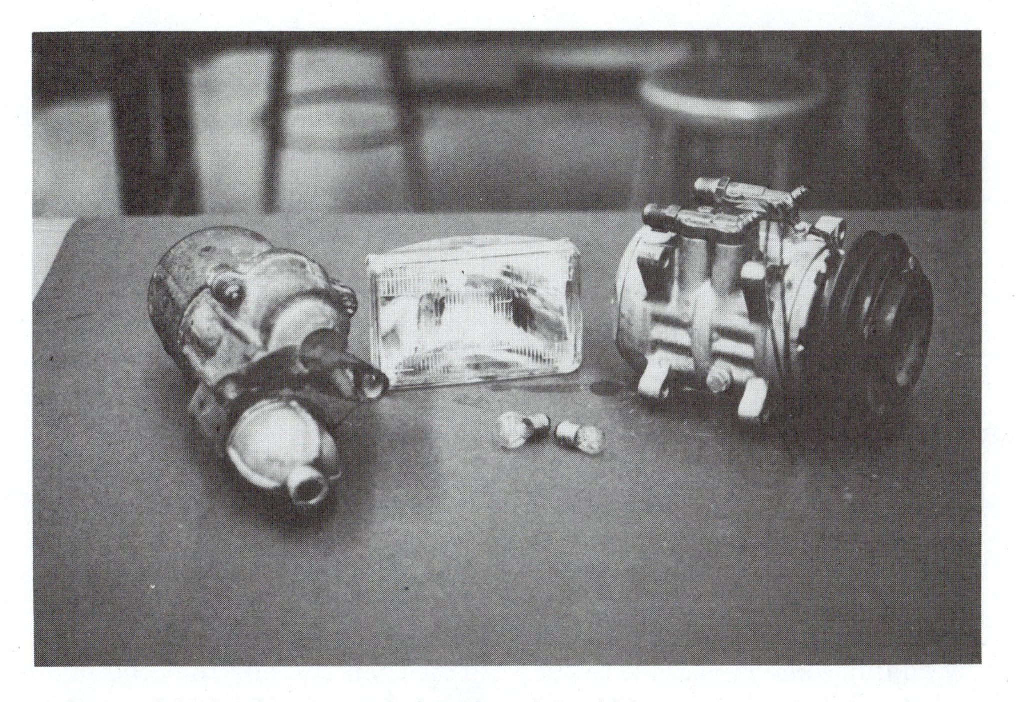

FIGURE 1-14 Some of the loads commonly found on the vehicle.

unwanted resistance of this load. Unwanted resistance is frequently referred to as **false loads**. Your diagnosis job on vehicles usually involves finding and eliminating false loads. These loads do produce work, but usually are not engineered into the vehicle. Instead they tend to appear after a few years and can cause some very funny and exasperating results.

The load designed into the circuit will control the current because of its resistance. Decreasing the resistance will increase the current flow and the amount of work. Increasing the resistance will decrease the amount of current flow and the amount of work.

CIRCUIT CONTROLS AND PROTECTION

Before we actually wire up some sample circuits, we have to take a look at two other very necessary components, which control and protect most automotive circuits. Without them, we would lose the ability to decide when our circuit does work and run the risk of an electrical fire in case of an overload. Our circuit can only handle a predetermined amount of current. Anything in excess is considered an overload.

Let's examine the controls first. Most circuits on the modern vehicle are not energized or live all of the time. A break, or "open", is designed into the wiring that can be closed whenever we need current to flow through the load. A switch is the best and simplest example of a control, Figure 1-15. We open the switch when we want the circuit to be off and close the switch when we want the circuit on. Keep in mind that a control is not a load and, therefore, should not have any resistance. If a switch does develop some resistance, it will become a false load and give off some heat. It might also have a detrimental effect on the rest of the circuit. If you ever feel a switch and it is warm to the touch, it should be replaced. The switch contacts have probably corroded inside from the arcing that occurs each time the switch opens and closes. This corrosion is actually resistance and is giving off heat as current flows through it. The exception to this is the headlight switch. It will normally develop some heat as it dims the

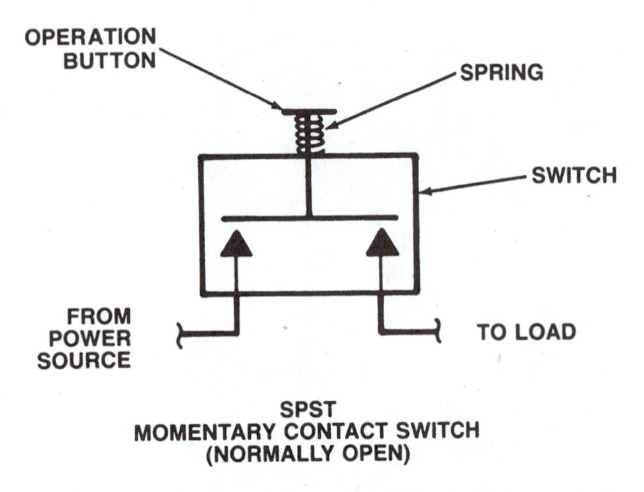

FIGURE 1-15 Single pole single throw switch. (Courtesy of Ford Motor Company)

dashboard lights. The dimmer function is accomplished by adding some resistance to the dashboard lights. This resistance causes a light amount of heat to be produced. The remainder of the switches used on vehicles should not be warm to the touch.

Relays are also a type of circuit control that allows us to control high current with lower current, Figure 1-16. We will spend additional time on relays in Chapter 3. Most modern vehicles utilize relays in their circuitry. The modern diagnostician must be capable of understanding how and why relays are used. For now, just remember that they are a type of control and should be treated as such.

If for some reason the total resistance drops too low within a circuit (such as when a bare hot wire touches the metal frame), the current flow will increase. All circuits have a limit as to how much current can flow through the controls and conductors. If we exceed this limit, we can turn the conductors into loads, producing heat and eventually causing a fire. For this reason, circuits are protected by fuses, fusible links, or circuit breakers. Let us examine each, but first the fuse, Figure 1-17. Probably the most common circuit protector is the in-line fuse. Made of a fusible or meltable material, it is designed to melt from a system overload of current and open the circuit before the conductors feeding the load are melted. The size of the fuse is indicated in amperage and is usually printed on the fuse body and sometimes on the holder or fuse panel. Never, under any circumstances, put a larger fuse into a circuit than the one that the manufacturer designed to protect it. If you install a larger fuse, the circuit might overload and be damaged or destroyed before the fuse melts. Remember, it is a circuit protector. It can only do its job if it is placed in a circuit that will have less current normally flowing than the rating of the fuse.

The second type of circuit protector is the fusible link, shown in Figure 1-18. It is made of meltable material like the fuse with special heat-resistance insulation. You will usually find it at a main connection or around the battery. It protects the main power lines before they are split into smaller circuits and run through the vehicle's fuse box. Most manufacturers use them because they are com-

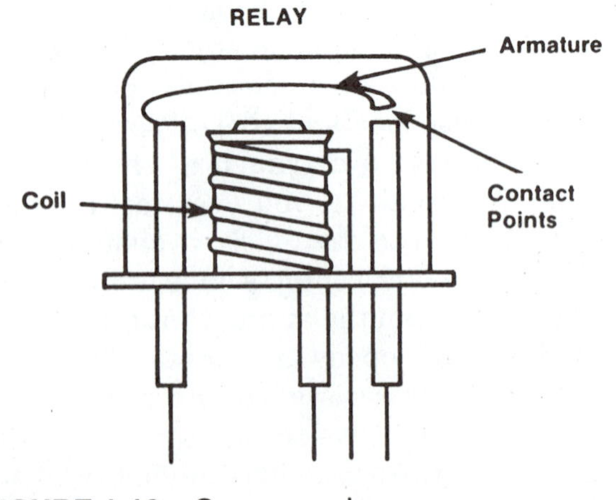

FIGURE 1-16 Common relay.

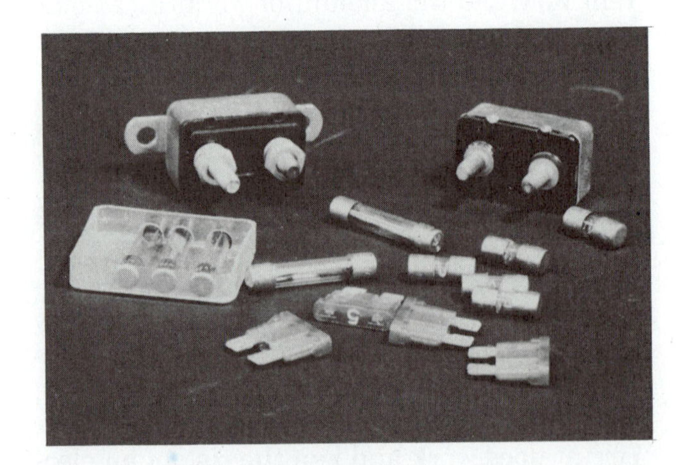

FIGURE 1-17 Types of fuses.

FIGURE 1-18 Fuse links located near the starting relay.

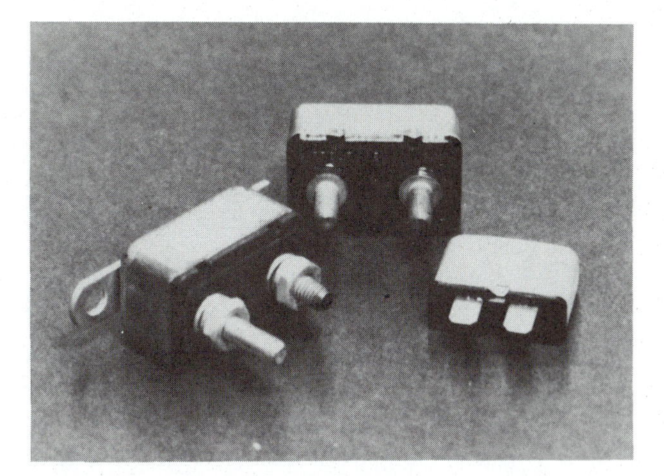

FIGURE 1-19 Circuit breakers.

pletely protected from the weather and road splash, which might contain salt or other chemicals. Fuses are usually not placed under the hood of the car because they are prone to corrosion. Instead they are usually found in a protected environment such as the passenger compartment. Underhood protection is most always done with fusible links. Their amperage capacity is determined by their size. They are usually four numbers higher than the circuit they protect. Four numbers higher is another way of saying two wire sizes smaller. The smaller the wire, the larger its number. In other words, a No. 14 circuit would be protected by a No. 18 fuse link, which is actually four sizes smaller. Using a No. 14 fuse link in a No. 14 circuit could result in the whole circuit burning up if a short occurred. Remember the object of circuit protection is just what the name implies—protect. Do not over fuse!

When manufacturers are designing a circuit that might experience an overload on a routine basis, they frequently use a circuit breaker as the form of protection, Figure 1-19. Similar in principle to a fuse, it is designed to open the circuit in case of a system overload. However, it has the advantage that it can be reset either manually or automatically once the overload is gone. A good example of the use of a circuit breaker is the power window circuit. In climates where ice is a problem, the window could jam or freeze shut. This could cause a current overload as the window motor fights the frozen window. The breaker should overload and open the circuit before damage to the motor occurs. Once the overload is gone, the breaker closes and is ready for the next try. There are other examples of circuit breakers, which we will look at later.

Now that we have identified the different components necessary for a practical circuit, let us look at the types available. Within the automotive field, we find the use of series, parallel, and series-parallel circuits. Let us look at series first.

SERIES CIRCUITS

If we take our insulated conductors, a switch (to control), a fuse (to protect), a battery (as a source of voltage), along with two light bulbs as loads, and wire them as in Figure 1-20, we have a simple **series circuit**. How does it work? The name implies the function—series. Each component is dependent on all the others if the circuit is to do any work. Bulb 1 is the source of B+ for bulb 2, and bulb 2 is the source of B– for bulb 1. The switch, conductors, and fuse are all dependent on one another. If bulb 2 burns out, the whole circuit will go dead. With only one path for current, an interruption or break anywhere along the path will turn off the whole circuit just as if the switch was opened. This dependence on all

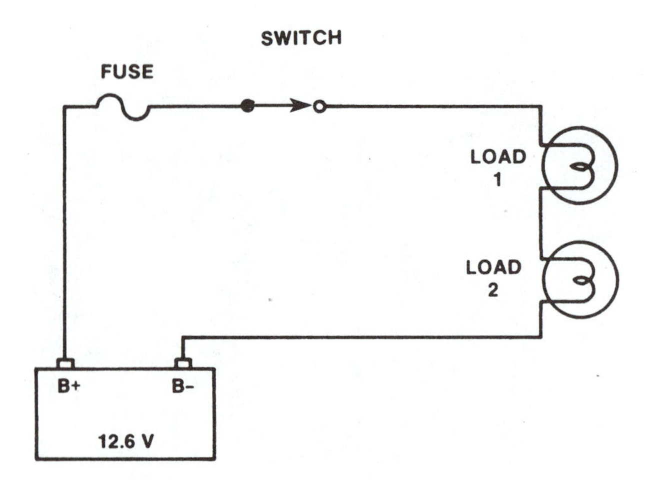

FIGURE 1-20 Two loads wired in series.

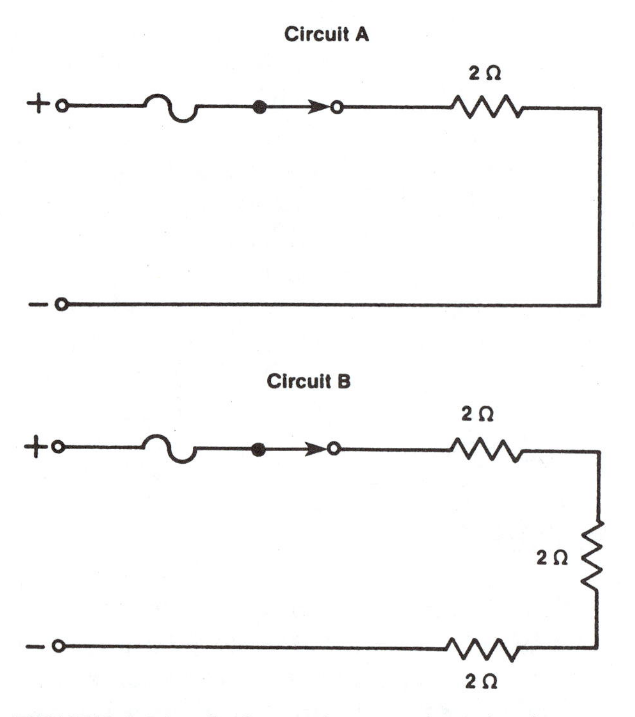

FIGURE 1-21 Series resistances in a circuit.

other parts is one reason why series wiring is not used too often in automobiles. We do not want all the lights to go out if the right front headlight burns out. Another characteristic of this type of circuit is the addition of resistances. Because the resistances are all in a row, they have a cumulative effect on the total resistance, Figure 1-21.

I. Procedure for Figuring Watts of Work

If we wire a circuit with three 2-ohm resistances and compare its action to that of a circuit with only one 2-ohm resistance, we can see the results in terms of work accomplished. Refer to Figure 1-22 for our Ohm's law formula and wattage formula. Circuit A will draw how much current? How many watts of work will be accomplished? How about the same questions for circuit B? Let us look at them together. First, in circuit A, Figure 1-23, cover up the amperage (the unknown value). This leaves voltage over resistance, or 12/2: 12 volts divided by 2 ohms equal 6 amps of current flowing. Therefore, 6 amps times 12 volts equal 72 watts of actual light, heat, and magnetic field produced. Now let us look at circuit B with three bulbs of 2 ohms each, Figure 1-24. The three resistances all add up (2 + 2 + 2). The battery is actually pushing current through 6 ohms of resistance now. Back to Ohm's law to find out the current: 12 volts divided by 6 ohms equal 2 amps of current flowing, then 12 volts times 2 amps equal 24 watts of work being accomplished.

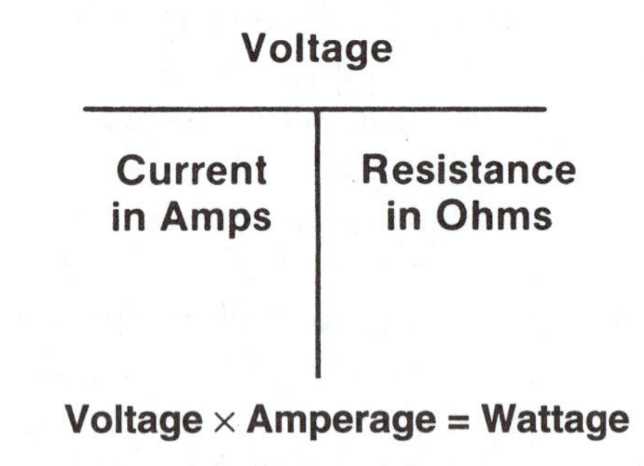

FIGURE 1-22 Ohm's law.

CIRCUIT A

$$\frac{\text{12 Volts}}{\text{2 Ohms}} = \text{6 Amps}$$

12 V × 6 A = 72 Watts

FIGURE 1-23 Figuring total work or wattage.

CIRCUIT B

$$\frac{12\ \text{Volts}}{6\ \text{Ohms}} = 2\ \text{Amps}$$

$$12\ \text{Volts} \times 2\ \text{Amps} = 24\ \text{Watts}$$

FIGURE 1-24 Figuring total work or wattage.

There is less total work because the resistances are wired in series. Series resistance will drop down the current within a circuit. If that resistance is unwanted, as in a corroded connection, the actual load will not receive as much current and do less work (false loads again).

The last trait of a series circuit is the voltage division or dropping that occurs. Battery voltage will have to divide or drop as it pushes current through each resistance. An example of this is seen in Figure 1-25. Notice that the voltage drops down as it pushes current through each bulb. You will see this for yourself when you wire up circuits in Chapter 3 and use the meters that we will study in Chapter 2. We have demonstrated mathematically what will happen in series circuits here. The opportunity to prove Ohm's law in Chapter 3 will help your overall understanding of circuits.

PARALLEL CIRCUITS

If we realize why most automotive circuits are not series wired, it becomes fairly obvious that certain characteristics are necessary or desirable for the next type of circuit we will look at: the **parallel circuit**. This circuit is named because usually the drawing style looks a little like a railroad track with its parallel rails, Figure 1-26. Notice in Figure 1-27 that some of the same components are used. We still have our voltage source, fuse, switch, and insulated conductors. The loads, however, are not wired one after another. Instead they each have their own source of B+ and B–. This will eliminate the dependence all loads have on one another that is characteristic of a series circuit. Independence is gained. Bulb 2 can now burn out without having an effect on bulb 1. Most of the vehicle is wired this way. If the right front headlight burns out, it will not turn off the rest of the lighting system. Independence is achieved because each load has its own path from and back to the source. In essence, it is the same thing as if the wiring were as in Figure 1-28. Parallel loads each receive full B+ voltage and therefore draw full current and produce full work. Current flow through a parallel circuit will divide down each leg. The water exam-ple pictured in Fig-

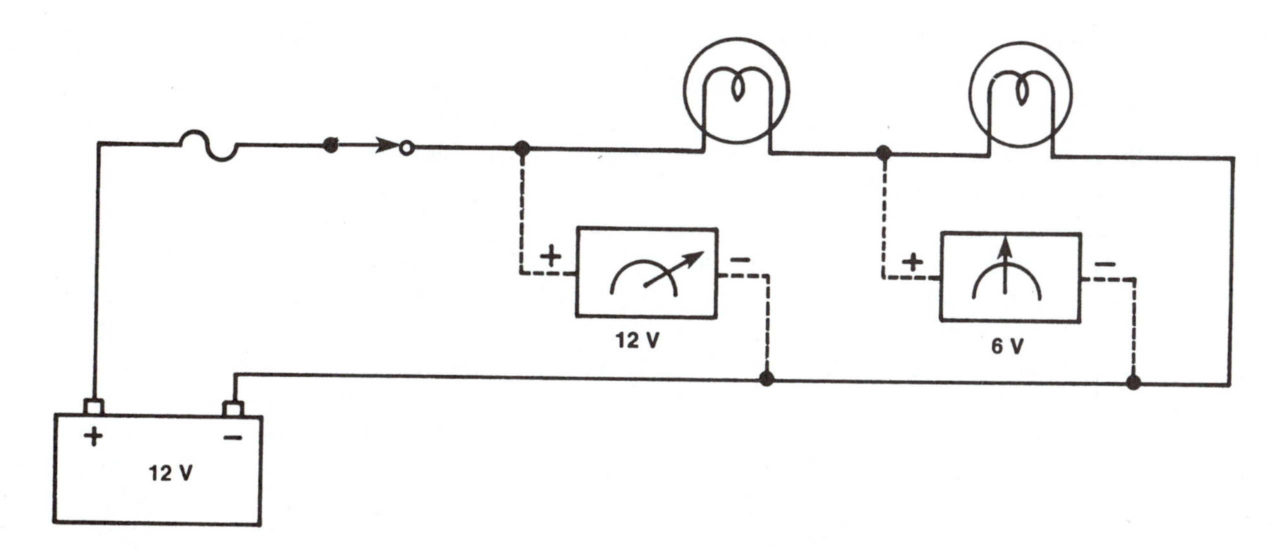

FIGURE 1-25 Voltage drops in series circuit.

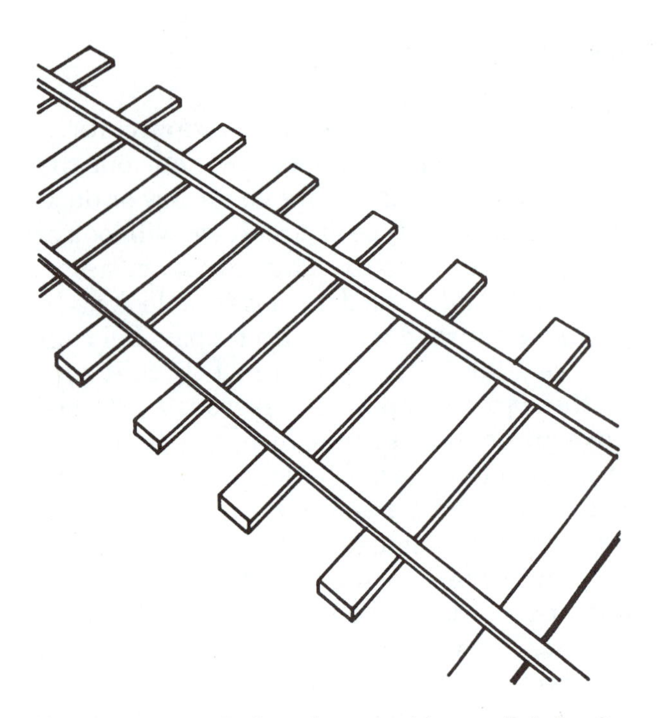

FIGURE 1-26 Railroads resemble parallel circuits.

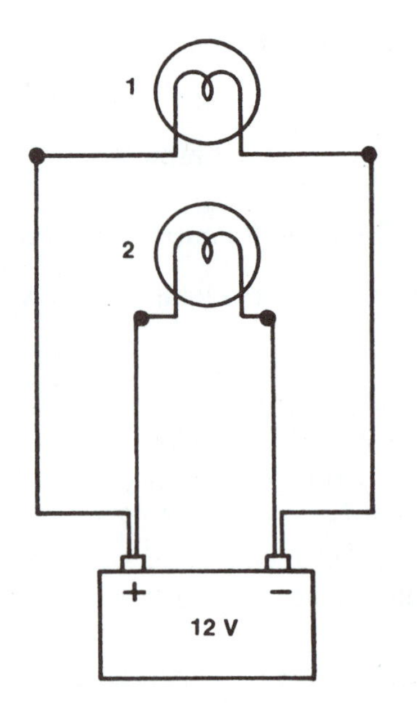

FIGURE 1-28 Two loads wired equal to 1.28.

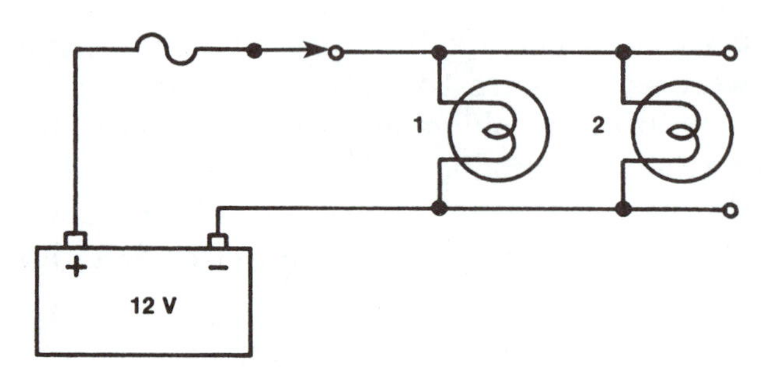

FIGURE 1-27 Two loads wired in parallel.

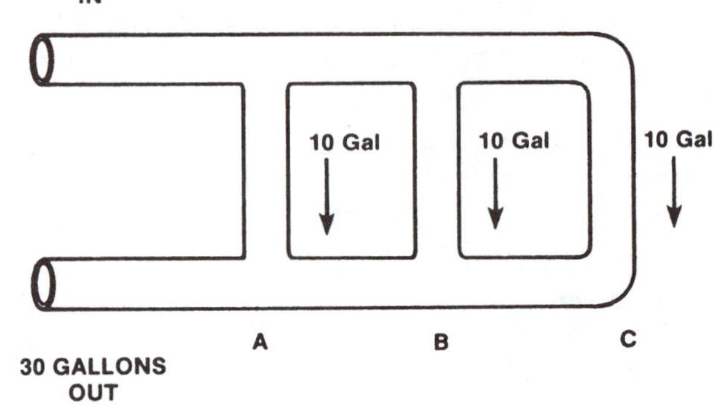

FIGURE 1-29 Water in pipes behaves like current.

ure 1-29 shows this best. Remember, current is like water. It is a quantity. If each leg of the water system has 10 gallons of water flowing in it, with three legs, the total water flow will be 30 gallons. The top main water pipe and the bottom main water pipe will have the 30 gallons flowing with each leg receiving its 10 gallons. How many gallons are flowing at point B? If you said 20 gallons, you are on the right track. How about at point C? Right, 10 gallons.

If we substitute loads in Figure 1-30 for each of the water legs and use Ohm's law and our wattage formula, we will be able to see the main differences between series and parallel circuits. Each leg has 12 volts pushing current through 2 ohms of resistance. Cover up amperage because it is the unknown: 12 volts over 2 ohms equal 6 amps; 12 volts times 6 amps equal 72 watts. Sound familiar? This is the same as our series example. But notice the difference. We have three separate legs to this circuit and if each leg draws 6 amps, the total current flow in the circuit will be 18 (6 + 6 + 6 = 18): 12 volts times 18 amps equal 216 watts of light, heat, or magnetic field produced.

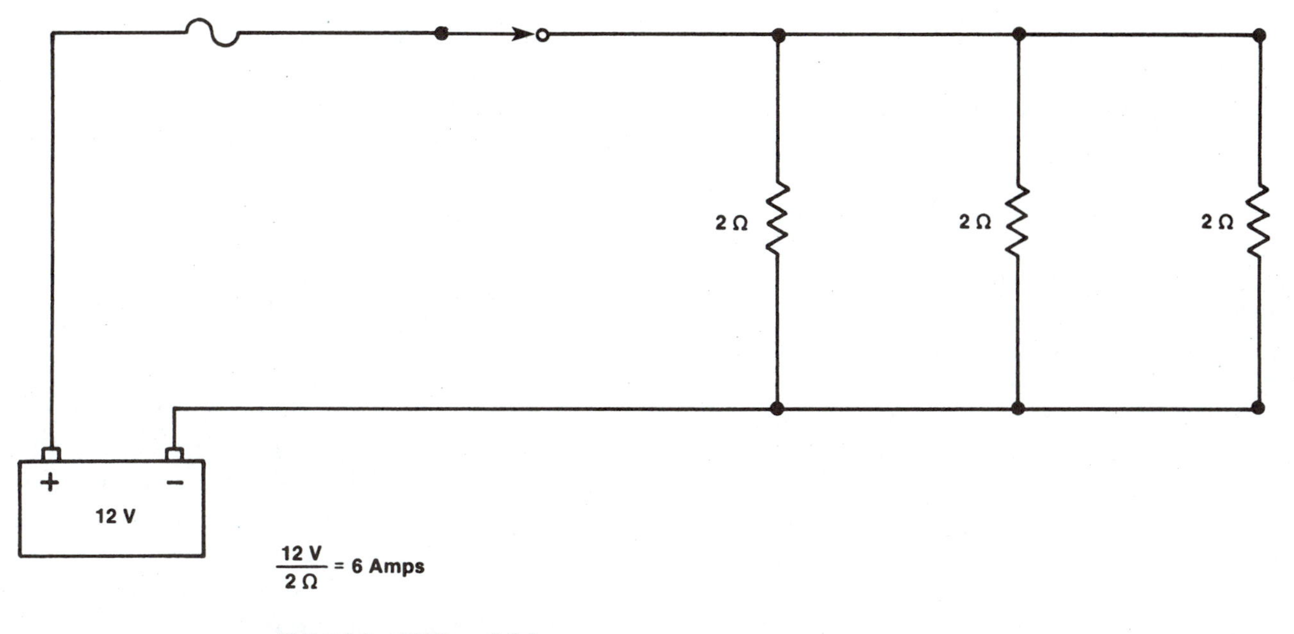

FIGURE 1-30 Figuring wattage in a parallel circuit.

Quite a bit more than the same loads wired in series! This also means that the main B+ and B– conductors must be capable of handling at least 18 amps and a fuse rating of probably 20 amps will be necessary.

II. Procedure for Figuring Current Flow

This example is with three equal resistances. What do you think will happen if the three legs have the resistances shown in Figure 1-31? You should realize first of all that each leg will draw different levels of current based on the actual resistance of the load in each leg. You should also realize that the total will be the total of each leg added together. Let us figure this one together, one leg at a time. The first load of 3 ohms will draw 4 amps of current: 12 volts divided by 3 ohms equal 4 amps. The second load of 4 ohms will draw 3 amps: 12 volts divided by 4 ohms equal 3 amps. The fourth load of 6 ohms will draw 2 amps of current: 12 volts divided by 6 ohms equal 2 amps. If we add up the individual amperages, we will know the total: 4 + 3 + 2 = 9 amps total. The important concept to understand here is that the current through any

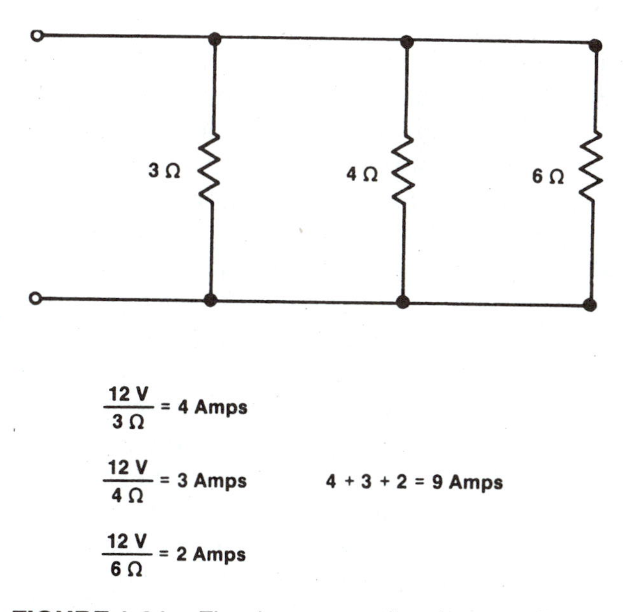

FIGURE 1-31 Figuring current flow in a parallel circuit.

parallel leg is dependent on the resistance of that leg, and the current will always be the total of all legs added together.

There is another way of figuring current flow or resistance within a parallel circuit. We do not have to look at each leg and add them up. We can, instead, figure the total resistance and use Ohm's law to figure the current. The only difficulty is that we cannot just add up the resistances as we did with the series circuit. Adding them together increases the total resistance, thereby reducing the total amperage. We have seen that the total current goes up, not down, so a different approach is necessary.

The formula for figuring out parallel resistance is as follows:

$$\text{Total resistance} = \frac{1}{\frac{1}{R1} + \frac{1}{R2} + \frac{1}{R3}}$$

If we plug in the actual resistance from our previous examples of 3, 4, and 6 ohms and then do the math outlined,

$$\frac{1}{\frac{1}{3} + \frac{1}{4} + \frac{1}{6}} = \frac{1}{\frac{4}{12} + \frac{3}{12} + \frac{2}{12}}$$

$$\frac{1}{\frac{9}{12}} = \frac{1}{\frac{3}{4}} = \frac{1}{.75} = 1.3$$

we come up with a total resistance of 1.3 ohms. Back to Ohm's law to find the unknown: 12 volts divided by 1.33 ohms equal 9 amps of current. Did you notice that we came up with the same results, 9 amps? Notice also that the total resistance of 1.33 ohms is lower than the lowest single resistance. R1 was 2 ohms, the smallest single resistance. Within a parallel circuit the total resistance will always be less than the smallest resistance. Remember, a parallel circuit always draws more current than a series circuit with the same components. More current flow must mean less resistance. Ohm's law proves this. Each time we add a parallel path for current to flow, we increase the total current and therefore decrease the total resistance.

Our experiments in Chapter 3 should help you see this relationship with volt, amp, and ohmmeters.

SERIES-PARALLEL CIRCUITS

The final type of circuit we will examine is the **series-parallel circuit**. You can tell from its name that it shares the characteristics of both series and parallel circuits. To be realistic, we must recognize that virtually all forms of parallel circuits have a series section. Figure 1-32 shows the circuit we have been working with. The loads are in parallel with a single control and a single fuse. The switch and fuse are series wired because there is only one path through them. No alternative or optional path here. Thus, if the switch or fuse were open, the circuit would be dead. The parallel sections are dependent on the series wired switch and fuse for their current path.

Another style of series-parallel is shown in Figure 1-33. Notice that an additional load has been placed in series with the three parallel loads. Each parallel load is independent from the other, but dependent on the series load. A burned out parallel bulb will have no other effect. But a burned out series bulb would shut down the whole circuit.

Your dashboard dimmer circuit is the best example of this type of wiring. A single variable resistance (to be discussed later) is wired in series with the parallel wired dashboard lights. Turning the dimmer increases the resistance in series. What effect do you think this has? Let us think about it. If we increase series resistance, voltage drops or divides and current goes down. This will reduce the available voltage to the parallel section and the lights will go dimmer. Reducing the resistance will brighten up the dash. The series load now has become a control because of its effect on voltage and amperage. Remember, Ohm's law has shown that as resistance goes up, current goes down, and as resistance goes down, current goes up. What about the bulbs in the dash? Why are they all dim? They are all in

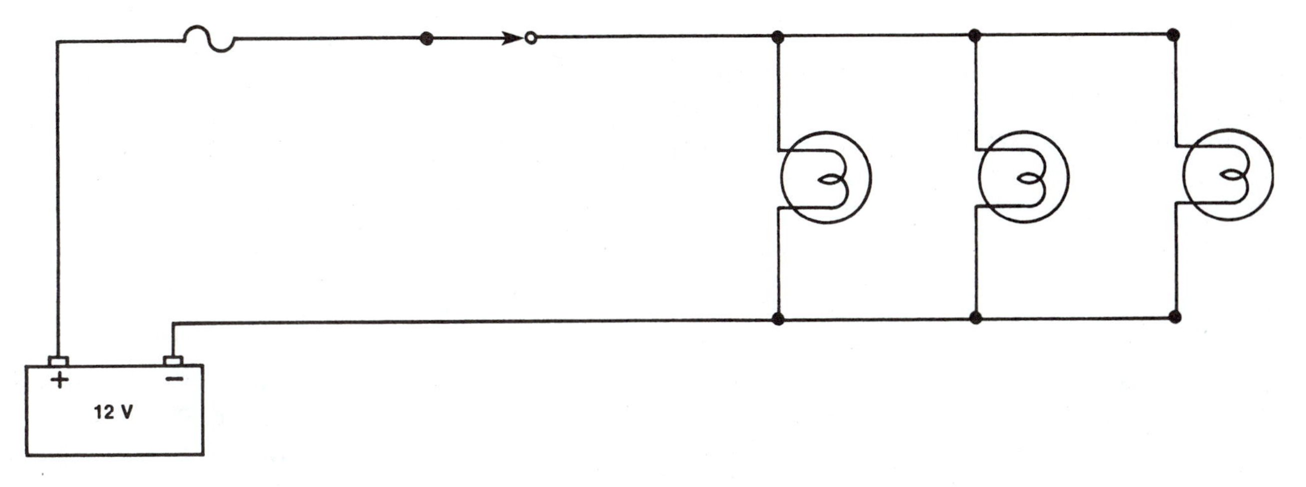

FIGURE 1-32 Light bulbs in parallel with a fuse and switch in series.

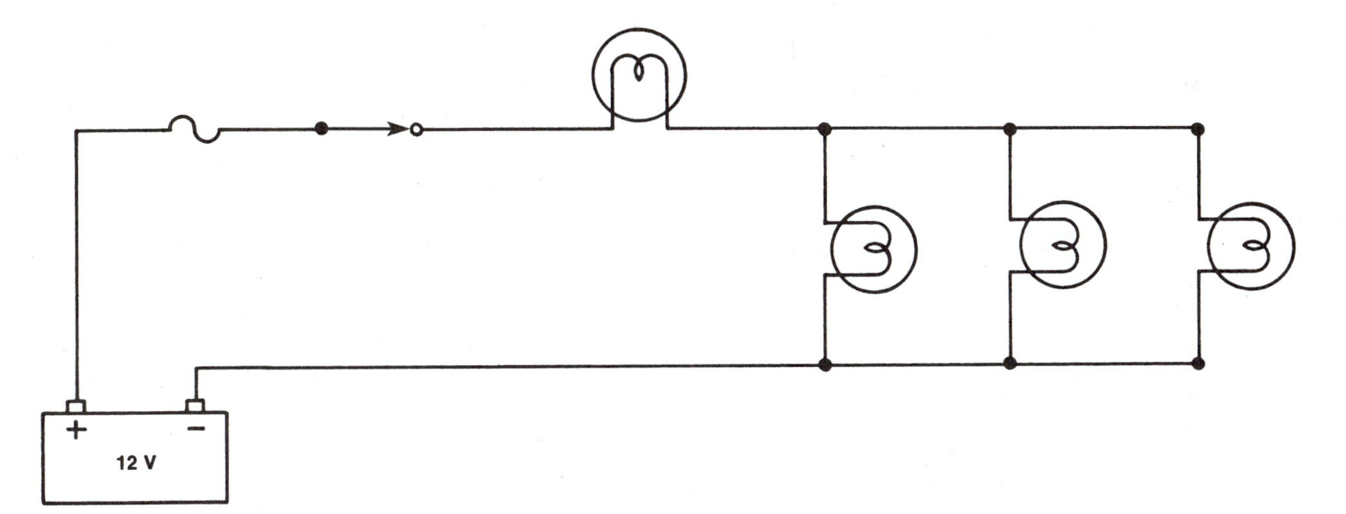

FIGURE 1-33 Series-parallel circuit.

parallel and have the same voltage across them.

The material presented within this chapter is the foundation of the rest of this book and quite possibly the foundation of your electrical career. Do not hesitate to reread it again. Make sure you understand the principles presented before you go on. It is easy to get lost from this point on if your background is incomplete. You should be able to easily answer the questions presented after the chapter summary. In addition, there are some sample problems dealing with Ohm's law for you to work through.

SUMMARY

In this chapter, we have looked at the atom and have seen that the valence orbit determines whether the element is a conductor or an insulator. Voltage was defined as electrical pressure; current, measured in amperage, was defined as moving electrons; and resistance, measured in ohms, as the force or opposition to the flow of electrons. The interrelationship between voltage, amperage, and resistance was expressed in the formula called Ohm's law, while the actual amount of work (defined as light, heat, or magnetic field) is expressed in the formula for figuring wattage. The use of Ohm's law and the wattage formula was then applied to series, parallel, and series-parallel circuits. In addition, we discussed the use of controls (switches or relays) and circuit protectors (fuses, circuit breakers, or fusible links) to make the circuit both functional and safe from overloads. Loads, both designed-in and false, were also discussed.

KEY TERMS

Valence orbit
Conductor
Insulator
Polarity
Current
Ampere
Amp
Ohm's law
Circuit
Short circuit
False loads
Series Circuit
Parallel circuit
Series parallel circuit

CHAPTER CHECK-UP

Multiple Choice

1. Which of the following is not needed for a complete circuit?
   a. source
   b. conductors
   c. controls
   d. load
   e. relay
2. Adding resistance to a series circuit will cause
   a. more current to flow.
   b. less current to flow.
   c. total resistance to go down.
   d. none of the above.
3. Voltage
   a. forces current through the circuit.
   b. resists current flow through the circuit.
   c. is measured in amps.
   d. none of the above.
4. Current is
   a. electricity flowing through the circuit.
   b. resistance in the circuit.
   c. voltage drop in the circuit.
   d. both b and c.
5. A short circuit could cause a
   a. fuse to blow.
   b. discharged battery.
   c. circuit breaker to blow.
   d. all of the above.

6. Voltage drop is the
   a. voltage used to push current through a resistance.
   b. current used to go through a resistance.
   c. electrons used to go through a resistance.
   d. none of the above.
7. Current is measured in
   a. ohms.
   b. volts.
   c. amps.
   d. none of the above.
8. Resistance is measured in
   a. volts.
   b. amps.
   c. ohms.
   d. none of the above.
9. An example of a load would be a
   a. motor.
   b. battery.
   c. fuse.
   d. diode.
10. Which of the following is an example of a power source?
    a. battery
    b. starter
    c. alternator
    d. a and c
11. Which of the following would be classified as a circuit protector?
    a. fuse
    b. circuit breaker
    c. fusible link
    d. all of the above

True or False

12. A closed circuit is a completed path for the flow of electricity starting from a source and returning to the source.
13. If a circuit is said to be open, current will be flowing through it.
14. In a parallel circuit, there is more than one path for electricity to follow.
15. The total resistance of all the parallel loads is always less than the smallest individual resistance.
16. Series resistances always add up.

17. To find the resistance of a circuit, divide the amperage by the voltage: A/V.
18. The current in a parallel circuit will divide among the resistances.
19. The voltage will divide among the resistances within a series a circuit.
20. The total amperage in a parallel circuit will be the sums of the individual circuits.

Matching

Match the definition on the right with the numbered word on the left.

21. series circuit
22. parallel circuit
23. conductor
24. insulator
25. amperage
26. voltage
27. relay
28. resistance
29. ohm
30. wattage

a. electrical pressure
b. resists flow of electrons
c. one path
d. unit of resistance
e. passes electrons easily
f. uses low current to control high current
g. more than one path
h. quantity of electrons
i. volts time amps

CHAPTER 

Meter Fundamentals

OBJECTIVES

At the completion of this chapter, you should be able to:

- test system voltage by connecting an analog voltmeter, observing correct meter polarity.
- determine the correct range for a voltmeter reading.
- test system amperage by connecting an analog ammeter, observing correct meter polarity.
- determine the correct range for the circuit being tested.
- interpret a scope trace and apply correct voltages.
- use a digital multimeter and correctly identify voltage and resistance values.

INTRODUCTION

A frequently heard comment from new automotive electrical students is "give me something I can see or hold and I can repair it. The trouble with electricity is that you can't see it." Sound familiar? This statement emphasizes a common misconception—that you cannot see electricity.

Using your test tools correctly will allow you to look inside a circuit and actually "see" the flow of electricity. The only difference is that we cannot directly see the electricity flow. Instead, we must rely on some instrument or meter to be our eyes, Figure 2-1. Knowing what you are looking at and being able to interpret various meter types is the scope of this chapter. We will look at four different meters:

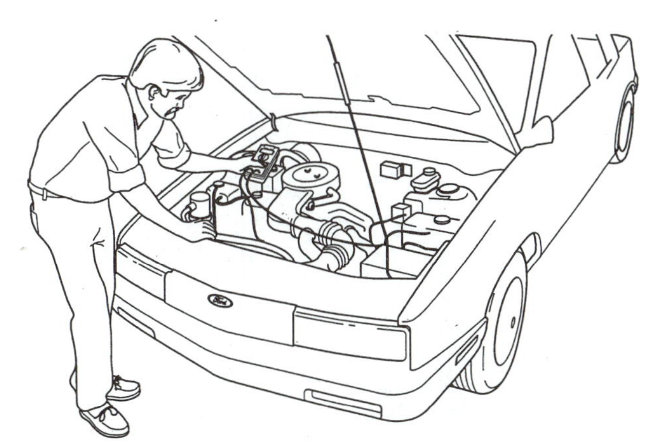

FIGURE 2-1 Electricity can be viewed through the use of a meter. (Courtesy of Ford Motor Company)

1. Voltmeters
2. Ammeters
3. Ohmmeters
4. Oscilloscopes (a type of voltmeter)

We will also look at digital multi-meters (DVOM) and briefly explain how we read them.

VOLTMETERS

First, let's look at the **voltmeter**. Probably the most used and useful meter in the shop is the voltmeter. Its purpose is to read voltage, which you know is the pressure behind the flow of electrons. Its simplest use is to measure system pressure or voltage, which is usually 10–14 volts. Let's look at this common voltmeter and let's analyze the front.The meter scale starts at zero on the left and goes up to 10 volts full scale on the bottom right end of the meter, Figure 2-2. This is called the **range** of the meter and will be important for both accuracy and for the protection of the meter. The range is the number of volts from lowest to highest that the meter can safely measure. Sometimes, a meter will have only one range, but most of the time multiple scales are available, Figure 2-3. Notice that this meter has another range or scale starting again at zero on the left and going up on the right to 20

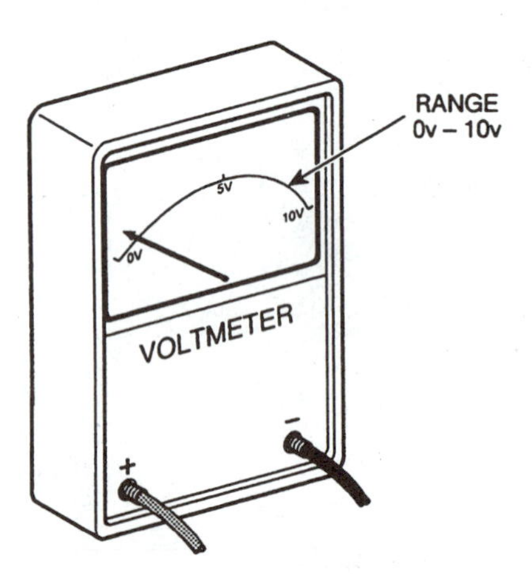

FIGURE 2-2 Single range voltmeter. (Courtesy of Ford Motor Company)

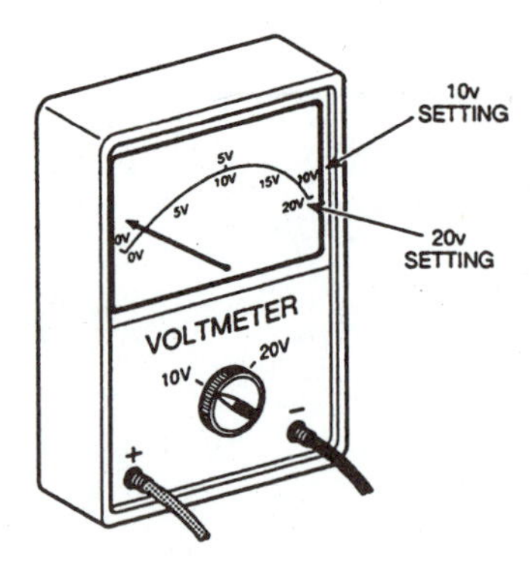

FIGURE 2-3 Multiple scale voltmeter; 0–10 volts and 0–20 volts. (Courtesy of Ford Motor Company)

volts. This is the second range and the highest voltage that this meter can safely measure: 20 volts. This meter is just an example. Most meters have more scales and can measure voltages much higher than 20 volts. With two different ranges available, we must think about what we wish to measure and set the meter in advance to the range we wish to use. This is done with the switch on the front of the meter. Note that it has the same numbers as the meter's full scale maximum voltages: 10 volts and 20 volts. It is extremely important that we always think before we hook up the meter and make sure that the voltage to be measured will not exceed the full scale reading. If, for instance, we were to attempt to measure a fully charged battery and had the meter set for a full scale maximum of 10 volts, we might damage the meter, Figure 2-4. To insure that the meter will not be damaged, always pick a range that is higher than the suspected system voltage. If you are uncertain of the voltage, set the meter on the highest range and, once you are sure that the next lowest range will not be exceeded, change down to the next lowest range. Ideally, you should try to measure with the range that gives you the most meter movement (lowest scale). This will insure that you will be reading on the most accurate scale and will not damage the meter movement as long as you observe meter and circuit polarity.

The two leads coming out of the front of the meter are color coded and labeled red for positive and black for negative, Figure 2-5.

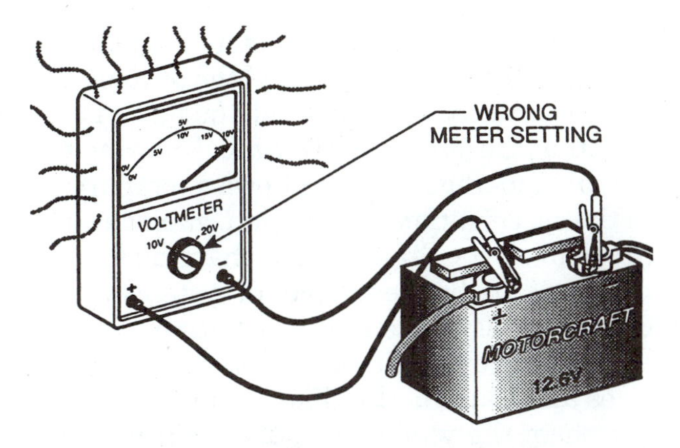

FIGURE 2-4 The battery voltage of 12.6 volts is 2.6 volts higher than the meter can accept, if it is set on the 10-volt range. (Courtesy of Ford Motor Company)

This is the most common color coding for meters found in the automotive industry. Remember the setting that allowed the meter to move up the scale within its normal range. Hooking the meter up with correct polarity will also allow the meter to move up the scale. Polarity can simply be described as the direction in which the electrons are flowing or, in the case of voltage, the direction of the pressure. This is significant especially if the matter can only measure in one direction. Our meter is capable of measuring this pressure only if the direction of the pressure and the direction of the meter match. Polarity is this matching of the meter to the circuit. All circuits have a negative and a positive. We must make sure that the positive and nega-

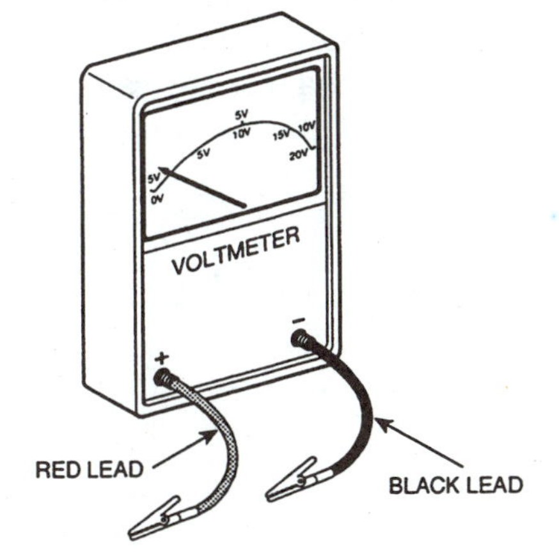

FIGURE 2-5 Voltmeter showing the two color coded leads. (Courtesy of Ford Motor Company)

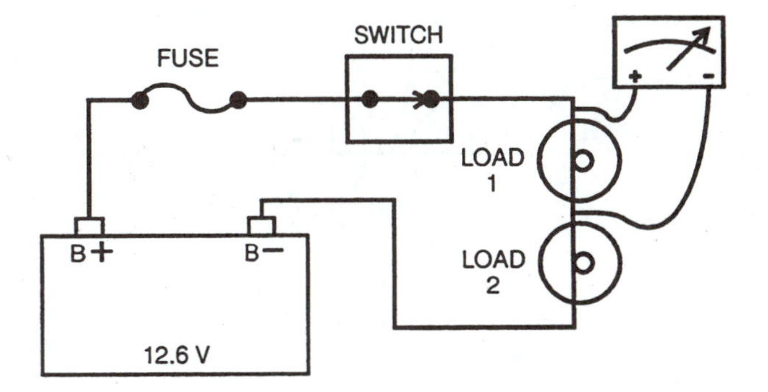

FIGURE 2-6 Circuit and meter polarity must always match. (Courtesy of Ford Motor Company)

tive of the meter match the positive and negative of the circuit as shown here, Figure 2-6. Sometimes in a circuit we wish to measure the pressure that has been lost. This is called **voltage drop** and it is in its measurement that we can get confused and quite possibly burn out a meter. Voltage drop is the voltage which is lost or reduced by pushing current through resistance. Voltage drop only occurs if current is flowing through resistance.

Refer to Figure 2-7 and notice that this simple series circuit has three loads. NOTE: To measure the voltage or pressure available from the battery, we would attach the positive meter lead to point A and the negative lead to point D in the circuit. Point A is the most positive part of the circuit, while point D is the most negative.

It makes sense that a single point is either positive or negative. However, to measure the voltage lost or dropped across each load, the situation becomes a little more complicated.

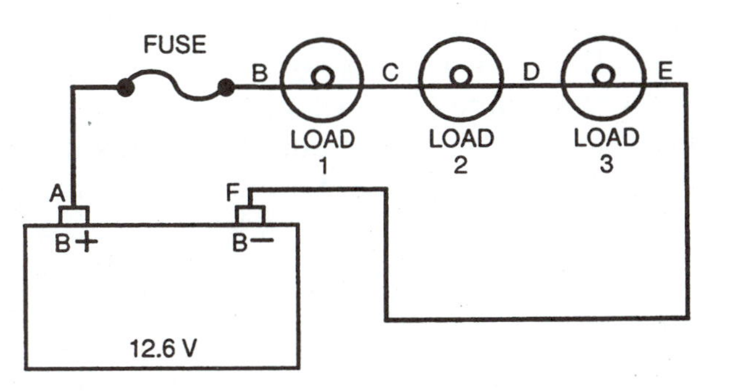

FIGURE 2-7 A point in the circuit can be either positive or negative, depending on what is being measured. (Courtesy of Ford Motor Company)

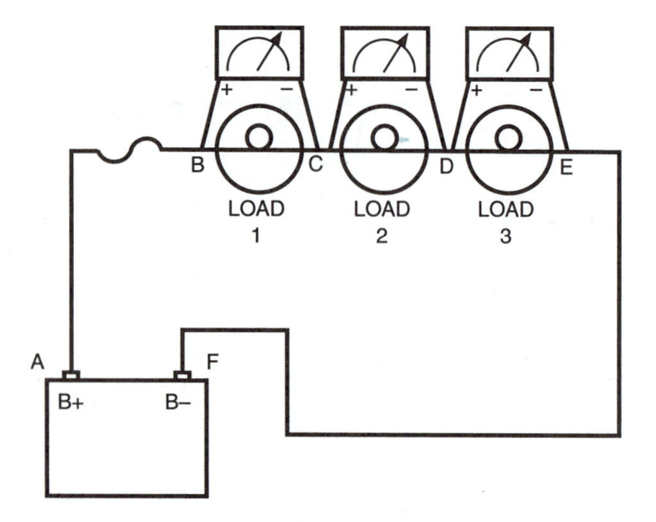

FIGURE 2-8 A point in the circuit can be either positive or negative.

A point in the circuit can be either positive or negative, depending on what we are measuring, Figure 2-8. If we want to know the voltage drop across load 1, we would connect the positive meter lead to point A and the negative lead to point B. The meter will now read the difference in pressure between these two points. This is the voltage drop across load 1. However, to measure the voltage drop across load 2, the polarity of point B changes from negative to positive. The negative connection point now becomes point C. The reason for this is that point B was the most negative point for load 1, at the same time that it was the most positive point for load 2. Keep this in mind and figure out how to measure the voltage drop across load 3. If you mentally connected your meter's positive lead to point C and your negative lead to point D, you are on the right track.

Always note both polarity and range before you use a voltmeter and the meter will be protected, thus insuring its accuracy for many years.

AMMETERS

The next meter we will discuss is the **ammeter**. Amperage or current that is being pushed through resistance will do work. The more work being done, the more current the circuit will draw. Frequently, the automotive technician needs to know the amount of current that the circuit has flowing. Knowing the actual amperage flowing allows a comparison to the specification and the determination if a problem exists. Starting and charging system diagnosis are both examples of systems where current measurement is of the utmost importance. Ammeters, like voltmeters, come in all sizes and shapes. Also, like voltmeters, if you use them the wrong way, they will be damaged. Correct care and use will insure the long life of a quality ammeter. Ammeters come in two varieties. A flow through or direct reading ammeter through which the total current actually flows, Figure 2-9. The second type, and probably more common style, is the inductive pick-up ammeter, Figure 2-10. An inductive pick-up ammeter does not have the total current flowing through it. Let's look at each type and discuss their connections and controls.

Direct Reading Ammeters

Refer again to Figure 2-9 and notice that the direct reading ammeter has heavy cables that are color coded and labeled just like the voltmeter: red for positive and black for negative. This should give you a clue to the connections. This meter is polarity sensitive just like the voltmeter. If you reverse the polarity, the meter's needle will be forced in the wrong direction and might damage the meter movement. Notice the control knob and meter scales available. As was the case with the voltmeter, you must pick a range that is higher

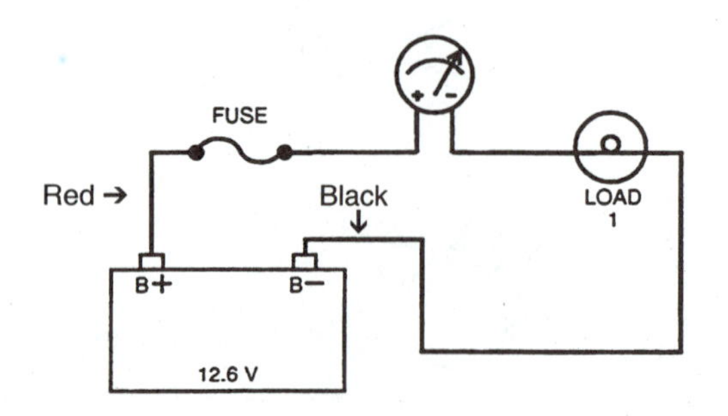

FIGURE 2-9 Direct reading ammeter. (Courtesy of Ford Motor Company)

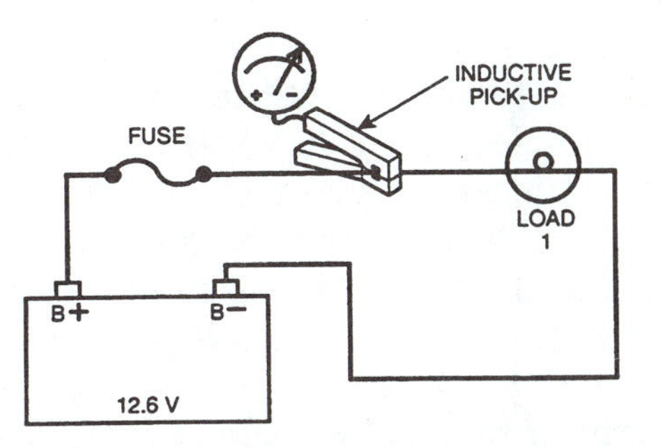

FIGURE 2-10 Inductive pick-up ammeter. (Courtesy of Ford Motor Company)

FIGURE 2-11 Never place an ammeter across a positive and negative potential. (Courtesy of Ford Motor Company)

than the maximum current expected or the meter needle will be damaged. When you "peg" the needle against the end stop, the meter's movement is usually damaged and accuracy becomes questionable. Up to this point, everything has been much like the voltmeter. There is an important difference, however.

NOTE: As we mentioned earlier, the voltmeter has a very high input resistance. This is frequently referred to as the **input impedance** of the meter. The input impedance of a meter is a measurement of the input resistance of the meter. Most modern computer circuits require a minimum input impedance of ten million ohms (10,000,000). This is what makes it a voltmeter and allows it to sample the pressure without actually having current flowing through it. A tire pressure gauge acts like a voltmeter in that it will measure the pressure without having to have all of the air flowing through it. The input impedance limits the current flow through a voltmeter. An ammeter, however, is designed to measure total current flow and will not limit the amount of current. It is more like a counter that will count the electrons as they flow through the meter movement. For this reason, they are designed with virtually no input resistance. No resistance is another way of saying a dead short circuit. If an ammeter is placed across a source of power, like a battery, the meter will act just like a dead short and possibly burn out because of the tremendous current which will flow through 0 ohms of resistance, Figure 2-11.

The correct method for connecting the ammeter into a three-load series circuit is to break open the circuit and wire the meter in series with the loads, Figure 2-12. This insures that the current flow will be determined by the resistance of the loads rather than by the resistance of the meter. Common methods for the use of this type of meter involve removal of a battery cable and series-wiring it between the battery post and cable, or replacement of a fuse with the meter. Remember, at no time should the meter be placed across the load or into a circuit that has no resistance. It is always connected in series with the circuit's resistance controlling current.

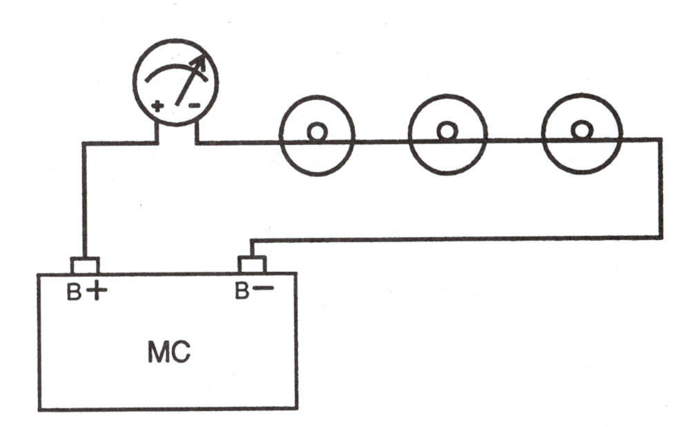

FIGURE 2-12 An ammeter correctly wired into a three-load series circuit. (Courtesy of Ford Motor Company)

Inductive Pick-Up Ammeters

A variation of the direct reading ammeter is the inductive pick-up design. The meter face and range setting is usually the same, and most of the meters are polarity sensitive. The difference is in the connection. This style of meter senses the amount of current flowing through a wire by measuring the magnetic field around the conductor, Figure 2-13. Any time current flows through a conductor, a magnetic field will be present. An inductive pick-up ammeter senses the amount of current by measuring the strength of the magnetic field surrounding the wire. By sensing the magnetic field around the wire, the meter is able to determine indirectly the amount of current flowing. Notice the large circle clamp on the end of the cable. This is the inductive pickup. It is usually a very sensitive electronic component and cannot be handled roughly. Dropping it or hitting it against a bench will usually impair the accuracy of the meter. An advantage to the inductive style is its simplicity of connection. No need to worry about connecting it in series, because current does not flow through it. This helps to prevent "pegging" the needle and destroying the meter movement. Don't forget, however, that many of these meters are still polarity sensitive. An additional advantage of an inductive ammeter is the ability to take the measurements without having to disconnect the vehicle's battery. Most vehicles today have multiple memory circuits that are powered off the battery. Disconnecting the battery to obtain an amp reading would allow the processor to lose its stored memory. Radio pre-sets, multiple driver's seat and mirror positions, stored diagnostic codes, in addition to basic vehicle operations information could be lost. Customers do not like to have to re-program their radio pre-set memories, so if at all possible, do not disconnect the battery. Use an inductive ammeter where possible. A non-polarity sensitive meter with equal amperage reading on either side of a center zero allows this meter to measure amperage with either a positive or negative polarity, Figure 2-14. This meter can measure amperage in both directions. It is especially helpful when doing starting and charging testing, because it can measure and tell the technician whether current is going into or coming out of the battery. Its use will be explained in the starting and charging section of this text. Keep in mind that this style of meter is still sensitive to overrange. If you are on the 100-amp scale and crank the vehicle over, the approximate 200 amps flowing will peg the needle and probably destroy the meter move-

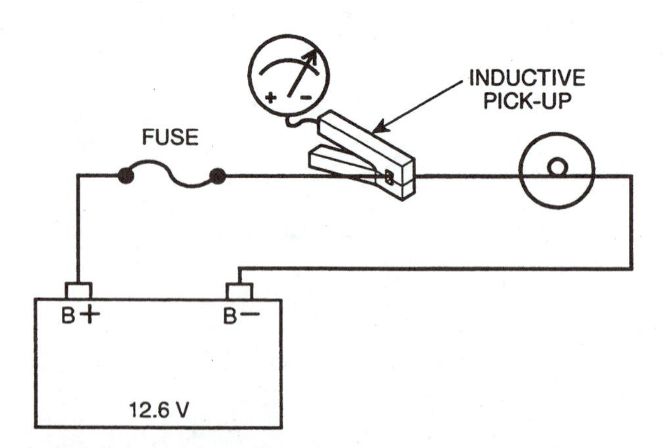

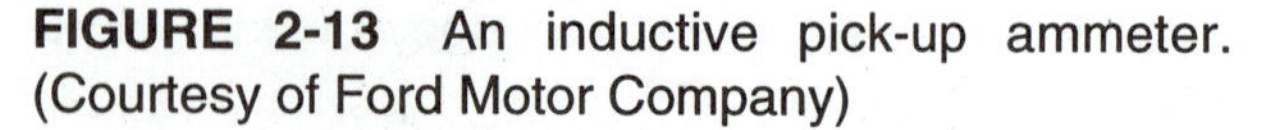

FIGURE 2-13 An inductive pick-up ammeter. (Courtesy of Ford Motor Company)

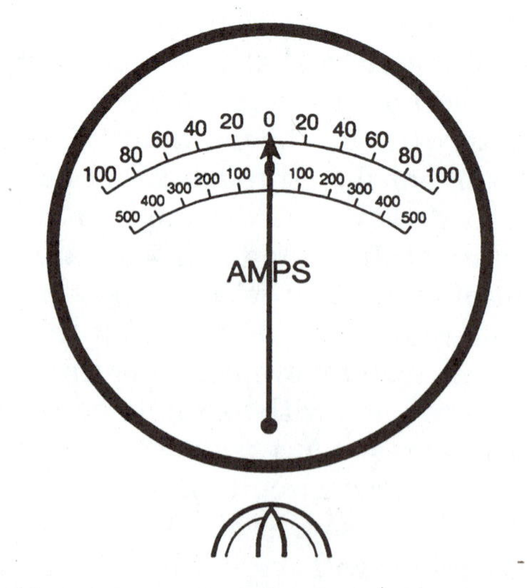

FIGURE 2-14 A non-polarity sensitive meter measures amperage with positive or negative polarity. (Courtesy of Ford Motor Company)

ment. Most quality meters can be repaired; however, you will be without its use during the costly repair. Remembering both polarity and range will prevent the need of most repairs.

OHMMETERS

The next meter we will discuss is the **ohmmeter**. This meter's use is slightly different from the voltmeter and ammeter because the circuit being tested must not be live, Figure 2-15.

The component or section of a circuit being tested must be isolated or disconnected from the power source. Ohmmeters have their own source of power, usually from an internal battery. If the component or circuit being tested has power from the vehicle's battery, the meter movement may be damaged.

Analog ohmmeter scales are usually backwards with zero being on the right end of the meter face and the higher numbers being on the left, Figure 2-16. Most analog ohmmeters are numbered in this manner, because actually the meter is measuring how much current the internal battery has been able to push through the component being tested. The more current that flows, the farther to the right the meter's needle will move. If you remember our discussion of Ohm's law, you should realize that more current converts over to less resistance. So as more current flows,

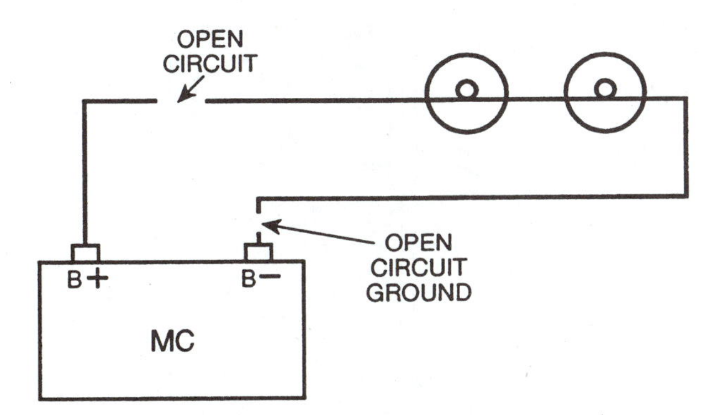

FIGURE 2-15 Component or circuit of an ohmmeter must be isolated from power source before connecting. (Courtesy of Ford Motor Company)

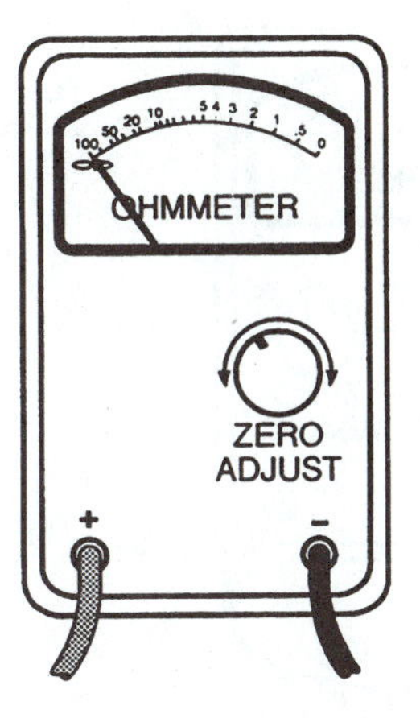

FIGURE 2-16 Analog ohmmeter scales are usually backwards from right to left. (Courtesy of Ford Motor Company)

the meter needle moves to the right where the meter face has been calibrated to read lower resistance. Maximum current flows when there is no resistance, hence 0 ohms is all the way on the right of the scale. No current flows when we have an open circuit (maximum resistance), so the left of the scale shows an ∞, which is the symbol for **infinity**. Infinity is for all practical purposes an open circuit, or more resistance than the meter can measure.

NOTE: Infinity on one scale might be a measurable resistance on a higher scale. For example, if the meter's scale is set to 1,000 ohms, and the component being measured has a 5,000-ohms resistance, the meter will show infinity. However, if the meter has a 10,000-ohm scale, the 5000 ohms can be read directly. Remember that infinity is more resistance than the meter can measure.

Ohmmeters require their own power source, usually batteries, inside the meter. The power source is needed because the meter will have to push current through the resistance being measured. Because of this, many analog meters have to be "zeroed" before they are used, Figure 2-17.

Zeroing the meter involves compensating for the condition of the batteries. As they age and lose their power, the meter is readjusted before it is used. After connecting the meter leads together, the ohms calibrate or zero ad-

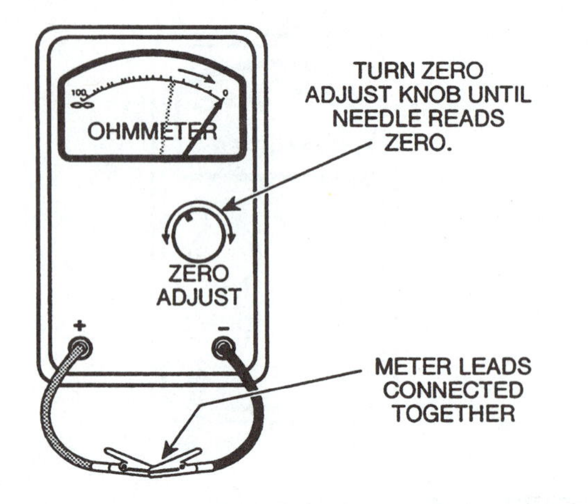

FIGURE 2-17 Analog ohmmeters must be zeroed prior to use. (Courtesy of Ford Motor Company)

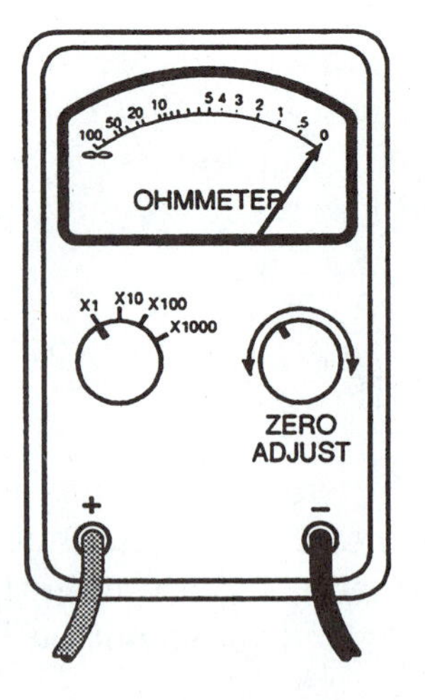

FIGURE 2-18 Most analog ohmmeters use a multiplier to figure higher resistances. (Courtesy of Ford Motor Company)

just knob is turned until the meter needle reads zero ohms. We are now able to accurately measure anything placed between the leads, and any meter reading will be the actual resistance. Notice again the scales that allow us to measure different components accurately. As is the case with other meters, we try to have the meter range set to approximately the resistance being measured. But unlike the ammeter or voltmeter, overranging or pegging the needle and damaging the meter movement is impossible if the circuit or component being measured has been isolated from power.

If the resistance is unknown, start on the lowest scale and work your way up to higher scales until a reading is obtained. This is especially important if you have an infinity reading. Infinity on the lowest scale can be around 500 ohms, and yet the component might be in the 10,000-ohm range. By changing the range to a higher scale, we would be able to read the actual resistance rather than the infinity, Figure 2-18.

The multi position switch on the meter front shows four ranges labeled R × 1, R × 10, R × 100, and R × 1K. Each meter reading obtained must be multiplied by the number to obtain the actual resistance. R stands for the reading of resistance, × means times, and the number that follows (1, 10, 100, 1 k) is the multiplier. Let's try some examples. The leads are connected together and the zero adjust is turned to zero on the meter.

Connect an unknown resistance between the leads and obtain a 22 R value with the meter on the R × 1 position: 22 times 1 equals 22; this then is 22 ohms. Figure 2-19. Here is another example.

After zeroing the meter, we obtain a 65 R value on the R × 100 scale. This reading is actually 6,500 ohms: 65 × 100 = 6,500 ohms, Figure 2-20.

How about an R value of 40 on the R × 1 k scale. If you said 40,000 ohms, you were correct. K is the abbreviation for kilo, or thousand. It is just like saying × 1000. 40 × 1000 = 40,000 ohms, Figure 2-21.

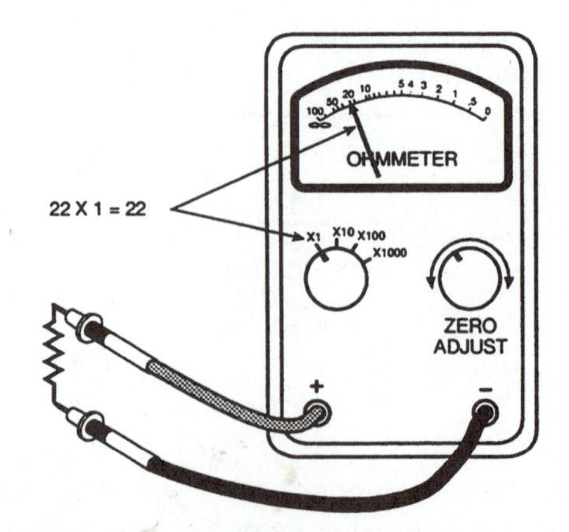

FIGURE 2-19 (Courtesy of Ford Motor Company)

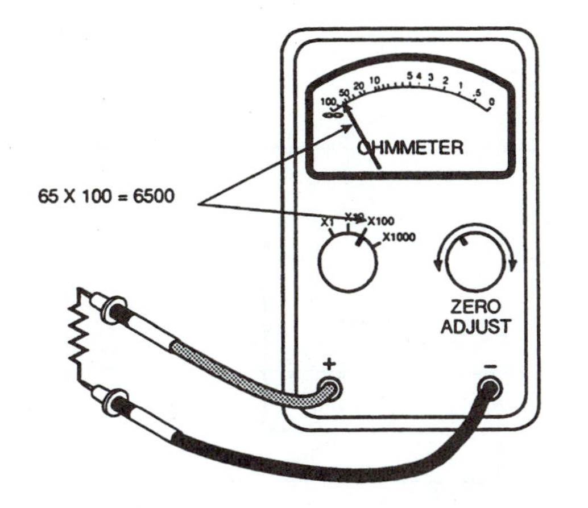

FIGURE 2-20 (Courtesy of Ford Motor Company)

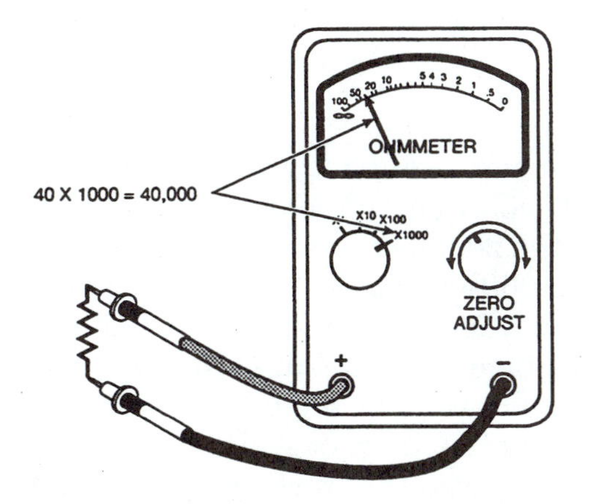

FIGURE 2-21 (Courtesy of Ford Motor Company)

Ohmmeters are useful to the technicians if they can compare the resistance found to some specification. These meters can also be useful when looking for a short circuit (0 ohms) or an open circuit (infinity). Their use is important to technicians. Practicing with one will allow you to gain the proficiency necessary. Always keep in mind the fact that the meter can only be used if the component or circuit being tested is isolated from power. The meter can be damaged by another power source or at least the reading will be wrong. Also, don't forget that most analog ohmmeters need to be zeroed. Some need to be zeroed each time you change scales. Keeping fresh batteries inside it will assure you of accurate readings.

OSCILLOSCOPES

The automotive **oscilloscope** is the next test instrument we will discuss. Many technicians do not recognize it as a meter, but it is a device to measure electricity, so we have included it in this section. The scope, as it is commonly called, is a major piece of test equipment found in many repair shops. It has been used for years to help diagnose ignition and charging problems. Additional uses for a lab type scope are being found every day with computerized circuits. We will cover the basics of how to read a scope and leave the diagnostic procedures until we cover different vehicle systems testing.

The scope is basically a voltmeter that has two very important advantages over a regular voltmeter: the ability to react extremely fast and the ability to display voltage changes over a period of time, Figure 2-22.

Some technicians refer to the scope as a voltmeter with a clock. The scope will sequence voltage changes over a period of time and display them for review and analysis. If we realize that a four-cylinder engine's ignition system might have to produce over 100 sparks per second at highway speeds, we can appreciate that no voltmeter needle, much less our eyes, could follow this many voltage changes. The scope will allow us to follow

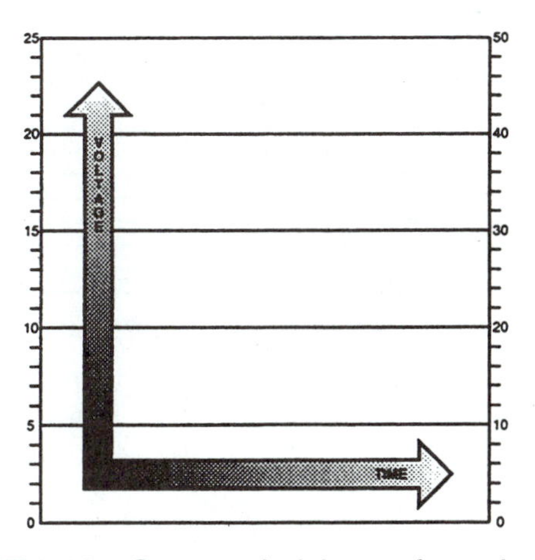

FIGURE 2-22 Some technicians refer to the scope as a voltmeter with a clock. (Courtesy of Ford Motor Company)

these 100 sparks per second. Another example of the speed necessary on today's vehicle systems is in crankshaft sensors which produce up to 35 pulses per engine rotation. At 2000 rpm (highway speeds) that equals 70,000 signals per minute or over 1100 signals per second! The only device that we have available to us is the scope, because it can react to voltage changes that occur in less than a 0.001 second (1 ms) and display them. This makes it a perfect diagnostic tool for high-speed voltage changes.

Refer to Figure 2-23: the screen, or CRT (cathode ray tube), is where all the important information will be displayed. As is the case with most meters, we must set the range. This scope has four ranges: 0 to 40 volts; 0 to 400 volts; 0 to 20,000 volts; and 0 to 40,000 volts. Notice the lines on the screen. This is called the radicule. Think of it as the numbers on a meter face. Find the zero line. Anything above this line will have a positive polarity and anything below will have a negative polarity. Each line above will represent a specific voltage, depending on the range set. The first line

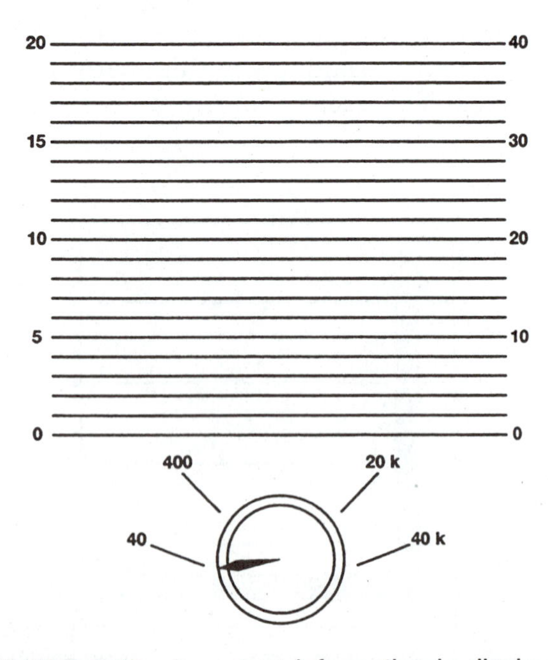

FIGURE 2-23 Important information is displayed on the oscilloscope screen. (Courtesy of Ford Motor Company)

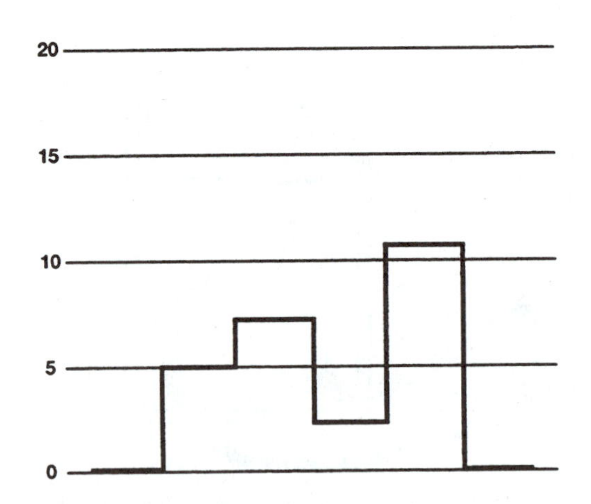

FIGURE 2-24 The scope will display voltage changes from left to right (just like reading a book). (Courtesy of Ford Motor Company)

above the zero will be 2 volts on the 40-volt scale, 10 volts on the 400-volt scale, 1,000 volts on the 20,000-volt scale, and 2,000 volts on the 40,000-volt scale.

Figure 2-24 shows a visual readout of the scope. Voltage lines or traces that are on the left of the screen occurred before those on the right. Notice the range or scale setting: 20,000 volts. Notice also the lines appearing on the CRT show 5 voltage changes that we can follow and sequence. First, the scope detected zero voltage, then a change from 0 to 5,000 volts. The 5,000 volts was maintained until a change occurred up to 7,000 volts, then a change down to 3,000 volts, then a change up to 11,000 volts until, finally, a change back down to zero. The scope has allowed us to note five voltage changes over a period of time much faster than any ordinary voltmeter could have reacted. Most oscilloscopes require one or sometimes two connections to the system being tested. Many utilize inductive pickups just like the ammeter, Figure 2-25. We will be more specific as to the connections when we test each different system.

Scopes are polarity sensitive just like other voltmeters. However, the polarity is important only because it gives the scope "trace," or line, reference, as to positive and negative. You need not be concerned about pegging or

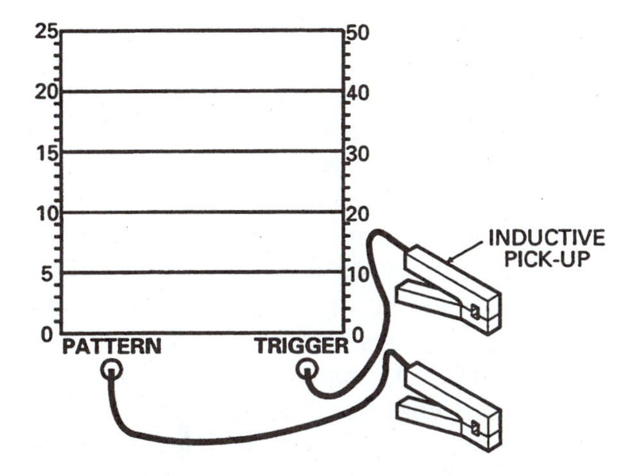

FIGURE 2-25 Scope utilizing multiple inductive pick-up. (Courtesy of Ford Motor Company)

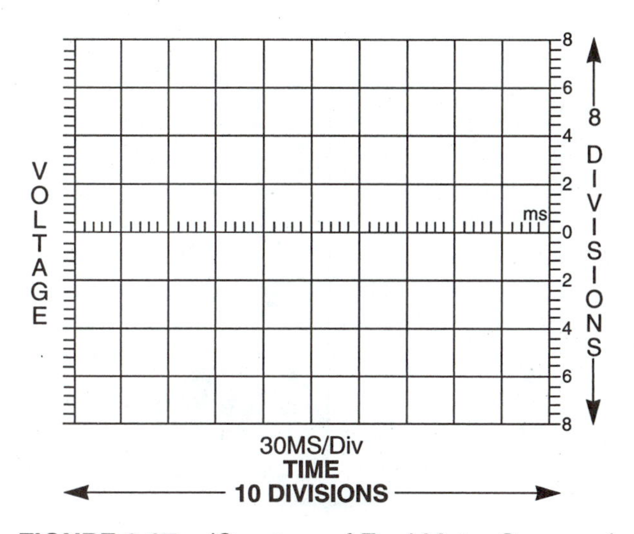

FIGURE 2-27 (Courtesy of Ford Motor Company)

burning out the scope by reversing the polarity or overranging it. It is almost impossible to burn out. This is one reason why it is used extensively as a diagnostic tool. Recent years have seen increased use of a lab type scope in the diagnosis of computer systems. The only difference between a lab scope and an ignition scope is the ranges available and the ability to be able to very accurately measure the time. Using a lab scope will involve one additional bit of information usually displayed as volts per division and time per division. Simplified, a box is drawn on the screen that signifies a specific period of time or a specific amount of voltage, Figure 2-26.

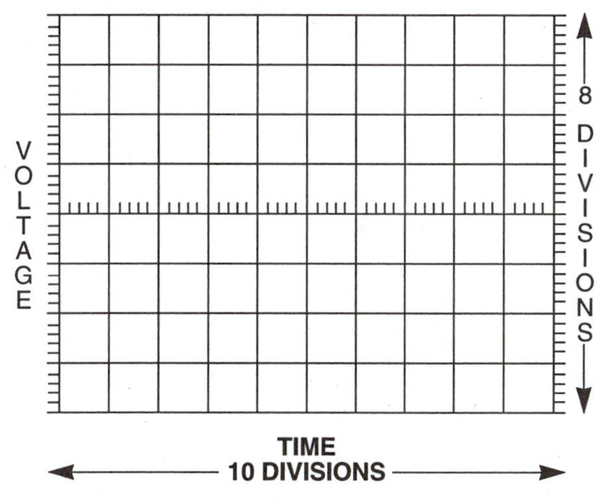

FIGURE 2-26 (Courtesy of Ford Motor Company)

The setting on the scope will now determine how long the time is from one edge of the division to the other edge (left to right) and how many volts from the bottom edge to the top edge. Let's use an example.

Refer to Figure 2-27. If the scope were set to .030 seconds per division and 2 volts per division, the time across the screen would actually be .300 seconds or 300 **milliseconds** (mS). A millisecond (mS) is a thousandth of a second. Ten divisions × 30 mS = 300 mS. This is sometimes referred to as the sweep rate. The eight divisions available up and down would result in a maximum voltage of 16 volts. Eight divisions × 2 volts = 16 volts Some of the lab scopes available today can measure 0–1 volt full screen with as little as 1 mS per division.

DIGITAL MULTIMETERS

The computer-operated systems on a vehicle today opened up the use in the early eighties, of the last meter that we will discuss, the **digital multimeter**, **DVOM** or **DMM**, Figure 2-28. First of all, realize that multimeters have been around for quite some time, even in automotive repair facilities. It is common to find a meter that measures voltage, resis-

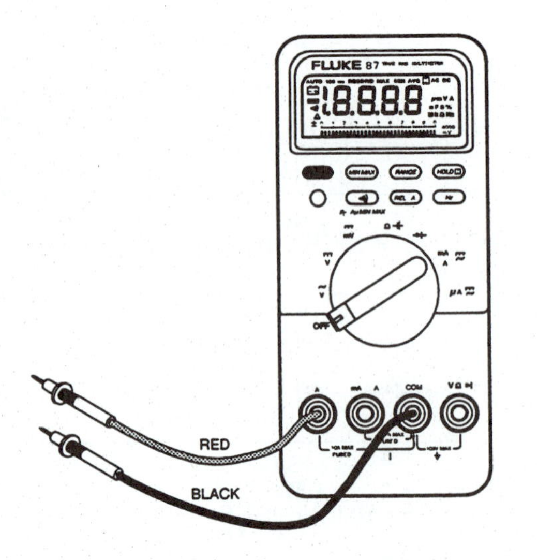

FIGURE 2-28 DVOM, DMM or digital multimeter. (Courtesy of Ford Motor Company).

tance, and low amounts of current in use rather than three separate meters. It should be realized, however, that typically this type of meter does not give technicians enough of a range in current measurement to be the only ammeter they use. Typically, multimeters have less than 1-amp capacity, with some going as high as 10 amps. Obviously, when testing starting and charging systems in excess of 100 amps, having only a 1-amp capacity meter is not adequate. Combination meters for voltage and resistance can be found in most shops.

The modern computer systems on vehicles today require the use of a DVOM, because they have integrated circuits that operate on very small amounts of current. These circuits can easily be "loaded down" by a meter with insufficient input impedance. Loading down is a way of saying changing the circuit. The old **analog meters** cannot do the job in the field. Enter the digital meter for automotive application. Its advantages are its very high input resistance: 10 million ohms, typically, with a digital display like a watch. This very high input resistance prevents the meter from drawing current when it is connected into a circuit. The accuracy of the reading and protection for the circuit being tested is maintained. Digital multimeters have other advantages over analog meters. They are not polarity sensitive, making their life expectancy quite a bit longer. Usually a + or – shows to indicate the polarity, with no needle to bend in the wrong direction. Many of the meters are also automatically zeroed on the ohms scale and are impossible to overrange. The only difficulty appears to be reading and interpreting what the display shows. In addition, there are two types of DMMs in use: manual ranged and automatic ranging. We will look at both.

I. Procedure for Reading the Digital Meter That Is Manually Ranged

Let's use the common four digit display and go through some examples of how to read the digital meter. Let's start with the ohms scales, Figure 2-29.

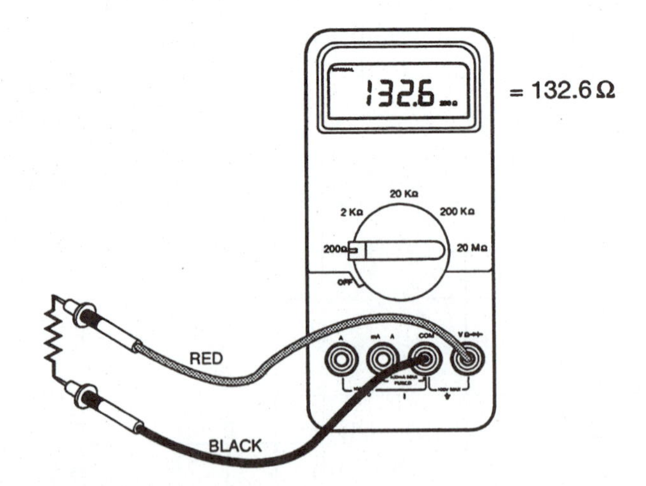

FIGURE 2-29 The decimal point between the third and fourth digit on the 200-ohm scale gives us an actual reading of 132.6 ohms. (Courtesy of Ford Motor Company)

The simplest method for reading a DMM is to look at the scale and put the decimal point after the number that appears on the scale. In other words, put the decimal point after 200 so that it becomes 200.0. The display will actually show 200.0 if the actual resistance being measured was 200 ohms. In our sample meter above, the actual resistance is 132.6 ohms. If we tried to measure something with an actual resistance of 500 ohms, the meter would be overranged and a 1 with three blanks, would show on the display, indicating that we must go to a higher range. Note: different manufactures use different indications of an overrange reading. "OL" or overload is displayed on some meters. Remember also that OL or overrange is another way of saying infinity, if you are measuring resistance. Consult the manual for the meter that you are using to determine just what an overrange condition will look like. The next range available on our example meter is 2 k ohms, Figure 2-30. Where will the decimal point be? Actually, 2 k is the exception to the decimal point rule because there are only four digits in 2,000. With four digits available, 2,000 ohms of resistance will show as 2000. The decimal point is not used on the display. Let's go back and pick up the 500 ohms that gave us an overrange OL on the 200-ohms scale. It will

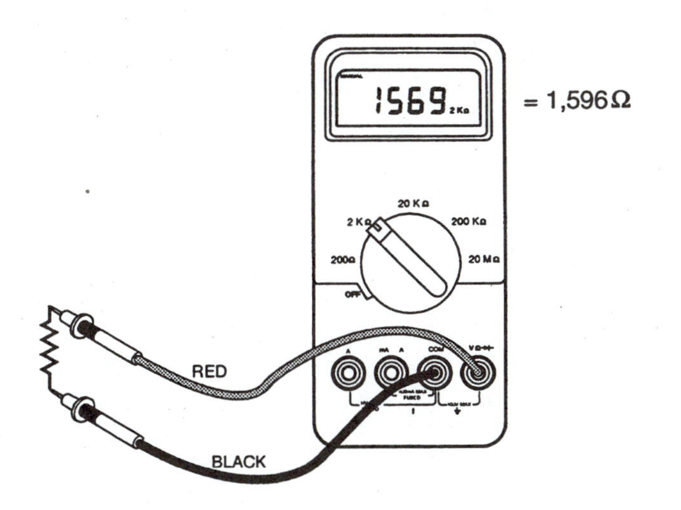

FIGURE 2-30 (Courtesy of Ford Motor Company)

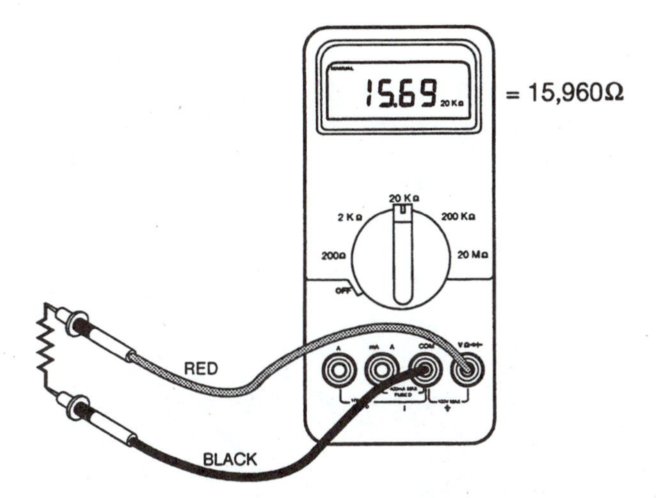

FIGURE 2-31 (Courtesy of Ford Motor Company)

look at the 2 k-scale as 500 with a blank in the first digit's place. How about 1,569 ohms? The meter will directly read the 1,569 ohms and all places will be filled. This range is usually the simplest because it is read directly. The next range available is 20 k ohms, Figure 2-31. An overrange OL on the 2-k range should push us into the next highest scale. The decimal point goes after the 20. So 20,000 ohms of resistance would show up as 20.00. Notice that we lost our last digit. When you are dealing with as high a resistance as this, 1 or 2 ohms can be quite insignificant, and so it is dropped. What will 500 ohms look like? The correct answer is .50. How about 15,693 ohms? The correct answer is 15.69 and the 3 is dropped.

200 k ohms range shows up as 200.0, while 20 M (million) ohms shows up as 20.00, Figure 2-32. Notice that the display looks the same for 20 k ohms as it does for 20 M ohms. Don't get these confused because there is a tremendous difference! Also remember overranging usually shows up as a 1 or as OL in the first place followed by blanks. Voltage reading on a digital multimeter is very simple, Figure 2-33. Again, look at the scale or range and place the decimal point after the number of the range. A 2-volt scale will be 2.000, 20-volt scale will be 20.00, 200-volt scale will be 200.0, and so on.

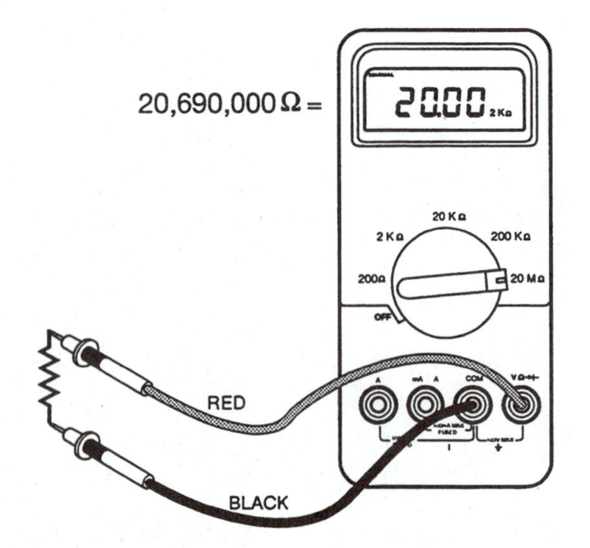

FIGURE 2-32 (Courtesy of Ford Motor Company)

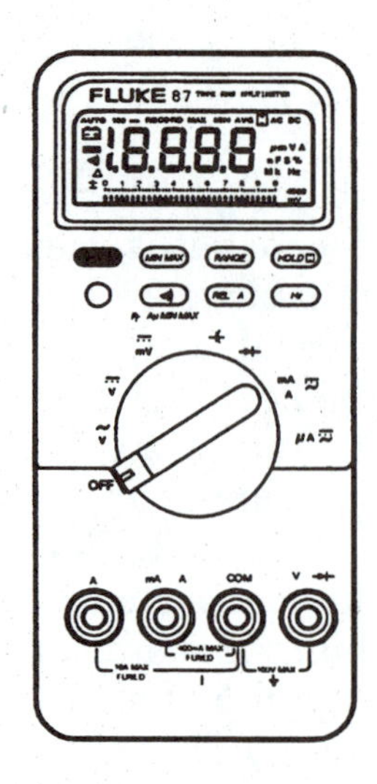

FIGURE 2-33 Modern multimeters have the capability of measuring volts, amps, and ohms. (Courtesy of Ford Motor Company)

II. Procedure for Reading an Auto Ranging Digital Multimeter

Many DMMs or DVOMs in use today are auto ranging. This greatly simplifies the front switch arrangement of the meter, but can involve some inaccurate readings if the technician does not understand what he is looking at. First of all we need to look at some of the abbreviations in use. There are only 3 that we need to be concerned with typically. A capital "M" means million. If an "M" appears on the right of the screen then the digits before any decimal point are millions.

For example, you are using a meter and it reads 45.8 M ohms, Figure 2-34. What does this really mean? 45,800,000 ohms.

How about a reading of 324.9 M ohms, Figure 2-35? If you answered 324,900,000 ohms—you are correct.

Don't forget, any digits before the decimal point are millions and anything after the decimal point is less than a million. The M scale is usually not encountered very often in typical automotive applications.

The next abbreviation that we will look at is the small "m". It stands for **milli** or a thousandth and is read as .001. A meter reading of 324 mV would actually be .324 volts, Figure 2-36. Three hundred twenty four thousandths of a volt—not much voltage is it? Actually, voltage readings this low are common on vehicle computer systems. For example oxygen sensors will generate between .200 volts (200mV) and .800 volts (800 mV) when they are functioning correctly. .200 is usually read as 200 millivolts. Don't get the "m" and the "M" mixed up. Obviously, there is a world of difference between a milli and a million.

The last abbreviation that we need to look at is "k" or **kilo** which is actually 1000 as we have seen before.

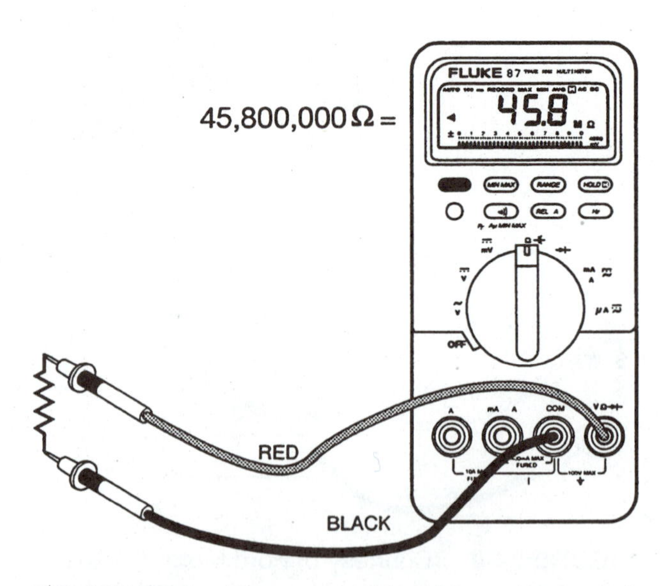

FIGURE 2-34 (Courtesy of Ford Motor Company)

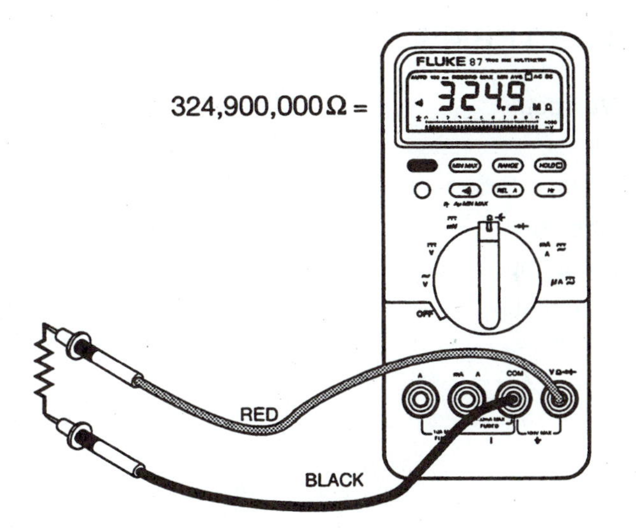

FIGURE 2-35 (Courtesy of Ford Motor Company)

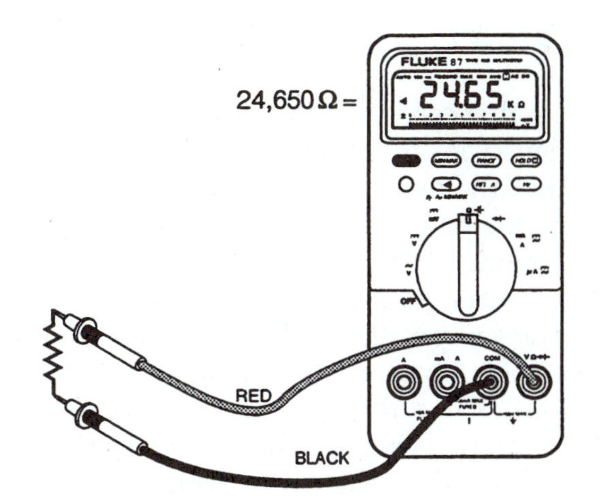

FIGURE 2-37 (Courtesy of Ford Motor Company)

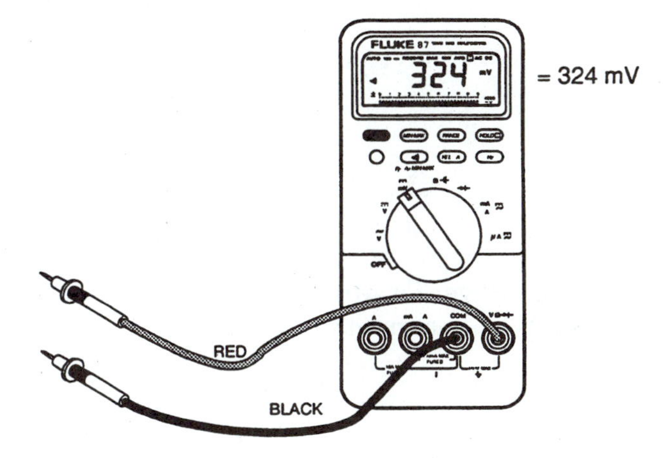

FIGURE 2-36 (Courtesy of Ford Motor Company)

A reading of 24.65 k ohms is 24,650 ohms, Figure 2-37.

A reading of .456 k Ohms is actually 456 ohms, Figure 2-38.

Most auto ranging meters can display readings as low as .001m and as high as 999,000,000 all automatically. Always note the abbreviation on the right of the meter—it is extremely important to the accuracy of your reading. In addition, most DVOMs have additional functions other than the reading of volts, amps, or ohms. Many include frequency, which is handy for Ford manifold absolute pressure sensor readings. Additional functions frequently are tachometer (engine speed), pulse width modulation, and percent of dwell.

Two recent features that are becoming more common are the touch/hold and the minimum, maximum, average recording feature. We will use these features in future chapters. Don't be fearful about using a digital multimeter. They are easy, accurate, and almost impossible to burn out from overranging or reverse polarity. Its use is important on computer circuits as its high input resistance allows an accurate reading without loading down the circuit. If a series of meters are available for your use, don't hesitate to take readings in an effort to become proficient in their use. The more experience you get with meters, the easier they will be to read, and the more accurate the information obtained.

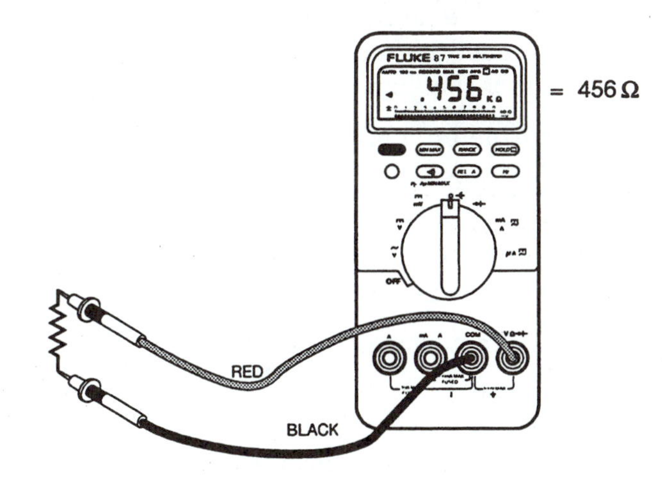

FIGURE 2-38 (Courtesy of Ford Motor Company)

SUMMARY

In this chapter, we have looked at the most common types of both analog and digital meters that the average technician must be able to use. Emphasis has been placed on the correct attachment of the voltmeter and ammeter, with attention to the polarity of both the meter and the circuit. In addition, correct meter range, both for accuracy and meter safety, was stressed. The ability to read voltages on an oscilloscope was introduced. Analyzing the digital readout on a digital multimeter was explained. The correct method of both attaching and interpreting ohmmeters was discussed and demonstrated.

KEY TERMS

Voltmeter
Range
Voltage Drop
Ammeter
Input Impedance
Ohmmeter
Infinity
Milliseconds
Oscilloscope
Digital multimeter (DVOM, DMM)
Analog meter
Milli
Kilo

CHAPTER CHECK-UPS

Multiple Choice

1. Technician A states that analog meters are polarity sensitive. Technician B states that digital meters usually do not require zeroing before use. Who is correct?
   a. Technician A only
   b. Technician B only
   c. Both Technician A and B
   d. Neither Technician A nor B
2. The object of the range setting on a voltmeter is to
   a. insure that the meter is correctly polarity connected.
   b. give the meter the ability to accurately measure a wide range of voltages.
   c. protect the meter from overranging the movement.
   d. all of the above.
3. A voltmeter is always wired in ___________ with the load.
   a. parallel
   b. series
   c. either series or parallel
   d. none of the above
4. An ammeter is always wired in ___________ with the load.
   a. parallel
   b. series
   c. either series or parallel
   d. none of the above

5. A scope will measure voltage over a
   a. period of time.
   b. limited number of vehicles.
   c. range of most digital meters.
   d. analog meter.
6. Technician A states that a digital meter is preferred to an analog meter when dealing with sensitive computer circuits. Technician B states that an analog meter might load down a computer circuit. Who is correct?
   a. Technician A only
   b. Technician B only
   c. Both Technician A and B
   d. Neither Technician A nor B
7. An ammeter is connected directly to the positive and negative of a battery. This will result in (a, an)
   a. accurate reading.
   b. indication of the state of charge of the battery.
   c. discharging the battery.
   d. burned out fuse or meter.
8. An ohmmeter is connected to a conductor and reads infinity. This indicates
   a. nothing.
   b. the wire is open.
   c. the wire is shorted.
   d. none of the above.
9. A digital reading of 2.370 k Ohms is actually
   a. 2.370 ohms.
   b. 2,370,000 ohms.
   c. 2,370 ohms.
   d. 23,700 ohms.
10. A digital reading of 1472 mA is actually
   a. 1.472 amps.
   b. 1,472 amps.
   c. 1472 amps.
   d. 147.2 amps.

True or False

1. Analog ohmmeters need to be zeroed.
2. Voltmeters have their own power source (usually an internal battery).
3. Digital meters react faster than oscilloscopes.
4. A reading of infinity on a digital meter is usually displayed as OL.
5. Inductive pick-up ammeters cannot be overranged.

6. Voltmeters are usually polarity sensitive.
7. Ammeters should be connected into circuits that already have resistive loads in them.
8. A digital reading of 14.62 on the 20,000 ohm scale is actually a reading of 1,462 ohms.
9. Voltmeters are normally connected across the load (in parallel).
10. Maximum current flows through a circuit that measures infinity.

Circuits That Do Work

OBJECTIVES

After completion of this chapter, you should be able to:

- wire series, parallel, and series-parallel circuits.
- measure current flow and voltage drops through circuits.
- figure wattage.
- analyze use of relays and diagnose them.
- understand the function of the starter solenoid.
- diagnose and repair ground (⏚)-related problems.
- diagnose and repair short circuits.
- use Ohm's law to figure voltage, current, and resistance in series, parallel, and series-parallel circuits.

INTRODUCTION

Now that you are familiar with the use of meters and have a basic understanding of voltage, amperage, resistance, and wattage, it is time to start wiring circuits and using meters to record electrical changes. Remember, the meter can be considered your "eyes into the circuit." Do not hesitate to use it. The information it will give you will be invaluable when analyzing modern vehicle systems or diagnosing circuit problems. In this chapter, we are first going to review the basics of circuits and work. Then we will wire up some simple circuits and use the meters to prove Ohm's law. Next, we will look at some simple control circuits in use today and walk through a procedure for open and short circuit diagnosis. Obviously, this is a lot of material to absorb. Reread it as many times as is necessary until the material becomes second nature to you. The lab experiments, which are optional, are designed to get you involved with meter readings and make the math of electronics easier. Make sure you ask your instructor questions as we proceed.

CIRCUIT COMPONENTS

Let us begin our review by looking at the common components of all circuits. Remember them? Look at Figure 3-1. This circuit should look familiar. It is the same one we looked at in Chapter 1. It has all the necessary components to make it a complete and functional circuit. First, we need power: **B+** and B–. Putting our voltmeter across B+ and B– will give us an idea if we have any potential or pressure difference. Remember, the voltmeter reads difference in pressure. Our voltmeter in this circuit has been drawn as a rectangular box with an analog meter face in it. It is reading a difference of 12.6 volts between the meter leads. Notice that the positive and negative meter leads are hooked into the circuit to match the polarity of the power source. The power source on most vehicles is either the 12.6-volt battery if the car is off, or the charging system (at around 14 volts) if the vehicle is running. The next thing we need is insulated conductors; insulated to ensure that the current gets to the right place. The conductors should have adequate size to handle the current for the entire length of the circuit. A switch to control the action of the circuit and a fuse to protect against overload brings us to the business end of the circuit: the load. Remember, it is the load that actually does the work of the circuit. Let us define work again as light, heat, or magnetic field, which is produced when current is pushed through the resistance of a load. It is the resistance of the load that will do the work. The current being forced through the resistance will also use up some of the voltage. This is called voltage drop and occurs anytime voltage pushes current through resistance. How about false or unwanted loads? These are the extra or unwanted resistances that might become part of the vehicle, especially as it ages. Corrosion is the most common cause of a false load. Putting a voltmeter across two points and seeing a reading with current flowing indicates

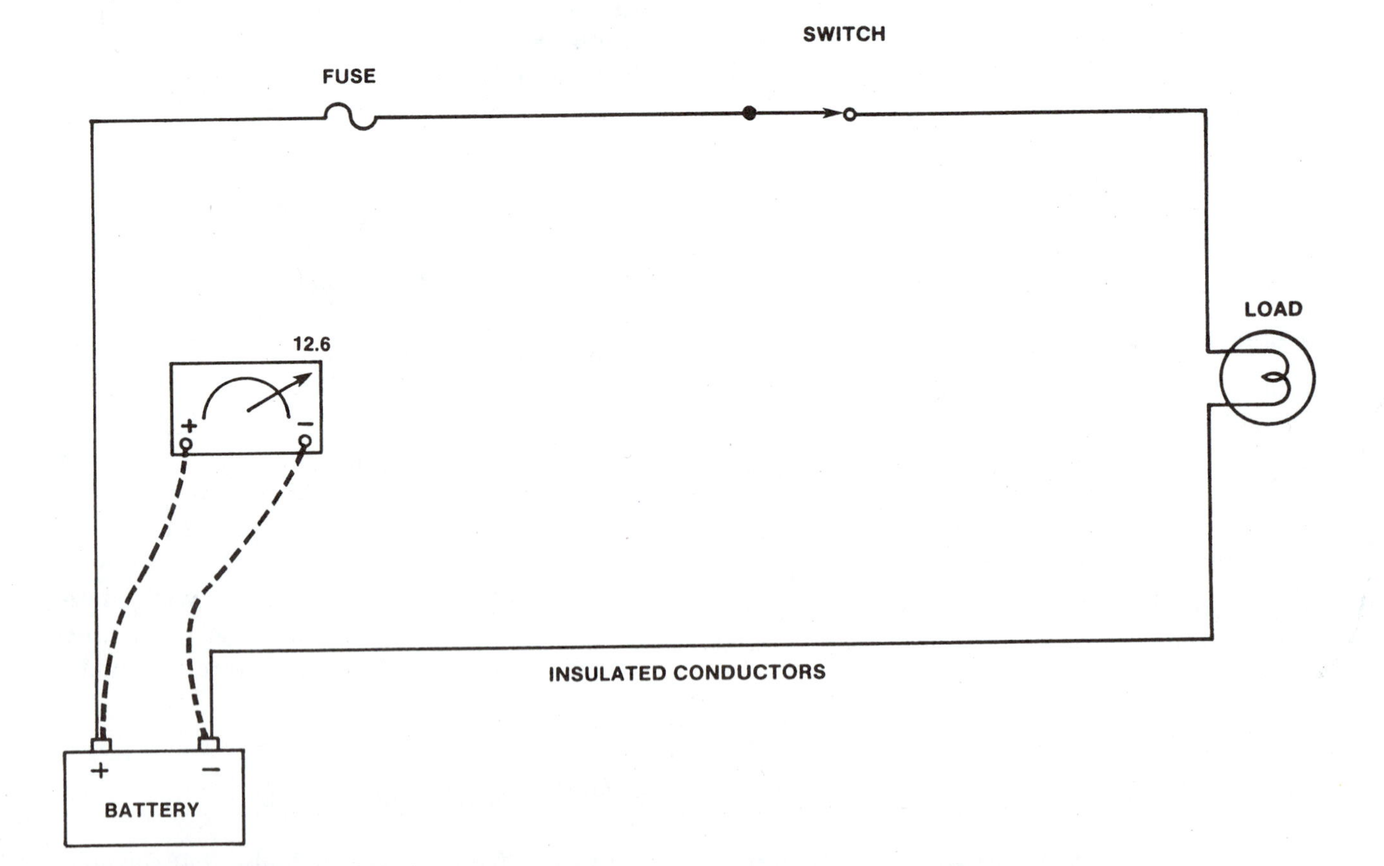

FIGURE 3-1 Common electrical components.

that resistance must be present in the circuit between the points where the voltmeter leads are connected. This is an important point to remember, so let us restate it a little differently. A voltmeter will read a voltage drop if it is placed across a resistance that has current flowing through it. We will see this with our voltmeter in the next section of this chapter.

This concludes the review of previous material. If you were confused, please go back and review Chapter 1. Remember, we are building a foundation for the entire automotive electrical system. The basics must be understood before you move on. If the material presented appears to be confusing, review it as many times as is necessary until it becomes second nature to you. In addition, make sure your instructor is aware of any questions you might still have.

I. Procedure for Wiring a Simple Circuit

Get a voltmeter, ammeter, ohmmeter, three automotive light bulbs or resistors, switch, fuse, and insulated conductors (preferably with clips on the ends). It is time to see electricity in action!

Wire up a simple **series circuit** as shown in Figure 3-2 and turn it on. The bulb should light, or the resistor should get warm. With the light on, use your voltmeter and record the pressure difference found in the blanks

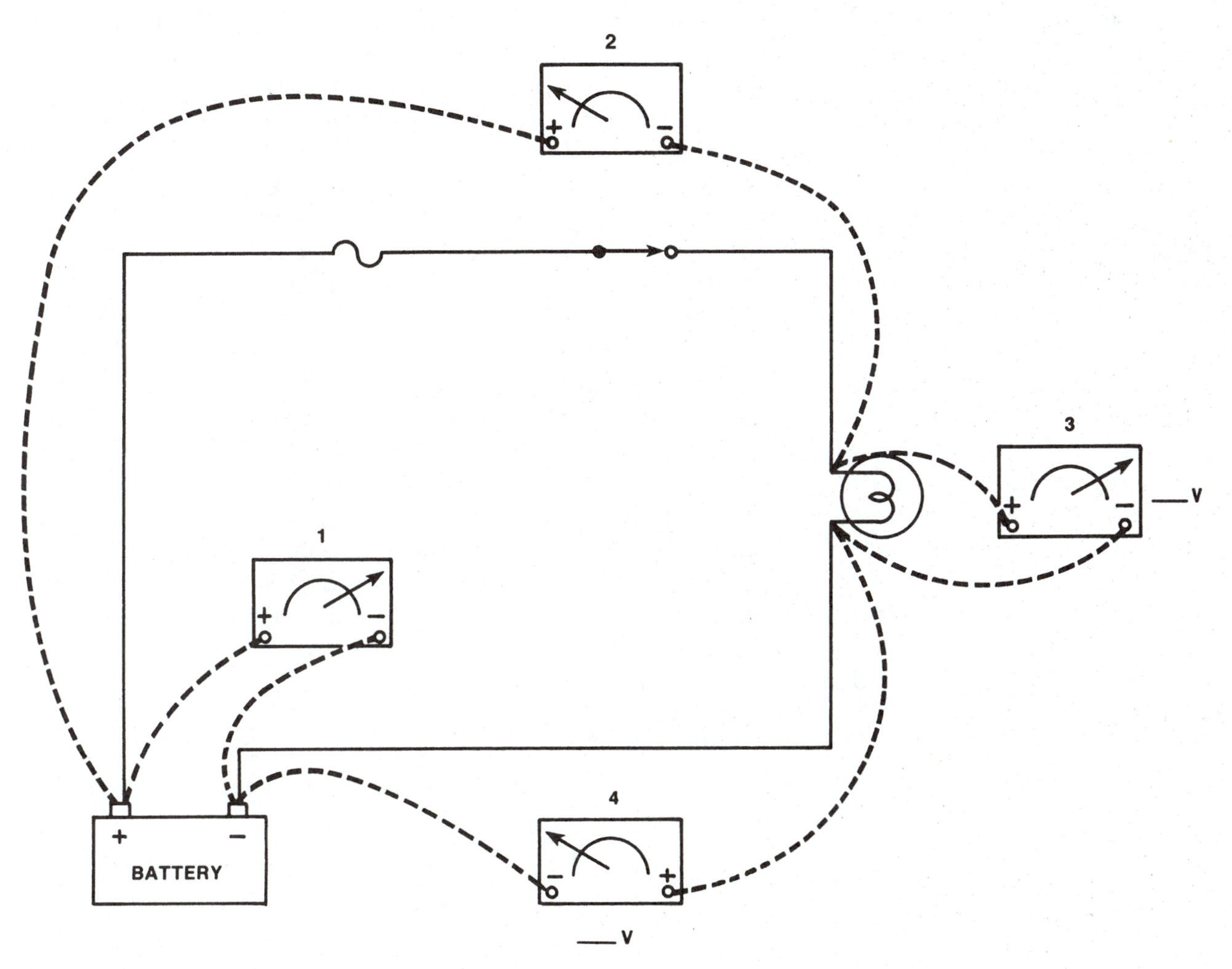

FIGURE 3-2 Voltmeter measuring voltage drops.

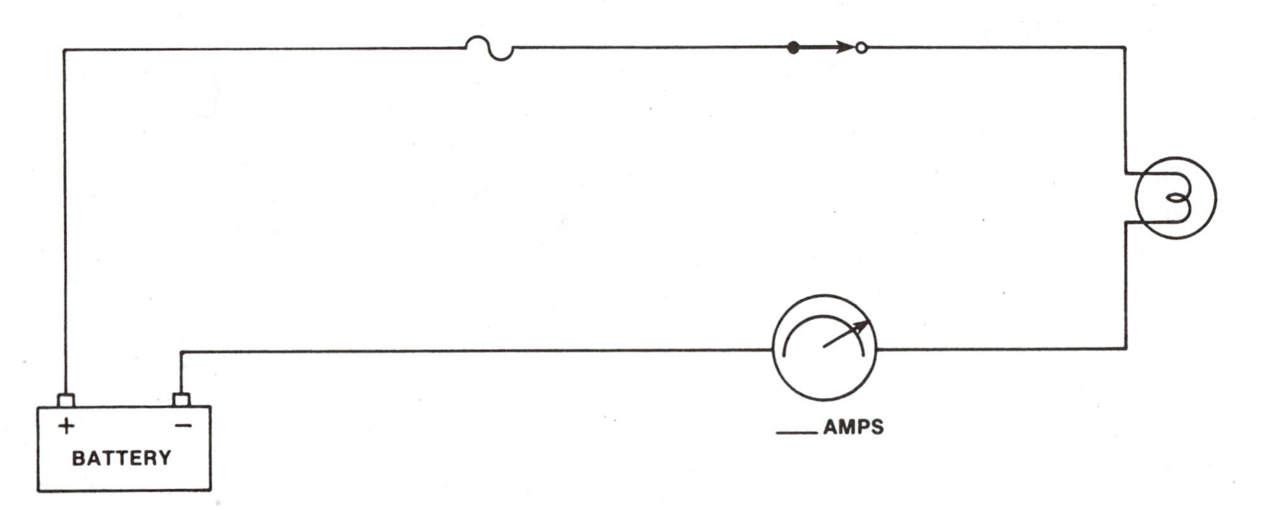

FIGURE 3-3 Ammeter measuring circuit current.

provided. Do not forget, the voltmeter (unless it is digital) is polarity sensitive. The positive lead should always be placed in the most positive part of the circuit, and the negative lead in the most negative part of the circuit. As you record your voltmeter readings, do not worry about what the meter is indicating, we will analyze what and why shortly, once all the blanks are filled in. Look at the reading on voltmeter 1. What is it indicating on? If you said the available pressure, you are correct. This meter tells us that with current flowing in the circuit (the bulb is on) we have a certain pressure. This is the load applied voltage reading. Notice that this number is the same as voltmeter 3. If it is not, rewire your circuit. The same reading indicates two different things. First, it tells us the voltage drop across the resistance of the load, and second, because it is the same as the applied voltage, it tells us that the load is the only resistance in the circuit. If, for instance, it was 2 volts lower than the applied voltage, we would know that somewhere in the circuit there must be additional resistance. This resistance was not designed into the circuit by you and is, therefore, unwanted. Remember that unwanted resistance is referred to as a false load, but more on this later. We can prove that there is no other resistance in the circuit by looking at voltmeters 2 and 4. They should both read 0 (zero). Why? Think about what causes differences in pressure: resistance. No difference in pressure equals no resistance. Our voltmeters are indicating that there is no difference in pressure and, therefore, as long as current is flowing, there must not be any resistance.

Open the switch and break the circuit open at the source (either positive or negative). Insert an ammeter in in series, as Figure 3-3, so that all current flowing must go through the meter. Do not forget the ammeter is polarity sensitive. The positive lead must be in the most positive part of the circuit, and the negative lead must be in the most negative part of the circuit. Remember also that an ammeter must only be used in a circuit that has some resistance in it. The resistance of the circuit *not the ammeter* must control the current flow. Our circuit has the resistance of the bulb to control current. Record the ammeter reading after you close the switch and reactivate the bulb. We will use this reading as a reference later.

II. Procedure for Determining Voltage Drop

Follow Figure 3-4 and wire up a simple circuit with two bulbs. Use two bulbs of the same size. The number printed on the two

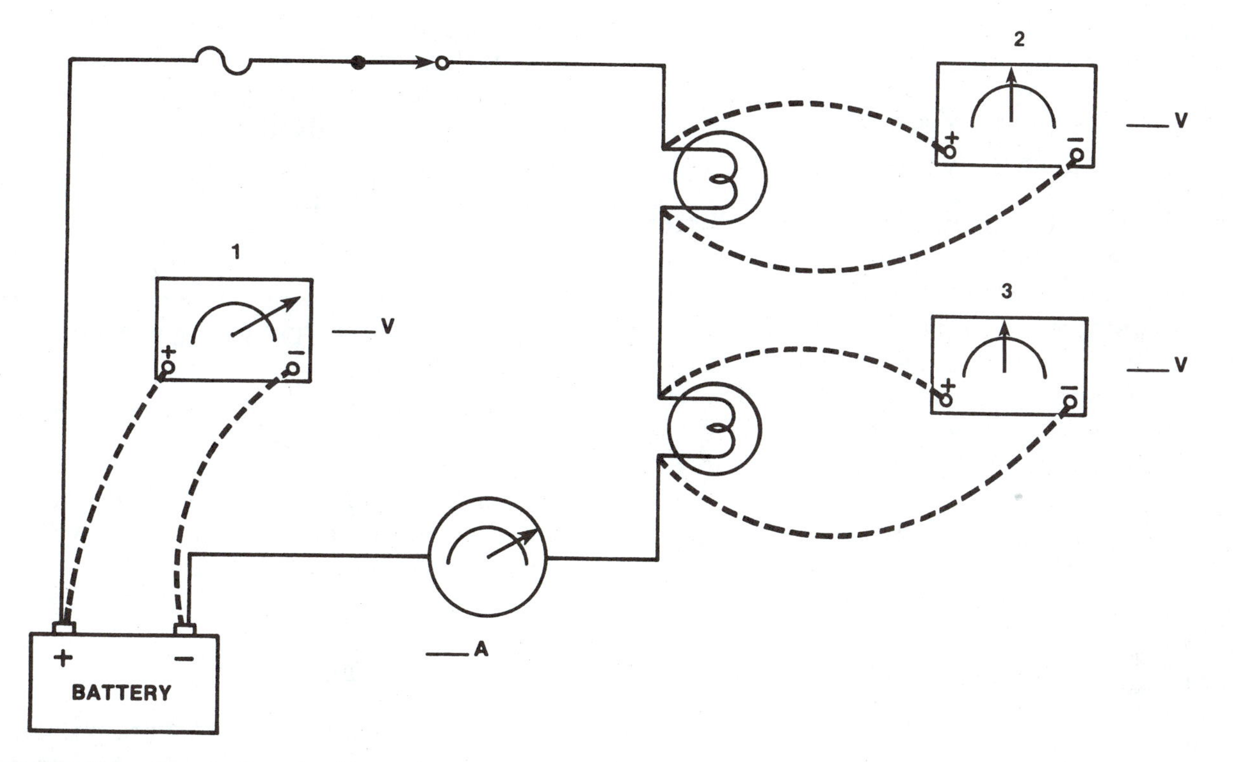

FIGURE 3-4 Two bulbs wired in series.

bulbs should match; 1156 is a good size to use as it has one filament. Record the voltmeter readings and the ammeter readings. Let us analyze the meter readings. What does the first one tell you (the first one being the applied voltage that pushes the current through the circuit)? This number will be a reference. How about number two, what does it show? The voltage drop across the first bulb will probably be a voltmeter reading very close to half of the applied voltage, and the same thing with voltmeter three. What actually happened here was the pressure dropped or divided equally as it pushed current through the resistances of the bulbs. The resistances wired in series are about the same number of ohms each, so the voltage divided equally. How is it that the ammeter reading is about half the reading of Figure 3-3? Hopefully, you realized that the current went down because the resistance went up. Another way of saying this is to say that the current had to be pushed through twice the resistance and, as a result, only half of the current was allowed through. The resistance in the circuit is what controls the current. Increasing the resistance decreases the current. Our ammeter reading dropped as the resistance increased. We will look at current flow more closely in a future section. If you are using resistors, you can measure the increased resistance of the circuit. Does this sound familiar? I hope so, because this is Ohm's law. You just proved it to yourself. We will use the math of Ohm's law in the future with some additional problems.

Add another bulb (the same size) in series and look at the current flow. Did it drop down like you thought it would? What is the meter telling you? The ammeter reading is probably about one third the value of the reading in Figure 3-3. Why? Because the resistance was tripled. Each time we raise up the resistance, we drop the current. Did you notice that the bulbs are getting dimmer each time you added another one? The current times the voltage will give you the wattage of work being accomplished. Dropping the voltage down by half and the current down by half reduced the wattage down to one quarter. You can prove this to yourself with the following simple formula:

Example 1 $V \times A = ?$ watts
Example 2 $V \times A = ?$ watts

This formula has shown you why the bulbs are progressively getting dimmer. Each time another load is wired in series, the voltage and the amperage drops down lower. Multiplying this reduced voltage times the reduced amperage gives us lower wattage levels. Remember that wattage is an indicator of the amount of work actually accomplished in the circuit.

Let us summarize. The voltage dropped or divided in the series circuits is based on the combined resistances of the loads. Equal resistances give us equal voltage division or drops. Each additional equal resistance wired in series causes the voltage to redivide or drop evenly throughout the rest of the circuit.

Wire up the circuit from Figure 3-4 again, but this time use different size bulbs (different numbers). Determine the voltmeter readings below and analyze the differences.

Voltmeter 1 = ? volts
Voltmeter 2 = ? volts
Voltmeter 3 = ? volts

What happened here? Obviously, the voltage dropped or divided, but this time it did not do so equally, as in the other examples. What was the difference? Hopefully, you have realized that the unequal resistances are the reason for the unequal voltage division. Unequal resistance equals unequal voltage division. The total voltage drop still equals the applied voltage, but the individual drops are unequal. Ohm's law is still at work here.

SERIES CIRCUIT

Let us look at the ammeter reading that we were able to obtain in the examples of series circuits wired. We have seen how the current was reduced by the addition of series resistance. Again, this is Ohm's law in action. Increasing the resistance reduces the current. When we doubled the resistance, we found the current was halved. Let us use a circuit with specific resistances and prove this to ourselves mathematically.

III. Procedure for Determining Current Flow

Figure 3-5 shows a series circuit with one single 2-ohm resistance and an applied voltage of 12 volts. The current in this circuit will be 6 amps. Ohm's law states that:

$$\frac{12 \text{ volts}}{2 \text{ ohms}} = 6 \text{ amps}$$

If, on the other hand, we were presented with the same circuit and did not know the total resistance, we could use a different form of Ohm's law.

$$\frac{12 \text{ volts}}{6 \text{ amps}} = 2 \text{ ohms}$$

and yet another way:

$$6 \text{ amps} \times 2 \text{ ohms} = 12 \text{ volts}$$

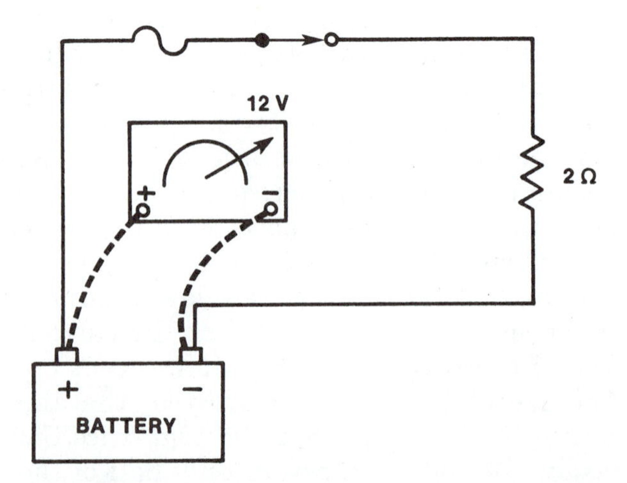

FIGURE 3-5 Two ohms across 12 volts.

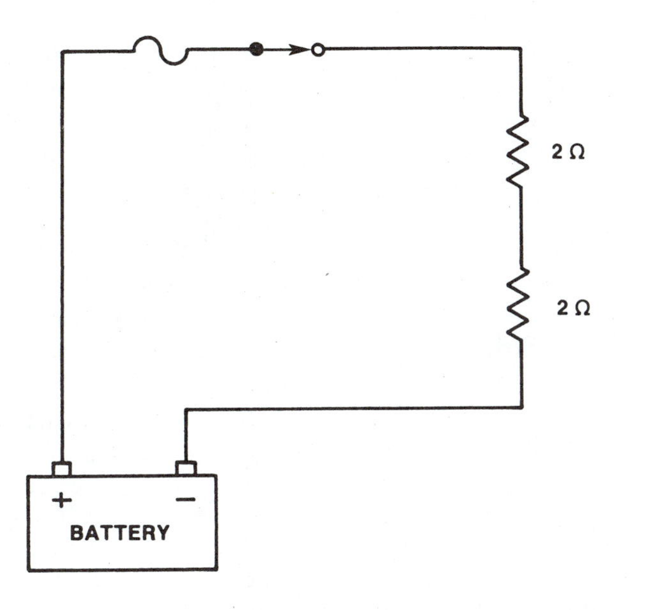

FIGURE 3-6 Two 2-ohm resistors in series across 12 volts.

Notice that Ohm's law give us the ability to find the unknown if we have the two other values. Now let us add more resistance to this series circuit. Figure 3-6 shows the addition of another 2-ohm resistor, giving us a total of 4 ohms wired in series. Total current flow will now be:

$$\frac{12 \text{ volts}}{4 \text{ ohms}} = 3 \text{ amps}$$

or again if we did not know the ohms but knew the current,

$$\frac{12 \text{ volts}}{3 \text{ amps}} = 4 \text{ ohms}$$

Again, the relationship between the known values and the unknown is Ohm's law.

Question: How much current will flow through a circuit that has 5 resistors wired in series: a 1-ohm, 2-ohm, 3-ohm, 4-ohm, and another 2-ohm as seen in Figure 3-7A. The approach to this is to first add up the resistances.

$$1 + 2 + 3 + 4 + 2 = 12 \text{ ohms total}$$

Now figure the current:

$$\frac{12 \text{ volts}}{12 \text{ ohms}} = 1 \text{ amp of current}$$

The total current flow in this circuit will be 1 amp.

Question: What will the individual voltage drops be across each resistor? This ques-

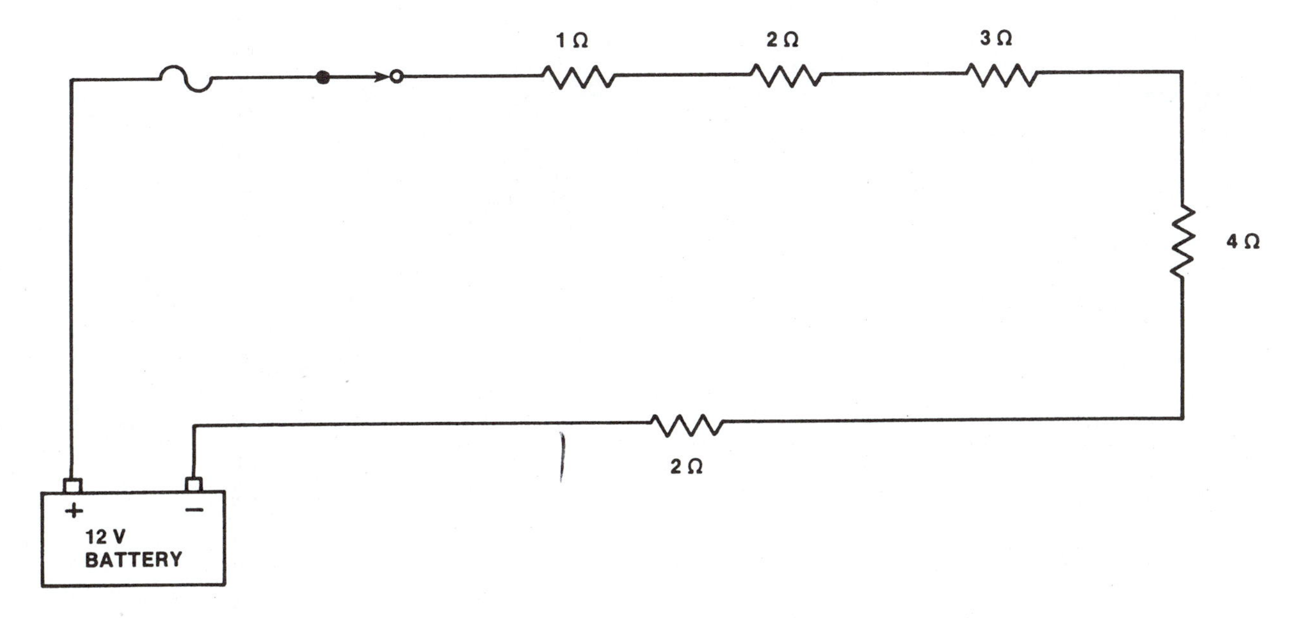

FIGURE 3-7A Five resistors wired in series.

tion raises a new point in our discussion of circuits and Ohm's law. It is easily answered, though, because we know two of three variables. Up to this point, we have only looked at the total resistance, voltage, and current. To look at individual drops is just as easy. We know the current flowing through each resistor is the same because this is a series circuit. Current flow through any and all parts of a series circuit is the same because there is only one path for current. If 1 amp is flowing into the first load, 1 amp is flowing out of the last load. We also know the resistance of each load. By applying Ohm's law to each resistor, we will be able to find the voltage drop across each resistor. If our numbers are correct, they will add up to the total applied voltage: 12 volts, in this example. Start with the 1-ohm resistor.

$$1 \text{ ohm} \times 1 \text{ amp} = 1 \text{ volt}$$

This 1 volt is the actual voltage drop measured across a 1 ohm resistor with 1 amp of current flowing. Now the rest of the resistors.

$$2 \text{ ohms} \times 1 \text{ amp} = 2 \text{ volts}$$
$$3 \text{ ohms} \times 1 \text{ amp} = 3 \text{ volts}$$
$$4 \text{ ohms} \times 1 \text{ amp} = 4 \text{ volts}$$

As a final check, add up the individual voltage drops. Do not forget that there are two, 2-ohm resistors.

$$1 \text{ V} + 2 \text{ V} = 2 \text{ V} + 3 \text{ V} + 4 \text{ V} = 12 \text{ volts}$$

The applied total voltage and the sum of the individual voltage drops was 12 volts, so our circuit follows Ohm's law perfectly. You will find additional series circuit problems presented at the end of this chapter. You may wish to go to them now and test your knowledge of voltage, current, and resistance in a series circuit before we go to parallel circuits.

PARALLEL CIRCUIT

Now that we have seen what current, voltage, and resistance do in series circuits, let us examine the same voltage, current, and resistance in parallel circuits. Wire up the simple two-bulb parallel circuit shown in Figure 3-7B. Use the same size bulbs. Take voltmeter readings and ammeter readings and fill in the readings. We will discuss the results after you have them filled in. Let us look at the ammeter reading first. Did you notice that it was about two times the ammeter reading you measured in

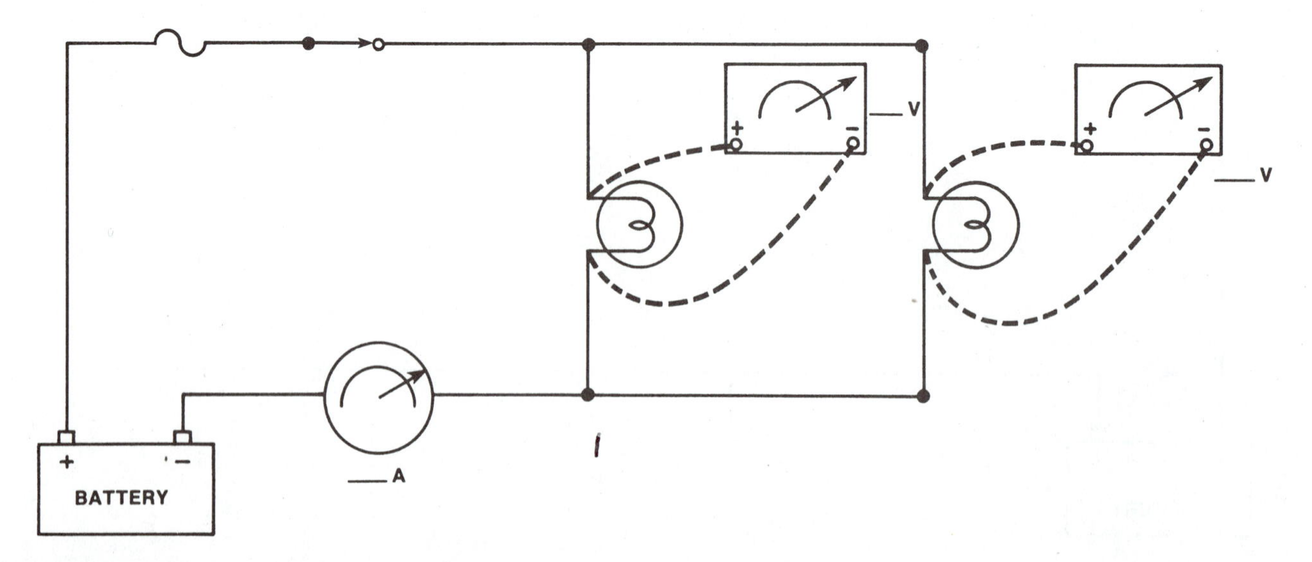

FIGURE 3-7B Two bulbs wired in parallel.

the first series circuit. Why is this? Simple. Each bulb has the full B+ and B– applied across it. Doesn't it? Do you remember the series circuit? With the maximum voltage applied across the load, the maximum current was pushed through the load. This means that each leg of the parallel circuit has full current flowing through it and that the total current flow is the addition of all legs together. This relates to the water system in Figure 3-8A. The top pipe has 4 gallons flowing because each downflow pipe has 2 gallons flowing: 2 gallons in each leg equals 4 gallons total. Current responds in much the same way. The current in the main leg will have the total of the two parallel legs. If each leg has full B+ and B– applied, it will draw full current, or the same current when it was all alone across the power source. Add the currents together because two loads are each drawing the current they want.

IV. Procedure for Determining Resistance

Let us think about the current in terms of the resistance. If the current went up or doubled, the resistance must have gone down by half. Does this make sense? I hope so. Let us do some resistance problems to prove that the resistance actually did go down. Do you remember the formula:

$$\frac{\text{Voltage}}{\text{Amperage} \times \text{Resistance}}$$

Let us say that we wire one bulb across a 12-volt power source and the ammeter reads 3 amps of current. Using Ohm's law, let us figure the resistance.

$$\frac{12 \text{ volts}}{3 \text{ amps}} = 4 \text{ ohms}$$

To keep the example going, if we were to wire up an additional bulb of the same resistance, it would draw 3 amps of current, also. The total current flow for the circuit is now 6 amps (3 amps for each leg). Figure the total resistance for the circuit. Use the total amperage and the applied voltage.

$$\frac{12 \text{ volts}}{6 \text{ amps}} = 2 \text{ ohms total}$$

How about that, the resistance actually is one half. This is the reason that the current doubled.

Now let us use a formula to add up the individual resistances. The total should equal our previous total. Do not let the math scare

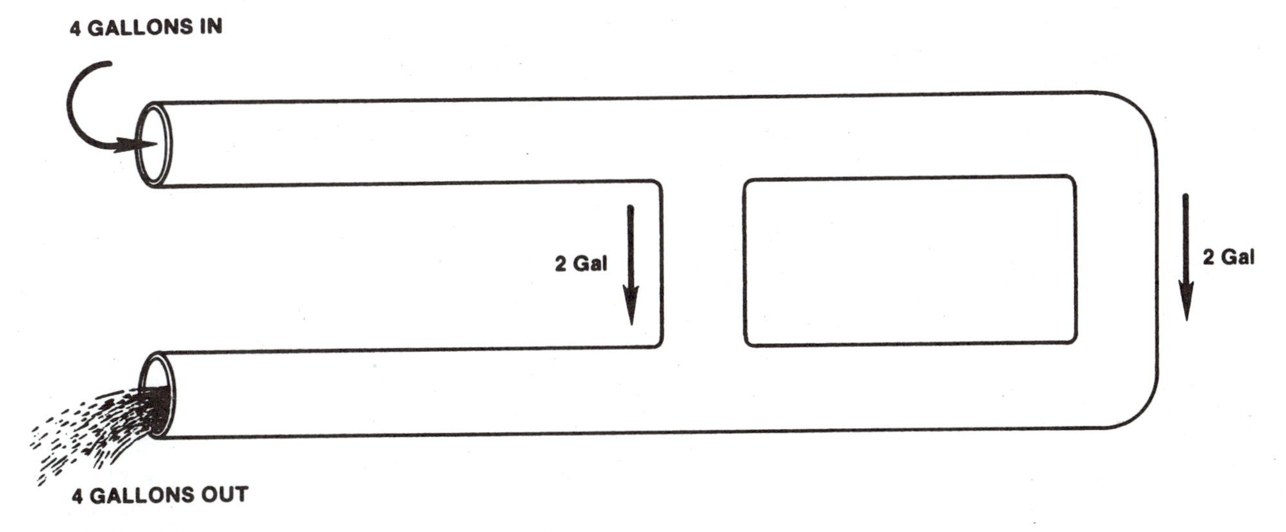

FIGURE 3-8A Water system is like a parallel circuit.

you. Just fill in the blanks. If you get lost, ask your instructor. We will be using a form of Ohm's law that states how resistances are totaled when they are wired in parallel. Keep in mind that you cannot just add them together as you did in a series circuit. Here is the formula.

$$\text{Total resistance} = \frac{1}{\frac{1}{R1} + \frac{1}{R2} + \frac{1}{R3} + \ldots}$$

For our example with two resistances of 4 ohms each:

$$Rt = \frac{1}{\frac{1}{4} + \frac{1}{4}}$$

The next thing we must do is add the 1/4 to the 1/4:

1/4 + 1/4 = 2/4, or 1/2. This now gives us:

$$Rt = \frac{1}{\frac{1}{2}}$$

Working again with the denominator 1/2, we can either divide it directly into the numerator and get 2 ohms total: (1 ÷ 1/2 = 2) or we can convert the 1/2 over to the decimal 0.5.

$$Rt = \frac{1}{0.5}$$

We then divide the 1 by 0.5 to reduce the fraction to its lowest form and find that the total resistance is 2 ohms. Notice that this is the same total that we found when we figured the total resistance by dividing the applied voltage by the total current flow in amps.

Figuring total resistance by looking at individual resistances and using the previous formula will show that as additional resistances are added in parallel, the total will always go down to a level lower than the lowest resistance. We introduced this to you in Chapter 1. Add an additional bulb to our two-bulb circuit. Measure the amperage and the applied voltage. Plug the figures into the formula below to find the total resistance.

$$\frac{\text{volts}}{\text{amps}} = \text{ohms of total resistance}$$

Did you notice that the total was less than the individual resistance we figured before? Notice that in parallel circuits, the current divides and the voltage remains the same throughout the circuit. This is opposite of what we found in series circuits. In series circuits, the voltage divided and the current was the same throughout the circuit.

Let us try an example with different resistances wired in parallel and try to determine the current flow. Figure 3-8B shows four resistances of 2, 4, 6, and 8 ohms wired in parallel with 12 volts applied. Try to follow along as we work through the math for total resistances in parallel.

$$Rt = \frac{1}{\frac{1}{2} + \frac{1}{4} + \frac{1}{6} + \frac{1}{8}}$$

First, to be able to add the fractions in the denominator, we will have to find the common denominator for all. This is the number that all fractions can be changed into. For our example, 24 is lowest common denominator. It is the lowest number that all of our fractions will evenly divide into. This now give us:

$$Rt = \frac{1}{\frac{12}{24} + \frac{6}{24} + \frac{4}{24} + \frac{3}{24}}$$

or

$$Rt = \frac{1}{\frac{25}{24}}$$

By dividing the 25 by 24 we get 1.04 as the denominator. Our fraction now looks like this:

$$Rt = \frac{1}{1.04}$$

By dividing 1 by 1.04, we see that the total resistance of a circuit with 2, 4, 6, and 8 ohms of resistance wired in parallel is 0.96 ohms.

Now that we know the total resistance we can figure the total current flow.

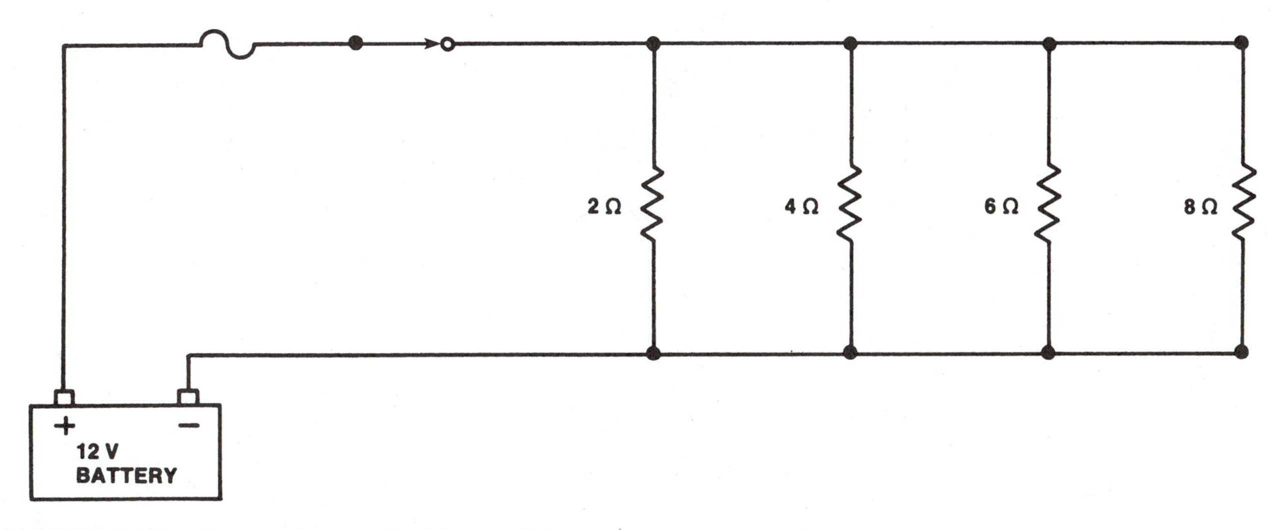

FIGURE 3-8B Four resistors wired in parallel.

$$\frac{12 \text{ volts}}{0.96 \text{ ohms}} = 12.5 \text{ amps}$$

Our circuit actually has 12.5 amps of current flowing through it. We can take this example one step further and look at the current flow through each of the legs of the circuit. If we do it correctly, the individual legs should equal 12.5 when added together.

As each leg of the circuit has 12 volts applied, all we have to do is divide the voltage applied by the individual resistances to figure current flow.

$$\frac{12 \text{ volts}}{2 \text{ ohms}} = 6 \text{ amps}$$

$$\frac{12 \text{ volts}}{4 \text{ ohms}} = 3 \text{ amps}$$

$$\frac{12 \text{ volts}}{6 \text{ ohms}} = 2 \text{ amps}$$

$$\frac{12 \text{ volts}}{8 \text{ ohms}} = 1.5 \text{ amps}$$

$$6 + 3 + 2 + 1.5 = 12.5 \text{ amps total}$$

There are additional problems for you to try at the end of this chapter. You may wish to give them a try now while the math of parallel circuits is still fresh in your mind. Feel free to use a calculator, because sometimes the numbers can be cumbersome to work with.

Next, we will look at series-parallel circuits and see that the characteristics of both series and parallel circuits are present.

SERIES-PARALLEL CIRCUITS

Now that we have seen how electricity behaves in a series and in a parallel circuit, let us examine a combination called a series-parallel circuit, S/P. From a practical standpoint, most circuits fall into this classification. With some sections of the B+ and B− limited to one or two main feed wires or cables before branching into parallel sections, the circuit can be classified as S/P. The battery cables are the series section and the individual loads are the parallel section. We will wire up a simple S/P circuit to see how voltage and current behave.

V. Procedure for Determining Resistance

Wire up the circuit in Figure 3-9. Take your voltmeter and ammeter readings and record

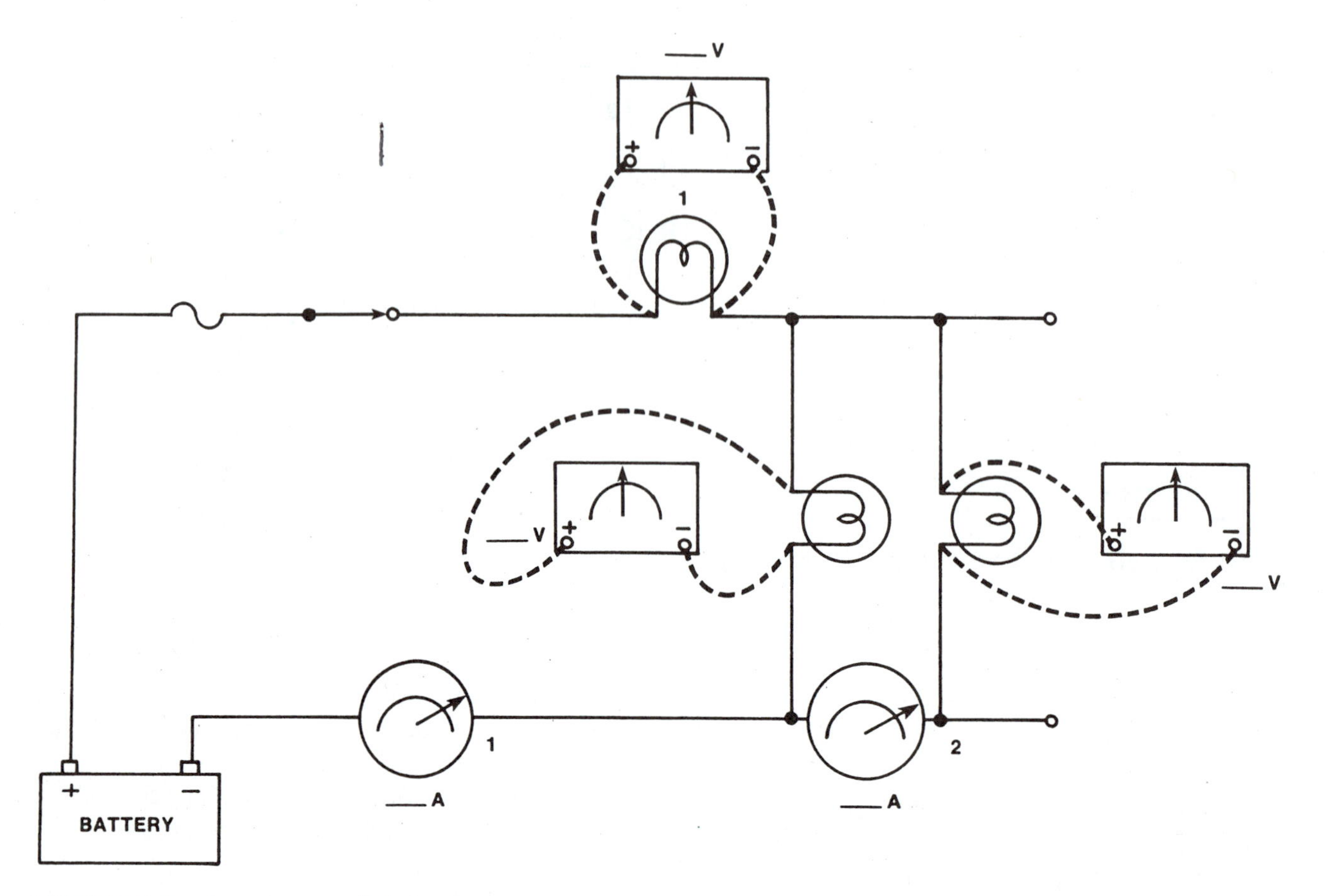

FIGURE 3-9 A series-parallel circuit.

them. Once you have completed this, move bulb 1 into the B– series section of the circuit and retake your voltage and ammeter readings. Do not forget to record them as we are going to refer back to them. Use the same size bulbs.

Okay. Hopefully, you have the numbers filled in and are ready to analyze them. Let us look at the parallel section first. Fairly straight forward here, isn't it? The current divided and the voltage to each leg remained the same. Notice though that the voltage across the loads is not the applied circuit voltage. This is because, as far as the applied voltage was concerned, it viewed the parallel two loads as one single resistance. How many ohms did it think was out there? Look at ammeter 2 and one of the parallel voltmeters. Use Ohm's law to figure out the total resistance of the parallel circuit.

$$\frac{\text{volts}}{\text{amps}} = \text{ohms (total parallel)}$$

If the total resistance of the parallel section looks like one single resistance of the value you have just figured, the voltage will divide between the first load (the series bulb) and the second series load (which is actually two loads in parallel). Look at Figure 3-10 for a schematic representation of this.

Back to the original question. Why did the parallel section not have the full applied circuit voltage across its loads? The voltage had to divide between the load in series and the two loads in parallel because the parallel loads were in series with the first load. Be sure to reread that sentence over a couple of times until it makes sense.

Notice in the second S/P circuit that the numbers remained the same. In other words,

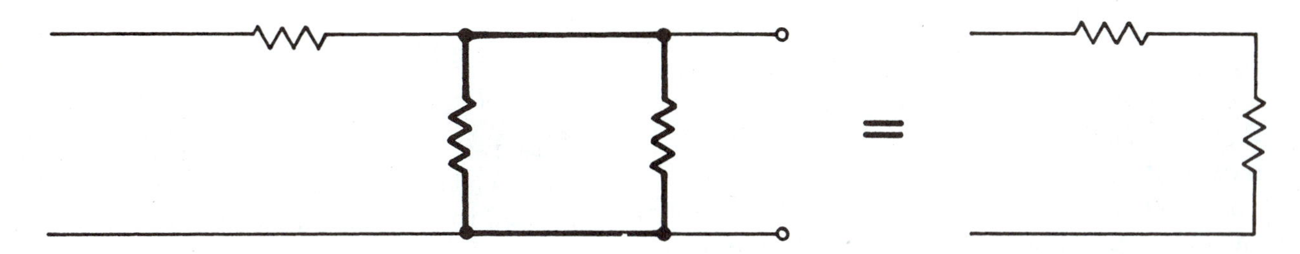

FIGURE 3-10 Two parallel resistances can be reduced to one equivalent series resistance.

moving the loads around had no effect. Resistance in series will cause the same voltage division if it is in the B+ section or the B– section. Remember the battery sees two series resistances to divide its voltage between. It could not care less what their position is. The voltage division takes place because of the resistance of each load as a percentage of the total resistance. If a resistance is 25% of the total, it will drop the applied voltage by 25%. As long as the total does not change, altering the position will have no effect on the results.

Prove this to yourself by wiring up Figure 3-11. Take the voltage readings and figure out what happened.

The voltage drop of the first load is now greater isn't it? What happened? By adding another resistance in parallel, we reduced the total parallel resistance and the total circuit resistance. At the same time, we did not change the resistance of the series section (bulb 1), so it became a greater percentage of the total resistance and experienced a greater voltage drop across it.

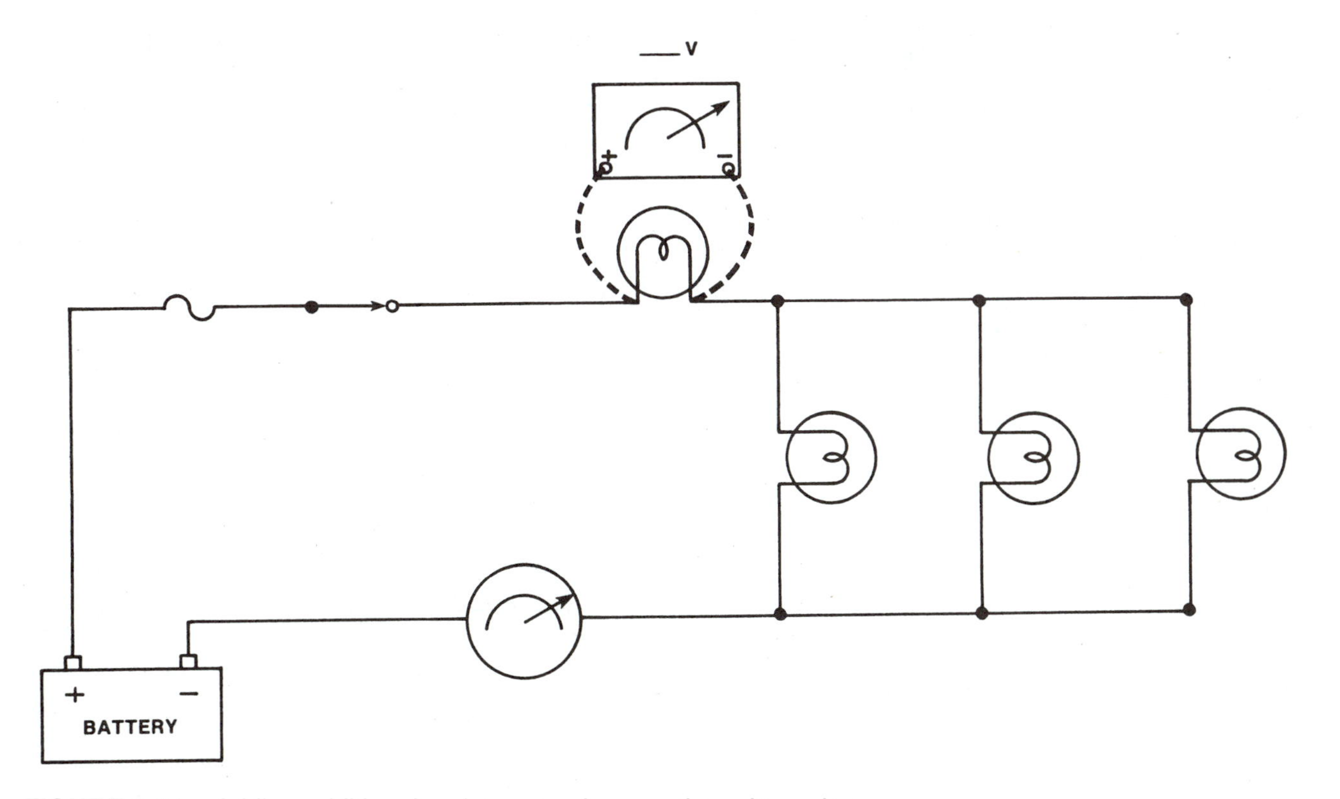

FIGURE 3-11 Adding additional resistances changes the voltage drops.

The important thing to remember is that resistance, when placed in series, will drop down all applied voltage to the parallel loads and reduce the current that the loads draw. If you experience a voltage drop before a parallel load, examine the wiring diagram and look for a similar problem on the other loads wired in parallel. If all loads have the same reduced voltage, the excessive resistance or false load must be in the series part of the circuit. This little, seemingly simple hint, can save you hours of diagnosis time. Many times we can diagnose the vehicle almost completely from the wiring diagram by being alert to how the circuit is wired and what the voltage reading is across the parallel loads.

Let us try another example with resistance values and see how Ohm's law would be applied. The circuit shown in Figure 3-12 has a 6-ohm load in series with 2- and 4-ohm loads in parallel; 12 volts is applied. *Question:* What is the total resistance of the circuit?

$$Rt = \frac{1}{\frac{1}{2} + \frac{1}{4}}$$

The lowest common denominator is 4.

$$Rt = \frac{1}{\frac{2}{4} + \frac{1}{4}} = \frac{1}{\frac{3}{4}}$$

4 is divided into 3 and equals 0.75.

$$Rt = \frac{1}{.75} = 1.33 \text{ ohms total (parallel section)}$$

The total resistance of the circuit is the sum of 1.33 ohms (parallel) + 6 ohms (series) or 7.33 ohms total in the circuit. Total amperage can now be figured.

$$\frac{12 \text{ volts}}{7.33 \text{ ohms}} = 1.65 \text{ amps}$$

The total resistance of 7.33 ohms allows 1.65 amps of current to flow. Do not forget this current will all flow through the series section of the circuit but will split or divide as it goes through the parallel part of the circuit. With the information we now have, we can figure the rest of the unknowns. *Question:* What is the voltage applied to the parallel circuit legs?

Figuring the voltage to the parallel will involve subtracting the voltage drop of the series from the total applied voltage. We will figure the drop again by Ohm's law using our previous 1.65 amps and the resistance of 6 ohms.

$$6 \text{ ohms} \times 1.65 \text{ amps} = 9.8 \text{ volts}$$

Next, we subtract the series voltage drop from the applied and we have our answer.

$$12 \text{ volts} - 9.8 \text{ volts} = 2.2 \text{ volts}$$

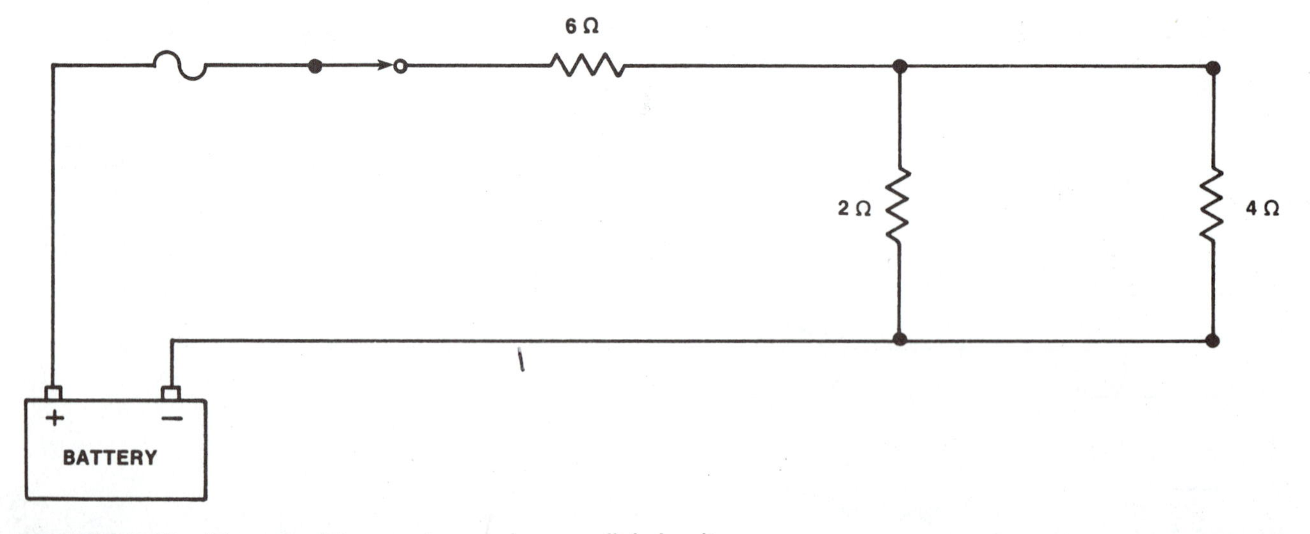

FIGURE 3-12 Three resistances in a series-parallel circuit.

The 2.2 volts is being applied to the parallel sections of the circuit because the series section dropped the voltage down by 9.8 volts.

Question: How many amps of current are flowing through each leg of the parallel section? This question will again rely on previous information. We now know that the parallel legs have 2.2 volts applied and we also know the individual resistances. Ohm's law again will tell us the individual amperages as follows.

$$\frac{2.2 \text{ volts}}{2 \text{ ohms}} = 1.1 \text{ amps}$$

$$\frac{2.2 \text{ volts}}{4 \text{ ohms}} = .55 \text{ amps}$$

1.1 amps + .55 amps = 1.65 amps total in the parallel section of the circuit. Again, our circuit is complete with all numbers matching.

There are additional series-parallel voltage, resistance, and current questions at the end of this chapter. By working through them you will become more confident in Ohm's law and will have a greater understanding of just what is happening within a circuit.

CONTROL CIRCUITS

Frequently, on modern vehicles, the current draw of the accessories is beyond the capacity of the ignition switch and sometimes even the normal wiring. Under these circumstances, the manufacturer will use a relay. Simply defined, a relay is a device that uses low current to control high current. Look at Figure 3-13.

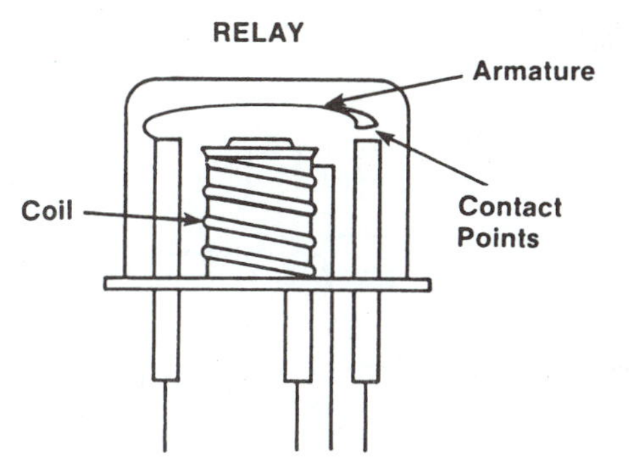

FIGURE 3-13 A relay can be used to control the flow of current.

It is a simple diagram of a common electromechanical relay. The relay coil has very high resistance, which means it will draw very low current. This low current will be used to produce a magnetic field. This field will close two electromechanical contacts. The contacts are designed to carry heavy current. When we apply B+ and B– to the relay coil, the contacts will close and direct heavy B+ to the main load that we want to control.

The simplest application of relays in vehicles is the horn relay circuit. It is quite common to find it wired as in Figure 3-14. Notice that B+ is applied directly to the coil. In the automobile, if you follow the B– side of the relay coil, you will notice that it comes through the fire wall, which separates the engine compartment from the passenger compartment, goes up the steering column to the horn button. One side of the horn button is grounded and pushing it in will apply a ground to the B– side of the horn relay coil. With both B+ and B–, current flows through the coil and develops a magnetic field. The field will close the contacts that will direct B+ to the horn. Because the other side of the horn is grounded, it will sound. Used in this manner, the horn relay becomes a control of the heavy current necessary to blow the horn. The actual control circuit can be wired with very thin wire as it will not have much current flowing through it. Horn relays might have as much as 50 ohms of resistance across their coils that will convert over to around 1/4 amp flowing through the control circuit. By contrast, horns might have less than 1 ohm and might have two wired in parallel that will convert over to 24 or more amps flowing through the relay coil contacts.

Another circuit that can utilize a relay is the blower motor circuit. Many blower motors use a series of voltage dropping resistors for the lower speeds and a relay for maximum speed. This is frequently done on air-conditioned vehicles where the blower motor has to move large quantities of air. Start with the re-

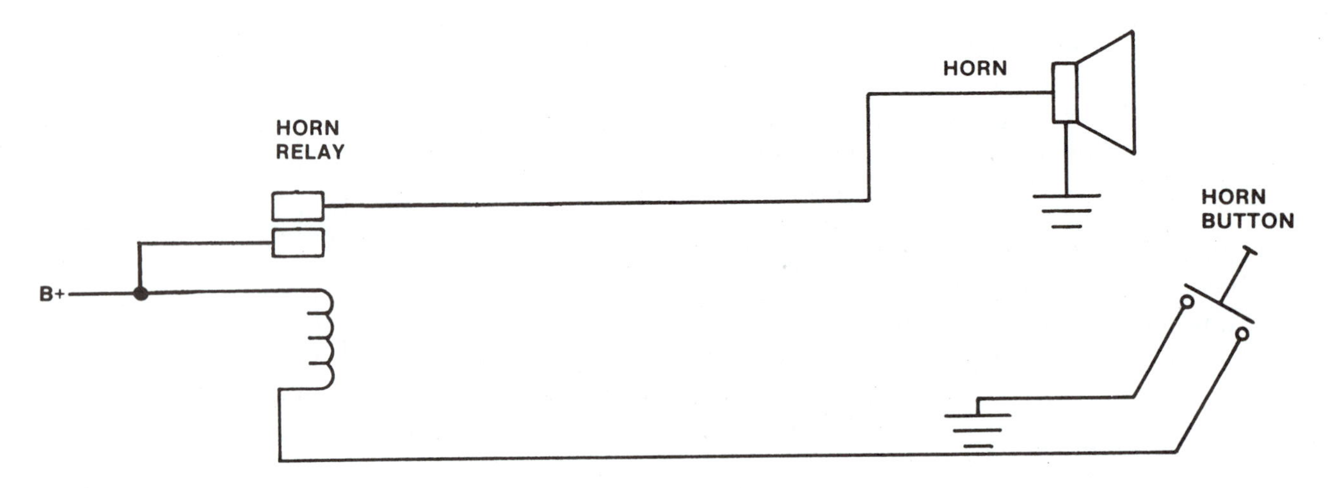

FIGURE 3-14 Common horn relay circuit.

sistor shown in Figure 3-15. It is normal for automotive manufacturers to use a heating coil wired in series to control the lower speeds of the blower motor. Remember that series resistance will drop the voltage and reduce the current to the motor. Lower current will cause the motor to go slower. We use the switch to control how much of the resistance is wired in series. Look at the diagram in Figure 3-16. Notice that we have now used the resistor and the switch in series to control the current. Moving the switch from OFF to LOW brings current to the motor through conductor A and all three sections of the resistor. This means that the three sections of the resistor will cause the maximum voltage drop before

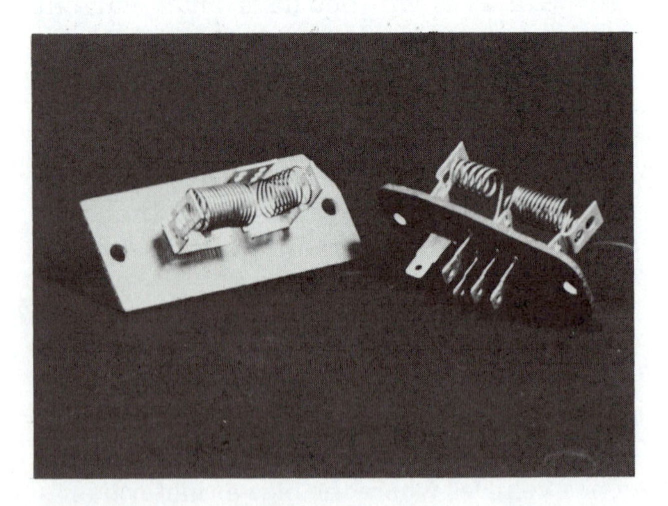

FIGURE 3-15 Common blower motor resistors.

the motor. Moving the switch up to MED (medium) brings current through conductor B and only two of the three resistors. With less resistance and more current flowing, the motor runs faster Moving the switch to HIGH brings current through conductor C and uses only one resistor in series. So far so good, you have three speeds by using three different series resistances before the motor. Now comes the relay. Look at Figure 3-17.

The conductor from the maximum position of the switch runs to the relay coil. The other end of the coil goes to B–. When the switch is placed into the maximum position, current flows through the coil, produces a magnetic field, and closes the contacts. Let us add a fused B+ conductor to one contact and the motor to the other one and you can see that when the contacts close, the motor runs full speed ahead, Figure 3-18. No series resistance equals maximum current flow.

The current through the switch is now lowest when the switch is in the maximum position. The relay coil has very high resistance, which reduces the current through the switch. Maximum current flow for the motor will go through the relay contacts when they are closed. What this effectively does is eliminate the need for the heavy and expensive contacts and conductors in the dashboard where the switch is located. The relay is usually placed under the hood in a direct line be-

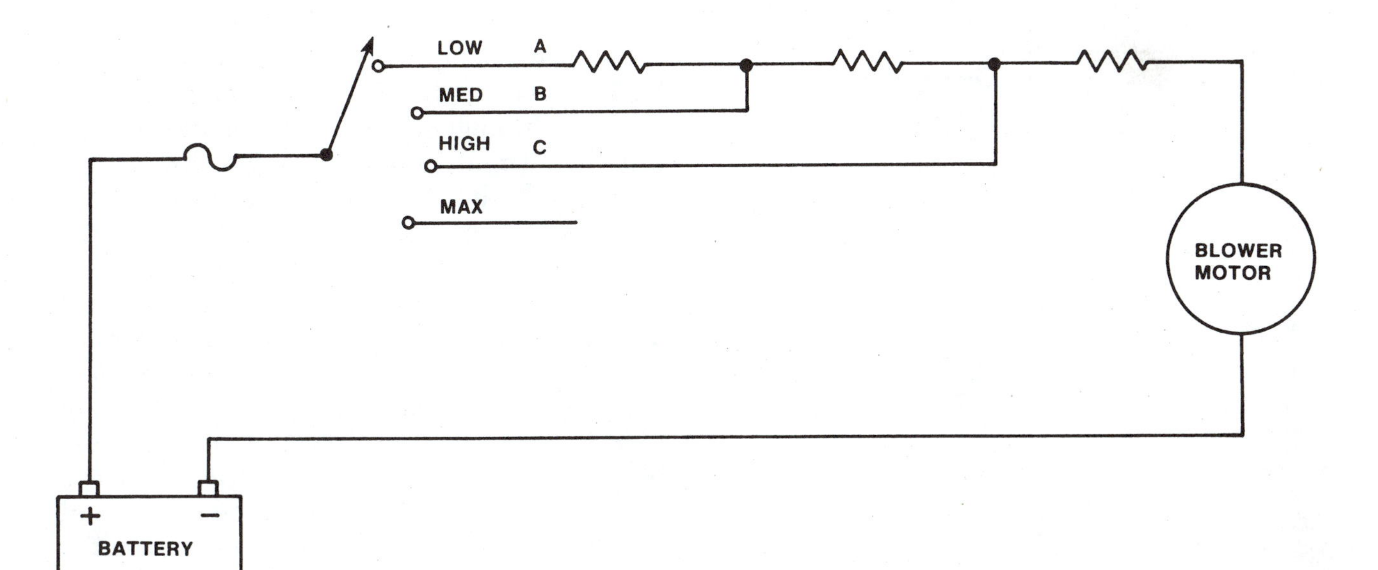

FIGURE 3-16 Common blower motor circuit.

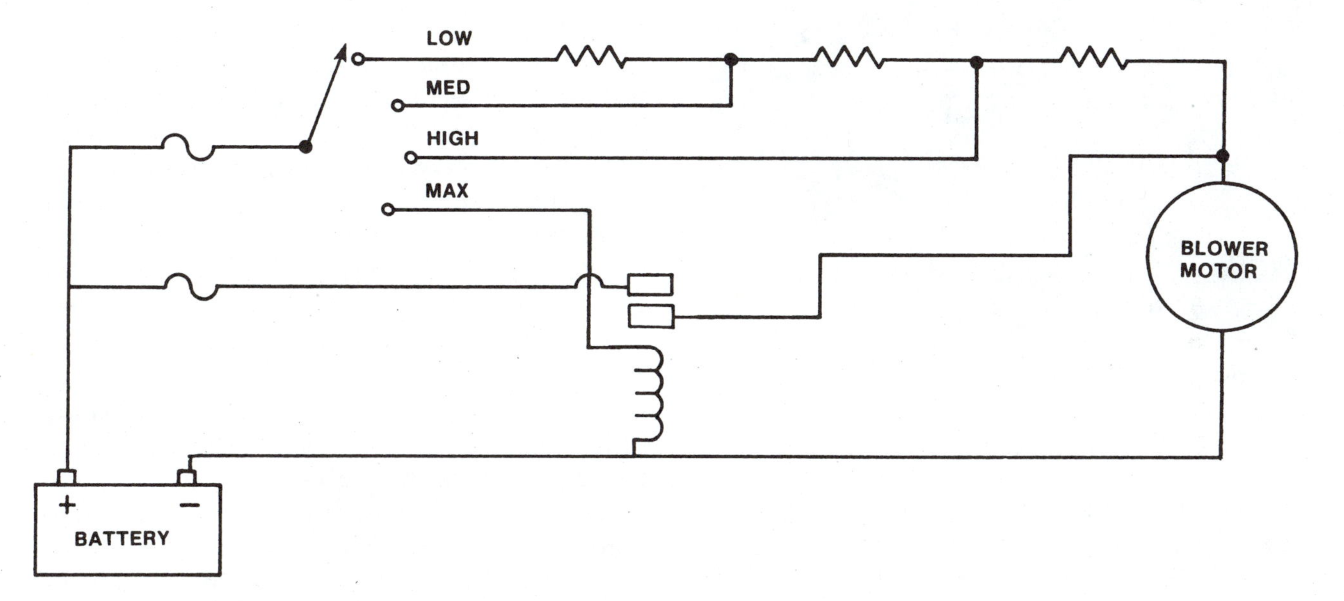

FIGURE 3-17 Blower motor circuit with a relay for maximum speed.

tween the battery and the blower motor in an effort to save on heavy connection wires, as in Figure 3-19.

Relays are in frequent use on today's vehicles. With all the heavy current devices being used, there is a move to eliminate some of the current flowing through the switches and dashboard. Look at a wiring diagram for a modern vehicle and you might be surprised at the number of relays in use.

In the last couple of years, we are finding much more reliance on electronics. In the next chapter we will analyze just what electronic components are found on vehicles and what their diagnosis procedure is. Let us look at a standard relay with some electronics added.

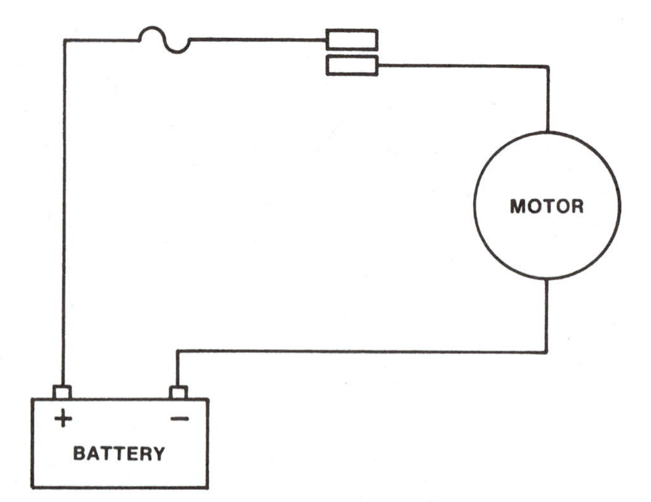

FIGURE 3-18 Closed contacts will deliver the maximum current.

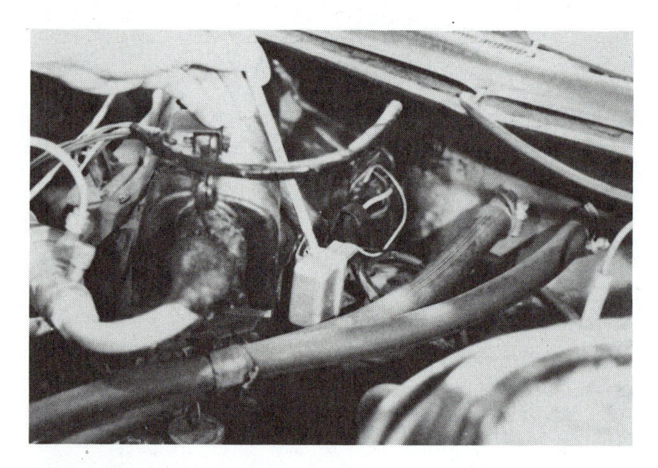

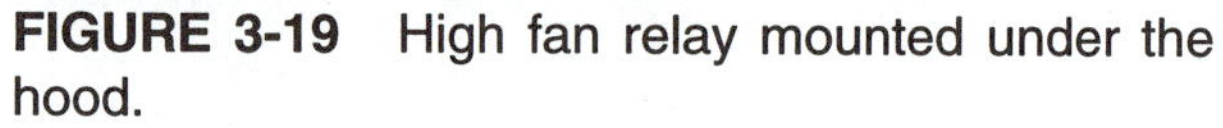

FIGURE 3-19 High fan relay mounted under the hood.

Typically, the rear window defogger found on vehicles today is a good example. Look at Figure 3-20, it is reproduced out of a Ford manual. Notice that the relay coil has a timer attached. This is basically a countdown electronic timer that will leave the rear defogger on for about 10 minutes before shutting it down. The relay also uses a momentary contact switch. A momentary switch is just what the name implies. The switch will not stay in the ON position, but rather spring back to a neutral position. If you follow the lead coming out of the contacts, you will see that the relay coil is powered off of the closed side of the contacts. The driver closes the contacts manually. This also applies B– to the relay coil. The relay coil's magnetic field now holds it closed until the timer removes the B–. Any interruption of either B+ or the timer running out will turn off the heating element at the rear window. Turning off the ignition switch, for example, will remove B+ from the relay coil and allow the contacts to open. The vehicle driver will have to restart the heating by closing the switch momentarily after restarting the car. This rather complicated setup is used because of the large current draw that the rear defogger has. Usually 10 minutes is all that is necessary to melt the snow or defog the inside window. It would be a senseless drag on the charging system to keep it on when it is not necessary. Vehicle mileage is the main consideration here. If the defogger is on longer than is necessary, it will represent a drag on the charging system that will reduce the overall vehicle mileage. Amperage that is being produced by the alternator makes it harder to turn. This equates over to a mechanical load. Some smaller engines will have their idle reduced by as much as 100 rpm when the rear defogger turns on—quite a mechanical drag.

Another type of relay, and quite possibly the first application ever found on vehicles, is the electrical solenoid, which is used in starting systems. From a definition standpoint, a solenoid is very similar to a relay because it uses low current to control higher current. It also will give some mechanical motion. Many starting systems use the solenoid to move the starter drive mechanism in to mesh with the flywheel before it powers the starter motor. In this manner, the solenoid times or sequences the starting action. First, the drive mechanism is meshed with the flywheel, then current is switched on to the starter. Let us look at a typical solenoid. Figure 3-21 is an electrical solenoid. Notice that the relay coil has two windings labeled pull-in and hold-in. The mechanical motion of pulling the starter drive in to mesh takes quite a bit of current, but it will take very little current to hold it in place. For this rea-

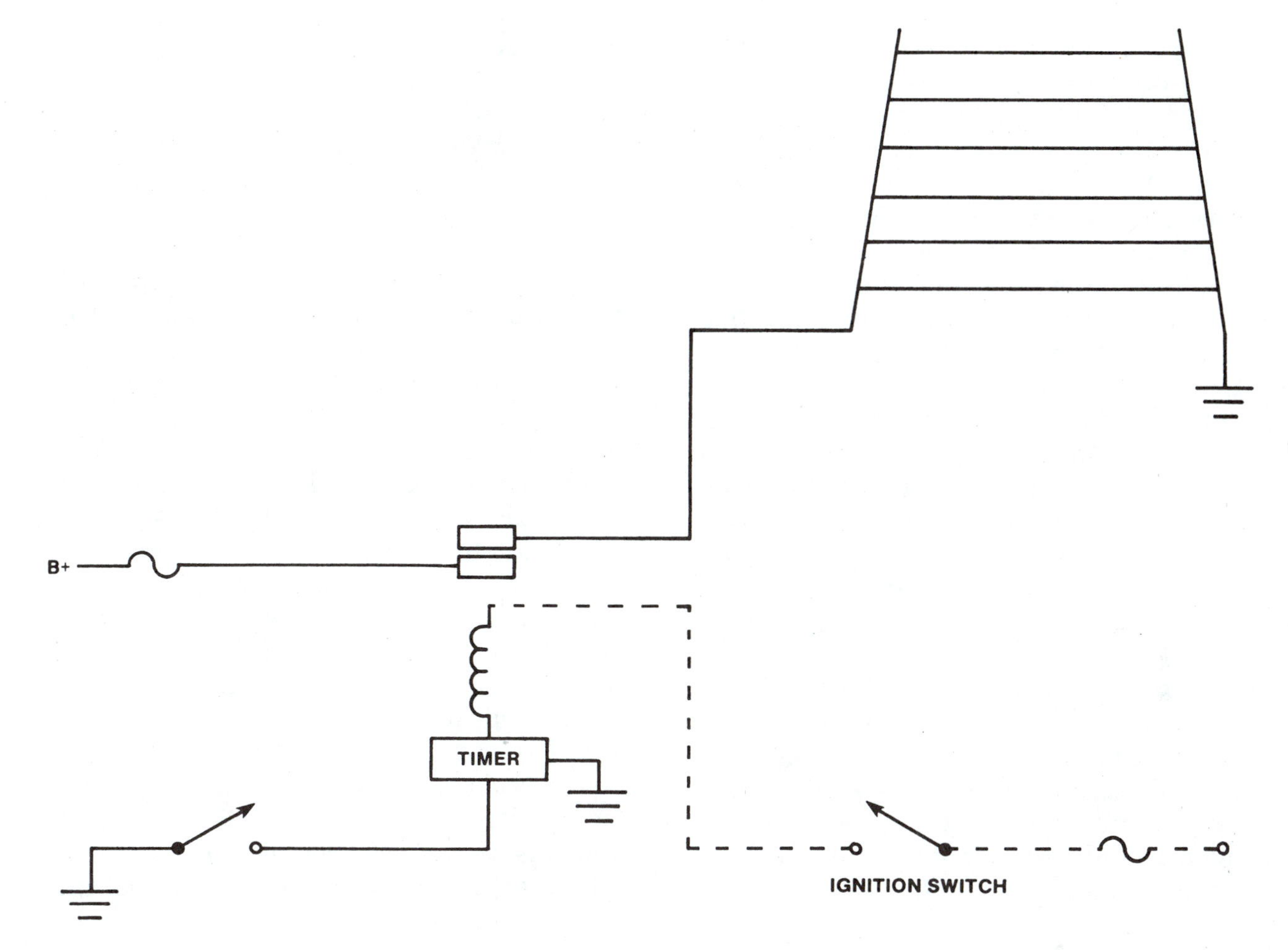

FIGURE 3-20 Common rear defogger circuit with a timer.

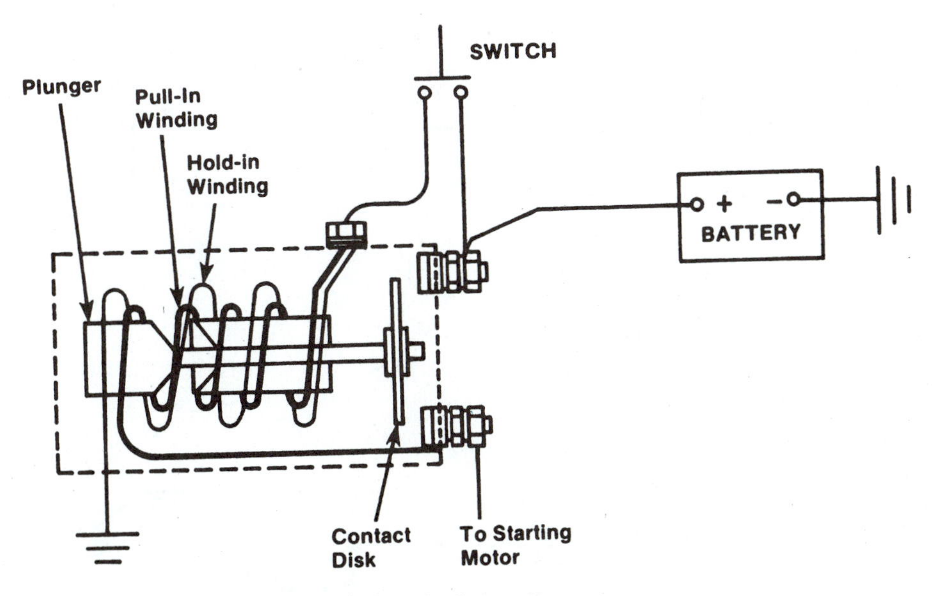

FIGURE 3-21 Solenoid with pull-in and hold-in windings.

son, two coils are used in parallel to pull the plunger into the coil. As the ignition switch is closed, B+ is brought to both coils. The pull-in winding is using the starter for its source of B–. As the plunger moves into the hollow coil, it will push the contact disc into the top and bottom contacts. The top one comes from the battery and the bottom one goes to the starter. The contact disc acts like a switch and brings the current down to the starter. At the same time, it will shunt out the pull-in winding by applying B+ to both sides of it. If the same potential is on both sides of the coil, no current will flow. The hold-in winding is still drawing current because it has B+ on one side of it and B– on the other. This holds the plunger and the starter drive in mesh, while starting current turns the engine over. This style of two windings also frees up additional current to run the starter. Any current necessary to move the starter in to mesh will be available to turn over the engine once the contact disc shorts out the pull-in winding. We will look at the solenoid and how it is wired into the starting circuit in a later chapter.

OPEN CIRCUITS

Open circuits are very common in the automotive repair field. Frequently, a vehicle comes in with the repair ticket stating that some electrical component is not working. Early testing to pinpoint the difficulty as an open circuit will be discussed. But before we do this, let us identify the most common wiring style on vehicles. This is often referred to as **single wire circuits**. Obviously, we cannot have current flow or a circuit with only one wire. However, most automobiles have a metal frame that runs the full length of the vehicle. This metal is used as one of the wires in a circuit. Refer to Figure 3-22. This simple diagram shows that the return path for electricity, or B–, will be the frame of the vehicle. This will allow us to run the one wire (B+) to the load, and then connect the load's negative side to the frame. This is called a negative grounded load. Do not assume the ground is okay, unless you have tested it and know it for a fact. Probably 25% of vehicle problems are on the ground side. Schematically, a ground is shown with a symbol like this ⏚ or this ⏛.

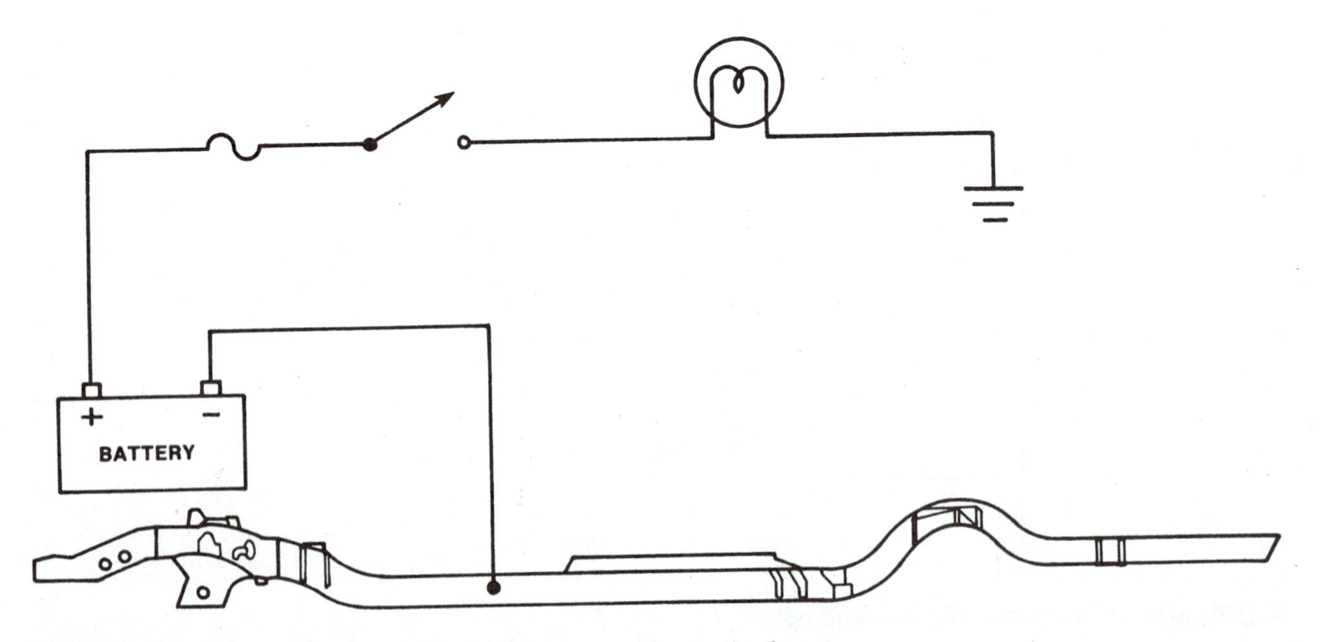

FIGURE 3-22 "Ground" refers to the battery connection to the frame.

VI. Procedure for Diagnosing Open Circuits

There are obviously many different methods for diagnosing open circuits. The one presented here works well if you follow it completely. Do not skip or jump around any of the items. Doing the same procedure for every open circuit will ensure that your own repair will not forget an item or two. Here is the procedure:

1. Verify the defect.
2. Jumper B+ to the load.
3. Walk through the circuit.
4. Isolate testing for power at connections.
5. Repair the open.
6. Retest.

Sounds simple doesn't it. Actually, it is quite simple as long as you take it by the numbers.

1. *Verify the defect.* Do not take the customer's word as to what the problem is or, for that matter, do not accept anybody else's diagnosis. Insist on doing your own work. Try to duplicate the condition that the customer is complaining about. Let us use an air-conditioning clutch as an example. Think about the circuit and identify the work that is supposed to be done. In this case, it is "engage the compressor clutch", Figure 3-23. When current will flow through the resistance of the clutch's coil, the strong magnetic field that is generated will engage the crankshaft to the outer pulley. Let us verify this defect. Open the hood, start the engine, and set the controls to maximum cooling. Is the clutch engaged? If not, you have verified the defect. Try all the air-conditioning positions that should engage the clutch. If the clutch will not function, continue to Step 2.

2. *Jumper B+ to the load,* Figure 3-24. Attach a jumper wire up to the positive battery connection and the other end up to the compressor clutch. Break open the circuit and jumper as close to the load as possible. If the load does in fact work when you have it jumpered, you have proven two very important things to yourself. First, you have eliminated the load from your list of possible problems, and second, you know that the circuit ahead of the load must be open or must contain excessive resistance. Keep in mind that if the load has s ground wire instead of relying on the vehicle frame, you will have to jumper it also. A break in the ground return circuit will appear to be a bad load if you only jumper the B+ and note that the load does not work. Items such as blower motors, headlights, taillights, or anything else mounted on an insulator will have an independent ground wire. Jumper the load on both sides.

3. *Walk through the circuit.* We do not mean walk through the vehicle wiring by hand. This is virtually impossible on modern vehicles. Instead, use the wiring diagram and look at the circuit or circuits affected. The di-

FIGURE 3-23 Compressor electromagnetic clutch.

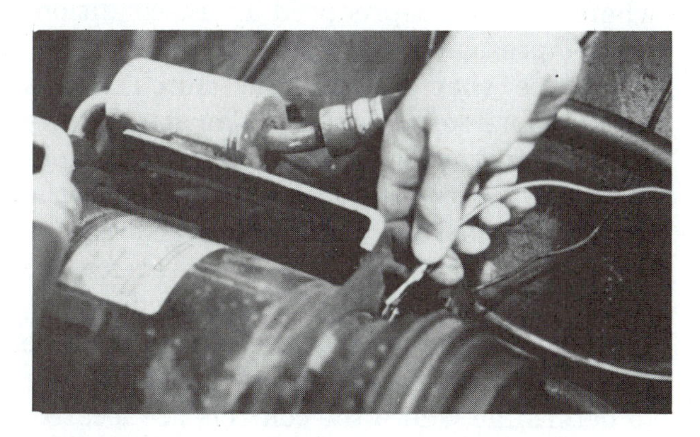

FIGURE 3-24 Jumpering the load by bringing B+ directly to it.

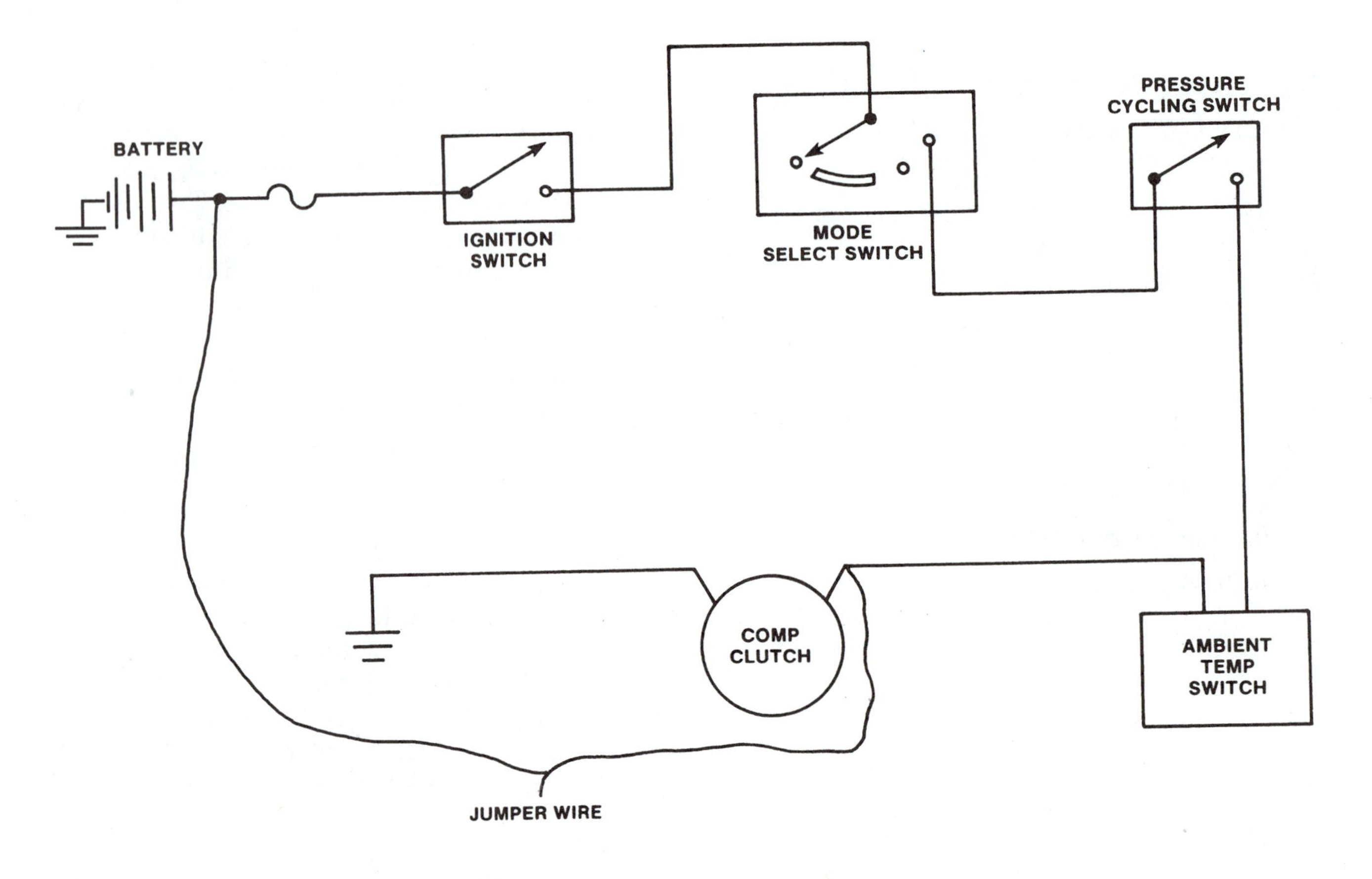

FIGURE 3-25 Air-conditioning clutch circuit.

agram will do two things for you. First, it will identify how the circuit is wired and with what color wires, and second, it will tell you where the connectors and components are located on the vehicle. You might have to use an independent component locator along with the wiring diagram. Refer to Figure 3-25 where we have reproduced an air-conditioning compressor clutch circuit.

Notice that B+ for the clutch comes through four series controls. Can you identify them? The ignition switch, the air-conditioning mode control, the cycling switch, and the ambient temperature sensor—these four controls or switches must all be closed for the clutch to engage. Without the wiring diagram, you would not know that these controls were wired in series. Now to the component locator to determine where the controls are located. Figure 3-26 shows you what a component locator for this circuit would look like.

4. Isolate testing for power at connections. This is the heart of the procedure, starting at the simplest end of the circuit (simplest or easiest to get to). No need to tear the dashboard apart, unless the underhood testing points to the mode switch. Locate the connector closest to the load, the same one you jumpered. Test for B+ with a 12-volt test light like Figure 3-27 shows. An open circuit will not pass any current. With the test light grounded and the other end in the feed line to the load, an open will be easy to spot. The test light will be off. No current available equals no light! If no light is present, we move up the B+ line to the next component or connector as identified by the wiring diagram. As we move toward the source, we are looking for B+ and a test light that lights. Once we find a connector that is hot, we know that the open is between the dead connector and the hot one, Figure 3-28.

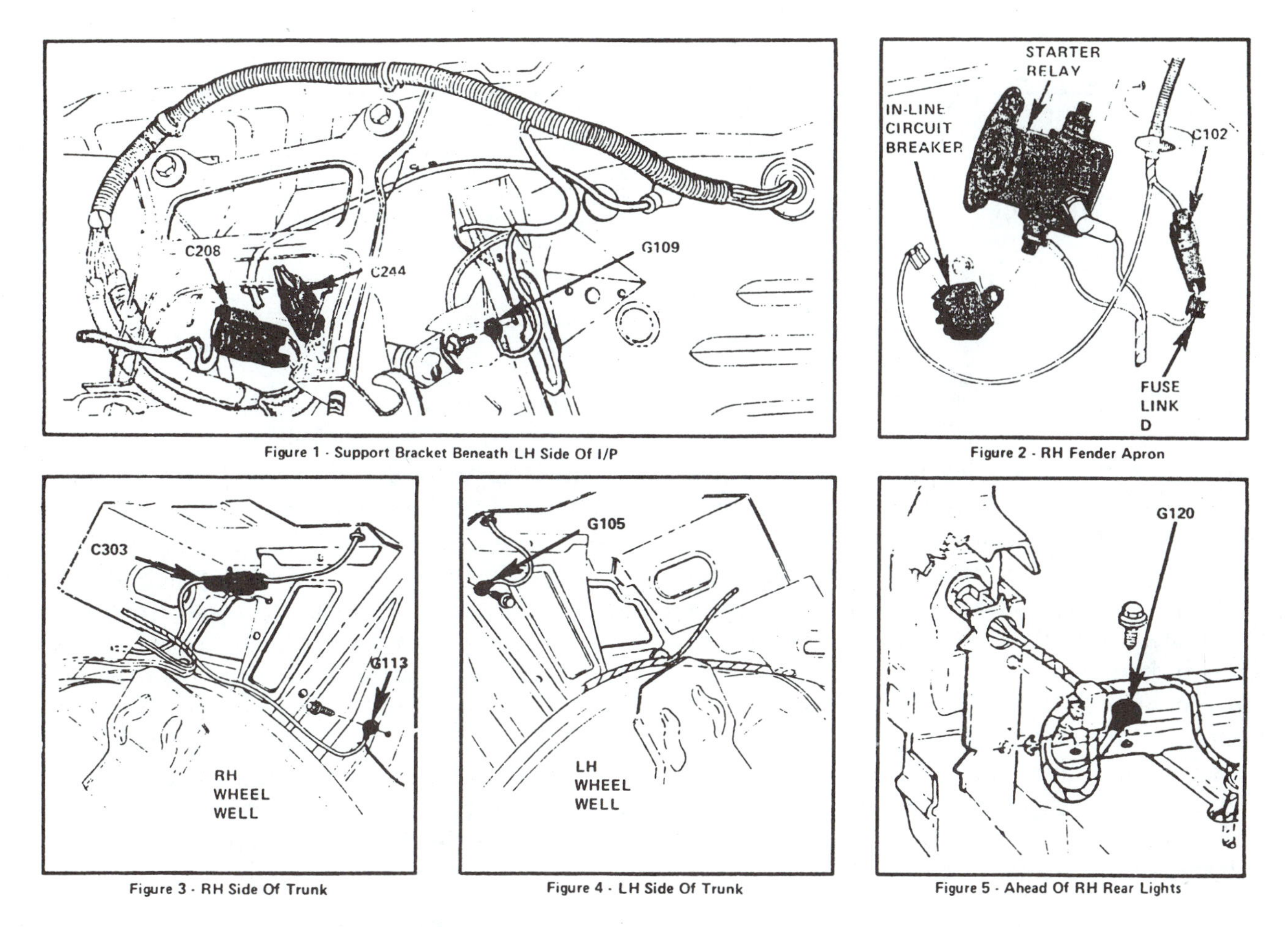

FIGURE 3-26 Component locator is a necessity on today's vehicles. (Courtesy of Ford Motor Company)

FIGURE 3-27 Testing for power at connectors.

5. *Repair the open either by replacing an open component or repairing a conductor.* This becomes the simplest part of the procedure. Identifying the open is sometimes time consuming because of the length of the circuit. Repairing the open is usually accomplished in a fraction of the diagnosis time. An open wire is repaired with a solderless connector or sometimes soldered and insulated with tape or heat shrink tubing Repairing the defect so that it will not reappear is important. Take your time and do it right. Remember, if the circuit went open because of a weakness in the harness or excessive movement of the wire, it can reoccur. Use wire ties or clamps and eliminate harness flexing so that your repair will last longer than the original.

6. *Retest.* Obviously, this is the easiest item to accomplish. Turn on the circuit. Make

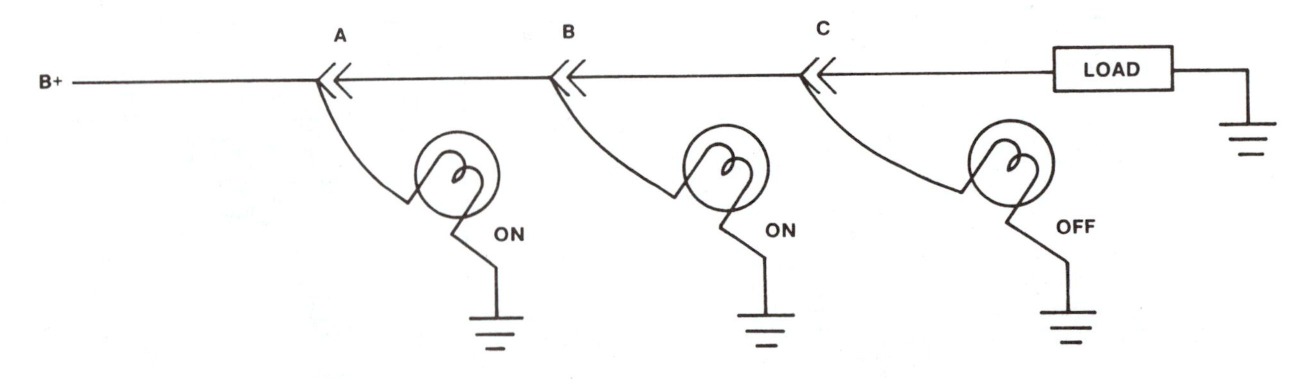

FIGURE 3-28 Open circuit identified between connectors B and C.

sure it works to specification. Also, measure the applied voltage to the load to make sure that no additional resistance is present, Figure 3-29. This is extremely important. Remember, series resistance will decrease both the voltage applied and the current used. This, in itself, can cause additional problems with some loads. Our example of a compressor clutch is in fact one of those loads that reduced B+ will burn out. Not enough current will cause it to slip especially at high speeds. This will greatly reduce its life. If the applied voltage is the same as the battery voltage, your job is finished.

The last thing to make note of here is that you should look for the primary cause of the open. If in your diagnosis you find an open wire, try to figure why the wire went open. Was there too much current flowing? Was the wire flexing too much? Was it pinched in a hinge? Answers to these and other questions will help to determine just how you will repair the defect. Do not leave the cause of the open circuit in the vehicle if you can help it.

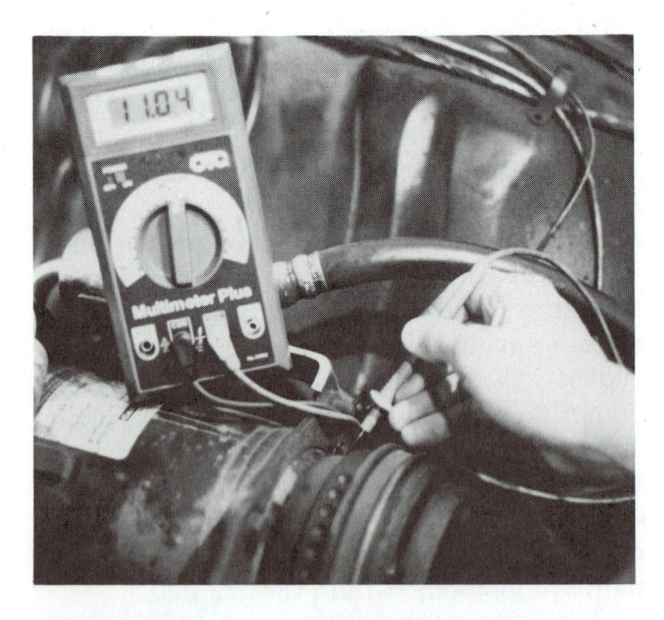

FIGURE 3-29 Measuring the voltage applied to the load.

Also, keep in mind that many open circuits are actually short circuits that have burned out the circuit protector. Finding and eliminating short circuits is our next topic.

SHORT CIRCUITS

Before we discuss the procedure for finding and eliminating short circuits, let us look at them in terms of resistance. Most short circuits will burn open a circuit protector. A fuse, fusible link, or a circuit breaker should open up the circuit and prevent more extensive damage. Why does the circuit protector open up? Hopefully, you realize by now that a momentary overload of current burns open the element inside. Think about resistance and its effect on current. Remember, the less the resistance, the more the current. So, actually, we are looking for a reduction in resistance that has caused the circuit protector to open. Two sides of the circuit must have come in contact with one another. This is a lot easier than it sounds. The entire frame and most of the body on a majority of vehicles is metal and at ground potential. If a B+ conductor touches this ground potential, we have a short circuit around the designed resistance of the load, a circuit protector opens up, and a customer has a dead component or accessory on the vehicle. Always think of short circuits as a reduction in resistance that causes an increase in the amount of current.

VII. Procedure for Diagnosing Short Circuits

Let us use a procedure very similar to our open circuit diagnosis.

1. Verify the short and identify the type.
   a. Dead
   b. Intermittent
   c. Cross
2. Walk through the circuit.
3. Control the current with a false load.
4. Separate connectors to pinpoint short.
5. Repair.
6. Retest.

You can see that our diagnosis procedure is very much like open circuit diagnosis with a few changes. Let us look at them one at a time.

1. *Verify the short and identify the type.* Plug in a fuse and try to make it blow. Turn on all items on the circuit protector one at a time. Try to pinpoint the circuit that overloads the protector and opens it. If you can make the fuse blow, you are dealing with a dead or constant short to ground. If you cannot make the fuse blow and the vehicle came in the door with a blown fuse, you are dealing with an intermittent short that is not present right now. Cross shorts, or shorts between two circuits, are easy to identify.

The short causes two circuits to be powered off of each other.

2. *Walk through the circuit using the wiring diagram.* Identify where the connectors are with the component locator. Notice what components are on the circuit protector, Figure 3-30. Many times the manufacturer will wire different circuits together. This can make the diagnosis either easier or more difficult, depending on just how many circuits

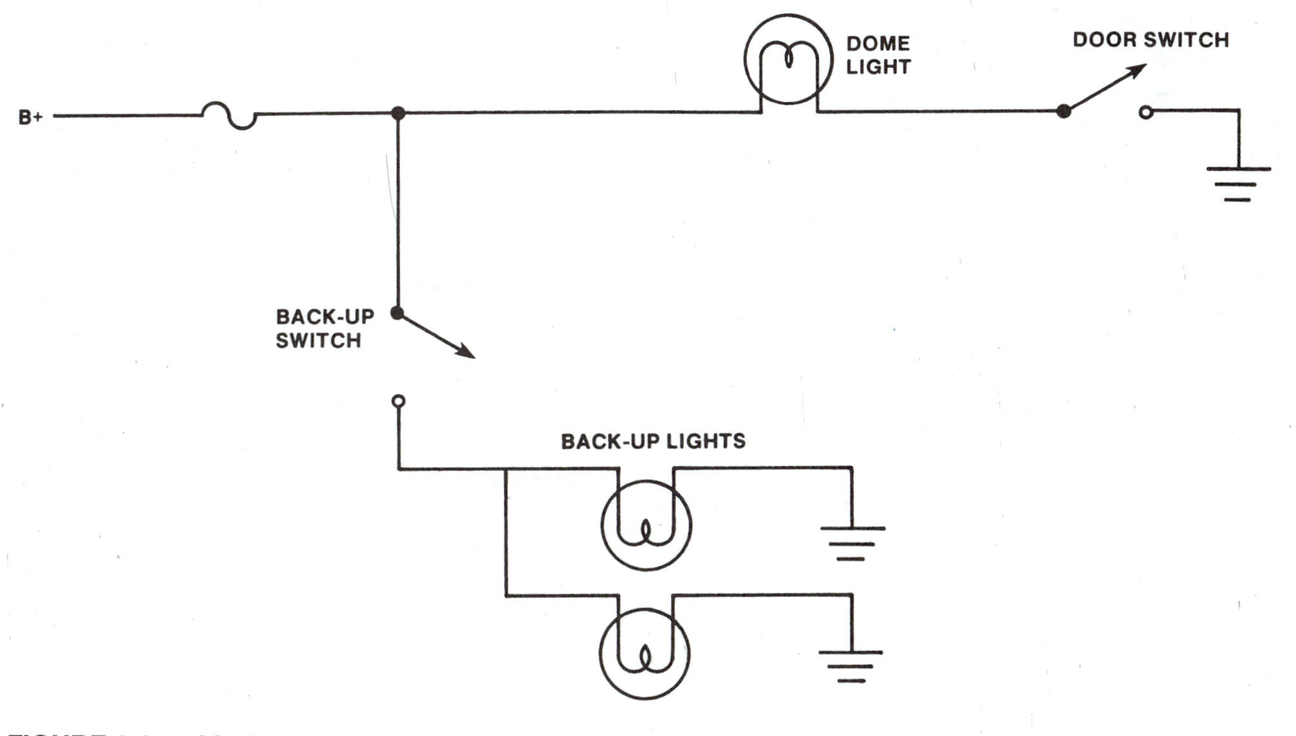

FIGURE 3-30 Multiple loads on one fuse.

are involved. Intermittent shorts on multiple load circuits can take a long time to trace down.

3. *Control the current with a false load.* Here is the most important concept. Whenever a circuit protector opens, it is because of an increase in current. This increase in current is because of the decrease in resistance. What we want is to be able to have time to diagnose our short while it is actually occurring. This is difficult to do because the short causes too much current to flow. Enter the short detector or loud test light. This device has a headlight and a key warning buzzer wired to a flasher. If it is connected to B+ and B–, it will buzz and light. More important than this, it will control the current if it is wired in place of the burned out fuse. The light and buzzer are forms of resistance that when wired in series will buzz and light because of the B+ available at the fuse box and the short to ground.

4. *Separate connectors to pinpoint the short.* Now comes the easy part. The excessive current has been controlled and we have a visual and audible indicator of the short. Now we separate the connectors in succession, moving away from the fuse. As you open the connector, if the loud test light stops, you have eliminated the short. In other words, it is beyond the connector you have opened. If the test light still is making noise, the short is ahead of the open connector. In this manner, we can pinpoint the short usually to a small section of the wiring harness and then physically examine the section looking for the short to ground.

5. *Repair the short in much the same manner you did in open circuit diagnosis.* Make sure you eliminate the cause of the short. Many times the causes of shorts are found at sharp corners or hinges. Make sure you tape up the wires to hold them away from the cause of the short or you will be looking at a repeat repair down the road.

6. *Retest.* Make sure the short is gone. Replace the loud test light with the normal fuse, turn on the circuit, and try to make it blow. If you have done your job, it will not!

This procedure really works and can save you tremendous amounts of diagnosis time. Just remember, the goal when dealing with dead shorts will be to turn the tester off. When you have turned the tester off, you have found the short. If the short is intermittent, you will hook up the short detector and try to move the wires until you turn on the tester. Turning on the tester means you have created the short. Now eliminate it in the same manner you did with the dead short and you are in business.

SUMMARY

In this chapter, we have taken a look at the three types of circuits: series, parallel, and series-parallel. Examples were given that showed the addition of resistances in series and parallel. Ohm's law was used to find any unknown given the two knowns. In addition, various forms of Ohm's law were used to find the voltage drops across loads wired into different circuits. The wattage formula was used to demonstrate loads in series and parallel and the actual work accomplished. The use of relays in simple circuits such as the horn were analyzed in addition to the solid-state type currently in use in rear window defoggers. Solenoids with both pull-in and hold-in windings were diagramed as they are used in starting circuits. The chapter ended with a procedure for diagnosing both open circuits and short circuits. The use of a short detector or loud test light was demonstrated.

KEY TERMS

B+

Open circuits

Single wire circuits

CHAPTER CHECK-UP

Multiple Choice

1. Adding additional resistance to a series circuit will
   a. have no effect on the current flow.
   b. reduce the current flow.
   c. increase the wattage.
   d. increase the current flow.
2. Technician A states that the resistance in a series circuit is added together to find the total. Technician B states that the total resistance in a parallel circuit is always lower than the lowest single resistor. Who is correct?
   a. Technician A only
   b. Technician B only
   c. Both Technician A and B
   d. Neither Technician A nor B
3. Three 10-ohm resistors are wired in parallel. What is the total resistance?
   a. 30 ohms
   b. 10 ohms
   c. 3.3 ohms
   d. 33.3 ohms
4. How much current will flow through the circuit in Question 3 if 12 is applied to it?
   a. 3.6 amps
   b. 39.6 amps
   c. 0.275 amps
   d. none of the above
5. A taillight bulb is glowing dimly. A voltmeter across the bulb reads 7.4 volts. Technician A states that there must be a voltage drop somewhere else in the circuit. Technician B states that resistance in parallel must be present. Who is correct?
   a. Technician A only
   b. Technician B only
   c. Both Technician A and B
   d. Neither Technician A nor B
6. A 2-ohm resistor is wired in series with two 7-ohm resistors in parallel. The total resistance is
   a. 3.5 ohms.
   b. not enough information is given.
   c. 5.5 ohms.
   d. 6.9 ohms.
7. To find the total current flow in any circuit
   a. multiply the amperage times the voltage.
   b. divide the voltage by the amperage.
   c. multiply the voltage times the resistance.
   d. divide the voltage by the resistance.

8. A 2-, 3-, and 4-ohm resistor are wired in parallel. Their combined resistance is
   a. 9.00.
   b. 0.93.
   c. 1.08.
   d. 2.16
9. How many amps will flow through the three resistances above if you apply 12 volts to them?
   a. 11.1
   b. 6.4
   c. 3.1
   d. 12.90
10. Open circuit diagnosis involves jumping the load. This tells the technician
    a. where the open circuit is.
    b. why the open occurred.
    c. whether the open is before the load or in the load itself.
    d. whether the open is intermittent.
11. A 12-volt test light is touched into a connector. The light lights. Following the conductor forward (toward the load), the next connector is probed with no light. This indicates
    a. nothing.
    b. open circuit between the two points.
    c. the load is open.
    d. there is an open before the first connector.
12. The object of using a false load in short circuit diagnosis is to
    a. control the current by adding resistance in series.
    b. reduce the total resistance.
    c. increase the current flow.
    d. decrease the current by adding resistance in parallel.

True or False

13. Voltmeters are always wired in parallel.
14. Ammeters are always wired in parallel.
15. The total amperage flowing in a parallel circuit is equal to the individual legs.
16. A short circuit will result when a hot wire is grounded before reaching the load.
17. A jumper wire is used to bypass a load.
18. Automobiles use single wire circuits.
19. The frame of the vehicle is normally attached to the negative battery post.
20. Circuit protectors are usually found on the negative side of the circuit.

Electronic Fundamentals

OBJECTIVES

After completion of this chapter, you should be able to:

- define the difference between a conductor, a nonconductor, and a semiconductor.
- measure the forward and reverse resistance of a diode.
- test a despiking diode.
- explain diode action.
- wire a PNP transistor or an NPN transistor into a simple circuit.
- explain the difference between a PNP and NPN transistor.
- determine temperature using a thermistor and an ohmmeter.
- identify position using a voltmeter and a throttle position sensor.
- identify pressure and pressure changes using a manifold absolute pressure sensor and a DVOM with frequency capability.
- determine air fuel ratios using an oxygen sensor and a DVOM
- outline the differences in mechanical switching, transistorized switching, and integrated circuit switching.
- list the precautions when working on vehicles equipped with solid-state components.

INTRODUCTION

The vehicle of today would be very different if it were not for electronics. For purposes of clarity, let's define electronics as the technology of controlling electricity. Electronics has become a special technology beyond electricity. Transistors, diodes, semiconductors, integrated circuits, solid-state devices, and micro-processors are all considered to be part of the electronics era rather than just electrical devices. Within this chapter, we will examine the history of electronics and then look at some of the common electronic devices in use today on vehicles.

Electronics didn't just begin a few years ago. Actually, scientists began experimenting in the early 1900s with gas-discharge tubes, Figure 4-1. These tubes contained nitrogen

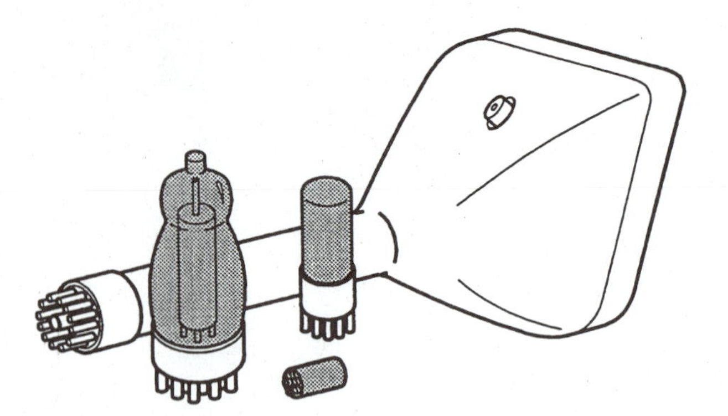

FIGURE 4-1 Electron tubes had many purposes in years past. (Courtesy of Ford Motor Company)

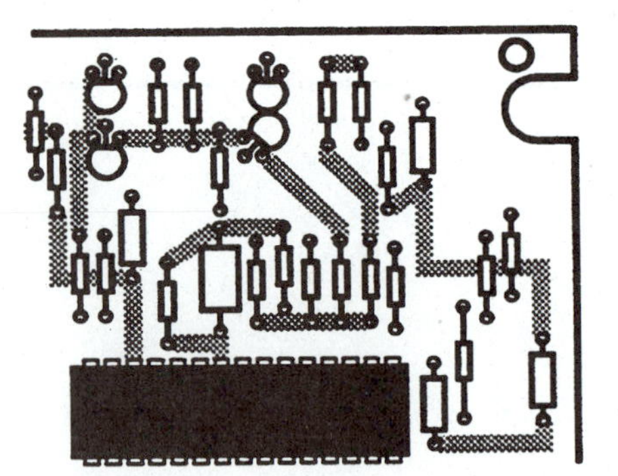

FIGURE 4-2 The integrated circuit or chip has changed the automotive electronics field. (Courtesy of Ford Motor Company)

and hydrogen and would do various things when electrical currents were passed through them. Switching on and off circuits or amplifying small signals were some of the common functions performed by these early tubes. By the 1920s, the same things were being done to vacuum tubes. Vacuum tubes aided in sending and receiving the first radio signals and ushered in the first phase of a three-phase development of electronics that lasted until the 1950s. The use of vacuum tubes in vehicle applications was limited to the dashboard radio, which became commonplace in the 1950s. The second phase of the electronics era began in the fifties with the introduction of solid-state devices. The term **solid state** is applied because of a comparison between vacuum tubes. Tubes did switching or amplifying when current was passed through a gas, or vacuum. Current flowing through certain "solid-state" materials could do the same switching or amplifying. The vacuum and/or gas that the tube had in it, gave the unit its characteristics. The vacuum was replaced with a solid, such as silicon or germanium. The solid-state phase of electronics has had a tremendous effect on the vehicles of today. Electronic ignition, alternators, electronic regulators, and am/fm radios are just some of the solid-state electronic components that found their way to the vehicle. Vacuum tubes were big, bulky, generated lots of heat, and generally were fragile. They also usually required very high voltages and were not well suited to most vehicle applications. Early solid-state devices were small by comparison and required low voltages. The late fifties and sixties found increased application for solid-state devices by automobile manufacturers.

This third phase of electronics came with the introduction of the **integrated circuit**, shown in Figure 4-2. The integrated circuit has really found a home within the vehicle of today. Controlling virtually every system on the vehicle is the computer, a micro-processor which is composed of many integrated circuits. Integrated circuits contain thousands of electronic parts on a paper-thin **chip**, Figure 4-3. The term chip is synonymous with integrated circuits. Many of the jobs performed by conventional electrical circuits can be done

FIGURE 4-3 The chip is usually made of silicon and can measure less than 0.15 square inches (4 millimeters). (Courtesy of Ford Motor Company)

with chips. But chips are hundreds of times smaller and, generally, less expensive to build and operate than conventional circuits. The earliest integrated circuits were designed in the early sixties and were for guided missiles and satellites. This led to the pocket calculator, digital watch, and automotive computer so common today.

What the next phase of electronics will be is not clear yet. Without a doubt the previous three have had a large impact on the vehicles of today. A working knowledge of their principles will help you in their diagnosis. In the next sections of this chapter, we will cover some of the common electronic components and chips that are in use on the vehicle. We will attempt to give you a working knowledge of what the component does, not necessarily how the item was designed or manufactured. We must have the diagnostic ability to find faulty electronics, but do not necessarily have to have an electronics degree to be able to do so.

CONDUCTORS AND INSULATORS

Let's begin with a review of a conductor and an insulator (nonconductor). We defined a conductor as a material whose outer electron orbit, the valence orbit, contained few, loosely held electrons, Figure 4-4. These one or two electrons easily moved to other atoms if an electrical pressure (voltage) was applied to them. We find cooper, aluminum, and so on in common use in electrical systems. These are conductors that will carry the electrical current to the load, where it will do the work for us.

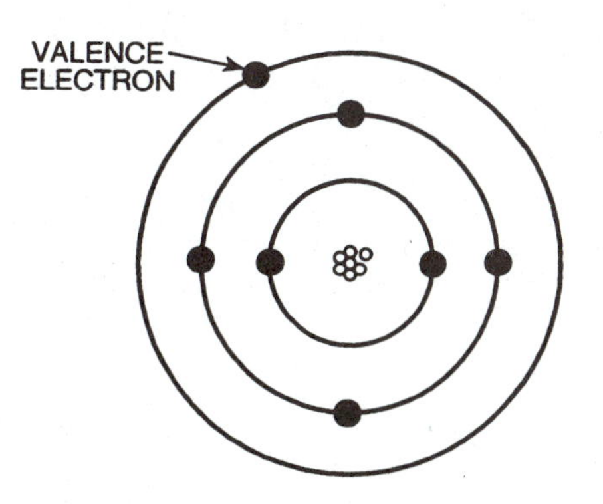

FIGURE 4-4 A conductor has one or two electrons in the outer orbit. (Courtesy of Ford Motor Company)

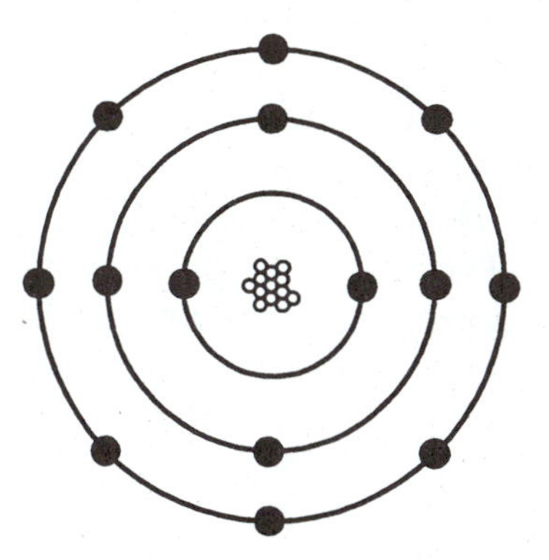

FIGURE 4-5 Nonconductors have many electrons in their outer orbit. The valence orbit (outer orbit) is usually full or nearly full in a nonconductor. (Courtesy of Ford Motor Company)

The electrical opposite of a conductor is a nonconductor. Nonconductors are most commonly referred to as insulators, Figure 4-5. It takes eight electrons to fill the outer orbit or shell. The electrons found in full valence orbits resist moving to other atoms when a voltage is applied to them This characteristic is what makes them have insulating properties. Plastic, rubber, and so on are common nonconductors or insulators found on vehicles.

What if a valence orbit is not completely full and yet contains more than the usual one or two electrons common to conductors? Their outer orbits with more than a conductor's number of electrons and yet less than an insulator's number make them not quite a conductor and not quite a nonconductor. They are referred to as semiconductors and usually have four valence electrons in their outer orbit. They are neither good insulators nor good conductors. Energy will be necessary to make them conduct, but more energy than is necessary for a conductor and less than is required for an insulator. The principal elements that exhibit the semiconductor characteristics are germanium and silicon. The most commonly used is silicon because of its availability, resistance to heat, and cost. Virtually every beach covered with sand is a potential solid-state device composed of semiconductors, because sand is silicon. To be of practical use, the silicon is treated or doped with impurities to give it certain electrical qualities. The treatment will make it either positive (P-type) or negative (N-type). Assembling various N-type and various P-types together, gives us electronic devices able to control, switch, and amplify current, among others.

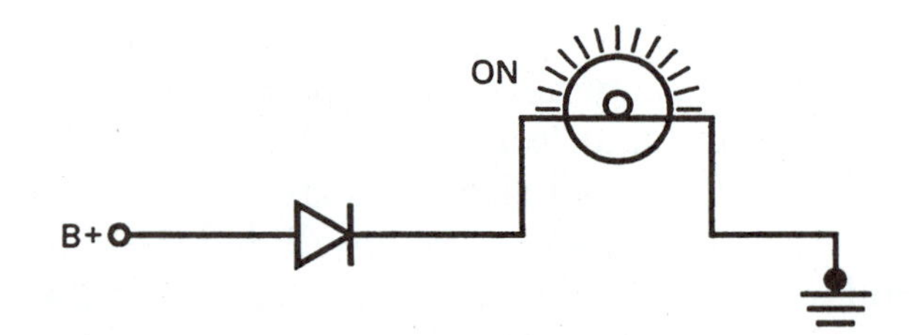

FIGURE 4-8 Schematic of a diode in a circuit with a power source and a small light bulb—bulb is on. (Courtesy of Ford Motor Company)

THE DIODE

One of the simplest semiconductor devices in use on the vehicle is the **diode**. It probably was also the first application of solid-state electronics to find its way into the automotive industry, Figure 4-6. A diode allows current to flow in one direction only. Current trying to flow in the opposite direction will be blocked. There are many different types of diodes, but they all operate on the one way check valve principle.

Figure 4-7 shows the schematic symbol for the diode. Notice that the left side of the diode (the side with the arrowhead) is identified with the notation anode. The flat line side of the diode is identified by the notation cathode. The arrow shows the direction that current will flow. With the anode toward the positive side of the circuit, the diode will conduct and the bulb will be on, as illustrated in Figure 4-8. Reversing the circuit will turn the bulb off, because the diode's cathode is toward the positive and will therefore not conduct, Figure 4-9. Diodes are frequently referred to as two-terminal devices.

Diodes are rated according to the job they must do. The amount of current they can pass easily is called the forward current rating. Exceeding this current level will damage or destroy the diode, usually burning it open, Figure 4-10. An open diode cannot pass current in either direction. The diode will get very hot if large amounts of current are passed through it. Most diodes in automotive use today will have a forward voltage drop of about 0.7 volt. We previously discovered that if a voltage drop is present, then light, heat, or a magnetic field is being produced. For this reason many diodes are mounted in finned heat sinks which are generally made out of aluminum. The aluminum fins will dissipate the heat produced by the diodes in much the same manner as a radiator will remove unwanted engine heat. Removing unwanted

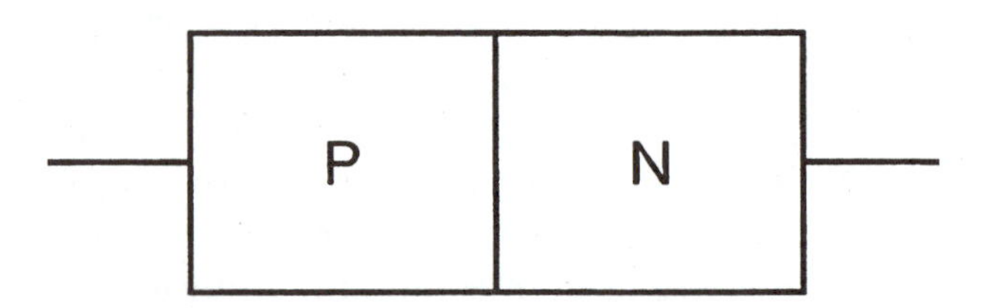

FIGURE 4-6 The diode is simply a one way check valve for electricity. (Courtesy of Ford Motor Company)

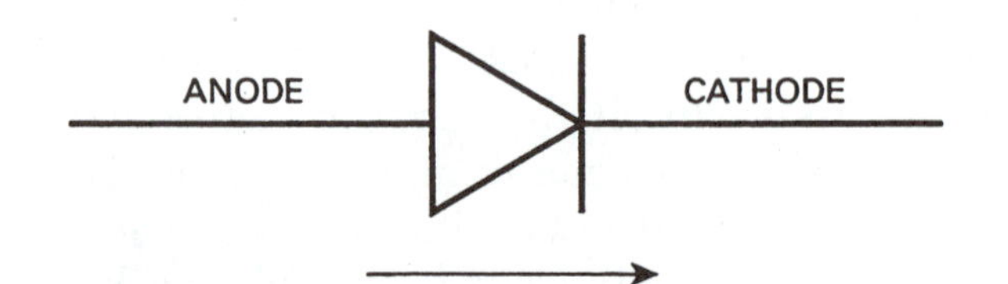

FIGURE 4-7 The schematic symbol for the diode. (Courtesy of Ford Motor Company)

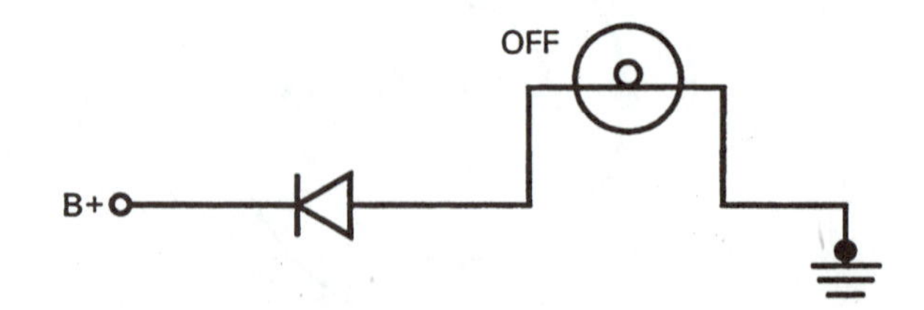

FIGURE 4-9 Schematic of diode with current blocked—bulb is off. (Courtesy of Ford Motor Company)

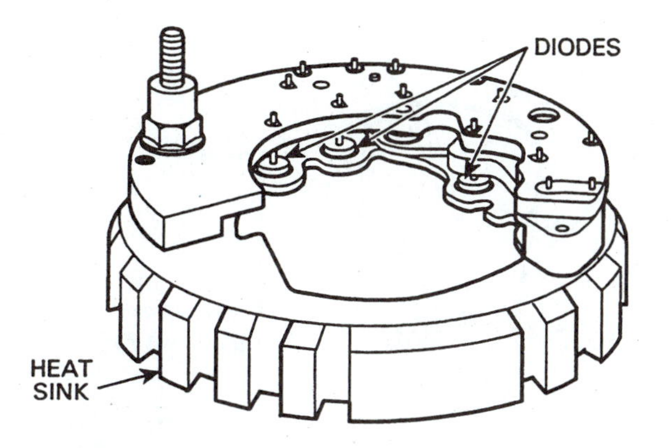

FIGURE 4-10 Diodes are usually mounted in heat sinks. (Courtesy of Ford Motor Company)

heat will protect the diode. Airflow around the heat sink is important if the forward current rating is to be reached without damaging the diode. The second method of rating diodes is their peak inverse voltage. This is the voltage that the diode can withstand safely in its blocking mode. Typically, diodes range from 50 to 1,000 volts or more. Exceeding the peak inverse voltage for even a split second can result in damage to the diode. Usually, diodes that have been subjected to peak inverse voltages in excess of their rating will short circuit themselves. A short-circuited diode will pass current in both directions.

Diodes are found in three common applications in the vehicle of today. The first application is in rectification circuits where they will take alternating current (current that flows first in one direction, then in the other direction) and change it into direct current. Figure 4-11 shows the use of six diodes in a rectification circuit. (We will see more of this in Chapter 10, Charging Systems.) This was most likely the first application of solid-state devices on a vehicle.

The second vehicle application is in isolation circuits. The blocking ability of the diode is used like a switch to prevent current from flowing where the manufacturer doesn't want it, Figure 4-12.

The third application of diodes is as a protection device for integrated circuits. A diode is located near large magnetic coils such as the air-conditioning compressor clutch or

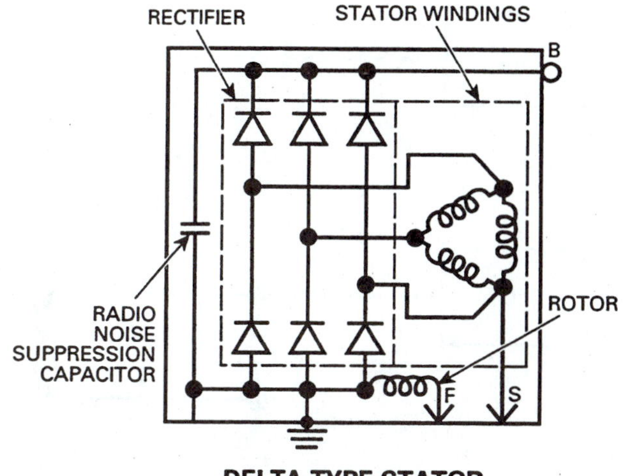

FIGURE 4-11 Six diodes in a rectification circuit. (Courtesy of Ford Motor Company)

where a relay coil is connected into a circuit that is integrated circuit controlled. When current flows through a coil of wire, it will generate a strong magnetic field. Turning off this field will generate a high voltage "spike" that, if allowed to reach the IC (integrated circuit), could destroy it. Locating the diode as in Figure 4-13, will prevent the spike from reaching back to the IC and yet will allow the IC to control the coil's magnetic field. The spike must not exceed the peak inverse voltage or the diode will conduct and destroy both

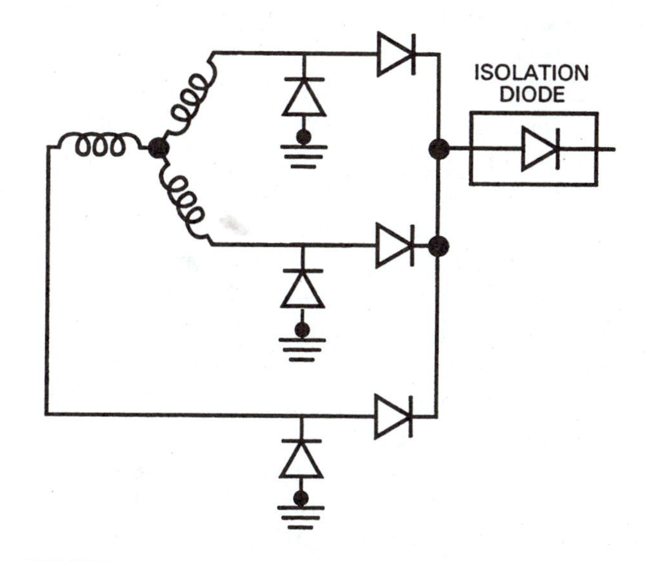

FIGURE 4-12 Isolation diodes. (Courtesy of Ford Motor Company)

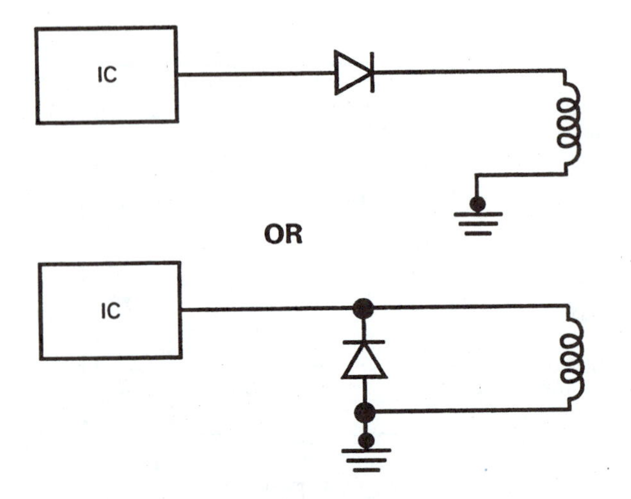

FIGURE 4-13 This diode protects the electronic device that controls it. (Courtesy of Ford Motor Company)

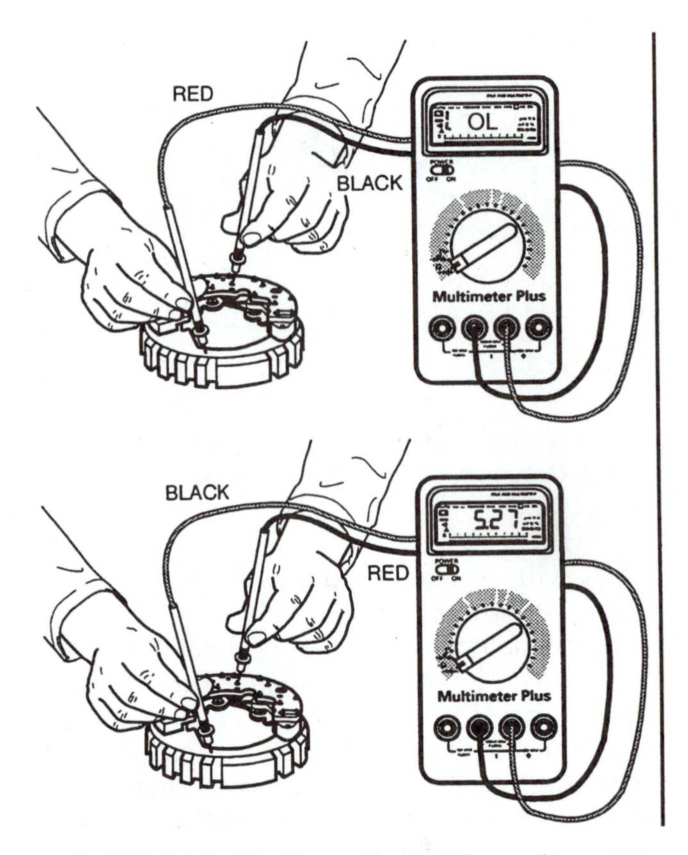

FIGURE 4-14 Testing a diode with an ohmmeter. (Courtesy of Ford Motor Company)

the IC and itself. Diodes used in this manner are called despiking or suppression diodes. When current is flowing through the coil in the normal direction, the diode is blocking. When the current to the device is turned off, a high voltage spike will be generated with a reversed polarity. This reversed polarity causes the diode to conduct, effectively giving the voltage spike a path in which to push current, other than to the IC.

We have found increased use of diodes in vehicles recently. The increased use of integrated circuits in vehicle microprocessors has increased the use of protection diodes. Some vehicles use as many as twenty separate diodes to protect the expensive and important chips. The diode can be tested by using an ohmmeter as shown in Figure 4-14, or a self-powered test light as shown in Figure 4-15. Both methods supply a DC current through the diode. If both ohmmeter readings are high or the light is off in both directions, the diode is open and should be replaced. If both ohmmeter readings are low or the test light lights in both directions, the diode is shorted and should be replaced Analog ohmmeters are generally preferred over digital ohmmeters because they will measure the resistance with a greater amount of current flowing. When using an analog meter, the × 1000 test position should be used. In recent years, however,

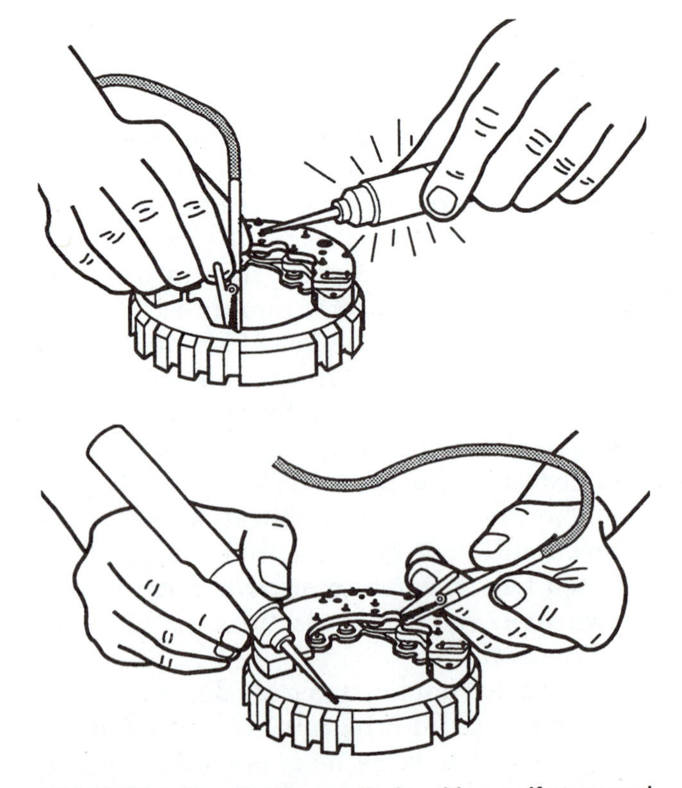

FIGURE 4-15 Testing a diode with a self-powered light . (Courtesy of Ford Motor Company)

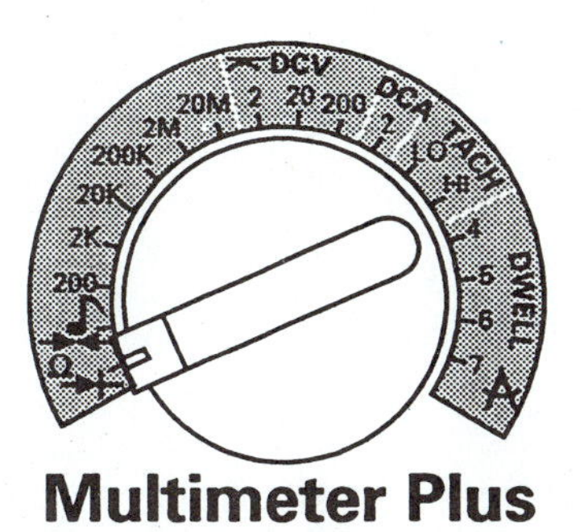

FIGURE 4-16 Digital meter with diode test position. (Courtesy of Ford Motor Company)

there have been an increased number of digital ohmmeters that have a diode test position, Figure 4-16.

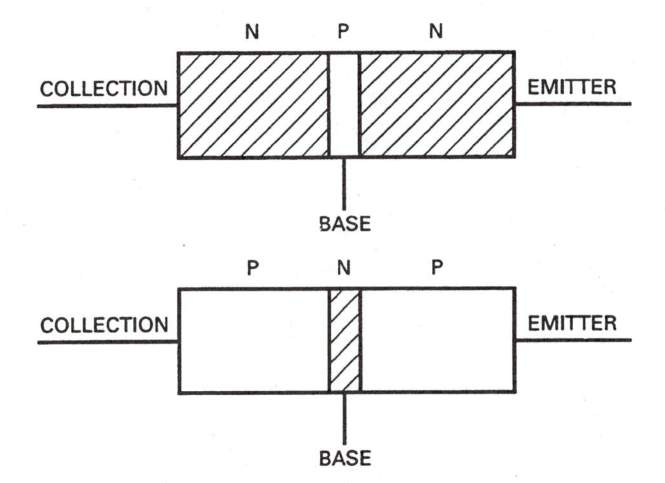

FIGURE 4-17 The two basic types of transistors, NPN and PNP. (Courtesy of Ford Motor Company)

TRANSISTORS

As was mentioned earlier, the 1950s began the age of solid-state devices. Diodes are made by joining some P-type material and some N-type material, as we have seen. Taking this one step further puts us into the world of the **transistor**. The two basic styles of transistors are **NPN** and **PNP**. Figure 4-17 shows the construction of both types. The diode previously discussed had two connections: one for the P material and one for the N material. Notice that Figure 4-18 shows three connections available, one for each layer of positive or negative material. Size and construction details determine the application of the transistor. The automobile manufacturer decides which transistor will function correctly for the circuit being designed. You will not be responsible for the designing and wiring of transistorized circuits as a service technician. However, a general working knowledge can help you in the diagnosis of failed circuits. In addition, certain precautions are necessary when working on transistorized circuits, which we will also discuss. Generally, the repair of a transistorized circuit involves the replacement of the entire circuit after diagnosis has pinpointed the failure. Individual repair or replacement of single transistors is usually not attempted in the field. The diagnosis is

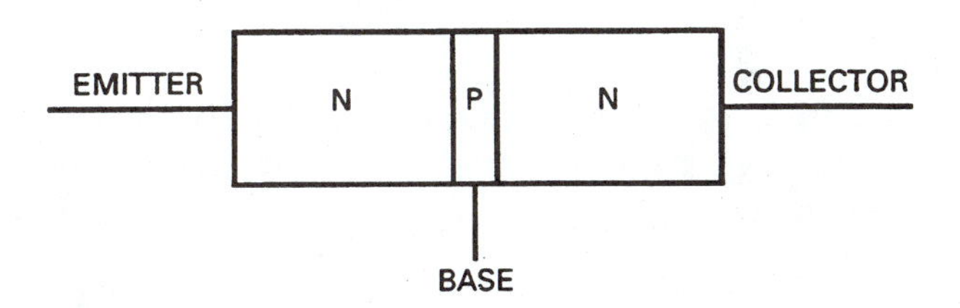

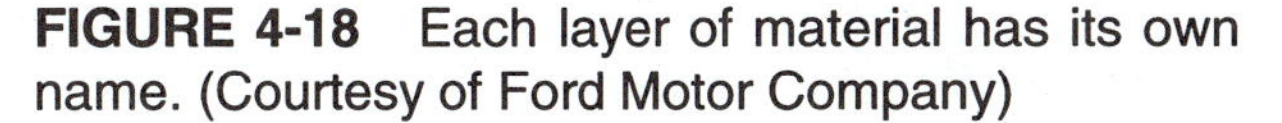

FIGURE 4-18 Each layer of material has its own name. (Courtesy of Ford Motor Company)

made simpler if you have the following working knowledge.

In Figure 4-19, notice that an NPN style transistor has the **base** and the **collector** both at positive potential, while a PNP has both the base and collector connected together at negative potential. The arrow drawn on the emitter always shows the direction of current flow (from positive to negative).

The **emitter** is the source (as in the water supply), Figure 4-20. The handle of the faucet is the equal to the base, while the end or opening where water will flow from is the equal of the collector. In a transistor, the collector is the output, while the emitter is the input. This leaves the base as the controller of current between the emitter and collector. Another way of looking and thinking about transistors is to assume that they are like two diodes with either their positive sides or their negative sides connected at the base, as in Figure 4-21.

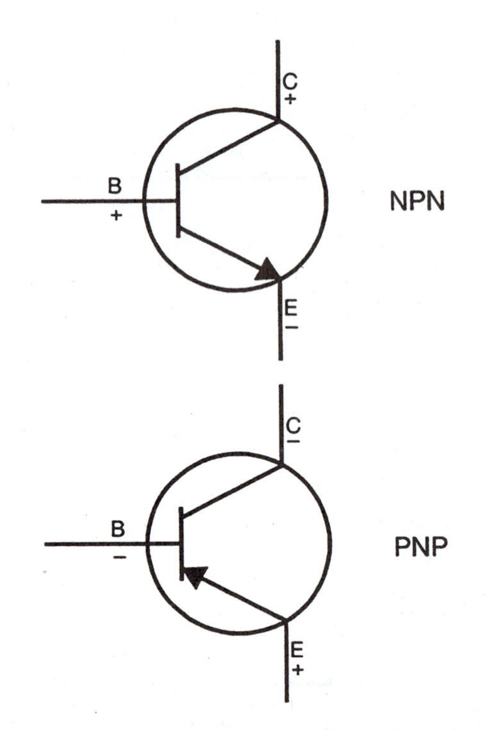

FIGUE 4-19 The schematic symbol of both an NPN and a PNP style transistor. (Courtesy of Ford Motor Company)

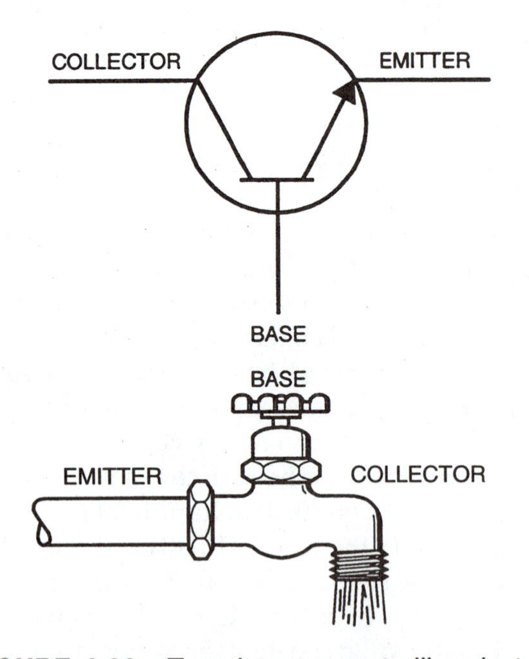

FIGURE 4-20 Transistors operate like electrical faucets. (Courtesy of Ford Motor Company)

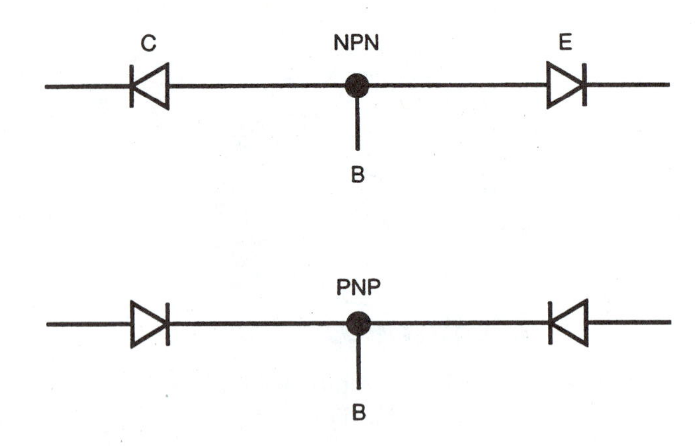

FIGURE 4-21 A transistor will test like two diodes. (Courtesy of Ford Motor Company)

The transistor is a controlling device. It should not be confused with a variable resistor or thermistor, which we will examine later in this chapter. It does, however, follow all the rules of Ohm's law. The simple diagram shown in Figure 4-22 will give you an idea of how a transistor works. Notice that voltmeter A will measure the emitter or input voltage. The collector circuit has a small light bulb to ground. We have connected a variable resistor between the base of the transistor and the positive potential. NPN transistors are positive base devices. Ammeter A will measure the base current. As the variable resistor is turned, the bulb will change intensity. This is because the variable resistance changes the current flow to the base. As current flow is added to the base (less variable resistance), more current flows through the collector and emitter. If base current is turned off completely, what will the collector emitter circuit do? With no base current, the transistor will shut down completely and no collector emitter current will flow. In this manner, the base current determines the amount of current flowing through the collector emitter. When a transistor is wired as shown, it is being used as an amplifier. A small amount of base current change can cause a larger amount of collector emitter current change.

When manufacturers are designing a circuit, their decision of whether to use an NPN or a PNP transistor is based on the design being negative or positive. PNP transistors

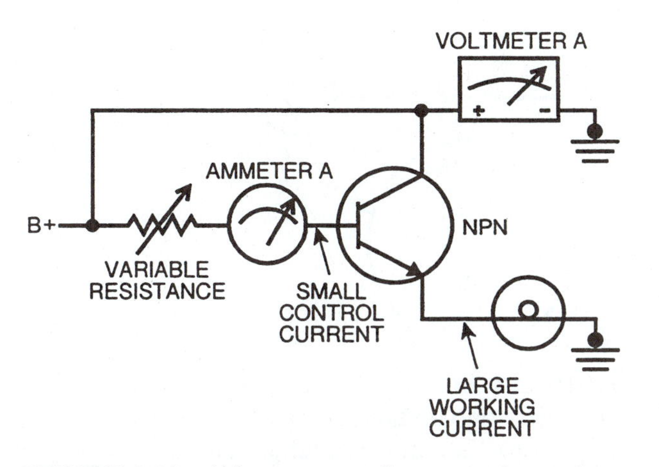

FIGURE 4-22 When current flows out the emitter, the bulb will go on. (Courtesy of Ford Motor Company)

are negative based, while NPN transistors are positive based, Figure 4-23.

Transistors are also used as switches, especially in electronic ignition. Their advantage is the fact that they can operate many thousand times a second without difficulty, while a mechanical switch can develop problems at a couple of hundred opening and closing cycles per second. We will examine their use in ignition systems in future chapters.

Our amplifier circuit had a variable signal or input to the base. Figure 4-24 shows a simple switching circuit that uses a transistor. When the base "sees" the positive current from the battery, it will conduct current through the collector and turn on the bulb. The resistor to the base limits the current that the base will see. In the example shown, the transistor can be switched on and off

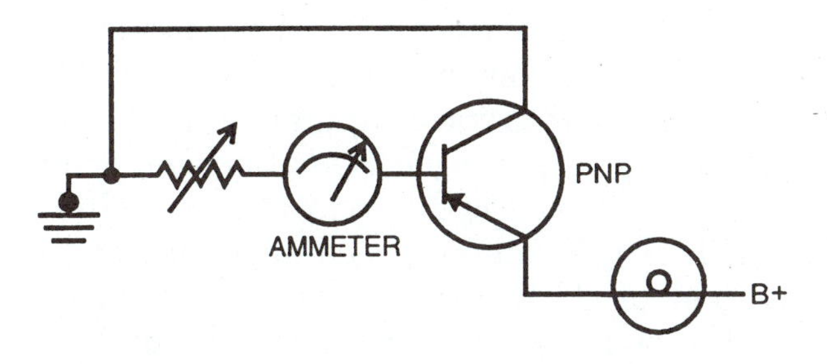

FIGURE 4-23 This is the same circuit as above, using a PNP transistor. (Courtesy of Ford Motor Company)

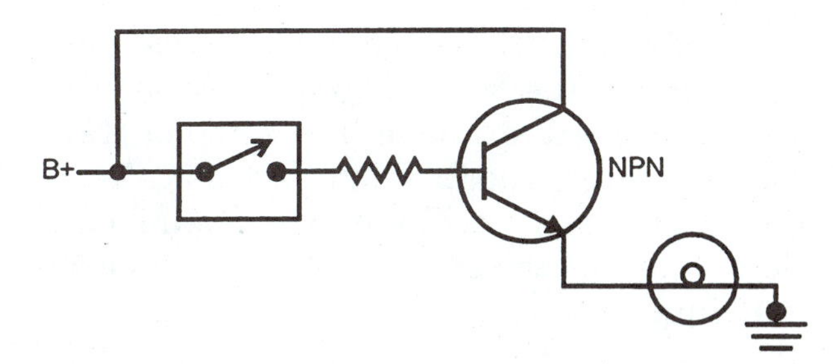

FIGURE 4-24 As a switch rather than as an amplifier, transistors are driven by an ON/OFF signal or input to the base. (Courtesy of Ford Motor Company)

many thousand times a second for extended periods of time. This advantage to a transistorized application results in greater reliability over a relay or mechanical switch. There is no arcing inside a transistor, as is the case with relay contact points. We have seen increased use of transistorized relays in automotive applications in recent years. In addition, some mechanical relays have transistor control circuits in them.

The precautions necessary when working around transistors are the same as for diodes. As a matter of fact, the precautions listed at the end of this chapter are necessary for the entire modern vehicle. Making a polarity mistake or allowing an over voltage condition to occur can cause extensive damage to most circuits. Usually, if a transistorized circuit works for a short period of time, it will function almost indefinitely. Failure of a circuit after a vehicle has been on the road for a number of miles can usually be traced to a problem outside of the semiconductors. Frequently, it can be traced to an over voltage condition caused by the charging system, or the use of too high a voltage during battery charging or jump starting.

ELECTRONIC CONTROL INPUT DEVICES

The age of electronic control of virtually everything has meant the increased use of input devices designed to give microprocessors information about temperatures, positions, pressures, and oxygen levels. In this section, we will examine how some of these common sensors do their job electronically. In later chapters we will see how the information is used by the processor to determine engine functions.

THERMISTORS

Thermistors are in use in most vehicles today. They are named because they are actually a temperature sensitive resistor. To be correct, thermistors are actually not part of the semi-conductor family. They are usually found connected to semiconductors, or integrated circuits, so we will look at their use here. Thermistors are used to detect various temperatures or changes in temperature. Their most frequent use involves the measurement of engine coolant temperature, or inlet air temperature.

The first use dates back to the mid sixties when electrical dashboard gauges first came on the scene. The thermistor was installed in the engine block with its end sitting in the engine coolant. As the coolant temperature changed, the resistance of the thermistor also changed. Ohm's law at work says that as the resistance of a circuit changes, the amperage will change also, assuming the voltage remains the same. As the temperature of the engine increases, the resistance of the thermistor decreases. This decreased series resistance increases the current flow through the ammeter and changes the needle's position. Rather than having the meter actually show amperage, the face is now redesigned to show temperature. Refer to Figure 4-25. Many dashboard temperature gauges operate on this principle. The thermistor is called the engine coolant temperature sensor (**ECT**) and usually has at least one wire attached to it from the gauge. By installing the sensor into the engine block, it is grounded, completing the circuit. Some ECTs in use today have two wires connecting them into a circuit. Their operation is the same with the ground being

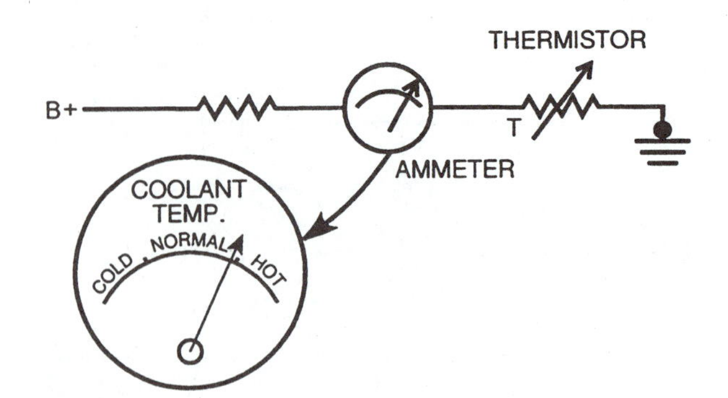

FIGURE 4-25 A common gauge circuit using a thermistor and an ammeter that has been calibrated in degrees Fahrenheit of temperature. (Courtesy of Ford Motor Company)

supplied by the additional wire rather than the engine block. This style of thermistor is a negative coefficient type. This means that as the temperature goes up, the resistance of the thermistor will go down.

When computers became responsible for air/fuel ratios and began to control ignition systems, the sensing of engine temperature and inlet air temperature became important. A cold engine requires a richer air/fuel ratio (more gas) and a hot engine requires a leaner air/fuel ratio (less gas). The engineers turned to the thermistor as the device that would relay engine temperature over to the computer. The engine computers in use cannot measure temperature directly. The input of engine temperature information must be in a language that the computer will recognize. This language is electricity. Remember, negative coefficient means that as the temperature increases, the resistance of the thermistor goes down. Notice that as shown in Figure 4-26, it is capable of inputting temperatures as cold as –40°F (below zero) and as hot as 248°F. This temperature range will become an input to the computer. "Input" is another way of saying information going into the computer. It is different from an output, which is the computer telling something to turn on or off. Everything associated with an on-board computer is either an input or an output. The thermistor signal is an input.

Many of the computer controlled fuel systems in use today utilize air temperature as

F	C	OHMS
-40°	-40	269,000
32°	0°	95,000
77°	20°	29,000
248°		1,200

FIGURE 4-26 A temperature to resistance chart for a negative coefficient thermistor-style engine coolant temperature sensor. (Courtesy of Ford Motor Company)

an input. Figure 4-27 shows an intake air temperature (**AT**) sensor. Thermistors are easily installed and wired into the computer and will have their resistance changes seen as temperature changes. The use of multiple thermistors is increasing.

The computer circuits using thermistors usually send out a small signal to the sensor. The resistance of the sensor varies the current flow and voltage drop through the circuit. The computer reads this voltage drop and knows the temperature. The small signal sent out from the computer is usually referred to as the reference voltage. Notice the wiring diagram in Figure 4-28. A reference voltage of 5 volts is common. The coolant temperature of 77°F (20°C) would put the thermistor's resistance as 29,000 ohms. In this way, the computer recognizes the voltage that has had a voltage drop caused by the 29,000 ohms as an indication of 77°F engine coolant temperature. It will now use this information (input) to help it make correct and accurate decisions or outputs. Even though it is not the intent of this text to cover computer systems completely, the input devices on many of the vehicles are relatively easy to understand and diagnose using a voltmeter and an ohmmeter. Knowing the approximate engine temperature and the actual resistance of the ECT (engine coolant temperature sensor) makes diagnosis a snap. For example, (refer to Figure 4-26) an engine temperature of 248°F and an ECT resistance of 95,000 ohms would tell the technician that he/she is looking at a bad sensor. The actual engine temperature and the temperature that the computer "sees" as an input are different by 178°F. An ECT resistance of 95,000 ohms should be an engine temperature input of 32°F (0°C) rather than the 248°F actual. In this manner, the technician can verify the information that the computer is accepting as an input is accurate.

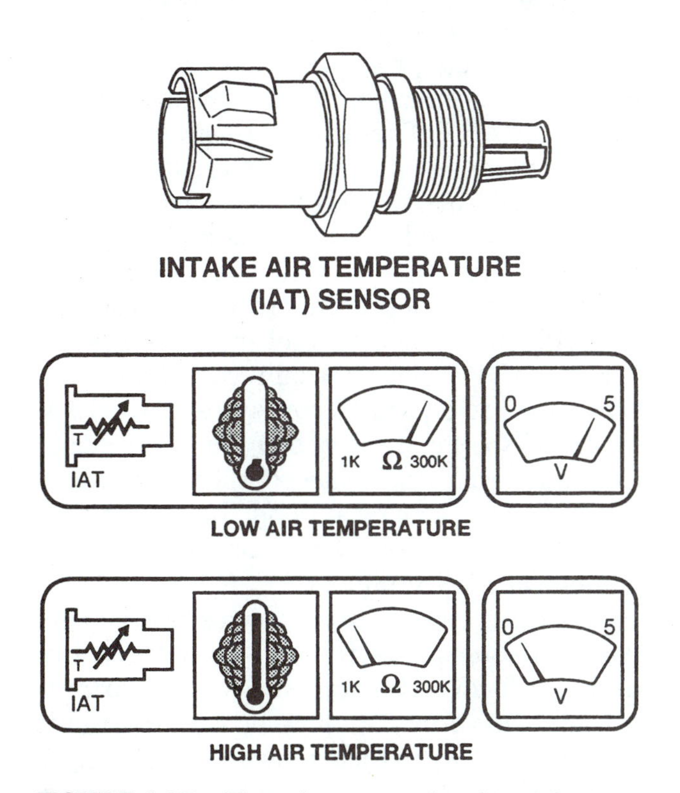

FIGURE 4-27 Thermistors can also detect the temperature of the air by using an air charge temperature sensor (IAT). (Courtesy of Ford Motor Company)

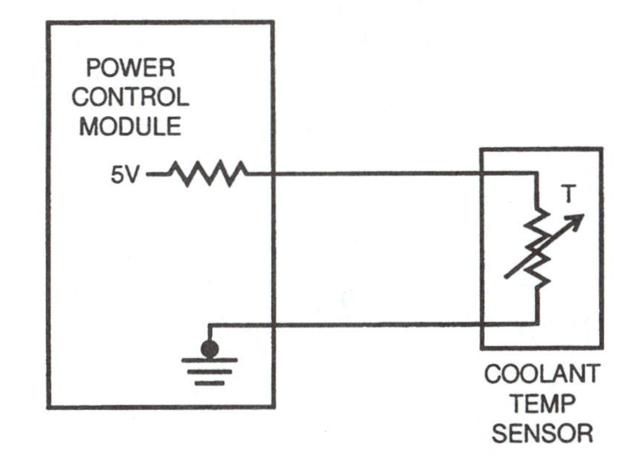

FIGURE 4-28 Diagram showing the computer sending out the 5-volt open circuit reference to the coolant temperature sensor. (Courtesy of Ford Motor Company)

Computer enthusiasts like to say "garbage in, garbage out," which translates to wrong inputs will result in wrong outputs and quite possibly driveability complaints. Thermistors are a frequently found input device anytime temperature is needed. This not only includes engine and air temperature, but transmission, transaxle, and air conditioning. It is not uncommon to have as many as 10 thermistors in use on today's vehicles.

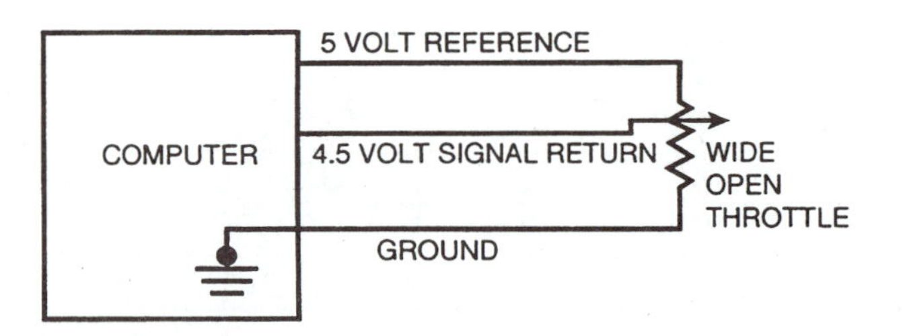

FIGURE 4-30 With a wide open throttle, the reference voltage will divide between ground and TP sensor return. (Courtesy of Ford Motor Company)

THROTTLE POSITION SENSOR

Another computer input is the **throttle position sensor** (TPS), Figure 4-29. As was the case with the coolant temperature, the computer must know what position the throttle is in to be able to determine air/fuel ratio, ignition functions, and various other outputs for which it is responsible. It uses the voltage division principle of a series circuit as its indicator of throttle position. As the throttle position changes, it will vary the resistance of the circuit. Notice that reference voltage is again applied to the TPS. With a fully closed throttle (idle), the reference voltage will divide between ground and TPS return. With the majority of the resistance between the return and the ground, there will be a small signal returned to the computer (typically about 1/2 volt). This 1/2 volt returned to the computer tells the computer that the throttle is at idle position. As the throttle opens, the voltage division changes. The majority of the division will occur between the reference and the return, Figure 4-30. This will leave little voltage difference between the return and ground. With the throttle at wide open position, the return voltage will be around 4.5 volts. Notice that at the two extremes of idle and WOT (wide open throttle) the voltage varied between 0.5 and 4.5 volts, Figure 4-31. What do you think the voltage will be at the TPS if the throttle is 50% open? The correct answer is 2.5 volts. Fifty percent throttle opening will put the TPS at its midpoint. Midpoint or 50% voltage division will put the return signal to the computer at 50% of reference: 50% of 5 volts is 2.5 volts. Most throttle position sensors operate on these principles. They can, however, operate in reverse: high voltage at idle and low voltage at WOT. They can also operate on a different reference voltage than this example, but in principal, they will be the same. Diagnosis of them is simplified if you understand what it is they are doing. Keep in mind that there is a mechanical connection between the throttle and the TPS. On a central fuel injection system (TBI), the TPS is mounted directly to the injection unit, and on a multiport system (MFI) it is mounted to the throttle body unit. In either case, as the throttle is moved, the wiper arm of the TPS is also moved. The three wires attached to it are the reference voltage, the return voltage, and the ground.

Another type of input device is a type of position sensor like the TPS and takes two different forms. The first one we will look at is the **pick-up coil**. The pick-up coil first made its way into the vehicle when electronic ignition was introduced in the mid-seventies. There are still some applications using pick-up coils on today's vehicles. You will sometimes also hear or read the term magnetic

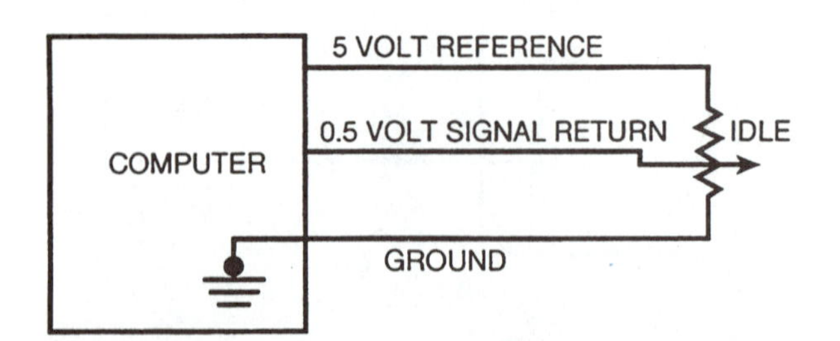

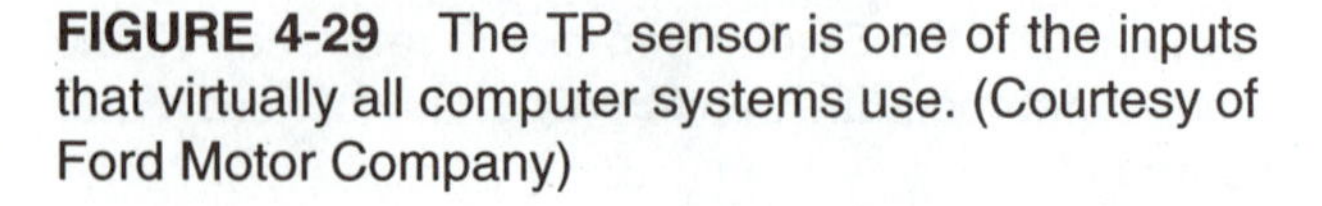

FIGURE 4-29 The TP sensor is one of the inputs that virtually all computer systems use. (Courtesy of Ford Motor Company)

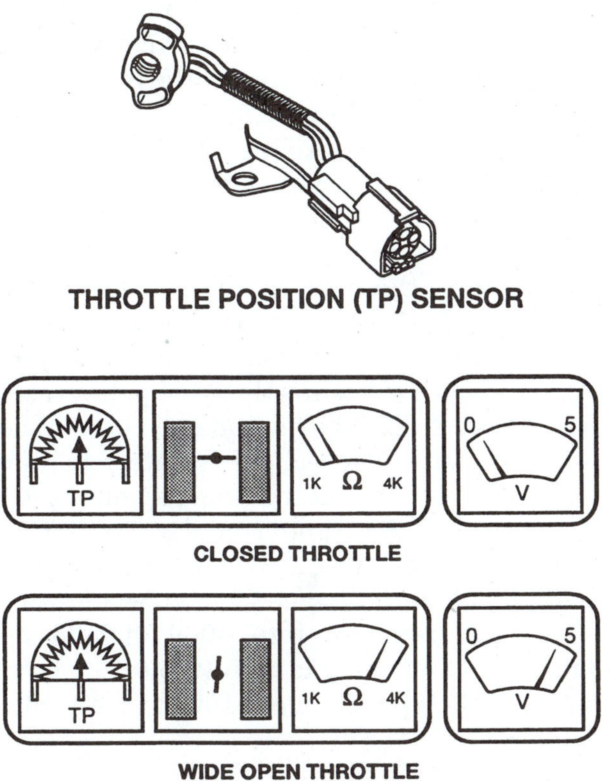

FIGURE 4-31 Varying voltage return is used as an input for the computer to determine not only the position of the throttle, but the rate of opening or closing. (Courtesy of Ford Motor Company)

generator or A.C. generator. All three terms are in common use and they all denote the same input device. Let's look at how it works

When lines of force move around coils of wire, a signal is generated in the coil. Figure 4-32 shows the relationship between moving magnetic lines of force and conductors or wires that are used in most vehicle subsystems. Starting, charging, ignition, fuel, etc., all use the generated voltages from moving magnetic fields. The pick-up coil will generate a signal because the magnetic lines of force from the permanent magnetic move as the distance between the armature tooth and the stator tip change Figure 4-33. In this example, the armature is spinning at 1/2 engine speed by the camshaft. The assembly is inside the distributor housing. When the stator tip and the armature are lined up and the distance begins to increase, the polarity of the

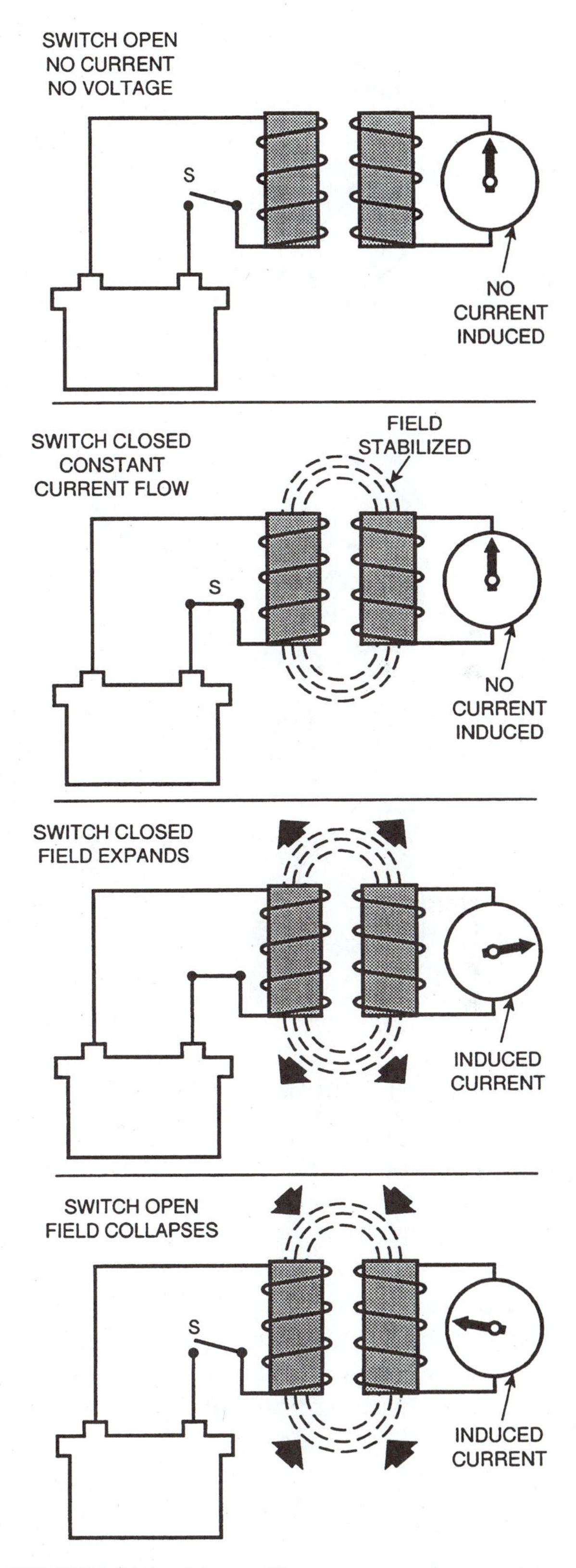

FIGURE 4-32 Lines of force surround magnets and are concentrated closest to either the North or South poles. (Courtesy of Ford Motor Company)

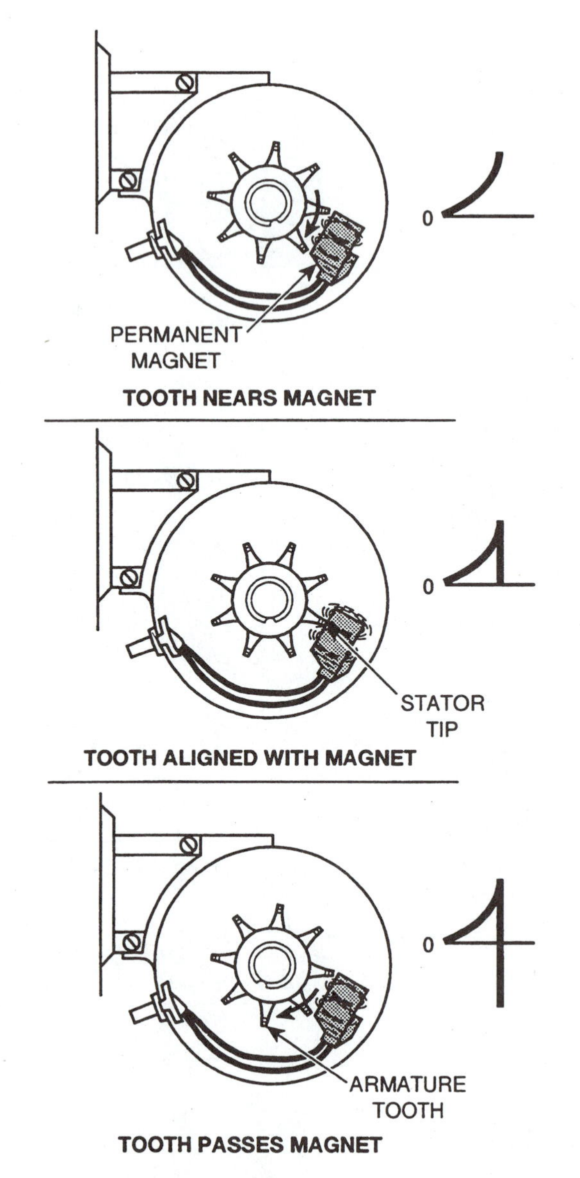

FIGURE 4-33 As the armature to stator distance gets smaller, the magnetic lines of force become more concentrated and a positive voltage signal is generated. (Courtesy of Ford Motor Company)

signal will change from positive to negative. It is this changing polarity that is used to signal the position of the distributor. This information is used by the ignition system to help determine ignition timing.

The second type of position sensor currently in use, employs a **Hall effect switch**, Figure 4-34. An example of this type of input device is a **CKP** (crankshaft position) sensor. The **CKP** sensor consists of a rotary vane cup and a Hall effect switch. A weak permanent magnet is located opposite the sensor with the vane cup in between. However, when a space is between the switch circuit and the magnet, the signal goes off or down, Figure 4-35. The amplitude or voltage doesn't change in this type of sensor. Instead, the number of times the on/off cycle is repeated is changed. This is called the frequency. We will see many uses for this type of sensor on the modern vehicle. The signal from a CKP sensor or any other sensor that utilizes a Hall effect switch is a digital on/off signal which can be directly read by the processor, Figure 4-36. Its signal may be used to determine position, speed, or both. Keep in mind that the faster the rotary vane cup moves, the more frequent the output signal will be. A CKP sensor for a 4-cylinder distributor will generate 4 high and 4 low voltage signals per rotation. This will translate into 4000 signals per minute at 2000 engine rpm (typical highway speeds) 2000 rpm engine is 1000 rpm distribu-

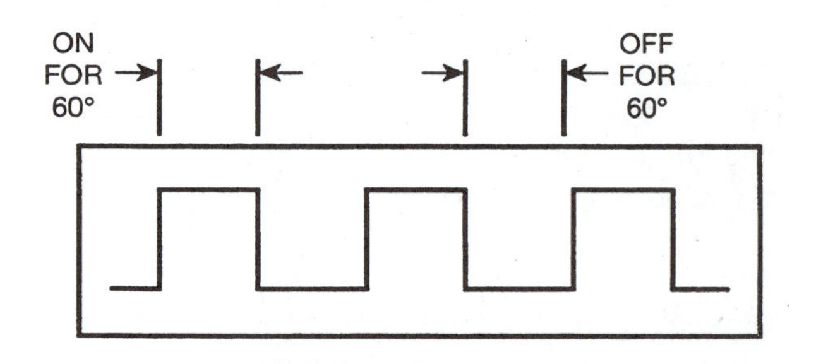

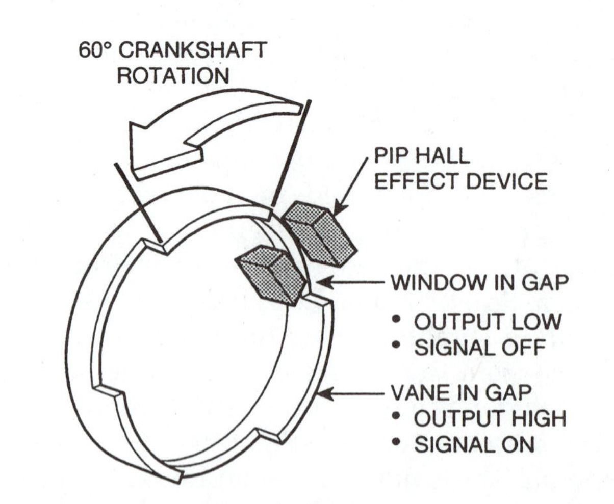

FIGURE 4-34 Voltage applied to the Hall effect switch effectively turns it into a device which will be able to sense magnetic fields. (Courtesy of Ford Motor Company)

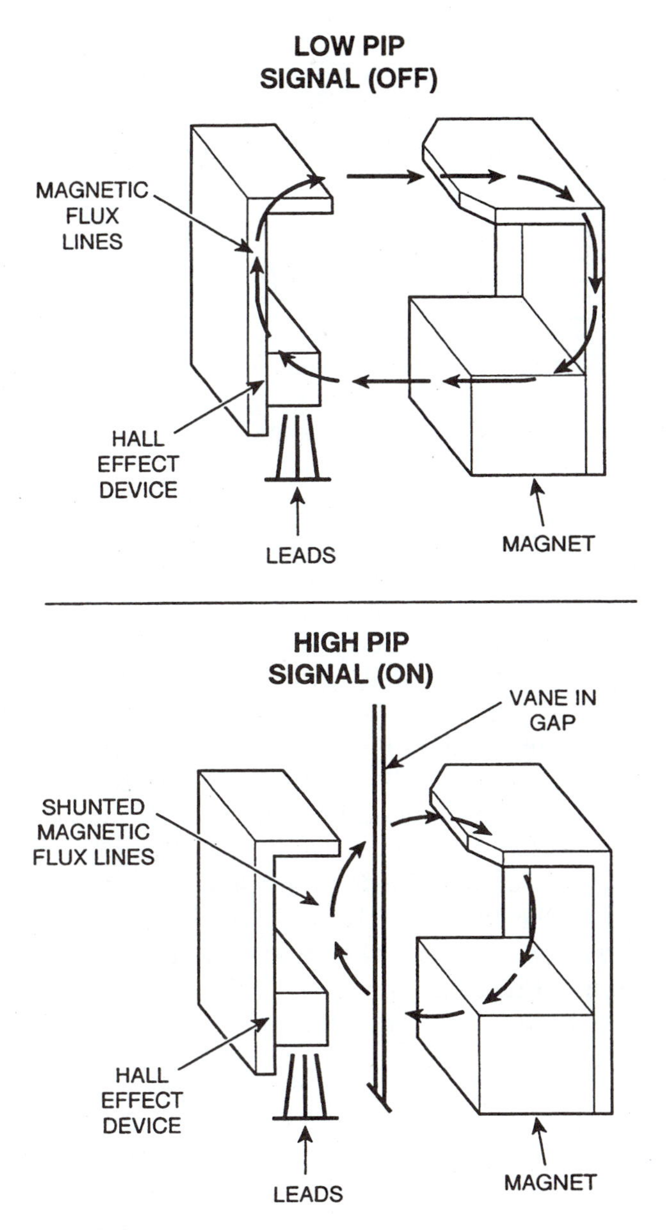

FIGURE 4-35 When the vane blocks the magnetic field from reaching the switch, the signal from the switch is on or high. (Courtesy of Ford Motor Company)

tor. With 4 signals per distributor rotation: 4 × 1000 rpm = 4000 signals per minute delivered to the processor.

INTEGRATED CIRCUITS AS INPUT DEVICES

The eighties saw the world of automotive electronics change by leaps and bounds. The changes are continuing during the nineties.

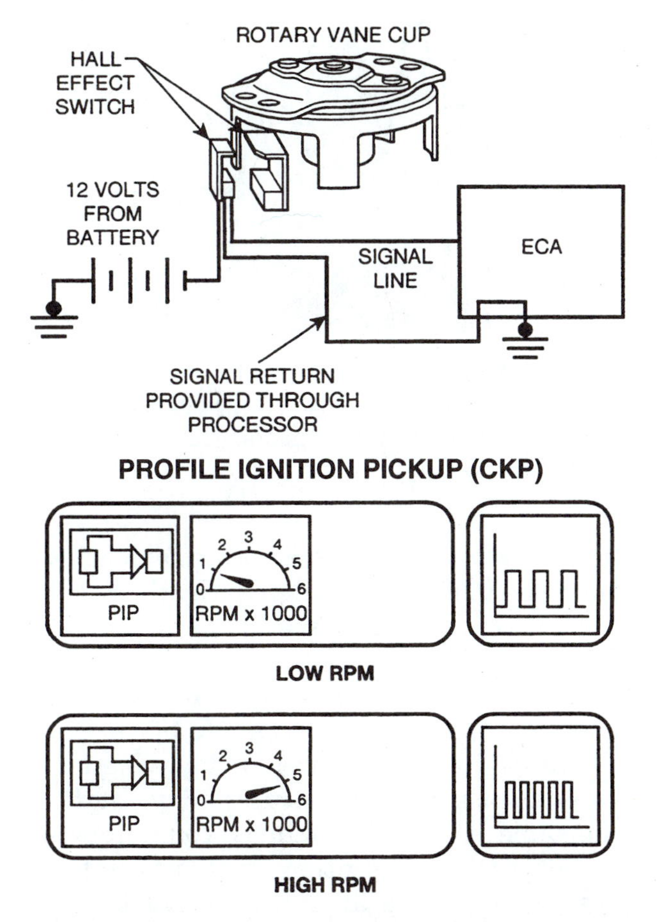

FIGURE 4-36 The processor may be the electronic control module (ECM) or the powertrain control module (PCM) or any other processor that needs information about speed and/or position. (Courtesy of Ford Motor Company)

The transistorized circuits that we have looked at have been replaced by the integrated circuit. What is an integrated circuit? Simplified, it is a paper-thin piece of silicon that can contain thousands of electronic parts as shown in Figure 4-37. Remember, the conventional circuit has all of its components wired together or soldered into a board called a printed circuit. In the early 1960s scientists were able to take the conventional circuit and reduce it down to a very small chip of silicon that can measure less than 0.15 square inch (4 millimeters). Chips have made their way into the automotive industry because of their low cost and reliability. The modern automobile contains hundreds of chips. Many times, they are part of a group and together form the radio,

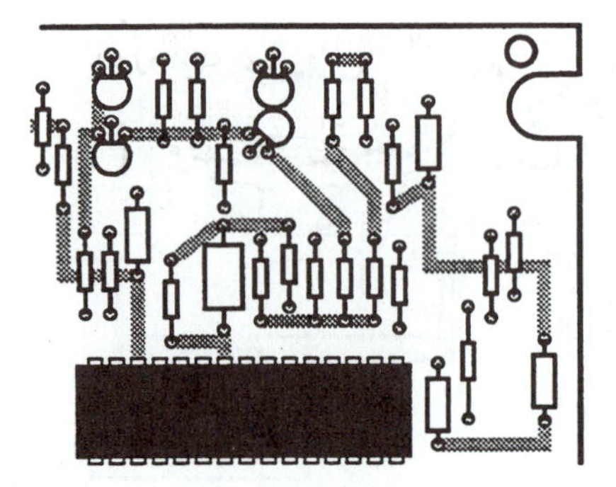

FIGURE 4-37 Integrated circuits can accomplish the same tasks as conventional circuits consisting of transistors, diodes, and resistors. (Courtesy of Ford Motor Company)

MANIFOLD PRESSURE (VACUUM) TO VOLTAGE CONVERSION		
Absolute Pressure	**Vacuum**	**Sensor Voltage**
14.7 psi	0"	4.5 V
10.5 psi	6.5"	3.75 V
7 psi	13"	2.5 V
3.5 psi	19.5"	1.25 V
0 psi	26"	0.5 V

FIGURE 4-38 The relationship between pressure and voltage that the IC should input to the computer. (Courtesy of Ford Motor Company)

the computer, and the dashboard, among others. Other times, they might be by themselves and act as input signal sources for another circuit.

The **MAP** sensor, or manifold absolute pressure sensor, is an example of an input signal device that utilizes a chip. There are two designs in frequent use today. Both do the same job; that is, tell the computer-controlled fuel and ignition systems what the pressure is inside the intake manifold. Let's look at their design and see what the IC (integrated circuit) or chip is doing.

The vacuum line attached to it goes to the intake manifold so that it sees the pressure in the manifold. As the throttle is opened and closed and the vehicle accelerates or cruises, the pressure inside the manifold will change. The computer needs to know this changing pressure if it is to change the air/fuel ratio (how much air and how much fuel) along with other functions. For example, when we ask for maximum power, as in accelerating up a steep hill, the computer needs to add more fuel. When our foot is to the floor, pressure is high (14 psi). This pressure inside the manifold is very close to normal atmospheric pressure because the throttle is not restricting the airflow. Obviously, the computer is an electrical device and cannot recognize pressure directly. The pressure must be converted over to a computer-recognizable voltage signal by the chip and inputted into the computer.

Figure 4-38 shows the relationship between pressure and voltage that the IC should input to the computer. Note: This chart is for a voltage style of MAP sensor in common use among many manufactures. When we idle or decelerate, the pressure in the manifold goes down. You have probably referred to this as a vacuum. It might go as low as 1 or 2 psi (which converts over to .5 volts at the computer). Notice from the chart that 7 psi is 2.5 volts to the computer. In this way wide open throttle will be approximately 4.5 volts and deceleration will look like 0.5 volts. MAP sensors also can compensate for altitude and weather conditions, Figure 4-39.

In this way, the computer can vary the air/fuel ratio as the vehicle climbs a mountain and the air gets thinner. Less air requires less fuel. The reduced air pressure changes the voltage signal to the computer by 0.1V for

Altitude	Volts
Below 1,000	1.7-3.2
1,000-2,000	1.6-3.0
2,000-3,000	1.5-2.8
3,000-4,000	1.4-2.7
4,000-5,000	1.3-2.6
5,000-6,000	1.3-2.5
6,000-7,000	1.2-2.5
7,000-8,000	1.1-2.4
8,000-9,000	1.1-2.3
9,000-10,000	1.0-2.2

FIGURE 4-39 MAP sensor inputs to the computer showing altitude compensation. (Courtesy of Ford Motor Company)

every 1,000 feet. We will see how to diagnose computer input devices in a future chapter. In the meantime, keep the purpose of this IC or chip in mind. It is an excellent example of integrated circuits on the automobile.

Another type of MAP sensor is shown in Figure 4-40. The information inputted to the computer will be the same: manifold pressure. Ford vehicles use this type of sensor. The manner in which this IC functions is different, however. Frequency refers to the number of voltage shifts during a period of time. For example, a frequency of 60 cycles, or hertz, is used in our household electrical systems. This means that the voltage shifts or changes sixty times a second. This is its frequency. Figure 4-41 shows the conversion of manifold pressure to frequency. Instead of being a voltage divider as was previously discussed, it is a frequency generator. The MAP sensor in this circuit will vary the frequency of its output (input to the computer) as the pressure changes. The principle is the same, but you will see that the diagnosis procedure is different. Our first example can be diagnosed with a vacuum pump (to lower the pressure) and a voltmeter to note the pressure changes, whereas the second example will require the ability to measure frequency using either a scope or a DVOM with frequency capability. As noted before, we will look at their diagnosis in detail in a future chapter.

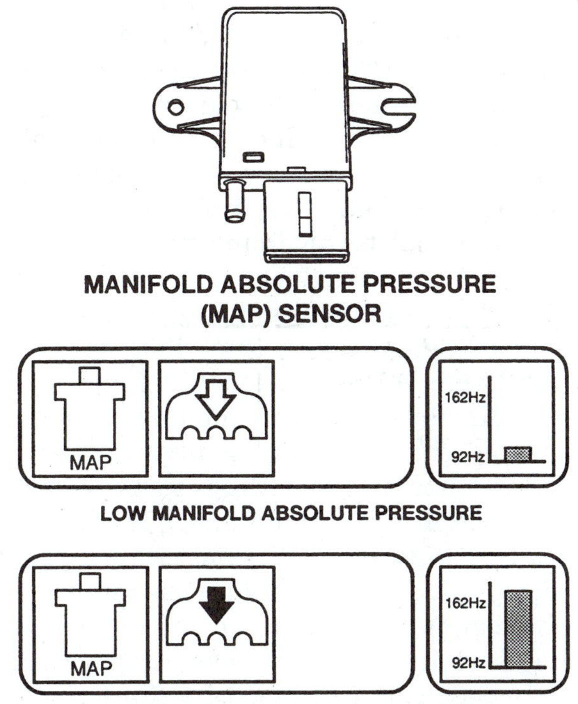

FIGURE 4-40 MAP sensor as a frequency generator. (Courtesy of Ford Motor Company)

APPROXIMATE SIGNAL FREQUENCIES OF A FORD MAP SENSOR AT SEA LEVEL	
Manifold Vacuum	**Map Frequency**
0 in HG	160 Hz
13 in HG	125 Hz
26 in HG	90 Hz

FIGURE 4-41 Conversion of manifold vacuum to frequency. (Courtesy of Ford Motor Company)

INTEGRATED CIRCUIT PRECAUTIONS

We have noted that the use of integrated circuits is increasing in the automotive industry. This fact makes precautions all the more important. The following are some general rules or precautions that should be followed when working on IC circuits.

1. Do not apply voltage directly to the IC unless the manufacturer's procedure calls for it.
2. Do not connect or disconnect an IC with the power on to the circuit.
3. Keep the battery voltage less than 20 volts during charging or jumping.
4. If a relay is added to the vehicle, as in a rear window defogger, use one that contains a de-spiking diode, as shown. This will prevent a relay surge from getting to an IC and damaging it. High voltage surges will destroy chips.
5. Do not run the vehicle without the battery being connected.

6. Always make sure that the charging system is not producing excessive A.C. A.C. can damage sensitive computer circuits, or at least change some of the input information.

OXYGEN SENSORS (O_2S)

When tailpipe emissions became as important as they are today, manufacturers looked to a type of electronic sensor called an oxygen sensor, Figure 4-42. The sensor is a zirconium dioxide-type voltage generator which typically can generate a voltage between 0 and 1 volt. Controlling air fuel ratios very tightly called for the processor (ECM or PCM) to have information about the effectiveness of the burn within the cylinder. An air fuel ratio of 14.7 parts air to 1 part fuel by weight has been determined to be efficient for most engine operations. This air fuel ratio is called a stoichometric ratio. It was virtually impossible to achieve and hold without the use of electronics, especially the O_2S. If a lean (less fuel) condition exists, less oxygen will be used to burn the fuel. This will usually leave behind in excess of 2% O_2 and the sensor will generate a low signal (generally lower than .450 V). However if a rich (more fuel) condition exists, more oxygen will be used to burn the fuel. This will leave behind less than 1% oxygen and the signal from the sensor will go higher than .450 V. This high signal = rich, low signal = lean, will be used by the PCM (powertrain control module) to determine if the air fuel ratio is stoichometric after the burn has taken place. In this way, it is a check on the actual burning within the cylinder. By holding the A/F ratio constant (1%–2% O_2) the emissions are controlled tightly. The oxygen sensor is an extremely important emission sensor. Its output is a pulsing direct current (DC) voltage between 0 and 1 volt.

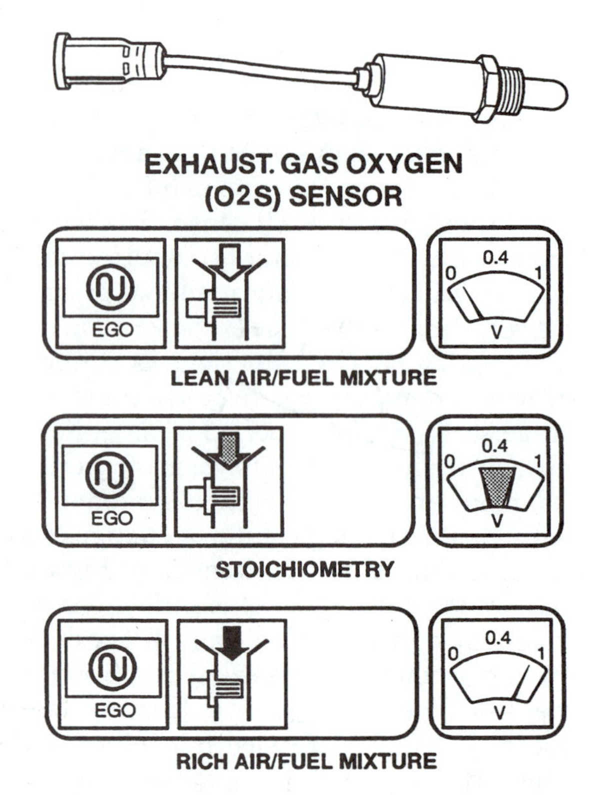

FIGURE 4-42 Oxygen (O_2S) sensor. (Courtesy of Ford Motor Company)

STATIC ELECTRICITY

While we are on the topic of protecting electronic components during service, it is a good spot to mention and discuss static electricity. You are all familiar with static charges in one form or another. As you slide your feet across a carpeted floor and then touch something, you might feel and see a slight static discharge. The action of sliding your feet across the carpet actually placed a slight charge on you. A change in the number of electrons put you at a different charge level than the objects around you. When you touched them, there was a discharge between you and the object. Although this discharge generally does no damage other than to perhaps surprise you, it can do potentially great damage to electronic circuitry. Static electricity has become a topic of many articles written with the automotive technician in mind. As more and more microprocessors are added to the vehicle, more and more concern is expressed by the manufacturers for the environment that we typically work in. Our shops are generally not as static free as an electronic repair shop, and yet we

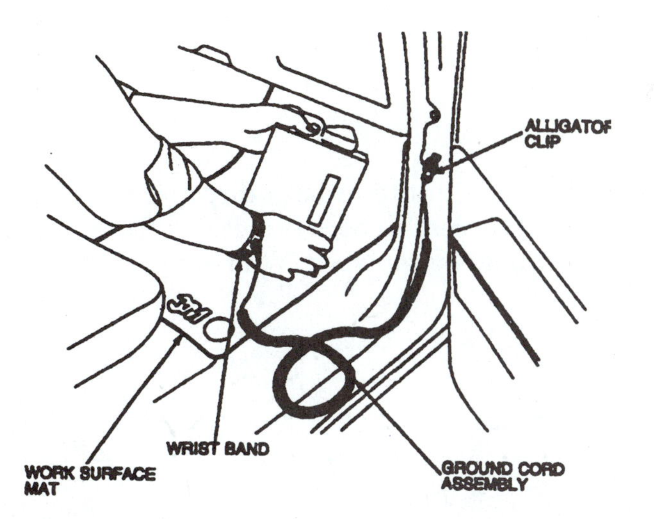

FIGURE 4-43 Grounding strap, used to protect components from static discharge. (Courtesy of Ford Motor Company)

work on very sophisticated integrated circuitry. Technicians of today must realize the static electricity that they have ignored for so long will have to be discharged safely before they begin working on an electronic component or processor. Dashboards, trip computers, memory circuits, processors for fuel and ignition control, anti-lock braking, body computers . . . the list grows each year. To effectively work on these circuits, some precautions are necessary. Generally these can be summarized by the statement that you must be at the same electrical potential as the component you are working on and the vehicle you are working in.

Sounds simple doesn't it? We are seeing increased use of grounding straps which connect the technician, the component, and the ground system of the vehicle all together, as in Figure 4-43. Again, the theory behind this is to place all things that will touch at the same electrical potential so that a discharge will not take place. Even if you do not have all the special grounding equipment, run jumper ground wires between the components and the vehicle, and ground yourself to the vehicle before you begin working. In addition, follow any precautions printed on the packaging materials of the replacement components. Frequently, special packaging protects the component from static discharge. Leaving the components in their original packaging until they are ready to be installed is the best insurance.

SUMMARY

In this chapter, we have looked at the history of electronics from the vacuum tube radios of the 1950s to the integrated circuit computer input devices in use today. We examined the difference between conductors, nonconductors, and semiconductors and saw how doping certain materials gave them either P or N type characteristics. By placing a P to an N, we were able to construct a diode, which we found out was an electrical one way check valve. We discussed the use of diodes in both rectification and protection circuits. The addition of a third doped material produced the transistor, which we saw could be used for amplification of a small signal or in a switching circuit similar to electronic ignition. Thermistor, throttle position sensors, manifold absolute pressure sensors pick-up coils, Hall effect switches and oxygen sensors were examined as sources of input information to the computer. The precautions necessary when working on the modern vehicle were outlined with emphasis on preventing either high voltage, static or spikes, from reaching the various devices.

KEY TERMS

Solid state	Emitter
Integrated circuit	Thermistor
Chip	ECT
Diode	IAT
Transistor	TPS
NPN	Pick-up coil
PNP	Hall effect switch
Base	CKP
Collector	MAP

CHAPTER CHECK-UPS

Multiple Choice

1. An ohmmeter is placed across a diode and indicates continuity in both directions. Technician A states that this indicates the diode is blocking current in both directions. Technician B states that this indicates the diode is open and should be replaced. Who is correct?

   a. Technician A only
   b. Technician B only
   c. Both Technician A and B
   d. Neither Technician A nor B

2. A despiking diode

   a. protects some solid-state circuitry.
   b. cannot be tested with an ohmmeter.
   c. is not used on modern vehicles.
   d. cannot be tested with a self-powered test light.

3. A negative coefficient thermistor will have its total resistance

   a. go up as its temperature increases.
   b. remain the same as its temperature increases.
   c. go down as the temperature increases.
   d. go down as the temperature decreases.

4. Thermistors are frequently used as

   a. despiking units.
   b. computer input devices.
   c. computer output devices.
   d. none of the above.

5. Transistors (either NPN or PNP) have three leads identified as

   a. anode, cathode, gate.
   b. neutral, positive, negative.
   c. base, emitter, collector.
   d. negative, positive, negative or positive, negative, positive.

6. At one half open throttle, the TPS return voltage on a 5-volt computer system is 0.5 volts. Technician A states that this is normal. Technician B states that this is not normal. Who is correct?

   a. Technician A only
   b. Technician B only
   c. Both Technician A and B
   d. Neither Technician A nor B

7. Manifold pressure is (a,an) ___________ type input on a Ford product.

   a. inches of vacuum
   b. pulse width modulation
   c. frequency
   d. voltage

8. An O_2 sensor is generating a voltage consistently below .450 V. Technician A states that this indicates normal operation of a lean vehicle. Technician B states that this indicates a vehicle that is not running ideally. Who is correct?
   a. Technician A only
   b. Technician B only
   c. Both Technician A and B
   d. Neither Technician A nor B
9. Battery voltage is jumped directly to the MAP sensor. Technician A states that this is incorrect testing procedure and might destroy the IC inside the sensor. Technician B states that this procedure will not give any valid information. Who is correct?
   a. Technician A only
   b. Technician B only
   c. Both Technician A and B
   d. Neither Technician A nor B
10. Grounding yourself to a vehicle will
   a. prevent static electricity from destroying electronic components.
   b. allow safe removal and installation of various components.
   c. place you and the electronic component at the same potential.
   d. all of the above.

True or False

1. Diodes are designed with high resistance in one direction and low resistance in the other.
2. Most digital ohmmeters in use today can test diodes accurately.
3. Proper testing of a thermistor requires both a resistance reading and a temperature reading.
4. Thermistors are frequently used as computer output devices.
5. TP sensors are variable resistors with two wires attached.
6. Reference voltage is the voltage signal that the PCM usually sends out to sensors.
7. An ohmmeter cannot be used to test a MAP sensor.
8. A PNP transistor needs the base and collector at negative potential to have current come out of the emitter.
9. Generally both IAT (intake air temperature) and ECT (engine coolant temperature) sensors are negative coefficient thermistors.
10. A voltage above .450 V from an oxygen sensor indicates a rich condition

CHAPTER 5

Reading a Wiring Diagram

OBJECTIVES

At the conclusion of this chapter, you should be able to:

- identify the common symbols used in many wiring diagrams.
- identify and use common point diagnosis.
- use location codes to find both components and connectors.
- trace circuits using a major automotive manufacturer wiring diagram.
- trace circuits using an aftermarket wiring diagram.
- trace through a printed circuit dashboard.

INTRODUCTION

The ability of the technician to follow the maze of wires on the modern automobile is next to impossible today. The vehicle of today is wired in sections called looms or harnesses. These harnesses are prewired for a particular section of the vehicle, such as the dashboard or the engine compartment. Figure 5-1 shows just such a harness. Notice that the wires are all tied or bundled together and tightly taped. this makes the task of following a particular wire through the vehicle both time consuming and very difficult. In addition, the wire might change color or go through a part of the vehicle that has nothing to do with the actual circuit being followed. The use of a wiring diagram allows the technician to follow the wire to some point where it can be conveniently

FIGURE 5-1 Typical wiring harness of the modern vehicle.

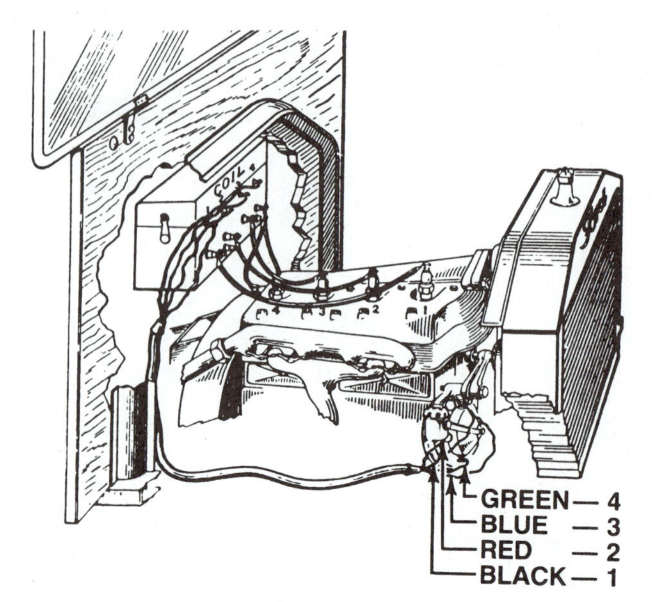

FIGURE 5-2 Model T wiring diagram. (Courtesy of Ford Motor Company)

tested. By eliminating the need to physically follow the wire throughout the vehicle, the technician will reduce diagnosis time resulting in reduced consumer cost. There is no doubt that the wiring diagram is essential to repairing an electrical system on the modern vehicle. As the vehicle has progressed through the years, so has the complexity of its wiring. Figure 5-2 shows the wiring diagram for the entire vehicle that put the Ford Motor Corporation on the map: the Model T. This diagram could be reduced down to billfold size and still show all the necessary information required to repair it. By contrast, a wiring diagram for a modern vehicle can be 15 pages long with hundreds of wires, connectors, and components. Times have changed.

WIRING DIAGRAMS: TYPES AND SYMBOLS

Wiring diagrams usually come in one of two different styles: pictorial and schematic. Figure 5-3 shows the basic difference between the two. Pictorials show actual pictures of the components, while schematics use symbols to illustrate components. Sometimes manufac-

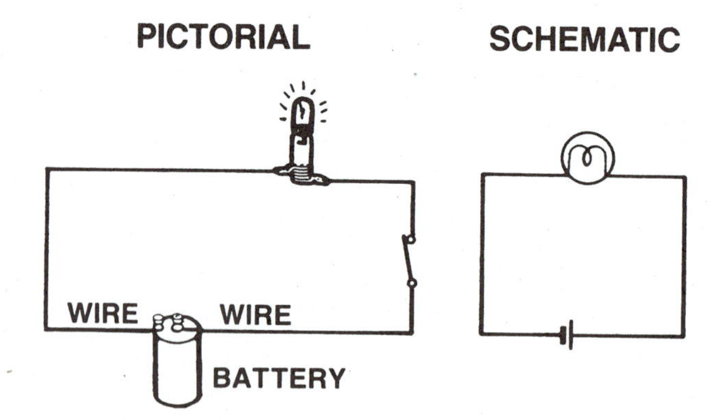

FIGURE 5-3 Pictorial versus schematic drawing style. (Courtesy of Ford Motor Company)

turers will use both styles on one diagram. Generally, schematic diagrams are chosen by the automotive manufacturers because they are considered more informative, show internal electrical paths, and take up less space than pictorial diagrams. To be able to understand the typical schematic wiring diagram, you will need to know the most common symbols. Probably the symbol most frequently encountered is that of ground. Figure 5-4 shows ground symbols. It can be represented in either a case or remote form. Figure 5-5 shows a section of a wiring diagram that has both **remote and case grounds**. Notice that the taillamps have a case ground. The black dot right on the light indicates that the ground for the bulb will be completed when the bulb's socket is installed in the housing. The housing must be made of metal, which will conduct electricity. The side marker lamps and the license lamp, however, do not have the same black dot attached to the bulb. Instead, notice that they have a ground wire, which eventually connects to a ground. The circle and ground symbol indicate that the wire has an

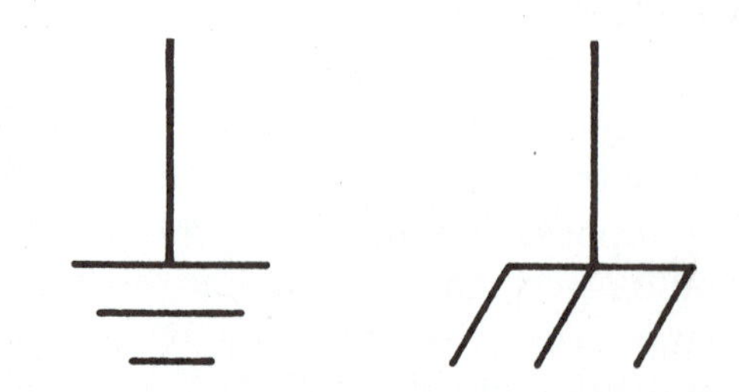

FIGURE 5-4 Common symbols for grounds.

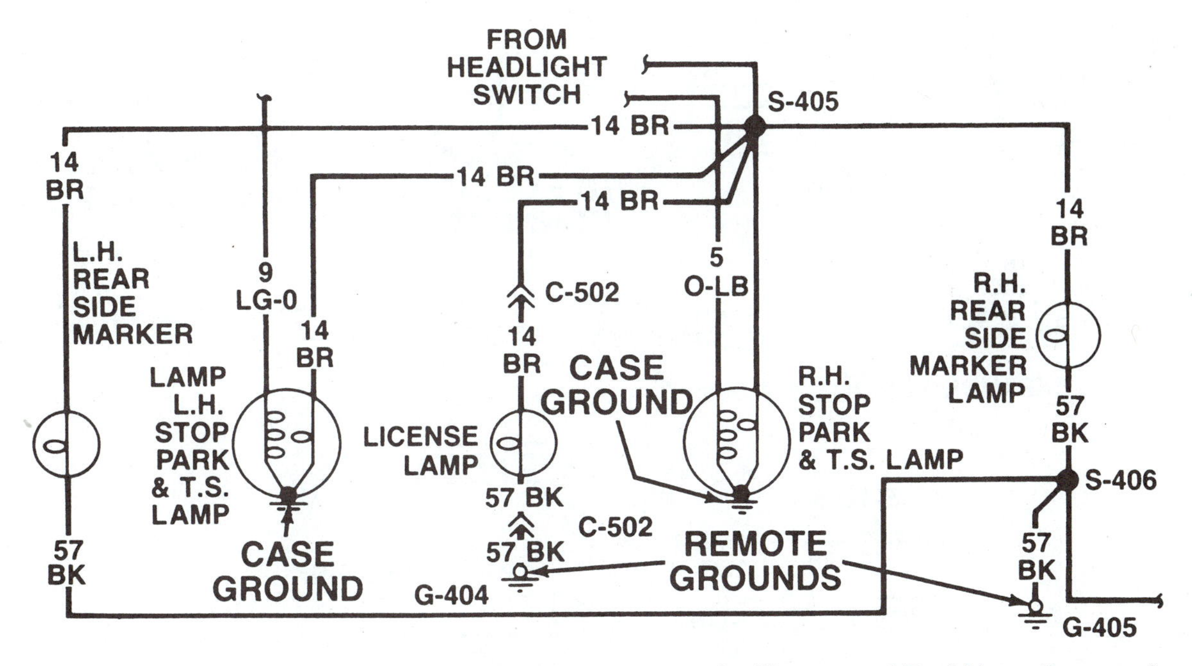

FIGURE 5-5 Schematics contain both case and remote grounds. (Courtesy of Ford Motor Company)

islet connected to it and a screw or bolt connects it to the metal of the frame. In this case, remote just means not attached. Notice also from the diagram that the side marker lamps are connected to the same ground, G-405. It is possible for a single ground to be the ground for more than one component. While we have this diagram to refer back to, let us discuss splices. A **splice** is a spot where two or more wires are connected. It is considered a **common point** or connection point. It usually is represented by a black dot. This section of the diagram shows two splices labeled S-406 near the righthand rear side marker lamp and S-405 near the top of the diagram. S-405 has six wires all connected together while S-406 has four. A wire can cross over another wire many times, but they are connected together only if they have the black dot indicating a splice. In addition, these splices all carry a number that we will look at later when we discuss **location codes**. Figure 5-6 shows some of the more common splice drawings, including something called an **expanded splice**. Remember, the object of any wiring schematic is to improve the ability of the technician to find their way around the vehicle. If the splice on the vehicle has more wires than can be easily shown on the diagram, an expanded drawing will be used with more than one splice drawn. To identify it as the same point or splice on the vehicle, both splices will have the same number. In this example, S-307 shows up at two splices with a wire labeled S-307 connecting the two together. The two splices and the wire connecting them together are actually the same splice drawn in two different spots to make the drawing easier to read. The S preceding the number indicates a splice. There can be only one splice on the entire vehicle with this number. If it appears more than one time, it is for drawing convenience. There will only be one S-307 on the vehicle.

Connectors are the next item we will discuss. A connector is a terminal where two wires are connected together. Figure 5-7 is the most common connector symbol. It represents the bullet connector shown in Figure 5-8. Refer back to Figure 5-5 (grounds). In the middle of the drawing is C-502, which is a bullet connector just ahead of the license lamp. As is the case with splices, each connector is numbered and carries the C designation. They are usually convenient places to test the circuit be-

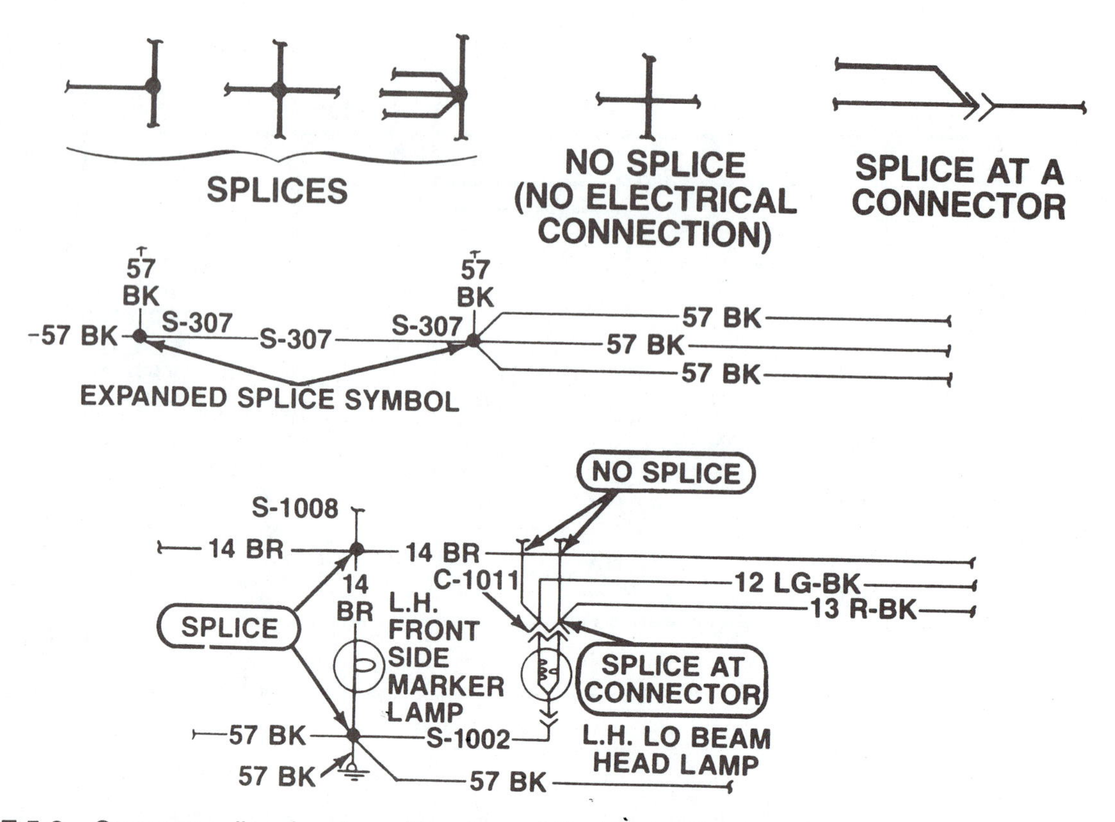

FIGURE 5-6 Common splice drawings. (Courtesy of Ford Motor Company)

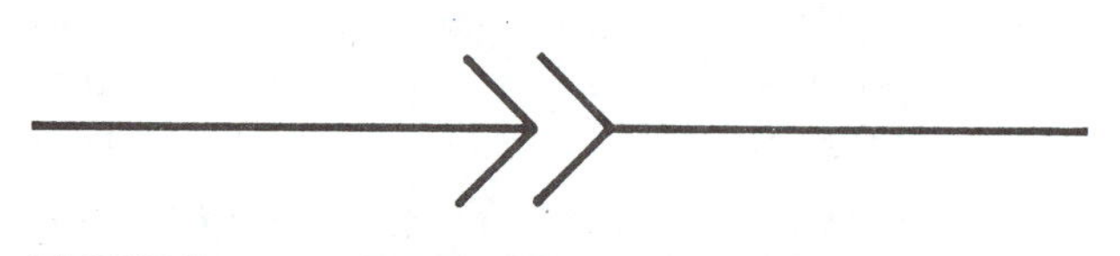

FIGURE 5-7 Schematic symbol for a bullet connector.

cause they are easily disconnected. Sometimes the manufacturer might wish to connect more than one wire together in a group. A multiple wire connector is used. Figure 5-9 shows that three wires are joined to three wires in a disconnected connector. This connector would be

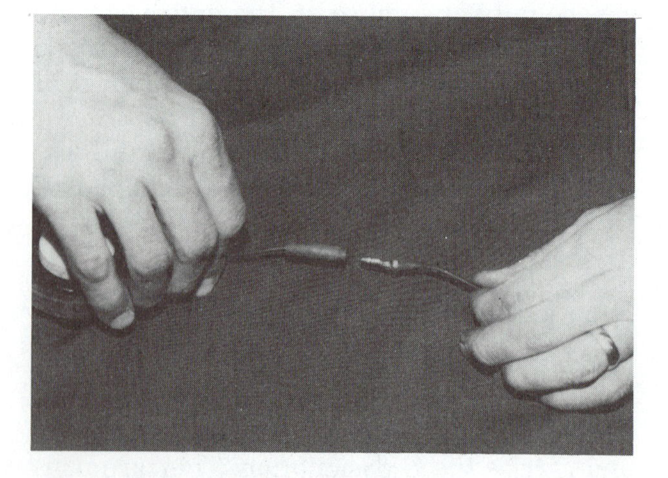

FIGURE 5-8 Bullet connector.

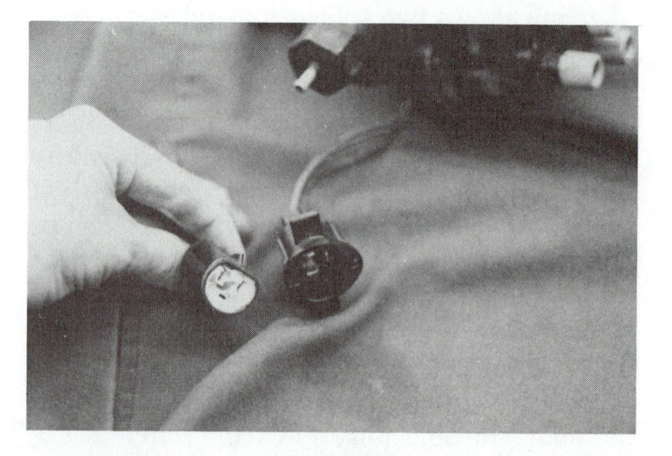

FIGURE 5-9 Electronic ignition three-wire connector.

FIGURE 5-10 Schematic diagram of electronic ignition three wire connector.

drawn schematically as in Figure 5-10. Notice that the general shape of the connector is shown alongside the schematic representation of the connector. Usually, both will not be present on the diagram. The dotted line through the three female wires and the dotted line through the male wires indicates that these three wires are in the same connector body all molded together. The closeness of the connections would probably lead you to figure this out even if it did not have the dotted lines. However, sometimes the wires in a connector might have to be drawn quite far apart from one another. When this occurs, the diagram will use a technique similar to that of the expanded splice. Dotted lines and connector numbering will indicate that the wires are all in the same connector. Figure 5-11 shows a schematic of a headlamp switch. Do not worry about the inside of the switch for now, but notice that the four wires coming in the top and the four wires coming in the bottom of the switch, all carry the same connector number, C-707, and are drawn with a dotted line. This indicates that this molded connector, C-707, has eight wires in it and seven female terminals (there are

○ FEMALE ROUND TERMINAL
⊗ MALE ROUND TERMINAL
○ EMPTY CONNECTOR CAVITY
▯ FEMALE ARCLESS TERMINAL
▮ MALE ARCLESS TERMINAL
▯ EMPTY CONNECTOR ARCLESS CAVITY

FIGURE 5-12 Connector shapes graphically displayed. (Courtesy of Ford Motor Company)

two wires connected to one terminal in the lower half). Remember again that there will be only one connector numbered C-707 on the vehicle and even though it is drawn as two, its number indicates that, in fact, it is one connector with eight wires attached to the headlamp switch. Sometimes the drawing might use graphic symbols, as shown in Figure 5-12. Sometimes large connectors with many wires will be shown with the graphic symbols rather than the arrow variety.

PROTECTION DEVICES

We have studied the three types of protection devices: **fuses**, **circuit breakers**, and **fusible links**. Now let us look at what they typically look like on a schematic diagram. Figure 5-13 shows two different styles of fuses. On the left is the common symbol for a fuse that is mount-

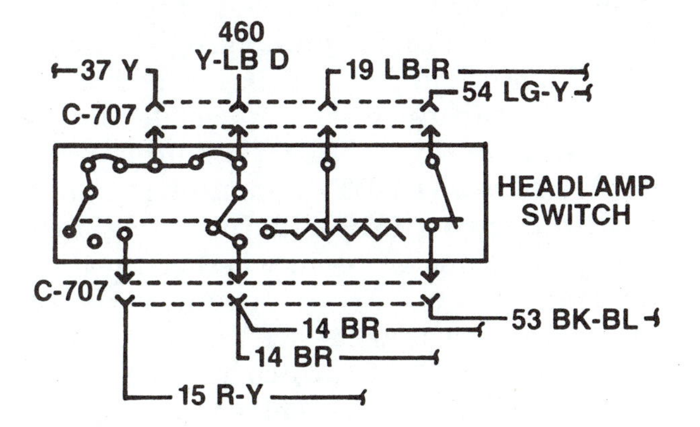

FIGURE 5-11 Headlight switch. (Courtesy of Ford Motor Company)

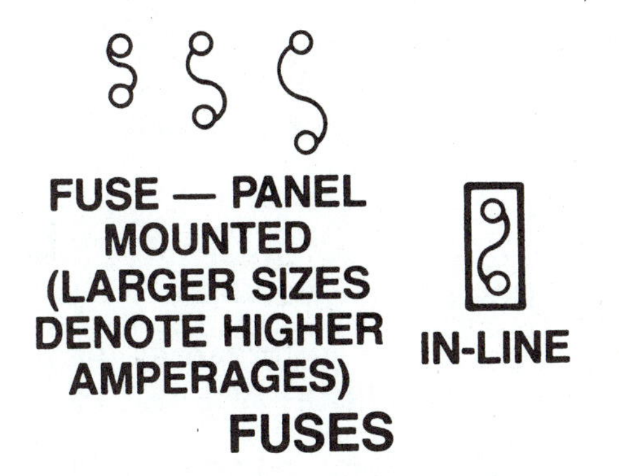

FIGURE 5-13 Schematic diagram of different amperage fuses. (Courtesy of Ford Motor Company)

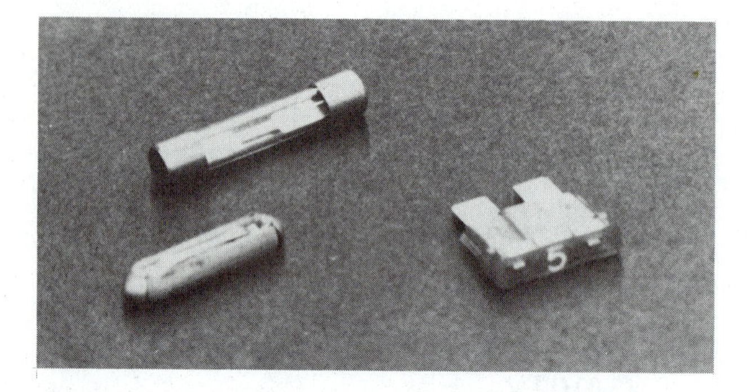

FIGURE 5-14 Three types of fuses are currently in use.

ed in a panel called the fuse box. The fuses are plug in and can be either glass, ceramic, or blade variety. On the right is the same symbol with a box drawn around it. This fuse is not in the fuse box. It is spliced into a circuit and is called an in-line fuse. Figure 5-14 shows these common types of fuses. Circuit breakers are utilized in place of fuses in circuits that are prone to temporary overloads such as windshield wipers that might ice to the window in the winter. The temporary overload caused by the ice will overheat the metal inside the breaker and open the contacts. Once the breaker cools down, it will close again restoring power to the circuit. Figure 5-15 shows

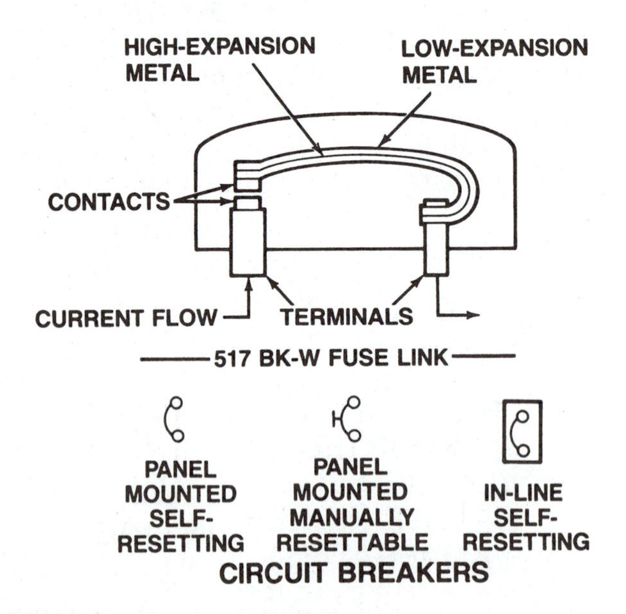

FIGURE 5-15 Pictorial diagram of a circuit breaker, with schematic symbols below. (Courtesy of Ford Motor Company)

what the inside of the breaker looks like and the schematic symbols for the most common circuit breakers. As was the case with fuses, a box drawn around the breaker symbol indicates that it is not in the fuse box but is located somewhere else. A placement drawing shown in Figure 5-16 indicates the placement for all of the fuses and circuit breakers that are panel mounted. In addition, a listing of what circuit is being protected is usually provided. Figure 5-17 shows the actual fuse box or panel that the drawing represented.

SWITCHES

The next group of schematic symbols to look at are switches. Switches control the on/off or direct the flow of current through various circuits. The contacts inside the switch assembly carry the current when they are closed and open the circuit when they open. Another term for opening is to "break" the circuit, while closing is considered "making" the circuit. You will frequently find the reference of pole and throw applied to a switch. This is a hold over from the days of the knife switch, Figure 5-18. In this type of switch, the conductors were attached to "poles" and the knife could be "thrown" to close the contacts. This knife switch would be considered a single pole, single throw switch. Simplified, this means that there is one input and one output. Pole refers to the number of inputs, while throw refers to the number of outputs. Therefore, a switch with one input and one output is referred to as **single pole, single throw (SPST)**. We find frequent use of additional styles of switches on the automobile: **single pole, double throw (SPDT); double pole, double throw (DPDT)**; and **multiple pole, multiple throw (MPMT)**. Switch contacts are also designated as to whether they are **normally open (NO)** or **normally closed (NC)**. The momentary switch drawn in Figure 5-19 is an example of a switch drawn NO. The contacts will not allow current through in their position. They are, therefore, normally open. Usually the wiring diagram will have

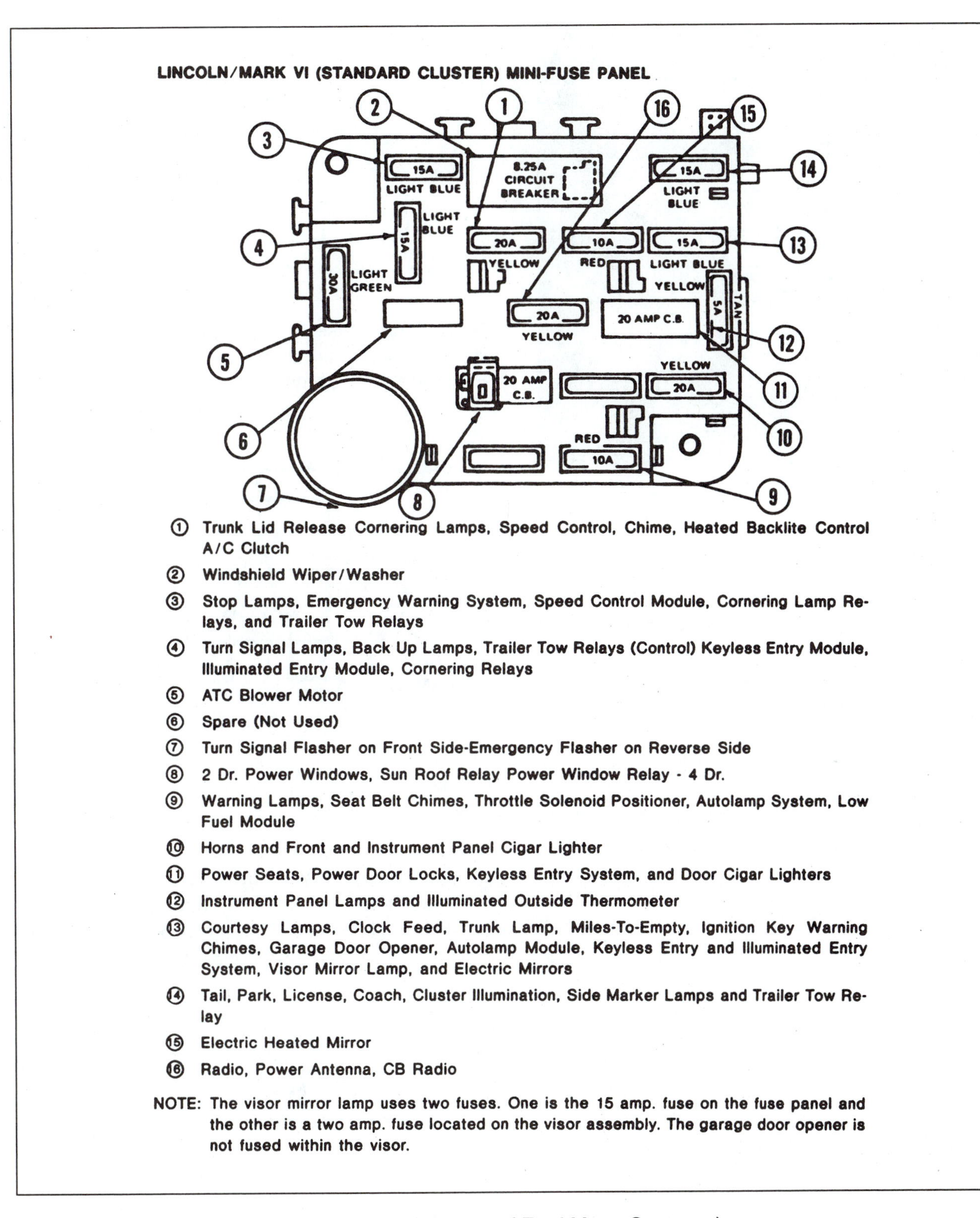

① Trunk Lid Release Cornering Lamps, Speed Control, Chime, Heated Backlite Control A/C Clutch

② Windshield Wiper/Washer

③ Stop Lamps, Emergency Warning System, Speed Control Module, Cornering Lamp Relays, and Trailer Tow Relays

④ Turn Signal Lamps, Back Up Lamps, Trailer Tow Relays (Control) Keyless Entry Module, Illuminated Entry Module, Cornering Relays

⑤ ATC Blower Motor

⑥ Spare (Not Used)

⑦ Turn Signal Flasher on Front Side-Emergency Flasher on Reverse Side

⑧ 2 Dr. Power Windows, Sun Roof Relay Power Window Relay - 4 Dr.

⑨ Warning Lamps, Seat Belt Chimes, Throttle Solenoid Positioner, Autolamp System, Low Fuel Module

⑩ Horns and Front and Instrument Panel Cigar Lighter

⑪ Power Seats, Power Door Locks, Keyless Entry System, and Door Cigar Lighters

⑫ Instrument Panel Lamps and Illuminated Outside Thermometer

⑬ Courtesy Lamps, Clock Feed, Trunk Lamp, Miles-To-Empty, Ignition Key Warning Chimes, Garage Door Opener, Autolamp Module, Keyless Entry and Illuminated Entry System, Visor Mirror Lamp, and Electric Mirrors

⑭ Tail, Park, License, Coach, Cluster Illumination, Side Marker Lamps and Trailer Tow Relay

⑮ Electric Heated Mirror

⑯ Radio, Power Antenna, CB Radio

NOTE: The visor mirror lamp uses two fuses. One is the 15 amp. fuse on the fuse panel and the other is a two amp. fuse located on the visor assembly. The garage door opener is not fused within the visor.

FIGURE 5-16 Fuse placement drawing. (Courtesy of Ford Motor Company)

FIGURE 5-17 Fuse box.

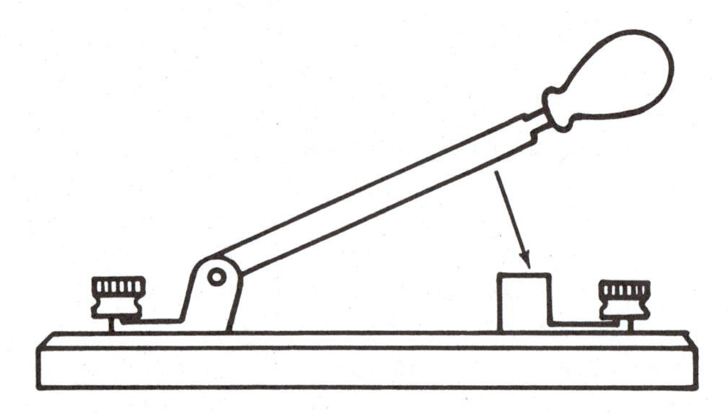

FIGURE 5-18 Knife switch. (Courtesy of Ford Motor Company)

the switch drawn in the position it will be in with the vehicle off.

An example of a single pole, double throw switch would be the typical headlight dimmer switch in Figure 5-20. It has one input and two outputs. In addition, one set of contacts will be drawn normally open, while the other will be drawn normally closed.

An example of a multiple pole, multiple throw switch would be the combination neutral safety switch, back-up light switch shown in Figure 5-21. It has two inputs and the possibility of six outputs or positions. Two moveable wipers move together. The dotted line between the poles usually indicates that the poles are "ganged", or connected together, mechanically. Based on the drawing, when will power be able to flow through the switch to the starter relay? The jumper between P and N indicates that current will be able to

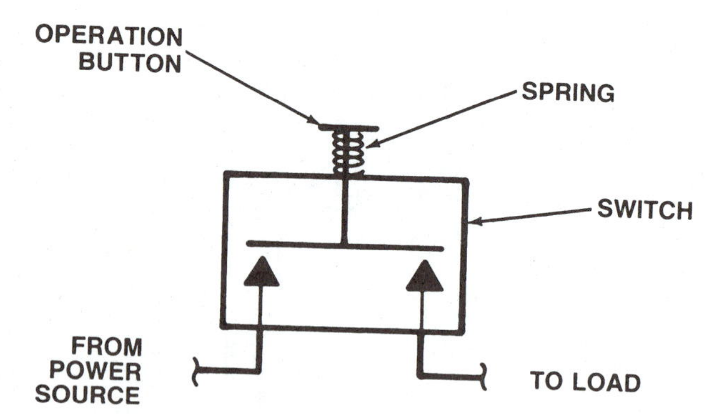

FIGURE 5-19 SPST (single pole, single throw) momentary contact switch (normally open). (Courtesy of Ford Motor Company)

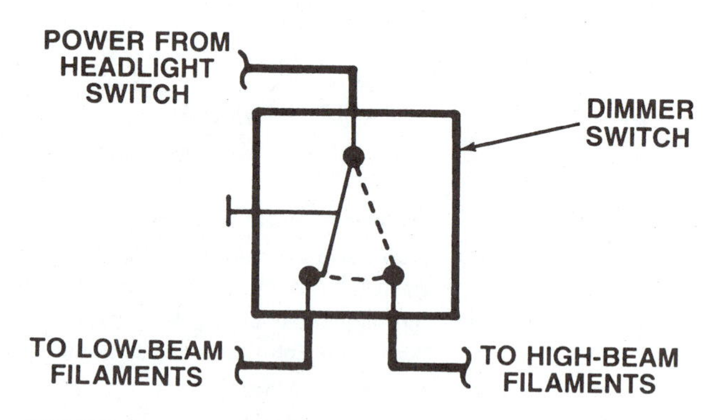

FIGURE 5-20 SPDT switch (single pole, double throw). (Courtesy of Ford Motor Company)

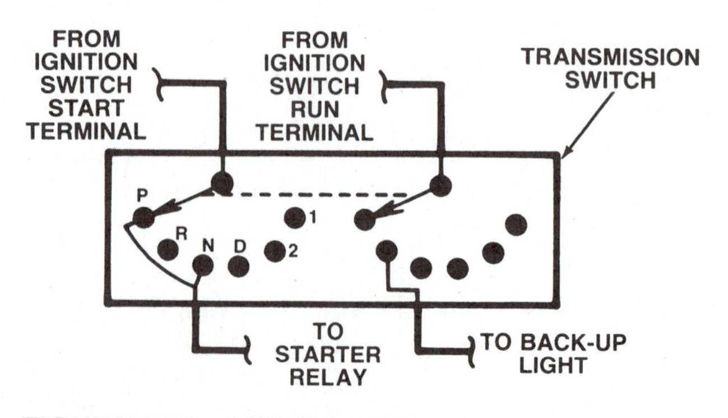

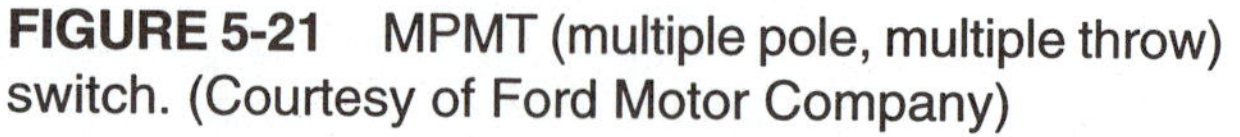

FIGURE 5-21 MPMT (multiple pole, multiple throw) switch. (Courtesy of Ford Motor Company)

flow through when the switch is in the P (park) or N (neutral) position. How about the back-up lights? When will they go on? Hopefully, you can figure out from the drawing that they will be able to go on only when the switch is in R or reverse. Ganged switches are frequently used in many applications where the motion of the switch is needed to control more than one circuit. In this example, the movement of the gear shift allows the starter to be engaged in park and neutral and the back-up lights to be on in reverse; thus, two functions off of the one switch.

Another example of a ganged switch is the ignition switch. It is probably the most complicated switch utilized on most vehicles. An example is reproduced from a wiring diagram in Figure 5-22. It has five wipers all ganged together. The dotted line extending to all wipers indicates that they will all move together. In addition, the legend printed in the lower left indicates the five different positions that the switch could be in. All wipers will move together and are currently shown in the L, or lock, position. This is the main switch for the entire vehicle and has many wires connected to it that perform many different functions. For example, the two wipers on the left are going to be utilized to turn the warning lights on during cranking to test the bulbs. The ground available on the two wipers (points E and D) will be applied to the two circuits and will turn the bulb on. This is the test of the circuit that most warning lights go through during cranking. The other three wipers all have B+ applied to them and they will switch it over to different circuits that require power. Let us test your understanding of the wipers. What positions will allow the turn signals to function? The last wiper has power to this circuit when it is in the R, or run, position. This makes sense, doesn't it? We want the turn signals to be functional only when the engine is running. How about the power windows? Wiper C shows a connection to wire 297, which is the circuit that will feed the power windows. Notice the jumper between position A and R. This means that the

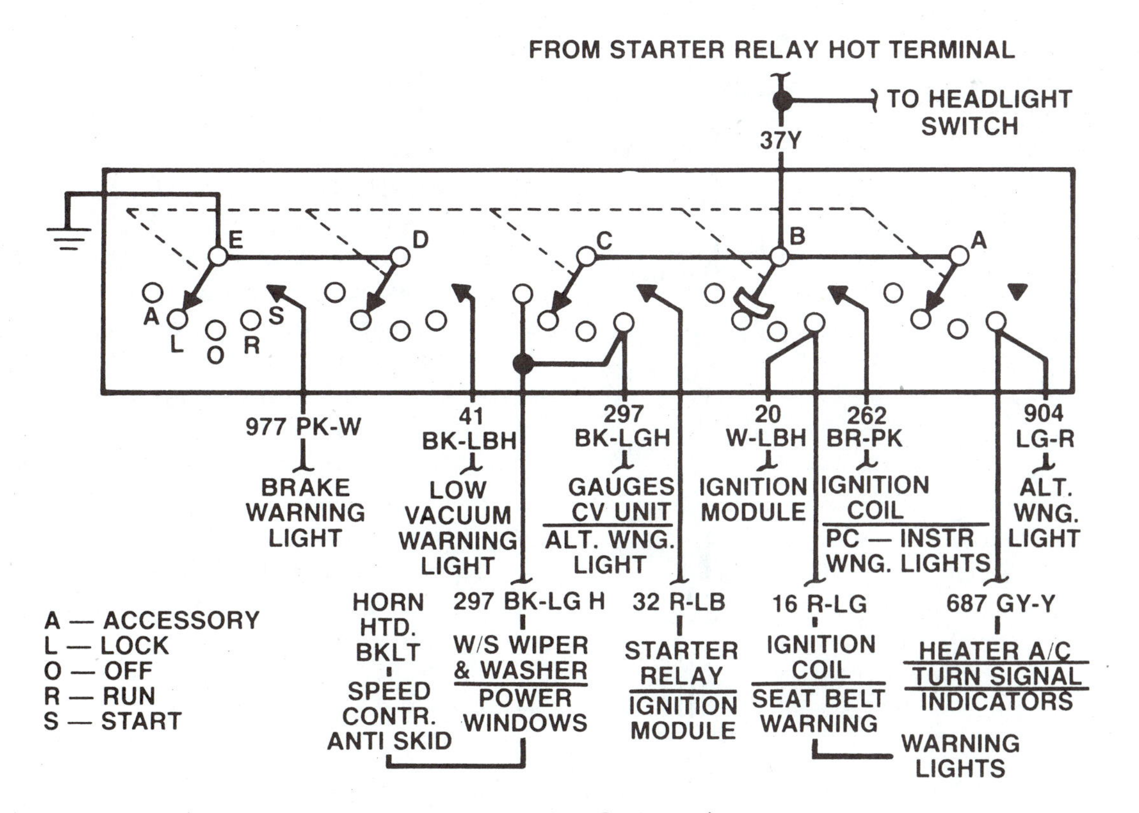

FIGURE 5-22 Ignition switch. (Courtesy of Ford Motor Company)

power windows, and anything else on this circuit, will be operational in the accessory and in the run position. You should be capable of following the current path through this multiple pole, multiple throw switch. You may encounter a situation where the contacts are not functioning as they should. Your understanding and ability to interpret the diagram will allow you to test the switch with a meter or test light.

SOLENOIDS

Frequently switches, especially the momentary contact type, control **solenoids**. Electric door locks, hood latches, and trunk latches can be solenoid operated. Figure 5-23 shows a solenoid-operated luggage compartment latch. Let us use all the information to date and follow the flow of current until it reaches ground. B+ is available through wire 296 to the deck lid release switch at connector C-1603. The momentary contacts allow B+ through this SPST switch and out C-1603. C-1603 must therefore be a molded two-wire connector. When the contacts close or make, B+ flows through three additional connectors: C-1604, C-1605, and C-1606, and reaches the solenoid coil. Current flowing through the coil creates a magnetic field and pulls the latch open. The solenoid is grounded with a case ground to the metal surrounding it. A pictorial drawing of the same circuit without the connectors is shown in Figure 5-24.

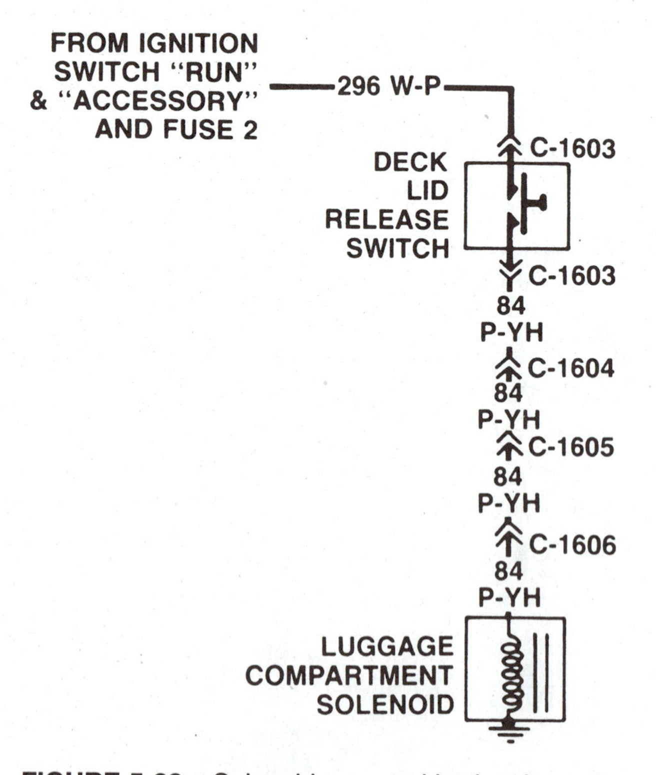

FIGURE 5-23 Solenoid operated latch schematic. (Courtesy of Ford Motor Company)

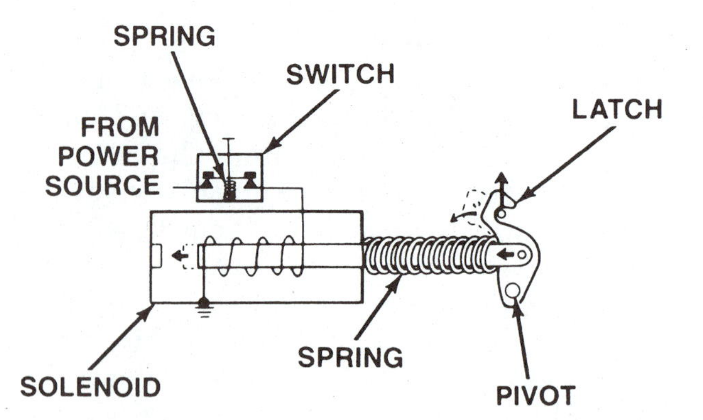

FIGURE 5-24 Pictorial diagram of solenoid operated luggage compartment latch. (Courtesy of Ford Motor Company)

RESISTORS

The next symbols that we will look at are those of the various types of resistors in use in the vehicle's electrical system. We have used the basic resistor symbol shown in Figure 5-25 in our discussion of circuits. You should realize that there are numerous applications using fixed resistors. For example, blower motor speeds or windshield wiper speeds are frequently controlled by using a resistor in series. Most automotive wiring diagrams will have the specified ohms of resistance printed near the resistor for reference purposes.

Sometimes the need arises for a variable resistance, such as a dimmer on the dash-

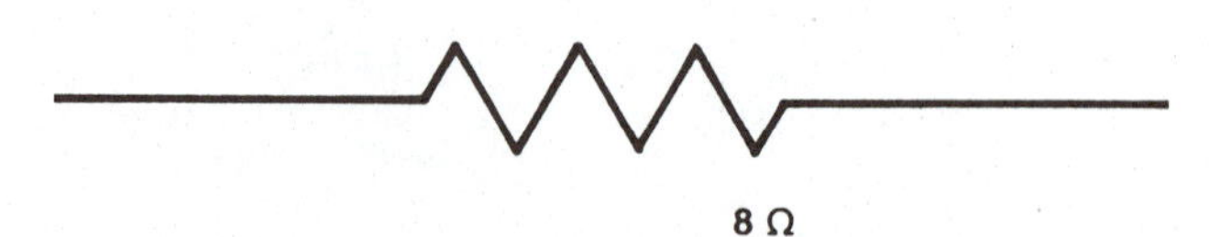

FIGURE 5-25 Schematic symbol for a resistor.

FIGURE 5-26 Schematic symbol for a variable resistor (rheostat).

board lights. Increasing the resistance dims the lights, while decreasing the resistance brightens the lights. This is usually done with a rheostat. The symbol shown in Figure 5-26 is most frequently used. Notice the arrow drawn through the center of the resistor. This is what indicates that it is in fact a variable resistor. Unlike the fixed resistor, it will usually not have a value printed near it on the schematic diagram.

The increased use of computers in vehicles has seen the introduction of two additional forms of resistors: the thermistor and the potentiometer. These two items are used as input signals for the computer. Let us look at the potentiometer first. Most throttle position sensors are potentiometers. They are drawn on schematic diagrams as Figure 5-27 shows, and they look like the photo in Figure 5-28. The three connections coming off of the sensor go to the computer and will indirectly tell it the throttle opening angle. From a drawing standpoint, it is similar to a rheostat and it does change resistance. The major difference comes in the number of wires. The potentiometer will have three while the straight variable resistor will usually only have two.

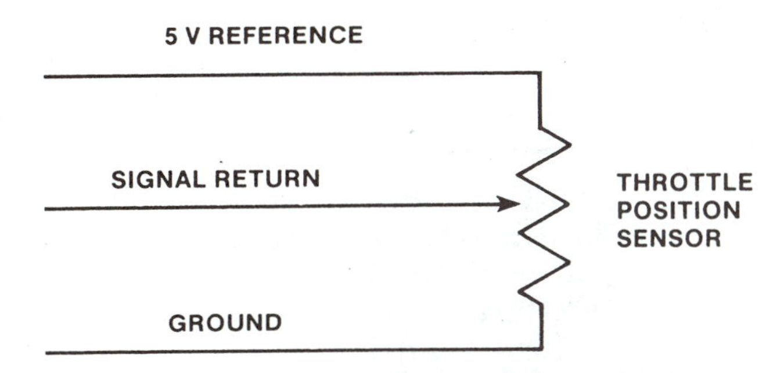

FIGURE 5-27 Schematic symbol for a potentiometer such as a throttle position sensor (TPS).

FIGURE 5-28 TPS mounted to a throttle body injection unit.

FIGURE 5-29 Schematic symbol for a thermistor.

Figure 5-29 is a drawing of another type of variable resistor called a **thermistor**. The capital letter T near one end of the resistor is the indicator of a thermistor. As is the case with the potentiometer and the rheostat, the thermistor is a variable resistor. The difference is that resistance change will be temperature related ("therm"). The most frequent use of the thermistor is to indicate engine termperature to the computer. It is usually installed in the coolant passages and can have either a negative or positive coefficient. Negative thermistors will have their resistance decrease as the coolant temperature increases, while positive thermistors will have their resistance increase as the coolant temperature increases.

The last form of resistance commonly found on the vehicle is the heating resistor. Figure 5-30 shows the symbol for a heater.

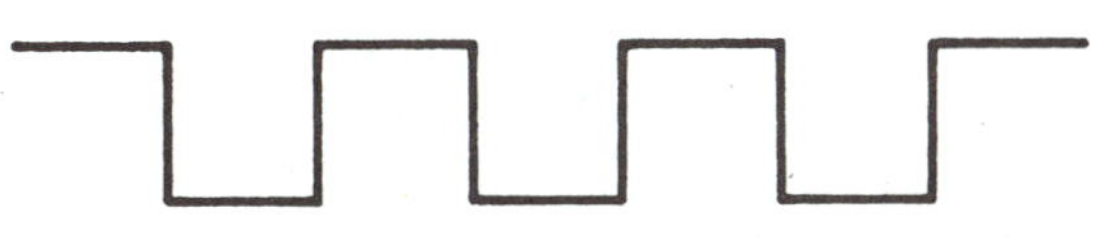

FIGURE 5-30 Schematic symbol for a heating element such as a rear window defogger.

The squared off resistor symbol will be found for the rear window defogger or a choke heater on the schematic diagram. It is a resistance heater that will develop sufficient heat when current is passed through it to melt the snow off of the window or heat the choke assembly to open it.

MOTORS

There are many electric motors in use on today's vehicle. Figure 5-31 shows the more common ones schematically. The differences between them is: how they are grounded, either case or remote; whether or not they are reversible; and if they are multispeed. The schematic for the single speed bidirectional motor would be found for power windows, antennas, or seats where the customer reverses the motor direction, or to control the window or antenna. It usually has two wires coming in and no ground. Reversing the current flow through the motor windings causes it to reverse its direction. The single-speed unidirectional (one direction) motor shown in the center is the type found on engine cooling fans or heater blower motors. Notice there is only one wire coming in and the unit is case grounded. The third motor shown is similar to the style used in windshield wipers, two-speed unidirectional. One direction but multispeeds either case or remote grounded. Keep in mind that the motors can have a circuit breaker installed and shown on the schematic as Figure 5-32 indicates. This will protect the motor if

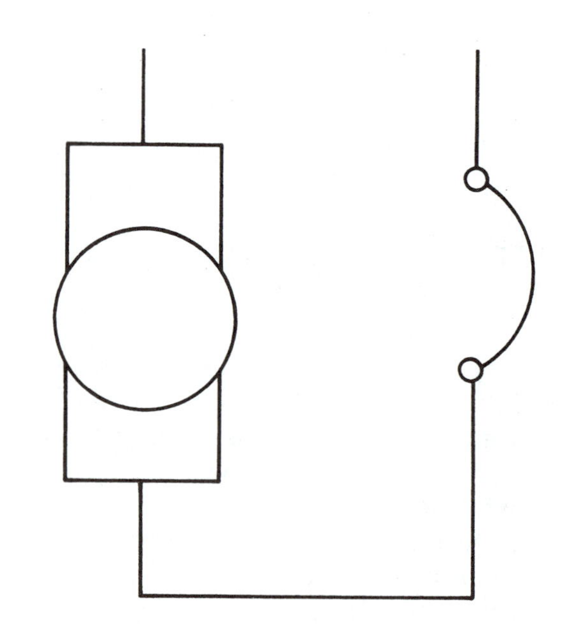

FIGURE 5-32 Motor with an internal circuit breaker.

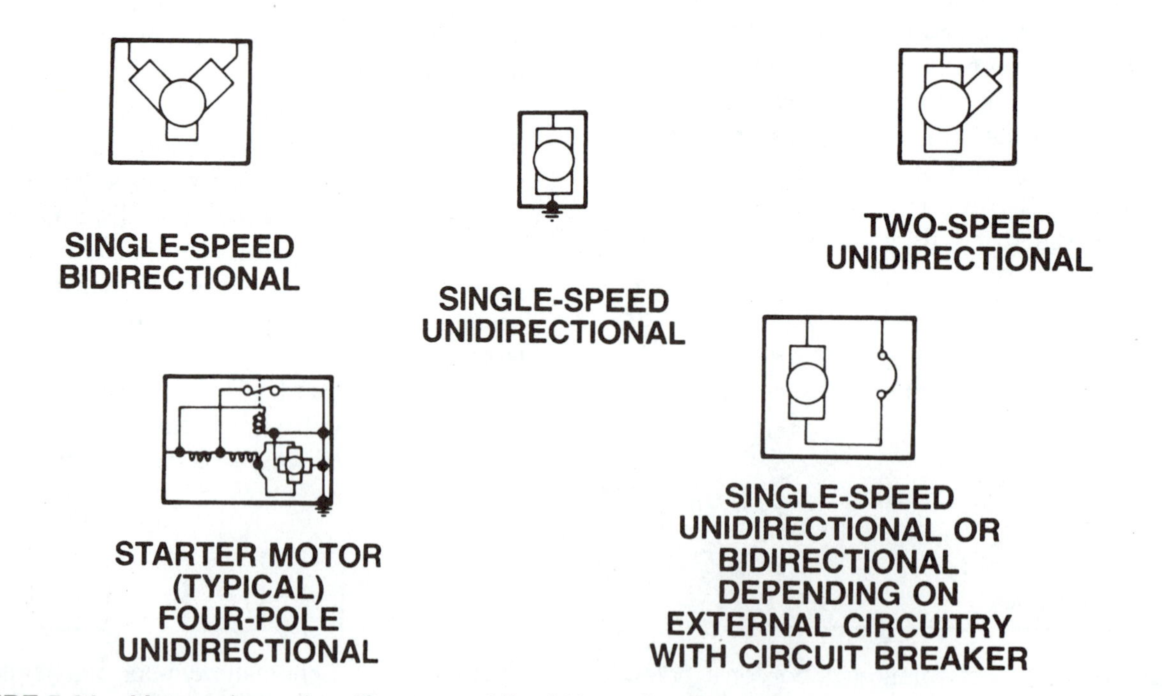

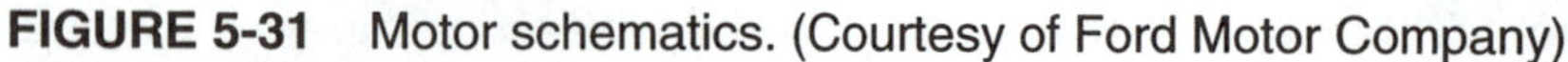

FIGURE 5-31 Motor schematics. (Courtesy of Ford Motor Company)

the mechanism is locked up and cannot move. The one shown is automatically resetable once the motor is able to turn again.

LOCATION CODES

Now that we have looked at some of the common schematic symbols and have a basic idea of how to trace through a wiring diagram, let us look at how one would actually follow a circuit on the vehicle. Remember that it is virtually impossible to actually follow a single wire through the vehicle. It might change color many times and go through sections of the vehicle that are totally unrelated to the circuit being followed. This is the main reason why the technician must rely on the wiring diagram. It will provide information that will be invaluable when actually trying to follow the circuit on the vehicle.

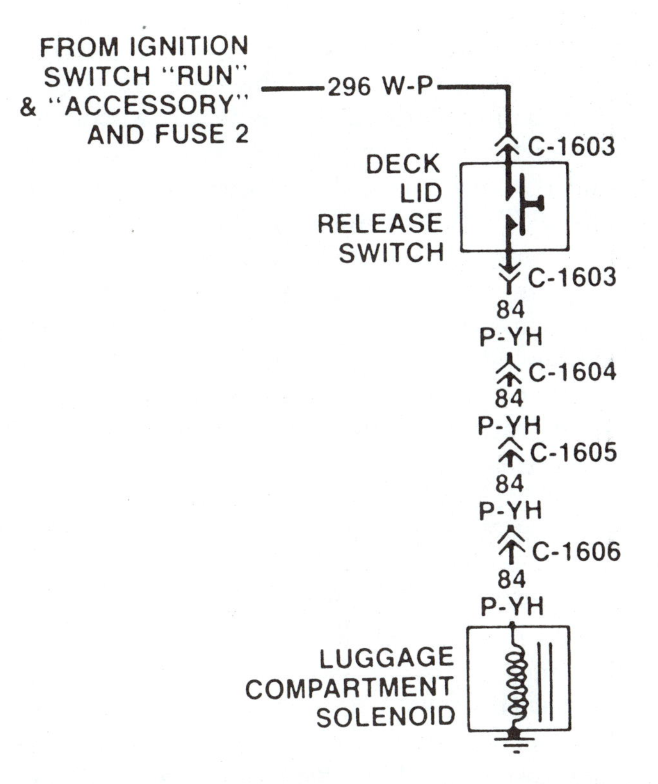

FIGURE 5-33 Solenoid operated luggage compartment latch. (Courtesy of Ford Motor Company)

I. Procedure to Determine an Open Circuit

Figure 5-33 shows the same circuit we used to describe solenoids. If a vehicle were to show up at your shop with an inoperative luggage compartment solenoid, your test procedure would probably involve a wiring diagram. Let us look at a diagnostic procedure that will be fast and efficient and use the capabilities that are built into the wiring diagram. The first step is always to verify the complaint. Try the release button a few times. Next go to the load, which in this case is the solenoid, and jumper B+ over to it getting as close as possible to the solenoid. The wiring diagram shows connector C-1606. But how do we know where this connector is? The location codes will tell the story. As we previously mentioned, every wire, connector, splice, and switch, among others, has a code that can be found in a chart telling you its location. A small section of the chart is reproduced here. Find C-1606 and determine its location.

C-1603 back of deck lid release switch

C-1604 lower left kickpanel

C-1605 lower left trunk over wheelwell

C-1606 back of luggage compartment solenoid

The chart indicates that C-1606 is the connector that will plug into the back of the solenoid. This will be an easy place to apply power and see if the solenoid actually works. It will also be the place where we should expect to find B+ with our test light. Assuming the solenoid is functional, our diagnosis of this circuit will involve working our way back through the circuit looking for B+ with our test light. The next most logical test point will again be a connector, C-1605, which is located in the trunk over the wheelwell on the left side. Notice how specific the description is. We know exactly where to look. This is the major advantage to location codes. They make testing far simpler because they eliminate the need to follow a wire. By tracing back through the diagram and looking for power at convenient test points like connectors, we will be able to quickly determine where the open circuit is. It will be between the connector, which had no power, and the connector, which had power. Testing systematically using the location codes makes the testing a snap. Wiring diagrams that do not have location codes are much more difficult to work with. Some aftermarket wiring diagrams do not include location codes on the diagram and instead have a separate book available, which lists the information.

This simple diagram also gives us additional information. It has identified the manufacturer's circuit number, 84, and also indicated that the feed wire to the switch is a white wire with a pink strip (W-P), which changes color as it comes out of the switch. It becomes a pink wire with a yellow hash mark (P-YH). This allows us to identify the wire easily should it be harnessed together with other wires for other circuits. Additionally, we know from the diagram that the circuit is case grounded at the solenoid. If you are able to "read" the diagram, it will prove to be a valuable and time-saving asset.

COMMON POINT DIAGNOSTICS

Many technicians refer to common point diagnosis as a tool or trick of the trade. It refers to the use of the wiring diagram to prediagnose, especially an open circuit. When the manufacturers design the vehicle they electrically tie different circuits together in a series-parallel arrangement. Figure 5-34 shows a simple example of this arrangement. The fuse is series wired to the three parallel loads. The common point in the circuit is just after the fuse where the three loads branch off. If the vehicle has a load A, which will not function in this circuit, the technician would take advantage of the information that the wiring diagram has supplied. Two additional loads are common point wired (B and C) to the fuse at S1. By being alert to this simple fact, technicians can save themselves valuable diagnosis time. If load B and C both function, the open must be beyond common point S1. If none of the loads function, the open is probably before the common point. The wiring diagram has just saved us from having to test at least half of the circuit and if the half we do not have to test is hidden behind panels, we have substantially reduced our diagnosis time.

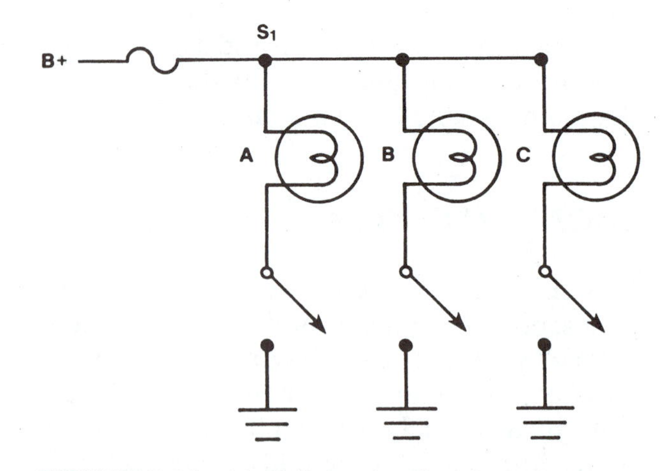

FIGURE 5-34 Multiple loads off a single fuse.

Let us look at another example and use common point diagnosis to decrease our diagnosis time of a short circuit. Figure 5-35 shows a vehicle with three loads running off of fuse number 4. Customer complaint is a blown fuse. Obviously, we would not want to crawl around this extensive circuit, because it covers almost the entire vehicle. The time involved to locate the short circuit would be hours. Instead, common point diagnosis can be used to identify the most likely location of the short before electrical testing begins. Remember, short circuits decrease the overall circuit resistance and, generally, blow the circuit protector, which in this case is fuse number 4.

II. Procedure for Isolating a Short

To isolate the circuit, we would first turn off all loads on fuse 4: air conditioning (AC) off, turn

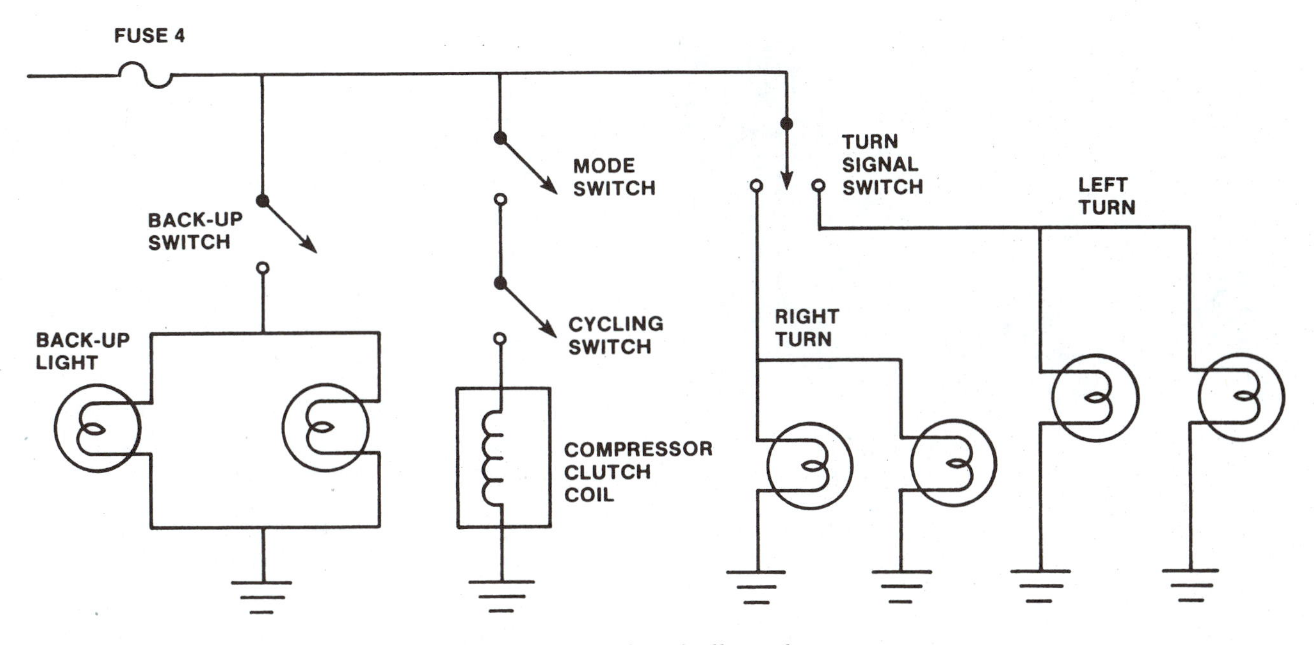

FIGURE 5-35 Three separate circuits common pointed off one fuse.

signal in a center position, and gear selector in park with the vehicle off. This should eliminate all current flow through the circuit. Next, we would plug in another fuse. Remember our goal here is to isolate the short to as small a section of the entire circuit as is possible. Turn on the ignition; did the fuse blow? If it did, where is the short? The common point after the fuse up to the switches is the only part of the circuit that is live. We would now know that our work will be in the dashboard area before the individual circuit switches. We would now separate the circuit at connectors after the common point and before the switches to further isolate the circuit. If the fuse did not blow, we know that our short is after the common point and the switches. By now you probably have figured out that the next step, if the fuse has not blown, would be to turn on each circuit separately until the fuse burns out. The short would be in the section of the circuit that, when energized, caused the fuse to burn out. Diagnosis like this can save you a tremendous amount of time. If, for example, the vehicle above came in with the complaint of inoperative AC and you did not at least walk through the circuit to determine if additional loads were on the same fuse, you might spend hours looking for a short in the AC circuit when, in fact, you might have a short in the back-up light assembly. Until the gear selector is placed in reverse and power runs to the back-up light assembly, the short would not appear. The customer complaint of inoperative AC or inoperative turn signals ignores the additional circuit on the fuse. Most customers are unaware of inoperative back-up lights because they cannot see them. They recognize only the circuits that have a direct bearing on themselves. Being aware of additional circuits on a fuse and utilizing common point diagnosis will pay the cost of the wiring diagrams on the first job.

PRINTED CIRCUITS

The last thing we will look at in this chapter is the use of printed circuit boards used on many dash assemblies. Figure 5-36 is a picture of a printed circuit board from the back of a dashboard. It is composed of a film of plastic with small strips of copper. The current for various sections of the dash will run through

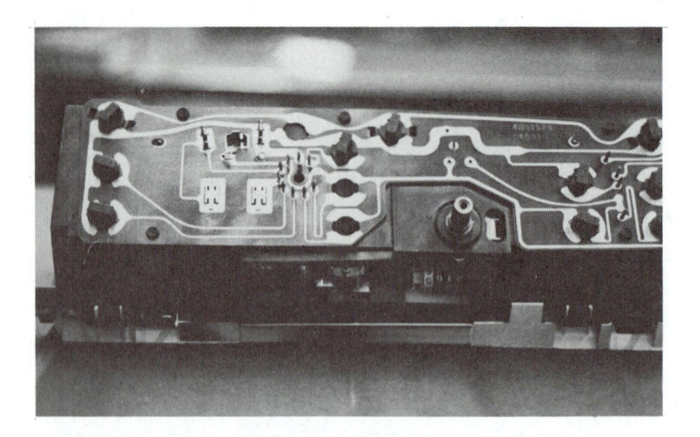

FIGURE 5-36 Printed circuit board.

these strips of copper. Lights, gauges, and indicators will plug directly into the printed circuit board and one harness connector will bring both B+ and ground to all the circuits. Printed circuit boards became popular as more and more of the manufacturers began to utilize plastic dashboards. The majority of the wiring diagrams will show the layout of the board as in Figure 5-37. It is possible to trace through the circuit as you would with any other wiring diagram.

PRECAUTIONS

There are some precautions that should be observed when working on a printed circuit.

1. Never touch the surface of the board. Dirt, salts, and acids on your fingers can etch the surface and set up a resistive condition. It is possible to knock out an entire section of the dash with a fingerprint.
2. The copper conductors can be cleaned with a commercial cleaner or an eraser lightly rubbed across the surface.
3. The printed circuit board is damaged very easily because it is very thin. Be careful not to tear the surface when plugging in bulbs. Replacement circuit boards are usually available from dealers, but the inconvenience and time necessary to obtain a replacement makes a mistake costly. Handle them with care.

One of the advantages of printed circuits is their high degree of reliability. In addition, they allow the dash to be pulled back and worked on with the entire circuit laid out and

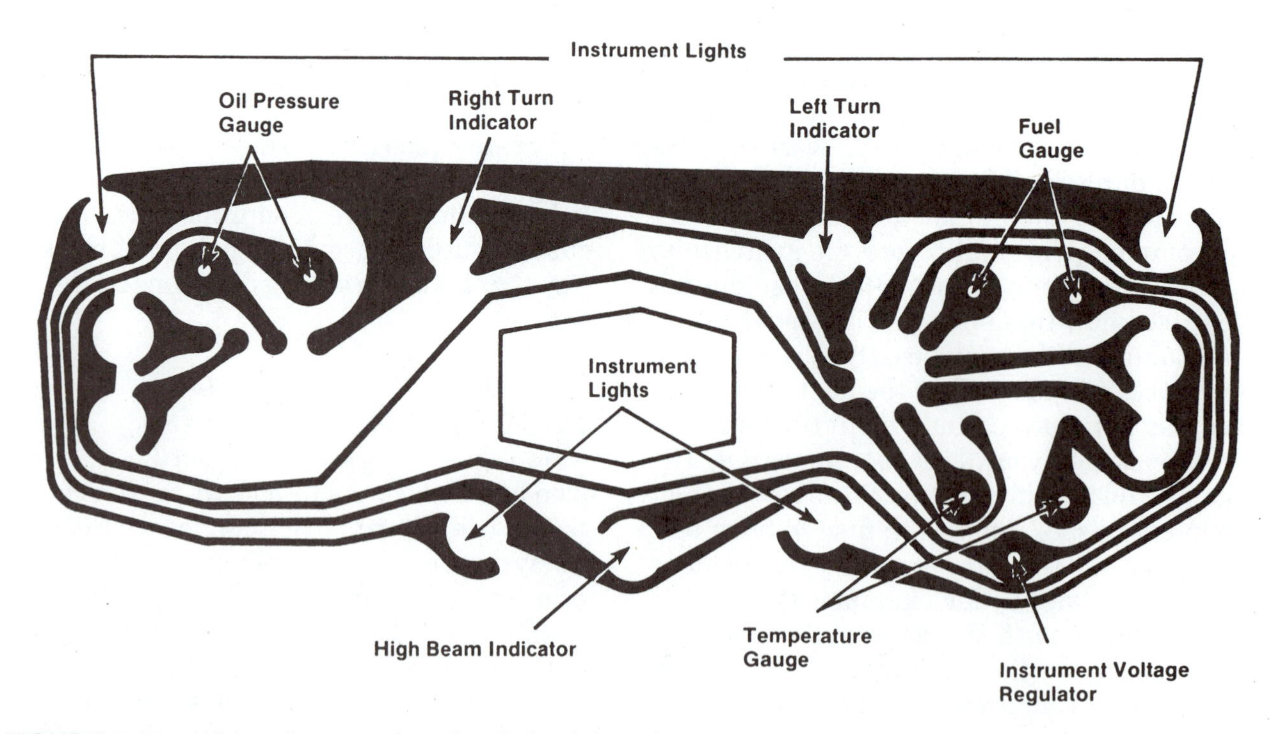

FIGURE 5-37 Wiring diagram of a printed circuit board.

functional. Because both B+ and ground come through the connector, the board will function not installed. This allows you to use your voltmeter or test light on the circuit with it functioning normally.

SUMMARY

In this chapter, we have looked at the most common symbols used in many wiring diagrams. Both pictorial and schematic symbols were shown. Splices, grounds (both remote and case style), and connectors were analyzed. The three types of circuit protections most commonly found were shown. Schematic diagrams were shown for a variety of switches, including single pole, single throw, and multiple pole, multiple throw, and ganged switches were looked at including an ignition switch. Variable resistances, such as those used in dimmer assemblies in addition to the various computer input symbols, were shown from manufacturers' wiring diagrams. The chapter ended with motor types and their symbols, and a discussion of the use of printed circuit diagrams. The use of the wiring diagram in common point diagnosis was examined and demonstrated.

KEY TERMS

Remote and case grounds
Splice
Common point
Location codes
Expanded splice
Fuse
Circuit breaker
Fusible link
Single pole, single throw (SPST)
Single pole, double throw (SPDT)
Double pole, double throw (DPDT)
Multiple pole, multiple throw (MPMT)
Normally open (NO)
Normally closed (NC)
Solenoid

CHAPTER CHECK-UP

Multiple Choice

1. Technician A states that a common point is represented by a black dot on the wiring diagram. Technician B states that a connector can be a common point. Who is correct?
   a. Technician A only
   b. Technician B only
   c. Both Technician A and B
   d. Neither Technician A nor B
2. —>>— is a symbol for a
   a. common point.
   b. splice.
   c. location code.
   d. connector.
3. ⏚ is a symbol for
   a. ground.
   b. connector.
   c. common point.
   d. none of the above.

4. Technician A states that if a fuse burns out when a switch is turned on, a short circuit is present before the switch. Technician B states that the short circuit is completed when the switch is closed. Who is correct?

   a. Technician A only.
   b. Technician B only.
   c. Both Technician A and B.
   d. Neither Technician A nor B.

5. A remote ground is

   a. where a ground connection is made at some point other than the component being grounded.
   b. where the ground is completed by placing the component into a metal housing.
   c. where two or more grounds are connected.
   d. none of the above.

6. A dotted line between connectors indicates that

   a. the connectors are electrically connected.
   b. the connectors are physically the same.
   c. the connector is a remote connector.
   d. none of the above.

7. Technician A states that two connectors with the same number are physically the same connector. Technician B states that the two connectors are drawn apart so that the drawing is easier to read. Who is correct?

   a. Technician A only
   b. Technician B only
   c. Both Technician A and B
   d. Neither Technician A nor B

8. Location codes are generally given for

   a. splices.
   b. connectors.
   c. components.
   d. all of the above.

9. Printed circuit boards are frequently used in

   a. dash assemblies.
   b. headlight switches.
   c. air-conditioning control circuits.
   d. all of the above.

10. A switch is drawn on a wiring diagram with NC printed beside it. This indicates

    a. the switch will be normally closed with the vehicle running.
    b. the switch will be open with the vehicle off.
    c. the switch will be closed with the vehicle off.
    d. none of the above.

True or False

11. A common point is a point where two or more wires are connected.
12. A splice will never be a common point.
13. SPST stands for single pole, single throw—a switch style.
14. Touching a printed circuit board with dirty hands might etch the surface and open circuit it.
15. Dotted lines between switch wipers indicate that all wipers will move together.
16. A location code is a code that will tell one where a component is on the wiring diagram.
17. A case ground is a remote point where two or more grounds are connected.
18. Printed circuit boards can be cleaned up with a good eraser.
19. A resistor drawn with an arrow through it is a symbol for a thermistor.
20. W-BH is printed on a wire. This is the color of the wire on the vehicle.

CHAPTER 

Batteries

OBJECTIVES

After completing this chapter, you should be able to

- define a primary cell.
- define a secondary cell.
- figure voltage for series and/or parallel cells.
- explain three common methods of rating batteries.
- use a temperature-compensated hydrometer to determine the state of charge.
- use a voltmeter to determine the state of charge of a sealed battery.
- determine the correct size battery for a vehicle.
- observe safety rules when working around a battery.

INTRODUCTION

An automotive battery is a device that produces electricity by means of chemical action. Technically, to be considered a battery, it must be composed of two or more **cells**, each cell being independent and containing all the chemicals and elements necessary to produce electricity. The term battery actually means groups. Our modern 12.6-volt battery used in vehicles is composed of six cells. Before we look at an automotive battery, let us examine the cell in detail.

TYPES OF CELLS

There are two common types of cells available today: **primary** and **secondary**. Their difference is in the ability of a secondary cell to be recharged. Primary cells cannot effectively be recharged and are the common A, C, D, and so on, cells that we use in flashlights and radios. When their ability to supply current at the correct voltage is gone, they must be replaced. Let us look at this primary cell closer. Knowledge of how it works, will help your understanding of a secondary cell.

Figure 6-1 shows a drawing of a carbon-zinc primary cell. The case of the cell is negative and is made of zinc and the center terminal is positive and is actually a carbon rod. Between the carbon rod and the zinc case is a paste called the **electrolyte**. It is the

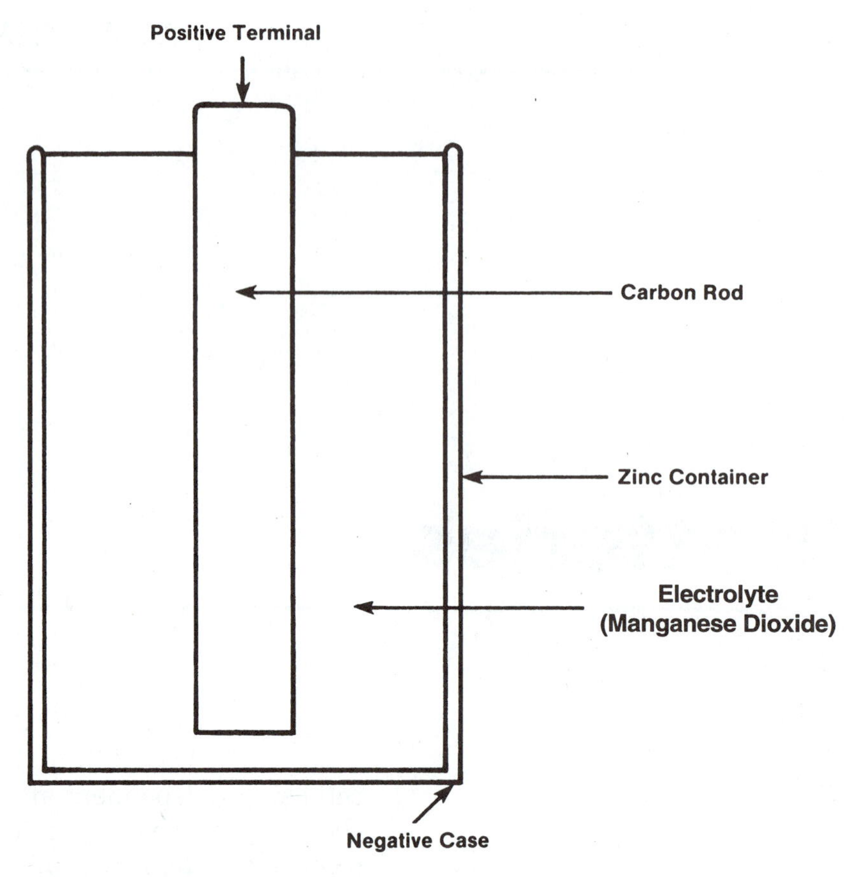

FIGURE 6-1 Carbon-zinc primary cell.

chemical action of the electrolyte with the carbon and zinc that produces the 1.5 volts and small amperage the cell has. As you run the flashlight or radio, the electrolyte wears away and the light gets dimmer and dimmer until the cell is dead. Using a battery charger on a zinc-carbon cell can extend its life, but will never recharge the cell back to its full potential.

There are many variations of the zinc-carbon cell. Among these are the alkaline, the mercury, and lithium, among others. The chemical action within these cells is slightly different and results in a greater current capacity when compared to carbon-zinc. Generally, their voltage also remains higher throughout their use.

CONNECTING CELLS

Have you ever noticed that cells are installed in flashlights with the positive of one cell in contact with the negative of another cell? Have you wondered about the 9-volt battery used in many transistor radios? Both of these are examples of series wiring of cells together. Figure 6-2 shows a pictorial diagram of our common two-cell flashlight. By wiring the cells in series (positive to negative), we are able to increase the voltage. Two cells in series produce 3 volts ($1.5 \times 2 = 3$ volts). How about four cells in series? If you think four cells in series produce 6 volts, you are correct: $1.5 \times 4 = 6$ volts. Now look at Figure 6-3. A 9-volt battery has six cells wired in series and stacked on top of one another ($1.5 \times 6 = 9$ volts). The

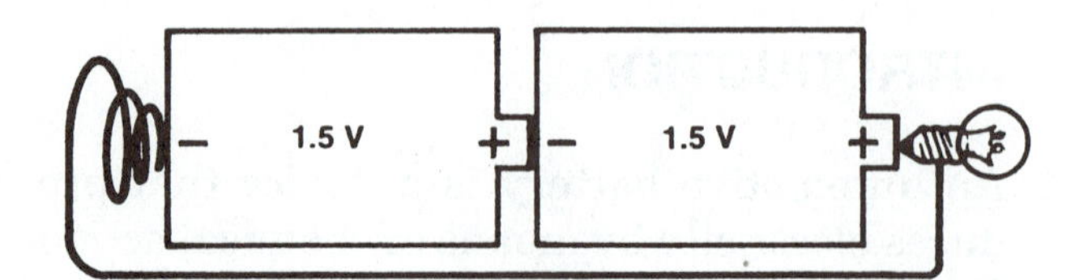

FIGURE 6-2 Two-cell flashlight operates on 3 volts.

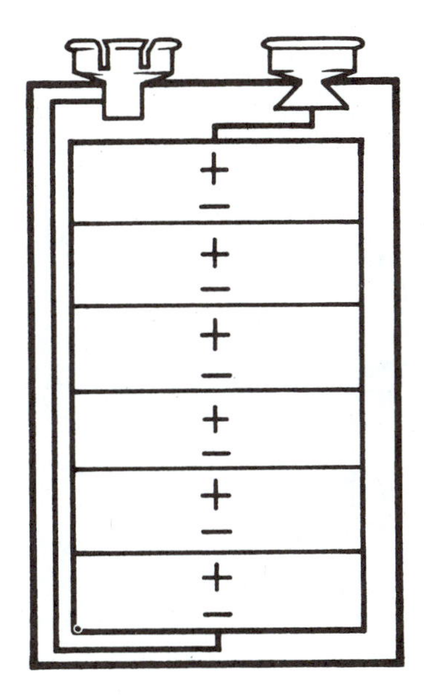

FIGURE 6-3 Six cells wired in series (+ to –) develops 9 volts.

conclusion here is that wiring of cells in series increases the voltage.

By now you have probably figured out what happens if we wire cells in parallel (positive to positive and negative to negative). The voltage will remain the same, but the ability of the cell to produce current will be doubled. If we wire a 1.5 volt to one cell and turn it on, the bulb will remain on for a period of time. If we wire two cells in parallel, the amount of time the bulb will stay on will be doubled. The conclusion here is that wiring of cells in parallel increases the available current, while keeping the voltage the same.

AUTOMOTIVE BATTERY CELLS

Now that we have an understanding of how a primary cell functions, let us turn our attention to the rechargeable cells that make up our automotive storage batteries. **Rechargeable** is the key difference that makes this type of secondary cell ideally suited for automotive use. Each morning the amperage necessary to crank and start the vehicle would quickly kill a primary cell. But, with a secondary cell, the charging system replaces the amperage used and brings the battery back up to full potential.

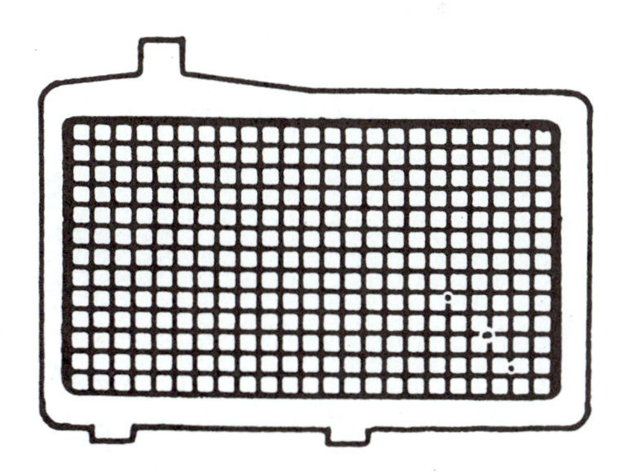

FIGURE 6-4 Typical battery grid.

Let us take the top off of a typical battery and look at the cell. A cell is composed of positive plates, negative plates, separators, and electrolyte. A plate, either positive or negative, starts out with a grid similar to a window screen. Figure 6-4 shows a typical grid. Grids are made out of lead alloys, usually either antimony or calcium. The active material (lead dioxide for positive, sponge lead for negative) is pressed into the grid in paste form. Notice that the two plates are chemically different, but both are forms of lead. Do you remember our primary cell? It was composed of two different materials also, zinc and carbon. We will look at what happens chemically in a future section. By taking a series of plates and wiring them together by a strap, we form either positive or negative plate groups. These plates are insulated from each other by separators made of plastic or glass. Older batteries were separated by wood. Figure 6-5 shows the assembled element that will be placed into the battery case. This element will sit in its own individual container, isolated from the other cells with a weak solution of sulfuric acid and water called the electrolyte.

Remember our discussion of series and parallel wiring of cells? Within the individual cell we have just assembled, all of the positive and all of the negative plates are wired together in series. This increases the amperage capacity of the cell, while the cell voltage remains the same (2.1 V). The more active material contained within the element, the greater the am-

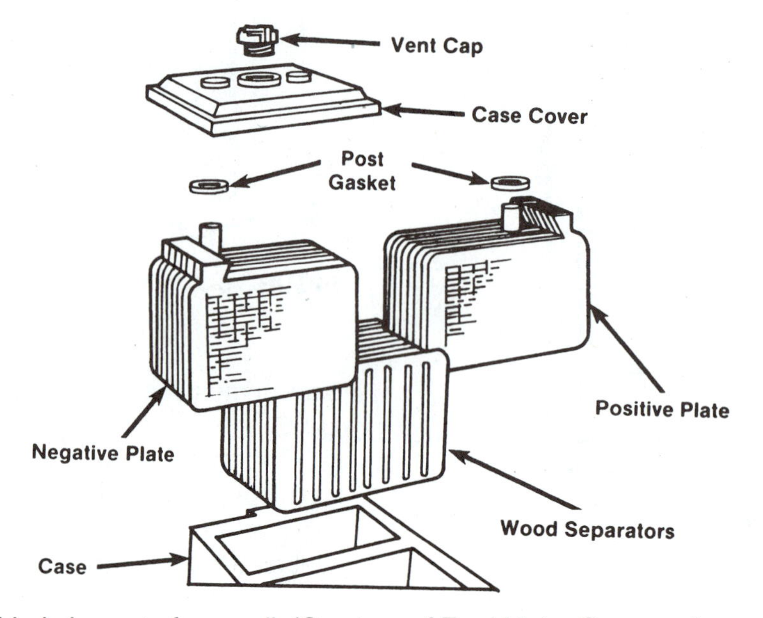

FIGURE 6-5 Assembled element of one cell. (Courtesy of Ford Motor Company)

perage capacity of the cell. The six cells will each produce 2.1 volts. They are wired in series to produce the 12.6 volts we need. Notice Figure 6-6A. It shows the connectors that connect the positive plates of cell one to the negative plates of cell two. Cell two's positive plates are connected to the negative plates of cell three, and so on. This series wiring of each cell to the other cells produces the required 12.6 volts from a fully charged battery. Notice also from Figure 6-6B that the element sits down in the case raised up slightly from the bottom on element rests. These spaces form the sediment chamber where active material that flakes off of the plates can collect without shorting out the cell.

Most batteries are produced with the plates sitting the length of the cell. There are, however, some batteries where smaller plates are placed the width of the cell as in Figure 6-7. This type of cell will typically have many more plates, with each one being smaller. They are wired with all the positives and negatives together as in our other example.

CHEMICAL ACTION IN A BATTERY

Now that we have an idea of how the battery is constructed, let us look at the chemical action during charge and discharge. You will not need a degree in chemistry to follow along, but a little understanding of chemical formulas will help. A chemical formula tells us what elements make up whatever we are looking at. For instance, H_2O is the chemical formula for water. The number 2 after the H indicates that 2 atoms of hydrogen (H) are in water. This leaves the O, which stands for oxygen. Without a number after the O, we must realize that one atom of oxygen is present. No number after the element indicates 1. By looking at the formula H_2O, we can see that 2 atoms of hydrogen and 1 atom of oxygen combined together form water. This formula will become very important during the charging of a battery.

The plates in a primary cell were made of different materials, carbon and zinc. A secondary cell must also have two different or

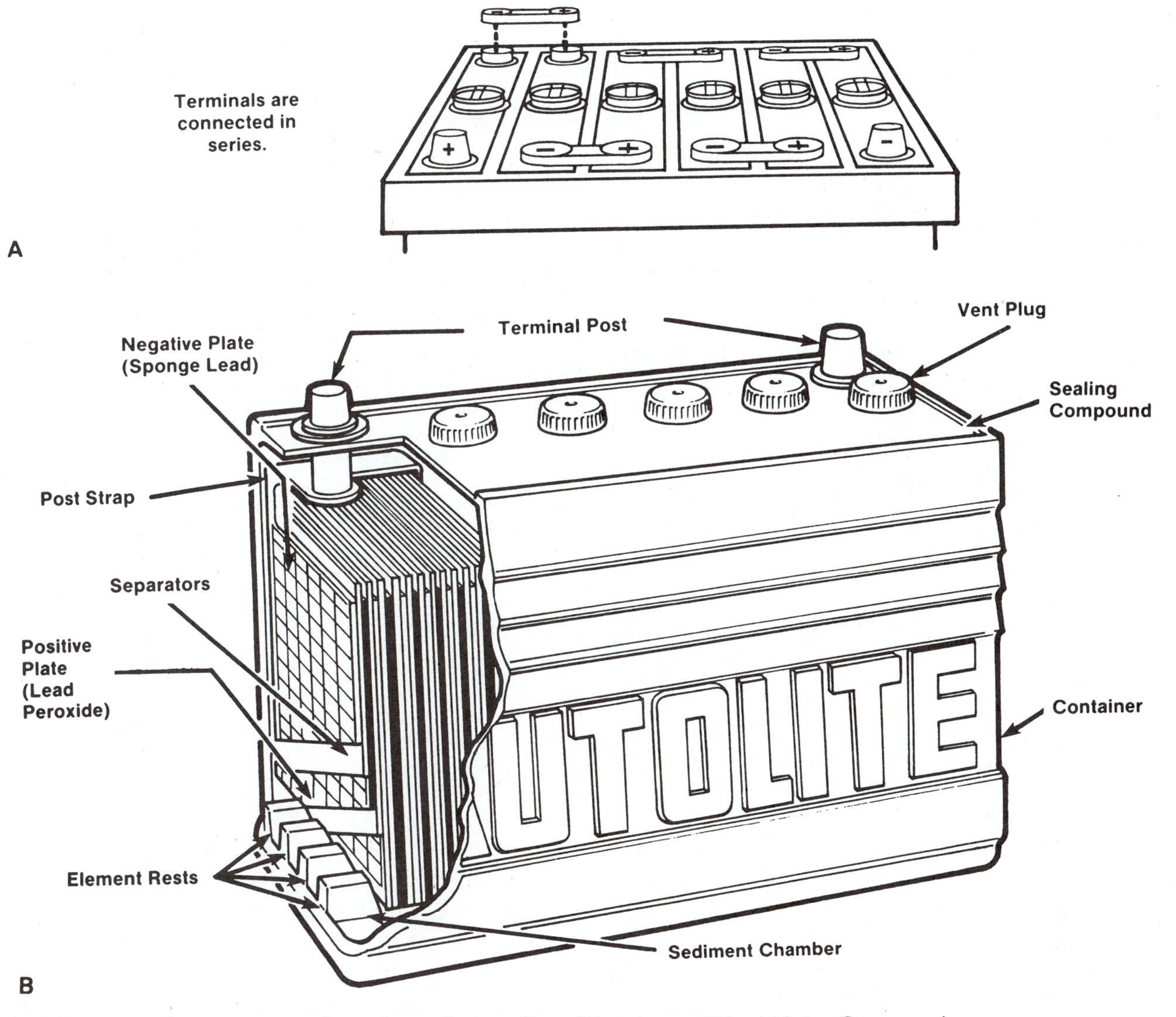

FIGURE 6-6 Connectors series-wire cells together. (Courtesy of Ford Motor Company)

dissimilar materials to function. The positive plate is a compound of lead and oxygen called lead dioxide (PbO_2), while the negative plate is composed of spongy lead (Pb). The electrolyte that the plates are in is sulfuric acid (H_2SO_4). Look closely at Figure 6-8. It simply shows what happens to the plates and electrolyte during discharge. The lead (Pb) from the positive plate combines with the sulfate (SO_4) from the acid forming lead sulfate ($PbSO_4$). While this is happening, oxygen (O) in the active material of the positive plate joins with the hydrogen from the electrolyte forming water (H_2O). This water dilutes or weakens the acid concentration.

A similar reaction is occurring in the negative plate. Lead (Pb) is combining with sulfate (SO_4) forming lead sulfate ($PbSO_4$). Notice that the end result of discharging is changing the positive plate from lead dioxide into lead sulfate at the same time that the negative plate is also being changed into lead sulfate. Remember that a cell will be able to produce voltage only if it is composed of two dissimilar materials. Discharging a cell makes the positive and negative plates the same. Once they

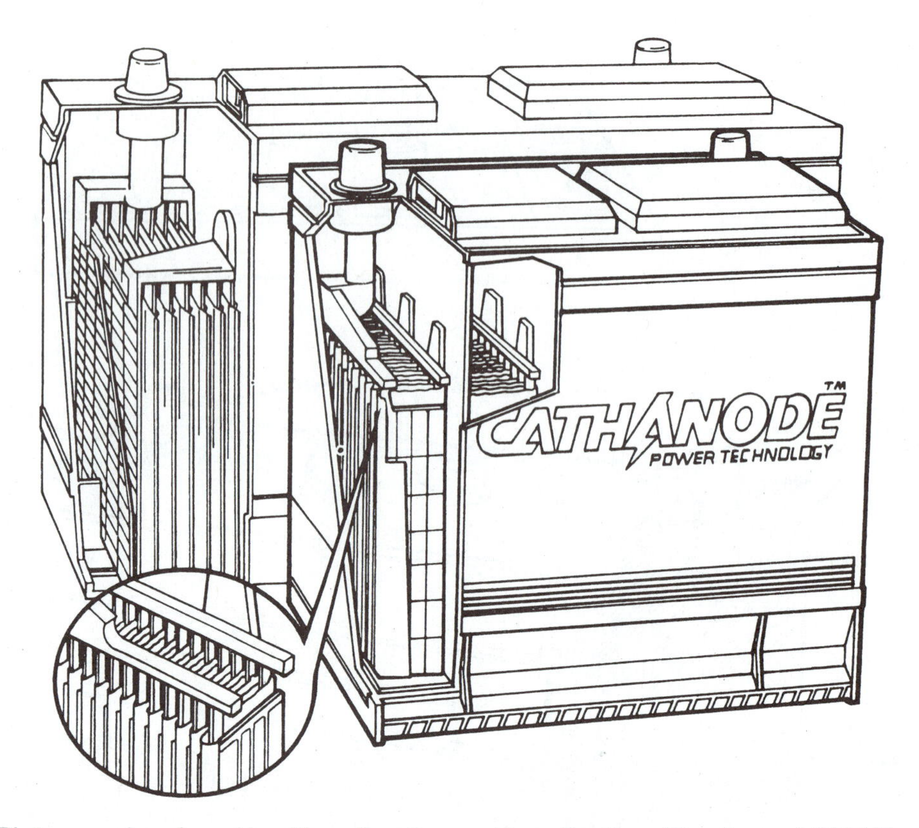

FIGURE 6-7 Plates can be placed in either direction and be effective. (Courtesy of GNB Technologies Inc.)

are the same, the cell is dead, and additional water has been added to the electrolyte diluting its strength. Look over Figure 6-8 again that shows the discharge cycle.

The charge cycle is exactly the opposite, which is shown in Figure 6-9. The lead sulfate ($PbSO_4$) in both plates is split into its original forms of lead (Pb) and sulfate (SO_4). The water in the electrolyte is split into hydrogen and oxygen. The hydrogen (H_2) combines with the sulfate (SO_4) becoming sulfuric acid again (H_2SO_4). The oxygen from the split water combines with the positive plate forming the lead dioxide (PbO_2). This now puts the plates and the electrolyte back in their original form and our cell is charged. In essence, this chemical

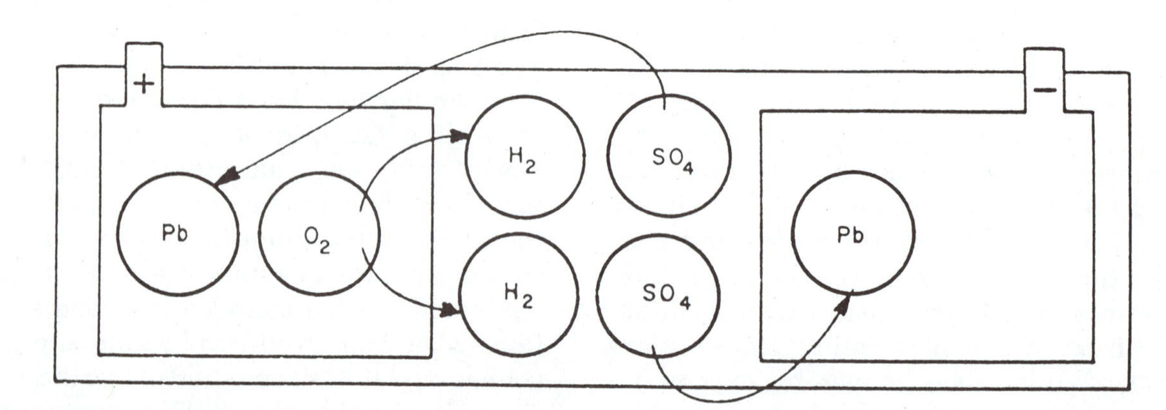

FIGURE 6-8 Chemical change: cell being discharged.

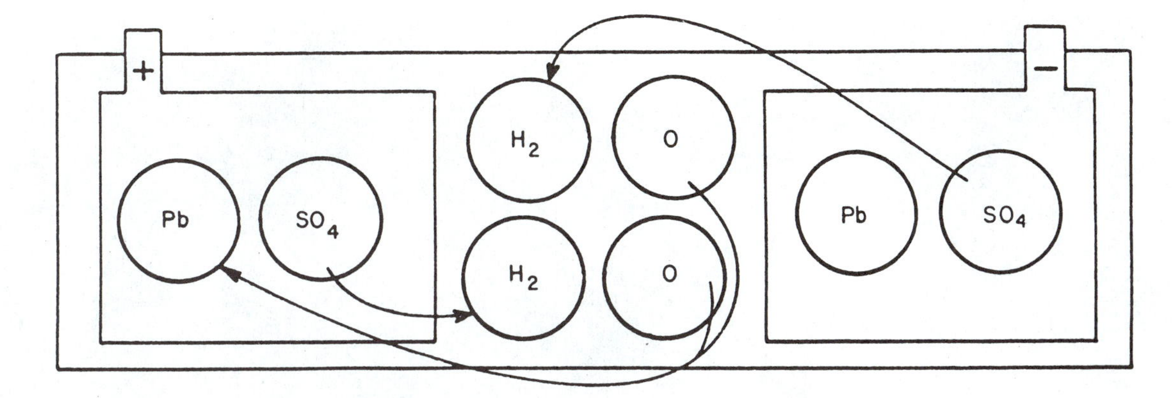

FIGURE 6-9 Chemical change: cell being charged.

action is a closed cycle. Nothing is left over; everything is used in the discharge and charge cycle.

If we zero in on the electrolyte, we can discuss something called specific gravity and see that it can give us information about each cell. Simply defined, **specific gravity** is the thickness of a liquid. Water at 68°F (20°C) is assigned the number 1000. Anything that is thicker than water will be assigned a higher number. Anything thinner, a lower number. In other words, a liquid with a specific gravity (SG) of 2 is twice as thick as water, while a liquid with a SG of 0.5 is half as thick as water.

Do you remember our electrolyte? It is composed of sulfur and water and is obviously thicker than water alone. As a matter of fact, it has a specific gravity of 1.265 (usually read as twelve sixty-five). In a little while, we will learn how to use a hydrometer. It is a device that measures the specific gravity of a liquid.

Let us get back to our chemistry of discharge. Hopefully, you have not forgotten that oxygen from the positive plate joins with the hydrogen in the electrolyte during discharge, forming water. This water dilutes the electrolyte and changes the specific gravity. Which way does it go? Well, if water is our standard of 1.000, discharging a cell will change the electrolyte more toward 1,000. As the cell is being discharged, the concentration of the electrolyte is changing from the high of 1.265 down to a low of about 1.100. By the time the SG reached 1.100, the cell is dead. Recharging the cell will again split apart the water with the oxygen going into the positive plate and the SG going up again. The chart reproduced here shows the relationship between specific gravity and the amount of charge the cell has. These numbers are approximate and are for electrolyte at 80°F, Table 6-1.

TABLE 6-1 Specific gravity versus state of charge.

State of Charge	Specific Gravity
100%	1.265
75%	1.225
50%	1.190
25%	1.155
Discharged	1.120

TESTING WITH A HYDROMETER

Examine Figure 6-10. It shows a **hydrometer.** The float inside the clear barrel is graduated from 1100 on the bottom to 1300 on the top. Usually, hydrometer floats are also color coded with green for the charged section, white for partially charged, and red for discharged. Notice that on the bottom of the hydrometer there is a tube to draw the electrolyte into the barrel. The electrolyte will enter the barrel and when the bulb is squeezed cause the hydrometer float to float. Toward the lower end of the hydro-

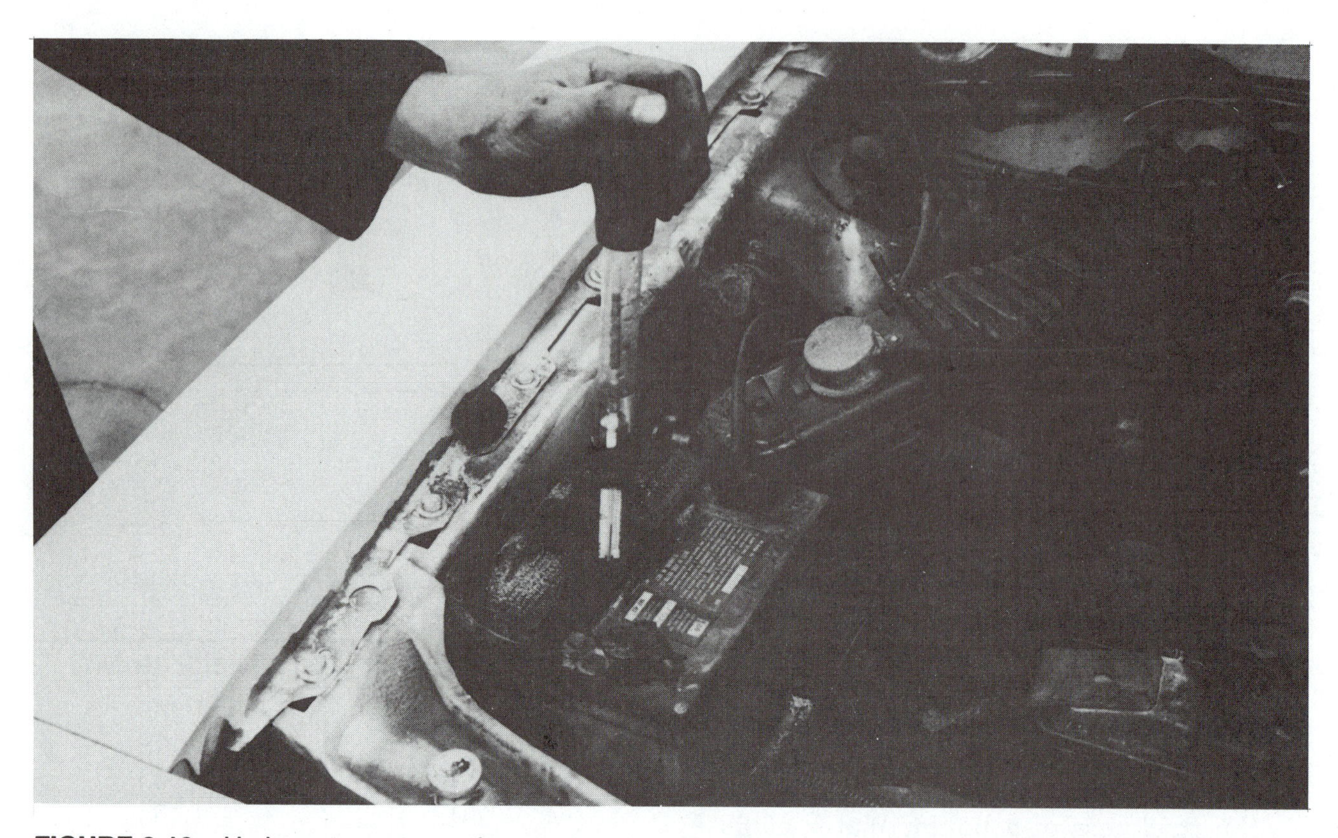

FIGURE 6-10 Hydrometers are used to measure a cell's specific gravity, for example, state of charge.

meter, there is a thermometer, which will indicate the temperature of the electrolyte. Temperature is very important as the specific gravity will change with temperature. The electrolyte's SG will be higher if the temperature of the liquid is lower. To be able to convert the SG over to a state of charge, the electrolyte must be at 80°F or be converted over to an adjusted SG, which has been compensated for temperature. For every 10° change, the SG reading must be altered by 0.004. Look at Figure 6-11. It shows this temperature compensation on a thermometer.

I. Procedure for Testing with a Hydrometer

Let us try a few examples. What is the actual SG if the temperature of the electrolyte is 100° and the float reads 1.240? Find 100° on Figure 6-11. To the right of 100 is +8. (This is actually +0.008.) The actual SG of the cell is 1.240 + 0.008 or 1.248. Got it? How about a battery that has been sitting out all night in the North and has a float reading of 1.220 at 10°F? If you first realize that this float setting indicates about 75% charged before compensation, you will see why it is so important. A battery that is 75% charged should be capable of cranking the engine. Let us temperature compensate to determine the actual SG and state of charge.

$$10° = -28 \text{ or actually } -0.028$$
$$1.220 - 0.028 = 1.192$$

Notice the difference? On Table 6-1, 1.192 indicates that this cell is actually 50% charged, *not* 75% like we first thought—quite a difference. It is possible that this cell will not be capable of supplying enough cranking power to

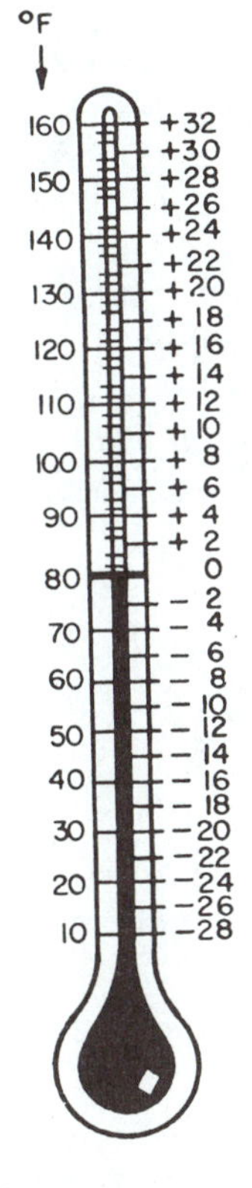

FIGURE 6-11 Temperature compensation chart from a battery hydrometer.

start the vehicle because it is only 50% charged. Temperature compensation is necessary if the electrolyte temperature is either above or below 80°F if an accurate reading is desired. Interpreting specific gravity readings will be covered in the next chapter.

SIZES AND RATINGS

When we look at automotive type batteries, we have to realize that there are different size batteries out in the field. Usually, their size is related to the job that they must do. A large V-8 requires a greater starting amperage than a 1.6 liter, 4-cylinder. Usually, the size of a battery is related to the amperage necessary to crank the engine. Automotive manufacturers will usually install larger batteries in vehicles with larger engines. Electrical accessories are also sometimes taken into consideration when figuring the required size of the battery. We will need to know the actual size and the recommended size battery when we test in the next chapter. Basically, there are three common methods of rating both American and foreign batteries.

Ampere Hour

Probably the oldest and still quite common method of rating batteries is ampere hour, or A/H. It is also referred to as the 20-hour method. Simply stated, it tells the amperage that a battery can produce until its terminal voltage falls to 10.50 volts. For example, if a battery can be discharged for 20 hours at a rate of 2.5 amps before its terminal voltage reaches 10.50 volts, it would be rated at 50 A/H.

$$2.5 \times 20 = 50 \text{ A/H}$$

In theory, a 50-ampere hour battery can produce a current level that when multiplied by time equals the number 50: 25×2 or 10×5, and so on. This low rate of discharge does not necessarily indicate the cranking capacity of the battery, but larger A/H batteries will deliver more current. Typically 40 to 70 A/H batteries are used by the automotive manufacturers as original equipment.

From a practical standpoint, the ability to produce current for 20 hours does not necessarily indicate the ability to crank an engine during a 20° below zero morning.

Cold Cranking Amperage

Cold cranking amperage is an indication of the battery's ability to supply current when cold. It is more difficult to supply current cold for short spans of time. The rating is done at 0°F and is an indication of the amperage the battery can supply for 30 seconds, while maintaining a voltage of 1.2 volts or more per cell (7.2 volts total). Cold cranking amps, or CCAs, usually fall in the range between 400 to 800 for passenger vehicles.

Reserve Capacity

The last rating we will discus is reserve capacity and it is defined as the ability of a battery to maintain vehicle load with the charging system inoperative. This minimum would probably include lights, ignition, wipers, and heater and be expressed in minutes. The reserve capacity is the minutes a new fully charged battery can be discharged at 25 amperes. A rating of 120 minutes means that the battery could be discharged at a rate of 25 amps for 2 hours (120 minutes).

MAINTENANCE-FREE BATTERIES

If we look at the typical lead acid battery, we would realize that the charging of it produces both hydrogen and oxygen in gas form. In the mid-seventies, the battery manufacturers began replacing the antimony in battery grids with calcium. The reason for this move was that a grid of antimony would gas under most charging situations. Even a slow charge would produce the gases through the electrolysis of the water. To maintain the appropriate amount of electrolyte meant that the technician or consumer had to replace the hydrogen and oxygen with water (H_2O). It was normal for a lead- antimony battery to use an ounce or more of water per cell per month. By reducing the antimony and replacing it with calcium, this water usage was reduced to around 10% of what it was. The term maintenance-free was attached to the calcium grid battery, because under "normal" conditions it would not require the addition of water in each cell during its life, Figure 6-12.

FIGURE 6-12 Calcium grid maintenance-free battery.

"Normal" should be defined at this point. A correctly functioning charging system with a voltage generally lower than 15 volts for a warm engine and battery is "Normal." Realize that if a maintenance-free battery is overcharged (over 15 volts) for extended periods of time, it will gas just as the lead-antimony battery did. If the manufacturer has sealed the cell tops to the point where you cannot add water, this gassing will eventually ruin the battery. Once the cell plates are not covered with electrolyte, they will begin to dry out and will be easily broken off from vibration. The life of the battery will be greatly diminished by this loss of water. At the same time, the concentration of electrolyte will be increased. It is important to note here that not all maintenance-free batteries are completely sealed. If the cell tops are removable, clean distilled or demineralized water should be added, if needed, just as we would with a lead-antimony cell. Never add water that is not pure. Distilled is preferred, especially if tap water has minerals suspended in it. Even calcium grid batteries will gas slightly under normal conditions. They will usually not require the addition of water during their lifetime, however.

Maintenance-free batteries with removable cell covers can have their specific gravity tested with a hydrometer. We do, however, find many nonremovable cell covers in use today. These batteries are tested in the same manner (next chapter), except that their state of charge is estimated based on their open circuit voltage (OCV). OCV is read by putting a voltmeter across the battery with no current flowing, Figure 6-13. All electrical accessories

FIGURE 6-13 Load testers are used to test batteries.

that draw current must be turned off. In addition, if the battery has been recently charged, a slight load (20 amps) must be applied for about a minute to drain off the surface charge before the open circuit voltage is noted. The OCV can be compared to specific gravity based on Table 6-2.

In this manner, the open circuit voltage can give us an indication of the battery's state of charge with a reasonable degree of accuracy.

NOTE: If a hydrometer can be used, it is the preferred method because it also gives us the comparison between cells that can lead us to a defective battery.

Remember, do not test the open circuit voltage just after the battery has been charged. Drain off the surface charge with a load before measuring OSV.

TABLE 6-2 OCV and specific gravity.

Charge	Specific Gravity	12V Bat	6V Bat
100%	1.265	12.7	6.3
75%	1.225	12.4	6.2
50%	1.190	12.2	6.1
25%	1.155	12.0	6.0
Discharged	1.120	11.9	5.9

FIGURE 6-14 Charge indicators are built directly into some batteries.

Some manufacturers are building a hydrometer into their cells before they seal the cell covers. Figure 6-14 shows a battery with a charge indicator. This indicator is accurate for the one cell it is in and should not be relied on to give total battery condition. If the one cell is charged, but the open circuit voltage is low, it would be an indication that other cells must not be up to full charge. This situation is similar to hydrometer readings that are more than the 50 points apart. Do not assume that the charging eye gives total battery information. Its information should be added to an OCV measurement and other battery tests, which we will examine in the next chapter.

TEMPERATURE AND BATTERY EFFICIENCY

Before we begin a discussion of battery testing in Chapter 7, we should look at battery efficiency and its relationship to temperature.

The activity of a battery is directly related to the temperature of its cells. The colder the cells are, the slower the electrochemical action is. The cell gets numb from the cold, just as we do. The chart that follows shows a comparison of cranking power available from a fully charged battery at 80°, 32°, and 0°. Notice that at 0°F, the efficiency of the battery is reduced to 40% of the capacity at 80°. In other words, a 50 A/H battery is really only the equal of a 20 A/H battery if it is at 0°, Table 6-3.

Table 6-3 assumes that the battery is fully charged and in good working order Obviously, if the battery has reduced capacity or is not fully charged, the percent of available power will be reduced even further. We are defining available power as the amperage that the battery can deliver during cranking.

Now let us think of what the vehicle battery is trying to accomplish. The hardest bit of work that this 12.6-volt battery must do is to crank the engine during starting. The chart to follow shows the amount of power required to crank an engine with SAE 10W-30 oil. Stiff engine oil is responsible for the increase in "drag" on the starting system, Table 6-4.

Notice that at 0°F, it will take over twice the effort to crank the engine. As is the case with the battery, this assumes that the engine and starting systems are in good condition. Realize that the engine requirements have doubled at 0°F while the battery's ability to supply current has been reduced to 40%. No wonder many vehicles do not start up in the North when the cold winters arrive. The importance of keeping the battery fully charged in the winter is shown by the fact that a half charged battery (SG of 1.190) will have only 21% of its power available at 0°F. Large batteries with cold cranking capacities that are adequate for the weather conditions expected have to be taken into consideration when you sell a customer a replacement battery—the bigger, the better! A larger battery will not have to work as hard cranking the engine in any temperature. Generally, the vehicle manufacturer will sell the vehicle with an adequate size battery. As the vehicle ages, its ability to start, especially in adverse conditions such as winter, might be diminished. With this in mind, it makes good sense to try to sell the customer the largest cold cranking amperage battery that will fit the battery case. Consider the original equipment size as the minimum required, not the ideal.

TABLE 6-3 Temperature and battery efficiency.

Temperature	Efficiency
80°F (26.7°C)	100%
32°F (0°C)	65%
0°F (-17.8°C)	40%

TABLE 6-4 Power needed to start the engine at various temperatures.

Temperature	Power
80° F	100%
32° F	155%
0° F	210%

One last thing to consider goes back to the chemistry of the cell. Remember that the concentration of the acid is weakened during discharge. The sulfur from the sulfuric acid is absorbed into the active materials of the plates. This leaves behind a larger concentration of water. Water freezes at a much higher temperature than fully charged electrolyte. Electrolyte at a SG of 1.265 will not freeze until its temperature is down to about 80° *below zero*! But, the same electrolyte at a SG of 1.100 will be frozen at 15° *above zero*. Under no circumstances should a battery be left discharged in below freezing weather. Once a cell has been frozen, the battery is usually worthless, because the ice has probably broken off the plates. Do not attempt to jump start a frozen battery either, because the possible open circuit condition of the cells can damage transistorized components once the vehicle starts. We will cover how this happens in the charging system chapters.

BATTERY SAFETY

The following are safety rules that must be observed when working around or on batteries.

1. Always remove the negative battery cable before doing any electrical work on the vehicle.
2. Never do anything that would cause a spark or arc near the battery. The hydrogen that the battery might be releasing is explosive.
3. Wear eye protection.
4. Wash your hands after you have handled a battery.
5. Rinse off the vehicle finish if battery acid has been spilled on it.
6. Do no attempt to jump start a frozen battery.

SUMMARY

In this chapter, we have looked at the various types of batteries currently in use in vehicles. The inability of a primary cell to accept a charge was examined and compared to the secondary cell in vehicle storage batteries. Both series and parallel wiring of cells were examined with the resulting voltage and amperage changes noted. The three most common methods of rating batteries: ampere, hour, reserve capacity, and cold cranking, were defined with examples. In addition, a discussion of the use of a hydrometer to determine the state of charge of an individual cell was discussed, or in the case of a sealed battery, the use of a voltmeter. Determining the correct battery for a particular vehicle and the safety rules for working around batteries ended this chapter. The information presented was designed to prepare you for the actual vehicle battery testing chapter that follows.

KEY TERMS

Cell
Primary
Secondary
Rechargeable
Specific gravity
Hydrometer

CHAPTER CHECK-UP

Multiple Choice

1. Technician A says that a primary cell can be recharged with a battery charger. Technician B says that a secondary cell produces a higher voltage than a primary cell. Who is right?
   a. Technician A only
   b. Technician B only
   c. Both Technician A and B
   d. Neither Technician A nor B
2. Six secondary cells are wired in parallel. Technician A says their combined voltage will be 2.1 volts. Technician B says their combined amperage capacity will be six times that of a single cell. Who is right?
   a. Technician A only
   b. Technician B only
   c. Both Technician A and B
   d. Neither Technician A nor B

3. Six secondary cells are wired in series. Technician A says their combined voltage will be 12.6 volts. Technician B says their combined amperage will be six times that of a single cell. Who is right?

   a. Technician A only
   b. Technician B only
   c. Both Technician A and B
   d. Neither Technician A nor B

4. A fully charged lead acid battery has plates made of

   a. lead dioxide and sponge lead.
   b. lead dioxide and lead sulfate.
   c. sponge lead and lead sulfate.
   d. calcium and antimony.

5. Technician A says that a fully charged battery will not freeze as easily as a dead one. Technician B says that the specific gravity of a dead battery is around 1.100. Who is right?

   a. Technician A only
   b. Technician B only
   c. Both Technician A and B
   d. Neither Technician A nor B

6. The process of electrolysis occurs

   a. if the battery is being overcharged.
   b. if the battery is charged after reaching full capacity.
   c. if the charging voltage exceeds 15 volts.
   d. all of the above.

7. A dead cell's positive and negative plates are both

   a. calcium.
   b. sponge lead.
   c. lead sulfate.
   d. lead dioxide.

8. Specific gravity readings should be temperature compensated because

   a. they are a more accurate indicator of the actual state of charge.
   b. electrolyte gets thicker as it gets warmer than 80°.
   c. hydrometers do not work on cold batteries.
   d. none of the above.

9. A 1.250 specific gravity reading is obtained with an electrolyte temperature of 60°. The actual SG is

   a. 1.250.
   b. 1.258.
   c. 1.242.
   d. 1.230.

10. A low-maintenance battery usually has grids made of
    a. calcium.
    b. antimony.
    c. glass.
    d. lead dioxide.

True or False

11. A 50-ampere hour battery is capable of delivering 10 amps for 5 hours before it is dead.
12. Engine size is the usual determiner of battery size.
13. Charging system voltage should not exceed 15 volts for a warm battery (80°).
14. Oxygen-recombination batteries frequently require water.
15. Distilled or demineralized water is the best to add if a cell requires water.
16. An open circuit voltage in the 12.6-volt range indicates an overcharged battery.
17. Battery efficiency is reduced as the temperature gets lower than 80°.
18. A 350 cold cranking amp battery is specified as original equipment. A replacement battery should be 175 CCA as a minimum.
19. A 120-minute reserve capacity battery can deliver 25 amps for 12 hours before it is dead.
20. As the outside temperature goes lower than 80°, the amount of current required to crank the engine goes down also.

CHAPTER 

Battery Testing and Service

OBJECTIVES

Upon completion of this chapter and the "on car" lab activities, you should be able to:

- jump start a vehicle safely.
- maintain a vehicle battery to trade standards.
- replace a battery.
- use a hydrometer to test the specific gravity of battery cells.
- use a load tester to test a battery for capacity.
- use a voltmeter on a sealed battery to determine relative state of charge.
- use a 3-minute test to test a battery.
- determine the correct load for a vehicle battery.

INTRODUCTION

Now that we have spent some time and have a good understanding of how a battery is constructed and how it works, it is time to get down to the important vehicle tests that are commonly done. The ability to be able to predict failure and determine if a battery is the cause of difficulty is the single goal of this chapter. The battery is often blamed for other system problems. It is probably one of the most replaced components in the vehicle, and its cost makes it all the more important that our diagnosis is accurate.

Before we get into testing, we have to discuss and understand the reasons why we test. If the reasons are clear, the testing will be all the clearer. We will also be better equipped to discuss with our customers what we are doing and why we are doing it. Communication with the customer is very important if we are to be successful. The best diagnostician must still have the ability to explain to a customer what he/she is doing. We sometimes have to talk customers into what is best for their vehicles—not necessarily the least expensive, but the best. We take into consideration how the vehicle is used, the weather it will be subjected to, and the age of the vehicle before we make any recommendations. Do not forget to communicate with your customers!

PURPOSES FOR BATTERY TESTING

Why do we test batteries? Simply stated, we usually test for one or more of three reasons. The first is to isolate a problem to one component. The second is for preventive maintenance purposes. The third is to predict failure. Let us look at each one of these reasons briefly.

1. *To isolate a problem to one component.* The age of replacing part after part on the vehicle until it is repaired is over! It is important to realize that we must make every effort to isolate the customer's problem to a single component. If the battery is bad, replace it. If the starter is bad, replace it, and so on. Just do not replace the entire system because the vehicle did not start. Isolate each area until you pinpoint the difficulty. A battery can hide or mask over other areas or components that might need repair. The starting system is a great example of this. A battery that does not have the ability to keep its voltage up during cranking will force the starter into drawing more current. As the technician looks at starting current and sees that it is too high, he/she might be tempted to replace the starter only to find that the new one cranks at the same amperage. The battery in this example has masked over the problem and caused our guess to be wrong. We never want to guess if we can help it. By isolating the battery, we are testing it independent of the vehicle. When we have finished, we are confident that the rest of the vehicle has not given us improper or false information. Never use the vehicle's starting system to test the battery. If you do, you will be one of those technicians who is frequently replacing both the starter and the battery and hoping that the customer complaint is cured.
2. *Preventive maintenance.* By testing and cleaning corrosion off battery cables and posts, we are ensuring that the current will have a low-resistance path and will be delivered where it will do the work. Preventive maintenance also means that we will do everything in our power to ensure that the vehicle will start. Does the battery appear to have enough power to make it through the winter? Is there something we can do to prevent breakdown? If we can prevent costly breakdowns, and quite possibly towing bills, our customers will have greater faith in us and our diagnostic abilities.
3. *Predicting failure.* This one goes hand in hand with preventive maintenance. If by our testing, we can predict that the battery will not be capable of starting the vehicle during the winter months, we are predicting failure. Explaining this to most customers will make them realize that now is the time for replacing the battery, not one morning when they are late for work, it is cold outside, and their vehicle's battery is dead. Certain tests can be used to determine if the battery will soon develop problems. The object of predicting failure and preventive maintenance is to alert the customer before the problem actually occurs. Do not wait for the vehicle to not start before you suggest to the customer that testing of the starting and charging systems is appropriate. Encourage a semi-annual battery starting and charging check-up with some preventive maintenance thrown in for good measure. This will prevent most problems that normally occur.

A REVIEW OF GENERAL SAFETY

Whenever we work around a battery, we must always keep three things in mind: acid, amps, and sparks. Let us be more specific. Batteries are filled with a solution of sulfuric acid. If this solution comes in contact with your eyes, they could be permanently damaged. Skin burns, clothes eaten away, and damaged vehicle finishes are also quite possible. Many bat-

FIGURE 7-1 Safety glasses should be worn while working around batteries.

teries are semi-sealed nowadays and some people believe that this eliminates the acid danger. Nothing could be further from the truth. It is true that gluing the top cell coverings down will help keep the acid in if you tip the battery over. But realistically, tipping the battery over is not a common problem. Always wear safety glasses with side shields as illustrated in Figure 7-1. These will prevent acid from accidentally being splashed into your eyes during battery servicing or filling. Immediately wash battery acid off your skin or the vehicle finish with lots of water. A mild soap will prevent the acid from irritating your skin. Always wash your hands after handling a battery as a small amount of acid might have remained on the battery and on your hands. Hands with a little acid on them will usually not be irritated. Our eyes will not be as lucky, however. Rubbing even a little acid residue into the sensitive eye can cause damage. If you accidentally get acid into your eyes, you must rinse them out and seek medical attention immediately. Many shops are purchasing eyewash fountains such as shown in Figure 7-2. The small expense involved can prevent blindness.

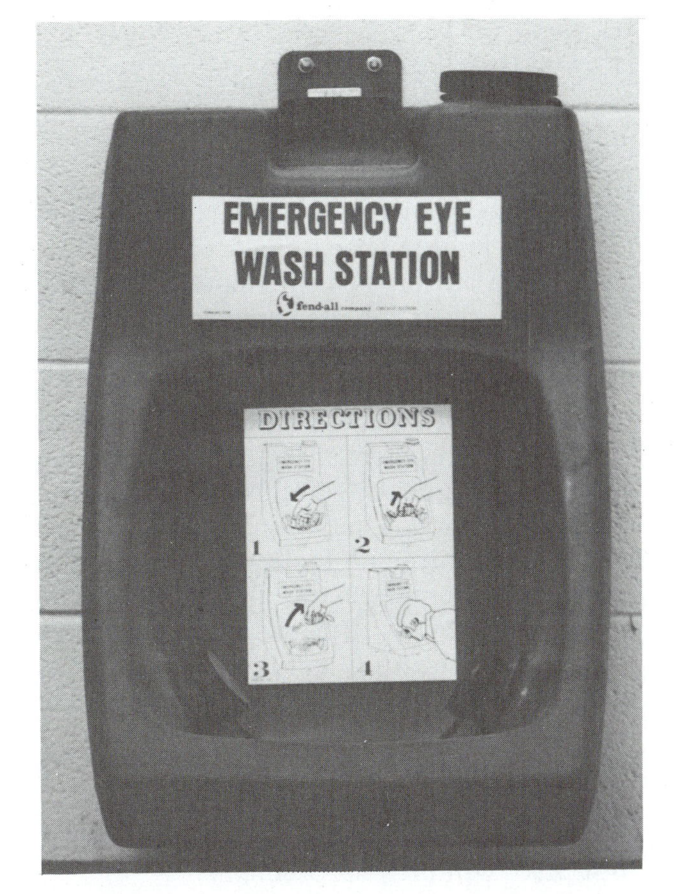

FIGURE 7-2 Eyewash fountain to wash electrolyte from the eyes.

Amps

This battery that we are servicing has tremendous potential to deliver very high amperage. Many batteries are capable of delivering 1,000 amps or more into a dead short. If we put something across the two posts with little or no resistance, like a wrench, sparks fly and damage to sensitive solid-state circuits could result. It is possible to short out a battery by accident and burn out a component that was on during the short. Always disconnect the negative cable (ground) before doing anything major under the hood of the vehicle. This precaution is especially important to observe if you are working on electrical components or items near the battery. This will ensure that damage will not be done if a short occurs. The battery cable is not fused and can cause extensive damage and quite possibly an underhood fire. Remember amperage flowing can give light, magnetic fields, and heat. Under dead short conditions, the amount of heat can be extreme.

Sparks

Do not forget that a charging battery can give off hydrogen and oxygen, which is explosive. Frequently, batteries explode right in the face of the person working on them. Never do anything that might cause a spark or flame near the top of a battery. This includes connecting or disconnecting a battery charger with the charger on. A common mistake is to charge a battery and then immediately attach the vehicle's cables with something on in the car. If, for example, the doors are open or the ignition switch is on, current will flow as soon as the cable is touching. This will cause a small arc that could ignite the hydrogen and oxygen still sitting around the top of the battery. The exploding battery will shower acid and pieces of plastic for quite a distance. The least of your worries is the customer who might have been watching the whole episode and quickly decided that you do not know what you are doing! The worst of your worries could be permanent blindness. Doing the job right in the first place will prevent problems.

CAUTION: Always disconnect the ground cable before doing major work on the vehicle, Figure 7-3.

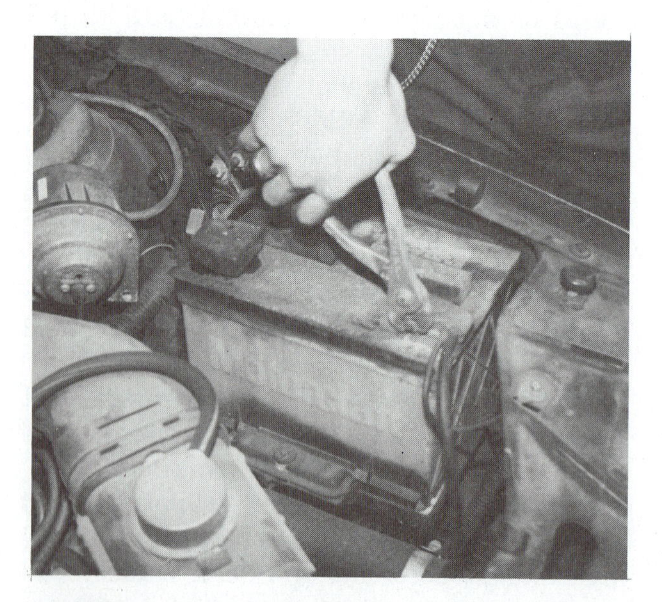

FIGURE 7-3 The battery's ground cable should be disconnected before most electrical work is begun.

Make sure that no sparks or arcs occur near the battery, especially if it is being charged. Wear safety glasses.

HYDROMETERS

In the previous chapter, we discussed a cell's specific gravity and defined it simply as the thickness of the electrolyte. Anything thicker than water has a specific gravity higher than 1.000 and anything thinner than water has a specific gravity lower than 1.000. Remember, we also discussed the fact that we could use the specific gravity to tell us the state of charge of a cell. A reading of 1.275 (usually read twelve seventy-five) indicates that the electrolyte in the tested cell is 0.275 times thicker than water. The thickness comes from the sulfuric acid in the electrolyte. As the battery discharges, the sulfuric acid is chemically absorbed into the plates leaving behind a greater concentration of water. This will lower the specific gravity of the cell. The opposite is also true: As the battery is charged, the sulfuric acid is driven (chemically) out of the plates and into the electrolyte, thickening it and raising its specific gravity. Interpreting the specific gravity of the cells in a battery give us two bits of important information. First, it tells us whether or not the cell is charged. Second, if we compare all six cell readings (12.6-volt battery), we can determine the overall battery condition. Think about the importance of this information. Knowing whether or not the battery is charged is very important to us. Further testing of a discharged battery does not prove anything and could result in our misdiagnosing a problem. Most batteries that are not charged will fail a load test tempting us to install a new battery. Two days later the customer is back with the same problem. If by contrast the battery is charged, we can expect it to pass a load test and will replace it if it does not. We have also received additional information about the charging system. A fully charged battery "usually" indicates a reasonable charging system. Comparison of all of the cells' state of charge or specific gravity gives us the second important bit of information. When the specific gravity of one cell is quite a bit different than the other five cells, it might have a tendency to pull the other cells down to its lower level. Remember, the cells in a battery are electrically connected and, therefore, one cell that is not charged will pull the other cells down to its level as it attempts to balance. Eventually, all cells will be equally discharged. This can occur overnight so the customer will have a dead battery in the morning. Battery manufacturers like to have less than a 0.025 difference from one cell to the others. The greater the difference, the greater the likelihood of self-discharge. By the time a battery reaches a difference of 0.075, it will most likely self-discharge during a short enough period of time to cause the customer some inconvenience. Use common sense when faced with different specific gravity readings. The greater the difference, the more you should encourage your customer to invest in a new battery. Keep in mind though that not all batteries will begin to self-discharge at a greater than 0.025-point spread. Some batteries with large point spreads last quite a long time and, yet some with very little difference fail the very next day! Explain the options to your customers. If they are hesitant, ask them to come back in a couple of weeks so you can retest the battery. If the specific gravity point spread is getting greater, you will not have as much difficulty convincing them to purchase a new battery.

Self-discharge of a battery overnight can often be traced back to a battery with a large point spread. Putting an ammeter on the battery and seeing no current draw with a vehicle that has a dead battery each morning usually indicates a self-discharge condition from cells whose specific gravity point spread is too great.

0.025 or less	Normal
0.025–0.050	Some batteries will discharge
0.050–0.075	Many batteries will discharge
0.075-0.100	Most batteries will discharge
0.100+	All batteries will discharge

Notice the relative and somewhat vague terms: Some, many, most. You will never be able to predict exactly which batteries will self-discharge. Look for the point spread and inform your customer. As the point spread gets greater so do the odds that battery failure is around the corner. Encourage your customer to purchase the battery before total failure takes place.

Look at Figure 7-4. It is a picture of a common type of hydrometer. Notice the float inside the barrel. The liquid electrolyte will float it up and allow us to read the liquid level

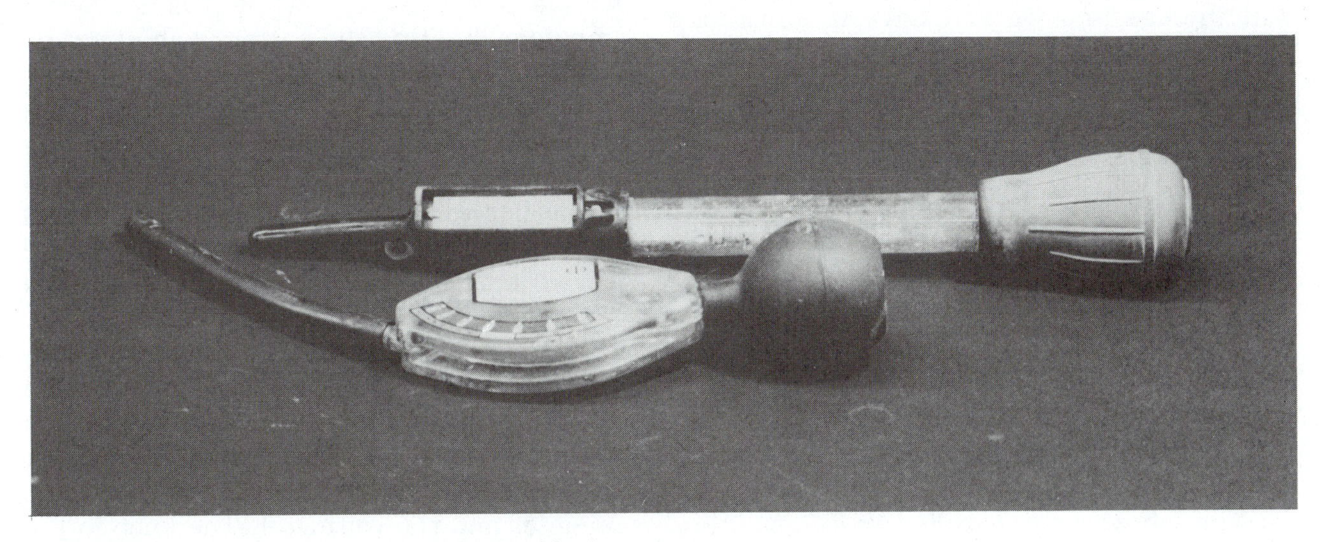

FIGURE 7-4 A temperature-compensated hydrometer.

TABLE 7-1 State of charge versus specific gravity.

State of Charge	Specific Gravity
Fully Charged	1.265
75% Charged	1.225
50% Charged	1.190
25% Charged	1.155
Discharged	1.120

against the scale printed on it. It sounds complicated but really is not. If you look at the liquid line on the float and it reads 1.250, what does this mean? It basically means that this cell is 75% charged or another way of saying it is one-quarter discharged. Table 7-1 will give you the percent charged from the specific gravity. Do not forget the reading 1.250 is usually referred to as twelve fifty. Recording the specific gravity of each cell will give us the information we want and allow us to make an intelligent recommendation to customers about their vehicles.

Most of the time when we test batteries with a hydrometer, we are especially interested in whether the battery is charged or discharged. We must realize that the hydrometer will not give us accurate information if the temperature of the electrolyte is either warmer or colder than 80°F. For this reason, most hydrometers are temperature compensated. As the electrolyte gets colder, it will thicken and we will have to subtract some points to get an accurate reading. For every 10°F, the specific gravity will change by 0.004 of 4 points. Refer to Table 7-1. What will the actual specific gravity of the cell be if you get a 1.220 reading at 50°. Hopefully, you said 1.208 because 50 degrees calls for a compensation of –12.

$$1.220 - 0.012 = 1.208$$

Simple, isn't it? It is also important because we are basing further testing and diagnosis on actual specific gravity readings. They should be temperature compensated.

The lab worksheet in Figure 7-5, should give you all the information you need to effectively analyze a battery with a hydrometer. Testing batteries with hydrometers is an invaluable source of information, assuming the top caps are not glued down. We will cover testing the batteries that we cannot use a hydrometer on later in this chapter.

LOAD TESTING

Now that you realize how and why we use a hydrometer, we can move on to the most important battery tester: the load tester. This tester has been around for quite a long time and its use is extremely important. Remember, battery size is directly related to the number of cubic inches of displacement. The ability to crank the engine is the most important consideration. The battery needs to be able to supply sufficient amperage to crank the engine fast enough to get it started. Cranking speed is important on many electronic fuel-injected vehicles. Without sufficient speed, the computer will not turn on the fuel metering system and the engine will not start. High-cranking speed also helps to prevent flooding of the engine. We will be measuring cranking speed in Chapter 8, "Starting Systems". Let us go back to the battery. A good supply of current with the voltage remaining high (typically 9.6 volts) is necessary. So this battery we are using must be able to keep its voltage above 9.6 volts while it is delivering amperage. Testing the battery's ability to do this is called **load testing**.

Do not confuse load testing with cranking the engine over. Load testing involves pulling a very specific amperage out of the battery. It is another example of independent testing rather than relying on the vehicle starting system to do the test. Look at Figure 7-6. It is a picture of a very common style of load tester. Actually, it is referred to as a VAT, or Volts-Amps Tester. With this device, we will be able to test the battery in addition to the starting system and eventually the charging system. It has a large carbon pile included with the ammeter and the voltmeter. The knob in the center of the tester will allow us to vary the

Specifications

Battery Size ______________ Engine Cubic Inches ______________

Hydrometer Readings

Cell Number	1	2	3	4	5	6	
Negative Post							Positive Post

Results (Check one.)

_____ All cells are within 25 points and are charged (continued testing).
_____ All cells are equal and are discharged (charge battery).
_____ Cell readings vary more than 25 points (skip load test and advise customer to replace battery.)

OR

Open Circuit Voltage Test

Load battery to 20 amps for about 1 minute before testing OCV.

______________ volts—open circuit (no load)

Refer to the chart and convert either the hydrometer reading or the OCV over to state of charge.

Charge	Specific Gravity	OCV (12.6)	OCV (6.3)
100%	1.265	12.6	6.3
75%	1.225	12.4	6.2
50%	1.190	12.2	6.1
25%	1.155	12.0	6.0
Discharged	1.100	11.9	5.9

Battery Load Test (for batteries at least 75% charged)

Set tester to the following:
1. Tester to STARTING
2. Variable LOAD off
3. Voltage—INTERNAL—higher than battery voltage

Attach the BST as follows:
1. Place large cables of tester clamp over the battery clamps, observing that polarity of the battery and tester are the same.
2. Attach inductive pick-up clamp around either of the BST cables *not* battery cables.

Determine the correct load:

Battery Amp/Hour × 3 = Load or
Cold Cranking Amps ÷ 2 = Load

NOTE: Compare the specific battery to the one actually installed. Load to the LARGER of the two.

Apply the calculated load for 15 seconds, observe voltmeter, and turn load OFF.

_____ GOOD—battery voltage remained above 9.6 volts during load test
_____ BAD—battery voltage dropped below 9.6 volts during load test (perform 3-minute charge test to check for sulfation)

3-Minute Charge Test (only for batteries that fail the load test or are less than 75% charged)

1. Disconnect the ground battery cable.
2. Connect a voltmeter across the battery terminals (observe polarity).
3. Connect battery charger (observe polarity).
4. Turn the charger on to a setting around 40 amps.
5. Maintain 40 amp rate for 3 minutes while observing voltmeter.

_____ GOOD—voltmeter reads less than 15.5 (battery is not sulfated).
_____ BAD—voltmeter reads more than 15.5 (battery is sulfated and should be replaced).

FIGURE 7-5 Lab sheet for battery testing.

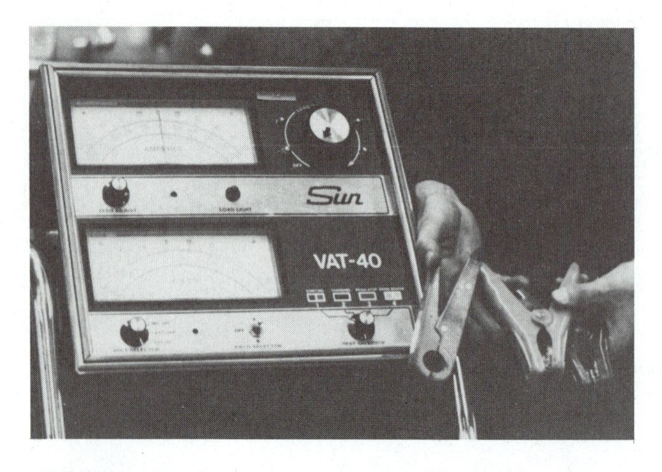

FIGURE 7-6 Volt amp tester with inductive pick-up.

pressure on the carbon pile. The more pressure we apply, the less resistance the carbon will offer to the flow of current. When we attach it to the battery as illustrated in Figure 7-7, the carbon will draw current out of the battery, producing heat. The ammeter will read the amount of current flowing. This then gives us an adjustable load for battery testing. This device is indispensable because it allows us to test each battery to its own level. The smaller the battery, the lower the load. The larger the battery, the more the load. Applying the load to the battery for 15 seconds and watching that the battery voltage remains above 9.6 volts is the key here.

FIGURE 7-7 Volt amp tester connected to a battery for a load test.

Load testers are normally connected across the battery to be tested, observing that the polarity of the meter is the same as the polarity of the VAT. In addition, the voltmeter must be on a range that is higher than the battery voltage. Our meter will have to be set on the 18-volt range. Once connected, the voltmeter should read battery voltage before we apply the load. The ammeter should read 0 (zero), indicating that no current is flowing through the load.

I. Procedure for Determining Battery Load

Now comes the actual loading of the battery. To determine the load, we first determine the size battery the manufacturer installed in the vehicle. This number will be our reference and the minimum battery that should be in the vehicle. At no time should you assume that the battery currently installed is the correct one: Compare and make sure. The actual load that we will apply will be determined by looking at the size battery installed and the manufacturer's original equipment. We load to the higher or larger battery. For example, let us assume that the manufacturer had originally installed a 50 A/H battery in the vehicle. The owner has since installed a 65 A/H battery. We would load it to the 65 level, not the 50 level. If on the other hand, the owner has installed a 35 A/H battery, we would load down to the 50 A/H level. Always load down to the manufacturers battery level as the minimum. Typically the manufacturer installs the minimum required size battery. We try to convince consumers, however, to install larger batteries when the original equipment fails. Load down to the larger of the two choices.

The actual load applied will be determined by the size of the battery as follows. The two most common methods of rating batteries are used in the following examples.

If the battery is rated in ampere hour, A/H, to determine the load to be applied, multiply the A/H by 3. In other words, a 50 A/H battery will be loaded down to 150 amps for 15 seconds with the battery voltage remaining above 9.6 volts during the load.

$$50 \times 3 = 150$$

If the battery is rated in cold cranking amps, divide the rating by 2 to get the load. A 300 cold cranking battery will be loaded down to 150 amps for 15 seconds with the battery voltage remaining above 9.6 volts during the load.

$$300 \div 2 = 150$$

In the previous examples, the batteries are obviously the same size. We would do the same procedure, however, with any size battery: 3 times the A/H or 1/2 the cold cranking amps.

The worksheet reproduced in Figure 7-5 shows that we first take our hydrometer readings. If they show that the battery is reasonably charged (75% or more), we load test. By looking at the results of the hydrometer and the load test, we have a real good idea of the condition of the battery. With the results, we can make an intelligent recommendation to the customer.

If the hydrometer readings are impossible to obtain because the battery does not have removable caps, perform an open circuit voltage test. An open circuit voltage test is read with no load on the battery, and then converted over to a state of charge. If the battery has been recently charged, the **surface charge** on the plates must be drained off. This is done by placinga load of about 20 amps on the battery for about 1 minute. Without draining off the surface charge, the battery will appear to be at a higher charge than it actually is. The conversion from OCV to state of charge is reproduced in Table 7-2. It is also on the worksheet in Figure 7-5.

TABLE 7-2 Conversion from OCV to state of charge.

Charge	Specific Gravity	12 V Bat	6 V Bat
100%	1.265	12.6	6.3
75%	1.225	12.4	6.2
50%	1.190	12.2	6.1
25%	1.155	12.0	6.0
0%	1.120	11.9	5.9

3-MINUTE CHARGE TEST

The load test we have just discussed is the industry standard. If the battery can pass the load test, the odds are it will be capable of cranking the engine under most conditions. This, of course, assumes that the battery is adequately sized for both the engine and the outside temperature. As a battery ages, a deterioration called **sulfation** might occur. Sulfation is a chemical action within the battery that interferes with the ability of the cells to deliver current and accept a charge. It is frequently found in batteries that have been sitting unused for extended periods of time.

The 3-minute charge test is a reasonably accurate method for diagnosing a sulfated battery, Figure 7-8. Usually, it is done if the battery is suspected of being sulfated or has failed the load test. When a battery fails the load test, it is not necessarily the battery's fault. It is possible that the battery has not been receiving an adequate charge from the vehicle's charging system over an extended period of time leading to sulfation. The 3-minute charge test determines the battery's ability to accept a charge.

Battery Cold Cranking Amperage (CCA) Testing

In recent years, some of the automotive equipment manufacturers have developed methods of testing the available capacity of a battery through the use of a computer. The majority of the testing systems apply a calculated load, such as 170 amps, to the battery for a specific

FIGURE 7-8 Battery charger and volt amp tester connected for a 3- minute charge test.

period of time while the voltage is graphically plotted. The resulting graph of the dropping voltage is compared to those in the computer's memory. The graphs in memory are the actual voltage changes that specific CCA batteries will experience. In this way the actual voltage of the tested battery compared to the stored voltage graphs allows the computer to accurately test the available CCA. This test is an example of computer technology being applied today to help the technician. Being able to tell the customer how many CCAs are available allows the technician to help guide the repair and predict failure. Typically, the CCAs of most batteries will drop as the battery ages, even though a standard load test will be passed. Now the technician can have additional information available to help predict failure rather than waiting until the failure has already taken place.

II. Procedure for a Battery That Has Failed Load Test

Connect the shop battery charger to the vehicle's battery. Remove the ground cable from the battery or at least make sure the computer, if the vehicle has one, has been disconnected. Most manufacturers have a quick disconnect or a convenient fuse for disconnecting the computer. On-board computers for fuel and ignition control are very sensitive to higher than normal voltages. Your battery charger might raise the system voltage above the level that the computer likes. Your customers will not be happy if your battery testing burns out their computers. Put a voltmeter across the terminals. Do not forget the meter is polarity sensitive. Turn the charger to a setting around 40 amps. Maintain this rate of charge for 3 minutes. Immediately after the 3 minutes, observe the voltmeter. If the voltmeter reads less than 15.5 volts, the battery is not sulfated and should be slowly charged before the next load test. If the voltmeter reads over 15.5 volts (7.75 volts for a 6-volt battery), the internal resistance of the battery is too high due to sulfation or poor internal connections to accept a full charge. The battery should be replaced. No further testing is necessary. The reason for doing the 3-minute charge test should be reviewed again. This test will pick out a sulfated battery that might be the result of an undercharge condition on the vehicle. A complete charging system test should be performed on the vehicle. The new battery might develop the same sulfation if the cause of the undercharge condition is not eliminated.

Now is a good time to make some comments regarding testing a sealed battery. Trying to figure out if a battery is fully charged is easy with a hydrometer or an open circuit voltage test. If you can get hydrometer readings, by all means do so. Indirectly, you will be getting information on the condition of the charging system and the general condition of the electrical system. A fully charged battery should be capable of passing a load test. The combination of the hydrometer or OCV and the load test is more than enough information on which to base a decision about the battery's future. Remember a load test on a discharged battery proves nothing! Something as simple as a loose fan belt can cause the charging system to never fully charge the battery. This battery will probably fail the load test. Replacing it will no doubt cure the no-start that brought the vehicle into your shop but will cause the customer to be displeased with your diagnosis when his/her *new* battery runs down and causes another no-start. Using the 3-minute test is not a substitute for charging a weak battery before load testing it for a second time. Also keep in mind that a hydrometer can be useful in predicting a future self-discharging battery. A large point spread is most always associated with a battery that runs itself down. Low open circuit voltage readings after the battery has been sitting a couple of hours are also indicative of a weak cell. The weak cell has a tendency to pull the rest of the cells down to its level. In some cases, you might be forced into charging a battery, disconnecting it, and letting it sit overnight. If in the morning the battery is dead, replace it. It is self-discharging. The procedure might not sound too convenient, but it certainly is conclusive.

BATTERY CARE

Now that we have seen how to test and evaluate batteries, let us spend some time discussing and doing the routine maintenance that is necessary even on a good system.

Cleaning the battery is probably one of the simplest and yet most important things that a technician can do to increase the life of a customer's battery. We should clean not only the cables but the top also. Notice the voltmeter reading in Figure 7-9. It shows that there is a difference in pressure across the top of the battery. Current must be very slowly draining across the top of the battery. Years ago, the tops of most batteries were made of a soft tar-like substance. As the battery charged and discharged and the gases were vented from the cells, the tar absorbed chemicals and dirt that

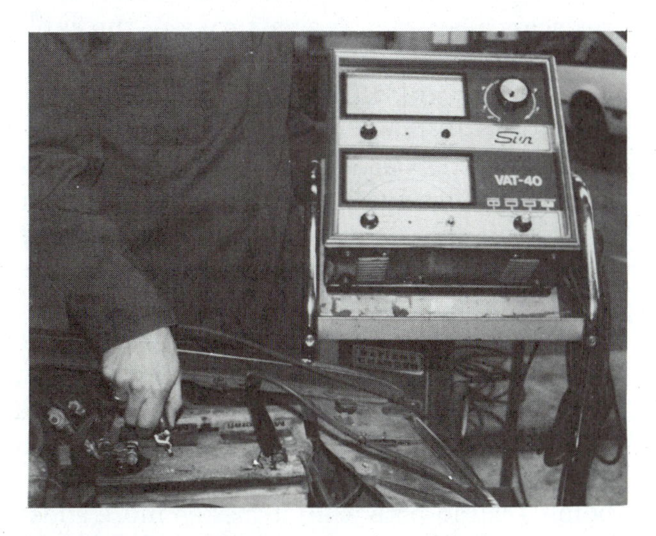

FIGURE 7-9 Voltmeter showing a discharge across a dirty battery top.

were conductive. This allowed small amounts of current to flow across the top of the battery and gradually discharge the cells. All batteries currently being produced have hard plastic tops covering the cells. This design greatly reduces the current drain across a dirty top, but it does not prevent it completely. For this reason, we should clean and neutralize any acid on the top of the battery. Make a paste of baking soda and water, and brush it on the top of the battery. After it bubbles and fizzes for awhile, rinse it off with large amounts of water. This procedure will ensure that current will not gradually drain off across the top. Our goal is to keep the battery fully charged once the vehicle is turned off, especially if the vehicle will not be run for awhile. Winter storage of a vehicle is a good example of the necessity of a clean top. If during a month or more of storage, a small drain occurs across the top of the battery, it is likely that the battery will sulfate and quite possibly be ruined by spring.

To check for a drain, place one voltmeter lead on the post (either one) and the other lead on the case top. A zero voltmeter reading is the goal here. Any other reading is not acceptable. A semi-annual cleaning will usually be sufficient.

Cable ends need more frequent care. Their connection is extremely important. You will see, in Chapter 9 "Vehicle Starting System Testing and Service," that any corrosion or resistance between the battery and the cable end will increase the amperage that will be needed to crank the engine. This increased amperage is harder on the entire system, including the battery. For this reason, we should remove the cable ends from the posts and clean the electrical contact surfaces as often as is necessary to ensure that the connection is tight and corrosion free.

We usually remove the negative cable first. The reason for this is quite simple. When we use a wrench on battery bolts, the wrench will be at the post's electrical potential. With the whole area around the battery usually being ground potential, it makes more sense to remove the ground cable first. If we accidentally come in contact with the frame or body with our wrench, no current will flow. If, however, we are removing the positive first and touch the surrounding metal, sparks fly! Neither the battery nor your expensive wrench will like having a couple of hundred amps flowing. Once the negative cable has been disconnected, the battery is now isolated from the vehicle and if the wrench comes in contact with the ground, current cannot flow. Always remove the negative cable first and you will save yourself from exploding the battery, embarrassing yourself in front of a customer, and no doubt, spend less money replacing tools that you have welded!

REPAIRING CABLE ENDS

In most cases when we find a badly corroded cable end, it can be cleaned and reinstalled with a new battery bolt and be good as new. However, we sometimes find that through repeated installations and removings, the lead in the cable end cracks or stretches to the point where it can no longer be tightened around the post. At this point, you have two options. One—replace the entire cable, or two—replace the cable end. Cables do not last forever and if a voltage drop test indicates excessive resistance inside the cable, do not hesitate to replace it. Starting system voltage drop testing is outlined in Chapter 9 "Vehicle Starting System Testing and Service." If the voltage drop across the cable is acceptable and there is sufficient length to the cable, replacing the end is a good solution to the problem, as long as it is done correctly.

First, cut back the insulation until no corrosion is present. You will have to cut off all the corroded section of the cable. Clean, corrosion-free copper is necessary if the repair is to last. Spread the cable strands out to inspect them, making sure that no corrosion exists. Corrosion that remains will spread to the rest of the cable end in a very short time, necessitating another repair. Once you are assured that the strands are clean, twist them together as in Figure 7-10. Solder the end together to make a really good connection, Figure 7-11.

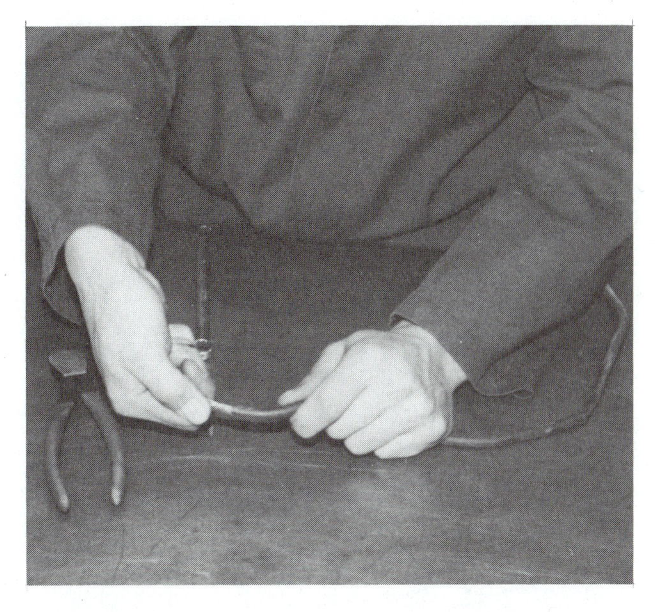

FIGURE 7-10 The cable strands should be twisted together.

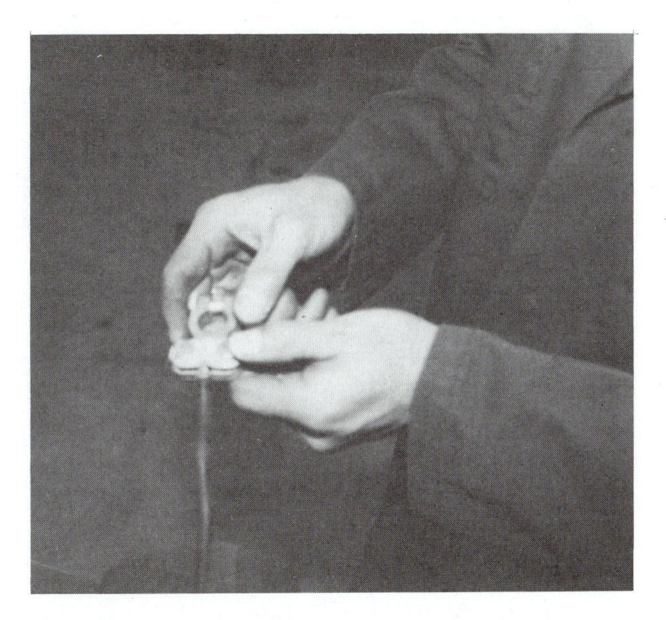

FIGURE 7-12 Install a new cable end.

A propane torch can be used to get the end hot enough to melt the solder. Once it is hot, allow the solder to flow up the cable about an inch. This soldering of the end is not always done out in the field. However, if done, the new cable end will be as good as the original equipment. Place a new cable end in place; do not forget any small wires that were originally part of the end, and tighten the two small bolts down until the cable end is tightly held in place on the cable, and the job is done correctly! (Figure 7-12.)

If the cable end does not have to be replaced, it should be removed, cleaned, and

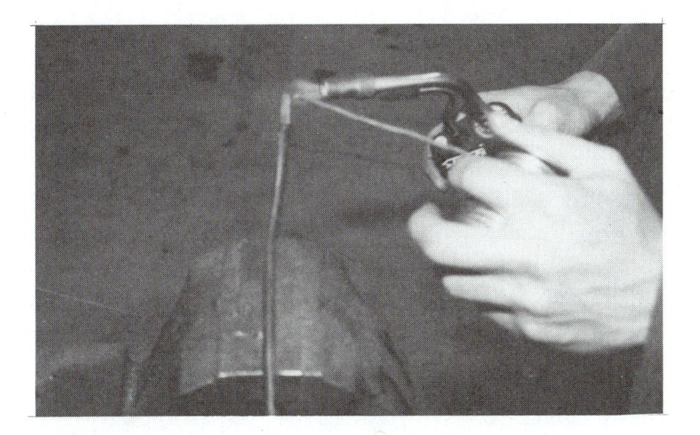

FIGURE 7-11 Heat and solder the twisted cable strands together.

reinstalled. The use of a battery cable remover is recommended, because it will push straight down on the post rather than applying pressure to the side. Side pressure from a screwdriver or wrench prying under the cable end might break off the post and ruin an otherwise good battery. Once the cable ends have been removed, clean both the post and inside cable clamp until they are shiny and bright. Figure 7-13 shows how this is done. Once the battery top, the post, and the cable clamp are clean, reinstall and tighten down the clamp bolt. If you have trouble fitting the cable end over the post, spread the clamp slightly, Figure 7-14. Just spread it enough to allow it to slip easily over the post. It is also a good trade practice to have extra battery clamp bolts on hand, because you will find, frequently, that they cannot be reused because of corrosion. Replacing them as necessary ensures a tight connection. All the current that the starter will draw must come through these connections. They *must* be tight and corrosion free!

Side terminal batteries are a little less likely to develop corrosion, but still must be cleaned every so often. They require a special cleaning brush shown in Figure 7-15. Follow the same procedure previously outlined with one exception: Always use a torque wrench on

FIGURE 7-13 Clean the cable and post with the special brush.

FIGURE 7-14 Spread the cable clamp slightly.

FIGURE 7-15 Side terminal cables require a special cleaning brush.

the cable end bolt. Over tightening will strip out the threads in the battery, ruining a good battery. The bolt threads into a lead strap that is easily stripped out. Be careful! A torque of 60 to 90 in. lb. is recommended. Notice—that is inch-pounds, *not* foot-pounds. If a torque wrench is not available, hand tighten the bolt and then tighten about one quarter of a turn more. With clean threads on the bolt, the one quarter turn past hand tight will get you pretty close to the recommended torque.

Filling batteries with distilled water is a simple task that should not be overlooked. Remember, a by-product of charging most batteries is hydrogen and oxygen. Water, or H_2O, replaces the liquid lost through electrolysis. Usually, we fill cells to the "split ring." This means that water is added until it just touches the bottom of the fill hole. Overfilling will cause the electrolyte to bubble out during charging.

Whenever we have to add water to a modern battery, we should be asking ourselves why? Why does the battery need water? Has it been in service for a very long period of time? Has it been overcharged by either a battery charger or the vehicle's charging system? We should realize that under most normal conditions, batteries of the calcium-lead variety require very little water during their lifetimes. Usually the water placed in the battery during manufacture will be sufficient for years, unless an overcharge condition exists. The old lead-antimony batteries by comparison, used large quantities of water. Adding water to a battery cannot be accomplished with a sealed battery. If the top is not sealed and the cells are a bit low on water, by all means refill with distilled or demineralized water. Then go look for the cause of this battery being low on water. Remember, overcharging is the most common cause.

Some batteries are shipped from the factory dry. These batteries must be filled with electrolyte before they are placed in service. The electrolyte is packed in plastic bags with a rubber pour spout. Fill each cell to the split ring, charge, and place in service. Notice the setup in Figure 7-16. It shows the correct method for filling a dry-charged battery. Remember, you are working with acid. Wear your safety glasses and be careful! A dry-charged battery is filled with electrolyte *not* water! Once in service, the cells are topped up with water *not* electrolyte.

Batteries are heavy and awkward to carry. Do not attempt to carry one if you have a bad back. Also do not carry one against your body as in Figure 7-17. This might leave some acid on your clothes, ruining them. Use a battery carrier or battery strap as in Figure 7-18.

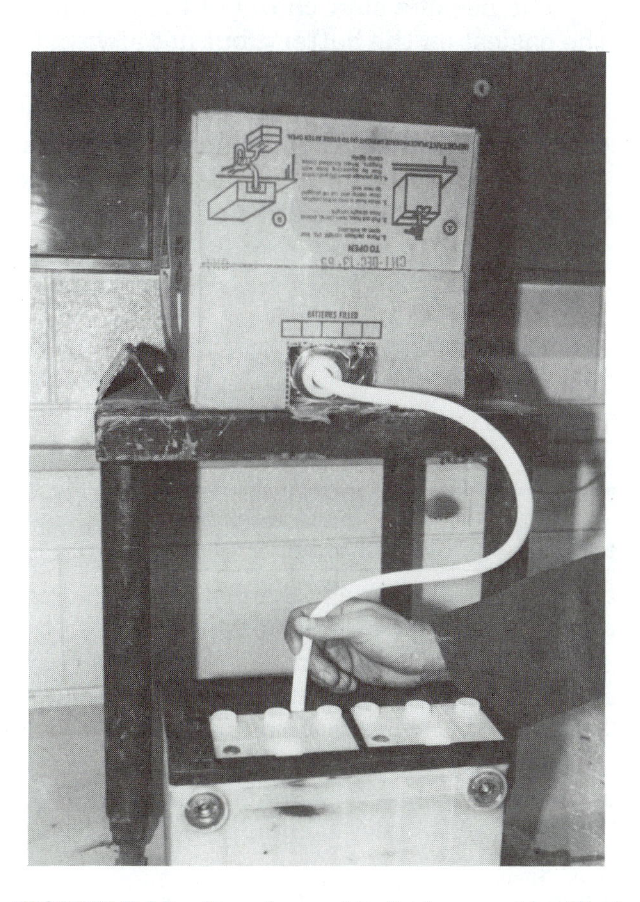

FIGURE 7-16 Dry-charged batteries must be filled with electrolyte before they are put into service.

FIGURE 7-17 Do not carry a battery against your clothes.

FIGURE 7-18 A battery strap is used to carry a battery safely.

Storing batteries that are not in use is important to the life of the individual cells. If your shop has a stock of batteries, make sure they are rotated with the first one stored being placed in service first. Use a hydrometer or measure the open circuit voltage on stored batteries to determine when they need to be charged. Slow charge (around 4 to 6 amps) as necessary. The worst thing for a battery is to be left in a discharged state extended periods of time. This accelerates the sulfation process and destroys the ability of the cells to deliver current. During storage try to avoid placing the batteries where they will be exposed to different temperatures because this accelerates the discharge rate. Store them on shelves, or on 2 × 4s. The colder, the better! A cold battery will retain more of its charge than a warm one over extended periods of time. The military frequently stores their batteries in the deep freeze! Even temperature, cold storage is considered to be the best alternative to actual vehicle use. Remember that this cold storage is for a fully charged battery. A dead battery will freeze if it is stored in too cold a temperature.

CHARGING

If an open circuit voltage test or a hydrometer indicates that a battery needs to be recharged, connect the charger positive to positive and negative to negative. It is always a good idea to remove the negative battery cable from the post before charging. This will isolate the vehicle from the charger and prevent the charger from damaging any microprocessor or electronic components. Battery chargers should be checked frequently to make sure they are not producing too high a voltage while they charge the battery. Many battery manufacturers recommend no higher than 15 volts be applied during charging of a warm battery. If the charging voltage is around 15 volts and the vehicle has been disconnected, you need not worry about damaging either the battery or the vehicle systems during charging.

The use of a slow charger (4 to 6 amps) is the easiest on the battery, but not always the most practical solution for the customer who needs his/her vehicle *now*! Under these conditions follow the recommendation of the Battery Council International, Table 7-3. Notice

TABLE 7-3 Battery Council International Charging Table. (Courtesy of Battery Council International)

Battery Capacity (Reserve Minutes)	Slow Charge	Fast Charge
80 minutes or less	10 hrs. @ 5 amps 5 hrs. @ 10 amps	2.5 hrs. @ 20 amps 1.5 hrs. @ 30 amps
Above 80 to 125 minutes	15 hrs. @ 5 amps 7.5 hrs. @ 10 amps	3.75 hrs. @ 20 amps 1.5 hrs. @ 50 amps
Above 125 to 170 minutes	20 hrs. @ 5 amps 10 hrs. @ 10 amps	5 hrs. @ 20 amps 2 hrs. @ 50 amps
Above 170 to 250 minutes	30 hrs. @ 5 amps 15 hrs. @ 10 amps	7.5 hrs. @ 20 amps 3 hrs. @ 50 amps

that the table is based on knowing the reserve capacity of the battery. Another method is to attach a voltmeter across the battery while it is being charged. The charging rate is low enough if the voltage does not exceed 15 volts (assuming the battery is at room temperature). If the voltage exceeds 15, simply reduce the charging rate until the meter is just below 15 volts. Keeping the voltage around 15 during charging will ensure the quickest charge for the customer and also be the safest rate for the battery. Many shop chargers have timers for fast charging that automatically reduce the charging rate to a slow charge once the time period (usually 1 hour) has passed. Do not overcharge a battery just because the customer is in a hurry! If you smell rotten eggs around a charging battery, it is being charged too fast. Reduce the charging rate to eliminate the gassing.

CAUTION: Hydrogen and oxygen are explosive. Do not disconnect the charging cables with the charger on or you might blow the battery up, and always wear safety glasses. You only have one pair of eyes.

JUMP STARTING

Jumping a dead battery from a live, charged battery is frequently done. Many times customers leave their lights on or crank their vehicles until the batteries are discharged to the point where recharges or jump starts are necessary.

III. Procedure for Jump Starting a Battery

Make sure the two vehicles are not touching. Connect the positive jumper cable to the charged battery's positive post, then to the discharged battery's positive post. Connect the negative jumper cable to the charged battery's negative post. Connect the other end to the engine block, as far away from the battery as possible. Make the final connection on the discharged battery's side of the circuit. This final connection will produce a spark. This spark should not occur anywhere near the charged battery as it might explode. Once the vehicle is started, disconnect the negative cable from the block immediately. Figure 7-19 shows this procedure. Under no circumstances should you make your final connection or your first disconnection at the charging battery.

FIGURE 7-19 Negative jumper cable being connected to the engine block.

Jumping batteries to get a vehicle started should not be attempted if the battery has been discharged overnight in sub zero temperatures. A discharged battery freezes at 19° above zero. Trying to jump start a frozen battery will sometimes result in the battery case exploding. The hydrogen and oxygen cannot get through the ice at the top of the cells and the resulting pressure can blow the side out of the case. Usually freezing the battery has ruined it anyway, and so it makes little sense to try to jump start it.

In this age of microprocessors and electronic gadgetry on the vehicle, you must be very aware of system polarity and system voltage. Reversing the polarity or using a starting unit whose voltage is too high could do thousands of dollars of damage to the vehicle systems. Under no circumstances should you jump start a computerized vehicle off anything except another vehicle. If you decide to charge a battery off another vehicle, disconnect the negative cable to isolate the vehicle systems from the charging battery. Many manufacturers void the vehicle warranty if it was jumped off anything other than another vehicle. We must be constantly aware of polarity and voltage when we work around modern vehicles.

SUMMARY

In this chapter, we have taken a look at the most common types of vehicle battery testing and service. Testing the battery involved either a hydrometer or an open circuit voltage test to determine the state of charge, followed by a load test, if the battery was at least 75% charged. The 3-minute charge test, as an indicator of a sulfated battery, was outlined. In addition, procedures for cleaning the top and side terminal cable ends was shown. Replacing worn or badly corroded cable ends was discussed along with jump starting a vehicle. The correct procedure for charging a battery by using a voltmeter to monitor the charge was shown. Emphasis on safety during all battery testing and service was discussed.

KEY TERMS

Load testing Surface charge Sulfation

CHAPTER CHECK-UP

Multiple Choice

1. Technician A says a 12.6-volt OCV reading indicates that a battery needs recharging. Technician B says a 1.265 hydrometer reading indicates that a battery needs recharging. Who is right?

    a. Technician A only
    b. Technician B only
    c. Both Technician A and B
    d. Neither Technician A nor B

2. A 450-CCA battery is being load tested. The correct load will be

    a. 1350 amps.
    b. 450 amps.
    c. 225 amps.
    d. 990 amps.

3. Technician A says that during a load test the battery voltage must not fall below 9.6 volts. Technician B says the load applied should be 3 times the A/H rating. Who is correct?

    a. Technician A only
    b. Technician B only
    c. Both Technician A and B
    d. Neither Technician A nor B

4. Hydrometer readings are between 1.260 and 1.275. Technician A says load test to determine the rest of the battery information. Technician B says a load test is not necessary. Who is correct?

    a. Technician A only
    b. Technician B only
    c. Both Technician A and B
    d. Neither Technician A nor B

5. Hydrometer readings of 1.200 to 1.220 are taken. Technician A says load test, then do a 3-minute charge test. Technician B says recharge, then load test. Who is correct?
   a. Technician A only
   b. Technician B only
   c. Both Technician A and B
   d. Neither Technician A nor B
6. Temperature compensation involves a ___ change in readings for every 10°.
   a. 0.004
   b. 0.040
   c. 0.400
   d. 4.000
7. Technician A says that the 3-minute charge test can help identify a sulfated battery. Technician B says a sulfated battery will usually fail a load test. Who is correct?
   a. Technician A only
   b. Technician B only
   c. Both Technician A and B
   d. Neither Technician A nor B
8. A 3-minute charge test is being done on a battery that has failed a load test. Technician A says set the charger to 40 amps. Technician B says the voltage should stay below 15.5. Who is correct?
   a. Technician A only
   b. Technician B only
   c. Both Technician A and B
   d. Neither Technician A nor B
9. Battery tops are cleaned periodically to
   a. look nice.
   b. prevent sulfation.
   c. prevent slow discharge across the top.
   d. eliminate excessive resistance that will increase starting amperage draw.
10. Negative battery cables are removed first
    a. to help prevent open circuits.
    b. only on side terminal batteries.
    c. to help prevent short circuits to ground.
    d. all of the above.

True or False

11. Jumper cables are connected to the vehicle's dead battery first when jump starting.
12. A load of 325 amps for a 650 CCA battery is the correct test load.
13. 10.6 volts during load testing indicates the need for a new battery.
14. A 3-minute charge test can be used instead of a hydrometer to test a cell's state of charge.

15. Having to add water to a low-maintenance battery can be an indication of a higher than specified charging voltage.
16. Acid is added to a battery that has a low electrolyte level.
17. A dry-charged battery is filled with water to activate it.
18. Fast charging a battery (higher than 15 volts) will increase its shelf life.
19. Sulfation is increased if a battery is left in a discharged state for extended periods of time.
20. Charging batteries can give off hydrogen and oxygen, which is explosive.

CHAPTER 8

Starting Systems

OBJECTIVES

At the completion of this chapter you should be able to:

- identify the magnetic principles necessary to understand starting motors.
- describe the main features of the permanent magnet starting motor.
- describe how a Bendix drive functions.
- describe how an overrunning clutch drive functions.
- describe the operation of a solenoid.
- describe the basic differences in a solenoid shift and a positive engagement style starter.
- describe how and why current draw is reduced as a starter armature spins faster.
- describe the operation of a neutral safety switch.
- describe the operation of a starting relay.
- trace the three different styles of starting systems in use today on a wiring diagram.

INTRODUCTION

Up to this point, we have spent considerable time studying electricity without looking at any specific electrical system. It is now time to put many of the principles discussed into practice as we analyze the starting system. Obviously, we would not get very far without the starting system. Many technicians think of it as the "first" system, because it is the system that must function correctly before any other system is called upon to do its job. Correctly functioning ignition, charging, and fuel systems would not be of much value if the starting system does not do its job first.

We will discuss some motor principles involving electromagnets and then discuss how we put magnetism to work in cranking the engine. We will end with a discussion of the most common American and foreign systems. The lab section will discuss and allow you the opportunity to test starting systems on the vehicle. Diagnosis and repair of current starting systems is not difficult, but it is very important. Most consumers take the system for granted because they are used to the instant response they get when the key switch is turned to the start position. High reliability is achieved with routine maintenance and diagnosis done at the professional technician's level. Keep these principles in mind and the rest is easy!

A TYPICAL STARTING CIRCUIT

Let us look at a typical cranking or starting circuit briefly before we zero in on the particulars. Refer to Figure 8-1. When the consumer turns the key switch, a small amount of current flows from the battery through a neutral safety switch or a clutch safety switch over to the starter solenoid. This current pulls a gear into mesh with the engine's flywheel and closes a large switch between the battery and the starting motor. Current now flows through the motor and cranks the engine. Releasing the key switch stops the flow of current and a large spring pulls the gear out of mesh thus allowing the engine to spin or run under its own power. This sequence is repeated each time the key switch is turned. We will look at each component of this system in turn. As we do this, keep in mind the underlying design principle. Starting systems are designed to deliver great amounts of horsepower for short periods of time. They have the ability to handle very large amounts of current. They are

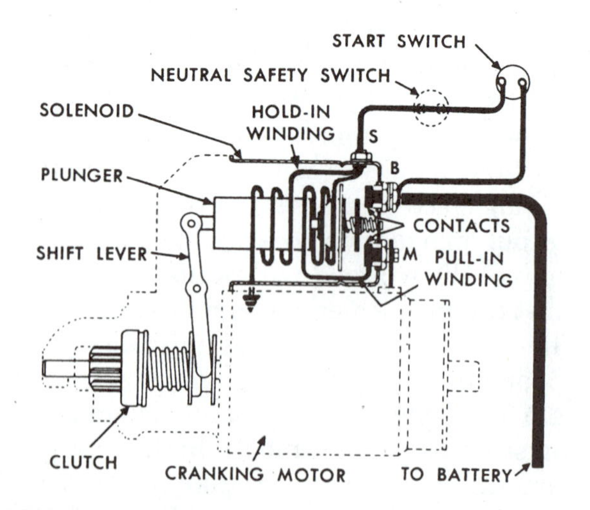

FIGURE 8-1 Cranking circuit. (Courtesy of Delco Remy Division)

sometimes referred to as "educated short circuits." Their resistance is so low that they are, for all practical purposes, a dead short to ground. Connections, cables, switches, solenoids, and batteries must be able to handle as many as a thousand amps under adverse conditions. How this works starts with magnetism.

MAGNETISM

One of the most familiar and yet misunderstood principles is magnetism. A magnet is a material that will attract iron and steel, plus a few additional materials. It will not attract plastic, aluminum, wood, and paper, among others. The magnetic attraction is strongest at the two ends, or poles. A magnet that is allowed to spin freely will align itself north and south. The end facing north is called the north pole, and the end facing south is called the south pole. A hiker's compass points to the north because the needle is actually a small magnet. As a child, you perhaps played with two magnets and discovered that like poles (two north poles or two south poles) would repel or push each other away and unlike poles (one north and one south) would attract each other. These principles are best illustrated in Figure 8-2, which shows the magnetic fields that surround a magnet. You can easily duplicate these fields by placing a piece of paper over a magnet and sprinkling iron filings over the paper. The filings will align themselves with the force fields of the magnet. Putting unlike poles together allows their fields to join, while putting like poles together forces the fields to split as in the illustration. This magnetic field attraction and repulsion is what will turn our engines over.

ELECTROMAGNETISM

The strength of a magnetic field can be greatly increased through the use of electricity. There is always a slight magnetic field around a conductor that is carrying current. The **inductive pickup** on our battery load tester

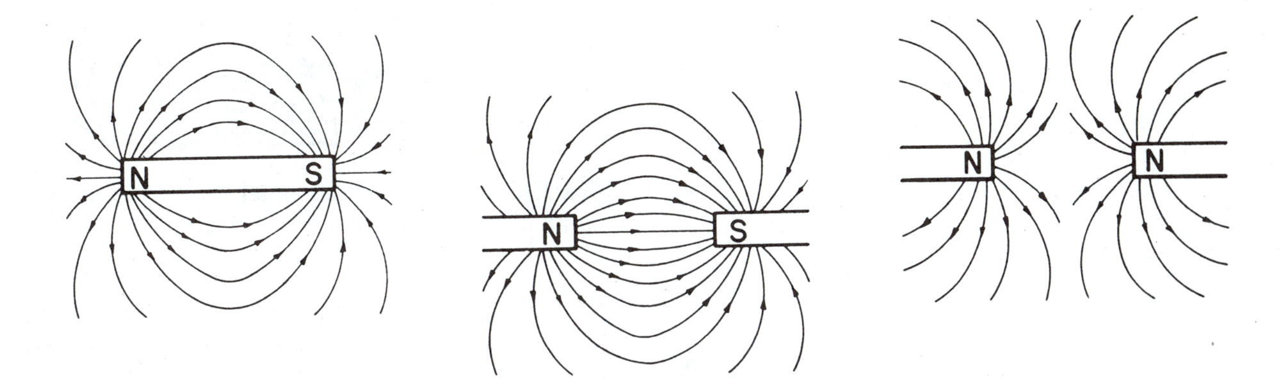

FIGURE 8-2 Magnetic fields around a permanent magnet.

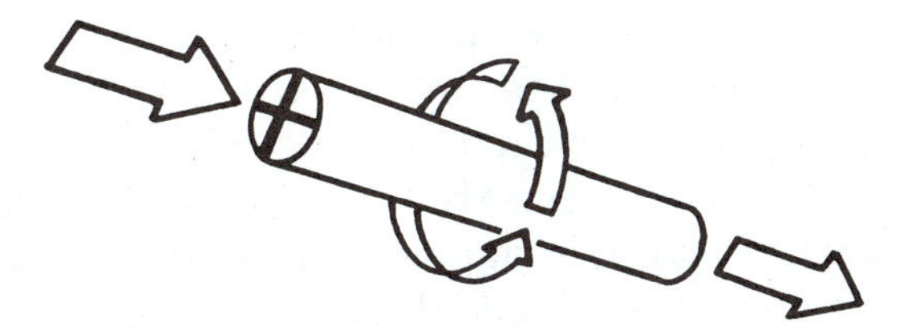

FIGURE 8-3 A weak field surrounds a conductor that has current flowing through it.

was able to convert this magnetic field over into an ammeter reading. Anytime current is flowing through a conductor, a magnetic field will surround it. With this principle in mind, we can make a simple, but powerful electromagnet. Look at Figure 8-3. It shows the weak magnetic field surrounding a single conductor. In other words, this conductor has a low flux density. **Flux density** is the concentration of the lines of force. If we place two conductors side by side, as in Figure 8-4, we will double the flux density and double the magnetic field. In a similar manner, increasing the current flow through the conductors will also increase the strength of the field. As we progress through the rest of this text, keep in mind two magnetic field principles:

1. The strength of a magnetic field is related to the current flowing through the conductor. Increasing the current increases the field strength.

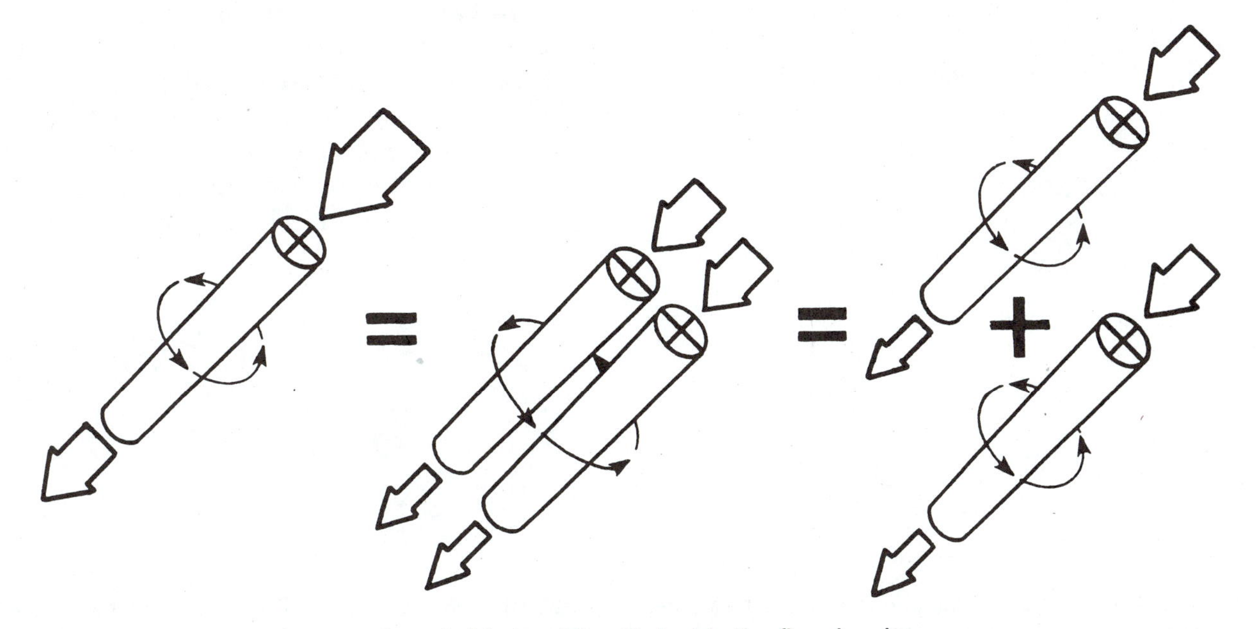

FIGURE 8-4 Two conductors placed side by side will double the flux density.

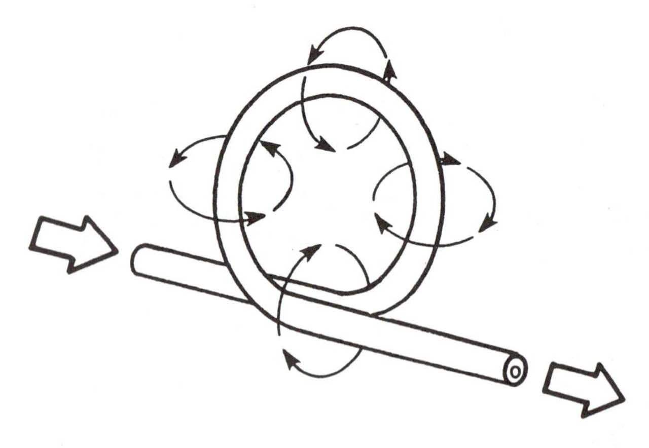

FIGURE 8-5 Looping a single conductor increases the flux density.

2. Adding more current-carrying conductors will also increase the strength of the field.

Let us look at principle 2 in detail. If we wish to increase the strength of a magnetic field, we can form a single wire into a loop as in Figure 8-5. Notice that looping the wire as shown will double the field (flux density) where the wire is running parallel to itself. The direction of the current flow is the same: from left to right in both sections of the wire. Magnetic lines of force cannot cross one another. The iron filings proved this, so the magnetic lines of force at the bottom of the loop will double in density. As we add more loops, the fields from each loop will join together and increase the flux density. Figure 8-6 shows how these lines of

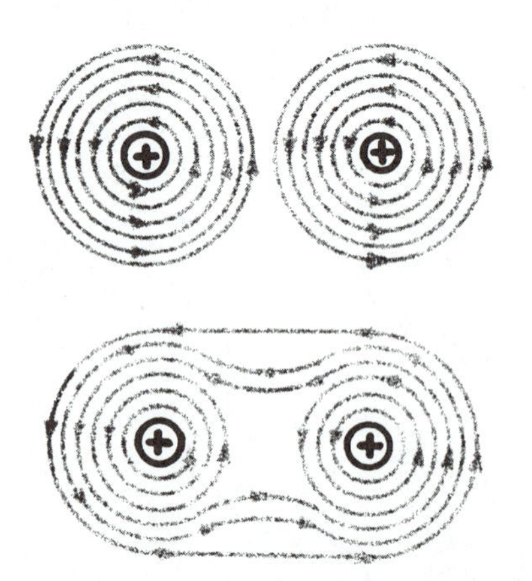

FIGURE 8-6 Lines of force join together if two current-carrying conductors are close together.

FIGURE 8-7 Insulators must be placed between conductors to prevent them from shorting out.

force will join and actually attract each other. This attraction is so strong that insulators must be placed between the conductors to prevent them from shorting out, Figure 8-7. The strength and thickness of the insulation will depend on the amount of current flowing. The more conductors with current flowing in the same direction, the greater the flux density and the stronger the magnetic field.

Let us examine this same principle with current flowing in opposite directions. Figure 8-8 shows this. Notice that the magnetic lines of force have a tendency to move apart because they have their force fields repelling each other.

Remember, like fields will repel one another. Current flowing in the same direction sets up fields that can join, while current flowing in opposite directions sets up fields that will repel each other. These principles are fundamental to your understanding of electromagnetism in the vehicle. By winding the current-carrying conductors into a coil, we greatly increase the flux density, Figure 8-9. One additional magnetic principle can be applied to increase the strength of the field. An iron core can be placed in the center of the coil. Iron will conduct or concentrate the lines of force easier than air. The iron frame used in

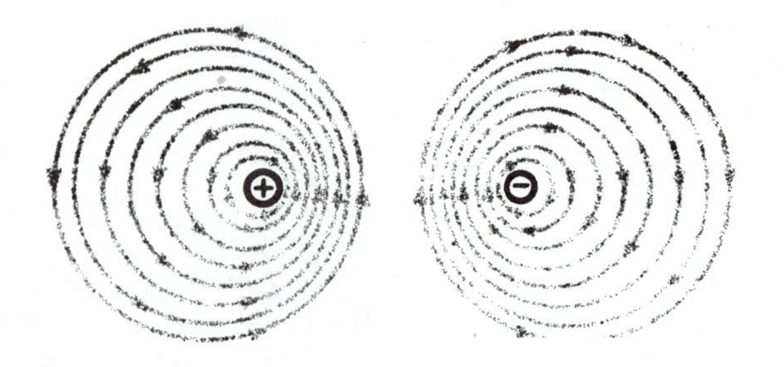

FIGURE 8-8 Opposite flux densities repel each other.

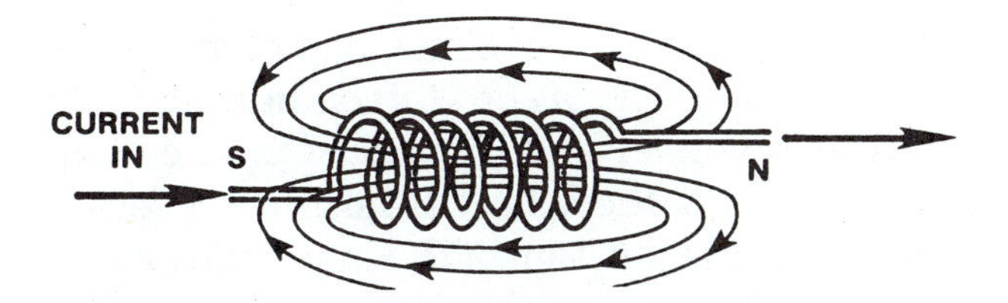

FIGURE 8-9 Winding a current-carrying conductor into a cell increases the flux density.

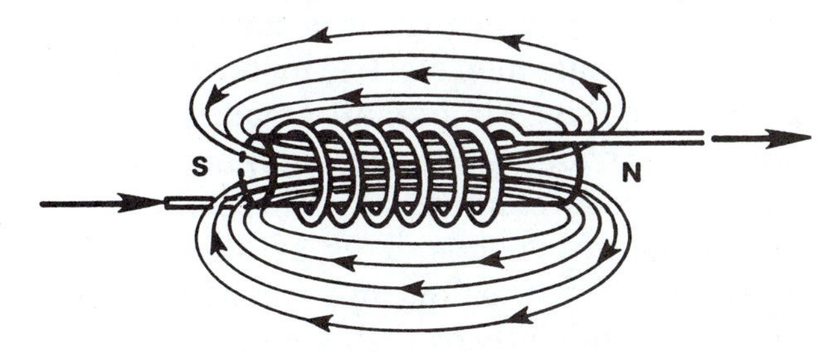

FIGURE 8-10 An iron core concentrates the flux density.

electromagnets allows both a place to assemble the windings, and greatly increases the strength of the field, Figure 8-10.

MOTOR PRINCIPLES

Examine Figure 8-11. It shows a horseshoe magnet with lines of force from the N pole to the S pole. This is its field. If a current-carrying conductor is placed within this field, we will have two force fields. On the left side of the conductor, the lines of force are in the same direction and will therefore concentrate the flux density. On the right side of the conductor, the lines of force are opposite those of the horseshoe magnet and will therefore result in a weaker field. With a strong, heavily concentrated field on one side and a weak field on the other, the conductor will tend to move from the strong field toward the weak field, from left to right as in Figure 8-12. The amount of force against the conductor is dependent on the strength of the magnet and on the amount of current flowing through the conductor. This movement can be summarized by the statement that a current-carrying conductor will tend to move if placed in a magnetic field. Refer to Figure 8-13. It shows a simple electromagnet style of motor. Trace the current as it flows from the battery around the two iron pole pieces. The direction of the windings will place the left pole at a south polarity and right side at a north polarity. Current continues through a half circle called a **commutator**, a loop of wire, and then out the other commutator. Current will now set up a

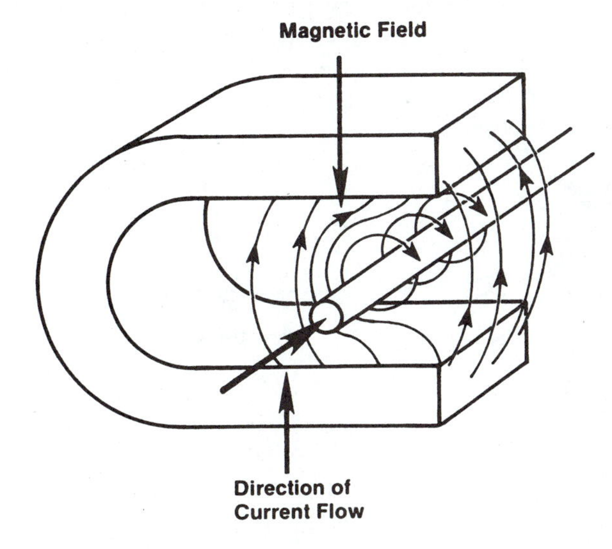

FIGURE 8-11 Two fields interact.

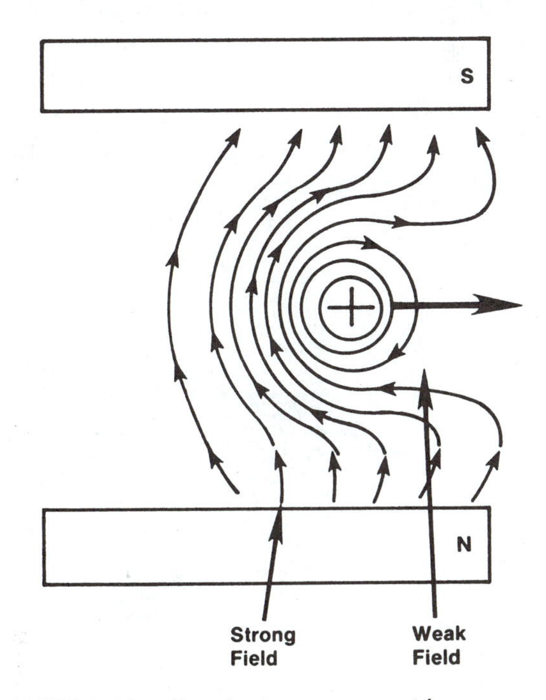

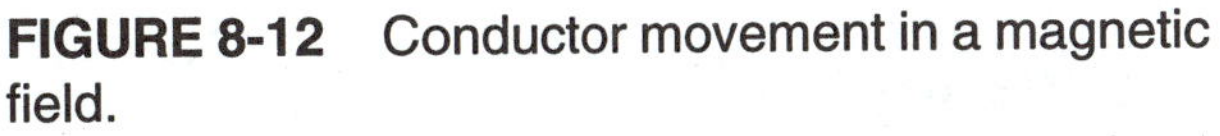

FIGURE 8-12 Conductor movement in a magnetic field.

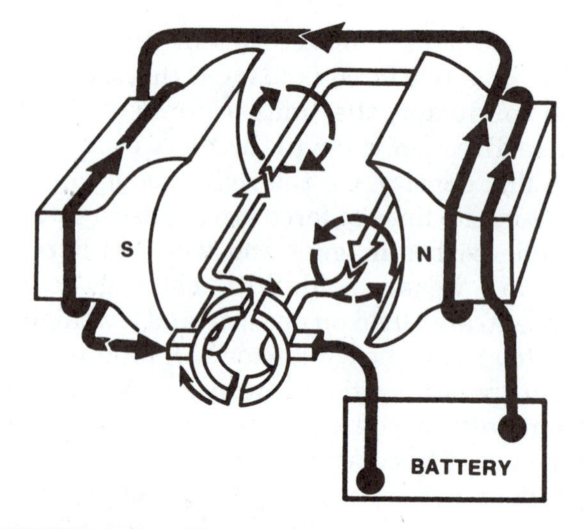

FIGURE 8-13 Electromagnet motor.

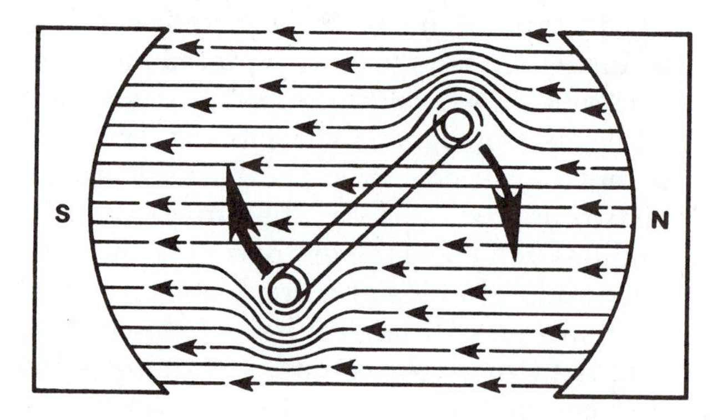

FIGURE 8-14 Fields force the rotation of the conductors.

magnetic field around the loop of wire that will interact with the north and south fields and put a turning or rotational force on the loop as in Figure 8-14. This force will cause the loop to turn. When the loop turns half of a turn, the position of the commutator bars will be reversed and the direction of current flow through the loop will also be reversed. Note that the position of the magnetic fields is the same, but the direction of current flow through the loop of wire has reversed. This will put the same clockwise rotational force on the loop as it had previously. This action will continue until the current is turned off. This two pole, single loop electric motor has the capability to develop a small amount of power and spin. Bringing the current into the loop of wire is accomplished with the commutator and a set of brushes. This arrangement allows the current to flow into something that is moving (the loop) through something that is stationary (the brushes). Each end of the loop will have a commutator bar soldered onto it.

Our single loop motor would probably not develop sufficient horsepower or torque (twisting effort) to actually turn an engine. The power it can develop can be increased simply by adding additional loops and/or pole pieces. Most cranking or starting motors use four pole pieces and many loops of wire in one assembly called an **armature**. Figure 8-15 shows a common armature. Figure 8-16 shows a common four-pole case and the interaction of the magnetic fields. There are many different methods of wiring the four poles and the armature together. The difference is dependent on engine speed, torque required, and battery size, among others. Technicians need not con-

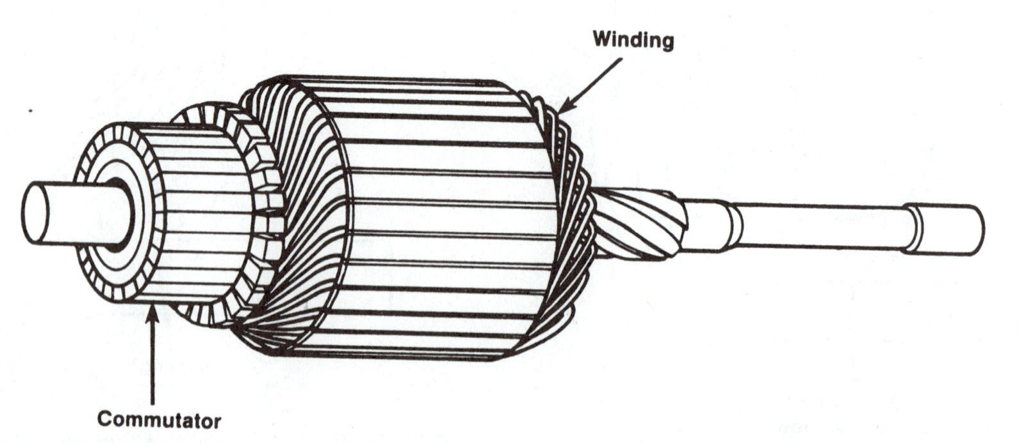

FIGURE 8-15 Armature.

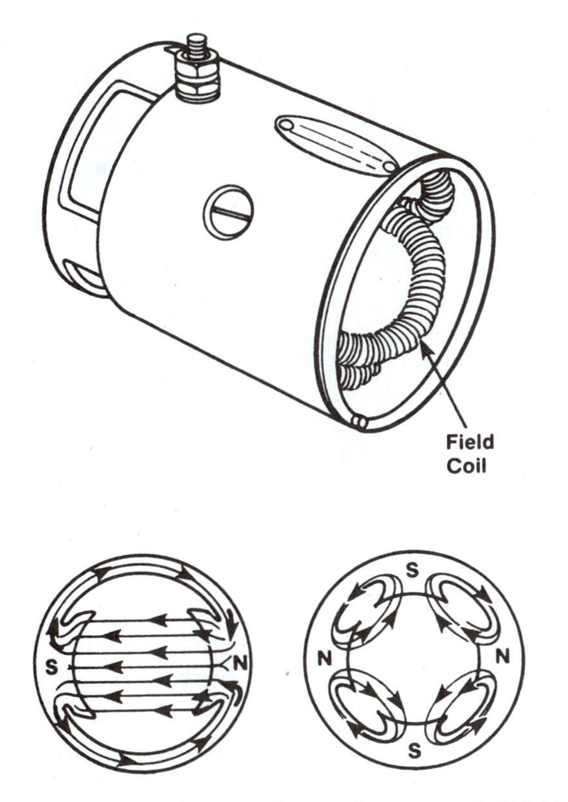

FIGURE 8-16 Two-and four-pole magnetic field interaction.

cern themselves with the design of the motor because options will probably not be available at the time of repair. The various worldwide manufacturers determine the motor design necessary for their application. Replacement or rebuilding in the field are the only options available to most technicians.

We have analyzed the current path through the typical motor and have seen that the current will develop magnetic fields that will cause the armature to turn. It is time to discuss the relationship between speed, torque, and current draw. This is a good time to discuss something called **counterelectromotive force**, or **CEMF**. We have all the conditions necessary for the generation of a voltage: a wire loop spinning in a magnetic field. In the future, we will be discussing charging systems and you will see that a conductor, a magnetic field, and motion between the two is all that is necessary for the generation of electricity. This is called **induced voltage** and will be used to charge our vehicle battery with the charging system. However, we will also have an induced voltage by the action of the armature spinning in the magnetic field of the pole pieces. This induced voltage will be opposite to the battery voltage that is pushing the current through the starter motor. The net result is less current flow through the starter as it spins faster. The faster the armature spins, the more induced voltage is generated. Remember this voltage is opposite, or counter, to the battery voltage and will actually reduce the motor's current draw. Figure 8-17 shows the relationship between speed and CEMF. Notice that CEMF is at 0 (zero) if the motor armature speed is at 0 (zero). This will allow the maximum current flow through the motor if the armature is stopped. Torque or twisting effort will also be highest with the motor armature stopped. As the armature spins faster and faster, the current and torque will be reduced proportionately. Keep this in mind as you begin your diagnosis of starting difficulties. Slow the armature down and the current draw will automatically increase. Speed the armature up and the current draw will be reduced. A practical example of this occurs

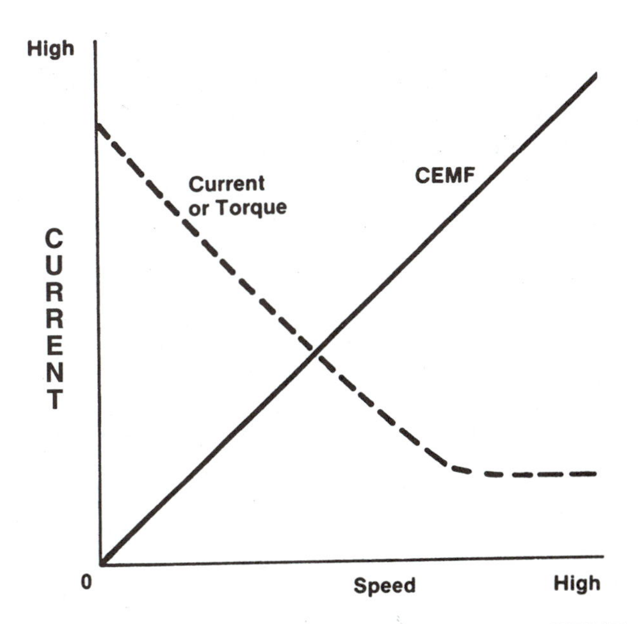

FIGURE 8-17 Armature speed increases CEMF and decreases current draw.

when a vehicle has oil that is too thick for the outside temperature. The additional drag that the lubricating oil puts on the engine results in increased cranking amperage.

Having a motor develop the horsepower we need and controlling it are two different problems. The control of the starting system can be divided into two different sections: electrical and mechanical. We will discuss the mechanical control first.

The action of the armature must always be from the motor into the engine. The engine must not be allowed to spin the armature. If the armature is spun by the engine, it could fly apart or "throw" solder. This means that the speed of the armature has exceeded its design limits and that the wires came off the commutator, thus, open-circuiting itself. It is the object of either the **Bendix drive** or the more common overrunning clutch to allow only this one-way action; the engine cannot be allowed to spin the armature. A somewhat mechanical diode or one-way check is necessary.

Bendix Drive

Although this style of drive is not as common as it used to be, there are still some applications on the road, and you should be familiar with the operation. It is frequently referred to as an inertia drive, because it depends on the inertia of the armature to move a small gear called a **drive pinion** in to mesh with the engine flywheel gear called the **ring gear**. Figure 8-18 shows the pinion mounted on the end of the armature in mesh with the engine flywheel ring gear. Refer to Figure 8-19 for a cutaway drawing of a Bendix drive. Notice the screwshaft threads (like a bolt). These will be a part of the armature and will spin at armature speed. On the end of the pinion and barrel is the gear that will mesh with the flywheel. It has internal threads (like a nut). When the armature is stopped, the barrel will be threaded onto the screwshaft and the pinion will not be in line or meshed with the ring gear. When current flows through the starter the armature will begin to spin. This spinning action will force the barrel to "unthread" itself much like taking a nut off a bolt. Once the barrel is unthreaded, the pinion will be in mesh and the flywheel will be turning by the armature. When the engine starts, the flywheel's increased speed will thread the barrel back along the screwshaft and disengage the pinion from the ring gear. In this manner, if the armature is trying to spin the flywheel, the drive pinion will be in mesh, but if the flywheel is trying to spin the armature, the drive pinion will be "unthreaded" from the ring gear. This design was in use for many years. It has, however, been replaced in most automotive applications by the overrunning clutch.

FIGURE 8-18 Starter drive will turn the engine's flywheel.

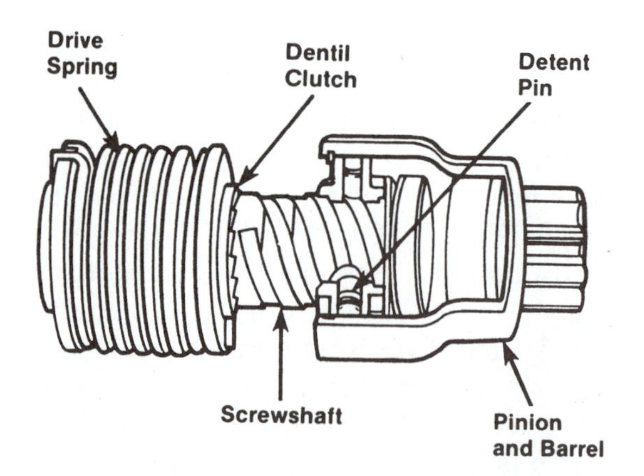

FIGURE 8-19 Bendix drive. (Courtesy of Delco Remy Division)

Overrunning Clutch

The **overrunning clutch** is a roller type clutch that transmits the torque of the armature in one direction only. In this manner, it functions like the Bendix style in preventing the engine from spinning the starter armature. Refer to Figure 8-20. Notice the spring-loaded rollers between the drive pinion and the roller retainer. The torque of the armature will be transferred to the drive pinion through these rollers. The clutch housing is connected to the armature and will spin at armature speed. This action will drive the spring-loaded rollers into the small ends of the tapered slots where they will wedge against the pinion.

The rollers will lock the pinion onto the armature and allow the motor to turn the flywheel. When the engine starts, the flywheel ring gear begins to drive the pinion faster than the armature. This spinning action releases the rollers and allows the pinion to spin faster than the armature. The name says it all; the unit allows the pinion to "overrun" the armature.

The overrunning clutch will require the addition of an electrical method of engaging the drive pinion that was not necessary with the Bendix drive. It automatically engages the pinion to the flywheel when the starter is energized. The overrunning clutch will require a shifting fork to move it in to mesh. Generally, this shifting fork is moved by an electrical solenoid or an open field coil.

Solenoids

In the section on the overrunning clutch, we mentioned that an additional item would be necessary to shift the pinion in to mesh with the flywheel. This item is the solenoid. It will have the job of pushing the drive pinion in to

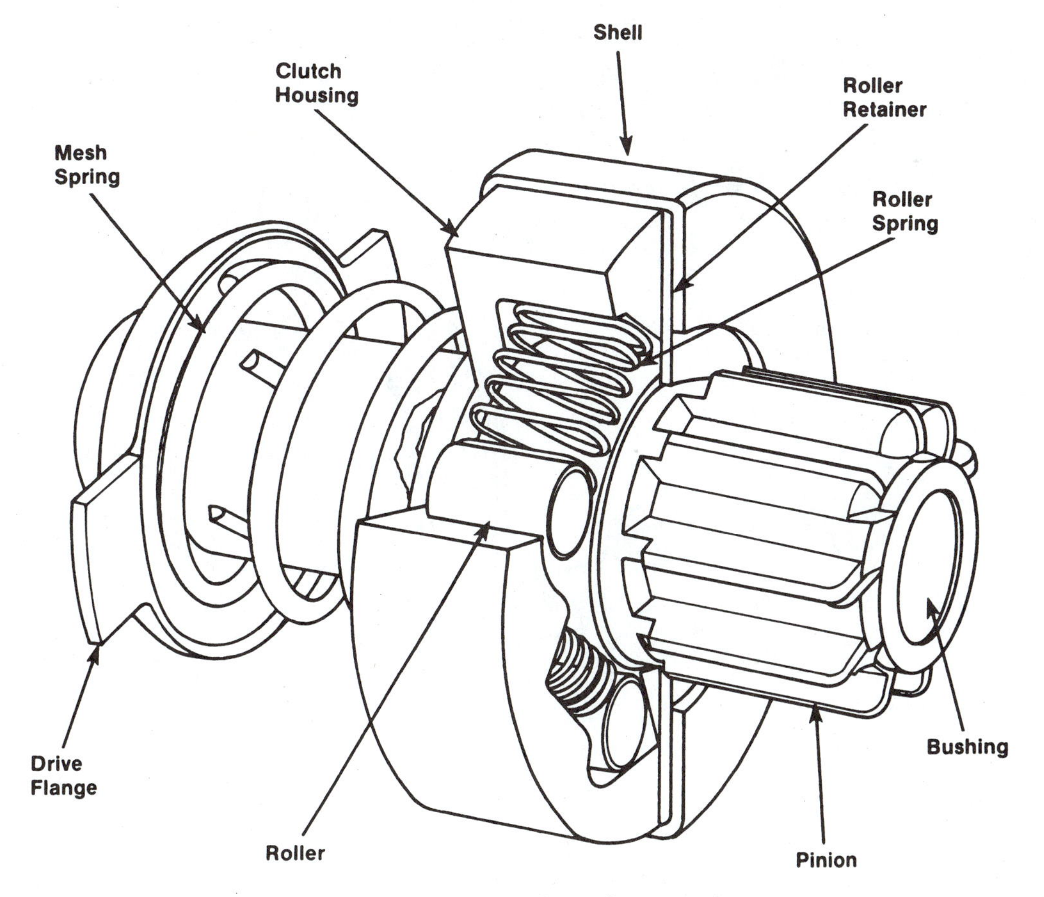

FIGURE 8-20 Overrunning clutch. (Courtesy of Ford Motor Company)

mesh when the key switch is in the start position. Refer to Figure 8-21 that shows a solenoid mounted on top of a starter motor. Notice the shift fork? It will pivot on the pivot pin and push the starter drive mechanism (overrunning clutch) in to mesh when the plunger is drawn into the solenoid windings. Current from the key switch will energize the solenoid windings and make it a very strong electromagnet. This magnetic field will attract or pull the plunger into the hollow windings. Electricity running through the windings will keep the overrunning clutch in mesh with the ring gear teeth. The first job or function of the solenoid is to engage the starter drive with the engine.

The second function is to electrically connect the battery to the starter motor. This is its relay function. The solenoid will act as a switch to energize the motor once the drive pinion is engaged. This sequencing of activity is extremely important. If the motor is energized before the pinion is engaged, damage to both the flywheel and the pinion results. The plunger contact disc located at the back of the solenoid will be pushed against two contacts by the solenoid plunger. One of the contacts is B+ and the other is the motor feed (M) or

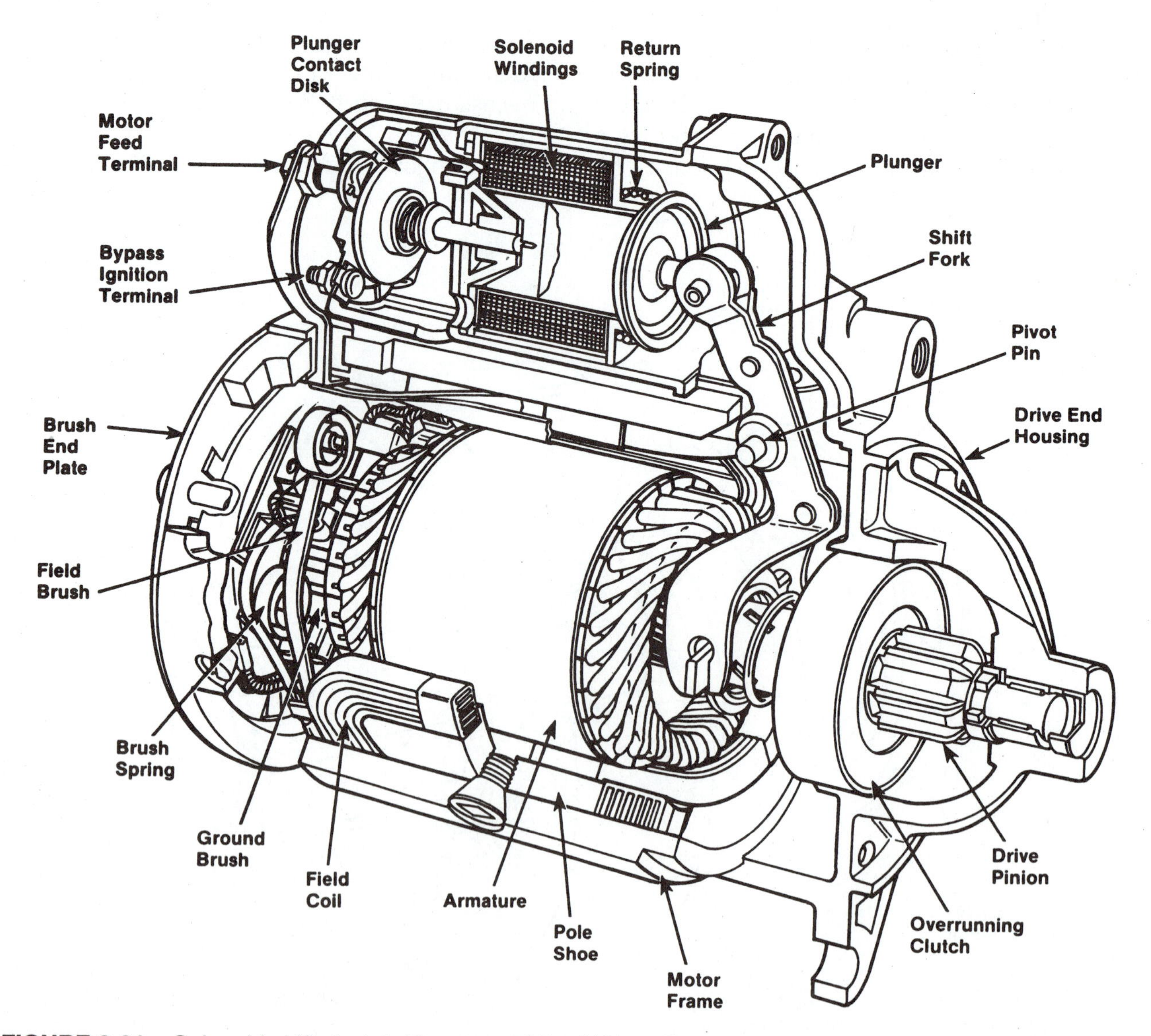

FIGURE 8-21 Solenoid shift starter. (Courtesy of Ford Motor Company)

starter input. Some solenoids also have an additional terminal that becomes B+ with the solenoid energized. This is called the **ignition bypass terminal**. It is used to supply current to the ignition system during cranking. Its use has gradually tapered off as certain styles of electronic ignition do not require a bypass circuit. We will cover the bypass circuit in Chapter 12 "Ignition Systems."

Most solenoids use two windings. They are called the pull-in and the hold-in windings. Their function is identified by their name. How they work is quite interesting, and a simple example of parallel loads. Refer to Figure 8-22, which shows a schematic of the two windings. Notice that the pull-in winding starts at the S terminal (start) coming from the ignition switch and ends or grounds at the M or motor terminal. This terminal will look like a ground because of the extremely low resistance of the stopped starter motor. Earlier in this chapter, we looked at how the motor functions. With the armature stopped, the windings will offer no resistance to low amounts of current. When the S terminal is energized with B+, the ground for the winding will be supplied by the starter motor and current will flow. By contrast, the hold-in winding has its own ground usually soldered directly to the case of the solenoid. It does not rely on the starter motor circuit for its path to ground. When B+ is delivered to the S terminal, both windings will be energized in parallel and a very strong magnetic field will be produced. This is because both windings are drawing current, sometimes as high as 50 amps. Once the plunger spring is compressed and the plunger is pulled in, a lot less current will be necessary to hold it in place. The contact disc will now be at B+ potential because

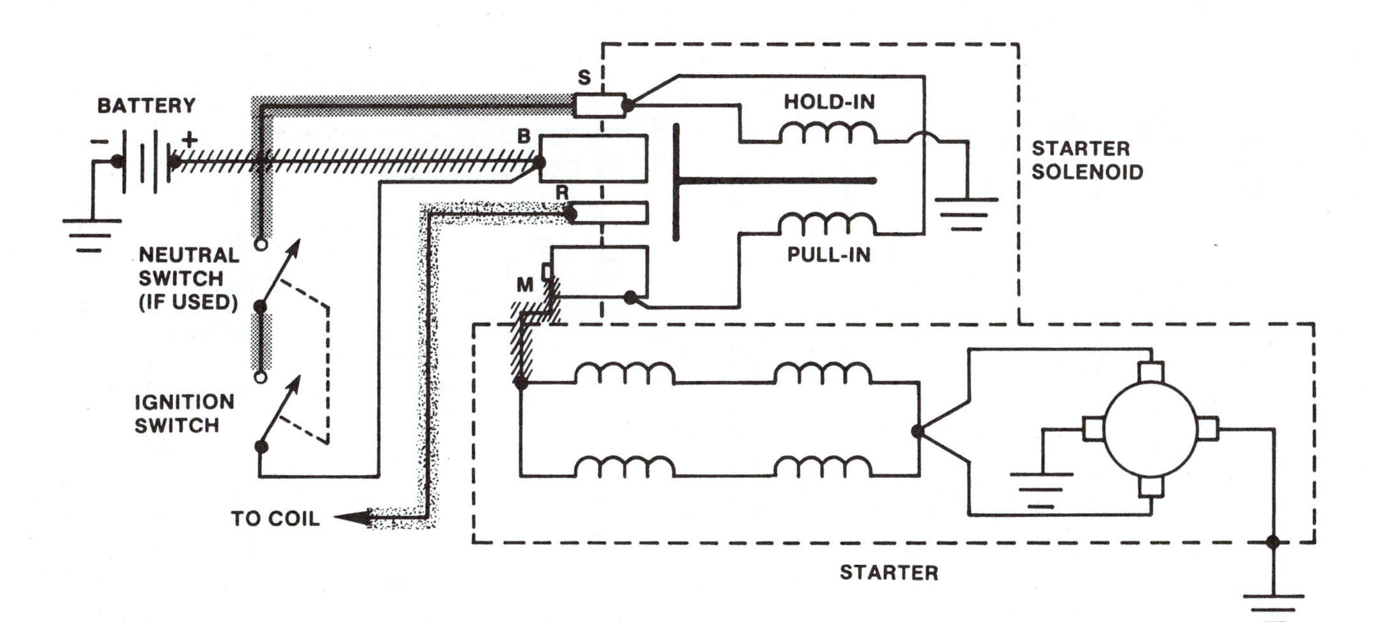

FIGURE 8-22 Solenoid shift schematic. (Courtesy of Ford Motor Company)

it will have been pushed against the battery terminal and the M terminal. Notice that the end of the pull-in winding is also attached to the M terminal. The M terminal has now become B+, and so has the S terminal. This places the same potential (B+) at both ends of the pull-in winding. With the same pressure or potential at both ends, all current flows through the winding stops. This design then allows the current, which the pull-in winding was drawing, to become available for use by the starter motor. The hold-in winding continues to draw current because the two ends of its coil have B+ and ground.

FORD'S POSITIVE ENGAGEMENT STARTER

The solenoid shift starter is by far the most used style worldwide. However, since 1960, Ford has used a modified design called the positive engagement type on many of its applications. It is used with a starting relay that we will look at in the next section. The simple schematic shown in Figure 8-23 illustrates the principle. The heavy contacts located between the two field coils are normally closed (NC) and will short the first coil directly to ground until they open. Opening them will allow current to flow through the rest of the starter. Remember the sequencing that the solenoid accomplished? First engaging the drive and then energizing the motor. The grounding contacts will accomplish the same thing. Refer to Figure 8-23 and Figure 8-24 together now and follow the action. Current arrives at the motor input from the starting relay (key switch in start position) and energizes the first open field coil. The switch is closed so most of the current goes through the first coil only and then to ground. The other path available is longer, as it involves two field coils and the armature. Remember, electricity takes the path of least resistance, or in this case, one field coil and ground. As much as 400 amps will flow through this one coil to ground and produce a tremendous magnetic field. This field will pull down on the moveable pole shoe forcing the drive pinion

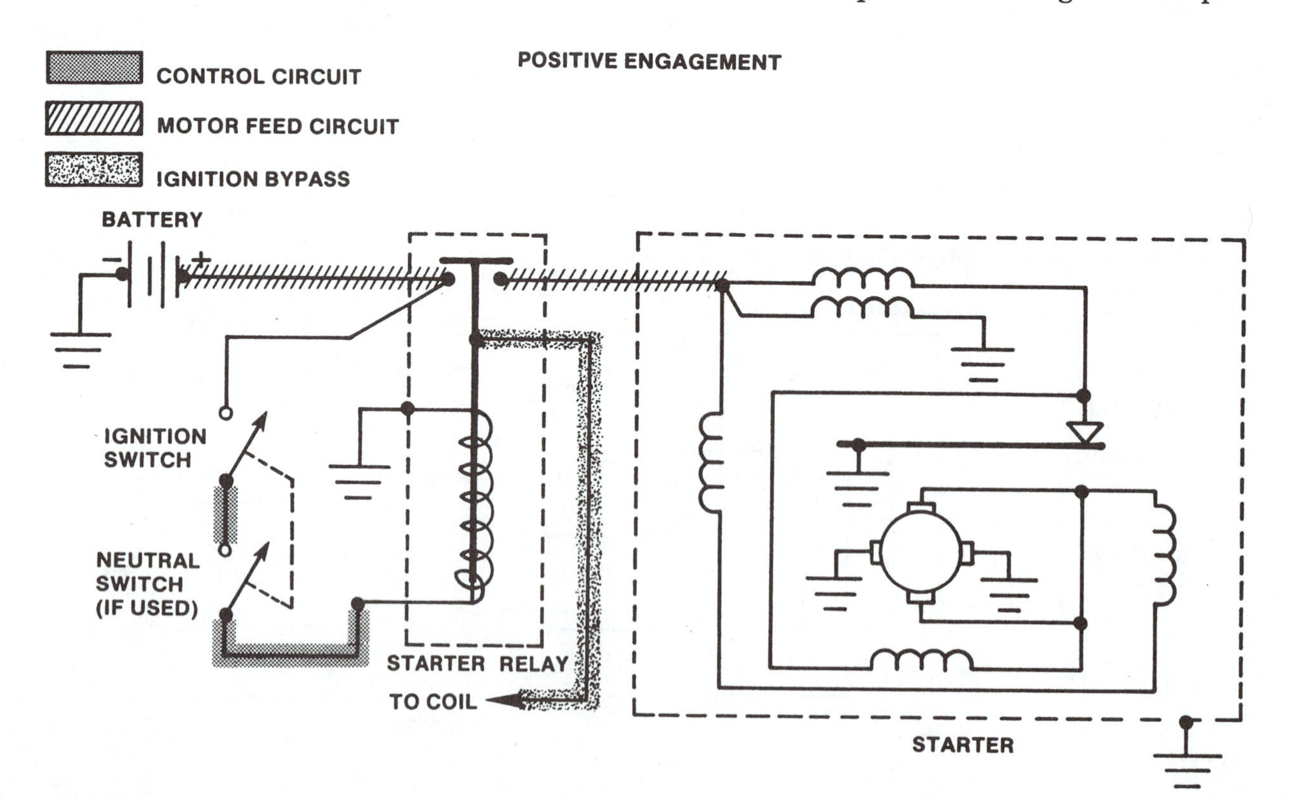

FIGURE 8-23 Positive engagement starter.

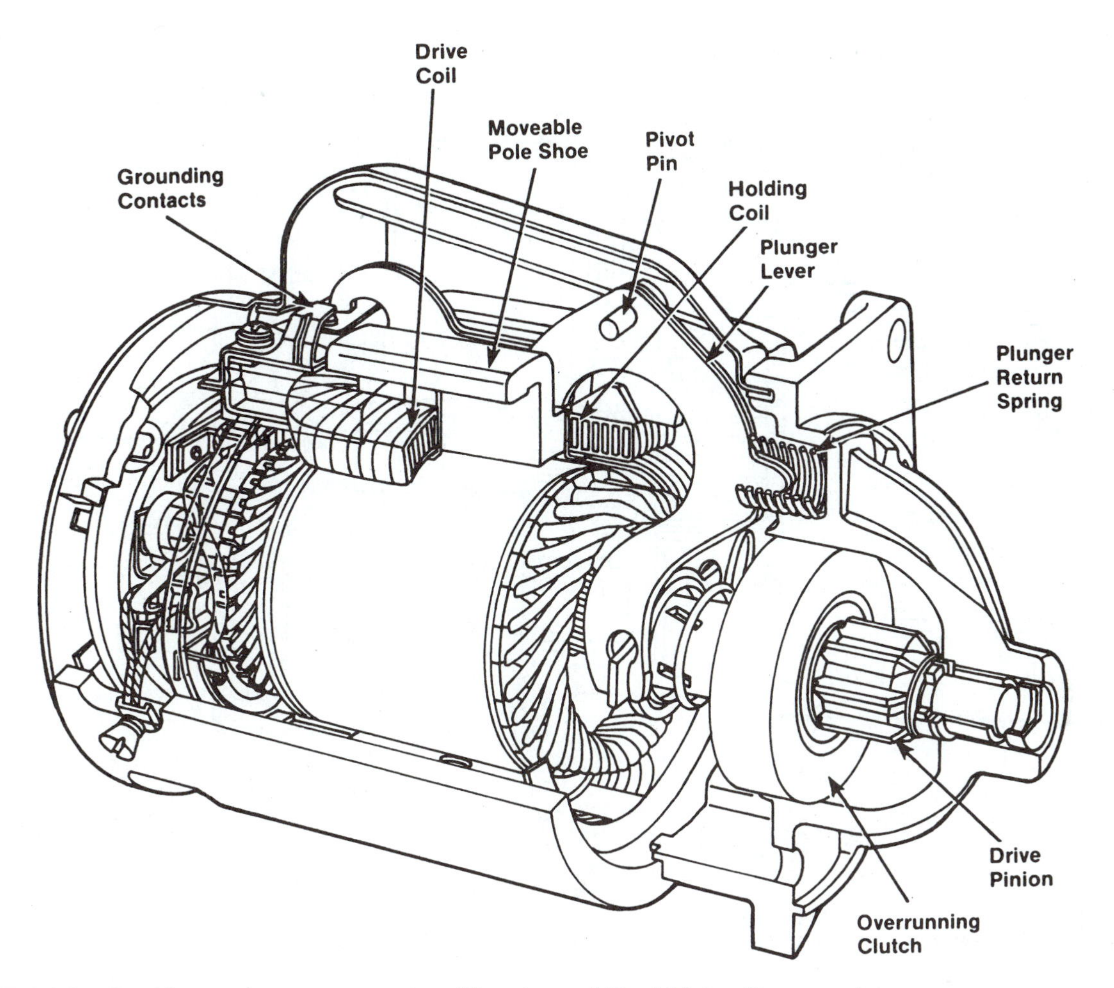

FIGURE 8-24 Positive engagement starter. (Courtesy of Ford Motor Company)

in to mesh with the flywheel. During this movement, the armature does not have any current flowing through it and is therefore stopped. On the end of the moveable pole shoe is a tab that will open the grounding contacts once the drive is in mesh. As the contacts open, the current now flows through the rest of the motor and the engine cranks. Many of these motors also use a smaller winding connected in parallel to the drive coil to assist in the holding down of the moveable pole shoe. This prevents the spring from disengaging the drive if battery voltage drops down low during cranking.

GEAR REDUCTION STARTERS

Another style of starter currently in use is the gear reduction type. Refer to Figure 8-25. It is a drawing of a common gear reduction type starter used by Chrysler Motor Corporation vehicles and some foreign vehicles. It differs in that the armature does not drive the drive pinion directly. Instead, a gear reduction assembly is used that allows a smaller motor running at a higher speed to crank with lower current. Many of these units use a 3.5 to 1 reduction ratio and greatly increase the torque available. The rest of the starter, including the solenoid, is the same as we have studied. Notice that an overrunning clutch is used to drive the flywheel ring gear.

PERMANENT MAGNET STARTING MOTORS

The most recent change in starting motors has been in the use of permanent magnets rather than electromagnets as field coils. Electrical-

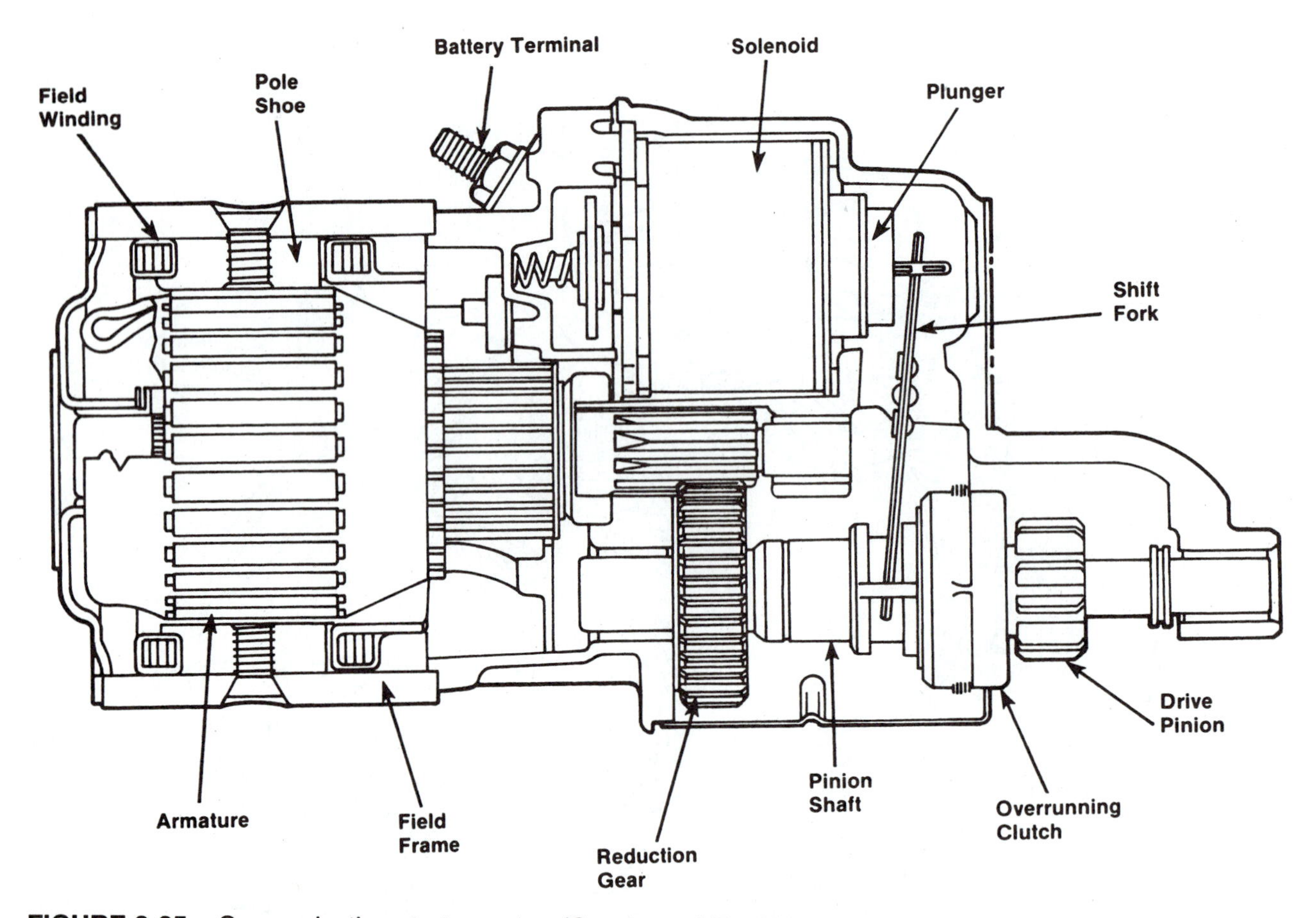

FIGURE 8-25 Gear reduction starter motor. (Courtesy of Ford Motor Company)

ly, this starter motor is simpler as it does not require current for field coils. Current is delivered directly to the armature through the commutator and brushes. Figure 8-26 shows this starter motor. With the exception of no electromagnets for the fields, this unit will function exactly as the other styles that we have looked at. Increased use of this style is expected in the future as production costs are greatly reduced. Maintenance and testing procedures are the same as for other designs. Notice the use of a planetary gear style gear reduction assembly on the front of the armature. This allows the armature to spin at greater speed to increase the torque delivered. In addition, notice the decreased size of the starter body. Reducing the overall size while keeping the torque high improves the adaptability of the unit in certain vehicles where space is at a premium. The unit still requires the use of a starting solenoid.

STARTING RELAYS

Many vehicles, both foreign and American, use relays. All positive engagement starters will use a relay in series with the battery cables to deliver the high current necessary through the shortest possible battery cables. Figure 8-27 shows a starter relay. Notice that it is very similar to the solenoid we have studied. It, however, will not be moving the pinion in to mesh like the solenoid. It is strictly an electrical relay or switch. When current from the key switch arrives at the ignition switch terminal, a strong magnetic field will pull the plunger contact disc up against the battery terminal and the starter terminal. This will allow the current to flow to the motor. Notice the use of the ignition bypass terminal to supply current to the ignition during cranking. Remember, not all vehicles make use of this terminal.

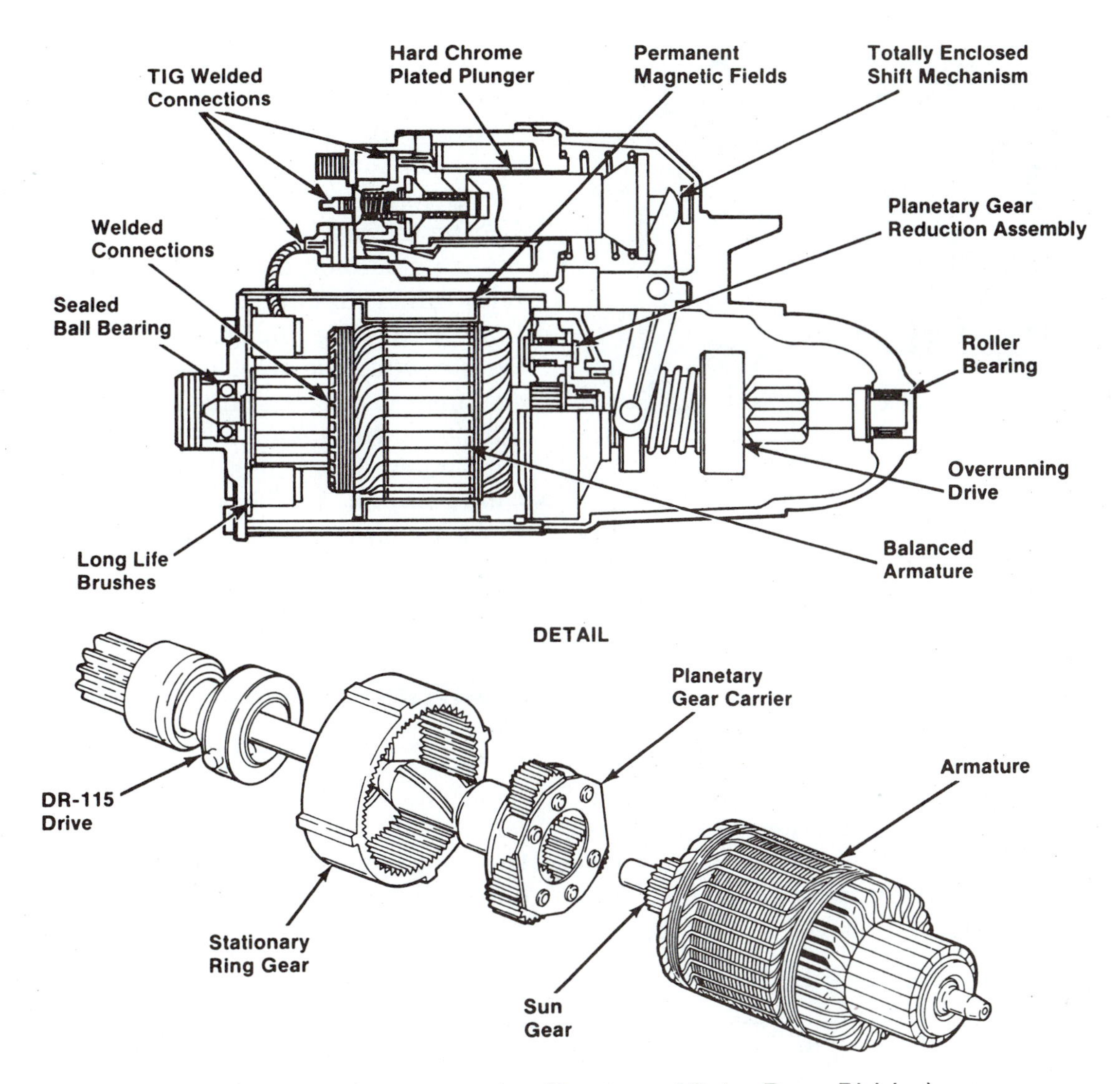

FIGURE 8-26 Permanent magnetic starter motor. (Courtesy of Delco Remy Division)

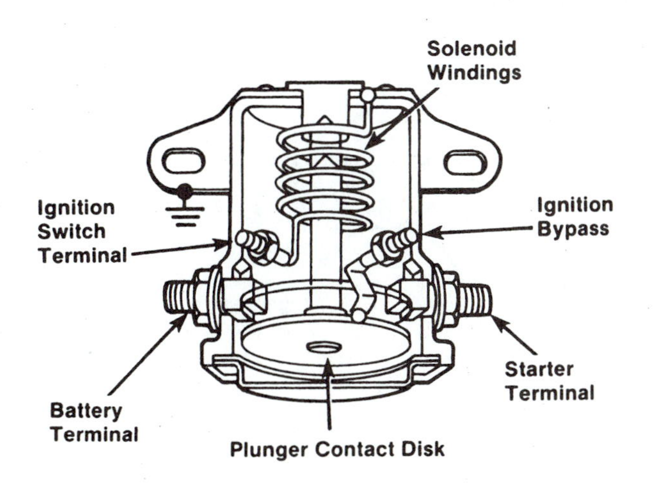

FIGURE 8-27 Starter relay. (Courtesy of Ford Motor Company)

Some vehicle manufacturers will use a starter relay with a solenoid. In this manner, solenoid current, which can be quite high, does not flow through the ignition switch. We will look at this type when we compare the wiring diagrams for the most common styles of starting systems.

THE NEUTRAL SAFETY SWITCH (NSS)

On automatic transmission vehicles, the starting system must be prevented from operating if the vehicle is in a drive or reverse position. Engaging the starter and having the vehicle

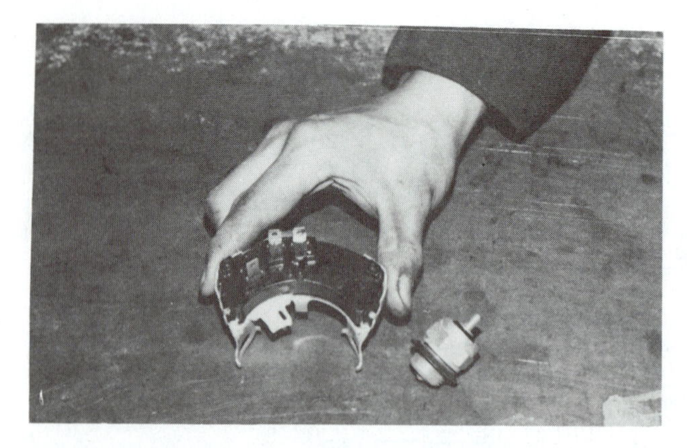

FIGURE 8-28 Neutral safety switches.

start in gear could have dangerous side effects. To prevent this, the manufacturers install an electrical switch in series with either the solenoid windings or the starter relay coil. The switch is closed if the transmission is in park or neutral and open in all other gear positions. If it is closed, the starting system functions normally and the vehicle cranks. If, however, the switch is open, no control or coil current can flow and the vehicle starting system is inoperative. Figure 8-28 shows two common types of **neutral safety switches** (NSS). The extra set of connections on the one on the left are for the back-up lights. Notice also that the mounting holes are elongated. This allows adjustment of the NSS position after installation. Replacement requires adjustment to prevent engagement of the starter in gear. The NSS on the right is a type that will thread into the transmission and supply a ground to the relay coil in park or neutral. It should be noted that preventing engagement of the starting system while the vehicle is in gear can be accomplished mechanically by a lockout at the ignition switch. In addition, some vehicles with manual transmissions will use a clutch safety switch. Pushing the clutch pedal in, closes the switch and allows the starting system to function. The new term applied to NSS is TR or **transmission range**.

STARTING SYSTEM TYPES

Basically, all the different systems in use today can be categorized into three types. Solenoid shift, solenoid shift with relay, and positive engagement with relay. Let us look at each one and see how they are wired. Refer to Figure 8-29, which shows a solenoid shift

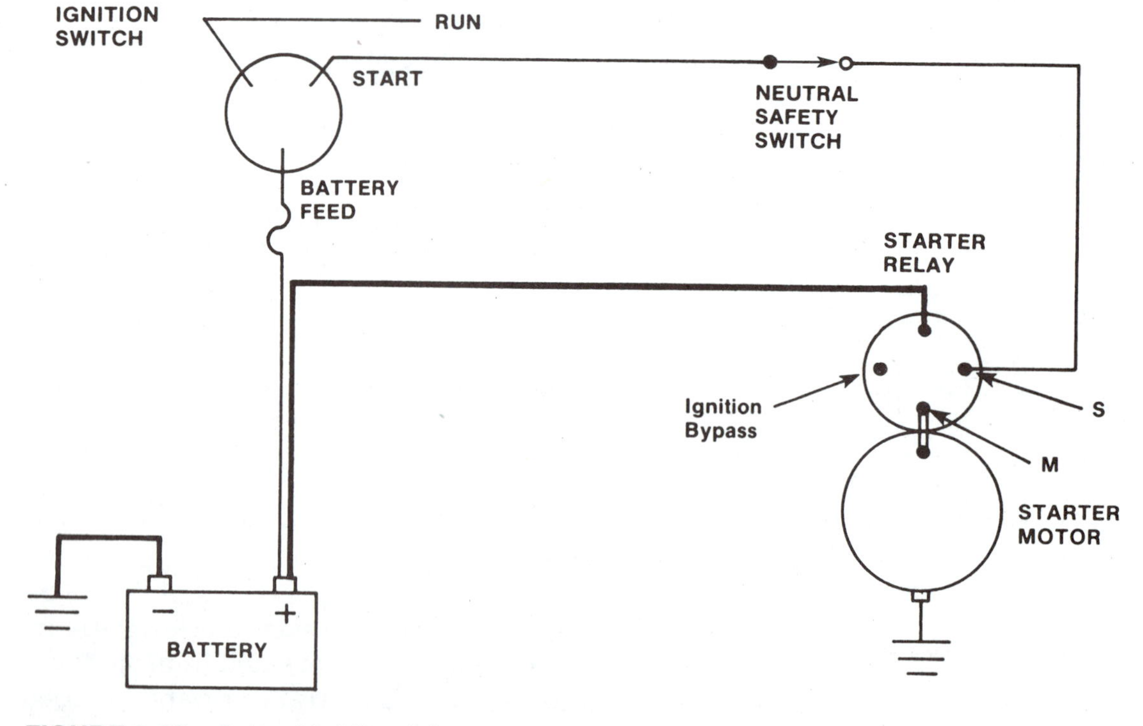

FIGURE 8-29 Solenoid shift starting system.

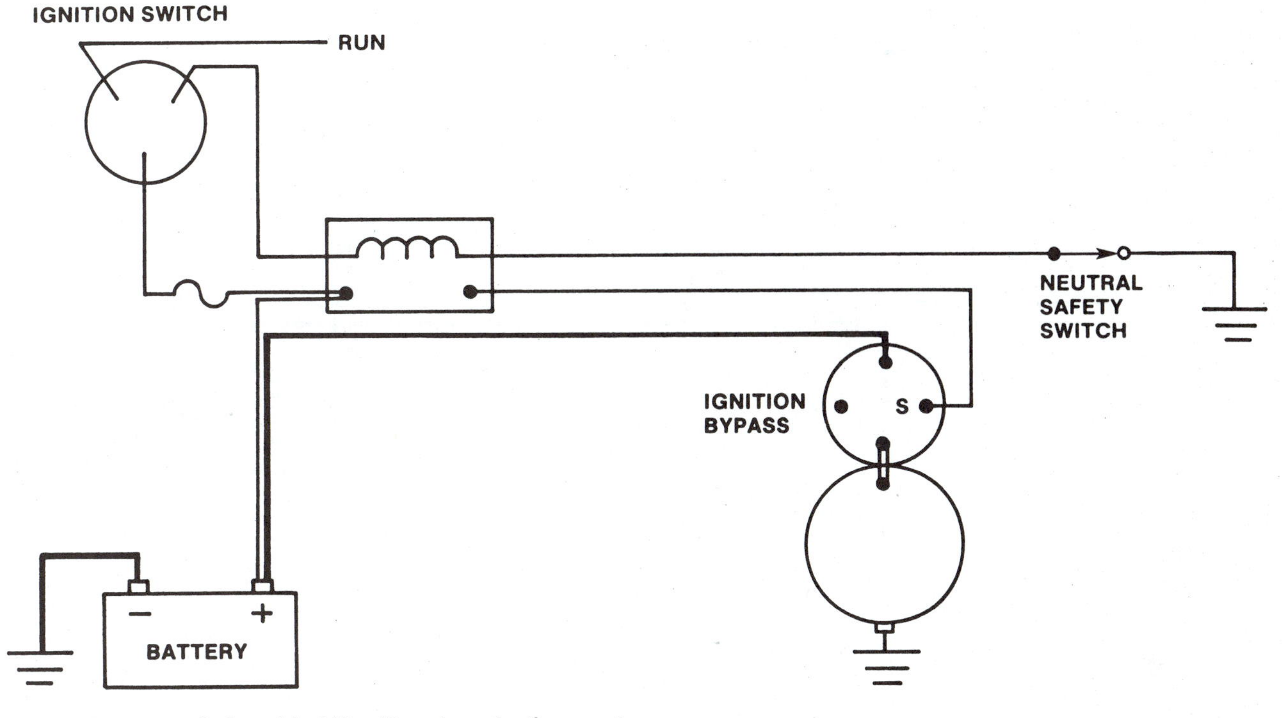

FIGURE 8-30 Solenoid shift with relay starting system.

starter motor with a neutral safety switch in the B+ feed to the solenoid. This is by far the most common type in use today. Again the ignition bypass connection may or may not be used depending on the style of ignition system in use. Notice that the solenoid is mounted to the starter. Let us trace through the function of this circuit. With the gear shift lever in either park or neutral, the ignition switch is turned to start. This brings B+ over to the S terminal of the solenoid. Both the pull-in and hold-in windings energize and develop the strong magnetic field necessary to pull the plunger in against spring tension and push the starter drive in to mesh with the flywheel. The plunger disc comes in contact with both the battery terminal and the motor terminal. This puts B+ on both sides of the pull-in winding at the same time that it energizes the motor. The engine is cranked by the armature driving the pinion, which is meshed with the flywheel. Once the engine starts, the pinion overruns the armature until the ignition switch is released and the plunger spring pulls the drive out of mesh.

Figure 8-30 shows the adaptation of the solenoid shift by the addition of a relay in the B+ feed to the solenoid pull-in and hold-in windings. Notice that the neutral safety switch is on the ground side of the relay coil rather than in the solenoid coil current. Let us trace through the relay as we did in the previous example. With the transmission in park or neutral, the NSS has applied a ground to the relay coil. When the ignition switch is turned to the start position, B+ is applied to the other side of the relay coil and current flows. The relay coil now develops a magnetic field that closes the large contacts, allowing B+ to flow through to the solenoid. Electrically, the rest of the circuit is as outlined for Figure 8-29.

The third style is illustrated in Figure 8-31. It uses the positive engagement starter with the open pole and a starting relay externally mounted. The neutral safety switch is in the B+ feed to the relay coil, and the pinion engagement function is performed by the moveable pole shoe and grounding switch. When the gear shift is in the park or neutral position, the NSS is closed and allows current

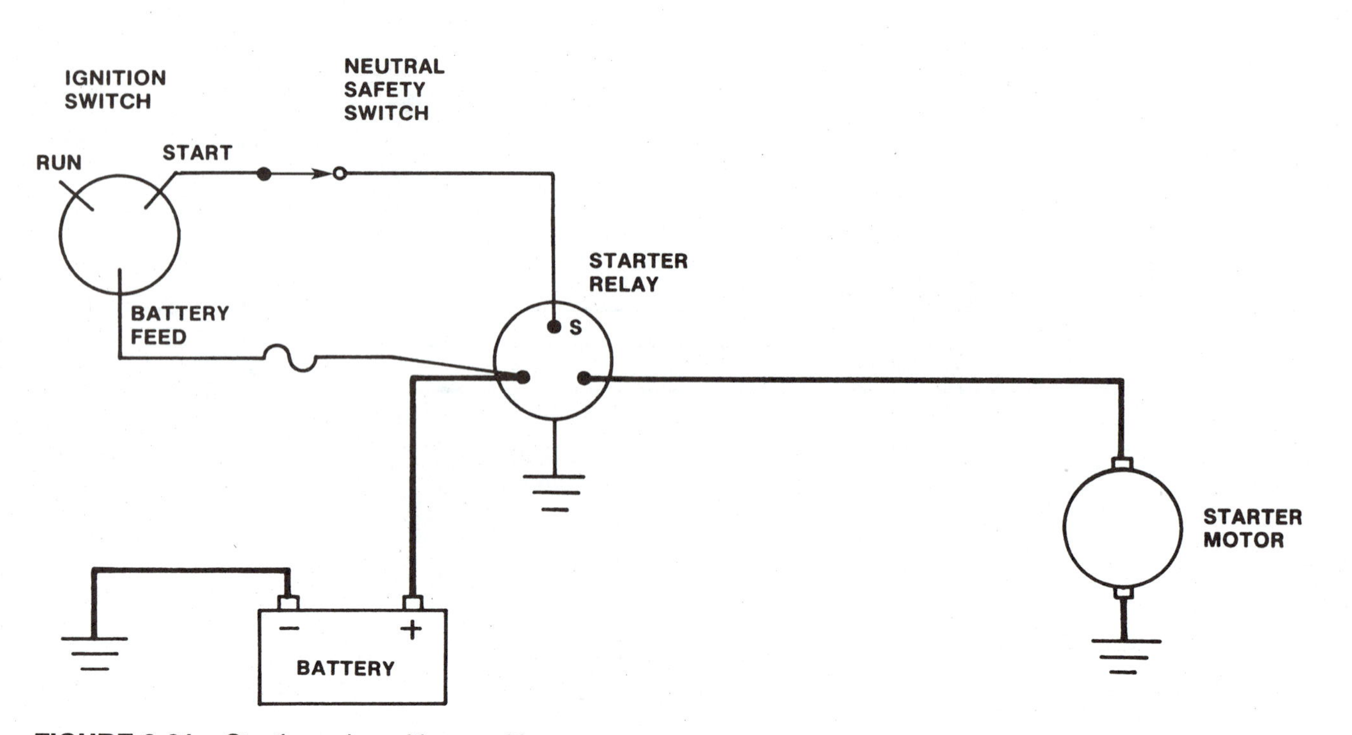

FIGURE 8-31 Starting relay with a positive engagement starter.

from the ignition switch to flow through it to the starting relay. The relay has one side of the coil grounded and will, therefore, develop a magnetic field to close the large contacts. The contacts allow current flow to the positive engagement starter motor. With the switch contacts closed, the heavy current flowing goes through the open pole to ground, developing a strong magnetic field that pulls the shoe down engaging the starter drive with the flywheel. Once engaged the contacts open and current flows through the remaining coils and the armature cranking the engine.

Note that all reference to an NSS would be deleted if the vehicle was equipped with a manual transmission. Many manual transmission vehicles have clutch safety switches that are closed by depressing the clutch.

SUMMARY

In this chapter, we have examined starting systems beginning with the principles of electromagnets and ending with actual vehicle systems. We examined the positive engagement style system with its relay and open pole field coil. We also discussed the solenoid shift style starting motor in addition to the gear reduction type. The permanent magnet starter was also examined. The two different methods of controlling the armature-flywheel action: the Bendix or the overrunning clutch were discussed. In addition, we examined current flow through a starter motor and saw how it was reduced as the armature speed increased. Neutral and clutch safety switch operation and use was also examined. We traced through the current flow on the most common types of starting systems.

KEY TERMS

Inductive pickup
Flux density
Commutator
Armature
Counterelectomotive force (CMF)
Induced voltage
Bendix drive
Drive pinion
Ring gear
Overrunning clutch
Igntion bypass terminal
Neutral safety switch
Transmission range

CHAPTER CHECK-UP

Multiple Choice

1. A current-carrying conductor will tend to do what when placed in a magnetic field?
   a. move
   b. remain stationary
   c. move toward the north pole
   d. move toward the south pole
2. Technician A states that coiling a wire around a soft iron core will increase the strength of its magnetic field. Technician B states that increasing the amount of current flowing through a conductor will increase the strength of its magnetic field. Who is correct?
   a. Technician A only
   b. Technician B only
   c. Both Technician A and B
   d. Neither Technician A nor B
3. An overrunning clutch protects the armature from
   a. spinning too slow.
   b. spinning too fast.
   c. electrical short circuits.
   d. starting the vehicle while it is in drive.
4. Technician A states that an open circuited pull-in winding will not allow the vehicle to start. Technician B states that the pull-in winding is grounded at the same point as the hold-in winding. Who is correct?
   a. Technician A only
   b. Technician B only
   c. Both Technician A and B
   d. Neither Technician A nor B
5. CEMF is
   a. voltage produced by the battery for starting.
   b. voltage produced by the armature spinning in a magnetic field.
   c. responsible for the increase in current draw as the armature spins faster.
   d. none of the above.
6. Technician A states that the neutral safety switch can also control the back-up lights on the vehicle. Technician B states that the neutral safety switch can apply a ground to the starter relay. Who is correct?
   a. Technician A only
   b. Technician B only
   c. Both Technician A and B
   d. Neither Technician A nor B

7. Technician A states that the positive engagement style starter will use the solenoid to move the starter drive in to mesh with the flywheel. Technician B states that the solenoid shift starter will use armature inertia to move the drive in to mesh with the flywheel. Who is correct?
   a. Technician A only
   b. Technician B only
   c. Both Technician A and B
   d. Neither Technician A nor B
8. You notice that the current draw on a starting system is reduced as the starter spins faster. Technician A states that there must be a weak battery in the system. Technician B states that this is due to the increased CEMF. Who is correct?
   a. Technician A only
   b. Technician B only
   c. Both Technician A and B
   d. Neither Technician A nor B
9. An open circuited hold-in winding will cause
   a. the battery to run down.
   b. a chattering (moving in and out) solenoid plunger.
   c. nothing.
   d. the starter drive to remain engaged after the engine is running.
10. The set of contacts found inside the positive engagement style of starter are
    a. to allow less current flow through the starter.
    b. short the first field coil to ground once the starter drive is engaged.
    c. replace the solenoid.
    d. short the first field coil to ground before the starter drive is engaged.

True or False

11. An open set of contacts in a positive engagement style starter will prevent engagement of the starter drive.
12. A neutral safety switch will not have continuity through it when the transmission is in drive or reverse.
13. A current-carrying conductor will tend to move when placed in a strong magnetic field.
14. A starter should draw less current once it is spinning.
15. A starter will develop the greatest torque (twisting effort) once it is spinning at normal speed.
16. The two windings in the solenoid are the pull-in and the hold-in.
17. Gear reduction starters will usually draw less current and develop less torque.
18. If an automatic transmission vehicle will start in any gear, the NSS is probably at fault.
19. Coiling a current-carrying conductor around a soft iron core will decrease its magnetic field.
20. The use of Bendix style of starter drives is increasing.

Vehicle Starting System Testing and Service

OBJECTIVES

At the completion of this chapter, you should be able to:

- measure amperage draw on a starting system.
- measure voltage drops on both the B+ and the ground side of the starting circuit.
- replace mechanically attached brushes.
- replace soldered in brushes.
- replace the starter drive in solenoid shift starters, positive engagement starters, and gear reduction starters.
- diagnose a no-crank condition with a test light.
- replace the solenoid in a solenoid shift starter and gear reduction starter.
- determine the pinion depth on a replacement starter.

INTRODUCTION

In the previous chapter, we discussed the operation of the starting system in small bite size pieces. It is now time to look at the entire system as it is commonly found in vehicles of today. It is interesting to note that of all the current systems on vehicles today, the starting system has experienced the least changes. On-board electronics, digital control, and integrated circuits have had a tremendous impact on most vehicle systems. The starting system has not seen these changes. It remains basically the same as it was in the thirties with the exception of the introduction of the permanent magnet motor. It should be considered part of the "bread and butter" systems that all technicians must know. ("Bread and butter" is just another way of saying basic and fundamental.) Employers expect their employees to be able to diagnose starting system problems quickly and get them repaired and back on the road.

In this chapter, we will look at a detailed on-the-vehicle test of the starting system for a no-crank condition and also a preventive maintenance diagnosis condition. We will also examine the replacement of relays, neutral safety switches, solenoids, and starters. We will also show the replacement of starter drives and brushes. The use of certain rebuilding tools will also be shown, even though starter rebuilding might not be attempted frequently in your geographical area. Local labor rates usually dictate whether a rebuilt starting unit will be purchased and installed or the existing unit rebuilt by the technician and reinstalled. In either case, the material will be presented here for both.

REVIEW OF THE VOLTS-AMPS TESTER (VAT)

Let us begin our discussion with a review of the volts-amps tester (VAT) that we used in battery testing. Refer to Figures 9-1 and 9-2, which show two common VAT load type testers. Both have an ammeter capable of measuring 500 amps on the high scale and about 100 amps on the low scale. In addition, both have voltmeters with two or more scales and a carbon pile that can load a battery or system down. This is the basic battery-starting-charging system tester. The information it will supply is necessary in our diagnosis. The VAT is a fundamental piece of diagnostic gear that most shops have. Many manufacturers offer it in their tester line.

This first section will be a testing series that a technician would do on a vehicle that does crank. The reason for doing it might be preventive maintenance, perhaps before a sea-

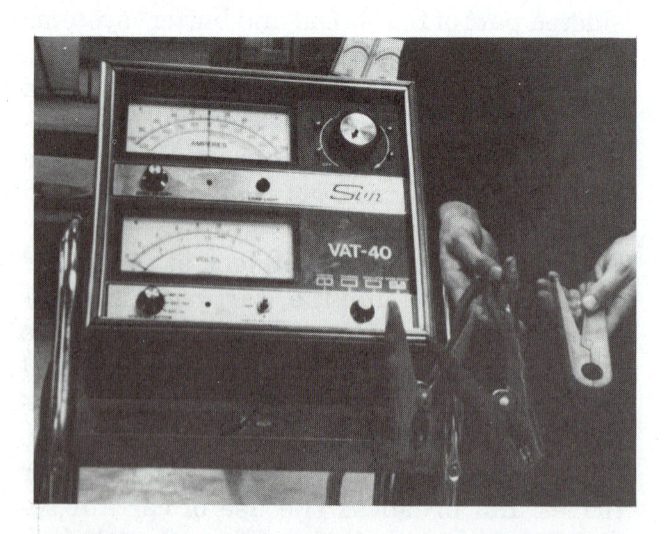

FIGURE 9-1 Volts-amps tester with inductive pickup. (Courtesy of Sun Electric Corporation)

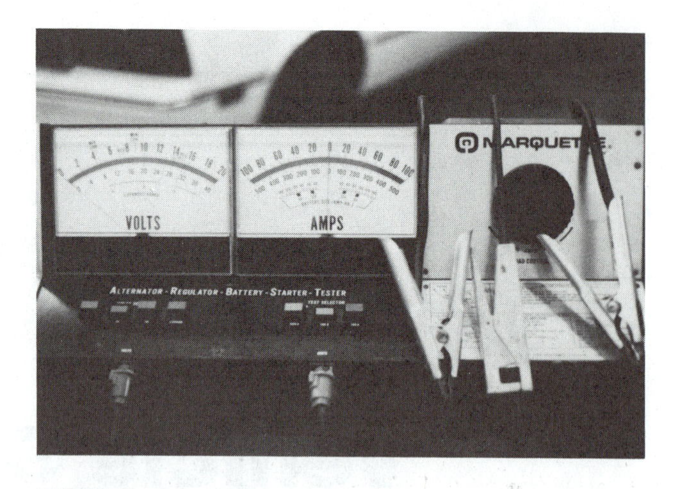

FIGURE 9-2 Volts-amps tester with inductive pickup.

son change like winter up in the North, or an intermittent cranking problem. This series of tests can be done on any vehicle. However, it should be noted that the specifications might be slightly different from the general guidelines printed. We must also realize at this point that a full battery diagnosis should be completed before looking at the starting system. The reasons for this are based in the interrelationship between battery performance and starting performance. Starting problems can be caused by the battery just as battery problems can be caused by the starting system. The battery must have the ability to keep its voltage above 9.6 volts and deliver the current necessary to crank the engine. For example, a starting system will draw twice the current if the battery voltage drops to half. To illustrate this example requires a review of wattage.

$$\text{Voltage} \times \text{Amperage} = \text{Wattage}$$

It is the wattage that will actually do the work of cranking the engine. Many V-8s crank at about 2,000 watts. At 10 volts applied to the starter, 200 amps will flow because.

$$10 \text{ V} \times 200 \text{ A} = 2{,}000 \text{ W}$$

However, if the battery cannot keep its voltage up, the starter will draw more current. More current drops the voltage lower, increasing the current which drops the voltage and so on. If 7.5 volts are delivered to the starter, it will draw 266 amps!

$$7.5 \text{ V} \times 266 \text{ A} = 2{,}000 \text{ W}$$

From a practical standpoint, it is important to note that many times the system is incapable of delivering extremely high amperage and the engine will just not crank. However, you can see how important it will be to deliver amperage at a reasonable voltage level. This is why we stressed the importance of correct battery size in previous chapters. Larger batteries will actually work less, and in theory, last longer in your customer's vehicle. Always install batteries that are at least equal to that installed by the original equipment manufacturer. Figure 9-3 is a reprint of the battery tests we outlined in Chapter 8. Do not hesitate to review them thoroughly. They must be accomplished before you begin to look at the starting system components. If the battery fails the load test and is fully charged, it must be replaced before going on. The battery that is weak will throw off your starting system testing.

I. Procedure for Testing the Starting System

With a fully charged, performance-tested battery installed, we can begin testing the starting system. We will want to crank the engine without it starting, so we should disable the ignition system in one of two ways. Either jumper the coil wire directly to ground at the distributor cap end, Figure 9-4, or remove the battery feed from the ignition system, Figure 9-5. Both of these methods will allow the engine to crank without starting. Under no circumstances should you crank the engine with the coil wire disconnected and open circuited as this could stress the coil and ignition module and might damage them.

The inductive probe of the VAT was around either the positive or negative cable of the tester for battery testing. It will now have to be moved over to the vehicle battery cable so that the VAT will record what the engine draws rather than what the VAT draws. The load should be off. After its installation, the meters should read battery voltage and zero amperage (unless there is a load on in the vehicle such as a light or the ignition switch). Turn on the headlights. The ammeter should move into the negative. If so, the cables are correctly installed. Turn off the headlights. If there is a zero adjust for the ammeter, adjust it until the needle rests or sits at 0 (zero). We are now ready to run a cranking amperage test. Either move the meters where you can observe them or ask someone else to crank the engine for about 5 seconds. Record both the voltmeter and the ammeter readings. Correctly functioning systems will crank at 9.6 volts or higher and will draw amperage relative to the cubic inches of the engine. Most V-8s will draw around 200 amps, six cylinders around 150 amps, and four cylinders around 125 amps. Remember these numbers are just guidelines. If manufacturers' specifications are available, use them!

Interpreting the numbers is not difficult but requires an understanding of the wattage principle that we discussed previously. If the specification on a particular vehicle was 9.6 volts and 200 amps and our results showed 300 amps at 6.8 volts, we would conclude that more testing is necessary to pinpoint the problem. A voltmeter reading lower than spec and an ammeter reading higher than spec is sufficient cause for detailed testing even if the vehicle does start. Preventive maintenance, if done correctly, will avoid future no-start conditions. Readings around specifications are considered good, but detailed testing should be completed if an intermittent problem is present. The detailed testing will attempt to pinpoint bad components. At this point, we should realize that our results were taken at the battery and that these readings might not be exactly representative of the actual voltage and amperage delivered to the starter. We frequently find voltage drops or losses through cables, con-

Battery Worksheet

Specifications

Battery Size ______________ Engine Cubic Inches ______________

Hydrometer Readings

Cell Number	1	2	3	4	5	6	
Negative Post							Positive Post

Results (Check one.)

_____ All cells are within 25 points and are charged (continued testing).
_____ All cells are equal and are discharged (charge battery).
_____ Cell readings vary more than 25 points (skip load test and advise customer to replace battery.)

OR

Open Circuit Voltage Test

Load battery to 20 amps for about 1 minute before testing OCV.

______________ volts—open circuit (no load)

Refer to the chart and convert either the hydrometer reading or the OCV over to state of charge.

Charge	Specific Gravity	OCV (12.6)	OCV (6.3)
100%	1.265	12.6	6.3
75%	1.225	12.4	6.2
50%	1.190	12.2	6.1
25%	1.155	12.0	6.0
Discharged	1.100	11.9	5.9

Battery Load Test (for batteries at least 75% charged)

Set tester to the following:
1. Tester to STARTING
2. Variable LOAD off
3. Voltage—INTERNAL—higher than battery voltage

Attach the BST as follows:
1. Place large cables of tester clamp over the battery clamps, observing that polarity of the battery and tester are the same.
2. Attach inductive pick-up clamp around either of the BST cables *not* battery cables.

Determine the correct load:

Battery Amp/Hour × 3 = Load or
Cold Cranking Amps ÷ 2 = Load

NOTE: Compare the specific battery to the one actually installed. Load to the LARGER of the two.

Apply the calculated load for 15 seconds, observe voltmeter, and turn load OFF.

_____ GOOD—battery voltage remained above 9.6 volts during load test
_____ BAD—battery voltage dropped below 9.6 volts during load test (perform 3-minute charge test to check for sulfation)

3-Minute Charge Test (only for batteries that fail the load test or are less than 75% charged)

1. Disconnect the ground battery cable.
2. Connect a voltmeter across the battery terminals (observe polarity).
3. Connect battery charger (observe polarity).
4. Turn the charger on to a setting around 40 amps.
5. Maintain 40 amp rate for 3 minutes while observing voltmeter.

_____ GOOD—voltmeter reads less than 15.5 (battery is not sulfated).
_____ BAD—voltmeter reads more than 15.5 (battery is sulfated)

FIGURE 9-3 Battery test worksheet.

FIGURE 9-4 Ground the ignition coil before doing starting testing.

nections, relays, or solenoids that diminish the delivered voltage to the starter. These voltage drops actually force the starter into drawing more amperage (wattage at work again). The detailed tests we will now discuss will allow us to look at the individual components that can have an adverse effect on the starting system.

CRANKING SPEED

The speed at which an engine spins is very important. It takes fuel, compression, and ignition to get an engine running. Compression and fuel will be dependent on cranking speed. The slower an engine cranks, the more compression loss there is. This loss will make it harder to draw in air/fuel and ignite it. General specs call for about 250 rpm for a standard starter and 200 rpm for a gear reduction starter. This speed can be measured easily with a tachometer. Most technicians do not actually measure them, however. A technician's experience and trained ear can be an accurate indicator of correct speed. A few years in the field and your ear will be "tuned" also. In the meantime, use a tachometer if there is a doubt in your mind that the engine is cranking over fast enough.

There is one set of conditions where we usually ignore the detailed tests and immediately pull the starter. This is the combination of a slower than normal cranking speed and a lower than normal cranking amperage. This set of conditions can only occur from internal resistance within the starter motor. Remember the relationship of speed to current draw?

FIGURE 9-5 Disconnect B+ from the electronic ignition.

Slow		High Cranking
Armature	=	Amperage
Speed		Normal

When an engine cranks slowly, we should expect to find high, not low amperage. Remember this combination. Lower than normal cranking speed and lower than normal amperage can only be caused by a problem inside the starter. Higher than normal amperage and slower than normal cranking speed usually requires some detail testing of the B+ and ground path. The theory behind detail voltage drop testing is quite simple. Voltage is lost or dropped when current flows through resistance. This is Ohm's law at work again just like we have seen in the past. Most of the manufacturers design their starting systems with cables, connections, relays, and solenoids that have very little, if any, resistance to the flow of starting current. Typically, it is less than 0.2 V on each side of the circuit (B+ and ground). This then means that the voltage across the starter input to the starter ground should be within 0.4 V of battery voltage. Figure 9-6 illustrates this principle. What we must do in the case of excessive current draw or in the case of an intermittent starting problem is to ensure that the voltage across the starter remains in the 0.4 V or less area when compared to battery voltage. For example, if the battery voltage during cranking was 9.8 V and the voltage delivered to the starter was 9.5 V, we would know that the resistance of the circuit has caused a 0.3 V drop. In most cases, this would be acceptable. If, however, the battery voltage was 9.8 V and the starter voltage was 8.8 V, we would know that a problem exists that will cause the starter to draw more current than is necessary. Do not forget that the current will go up if the voltage goes down (wattage again). Our next step will be to locate the resistance that is causing the 1 V drop.

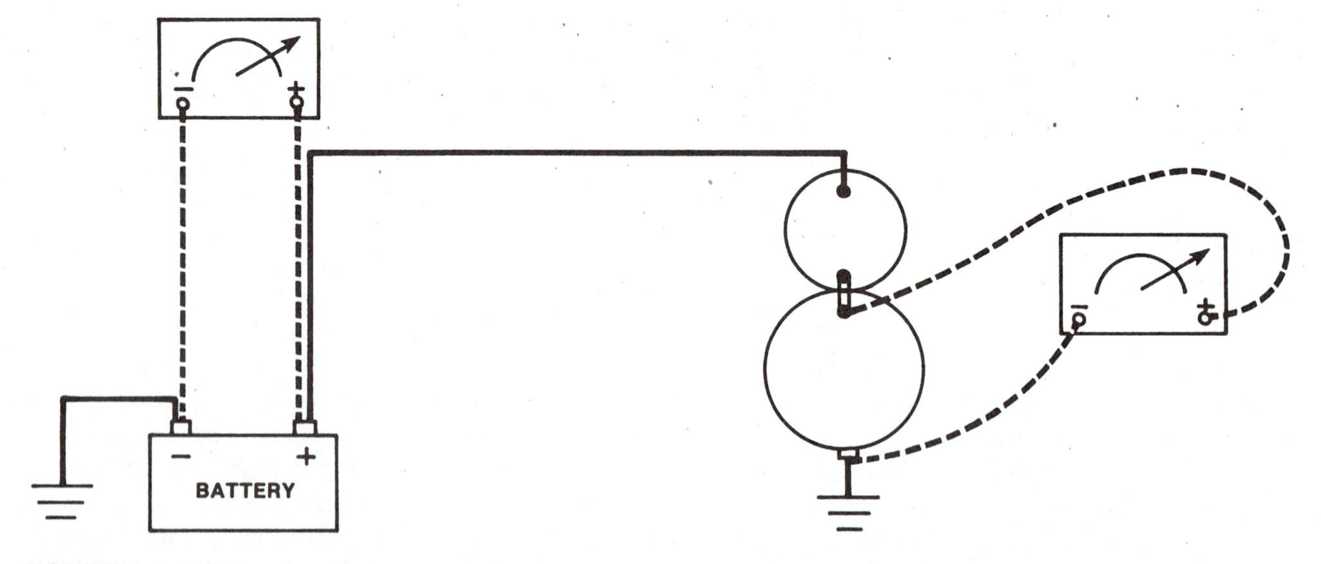

FIGURE 9-6 Voltage readings across the starter are compared to those at the battery.

Here is where we will use the lowest scale of the voltmeter, typically 2 to 4 volts. Keep in mind that current flowing through resistance causes voltage drops.

II. Procedure for Locating Voltage Drops with an Engine Cranking

A resistance will have a different pressure or voltage on each side. In Chapter 3, we measured the voltage drops across the resistance of a load. We will do much the same thing now while the engine is cranking. The voltmeter will be placed across either the B+ or the ground circuit as in Figure 9-7. This setup allows us to pinpoint the circuit half that has the excessive resistance. Once the circuit half is identified, it becomes a simple task to divide the circuit into small pieces and place the voltmeter leads across the pieces, looking for voltage drops with the engine cranking. Figure 9-8 illustrates this. The spec for the B+ or ground side of the circuit is 0.2 V maximum with normal current flowing. If higher than normal current is flowing, adjust the specification upward. If, for instance, 50% more current is flowing, the specification will be adjusted upward by 50% also to 0.3 V. We will be able to locate our problem when the voltmeter reads the voltage drop. As you do detailed voltage drop testing, remember that the positive voltmeter lead will be connected into the part of the circuit that is most positive and the negative lead into the most negative. For preventive maintenance checks, both the positive and negative paths should be tested and anything in excess of the 0.2 V B+ or 0.2 V ground specification should be replaced. If a connection appears to be where the resistance is, it should be cleaned, retightened, and retested. Corrosion and/or loose connections will cause voltage drops and increased current draw. The only part of the circuit that is designed to have over the 0.2 V drop is the starter motor. If the solenoid is part of the starter, it is tested along with the rest of the B+ path cables. It is part of the total allowable 0.2 V drop.

By doing the detailed voltage drop testing outlined here, you will be able to pinpoint the resistance. Repairing it is simply a matter of replacing the cable, solenoid, or whatever had the voltage drop. Voltage drop testing should be done at least once a year to pick out a resistive component before it causes a no-start condition. Preventive maintenance before a tow truck is needed is usually appreciated by most customers. After the voltage drops have been eliminated or at least reduced to an acceptable level, cranking amperage is used to condemn a

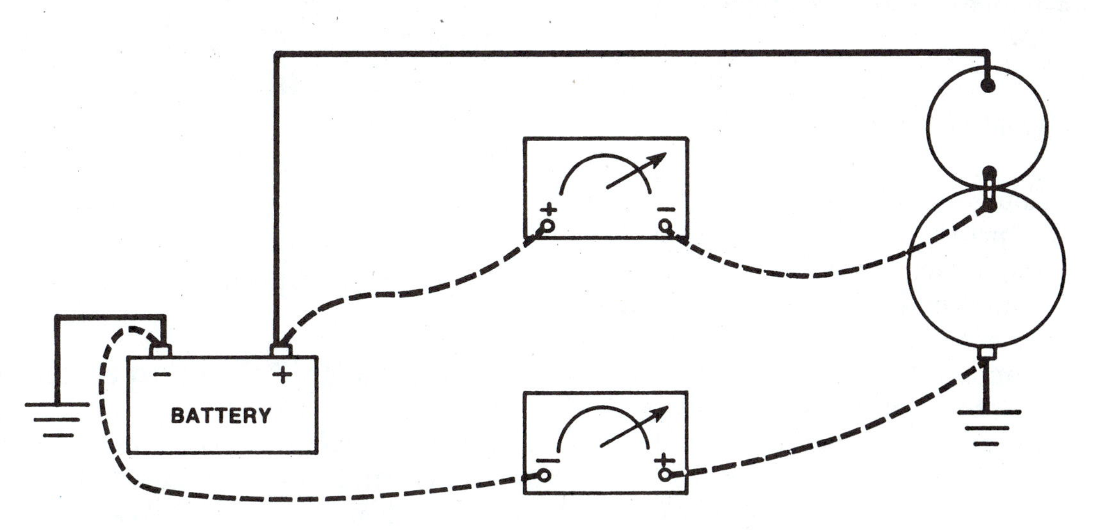

FIGURE 9-7 Voltmeters are used to find which circuit half has resistance.

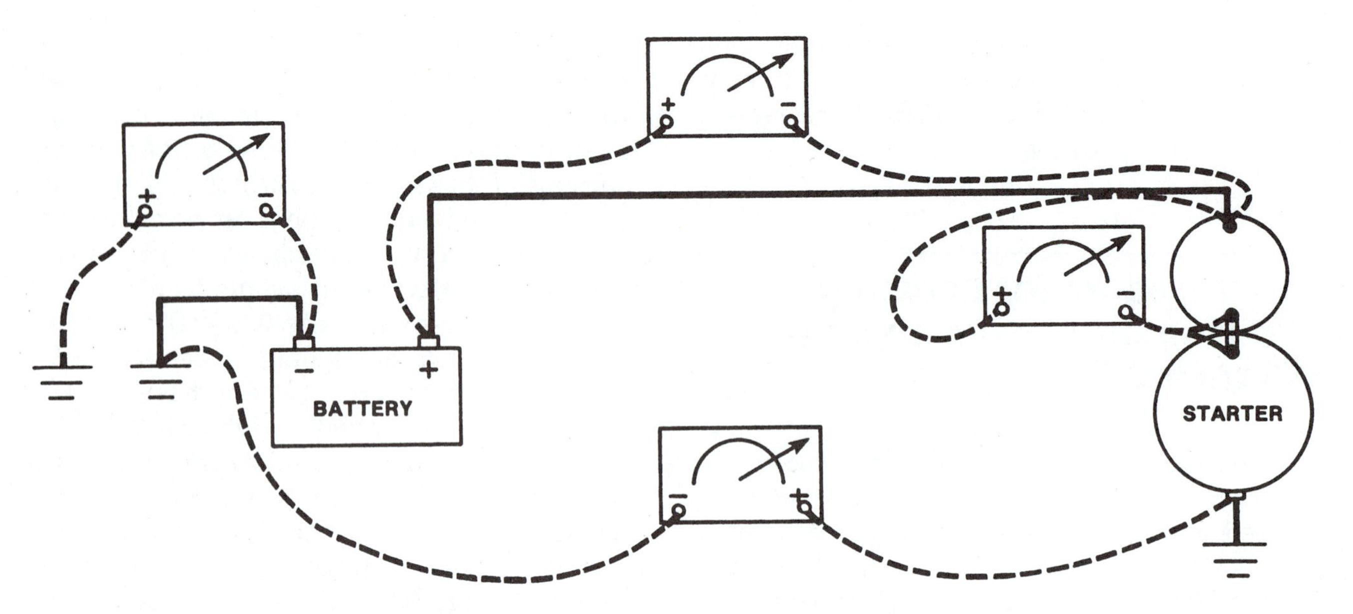

FIGURE 9-8 Voltmeter set up to measure entire system's voltage drops.

starter motor. Higher than normal amperage without excessive voltage drops are an indication of starter problems or an overly tight engine. The usual procedure, at this point, is to pull the starter off of the vehicle and disassemble it. The technician will look for obvious causes of excessive amperage draw, such as worn bushings or shorted windings. The key concept here is to not automatically pull the starter every time excessive amperage shows up. Eliminate voltage drops and test batteries first. Pulling the starter should be the last, not the first, step you do. Keep in mind that a tight engine will increase the amperage draw of the system, also.

NO-CRANK TESTING

The previous preventive maintenance approach to starting systems works well if the vehicle does, in fact, crank. Voltage drop testing is important and should be done frequently. However, if the vehicle will not crank over, a different approach to the situation is necessary. Most no-crank situations are the result of open circuits and can be diagnosed quite easily with a 12-volt test light. Let us look at the three most common types of starting systems and go through a no-crank situation in each.

The first step you always do in any no-crank situation is test the battery. Open-circuit voltage or hydrometer readings and then a load test will tell the story. Obviously, a dead or very weak battery can cause a no-crank situation. Battery testing *must* be the first approach in all types of systems. Once we are assured that the battery has the ability to supply current at a reasonable voltage level, we can begin our search for what will usually be an open circuit.

III. Procedure for Determining Open Circuit in No-Crank Condition

The procedure will be to start at one end of the circuit and use our 12-volt test light with the key in the start position.

Solenoid Shift Starting System

Let us use as our first example, the solenoid shift starting system, Figure 9-9. Notice the M terminal on the solenoid. This is the end of the circuit. If a 12-volt test light lights or a volt-

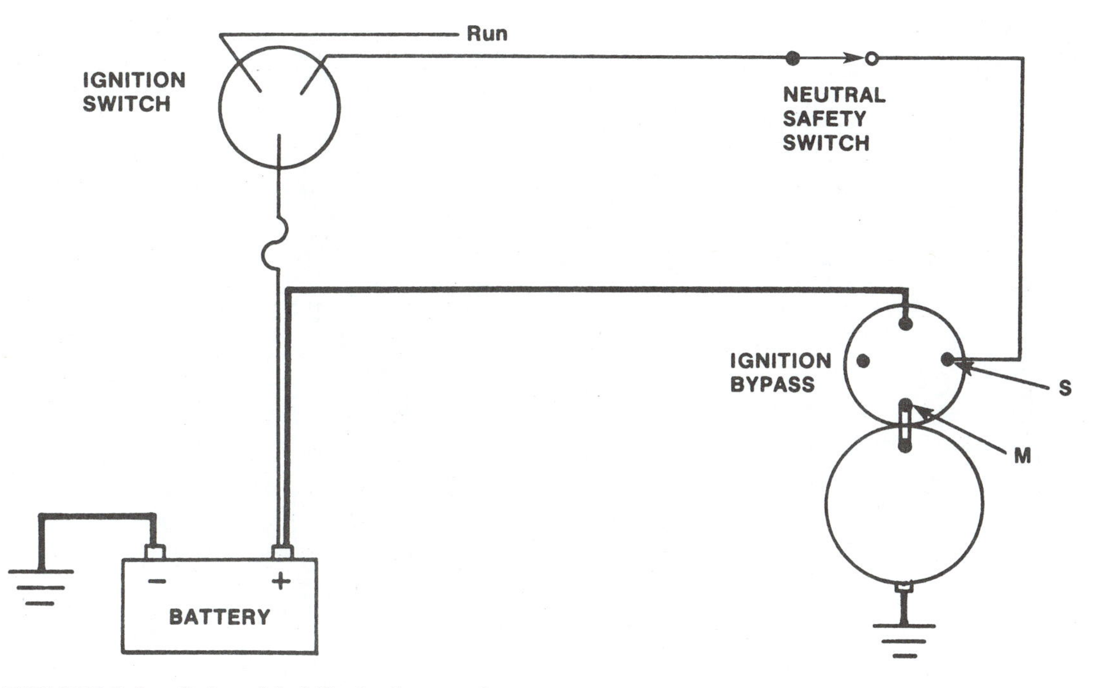

FIGURE 9-9 Solenoid shift starting system.

meter shows voltage when the ignition key is in the start position, the B+ feed to the motor is okay and we would move on to a ground circuit test. B+ to the motor is the object of the connections, solenoid, switch, and neutral safety switch (or clutch safety switch). Further testing of these individual components and wires is not necessary at this time because they must be passing current for the light to light. Keep in mind that the test light must be brightly lit. A very dim light indicates excessive resistance up circuit. Do not hesitate to use a voltmeter to measure the light bulb's voltage if you have difficulty recognizing voltage levels by looking at bulb brightness. In excess of 9.6 V and, in actuality, just about 12.6 V will usually be delivered to the M, or motor, terminal of the starter We now know that we have one of the three necessary factors to crank the engine: B+. The two unknowns at this time are the starter and the ground return path.

Let us look at the ground return path next. This is a simple test just like our first one was. Place the 12-volt test light's ground connection on the starter body (clean off the grease first) and the probe on the "known hot" M terminal. Turn the key to the start position. If the light lights, the ground return path is okay as in Figure 9-10. This would leave only the starter as our open circuit component and at this point we would pull it and bench test it, then repair or replace it.

The trick to starting diagnosis is to approach each section at the starter motor end and verify that B+ is present and a good ground return path is available before pulling the motor. Too often the starter is pulled and replaced when it is not the open component.

Let us go back to the B+ test and assume that the light did not light. Now what? Simply backtrack up the current path toward the battery until we get the test light to light. You will have to test the two sections separately. First, test the heavy current battery cable circuit as in Figure 9-11. The light should light with the key off (this is actually just like being on the battery positive post) and should remain lit when the key is placed in the start position. If it does not stay lit, repair the cable or end connections. The second circuit to look at is the control circuit and again all we need to do is move the test light toward the battery until we get it to light with the key in the start position. The point where it lights is a

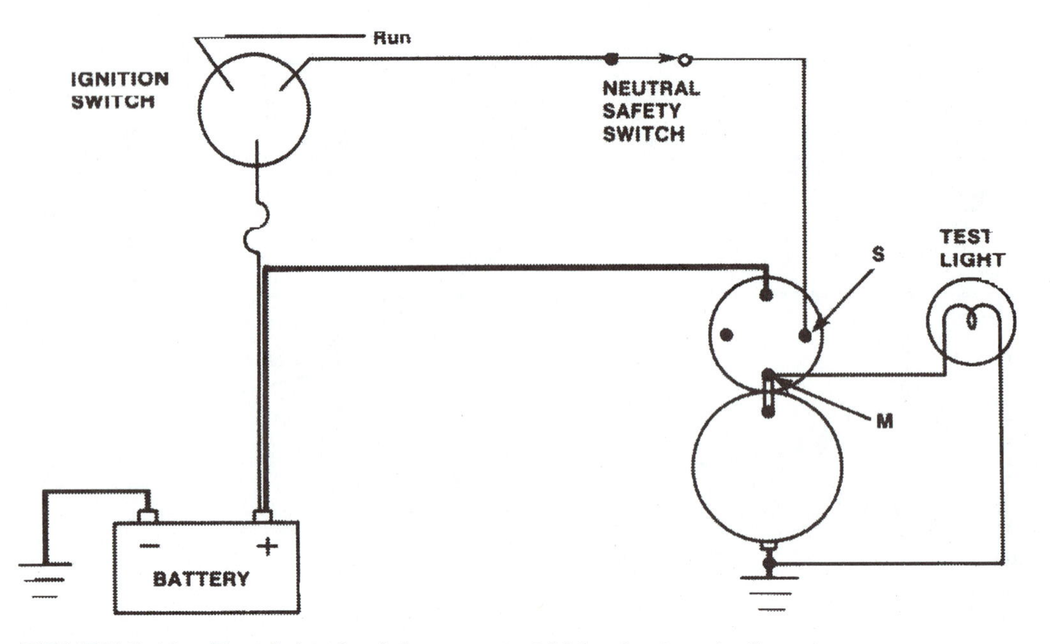

FIGURE 9-10 Test light check for power at M (motor terminal).

live section and the point where it does not light is dead. The component or cable between the light-on and light-off points is where the open is. If this sound simple, good! Actually, it is simple and usually does not take much time. Look at Figure 9-12, which shows that the light was on at the S terminal of the solenoid. What does this mean? Hopefully, you have been able to think your way through this common problem in the following manner: B+ to the S terminal, B+ to the B terminal of the solenoid, and nothing at the M terminal equals an open solenoid. Can you verify that it is open before you remove it? Sure you can.

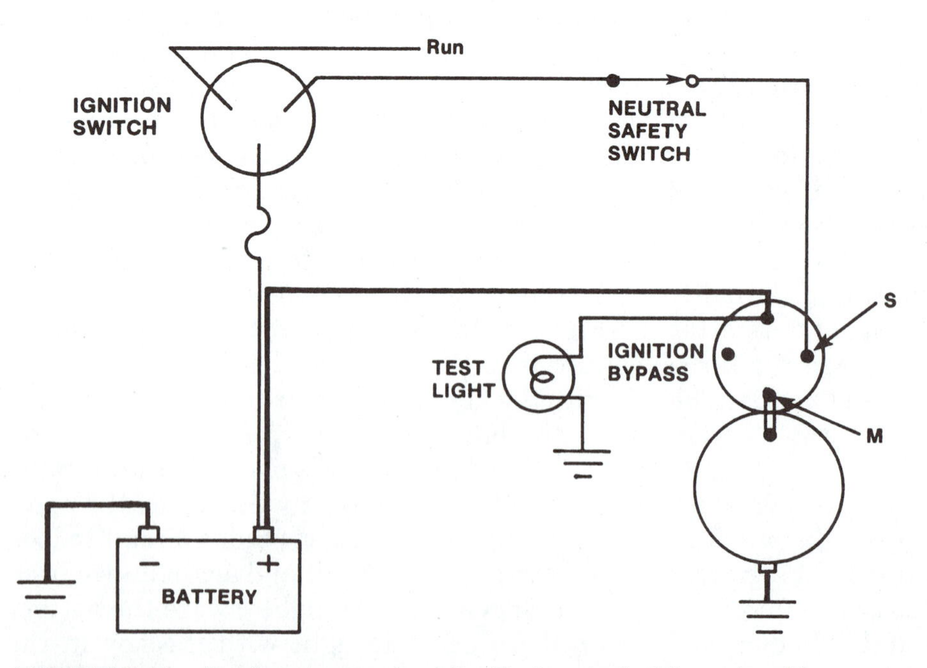

FIGURE 9-11 Testing the power at the B+ terminal of the solenoid.

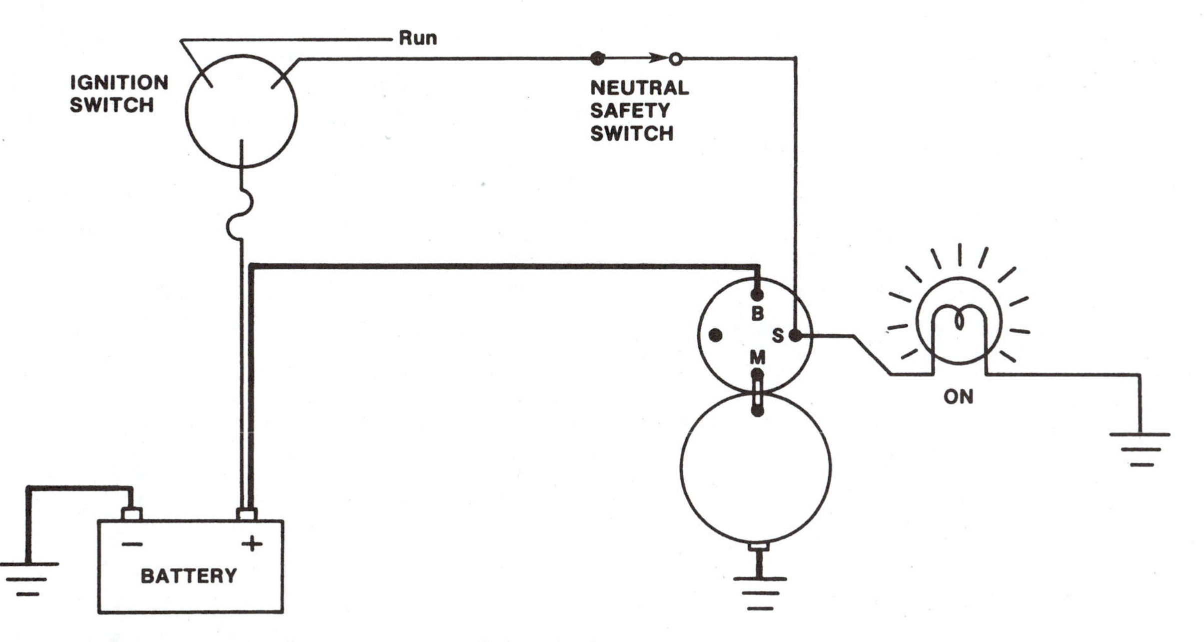

FIGURE 9-12 Testing for power at the S terminal.

Just jumper it with a jumper cable. Does the starter spin? If so, your diagnosis was correct. Pull the starter and replace the solenoid.

Let us try another example. See Figure 9-13. What do you think is the problem? With B+ available at the input of the neutral safety switch and nothing at the output, it is logical to conclude that the NSS is open. How will you verify your diagnosis? Jumper the NSS connector and see if the engine will crank. If the engine cranks with the NSS jumpered, it must be open and should be replaced.

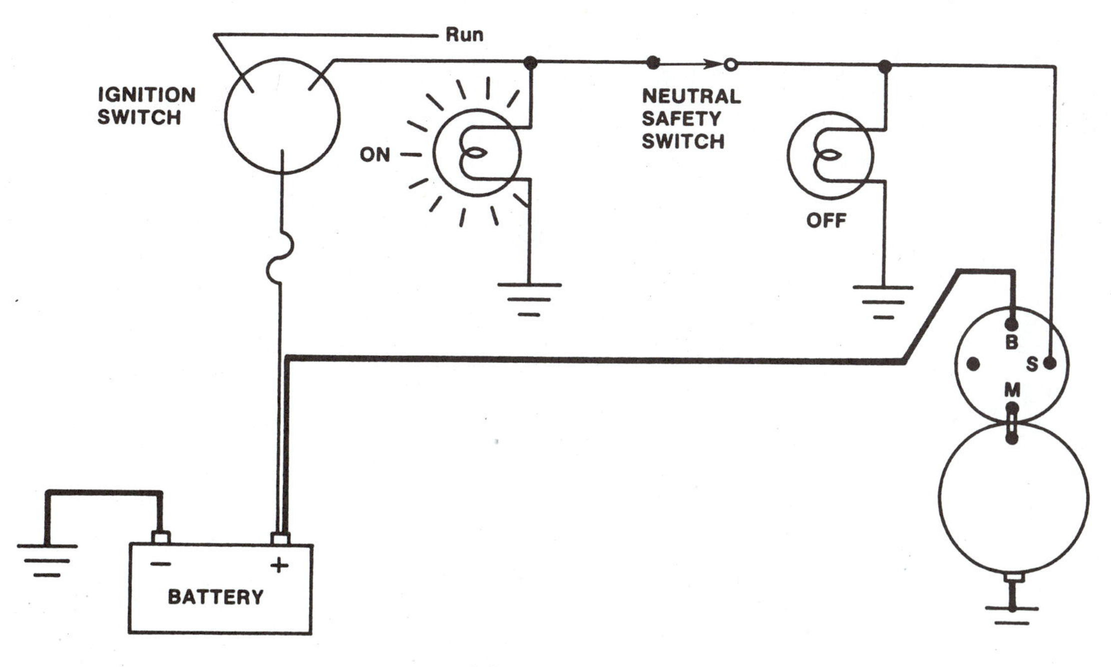

FIGURE 9-13 Testing for an open NSS.

Ford Starting System

Figure 9-14 shows the Ford starting system, which we looked at before. It will be tested in exactly the same manner as the solenoid shift system that we just looked at. Start looking for B+ at the motor input with the key in the start position, and work your way back along the circuit until the test light is on. Your testing procedure should include the additional battery cable from the solenoid output to the motor input. The remaining sections of the testing procedure remain the same. Do not forget to record your results on paper, as you might wish to refer back to them when discussing repair options with your customer. A copy of a report form has been included in Figure 9-15.

Solenoid Relay Circuit Starting System

We should spend time looking at a starting system that uses a relay in the solenoid circuit. Remember, a solenoid is a control circuit. It controls the higher starting current. Relay circuits like that shown in Figure 9-16 are also **control circuits** and must be tested as a separate parallel circuit. We again would start at the motor input and work our way back until we find B+ with the exception of the starting relay and neutral safety switch. Look at Figure 9-17. It shows the sometimes confusing testing of the NSS, which is on the ground side of a relay coil. Figure 9-17A is the test light conditions if the NSS is closed or the relay coil is open and in Figure 9-17B is the condition if the NSS is open. Can you figure out why the test light behaves in the manner shown? The light will be off if the NSS is closed because the wire from the relay coil to ground through the NSS is all at ground potential. Placing both sides of the test light at ground potential will not force any current through the bulb and it will be off. Opening the NSS by putting the gear shift lever in drive or reverse (with the engine off) removes the ground and the light glows dimly. The reason for this is that the bulb from the tester and the relay coil are now wired in series be-

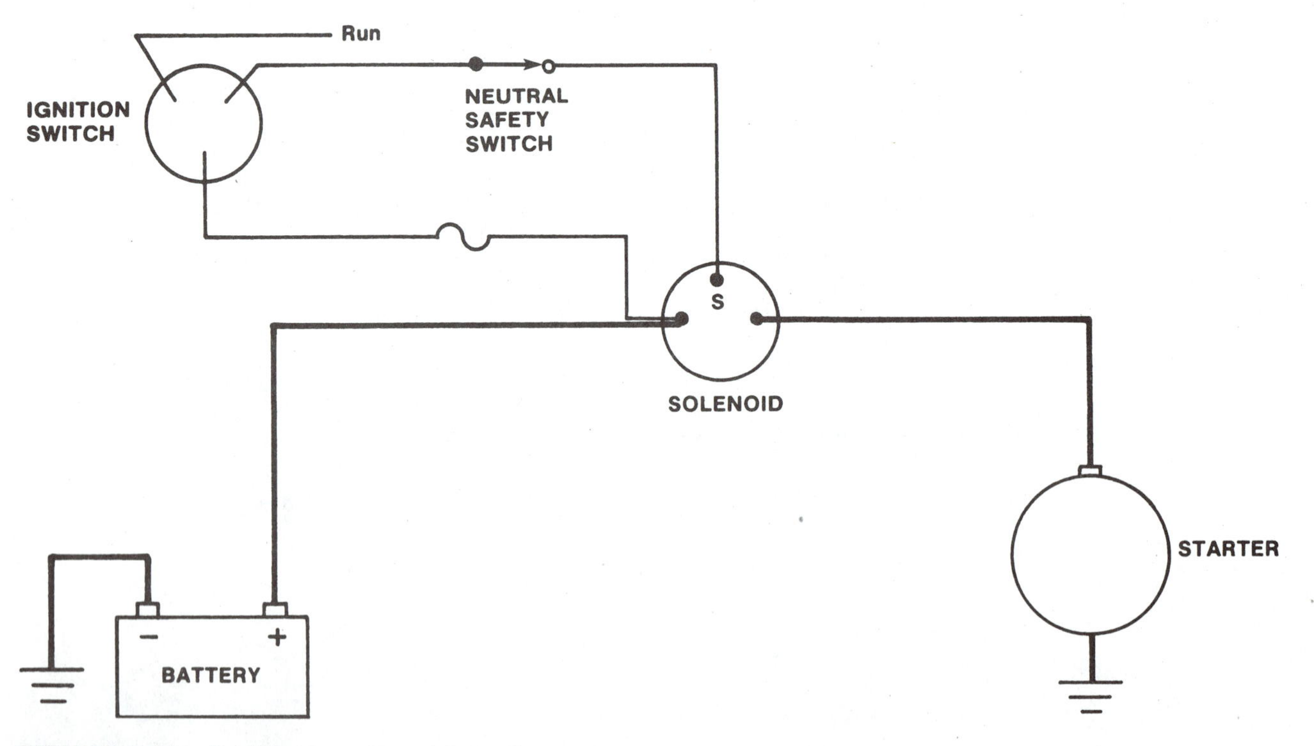

FIGURE 9-14 Positive engagement starting system.

Starter Amperage Draw

1. Place the inductive pick-up clamp around either the positive or negative battery cable. Note: If the battery cable has a small wire attached at the post clamp, it should be placed inside the inductive pickup.
2. Make sure the engine will not start by either removing B+ from the coil (electronic ignition) or by pulling and grounding the coil wire at the distributor cap. Do not allow the ignition system to run open circuit.
3. Crank the car over just long enough to stabilize the amp meter needle; observe and record volt and amp meter.

Voltage ______________ (while cranking)
Amperage ______________ (while cranking)

_____ Good—Voltage above 9.6 V
_____ Amperage—200 A V8
175 A 6 cylinder
150 A 4 cylinder

NOTE: Amperage readings are approximate and will vary somewhat from one manufacturer to another.

Starter Circuit Voltage Drop

Ground Path

1. Use external voltage setting and the external voltmeter leads.
2. Place the positive voltmeter lead on the starter body (ground).
3. Place the negative voltmeter on the negative *post* of the battery.
4. Crank the car over.

Record Your Reading

_____ Less than 0.2 volt reading while cranking; continue with positive voltage drop test
_____ More than 0.2 volt reading indicates excessive resistance in the ground return path. Isolate each section of the return path and record the voltage drop with the voltmeter leads placed as follows:

- _____ Bat negative post to bat clamp
- _____ Bat clamp to cable ground lug
- _____ Ground lug to engine block
- _____ Engine block to starter motor

NOTE: The bad section of the circuit will have the voltage drop.

Positive Path Voltage Drop

1. Hook the negative voltmeter lead on the starter input terminal (after solenoid).
2. Hook the positive voltmeter lead to the positive battery post.
3. Crank the engine over and read the voltmeter. Record below.
   _____ Less than 0.2 volt while cranking reading; continue with charging system testing.
   _____ More than 0.2 volt reading indicates excessive resistance in the B+ path.

Isolate each section of the path and record below:

_____ V positive post to positive clamp
_____ V positive clamp to cable lug
_____ V solenoid input to solenoid output
_____ V solenoid output to cable lug (if solenoid is not mounted on starter)
_____ V cable lug to starter motor input

NOTE: The bad or excessive resistance part of the circuit will have the voltage drop.

FIGURE 9-15 Starter on car test.

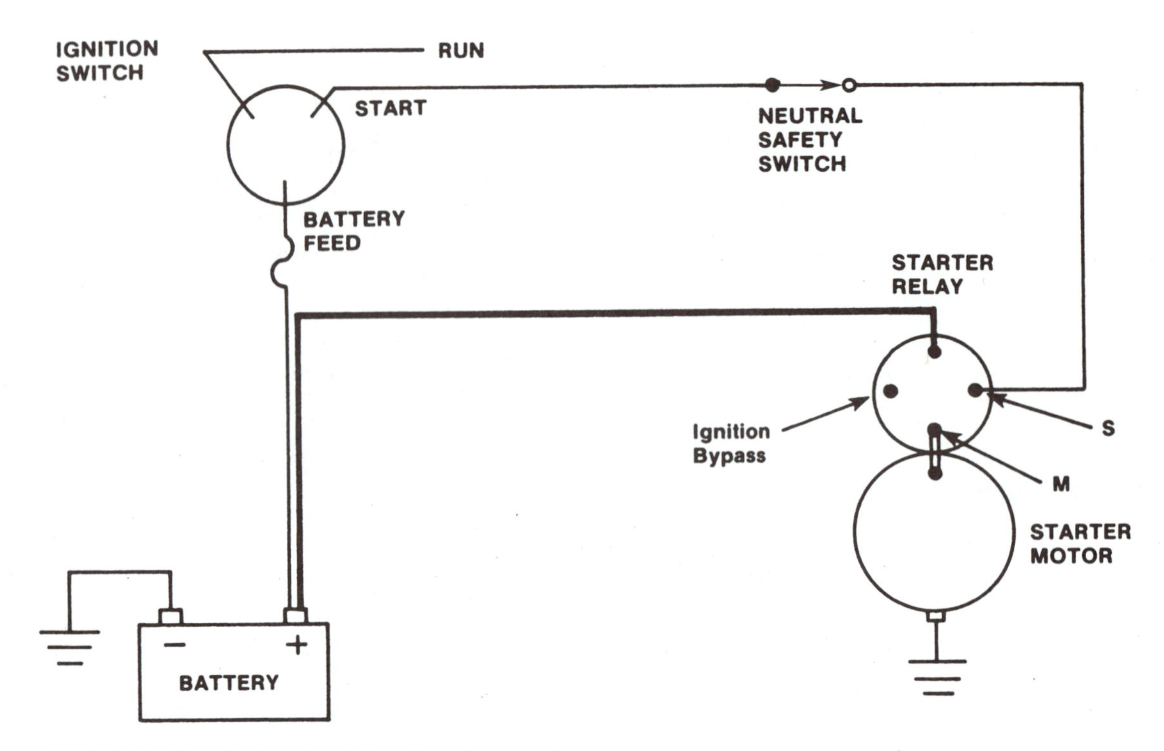

FIGURE 9-16 Solenoid shift with relay starting system.

tween B+ from the ignition switch and ground of the test light. In our testing of a no-crank condition, we are typically looking for B+ and a glowing or NSS side of the relay where we are looking for an off test light. Off will signify ground, which is desired, rather than off signifying open as in the rest of the circuits. This can be confusing, so do not hesitate to reread this section again. Also note that an open relay coil will give the same results, an off test light. To verify that the coil is or is not open, put the gear shift lever in drive or reverse. The light should glow. If it does not, the relay coil is open. This assumes that B+ is being delivered to the input of the relay coil from the ignition switch. Look at our next example of a no-crank. Use the test light indicators to diagnose the problem in Figure 9-18.

Hopefully, you have realized that we have an open NSS. The glowing test light on the NSS side of the relay coil indicates an open circuit. How will you verify the condition? Jumper the relay output to ground. The engine should crank. If it does, you have verified the problem. Replace the NSS.

CAUTION: Never allow a vehicle to leave your shop with a neutral safety switch or clutch switch jumpered. You could be held liable if the vehicle starts in gear and injures someone or does property damage!

Notice also that in each example we have attempted to verify the open by jumpering around the suspected open section. Whenever possible, verify in this manner before ordering the replacement part. It frequently will save embarrassment and ensure the accuracy of the diagnosis. Once the engine will crank, perform the voltage drop testing outlined in preventative maintenance testing to prevent further difficulties.

STARTER REPAIR

As noted in the early section of this chapter, starter repair or rebuilding is not always done in the trade. Various conditions, not the least of which is wages, dictate whether a starter will be rebuilt by the technician or a starter will be purchased and installed into the cus-

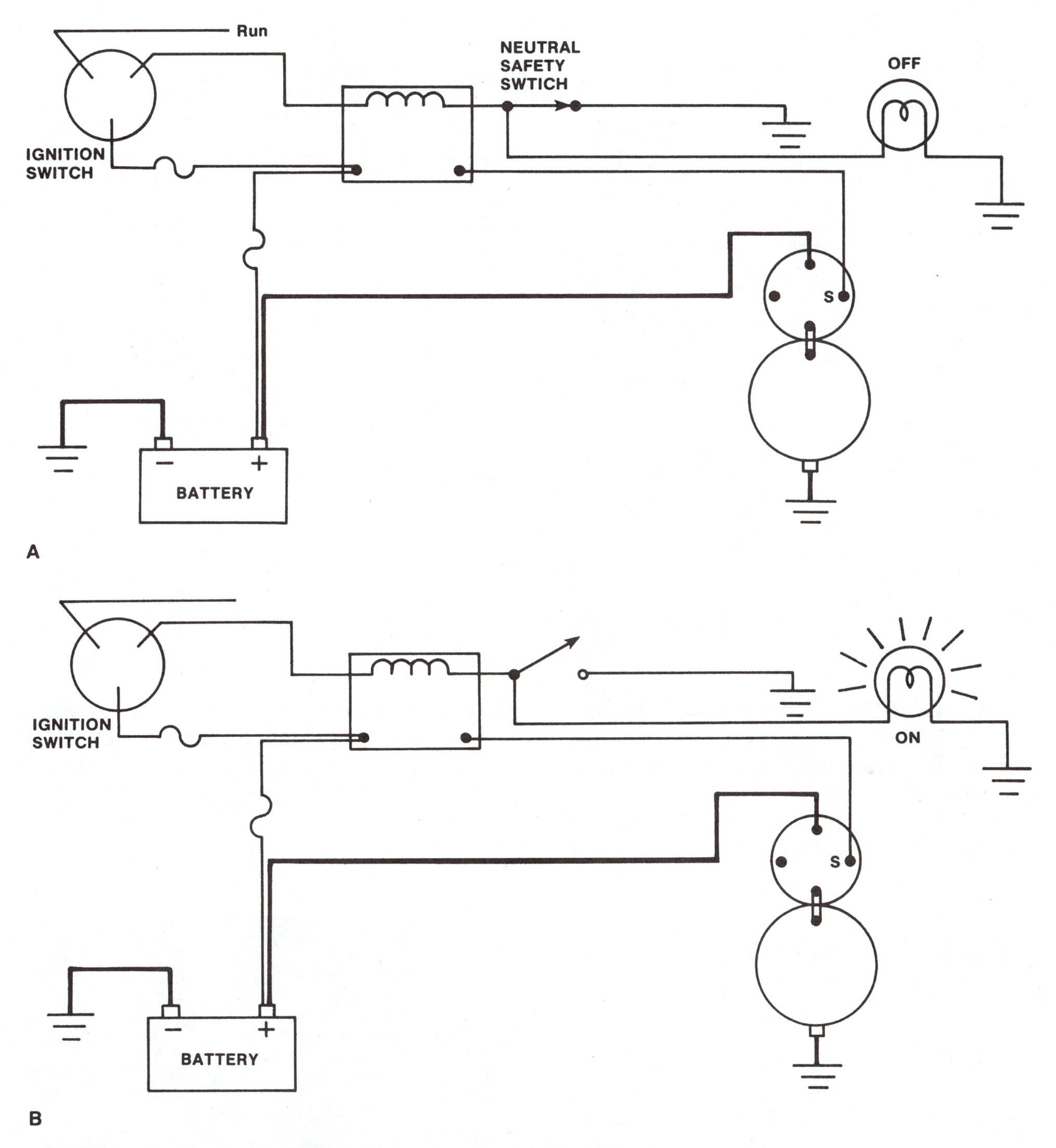

FIGURE 9-17 Testing an NSS that is on the ground side of a relay.

tomers's vehicle. As the hourly wage of the technician goes up, it becomes less advantageous to attempt to rebuild existing starters. We will, however, cover some of the operations of rebuilding even though local technicians might not attempt them.

Once the starter has been removed from the vehicle, clean the case, disassemble com-

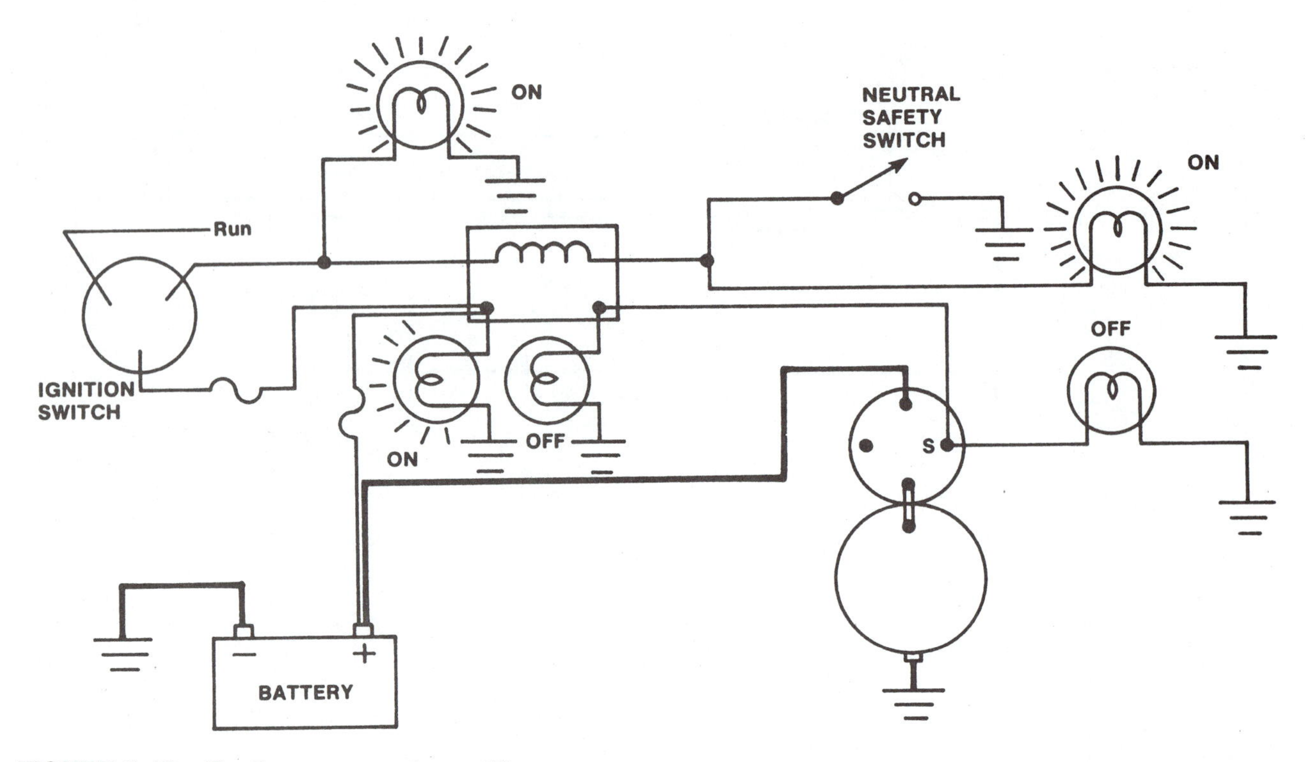

FIGURE 9-18 Testing a no-crank condition.

pletely, and follow the procedure outlined here to electrically test both the armature and case. We would not want to attempt to rebuild a unit that has electrical shorts or opens in either the case or armature. The testing procedure should identify a rebuildable unit.

IV. Procedure for Testing an Armature

Use either a self-powered test light or an ohmmeter. Remember, if using an ohmmeter, use the × 1 scale when looking for continuity and the × 1,000 scale when checking insulation.

1. Place the test probes on any two commutator bars. Continuity should be present (0 ohms or a test light on), Figure 9-19.
2. Place one test probe on the commutator and the other probe on the armature shaft. An open circuit should be present (infinity or no test light), Figure 9-20.
3. Place one test probe on the motor input and the other on the insulated brushes. Continuity should be present (0 ohms or a test light on), Figure 9-21.

FIGURE 9-19 Testing for continuity across a commutator.

FIGURE 9-20 Testing for a grounded armature.

FIGURE 9-22 Testing for a grounded field coil.

4. Place one test probe on the motor input and the other one on the case. An open circuit should be present (infinity or no test light), Figure 9-22.

Figure 9-23 shows the use of a growler to test both the armature and the case. **Growlers** are so named because they produce a very strong magnetic field that produces a buzz or growl when testing. They are used almost exclusively by professional rebuilders and, thus, are not too frequently encountered.

Now that the case and armature have been tested, let us start with brush installation. There are two different styles of brushes currently in use: soldered in place or screwed in place. When either of the two styles get too short, they do not make adequate contact with the commutator and have difficulty passing current. Notice Figure 9-24. This starter needs to have the brushes replaced. On this style, the replacement is very simple. Remove the four screws that attach the brushes to the holders and install the new brushes in place of the old, Figure 9-25. Figure 9-26 shows brushes that are being soldered in place. Generally, the old brushes are removed by cutting away the attachment wires, crimping the new connector over the old connection, and soldering them in place as in Figures 9-27 through 9-29. Note the large soldering iron. A 40-watt or larger iron is the

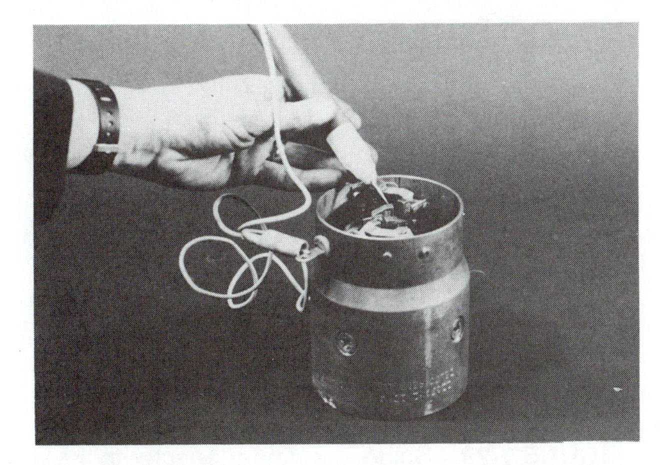

FIGURE 9-21 Testing the B+ path for continuity.

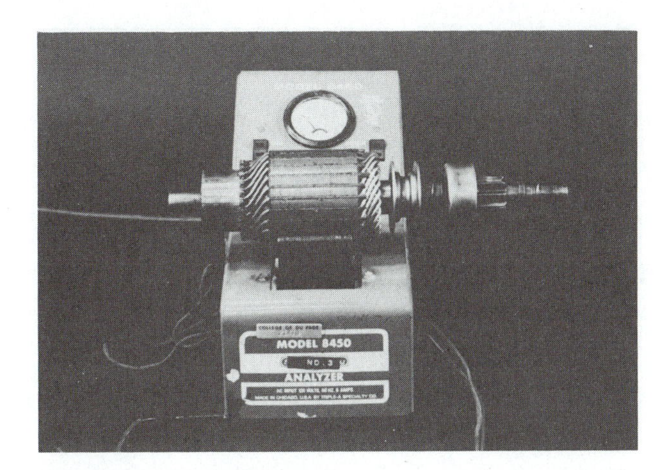

FIGURE 9-23 Growler.

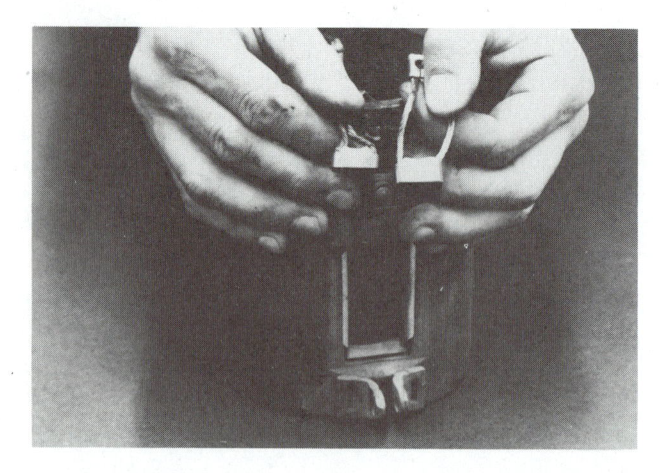

FIGURE 9-24 Worn brush compared to a new one.

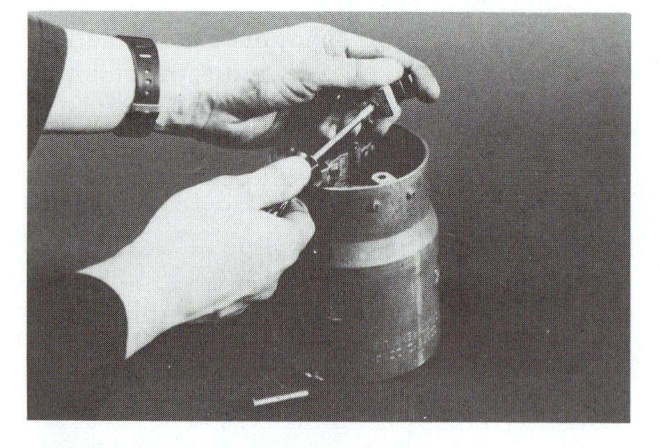

FIGURE 9-25 Replacing brushes.

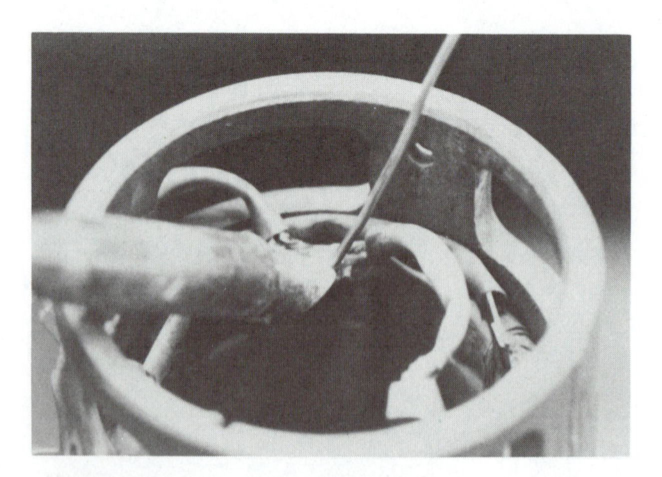

FIGURE 9-26 Soldering brushes.

FIGURE 9-27 Cut off old brushes and wires.

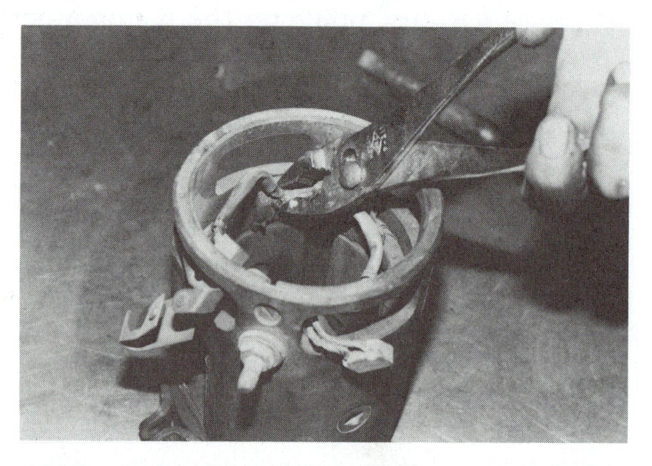

FIGURE 9-28 Crimp the new brush connector over the old.

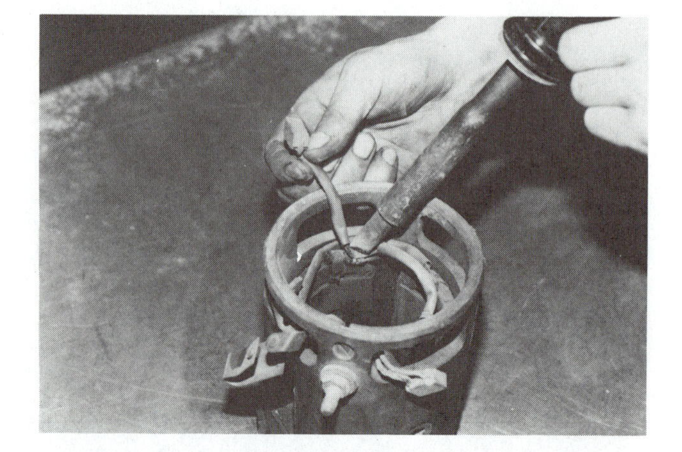

FIGURE 9-29 Solder the new connector to the old one.

minimum required. The use of rosin core solder is also required.

The new brushes should have a nice reconditioned surface to ride on . A commutator that is worn or partially burned can be cleaned up with an armature lathe and the insulation between the commutator bars can be undercut. Crocus cloth can also be used to clean up the commutator surface if it is not too badly worn. The armature is supported on each end by bearings that are also replaceable. They are, generally, pressed out with either an arbor specifically made for the armature or the new bearing as in Figure 9-30. Worn bearings will allow the armature to move toward the field coils. A very worn bearing might even allow the armature to contact the coils.

The next section of the starter we will replace is the starer drive mechanism. This is probably the most replaced internal component. It has a very high failure rate and will generally not last the life of the actual starting motor. Most noise that comes from the starter can be isolated to the drive mechanism. We discussed just how the overrunning clutch type operated in the last chapter. Replacement is the only cure available here. The units found on passenger vehicles are not internally rebuildable. Their replacement is outlined here with illustrations to guide you through, Figures 9-31 through 9-36.

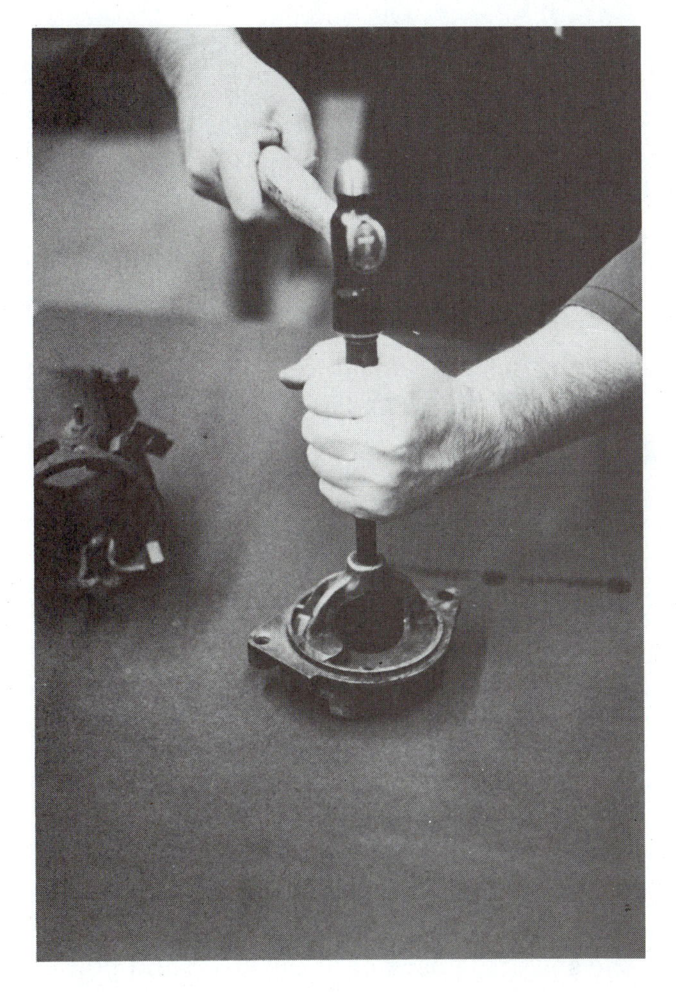

FIGURE 9-30 Pressing out a starter from bushing.

FIGURE 9-31 Knock the lock collar down.

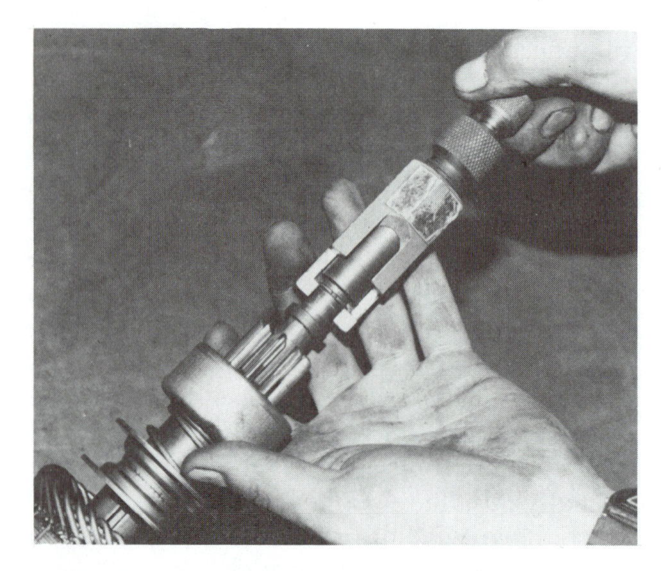

FIGURE 9-32 Pull the split ring off of the armature shaft.

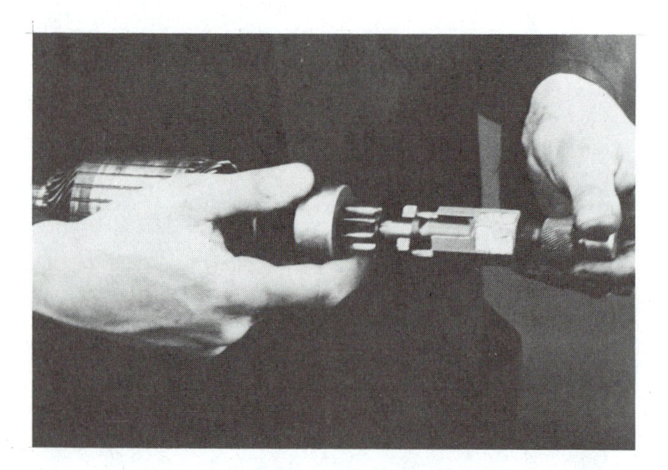

FIGURE 9-33 Remove the starter drive.

FIGURE 9-34 New starter drive and lock collar.

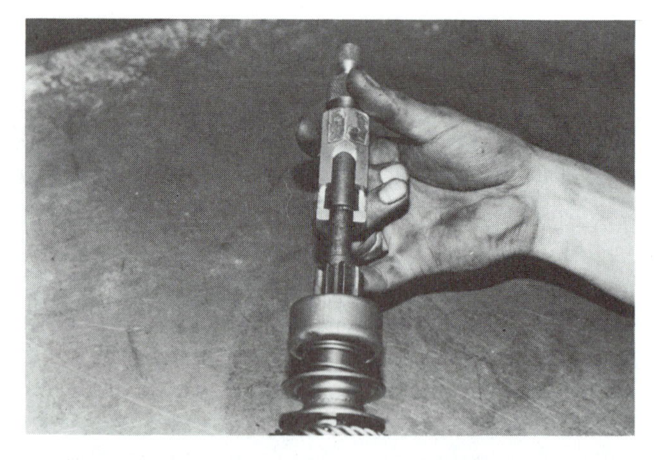

FIGURE 9-35 Slide snap ring down and into the groove.

FIGURE 9-36 Pull the back collar over the split ring.

V. Procedure for Replacing the Starter Drive

Remove the snap ring by first knocking down the collar with a socket or the special tool shown. Remove the snap ring, being careful not to spread it too far or distort it. It will be reused. Slide off the old overrunning clutch. Replace with a new one after first ensuring that the number of teeth, shape, and size are the same on both the old and new one. Push the snap ring back onto the shaft and pull up the retaining collar. This completes the replacement of the starter drive. This procedure is applicable on many different brands of starters.

A few manufacturers use a gear reduction starter. The procedure for the drive removal is outlined in Figure 9-37 through 9-43.

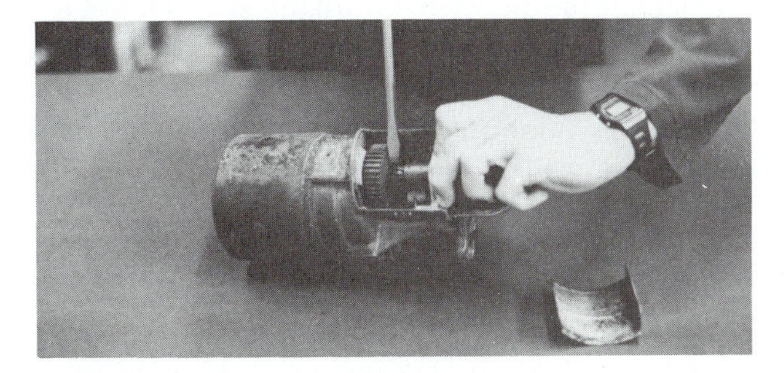

FIGURE 9-37 Pry off the dust cover and remove the driven gear snap ring.

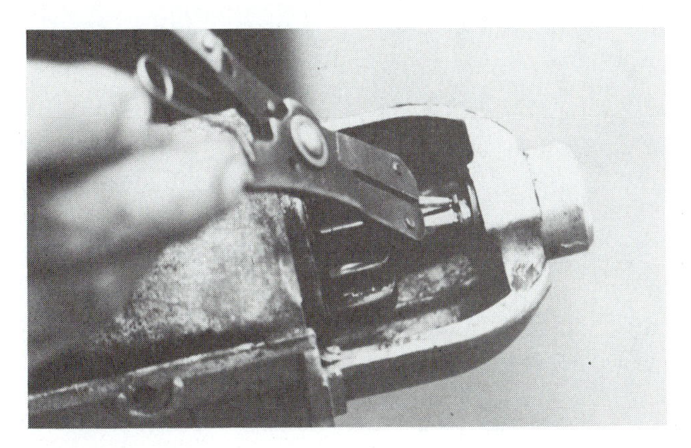

FIGURE 9-38 Remove the C-clip on the end of the pinion shaft.

FIGURE 9-39 Drive the pinion shaft toward the rear of the housing.

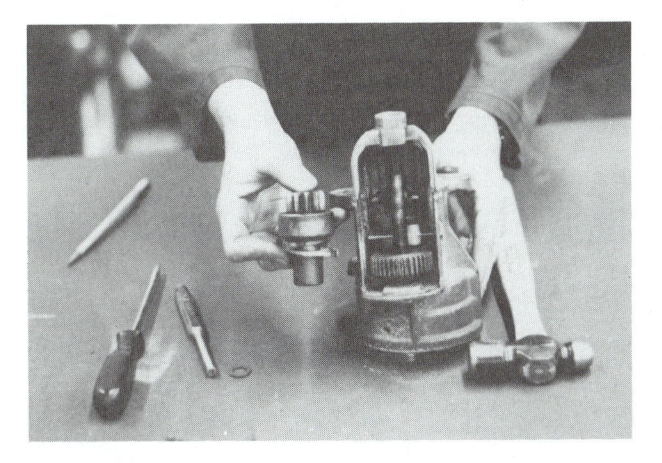

FIGURE 9-40 Remove the pinion.

FIGURE 9-41 Replace the pinion.

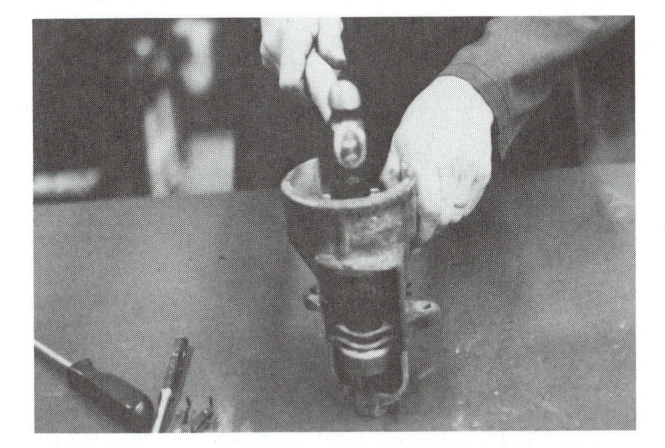

FIGURE 9-42 Drive the shaft through the drive pinion.

A

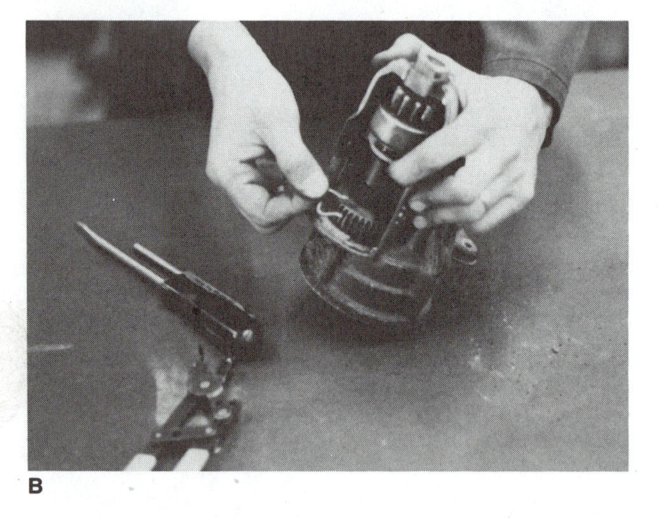

B

FIGURE 9-43 Replace the (A) snap ring and (B) C-clip.

Solenoids are also considered by most technicians to be a replacement item. Figures 9-44 through 9-47 show the replacement procedure for a solenoid shift starter motor commonly used. First, remove the screw that attaches the motor connection to the solenoid output. By loosening the two mounting screws, the solenoid assembly can be turned and released from the housing. The new assembly is replaced by pushing the plunger into the solenoid core with the return spring in place. Once in place, it is turned to lock it down. The two screws are replaced and the motor input reconnected to the solenoid output.

FIGURE 9-44 Unbolt the M terminal connector bolt.

VI. Procedure for Removing Drive Mechanism

Remove the dust shield to expose the drive mechanism. Separate the starter motor from the nose assembly. Remove the snap rings that hold the pinion assembly. Drive the pinion gear toward the back of the motor and remove the shaft. Remove the starter drive and discard. Install the replacement drive into the housing and drive the shaft into place. Replace the C-clip and snap ring.

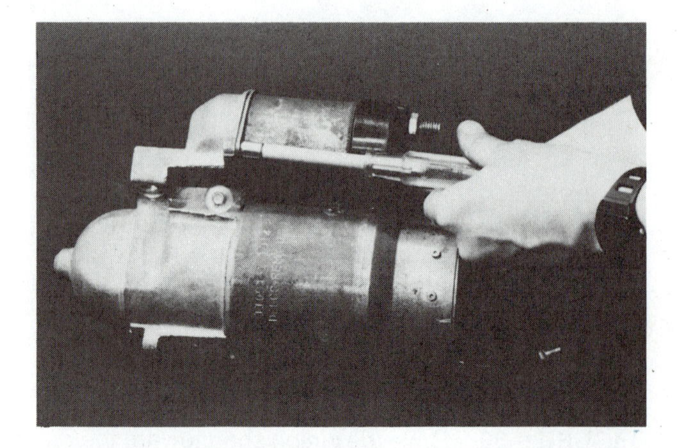

FIGURE 9-45 Unbolt solenoid hold-down screws.

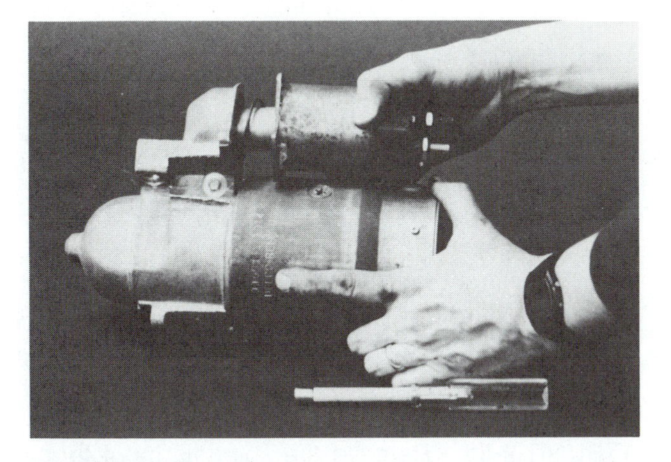

FIGURE 9-46 Turn the solenoid to unlock it.

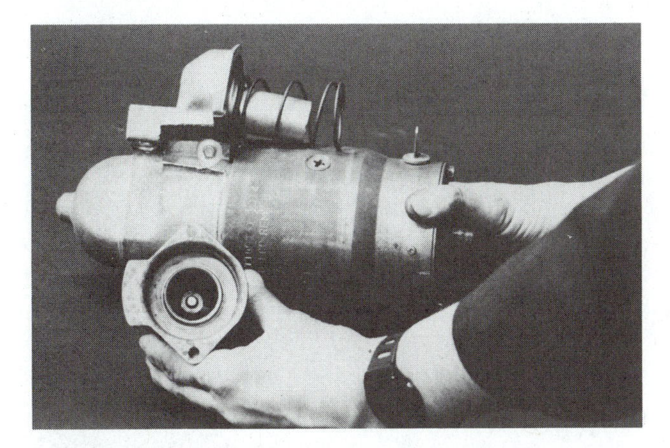

FIGURE 9-47 Remove the solenoid.

A gear reduction starter such as the Chrysler model will require that the starter be completely disassembled as follows in Figures 9-48 through 9-62. Remove the through bolts. Pull the end frame apart enough to expose the screw that attaches the field. Remove the screw and separate the two housing halves. Remove the nuts and washers from the solenoid input terminals. Unsolder the pull-in coil lead from the brush terminal and unwind the wire. Remove the brush holder and solenoid contact disc. Remove the solenoid winding and battery terminal from the brush holder. To reassemble, install the return spring into the solenoid core. Install the solenoid plunger and core into the brush holder. Wrap the pull-in coil wire around the ground brush holder and solder. The brushes are held against spring tension by the thrust washer. Attach the field coil screw. Place the armature into the housing. Replace the end plate and the two through bolts.

Most starter motors are disassembled and reassembled with one of the two procedures shown.

Once the motor has had any bad components replaced and is ready for reinstallation, a pinion contact or depth check should be made. It is also a good idea to check pinion contact if a rebuilt or new starter is going to be installed in the vehicle. Simply stated, pinion depth ensures that the gear on the flywheel and the pinion will match. If the pinion to flywheel clearance is too large, the starter might be noisy and could do damage to itself or the flywheel ring gear teeth. If there is too little clearance, the starter drive might stick or bind. In addition, amperage draw would most likely be increased forcing the entire starting system to work harder. Figure 9-63 shows pinion depth being measured on a starter that has the solenoid removed. The shims will be placed between the starter body and the engine to increase clearance. Many GM starters and a few foreign starters use shims to control pinion depth.

If in the diagnosis of starting difficulties, you find a broken tooth on the ring gear or a badly worn section, it should be replaced. Keep in mind that correct repair procedure will include a new starer drive mechanism to go along with the new flywheel.

FIGURE 9-48 Remove the through bolts.

FIGURE 9-49 Pull the end frame apart.

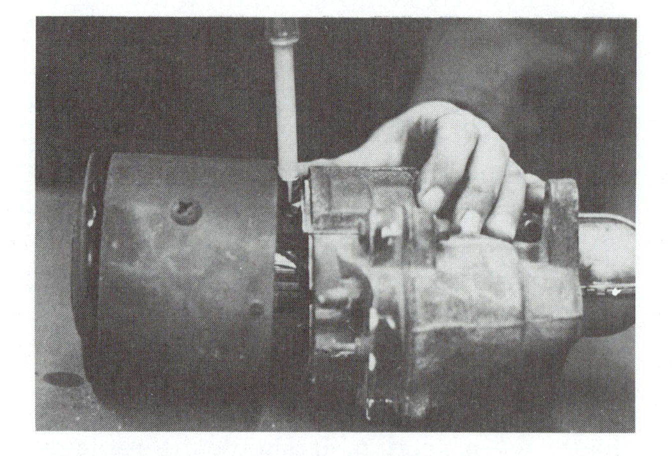

FIGURE 9-50 Remove the field screws.

FIGURE 9-51 Separate the housing halves.

FIGURE 9-52 Remove the solenoid input terminals.

FIGURE 9-53 Unsolder the pull-in coil lead.

FIGURE 9-54 Remove the brush holder and contact disc.

FIGURE 9-55 Remove the solenoid winding.

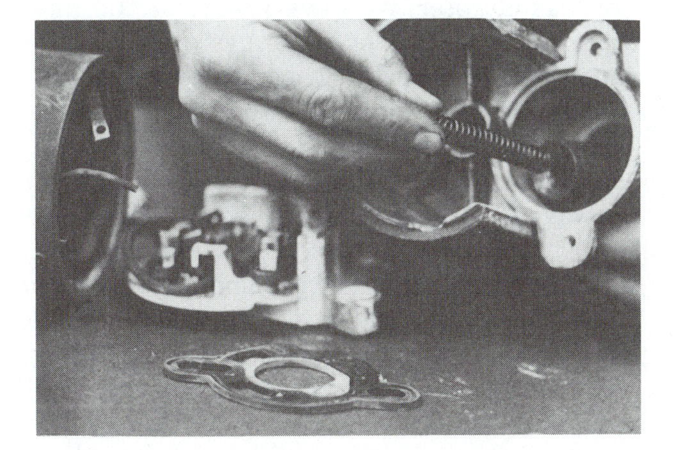

FIGURE 9-56 Install the return spring.

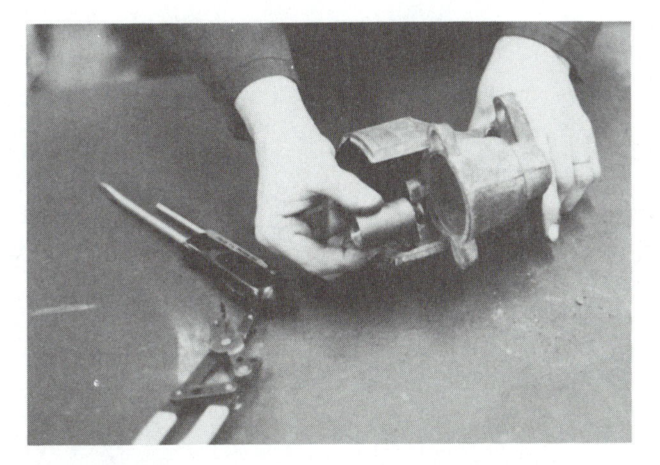

FIGURE 9-57 Install the plunger and core.

FIGURE 9-58 Wrap the pull-in coil wire and solder it.

FIGURE 9-59 Hold the brushes against their springs with the thrust washer.

FIGURE 9-60 Attach the field coil to the brush assembly.

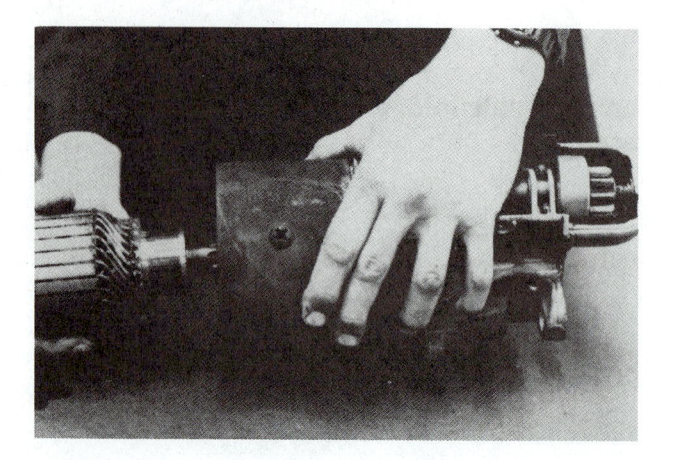

FIGURE 9-61 Place the armature into the end frame.

FIGURE 9-62 Replace the two through bolts.

FIGURE 9-63 Check pinion depth.

SUMMARY

In this chapter, we have looked at the most common types of starting system diagnosis and repair. Included was step-by-step procedures for locating excessive starting resistance in either the B+ or ground side of the circuit. In addition, a no-start procedure for locating a faulty component was outlined. Repair procedures for replacing brushes, starter drives, and solenoids were shown, as well as determining pinion depth on a shim style starter. Battery testing was reviewed and emphasized as being the first place to start in all starting difficulties.

KEY TERMS

Control circuits

Growler

CHAPTER CHECK-UP

Multiple Choice

1. Technician A states that slowing down the starter armature will result in increased amperage draw. Technician B states that speeding up the armature will result in decreased amperage draw. Who is correct?
   a. Technician A only
   b. Technician B only
   c. Both Technician A and B
   d. Neither Technician A nor B

2. A 0.7 volt drop is observed across the starter solenoid with the engine cranking. What should be done to the system?

   a. nothing—it is acceptable
   b. install a larger battery
   c. replace the cable or component between the voltmeter leads
   d. none of the above

3. If an engine requires 2,000 watts to crank, how many amps will the system draw if the battery voltage is 5 volts?

   a. 400 A
   b. 1,000 A
   c. 200 A
   d. not enough information is given

4. A starter is very noisy. Technician A states that the starter drive is probably the cause of the excessive noise. Technician B states that excessive current draw could cause the excessive noise. Who is correct?

   a. Technician A only
   b. Technician B only
   c. Both Technician A and B
   d. Neither Technician A nor B

5. The starter pinion to flywheel clearance is measured and found to be too small. What problem is likely to occur?

   a. decreased amperage draw
   b. increased amperage draw
   c. decreased voltage drops

6. An open NSS will be categorized by

   a. input test light on, output test light on.
   b. input test light off, output test light off.
   c. input test light off, output test light on.
   d. input test light on, output test light off.

7. Technician A states that brushes should be soldered in with rosin core solder. Technician B states that brushes should be replaced if they are worn to less than half their original length. Who is correct?

   a. Technician A only
   b. Technician B only
   c. Both Technician A and B
   d. Neither Technician A nor B

8. A starter amperage draw test is done and results in less than specification cranking speed and lower than normal amperage. What should be done next?

   a. remove the starter—it is bad
   b. run voltage drop testing of B+ and ground
   c. test the battery
   d. test for tight engine

9. The usual specification for voltage drop across the B+ circuit up to the starter is
   a. 0.1 volt.
   b. 0.2 volt.
   c. 0.3 volt.
   d. 0.4 volt.

10. A voltage drop of 1.8 volts is found and corrected in the starter solenoid. Technician A states that this will result in lower cranking amperage. Technician B states that this will result in decreased cranking speed. Who is correct?
    a. Technician A only
    b. Technician B only
    c. Both Technician A and B
    d. Neither Technician A nor B

True or False

11. Slower than normal cranking speed will result in lower than normal amperage.
12. Most noise from starters can be traced to starter drives.
13. Excessive voltage drops will increase starting amperage draw.
14. Commutators can be cleaned with crocus cloth if they are not too badly scored.
15. Slower than normal speed and lower than normal cranking amperage is the result of starter internal problems.
16. A tighter than normal engine will decrease the amperage draw of the system.
17. A ground NSS (Chrysler) has its input lead hot (test light is on) when it is functioning correctly.
18. A self-powered test light should not light when its probes are between the commutator and the armature shaft.
19. A weak battery will increase the amperage draw of the starter.
20. Continuity should be present between the case input and the case ground of a solenoid shift starter.

Charging Systems

OBJECTIVES

At the conclusion of this chapter, you should be able to:

- identify the major components of the charging system.
- identify the major alternator components and be able to explain their function.
- recognize the four styles of regulation currently being used by the automotive industry.
- relate regulator resistance to field current and relate field current to alternator output.
- identify the two different styles of stators currently in use and note the differences.
- identify and detail the operation of the vehicle indicator lamp, the ammeter, and the voltmeter.
- identify and block diagram an A circuit, B circuit, and an isolated A or B circuit.

INTRODUCTION

Now that we have the vehicle started with the battery and starting system, we must restore the battery to a full state of charge and get ready for the next start up. In addition, any accessories that are being used by the vehicle owner must have current available to them. These two functions, keeping the battery fully charged and supplying current to keep the vehicle running, will be the job of the charging system. In this chapter, we will look at how this system functions effectively. We will also look at just how we regulate or control the charging system. In the next chapter, we will look at servicing the charging system on the vehicle.

CHARGING SYSTEM OVERVIEW

The charging system has evolved right along with the rest of the electrical system. Starting with big bulky generators that generally could not effectively handle the high current demands of the modern vehicle and ending with computer regulation of the charging system of today, this evolution has kept this system in pace with the rest of technology. We will not cover generators within this text as their use has gradually diminished to the point where you will probably never encounter one in the trade. The purpose of the charging system will be to convert the mechanical motion of the engine to electrical power to recharge the battery and run the accessories. The same principles were in use in generators as in the modern charging system. This principle states that a voltage will be produced if motion between a conductor and a magnetic field occurs. This principle is exhibited in Figure 10-1. The amount of voltage developed will be dependent on the strength of the field and the speed of the motion. With this basic principle in mind, let us look at the modern charging system beginning with the alternator.

The charging system can be divided into three major components: the battery, the alternator, and the regulator. We have spent sufficient time discussing the battery, so we will not say too much about it here. We will, however, restate that the battery can be responsible for many charging system problems just as it can be for starting difficulties. The charging system cannot function correctly if the battery will not hold a charge or has limited capacity. In the next chapter, we will again emphasize to test the battery first! Our discussion of the rest of the system within this chapter will assume that a correct-capacity, well-functioning battery is in the vehicle.

THE ALTERNATOR

Let us look at the **alternator**. It is the device that turns the mechanical motion of the engine into the electrical power needed, Figure 10-2. Basically, it consists of three major components: the **rotor**, the **stator**, and the **rectifier** bridge housed in two end frames. As we look at each component, keep in mind that the basic principle of electrical generation is motion between magnetic lines of force and conductors. On alternators, the magnetic lines of force will be supplied by the rotor that will be spun by the engine.

Rotors

The object of the rotor will be to supply a spinning magnetic field. Examine Figure 10-3. It shows a rotor assembly that has a cooling fan and a pulley attached. The rotor is actually a coil of wire sandwiched between metal rotor halves. When current is passed through this coil of wire, a magnetic field will be produced. The strength of this field will be dependent on the amount of current flowing through it. This current is called **field current** and by controlling its amount, we will control the output of

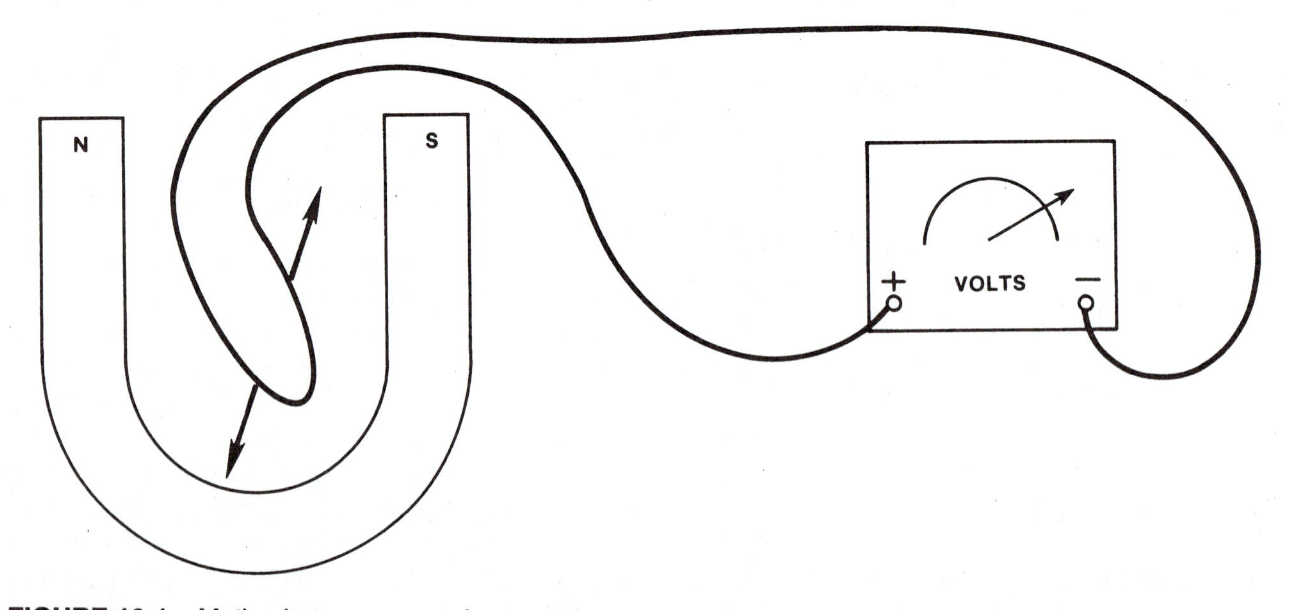

FIGURE 10-1 Motion between a conductor and a magnetic field produces a voltage.

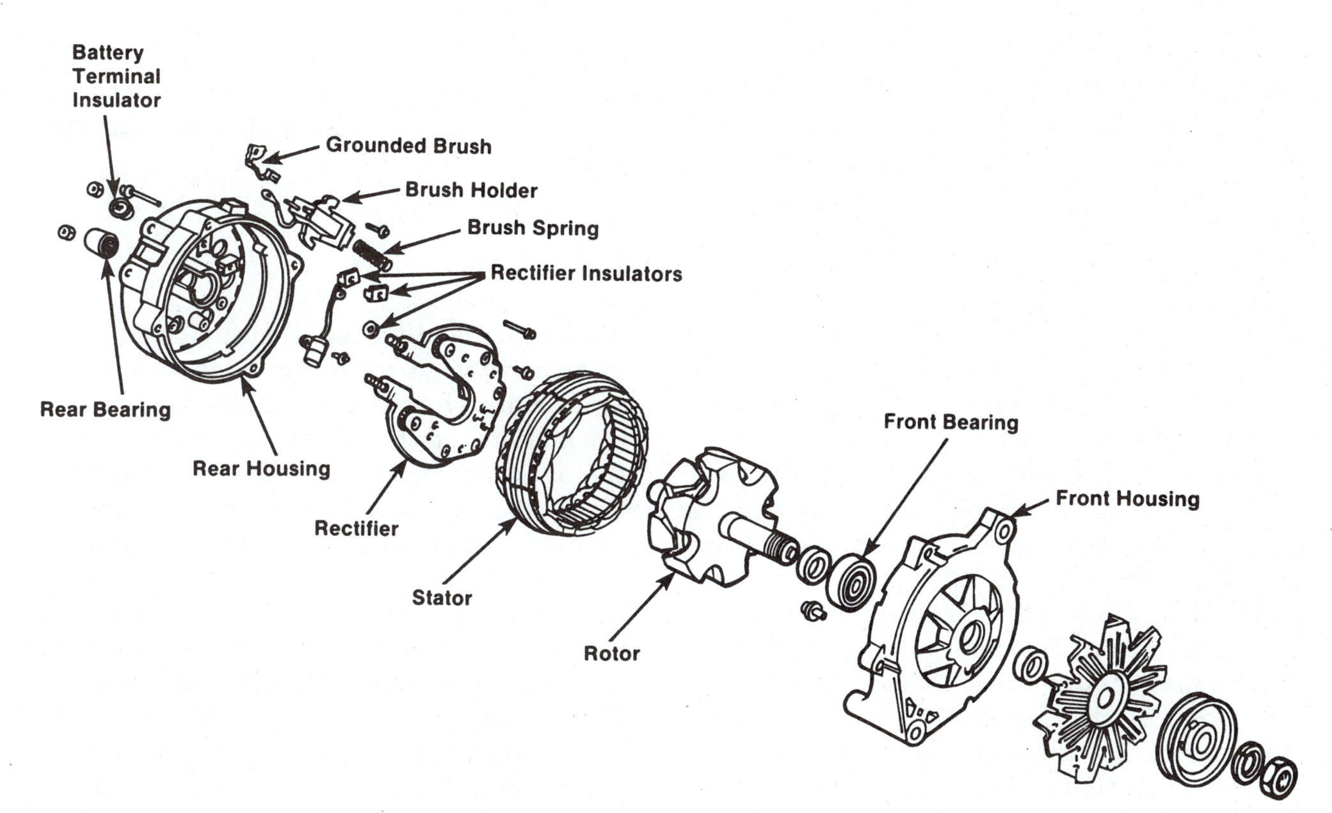

FIGURE 10-2 Exploded view of an alternator. (Courtesy of Ford Motor Company)

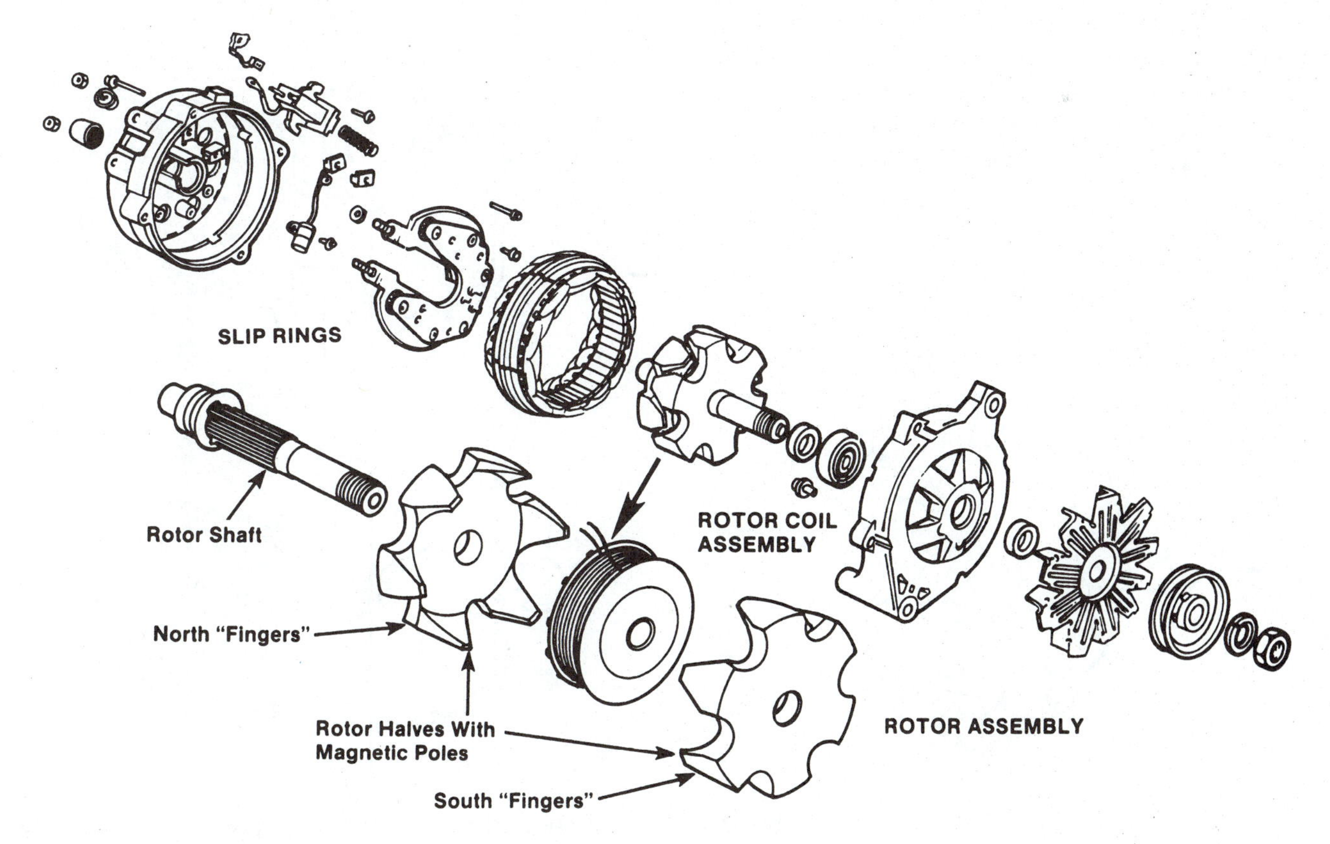

FIGURE 10-3 Rotor coil and case assembly. (Courtesy of Ford Motor Company)

the alternator. If a large amount of field current is flowing through the rotor, alternator output will be high. If a small amount of field current is flowing, alternator output will be low. It should also be realized now that many charging systems do not vary the amount of field current but instead vary the length of time that the field current flows. Full field current flowing for half the time will result in the same output as a constant half field current. The end result is the same: the control of alternator output to the level required.

Notice that the rotor coil is between two sets of interlacing fingers that will take on the polarity (north or south) of the side of the coil that they touch. In this way, we will spin a magnetic field that will first be north and then south polarity. Have you wondered how we can bring current into something that is spinning? Notice Figure 10-4. It shows how the wires from the coil are attached to two smooth rings that are insulated from the rotor shaft. These are called slip rings and their function is similar to an armature's commutator. The stationary brush passes field current into a slip ring. From here, it flows through the field coil, back to the other slip ring, and then out the stationary brush. This gives us a method of getting an adjustable or controllable magnetic field spinning. The rotor is supported by bearings on each end, which will allow it to spin at speeds in excess of 15,000 rpm.

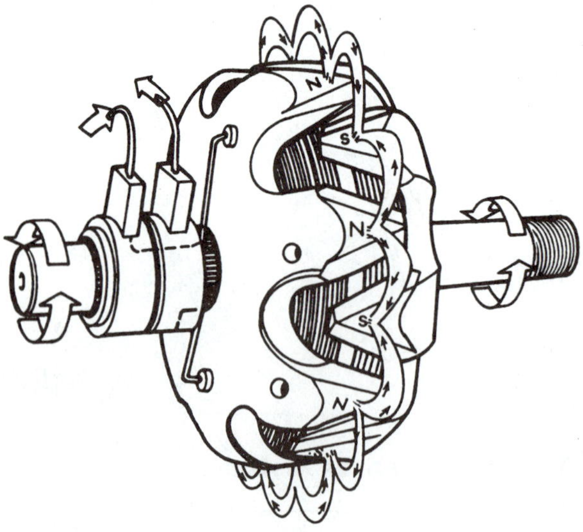

The Stator

Remember, generation of electricity involves motion between magnetic lines of force and conductors. Our rotor was able to supply both the motion and the variable magnetic field, so the only components left are the conductors. Remember the wire moving through the permanent magnet of Figure 10-1. The wire was having its valence electrons pressurized by the lines of force. Figure 10-5 shows the simplest one-wire stator. Notice that the loop of wire has its two ends attached to a load (A and B on the illustration). This setup shows that if a conductor is energized by a north polarity magnetic field, the electrons will be pressurized in one direction and if they are energized by a south polarity field, they will be pressur-

FIGURE 10-4 Current flows through the brushes and slip rings. (Courtesy of Delco Remy Division)

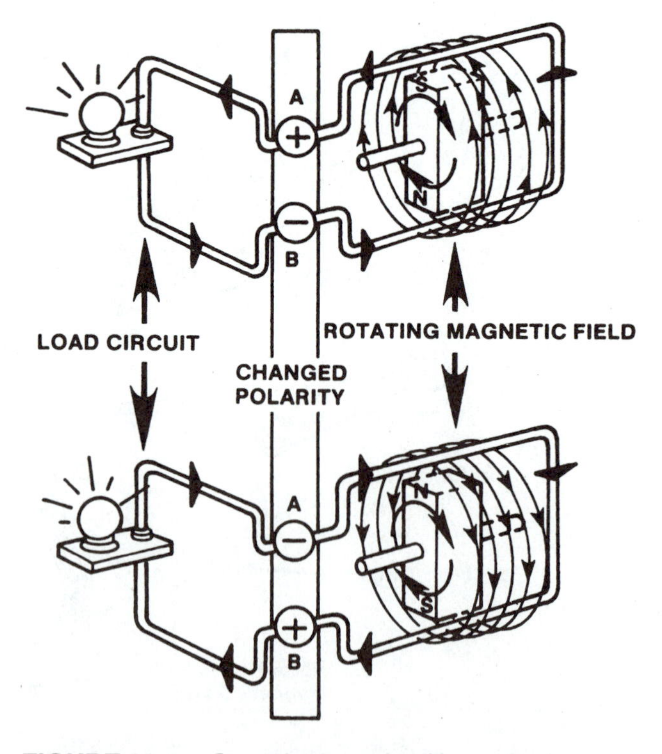

FIGURE 10-5 One wire energized by a rotor's magnetic field. (Courtesy of Delco Remy Division)

ized in the opposite direction. Electrons (or current) that move in one direction, stop, and then move in the opposite direction are called **alternating current**, or AC. Remember that the rotor is spinning with interlacing fingers of north and south polarity. This will generate AC in our simple one-loop stator.

The amount of current generated in one single loop of wire is too small to be practical. Many vehicles today require over 50 amps of current to maintain the battery charge and run the required accessories. To be able to generate this much current, the manufacturers wind lots of wire into an iron-laminated core or frame. Most alternators use three windings to generate the required amperage. They are placed in slightly different positions so their electrical pulses will be staggered and are wired in either a delta configuration or a wye configuration as seen in Figure 10-6. The **delta winding** received its name because its shape resembles the greek letter delta, Δ. The **wye winding** receives its name because its drawing resembles the letter "Y". Notice the difference in the windings. Alternators will use one or the other. Usually, lower output alternators use wye wound, while higher output alternators use delta. The reason for this will be explained in the diode section of this chapter.

The rotor fits inside the stator as seen in Figure 10-7 with a small air gap. This allows the magnetic field of the rotor to energize all of the stator winding at the same time so the generation of AC can be quite high if needed.

Before we move on to diodes, let us look at AC and examine something called the sine wave. Alternating current produces first a positive pulse and then a negative pulse. This can be represented on an oscilloscope, Figure 10-8. Notice that the complete wave forms start at

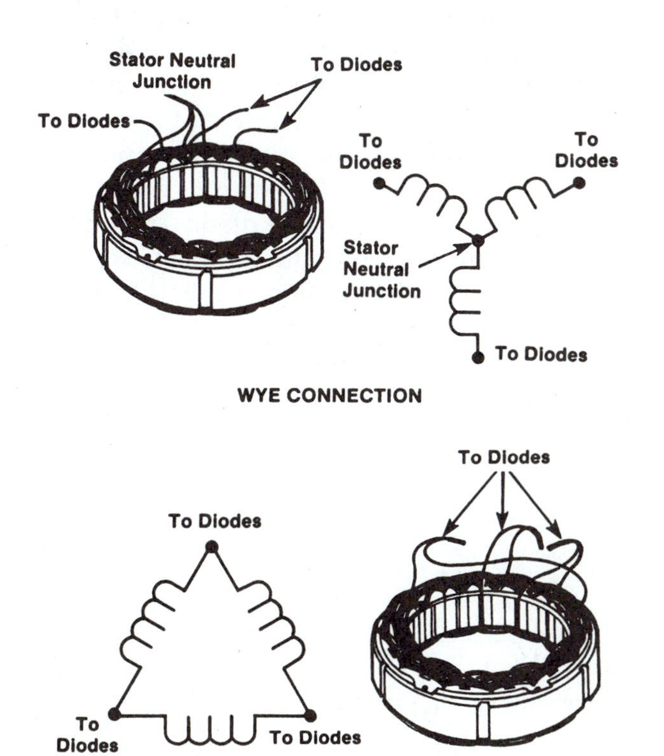

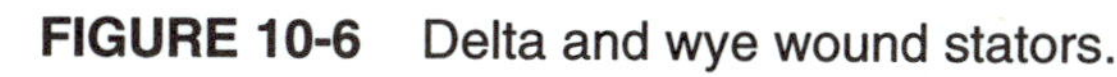
FIGURE 10-6 Delta and wye wound stators.

FIGURE 10-7 A small air gap exists between the spinning rotor and the stationary stator.

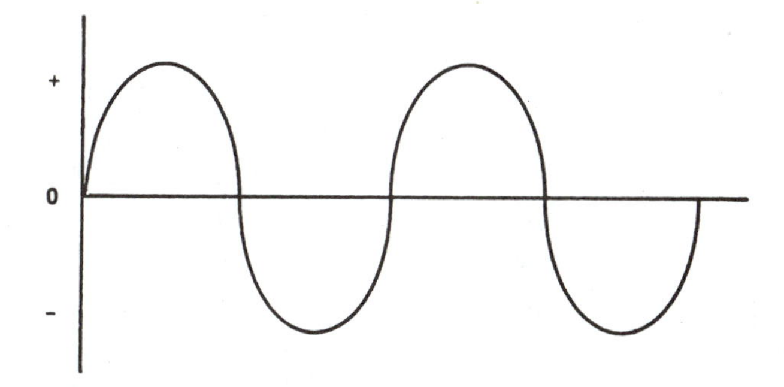

FIGURE 10-8 Sine wave from a stator.

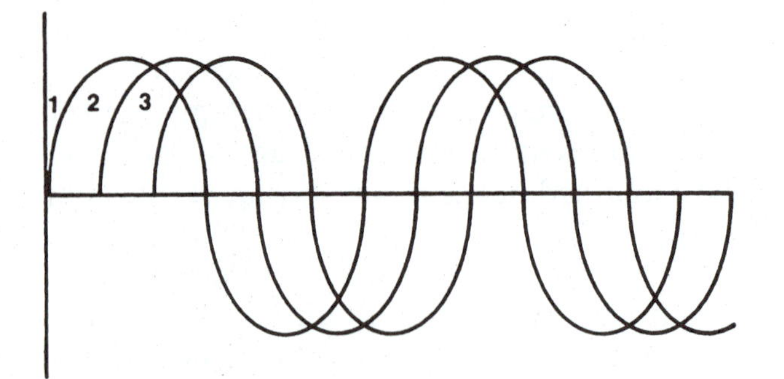

FIGURE 10-9 Three-phase sine wave.

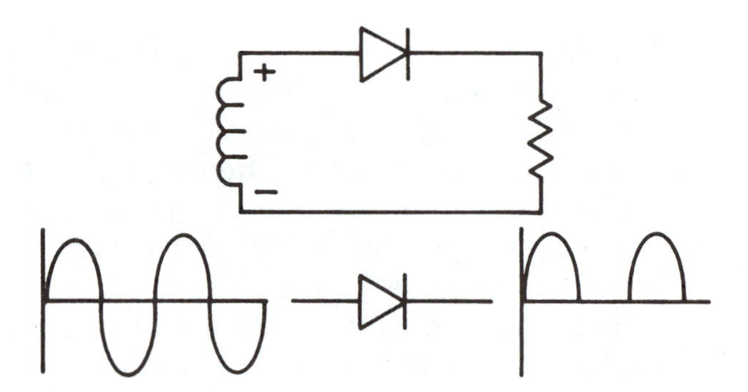

FIGURE 10-10 Half wave rectification diode positively biased.

zero, go positive, then drop to zero again before turning negative. The angle and polarity of the field coil fingers are what cause this sine wave in the stator. When the north pole magnetic field cuts across the stator wire, it generates a positive voltage within the wire. When the south polarity magnetic field cuts across the stator wire, a negative voltage is induced in the wire. A single loop of wire energized by a single north then a south results in a single-phase voltage. Remember that there are three stator windings overlapped. This will produce the overlapping sine wave shown in Figure 10-9. This voltage, since it was produced by three windings, will be called **three-phase voltage**. An important point to remember is that AC voltage and current comes off the stator. AC will not charge a DC battery, however. It will have to be changed or rectified into DC before we can efficiently use it.

The Diode Rectifier Bridge

Let us review where we are right now. Field current is flowing through the rotor generating a magnetic field. The engine is turning the rotor with a fan belt, thus giving us a spinning magnetic field. The magnetic lines of force are cutting across the windings of the stator, generating AC in its windings. If this brief overview does not make sense, do not hesitate to review the last few pages covering the generation of AC.

As we mentioned before, AC will not charge a battery because it is a DC source of current. To charge it, we need a direct current source at higher than battery voltage level. The AC coming out of the stator will have to be rectified by a set of diodes.

We discussed diodes in Chapter 5 and found that they were an electrical one-way check valve that would allow current to flow through them only in one direction. When the current reverses itself as it does in AC, the diode blocks and no current flows. With this in mind, look at Figure 10-10. It shows that if we take AC and run it through a diode, the diode will block off the negative pulse and produce the scope pattern shown. What would the pattern look like if the diode was reversed? Hopefully, you were able to figure out that it would look exactly the same, except that it would be blocked during the positive pulse and allow the negative pulse to flow through it, Figure 10-11. Because the single diode was able to block only half of the pulses (either the positive or the negative) and was able to pass half of the pulses, we would call this half-wave rectification. We have taken the AC and changed

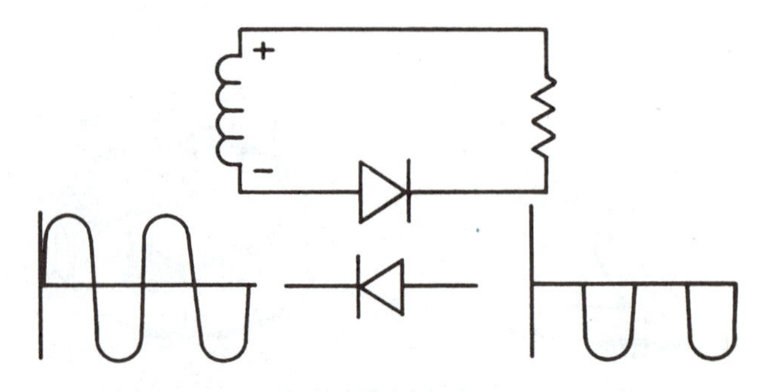

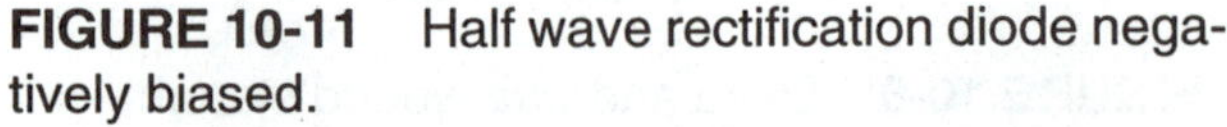

FIGURE 10-11 Half wave rectification diode negatively biased.

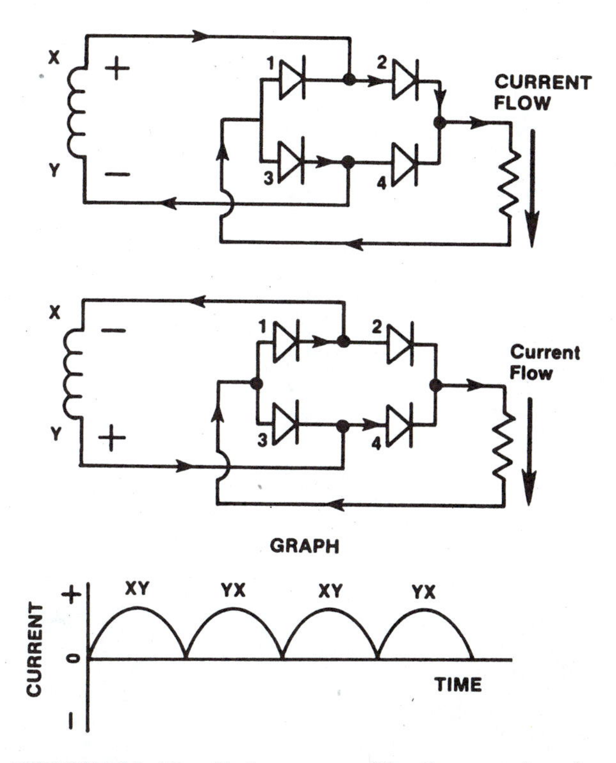

FIGURE 10-12 Full wave rectification requires four diodes.

it into a pulsing DC, but we are wasting the other half of the AC wave. To be able to have full wave rectification, we need to add more diodes as in Figure 10-12. Notice that in either direction we get DC out of the four diodes. Follow the arrows through the coil on the left. When current is from bottom to top, diode 1 is blocking because it sees pressure. Diode 2 is passing current out to the load. Diode 3 is passing current from the load, and diode 4 is blocking. When we reverse the direction of current through our coil of wire, all diode action is reversed so that 1 and 4 are passing current, while 2 and 3 are blocking current. The end result of both directions through the coil was the same. In this example, AC changes into DC and produces a pattern like that shown. This is called **full-wave rectification**.

Remember our stator winding? Let us examine diode action through six diodes and three stator windings. Figure 10-13 shows a wye wound stator with each winding attached to two diodes. Each pair of diodes has one grounded, or negative, diode and one insulated, or positive, diode. Notice also that the center of the wye contains a common point for all windings. This is called the **stator neutral junction** and can have a connection attached. At any time during the rotor movement, two windings will be in series, while the third coil will be neutral and doing nothing. Follow the arrows through the two windings and you will realize that the rectification of the AC into DC will be the same as before. As the rotor revolves, it will energize a different set of windings and in different directions. The end result will be the same, however. Current in any direction

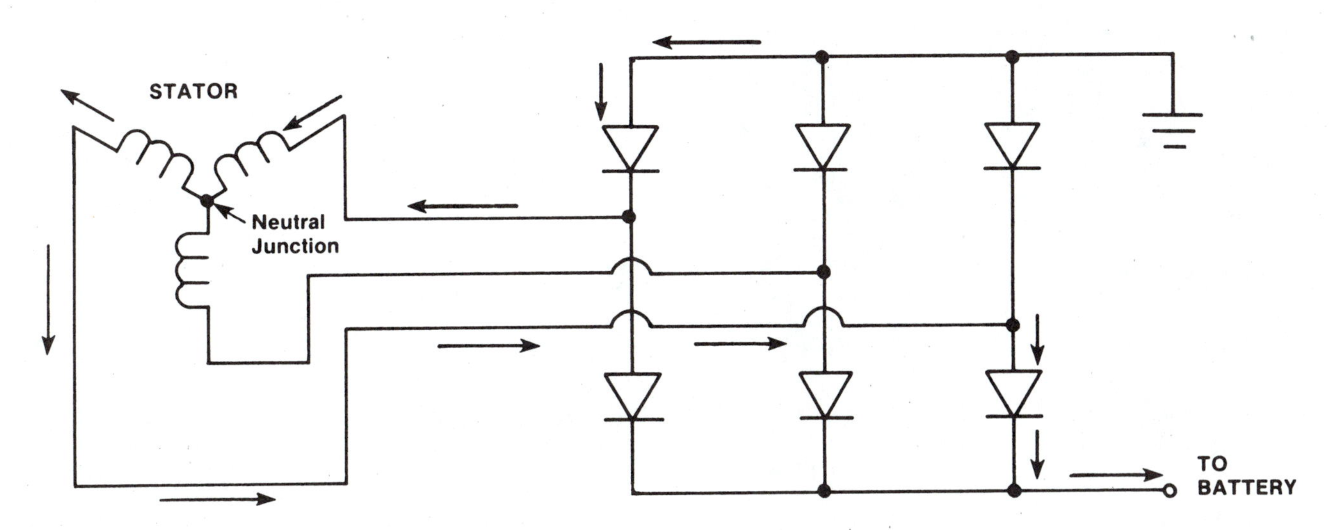

FIGURE 10-13 Wye stator wired to six diodes.

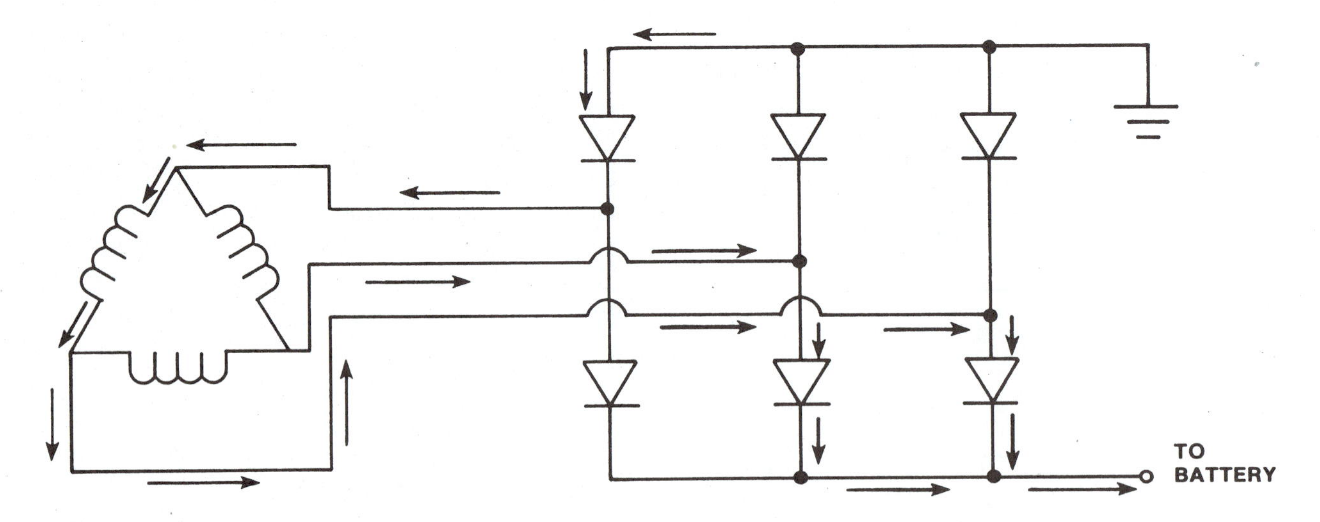

FIGURE 10-14 Delta stator wired to six diodes.

through two windings in series will produce the DC required by the battery.

Wiring the stator and diodes into a delta pattern does not change around the diode action. You can see from Figure 10-14 that the major difference is that two windings are in parallel rather than in series. This accounts for the increased current that is available from a delta wound alternator. The parallel paths will allow more current to flow through the diodes. The action of the diodes is the same, however.

In some alternators, the diodes are in a bridge assembly, as in Figure 10-15. In other applications, they are individually pressed into either the case (negative) or into a heat sink (positive), Figure 10-16. They must be electrically connected to the stator and either ground or positive, and mechanically connected to something that will dissipate the heat they will generate.

FIGURE 10-15 Diodes in a heat sink bridge assembly.

FIGURE 10-16 Individual negative diodes pressed into the alternator end frame.

End Frames

The end frames of the alternator provide the support for the rotor spinning over 15,000 rpm. Usually, a roller bearing is used in one end and a ball bearing in the other. They ordinarily have no method of being lubricated as they are packed with lubricant and sealed by the manufacturer. In addition, the rear end frame usual-

TABLE 10–1 COLOR CODING FOR AMPERAGE OUTPUT

Amperage	Color
38 amps	Purple
42 amps	Orange
55 amps	Red
61 amps	Green
65 amps	Black
70 amps	Black
85 amps	Red
90 amps	Red

Courtesy of Ford Motor Corporation

ly provides the insulated through connectors necessary to bring field current to the rotor, B+ out, and any other stator, diode, rotor connections. Different alternators found in the field will have different connections for various functions. Maintenance manuals will detail exactly what the different connections are actually for.

Alternators are generally made of aluminum and are therefore not to be pried on. Any excess force could easily bend the case making an otherwise electrically good alternator worthless. In addition, the case might have imprinted information about the type, style, or amperage output of the unit. Ford color codes their alternators as can be seen in Table 10-1. This can save the technician quite a bit of time when diagnosing charging system difficulties.

REGULATORS

Now that we have a functioning alternator, we need to realize that the system must be regulated to be efficient. Regulation of the charging system is extremely important for a variety of reasons. The first is the most obvious: keeping the battery fully charged. If the charging current is lower than the amount of current needed to run the vehicle, the battery will gradually run down. We must try to have the charging system supply enough current to run the vehicle, including the accessories, and also charge the battery. Once the battery is fully charged, the charging rate must decrease to a low enough level so as not to overcharge the battery. In addition, the advent of electronic accessories on the vehicle has made the charging voltage level critically important. On-board computers and other digital equipment can be damaged if the charging system is allowed to raise up the system voltage too high. You can see just how important the regulation of the charging system is. The modern charging system relies on the regulator to keep the battery fully charged, but not overcharged, run the vehicle electrically, and protect the system from an over-voltage condition. Quite a job!

Regulation is achieved by varying the amount of field current flowing through the rotor. If field current is high, output will be high. If field current is low, output will also be low. The simplest way to think of the regulator is to imagine it as a variable resistor in series with the field coil, Figure 10-17. If the resistance is high, field current will be low and if the resistance is low, field current will be high. By controlling the amount of resistance in series with the field coil, we will be able to control field current and alternator output. Some regulators function exactly as variable resistors, controlling field current based on charging system demands or needs. These needs are determined by the regulator by looking at two inputs: voltage and temperature. Looking at each one separately, first let us look at voltage. Charging voltage is critically important if we are to expect a fully charged battery. If the charging system is functioning at 11.5 volts, the battery will never come all the way up on charge. To ensure full charge, most regulators are set for a system voltage between 13.5 to 14.5 volts. To simplify this concept, look at Figure 10-18. This block diagram shows what would happen to the vehicle system if the voltage was at 11.5 volts. First,

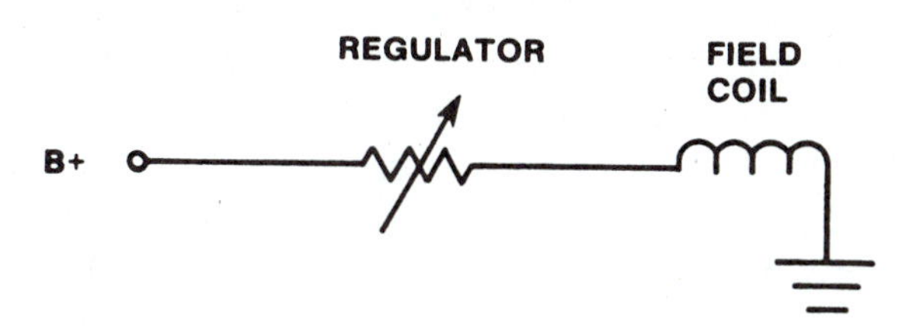

FIGURE 10-17 The regulator acts like a variable resistor.

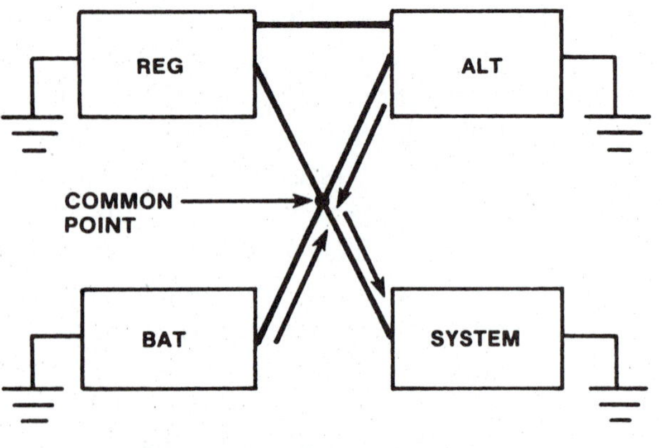

FIGURE 10-18 11.5 volts with current flowing out of the battery.

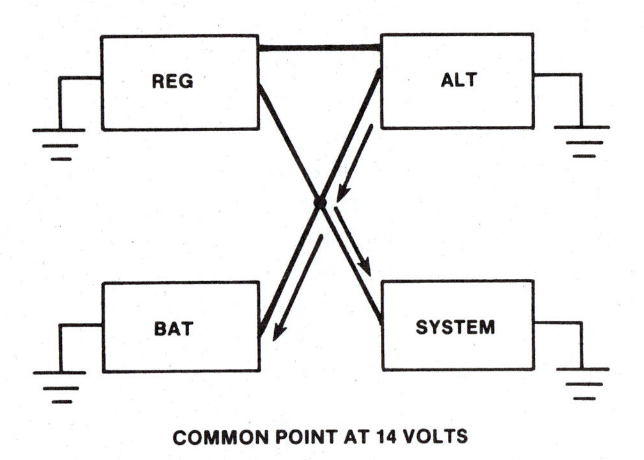

FIGURE 10-19 14 volts with current flowing into the battery.

the entire vehicle would be operating at 11.5 volts and in all likelihood current would be coming out of the battery, instead of going into it. By raising the voltage of the alternator above the voltage of the battery as in Figure 10-19, the alternator now becomes the source of current for the vehicle systems and current will now flow into the battery charging it. By keeping the charging voltage slightly higher than the 12.6 volts of the battery, we ensure that current will flow into rather than out of the battery. The regulator must have system voltage as an input if it is to regulate the charging system. This is sometimes referred to as **sensing voltage**, because the regulator is sensing system voltage and interpreting it to determine need for charging current. Low sensing voltage (less than regulator setting) will cause an increase in charging current by increasing field current. Higher sensing voltage will cause a corresponding decrease in field current and system output. The regulator is therefore responding to changes in system voltage by increasing or decreasing charging current. For example, you are running down the road with no accessories on and a fully charged battery. The regulator senses a high system voltage because the battery is fully charged and reduces the charging current until it is at a level to run the ignition system and trickle charge (2 to 4 amps) the battery. If you turn on the headlights, you will begin a sequence of events like this. First, the additional draw will drop or load the battery voltage down. This reduced voltage will be sensed by the regulator through the sensing circuit. It will, in turn, reduce the field circuit resistance and allow more field current to flow. This increase in field current will produce a stronger magnetic field and increase the alternator output in an effort to keep the system voltage within the specified range. This will all happen within a split second. Turning off accessories causes the opposite to occur. System voltage rises and the regulator cuts back on field current and alternator output.

The second input to the regulator is temperature. All regulators are temperature compensated as Table 10-2 shows. Notice that as the temperature goes down, the voltage regulator setting increases. This is to compensate for the fact that the battery is less willing to accept a charge if it is cold. The regulator will therefore raise up the system voltage until it is at a level that the battery will readily accept. Bringing the battery up on charge quickly is the goal here, even in very cold climates.

TABLE 10–2

	Volts	
Temperature	**Minimum**	**Maximum**
20°F	14.3	15.3
80°F	13.8	14.4
140°F	13.3	14.0
Over 140°F	Less than 13.8	—

FIELD CIRCUITS

Regulators fall into different categories of operation and also different styles. Within the automotive industry, we find three styles in use: mechanical, electronic, and digital. While the electronic or digital regulator is exclusively used on current American vehicles and most foreign, we still find a large number of older American and foreign vehicles that utilize mechanical styles of regulators. We will examine both mechanical and electronic regulators currently in use later in this chapter. Before we do, let us look at regulator placement and identify types of field circuits. Frequently, manufacturers will identify in their service literature the type of field circuit. It will be helpful if their identification system can be interpreted by you before you begin to service the various units on the vehicle.

The A Circuit

The first one to look at is the A circuit. Figure 10-20 is a block diagram of a typical A field circuit. Notice that the regulator is on the ground side of the charging system and that B+ for the field coil is picked up inside the alternator. Remember that resistance can be placed anywhere in a series circuit and it will have the same effect—that of reducing current flow. Having the variable resistance of the regulator on the ground side of the field coil will allow the control of field current just as easily as having the resistance on the B+ side.

The B Circuit

You have probably figured out that a B circuit has the regulator variable resistance between the B+ feed and the field coil as in Figure 10-21. The same results will occur,

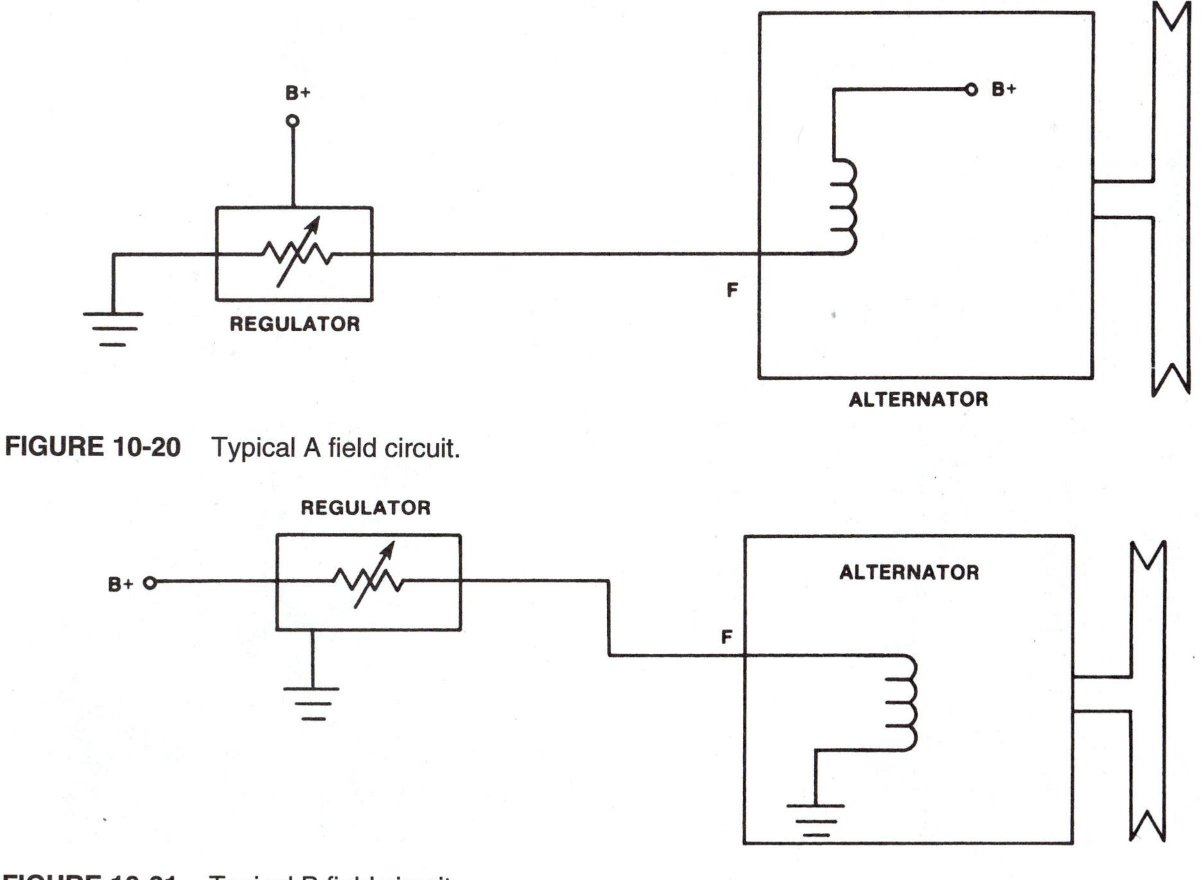

FIGURE 10-20 Typical A field circuit.

FIGURE 10-21 Typical B field circuit.

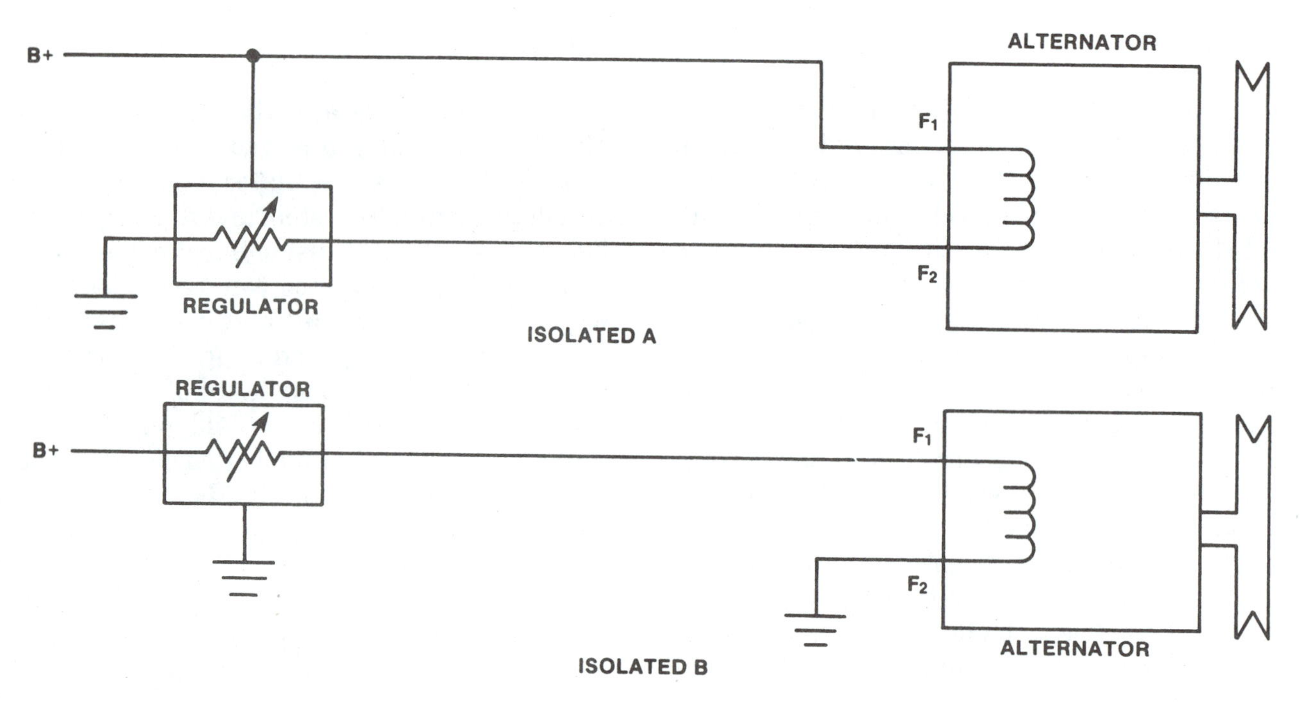

FIGURE 10-22 Isolated A and B circuits.

however, in either case. Notice that the field coil is grounded inside the alternator.

ISOLATED FIELD ALTERNATORS

Isolated field alternators pick up B+ and ground externally. They are usually easily identified because they have two field wires attached to the outside of the alternator case. Isolated is another way of saying insulated or not connected. Isolated field alternators can have their regulator placement on either the ground (A circuit) side or on the B+ (B circuit) side. Figure 10-22 block diagrams both an isolated A and an isolated B circuit.

It will be helpful if you understand the different types of field circuits before you begin to diagnose charging systems.

EXTERNAL REGULATORS

Let us continue our discussion of regulators with an external mechanical style. The name tells quite a bit about the style. It is mounted

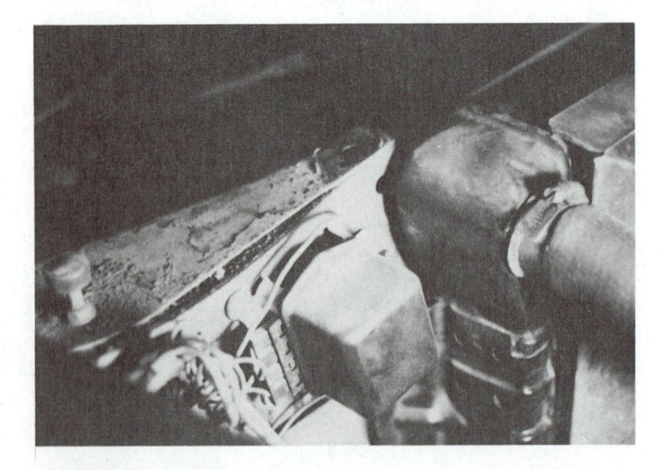

FIGURE 10-23 External mechanical regulator mounted in the engine compartment.

FIGURE 10-24 External mechanical regulator with the cover off.

on the fender or fire wall rather than being mounted inside the alternator. Figure 10-23 is an external mechanical regulator mounted to the fire wall of a GM product. Removing the cover of the same regulator as in Figure 10-24 reveals what looks like two small relay coils. Notice also in Figure 10-25A that the bottom of the regulator has some small wire wound resistors. We will discuss these resistors later. Before we analyze just how this regulator will control alternator output, we should review some principles of magnetism, because the two small coils inside the regulator are actually electromagnets. You should remember from previous lessons that when we have current flowing through a conductor, a weak magnetic field will be present around the conductor. The strength of the field can be increased by looping the wire and forming a coil. Its strength can be further increased by placing a soft iron core into the center of the coil.

Our regulator will utilize the electromagnet principles to move contact points or switches. Inside the regulator are usually two coils: the field relay and the voltage regulator. Let us look at the switch operation of the field relay

A

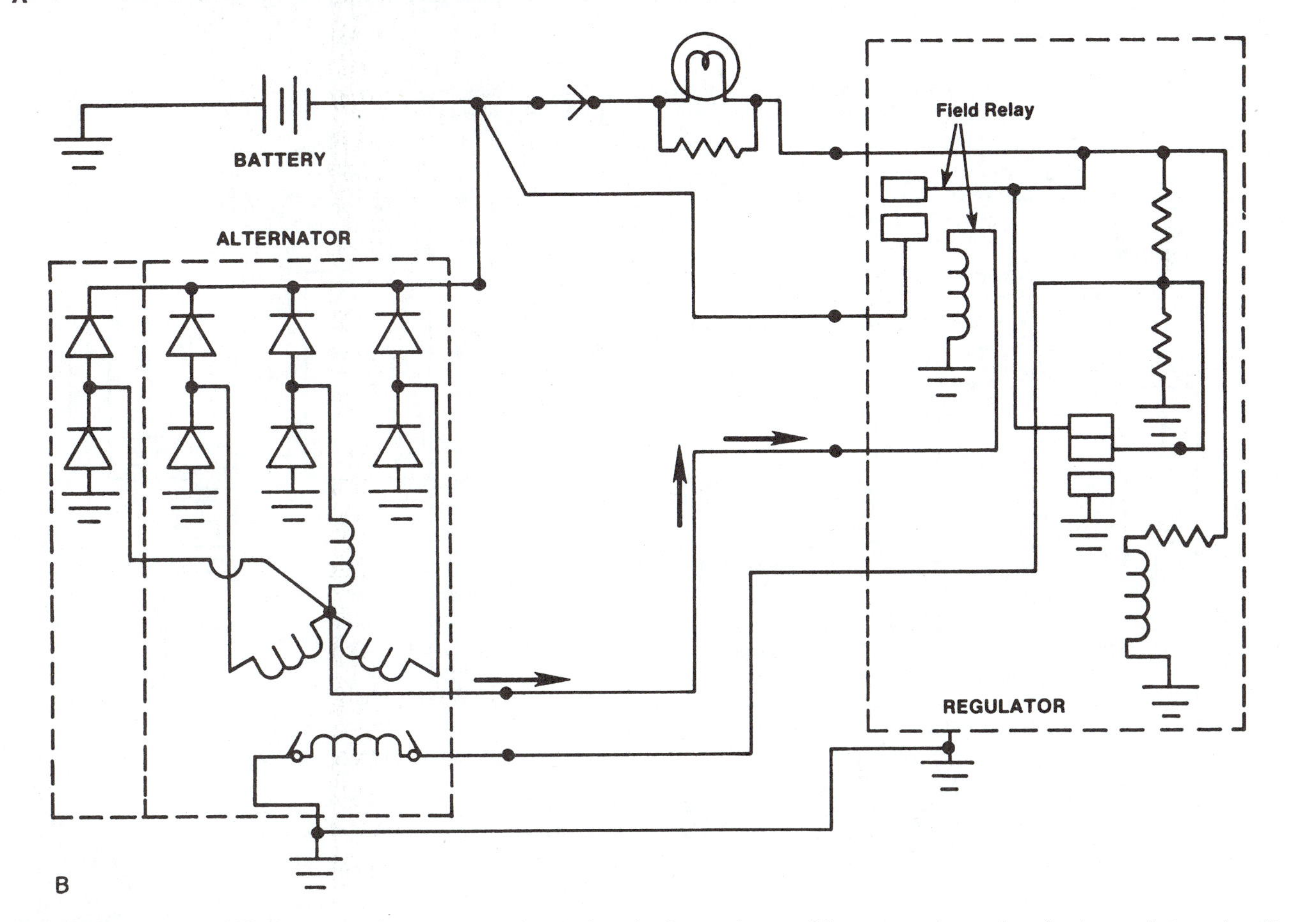

B

FIGURE 10-25 (A) Bottom of an external mechanical regulator; (B) external mechanical regulator circuit. (Courtesy of Ford Motor Company)

first. Figure 10-25B shows the relay coil and contact with no current flowing through the coil on the left. Notice that the contact switch is open; no current will flow through the open contacts. When current is allowed to flow through the relay coil, the magnetic field will pull the moveable contact arm down, as in Figure 10-26. This action will close the contacts and allow current to flow through. This field relay will have two positions only; open with no current flowing through the contacts and closed with current flowing through both the coil of wire and the closed contacts.

Our voltage regulator is slightly more complicated because it must be designed to recognize various levels of voltage. It will usually have two sets of contacts and a calibrated spring. The spring in Figure 10-27 is trying to pull the moveable arm up against the coil's magnetism that is trying to pull the arm down. This is a balancing act between magnetism and spring tension. Figure 10-28 shows the three positions that the moveable arm could be in. The voltage applied to the coil will determine the strength of the field and the position of the arm. Less than 13.5 volts will not have

FIGURE 10-27 Voltage regulator tension spring.

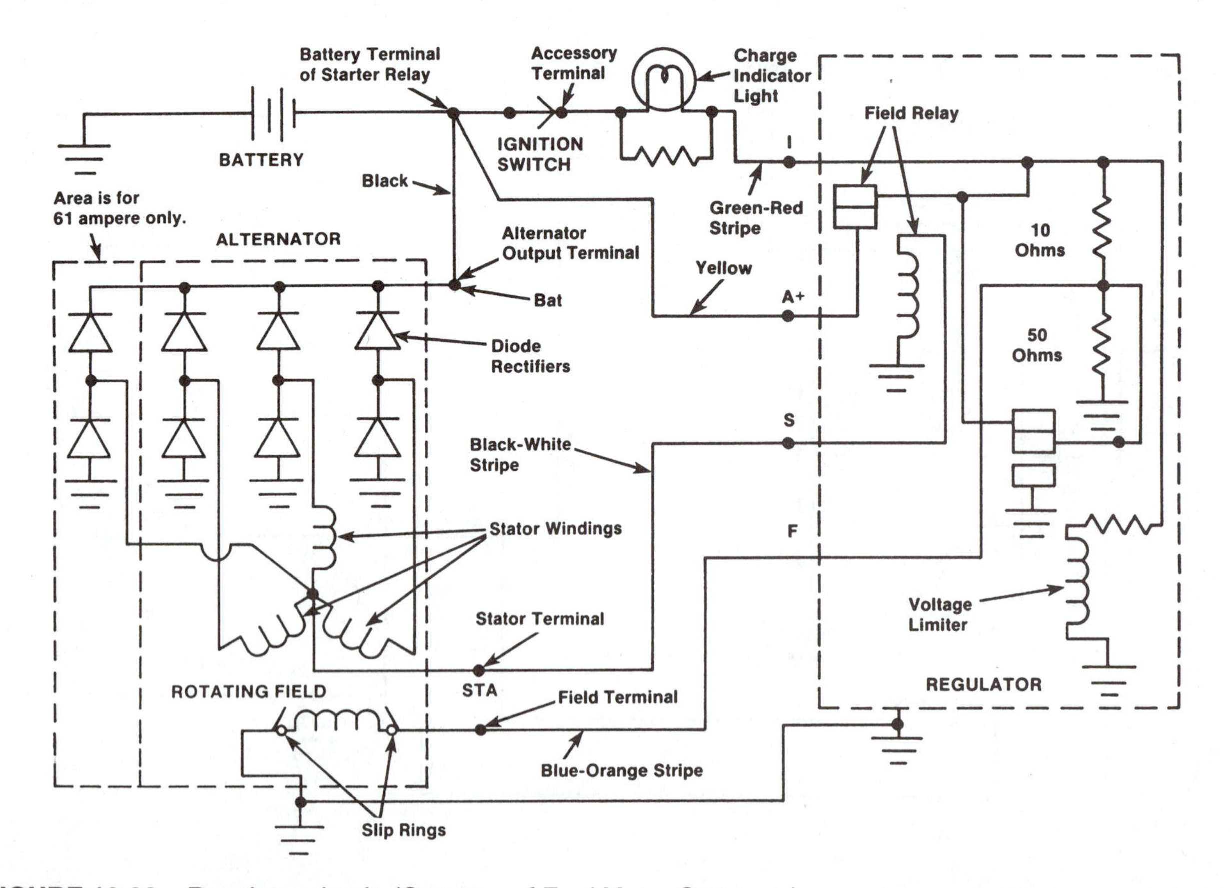

FIGURE 10-26 Regulator circuit. (Courtesy of Ford Motor Company)

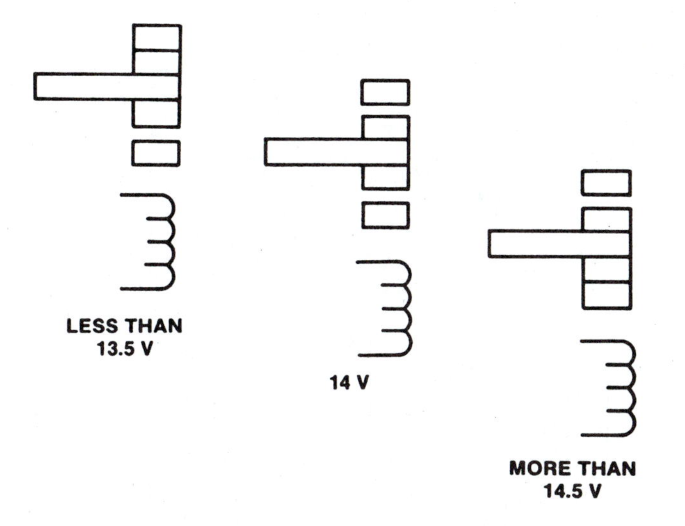

FIGURE 10-28 Regulator contact position.

sufficient magnetic strength to pull the arm against the spring tension. Around 14 volts will be capable of pulling the arm into the center position as shown, and more than 14.5 volts will pull the arm all the way down. Converting this over to contact position shows that less than 13.5 volts applied will allow current to flow through the top set of contacts. Whereas, 14 volts will open the top contacts and not close the bottom set of contacts resulting in no current flow through either set. In addition, 14.5 volts or more will close the bottom set of contacts and keep the top set open. If you can visualize contact position with different voltages, you will have no difficulty understanding electrically how the regulator will be capable of controlling the alternator. Do not hesitate to review this information if it is unclear to you.

Figures 10-29 through 10-32 show the vehicle in four different operating conditions. We will trace the current flow through the field circuit and show the effects on field current. Remember, if field current is high, alternator output is also high, and if field current is low, alternator output will also be low. How about if no field current is flowing? No output is the correct answer here. Without field current, there can be no magnetic field in the rotor and no output. This is also a good point to mention that most charging systems require about 2,000 engine rpm before they are capable of producing rated output. At idle, most systems will only be capable of about half rated output even with full field current flowing. Customers can sometimes even see this reduced output in the dimming of their headlights at idle with many accessories on. The headlights dim because the output of the alternator is less than the system calls for. The battery is asked to supply the difference between what is produced by the alternator and what is used by the system. The battery will supply current only if the system voltage drops below 12.6 volts. The drop from 14 volts down to 12.6 volts is the dimming of the headlights that your customer might see.

Let us turn our attention over to Figure 10-29, which shows the regulator with the vehicle off and the key on. Let us identify the regulator terminals first. F goes directly over to the field input terminal of the alternator. Notice that the field is grounded inside the alternator. This makes a B field circuit, because the regulator is between B+ and the field coil. Terminal 2 comes from the stator to the field relay coil. This will bring a very small amount of output current over to the coil when the alternator is producing current. Terminal 3 is the B+ we need for field current. Notice that it comes directly from a junction block off of the battery. This terminal is hot or live all the time. This junction block is a common point for a B+. It can also be the same point that we power the vehicle systems off of from the alternator or battery. Terminal 4 is the ignition switch input to the regulator. Turning the alternator on or off will be accomplished through this terminal. With the key on, current flow for the field will follow the arrows out the ignition switch through the resistor and bulb (wired in parallel), through the closed lower contacts of the voltage regulator, and out terminal F into the alternator through the field coil to ground. This puts the resistor/bulb in series with the field coil The small amount of current that can flow through the resistor/bulb will produce a very weak magnetic field in the rotor and turn the bulb on. This bulb is the charging indicator light in the dash. The resistor in parallel with the bulb allows some field current to flow

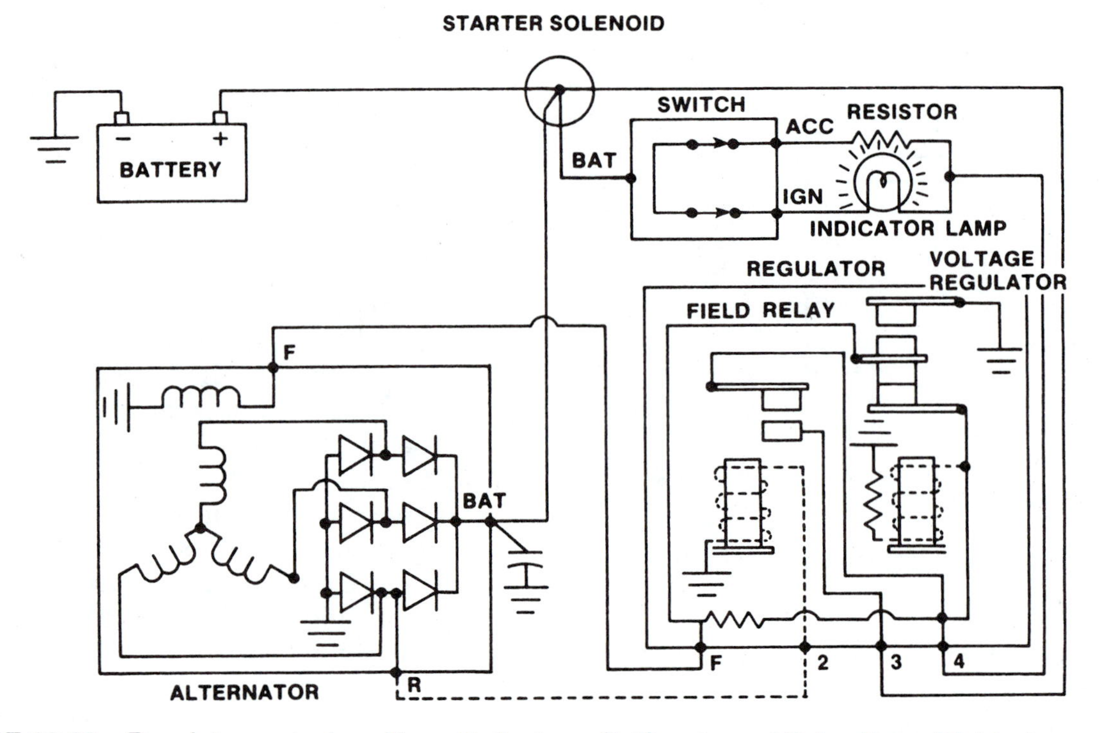

FIGURE 10-29 Regulator contact position with the key off. (Courtesy of Delco Remy Division)

even if the bulb burns out. With a burned or missing bulb, the charging system would not function at all without this resistor. Once you understand what is occurring with the key switch on and the engine off, move on to the next figure. Do not move on until you understand the current flow in Figure 10-29.

Figure 10-30 shows the charging system with the engine running and the alternator producing current at a voltage level below 13.5 volts.

NOTE: 13.5 volts is usually the beginning level at which regulation of the charging system becomes necessary. Lower than 13.5 volts should signal the regulator to allow full field current to flow through the field coil.

The arrows indicate current flow through two different circuits. First, it flows through the field relay coil. If the alternator is spinning, it will be producing some current. This puts B+ out the R terminal of the alternator and energizes the field relay closing the contacts. Closing the contacts does two things. First, it puts B+ from terminal 3 over to terminal 4 to turn off the indicator lamp. With B+ on both sides of the lamp, no current will flow through it and it will go out. Second, B+ directly from the battery becomes available for the field coil through the voltage regulator. With the voltage regulator coil common-pointed also to the battery (above terminal 4), the lower than 13.5 volts will not produce sufficient magnetic field to pull the contacts open. You should be able to follow B+ all the way from the junction block to the field coil in the alternator. No series resistance will allow maximum alternator output. Do not go on until you understand this full field current condition.

As the battery comes up on charge the voltage of the system will gradually climb into the regulator range (13.5 to 14.5 volts typically). Once this occurs, the voltage regulator coil will have a stronger magnetic field as in Figure 10-31. The magnetic field will overcome the spring tension and pull the lower set of contacts open. Follow the arrows as they trace the current flow through the resistor in the lower left corner of the regulator. We have now added more series resistance to the field coil in the alternator and obviously reduced the field current. Reducing the field current

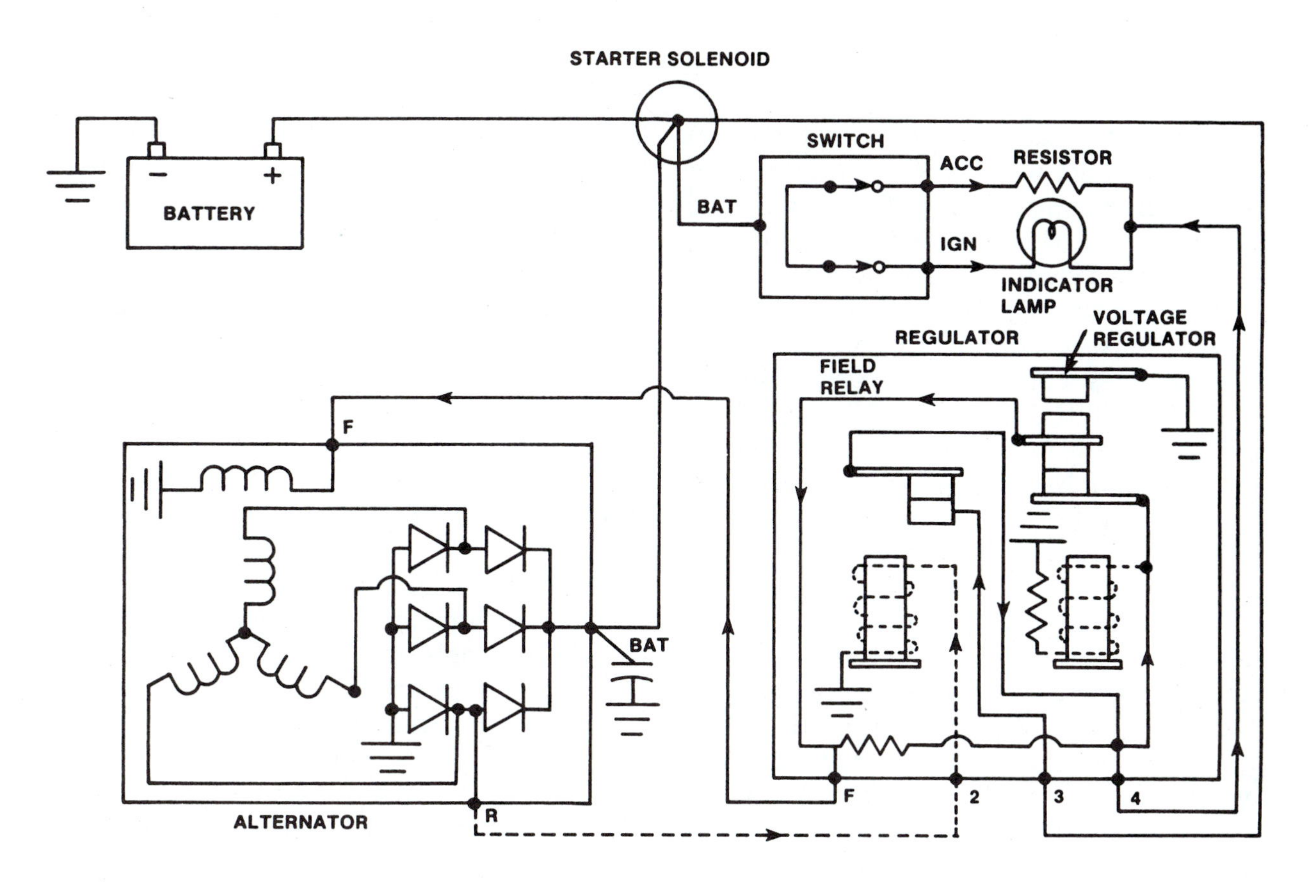

FIGURE 10-30 Regulator contact position with system voltage less than 13.5 volts.

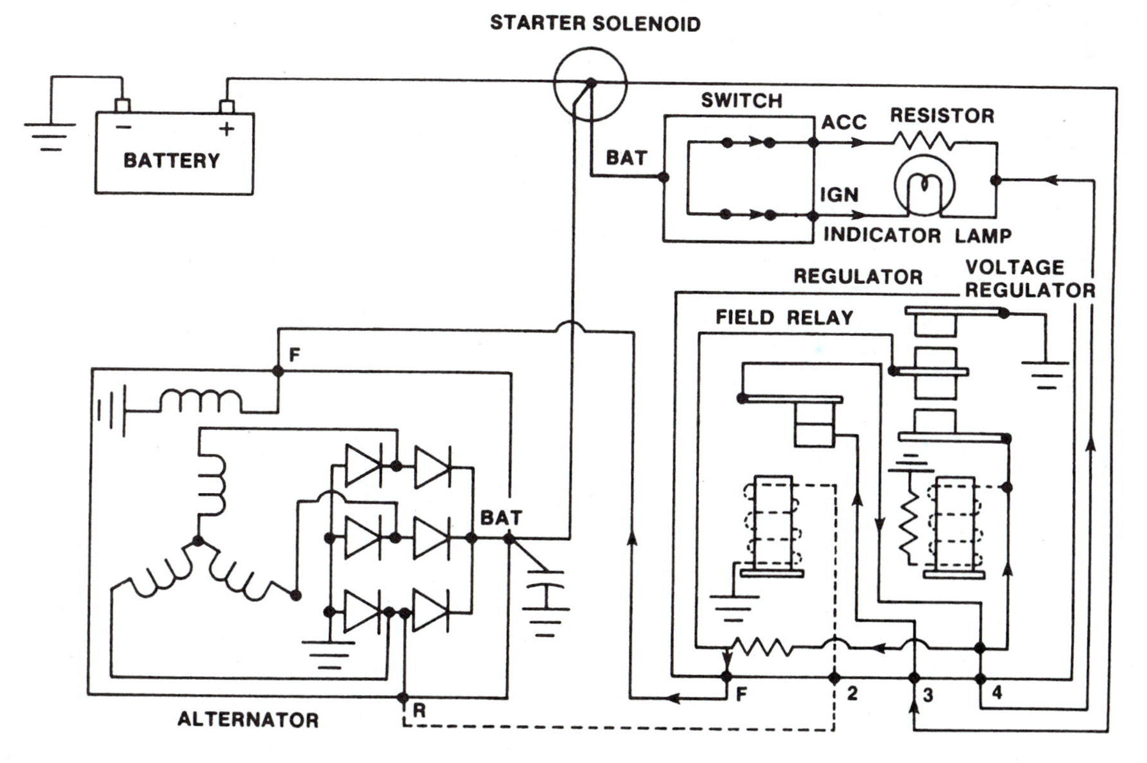

FIGURE 10-31 Regulator contact position with system voltage 13.5 to 14.5 volts.

will reduce the alternator output usually to about half its rating. Do not go on until you understand this half field current condition.

If the system draw is less than the alternator is producing, both battery and system voltage will continue to climb. As this voltage climbs, it will eventually cause sufficient magnetic field in the voltage regulator coil to pull closed the top set of contacts as in Figure 10-32. This will apply a ground to the F terminal and the field coil in the alternator. No current will flow through the field coil as both ends of it are now grounded. No field current equals no output and the voltage stops rising. This usually occurs around 14.5 volts (slightly higher if it is cold outside).

The regulator will vibrate between full output, half output, and no output as it maintains the voltage levels specified by the manufacturers. Reducing the system draw will place more resistance into the field circuit or shut it off completely. Increasing the system draw will reduce the field circuit resistance. If maximum output is desired, field current will be at maximum. This condition is called **full fielded** and it occurs when the voltage of the system drops below the level determined by the manufacturer (typically 13.5 volts).

Turning the ignition key off will shut off the engine. Once the rotor is no longer turning, the R terminal will go to 0 volt and open the field relay, effectively turning off the charging system.

EXTERNAL ELECTRONIC REGULATORS

Let us turn our attention to what is probably the most common regulator on the road today: the electronic regulator. It can be mounted externally or, in some cases, inside the alternator. There are some instances where it will be bolted to the outside of the alternator case. It can be transistorized or an integrated circuit design. Long life without maintenance,

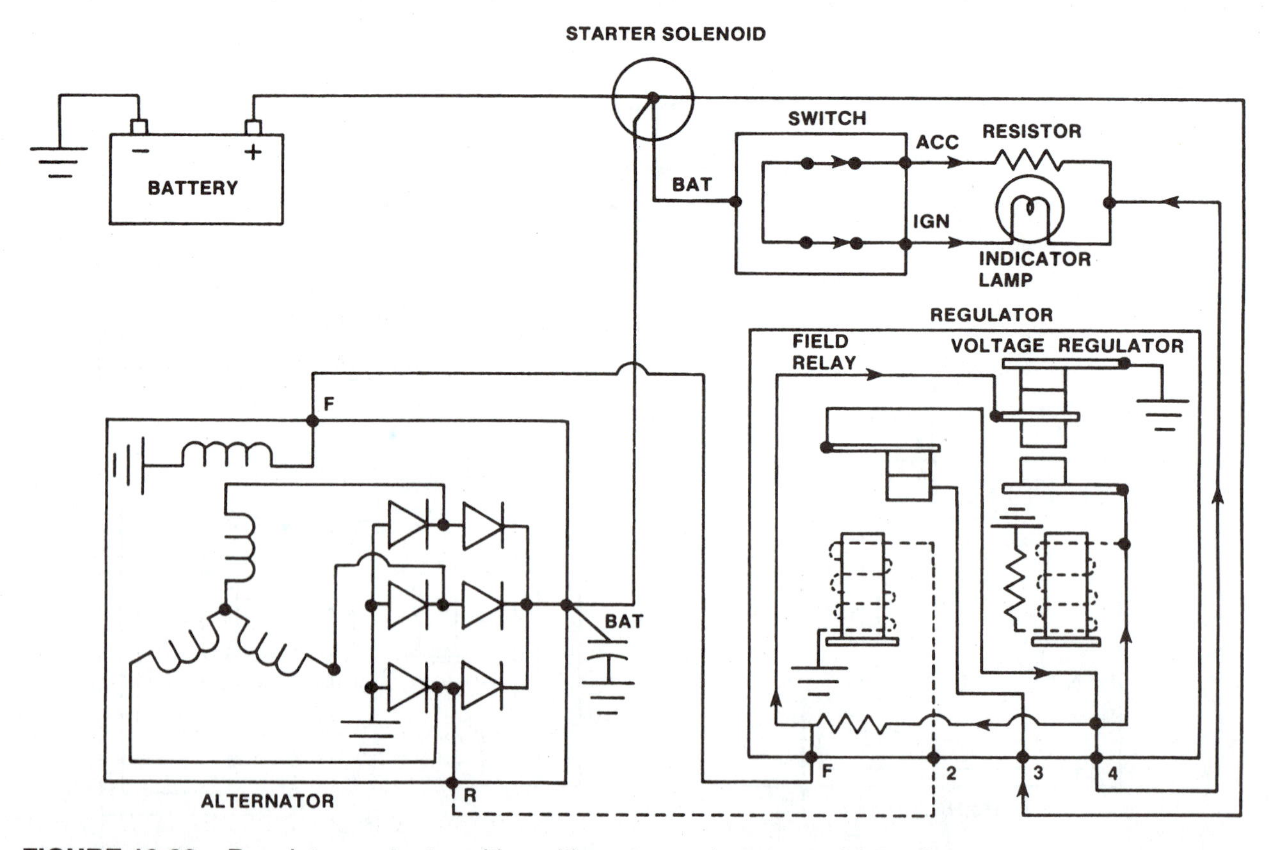

FIGURE 10-32 Regulator contact position with system voltage over 14.5 volts.

close voltage limiting, and inexpensive production are its main claims to fame. With no moving parts and contact points to burn from arcing, the unit will virtually last the life of the vehicle. The mechanical regulator will gradually lose both the ability to full field the alternator and the ability to hold the voltage within the specified range.

The heart of the electronic regulator lies in a component that we have not studied called the **zener diode**. It is much like the diodes in the alternator however, it has a significant difference—the ability to conduct in the reverse direction without damage. The zener diode used in regulators is doped so that it will conduct in reverse once a predetermined voltage has been achieved. You have probably guessed that this voltage level will be between 13.5 and 14.5 volts. Figure 10-33A and B shows a typical transistorized regulator circuit in simplified form. The key to the circuit is the zener diode, which has B+ on the anode (the side with the arrow). When the voltage of the vehicle system rises above the specification, the diode will conduct and permit current to flow to the base of transistor 1. This turns the transistor on, which will in turn switch off transistor 2. Transistor 2 is in control of field current for the alternator. With it off, no field current can flow, thus shutting off the alternator until the voltage level drops below specification. Transistor 2 is acting like a switch, turning field current on then off as the system voltage rises and falls above and below the specified voltage. This occurs many times a second and cannot be measured with a standard voltmeter. The thermistor in the upper left of the diagram will give the temperature voltage change necessary to keep the battery charged in cold weather. With no contact surfaces to pit and wear, the voltage level, which the zener diode conducts, will become the regulated voltage of the alternator for the life of the regulator.

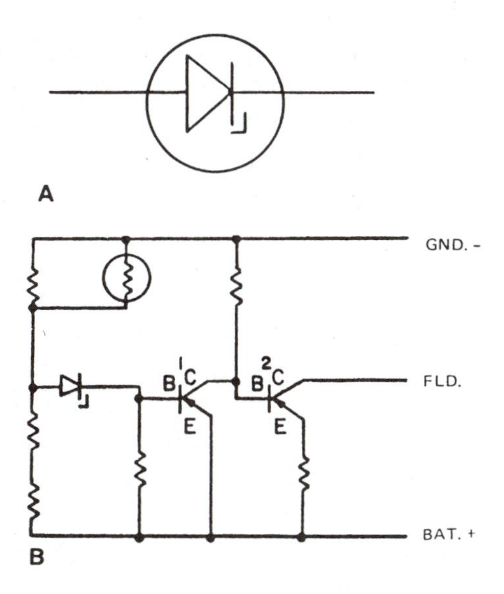

FIGURE 10-33 Zener diode and a simplified transistorized regulator.

This on again off again operation occurring many times a second, is an example of something called pulse width modulation (PWM). Pulse width modulation is used in many applications on today's vehicles, and started with regulators and charging systems in the 70s. Now, with computer control of many systems, it is the accepted method of control. It is really quite easy to understand. Let us use a 100 amp alternator as an example. If the demands of the vehicle's electrical system and the battery requirements are 50 amps, the regulator will turn the alternator's field on for 1/2 (50%) of the time, then off for 50%. This will occur very rapidly. This 50% on then 50% off will give 50 amps out of a 100 amp alternator. This is shown in Figure 10-34A. If the system's requirements change so that 75 amps are needed, the regulator will increase the on time of the field to 75% and decrease the off time to 25%. This is shown in Figure 10-34B. Another example of PWM is fuel injection systems. If the vehicle requires 50% fuel, the computer will send a pulse width to the injector with an on time of 50% and an off time of 50%. The extremely fast processors that are in use in today's vehicles will allow pulse width modulation of many thousands of times a second. Do not forget, the period of time usually does not vary. Only the on time of the device will vary.

Either transistorized or integrated circuit regulators are usually sealed with no means of setting voltage. They are serviced by complete replacement, which we will cover in the next chapter. Some electronic regulators use a mechanical field relay along with the electronic voltage regulator. Remember also that whether it is mechanical or electronic, the function is

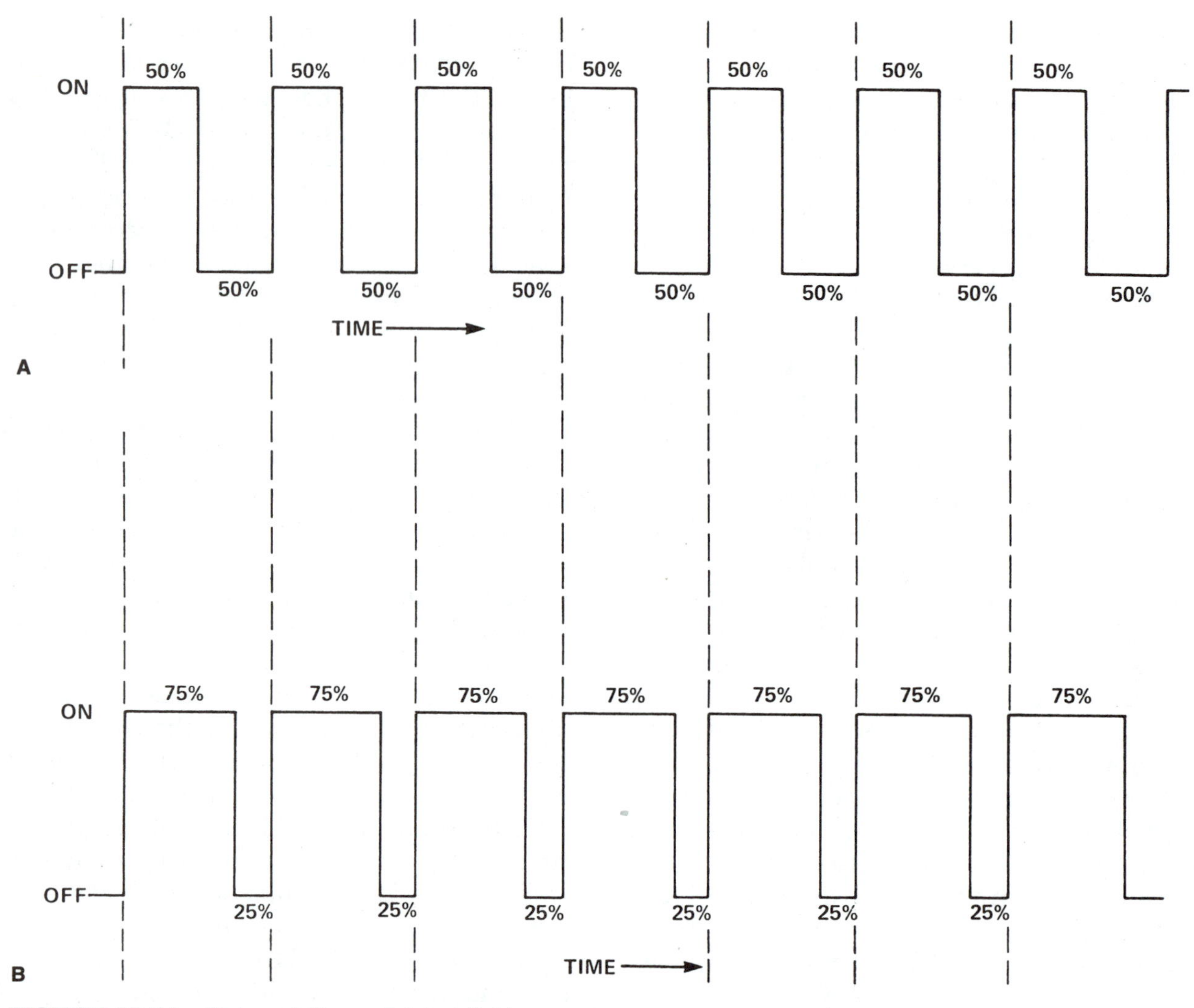

FIGURE 10-34 Pulse width modulated field current.

the same. Control field current to keep the system voltage within the specification. The only real difference is the sharp cutoff voltage of the electronic regulator. There is usually less of a range in voltage and actual can vary less than 0.5 volt, whereas the typical range with a mechanical is 13.5 to 14.5 volts or more. You will realize in the next chapter that their testing procedure is exactly the same.

INTERNAL ELECTRONIC REGULATORS

You have probably realized that the mechanical regulators that we started with are actually harder to understand than most electronic ones. Many technicians agree with you. Electronic regulators eliminated much testing and many intermittent failure problems. Typically they will either work completely or fail completely.

The next group of regulators to look at is exactly the same electrically as our last group. The difference is in their mounting and in the extra circuitry necessary. Putting the regulator into the alternator is a logical choice. It eliminates the interconnecting wires and greatly simplifies the vehicle building process. It might, however, introduce another variable—the diode trio—into our diagnosis. Let us look a the diode trio first.

When the manufacturers decided to install the regulator inside the alternator, they had to decide where they would get field current from. Their choice was to either bring it in from the outside or pick it up inside the alternator. GM decided to pick up field current inside the alternator right off the stator. This presents a problem, however, as this is AC, or alternating current. The field coil requires DC, so enters the rectifier of AC into DC—the diode. You should note on Figure 10-35 (10SI) the additional three diodes in the upper right. This is the diode trio. Figure 10-36 shows an actual diode trio taken out of an alternator alongside of the assembled trio sitting on top of the stator leads. The three diodes that make up the diode trio have their anodes attached to the stator and their cathodes common-pointed together at a regulator connection. Field current, once the alternator is producing some current, will come from the stator and be rectified by the diode trio. This B+ will also be used to turn off the indicator lamp.

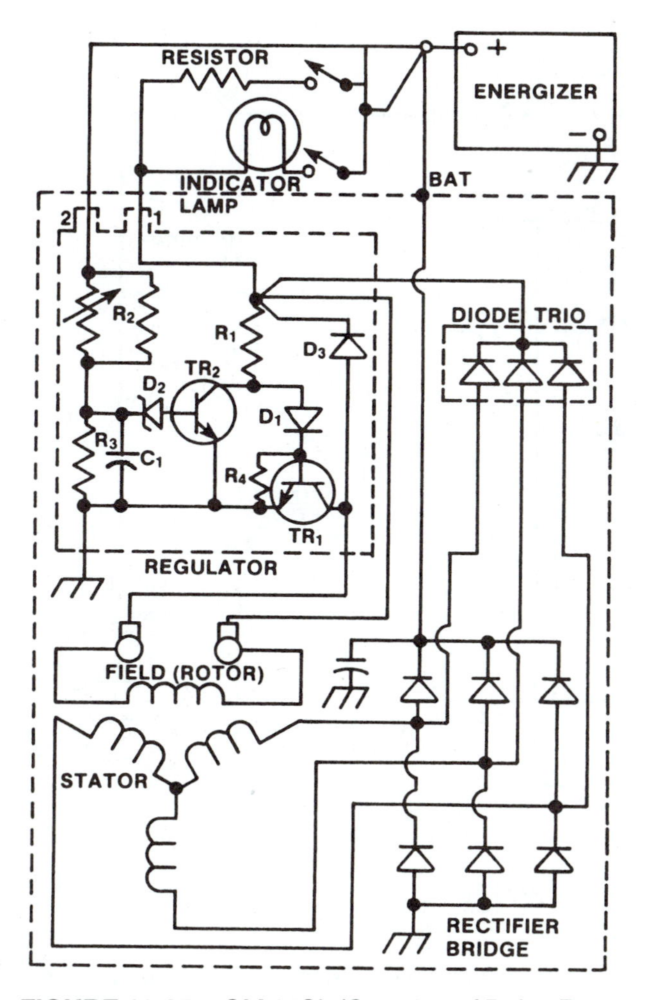

FIGURE 10-35 GM 10SI. (Courtesy of Delco Remy Division)

FIGURE 10-36 Diode trio.

Let us go through the operation of the charging system with an internal electronic regulator and a diode trio. Follow along on Figure 10-35. Take it step by step and do not go on until you understand the previous step. Rereading the section again will also help you in your understanding. This is a very common type of charging system in use today. Notice the two connections in the upper left of the figure. These are the only inputs to the system. Input 2 is the sensing circuit applied directly to the regulator. It has B+ available anytime the battery is installed and charged. Do you notice the zener diode inside the regulator? The sensing circuit is applied to the zener. Input 1 comes from the ignition switch through the dashboard indicator and resistor wired in parallel. Turning on the ignition switch applies B+ to input 1 and to the field through the common point above R1. TR1 conducts the field current coming from the field coil and a weak magnetic field is produced. It is weak because the current has been reduced by the indicator lamp and resistor. Once the alternator begins to produce current, the diode trio will conduct and B+ will be available for the field and terminal 1 at the common point above R1. Placing B+ on both sides of the indicator lamp will turn it off. It will remain off as long as the diode trio is conducting and supplying field current. The sensing circuit from terminal 2 will now be allowed to turn the switching transistor, TR1, on and off, as the voltage rises above and falls below the specified voltage. Actual regulator operation is the same as the previous electronic regulator with the exception of the diode trio. Do not hesitate to reread this section until it makes sense to you. Diagnosis of this style of charging system will be much easier if you have a working knowledge of it.

In recent years, General Motors has replaced the 10SI and 27SI with an alternator which cannot be easily full fielded. In addition, the diode trio has been eliminated. Field current comes from after the rectification diodes and is therefore DC. This source of DC eliminates the need for the trio. A special harness is available to allow full fielding. However, with the high cost of labor today, many of the repair facilities treat the unit as a non repairable, replace as a unit only, charging system. This eliminates the need to full field the alternator on the vehicle. If the facility you work for does repair alternators, you will need to purchase one of the harnesses, which will allow you to bypass the regulator. In addition, this alternator is a B circuit with the regulator supplying B+ rather than ground as in the SI series.

Ford Motor Company has an interesting internally regulated alternator on many of their vehicles. Figure 10-37 shows the IAR alternator and regulator. If you remember the GM 10SI you will note that field current came from inside the alternator. The Ford unit differs in that field current comes directly from the battery. This is the A terminal on the diagram and is the source of field current. The I or indicator terminal comes from the in-

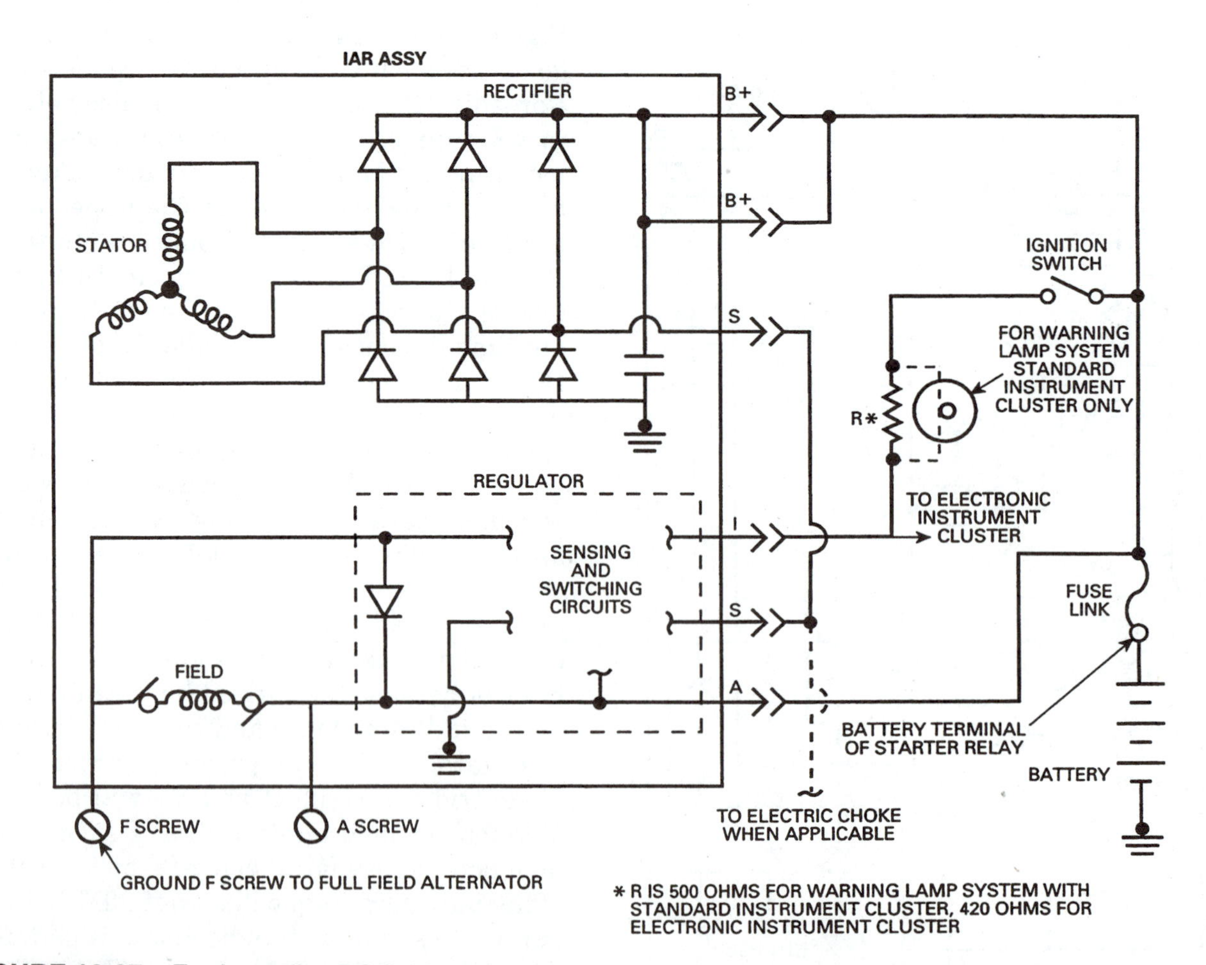

FIGURE 10-37 Ford supplies DC field current directly from a fuse link at the starter relay. (Courtesy of Ford Motor Company)

strument cluster and is ignition switched. Within the regulator is a circuit that will ground the I terminal if the alternator is not producing any current. Grounding this terminal will turn on the indicator lamp in the instrument cluster. Notice also that there are two B+ terminals to split the output until outside the alternator. Before the wiring reaches the starter relay battery terminal, the two conductors will join together.

This is a unique alternator because it allows the technician to easily full field the circuit. Full fielding is a process whereby the technician can take over the function of the regulator.

If a vehicle were to show up in a garage with no alternator output, the technician would ground the F screw and monitor alternator action as shown in Figure 10-38. Grounding this screw eliminates the regulator control and action. If the alternator produced full output with the F terminal grounded, and nothing with it ungrounded, the technician would replace the regulator. However, if the alternator does not produce output with the F terminal grounded or ungrounded, the technician would replace the alternator. Note: Frequently, especially at high labor rates, technicians will not separate the alternator and regulator, but replace the entire unit if either has malfunctioned.

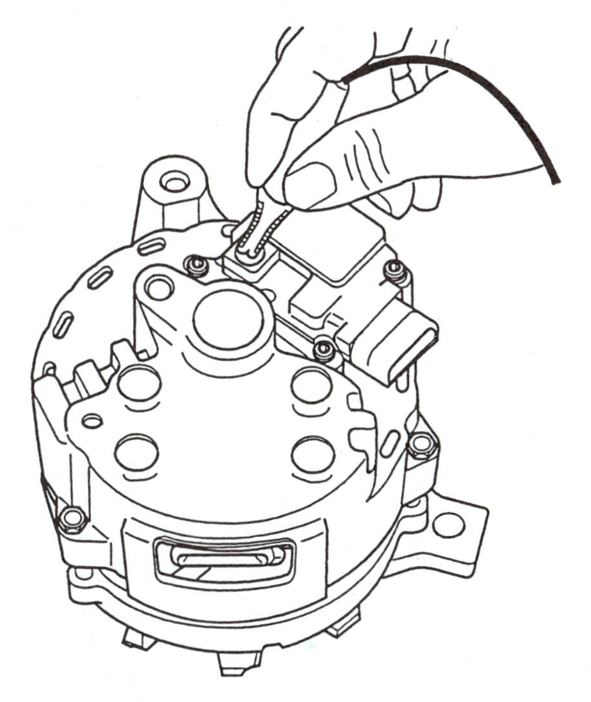

FIGURE 10-38 The back of the regulator has a screw which is labeled "ground here to test". (Courtesy of Ford Motor Company)

COMPUTER REGULATION

On many of today's vehicles, the regulator function has been taken over by the vehicle's computer. The operation is exactly the same as the integrated circuit electronic regulator we have studied. The field circuit will be grounded (A circuit) intermittently to keep the system voltage within the specified limits. Figure 10-39 shows a current computer regulator circuit. As the computer has gradually taken over various electrical operations on the vehicle, it is an obvious advantage to eliminate the regulator and turn its function over to the computer. The circuit is just as easy to diagnose as those we have previously studied. There is an obvious difference, however. Technicians, who are not the good diagnosticians they should be, probably are used to replacing the regulator anytime a vehicle comes in with a no-charge condition. This same procedure, when applied to a computer-regulated vehicle, will involve the replacement of a very expensive major component. The importance of diagnosing the exact problem and repairing it rather than replacing components until the vehicle charges, cannot be over stressed.

CHARGING INDICATORS

In vehicle charging, indicators usually take one of three available forms: the indicator lamp, an ammeter, or a voltmeter. We have looked at regulator circuits, which have used an indicator lamp, and hopefully you have realized that the lamp goes out when equal voltage is applied to both sides of the circuit. Usually one side of the bulb is connected to B+ coming from the ignition switch and the other side is con-

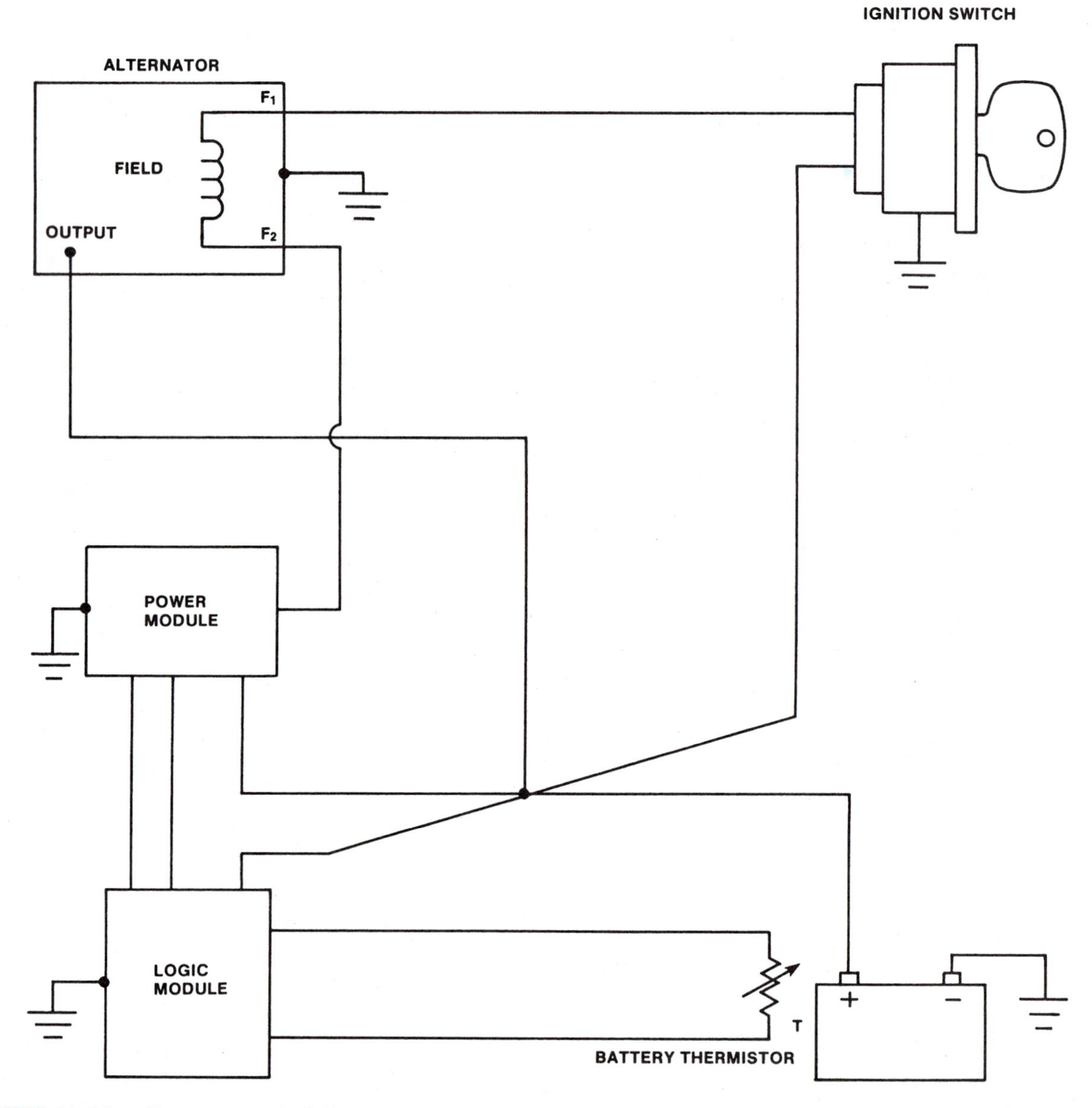

FIGURE 10-39 Computer regulation.

nected to alternator output that will become B+ once some output is present. It is important to note that the typical indicator lamp circuit does not recognize how much current is being developed or what the system voltage is. It will be off if there is any amount of current being produced and, therefore, should not be relied on to give much diagnostic information regarding system performance. It is, however, a pretty good consumer indicator as it will usually glow or be on brightly if there is no charging current.

The ammeter or charge indicator mounted on the dash gives the consumer much additional information over the indicator lamp. Many consumers, however, do not know what to do with the information. Correctly functioning, it will show a high charging rate with a recently started vehicle, Figure 10-40. Once the battery comes up on charge, it will gradually taper down until it is only slightly higher than zero. Occasionally, it will read discharge. A discharge condition will be present on most

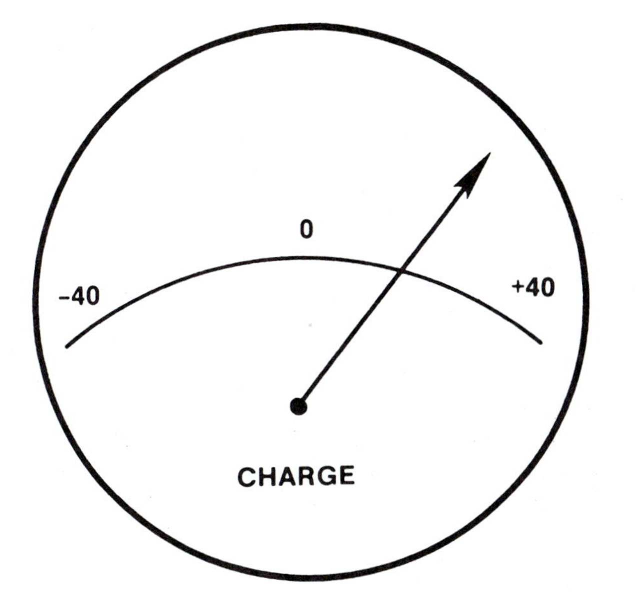

FIGURE 10-40 Ammeter will indicate charging current.

vehicles at idle with a lot of accessories on. The condition should change once the vehicle speed goes above idle. The wiring for most ammeters places them between the battery and a junction where the B+ for all systems is picked up. This is illustrated in Figure 10-41.

Ammeters can be quite useful to the knowledgeable technician as they can indicate battery, alternator, or regulator difficulties before a no-start condition exists.

Voltmeters mounted on the dashboard are also found on some vehicles. Like ammeters they can be useful to the knowledgeable consumer or confusing and ignored by others. Usually they are fed off the ignition switch so that they measure system voltage with the key on. They can be helpful in picking out bad regulators, alternators, or batteries. Figure 10-42 shows a typical in-dash voltmeter setup.

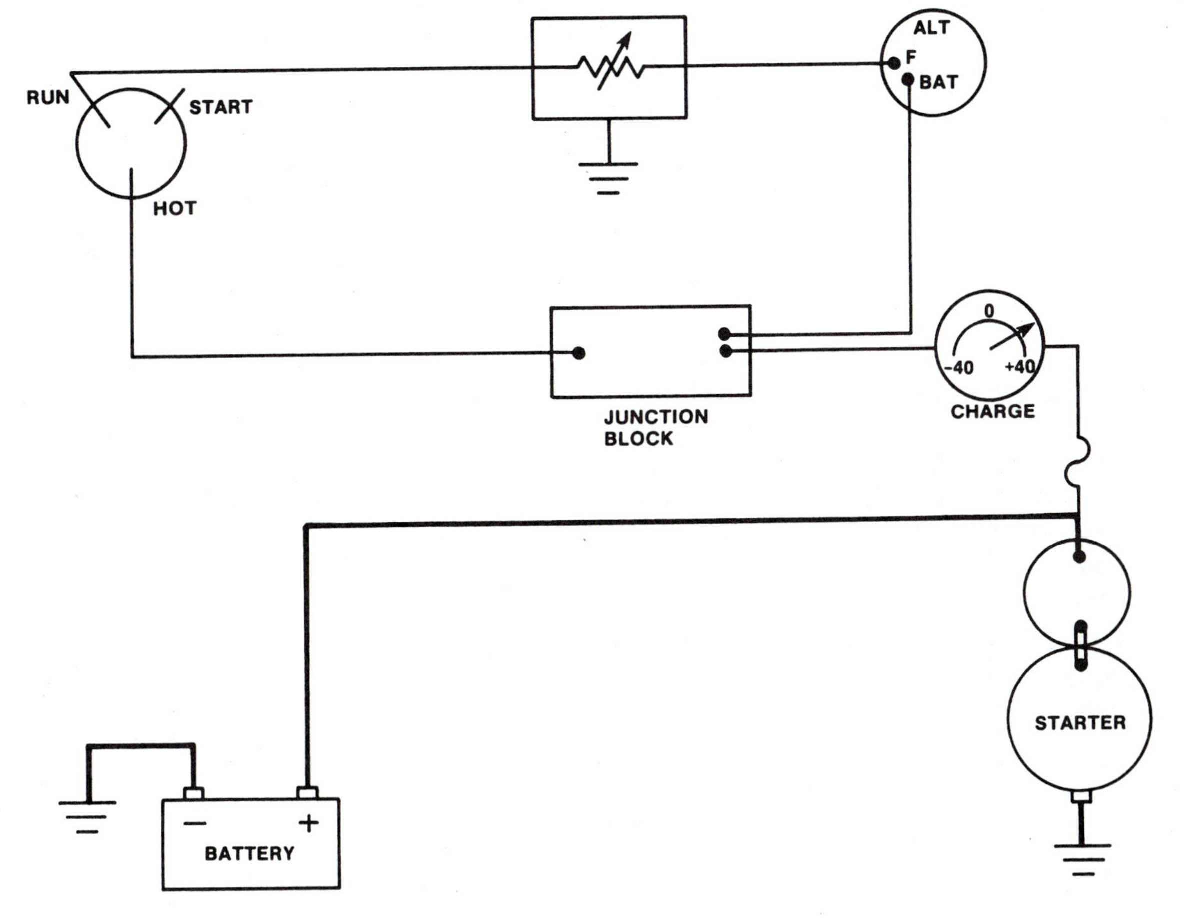

FIGURE 10-41 Ammeter installed in a typical system.

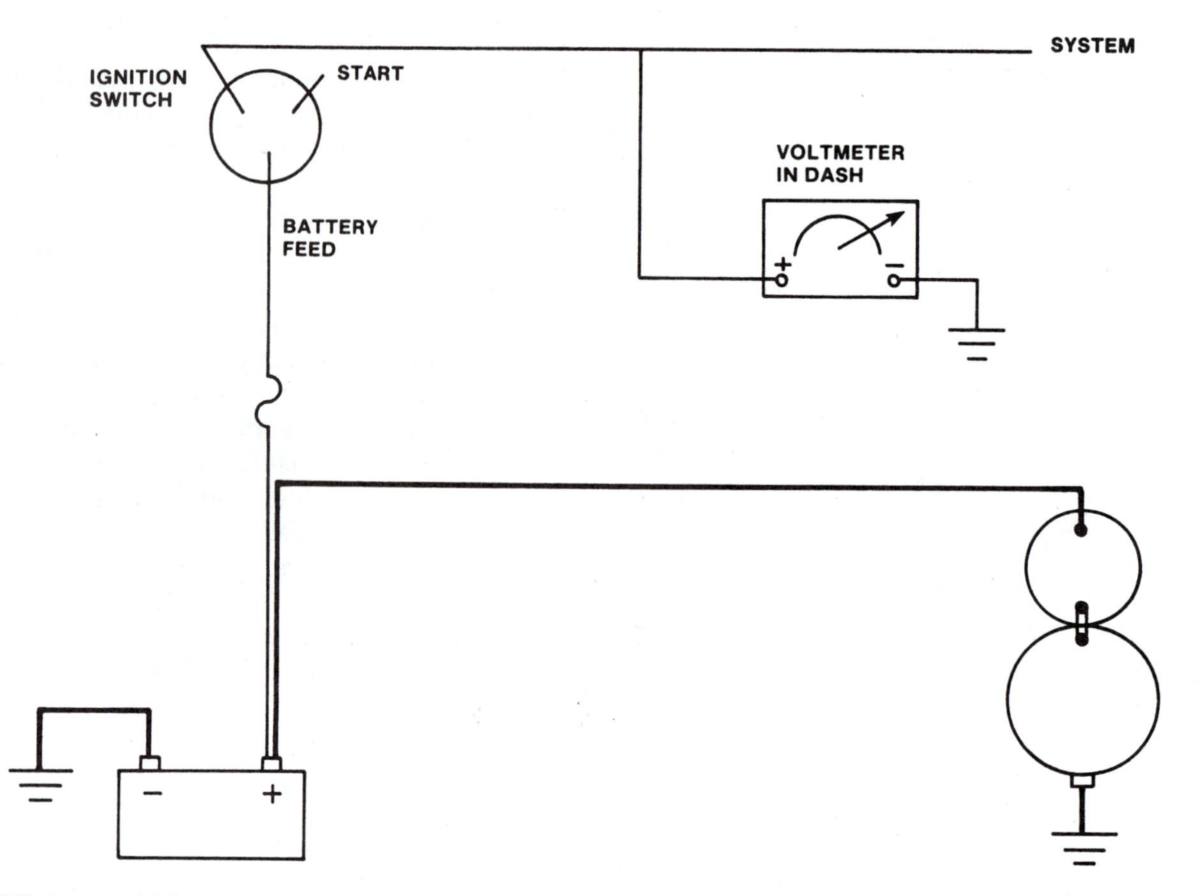

FIGURE 10-42 Voltmeter installed in typical system.

SUMMARY

In this chapter, we have looked at the charging circuit and its control. We began our discussion with the alternator and saw how the rotor, stator, and diodes produced the current necessary to both run the vehicle's electrical system and charge the battery. We then looked at the different types of regulators, starting with the mechanical style, the electronic type, and finally, computer regulation. The last part of the chapter covered the different types of instrumentation commonly found in vehicles: indicator lights, ammeters, and voltmeters.

KEY TERMS

Alternator
Rotor
Stator
Recifier bridge
Alternating current
Delta winding
Wye winding
Three-phase voltage
Full-wave rectification
Stator neutral junction
Sensing voltage
Full fielded
Zener diode

CHAPTER CHECK-UP

Multiple Choice

1. Alternator output at the Bat terminal of the alternator is
   a. DC.
   b. AC.
   c. three-phase unrectified.
   d. none of the above.
2. The magnetic field current of an alternator is carried in the
   a. stator.
   b. rotor.
   c. housing.
   d. diodes.
3. An alternator stator with three windings usually has ___ output diodes.
   a. 3
   b. 4
   c. 6
   d. 8
4. Technician A says that the alternator cannot produce alternating current until the stator windings are energized by battery current. Technician B says that the alternator cannot produce alternating current until battery current flows through the field coil of the rotor. Who is correct?
   a. Technician A only
   b. Technician B only
   c. Both Technician A and B
   d. Neither Technician A nor B
5. The output of an alternator is created in the
   a. stator.
   b. rotor.
   c. housing.
   d. brushes.
6. Technician A says that the drive pulley is attached to the rotor. Technician B says that the magnetic field cuts across the stator conductors. Who is correct?
   a. Technician A only
   b. Technician B only
   c. Both Technician A and B
   d. Neither Technician A nor B
7. A and B refer to ___ circuits.
   a. field
   b. stator
   c. diode
   d. rectifier

8. Technician A says that an internally regulated alternator will not have rectifying diodes. Technician B says that externally regulated alternators will not have rectifying diodes. Who is correct?
   a. Technician A only
   b. Technician B only
   c. Both Technician A and B
   d. Neither Technician A nor B
9. Increasing the regulator field resistance means that
   a. more output current is desired.
   b. the rotor is turning too slowly.
   c. less output current is desired.
   d. B+ sensing voltage is lower than specified.
10. The alternator field circuit is composed of
    a. battery, stator, and diodes.
    b. battery, regulator, and stator.
    c. battery, rotor, and regulator.
    d. regulator, stator, and rotor.

True or False

11. The magnetic field is carried by the stator.
12. The field current flows through the stator.
13. Brushes and slip rings carry the full alternator output.
14. Field current for the alternator comes from the battery.
15. Alternator output decreases as field current increases.
16. Alternator maximum available output increases as rotor speed increases.
17. Field relays are generally used with internal regulators.
18. Larger alternators generally have delta wound stators.
19. If output current is greater than needed, system voltage will fall below specification.
20. Sensing voltage is the voltage that tells the regulator just how much field current is necessary.

CHAPTER 11

Vehicle Charging System Testing and Service

OBJECTIVES

At the conclusion of this chapter, you should be able to:

- test the battery for charging system related-problems.
- test alternator output and compare to specifications.
- test regulator operation for full-fielding ability and for voltage regulation and compare to specifications.
- full-field alternator.
- replace diode trio.
- replace brushes.
- replace bearings.
- replace diodes.
- diagnose charging light problems.
- test diode function with an oscilloscope.

INTRODUCTION

Now that we have taken a look at the various components that make up the typical charging system, it is time to begin our discussion of actual charging system problems. As reliable as the modern system is, it is not perfect, and over time it will experience failure. Like the starting system, it cannot be looked at alone. The battery must be considered part of the system. Diagnosis of the battery must precede any evaluation of the alternator or regulator. The most perfect charging system will not be able to charge a faulty battery correctly. Sometimes technicians are tempted to shortcut the testing procedures and perhaps bypass a step or two. Eventually, these shortcuts produce a misdiagnosis and the technician has a customer who is not very happy. For this reason, it is extremely important that a complete diagnosis of what appears to be a charging system difficulty should precede any parts replacement. An example of the importance of this could be a no-start with a three-year-old battery. The customer calls for a new battery. As the technician, you should start with a state of charge test and then do a load test before condemning the battery. If you skip the state of charge test, perform the load test, and have the battery fail, replacement of the battery might not cure the problem. In this example, let us say that the fan belt is loose and it is allowing the alternator to slip under heavy electrical load. In a few days, your customer is back with a dead battery, or worse yet, someone else discovers the loose belt and tightens it, solving the no-start problem. What do you think your customer's (or perhaps ex-customer's) opinion of your mechanical skill is? Shortcutting correct procedure will result in misdiagnosis many times. Even though time is valuable, take the necessary time to completely evaluate the situation.

THE STATE OF CHARGE

Beginning with the battery, first consider the state of charge. Indirectly this will give us information regarding the charging system. A fully charged battery probably has a charging system that is producing some current. Do not assume the opposite, however. A partially discharged battery does not necessarily indicate a charging problem. The battery could have an internal drain or the vehicle could be drawing excessive current when the engine is off. If the battery passes the state of charge (80% or better), a load test should be performed to determine the capacity. Failure of the load test should lead you right into the three-minute charge test to test for sulfation. These three tests are again reproduced for you in Figure 11-1. If you have forgotten any of them, do not hesitate to review the chapters on the battery. Remember also that sulfation of a battery will occur if the battery is never brought up to full charge. It is possible for a charging system difficulty to produce a premature sulfation condition. Replacing a sulfated battery with a new one will only cure the immediate vehicle difficulty and not cure the cause. Weak, sulfated, or self-discharging batteries should be replaced before you try to diagnose the charging system. Substitute a known good battery for the testing procedure if you have to, but do not proceed until the battery is good or your results might not be accurate.

CHARGING SYSTEM TESTING

Once we are assured that the battery in the vehicle is functioning correctly, we can continue with test procedures. We will now look at alternator maximum output and see if it matches the specification for the vehicle. Keep in mind that the specifications listed in various manuals might list a couple of different alternators for the vehicle. When in doubt, look on the actual alternator case. Most manufacturers will print the rated output or color code it to indicate size. Usually, manufacturers will increase the size of the alternator as they add more current drawing accessories. Air-conditioned vehicles or vehicles with rear window defoggers, for example, usually will have a larger alternator than the same vehicle without air-conditioning. Make note of the specifications as you will refer back to them while you complete the testing procedure.

The volts-amps tester (VAT) should be hooked up to the vehicle as you did for starting testing: large cables over the battery posts and the inductive pickup around the vehicle's battery cable. Do not forget to include any small wires that are attached to the battery cable inside the inductive pickup, because this might be a charging wire. This is illustrated in Figure 11-2. Make sure that you are close to the battery with the inductive pickup or it might sense a magnetic field from some component under the hood and give you a false reading. Remember also, that during starting the meter range must be set high enough so as not to damage the meter movement. Once the vehicle is running, the range can be lowered for a more accurate reading (usually around 100 amps). Also, at this time, make sure all accessories on the vehicle are off.

The next thing to consider is engine and alternator speed. Most alternators will not be capable of producing their maximum output at idle. Around 2,000 engine rpm will be necessary. Use a tachometer and try to block the throttle open to 2,000 rpm while you perform an output test.

Specifications

Battery Size ______________ Engine Cubic Inches ______________

Hydrometer Readings

Cell Number	1	2	3	4	5	6	
Negative Post							Positive Post

Results (Check one.)

_____ All cells are within 25 points and are charged (continued testing).
_____ All cells are equal and are discharged (charge battery).
_____ Cell readings vary more than 25 points (skip load test and advise customer to replace battery.)

OR

Open Circuit Voltage Test

Load battery to 20 amps for about 1 minute before testing OCV.

______________ volts—open circuit (no load)

Refer to the chart and convert either the hydrometer reading or the OCV over to state of charge.

Charge	Specific Gravity	OCV (12.6)	OCV (6.3)
100%	1.265	12.6	6.3
75%	1.225	12.4	6.2
50%	1.190	12.2	6.1
25%	1.155	12.0	6.0
Discharged	1.100	11.9	5.9

Battery Load Test (for batteries at least 75% charged)

Set tester to the following:
1. Tester to STARTING
2. Variable LOAD off
3. Voltage—INTERNAL—higher than battery voltage

Attach the BST as follows:
1. Place large cables of tester clamp over the battery clamps, observing that polarity of the battery and tester are the same.
2. Attach inductive pick-up clamp around either of the BST cables *not* battery cables.

Determine the correct load:

Battery Amp/Hour × 3 = Load or
Cold Cranking Amps ÷ 2 = Load

NOTE: Compare the specific battery to the one actually installed. Load to the LARGER of the two.

Apply the calculated load for 15 seconds, observe voltmeter, and turn load OFF.

_____ GOOD—battery voltage remained above 9.6 volts during load test
_____ BAD—battery voltage dropped below 9.6 volts during load test (perform 3-minute charge test to check for sulfation)

3-Minute Charge Test (only for batteries that fail the load test)

1. Disconnect the ground battery cable.
2. Connect a voltmeter across the battery terminals (observe polarity).
3. Connect battery charger (observe polarity).
4. Turn the charger on to a setting around 40 amps.
5. Maintain 40 amp rate for 3 minutes while observing voltmeter.

_____ GOOD—voltmeter reads less than 15.5 (battery is not sulfated).
_____ BAD—voltmeter reads more than 15.5 (battery is sulfated).

FIGURE 11-1 Battery test should precede charging tests.

FIGURE 11-2 Volts-amps tester connected for battery tests.

The regulator will have to full-field the alternator for us to be able to read output. The easiest method for us to force this full-field operation is to use the carbon pile load to reduce battery voltage. If you remember our discussion of regulators in the last chapter, reduced voltage to the regulator is the signal for more alternator output. Usually, the load will have to reduce system voltage down to 12 or 13 volts before full-fielding will occur and alternator maximum output is achieved. Our ammeter will now register the current that is going into the battery. Record this level, turn off the load, reduce the vehicle speed to idle, and turn the engine off. To obtain the actual alternator maximum output, we must add to this number the current draw that the vehicle required to keep itself running. The ignition system and any full time accessories were drawing current along with the field coil in the alternator. The current they were using was produced by the alternator but never delivered to the battery for our inductive pickup to register. Turning the ignition key to on and recording the draw will give us the current we must add to our previous number to obtain actual alternator output. The number we now have should be very close (within 2 to 3 amps) of the rated output of the alternator. If it is, we now know that the alternator can produce its rated output and that the regulator does have the ability to full-field the alternator when it sees reduced system voltage.

If the total current available does not match the specification, we have identified a problem and will have to isolate the alternator, regulator, and interconnecting wires to determine just what is wrong. Trade standards are such that we will want to replace or repair the cause of our reduced output. Replacing both

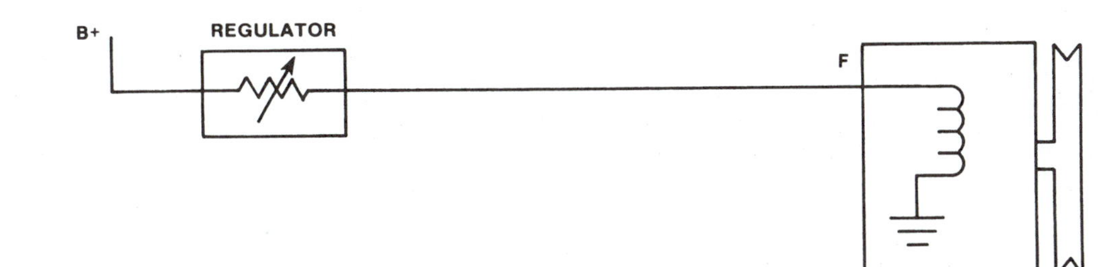

FIGURE 11-3 B circuit.

the alternator and the regulator will no doubt cure the problem but will leave you open to criticism regarding your diagnosis and possible over-repair. Isolation of the various components is best done by the process of full-fielding the alternator with the regulator disconnected from the circuit. If we supply full-field current to the alternator and still get reduced output, we have proved that the cause must be inside the alternator. If, however, full output is received when we full-field the alternator, we know that the regulator must have been the cause of the problem. Let us take it step by step with an externally regulated alternator of the B circuit variety. Hopefully, you remember that a B circuit indicates that the regulator is between the rotor and B+ and that the rotor is grounded inside the alternator. Figure 11-3 shows a block diagram of this style.

I. Procedure for Full-Fielding the Alternator

First, we disconnect the regulator with the ignition key off.

> **NOTE:** Disconnecting or reconnecting an electronic component with current flowing through it might cause it to burn out. A good rule to follow is never disconnect or reconnect anything with the key on.

By disconnecting the regulator, we have lost our source of B+ for the alternator's field circuit. Next, we start the engine and set it to the same approximate 2,000 rpm that we used before. Load down the battery with the carbon pile and then apply full B+ directly from the battery to the field terminal of the regulator connector as in Figure 11-4. If you are doubtful as to which connection is the field lead, consult the wiring diagram. Pin location and wire color should make the identification easy. By applying B+ to the field directly, we are bypassing the regulator circuit completely. The load can now be turned until the voltage is about the same as it was when we were asking the regulator to full-field the alternator. This is important because we will want to compare the amperage available with the regulator in full-field operation and with the regulator bypassed. The amperage should be compared only if the voltage is the same. In addition, a full-fielded alternator with the regulator bypassed will raise the system voltage up quite quickly. There is the risk that the unregulated voltage could climb high enough to

FIGURE 11-4 Apply full B+ to the regulator F connector.

cause some damage to a sensitive electronic component. Do not let the system voltage rise above typical regulator setting (around 15 volts) and you will not go wrong. If after you full-field the alternator it produces full specified output, the regulator is at fault. But if after full-fielding it produces less than rated or the same as when the regulator was in control, the alternator is at fault. What would you do if the following situation presented itself?

Specification = 65 amps

Regulator in control 51 amps @ 12.8 volts

Bypassed full-fielded 63 amps @ 12.8 volts

Hopefully, with the above information, you would concentrate your efforts on the regulator. It apparently does not have the ability to full-field the alternator. Field current must be reduced as it flows through the regulator. Regulator resistance in series with the field coil is the most common cause, especially with mechanical regulators. As the contact points pit and burn, they develop resistance, which reduces field current and alternator output.

How about the following situation?

Specification = 65 amps

Regulator in control 51 amps @ 12.8 volts

Bypassed full-fielded 51 amps @ 12.8 volts

This situation is the opposite set of circumstances. The regulator is doing all it can. It must be supplying full B+ to the rotor because the output is the same as when we supply B+ directly from the battery. It looks like the alternator is at fault here, doesn't it? This condition will require us to perform one additional step before we condemn the alternator. This step will be to full-field the alternator directly at the field terminal of the alternator. By doing this, we will be eliminating the wiring from the regulator to the alternator as a possible cause. Excessive resistance along this wire could cause the previous situation. If after directly full-fielding the amperage is the same, it is a safe bet to assume that the alternator or drive belt is faulty and needs to be serviced or replaced.

The key to this procedure is the use of the carbon pile to maintain the same voltage levels with the regulator in control and when you are full-fielding. In addition, do not disconnect or reconnect the regulator with the engine one.

This is a good place to mention the use of the belt tension gauge. If the drive belt is loose, the rotor might slip and alternator output reduce to the level of our second example. More than one alternator has been replaced when the real reason for the reduced output was a loose drive belt. Figure 11-5 shows a common style of belt tension tester that lists sizes of V-belts and common tensions for both new and used. The belt is placed between the guides and the tension is read directly off the scale.

Figure 11-6 shows the block diagram of an externally regulated A circuit alternator. Full-fielding it will be as easy as our B circuit example. The only difference will be that we will supply a ground to the field terminal rather than B+. The same principles apply to the results.

If you are not sure whether you are working on an A or a B circuit, try both B+ and ground. You will not do any damage if you apply the wrong test procedure in this case. A B circuit has one brush grounded in the alternator and requires B+ for correct full-fielding procedure. If you apply a ground to the F ter-

FIGURE 11-5 Belt tension gauge installed on a V-belt.

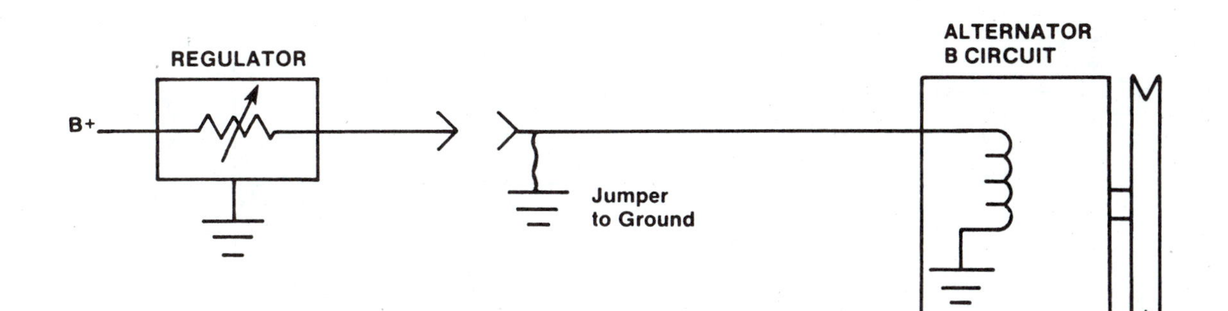

FIGURE 11-6 Externally regulated A circuit alternator.

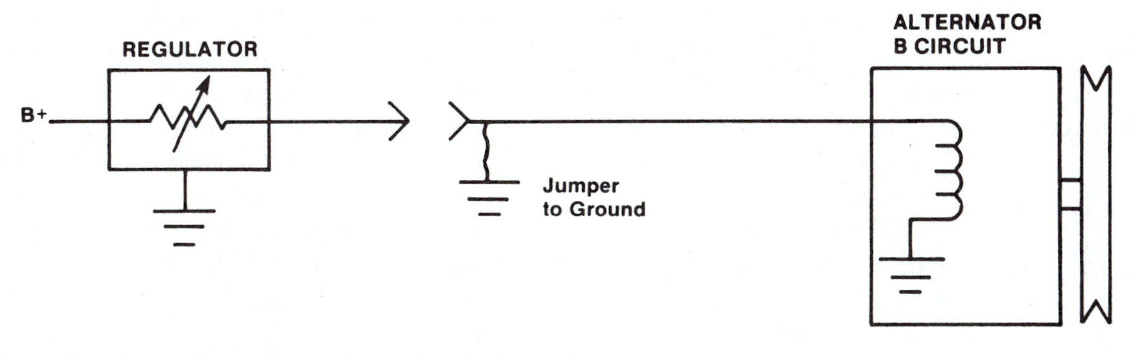

FIGURE 11-7 Grounding a B circuit produces 0 output.

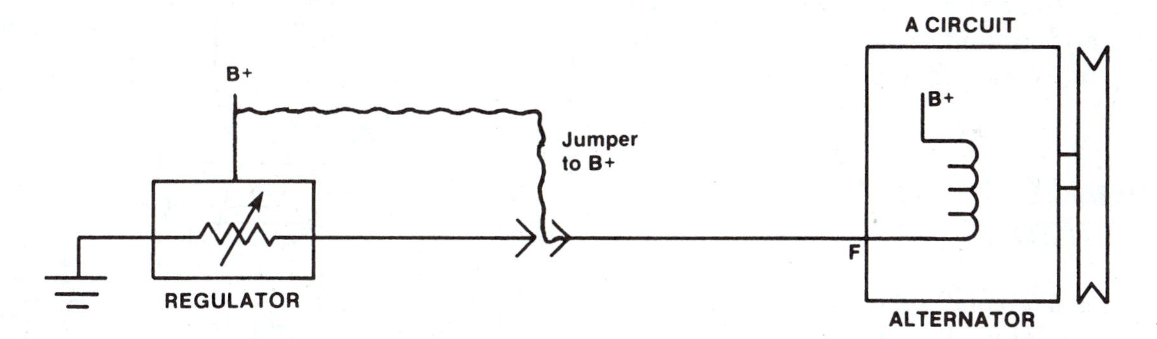

FIGURE 11-8 B+ to an A circuit produces 0 output.

minal thinking it to be an A circuit, your circuit looks like Figure 11-7. No current flows because you have a ground on both sides of the field coil. Similarly, B+ to an A circuit looks like Figure 11-8, and results in no current flow. Always try both B+ and ground if you have not identified the type of field circuit.

Full-Fielding an Isolated Field

The full-fielding procedure previously outlined will work well for either an A or a B circuit but must be modified very slightly for an isolated field. To review, an isolated field can be either an A or B circuit but will pick up B+ and ground outside of the alternator. Isolated field alternators have two field leads rather than the one usually found on other styles. Figure 11-9 shows a common wiring diagram for an isolated field Chrysler alternator. Notice that it is an A circuit because the regulator is on the ground side of the field coil. B+ for the circuit comes off the ignition switch and enters the alternator at the top field connection. The circuit continues through the rotor and out the bottom field connection over to the regulator. Notice also that the regulator receives its sensing voltage from the ignition switch off the same lead that is the field B+ source.

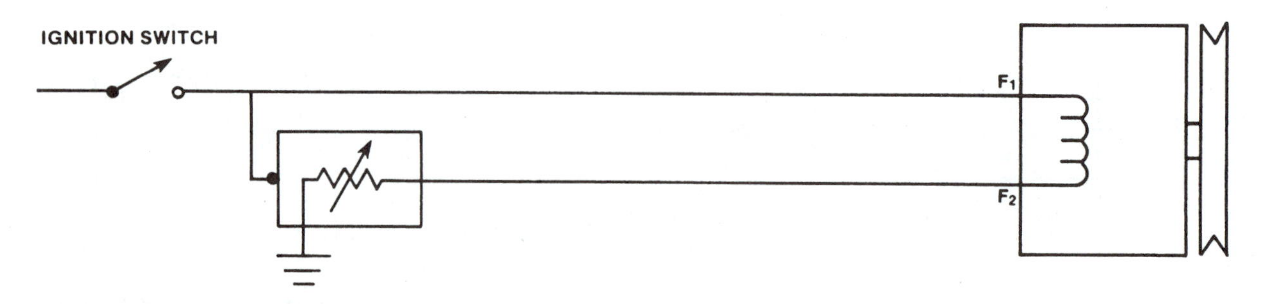

FIGURE 11-9 Common wiring diagram for an isolated field.

Full-fielding requires that we apply full B+ and ground to the rotor. On isolated field alternators, you will have to jumper one lead to B+ and jumper the other field terminal to ground. By doing both, you will be assured that you have truly separated the alternator from the rest of the circuit. If you were to just jumper the lower field terminal to ground (with the regulator disconnected) and the alternator were to produce reduced output, you still could have a good alternator. If reduced B+ is available for the field coil, reduced output will be the end result. Always verify that both B+ and ground are available when full-fielding an isolated field alternator.

INTERNALLY REGULATED ALTERNATORS

Let us look at a couple of different styles of internally regulated alternators in common use. Keep in mind that there are a few units on the market that cannot be full-fielded on the vehicle. These alternators cannot be isolated from their regulators. Their testing procedure involves removing the unit from the vehicle and disassembling it completely. The individual components can then be tested, usually with an ohmmeter. Some of the Robert Bosch charging systems are of this style. Maintenance manuals will outline the test procedure for the individual year, make, and model. There are however, many internally regulated alternators where a full-fielding procedure is applicable. Let us again look at the GM 10SI as there are millions of these units on the road. The 27SI, GM's larger output design, is tested in exactly the same manner. With the regulator inside the alternator we need not be concerned with the wires usually connecting the alternator to the regulator. Review Figure 11-10. It is the same diagram we used in the previous chapter. Remember that B+ for the field will come through the diode trio once the alternator is producing current. The back of the case of the alternator has a small D-shaped hole shown in Figure 11-11. This hole will line up with a small

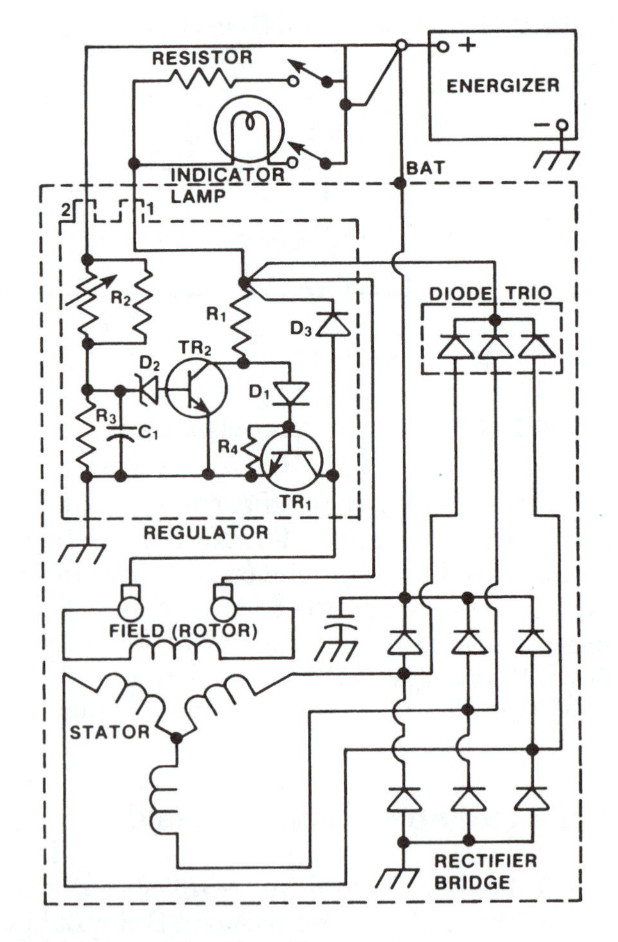

FIGURE 11-10 GM 10SI.

FIGURE 11-11 D hole location of Delco 10SI.

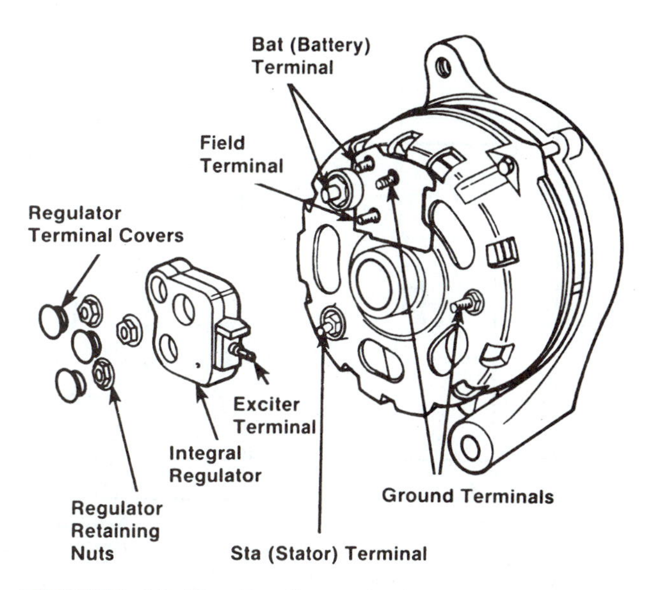

FIGURE 11-12 Ford's early integral regulator. (Courtesy of Ford Motor Company)

tab attached to the negative brush. Full-fielding this alternator is therefore quite simple because all you will have to do is place a small metal ground probe like a 1/8" allen wrench into the hole about 1/2" into the D-shaped hole, contact the tab, and touch the probe to the case of the alternator. This procedure is called grounding the D hole and it will force the alternator into full-field operation and allow you to observe an ammeter for maximum output.

CAUTION Make sure you have a load on the battery before you full-field the alternator to prevent system voltage from climbing too high and damaging something.

You will not disconnect the regulator, however, on this unit. It has been designed to allow full-fielding without damaging the sensitive electronic circuitry inside.

Ford Motor Company has two different designs of an integral regulator-alternator. Figure 11-12 shows an early design that had the regulator bolted onto the outside of the case with one terminal, called the **exciter**, connected to the outside of the regulator. Figure 11-13 shows the wiring of this alternator and regulator. Notice that the field receives its B+ from inside the alternator and that the regulator will supply the ground. By removing the protective cover from the field terminal (always the terminal closest to the rear bearing), you will be able to ground the field and bypass the regulator just as we have been able to do previously. As was the case in the GM 10SI, it is not necessary to disconnect the regulator before full-fielding it.

In the mid-eighties, Ford changed the regulator position, Figure 11-14. This unit has the connecting wires for the regulator external. Field current B+ comes form a connection ahead of the ignition switch. The ground for the circuit is supplied by the regulator. Notice the F terminal on the bottom of the diagram. This is the terminal that you will ground when you wish to full-field the alternator. The F terminal is usually protected by a cover but is easily accessible for testing purposes.

REGULATION

Figure 11-15 is a photo of this type of Ford alternator and regulator. Notice that the regulator has been removed and that the brush assembly is an integral part of the assembly. This regulator and brush assembly is very easy to replace out in the field. One of the mounting screws is the insulated F terminal which you will ground if you want to full-field this alternator.

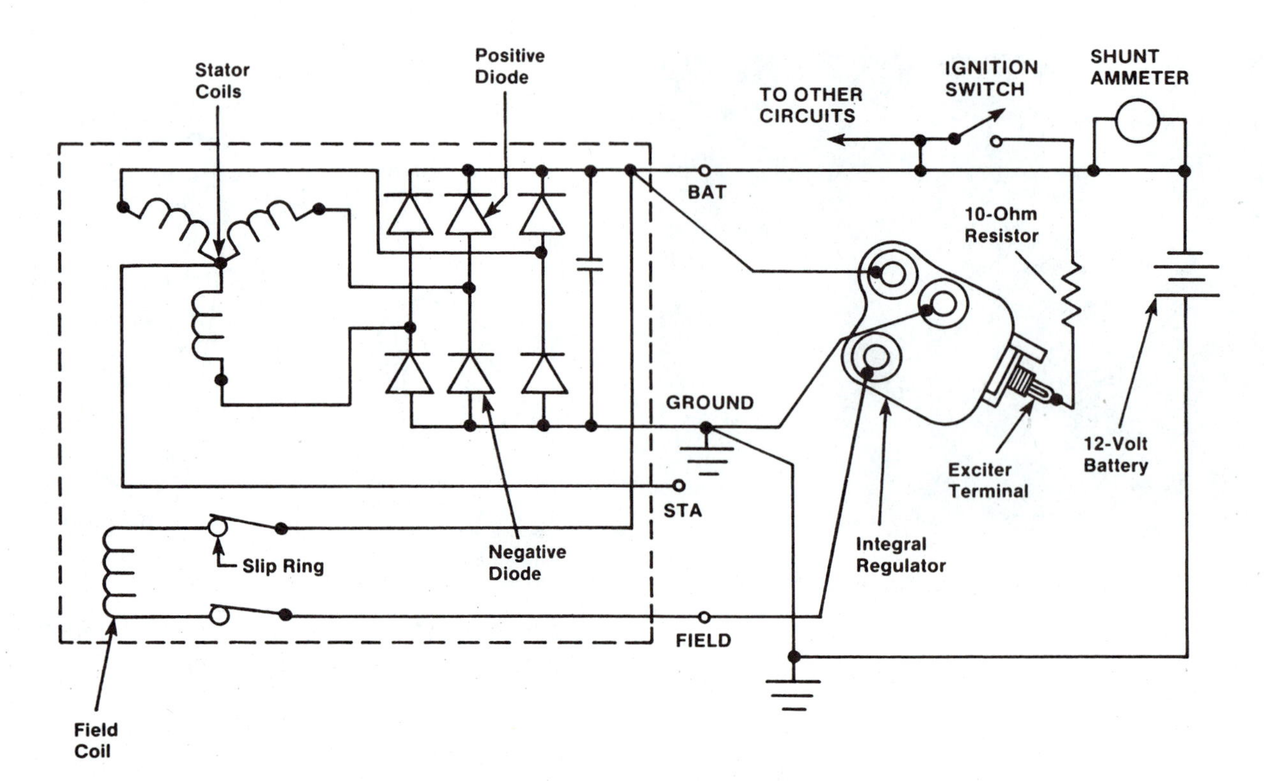

FIGURE 11-13 Schematic of Ford charging circuit. (Courtesy of Ford Motor Company)

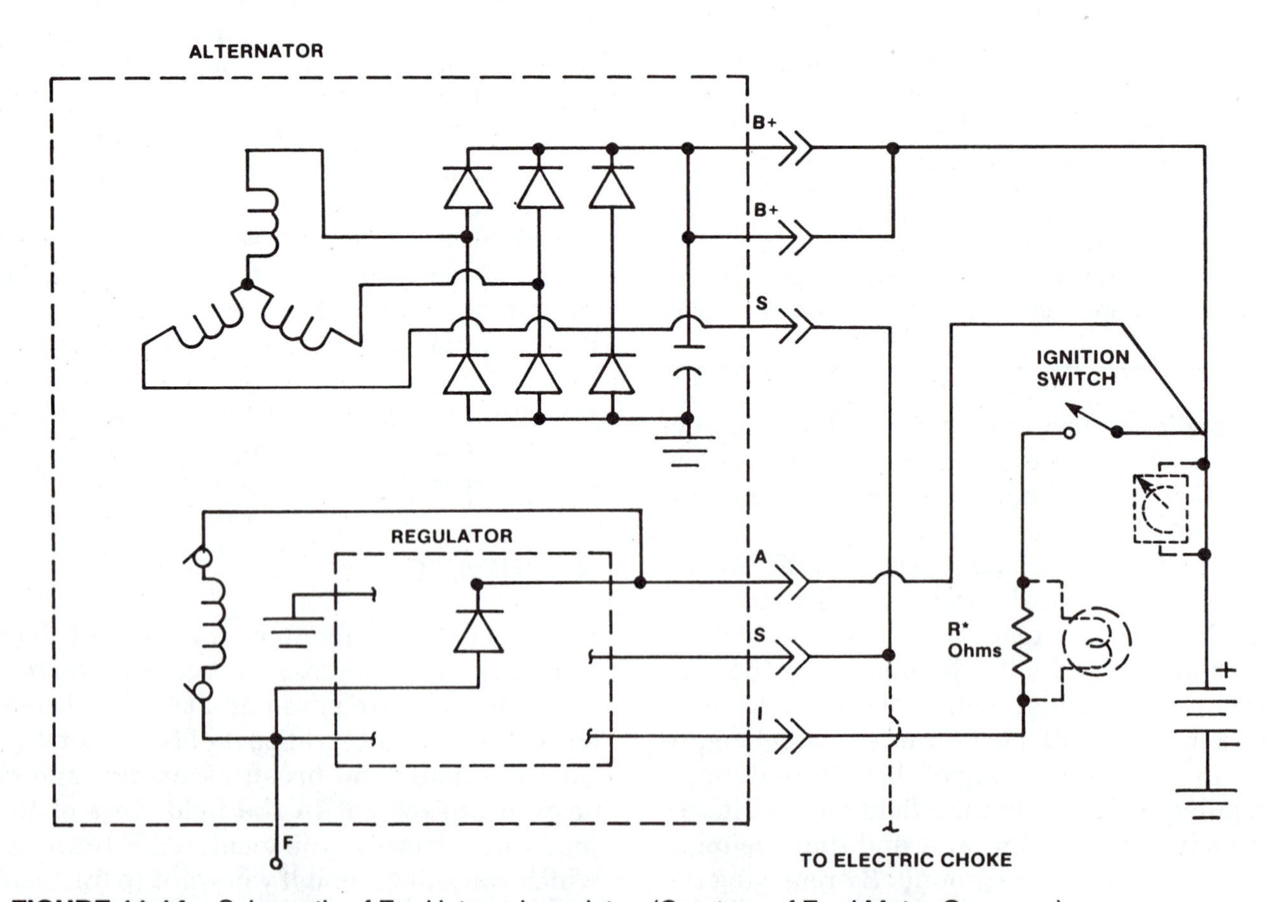

FIGURE 11-14 Schematic of Ford integral regulator. (Courtesy of Ford Motor Company)

FIGURE 11-15 Ford alternator with regulator removed. (Courtesy of Ford Motor Company)

COMPUTER REGULATION

As we studied in the last chapter, some of the manufacturers are using the vehicle computer to control the charging system. It is tested in exactly the same way as an external electronic regulator circuit is. With the key off, disconnect the field lead at the alternator. Start the vehicle and raise the engine speed up to 2,000 rpm. Load the system down with the carbon pile and ground the field terminal on the alternator. Vary the load until the voltage is the same as during your maximum output test and read the amperage. You will not find it necessary to treat a computer regulation circuit much different but should realize that regulator failure on this vehicle will require a new computer. Make sure of your diagnosis as this component is quite costly.

SYSTEM VOLTAGE REGULATION

As we have stated previously regulators really have two major functions. The first is to have the ability to full-field the alternator on demand. You have seen how this function is tested and compared to the bypassed regulator output to determine regulator condition. Regulators are replaced if they cannot full-field the alternator. The second function of the regulator is to keep the system voltage within a predetermined range, typically, 13.5 to 14.5 volts at normal operating temperature. As the temperature goes down, the regulator will raise up the operating voltage to help charge the cold battery. The next process we should check on our charging system is its operating voltage. This is very simply accomplished with the same volt-amp meter connections. As the battery comes up on charge, you will see a gradual decrease in the charging current. This decrease should occur within the specified operating voltage. Even with a fully charged battery and a charging rate of a couple of amps, the system voltage should not go above the regulator setting.

If the voltage setting is too low, the battery will not receive sufficient current to become fully charged and will gradually sulfate. If in testing regulator voltage it falls lower than the manufacturers specification, the regulator will have to be replaced. Years ago, regulators were adjustable. Most current regulators, however, cannot be readjusted to bring their voltage within specs and are replaced as a unit.

A higher than specification voltage level might not be the regulator's fault, however. Figure 11-16 shows how a 2-volt drop on the ground side of the regulator will raise up the system voltage by 2 volts. Excessive resistance within the sensing circuit will reduce the actual voltage that the regulator sees as system operating voltage. It will operate the alternator at a higher than normal output until it reaches the specified voltage that it sees. Any resistance causing voltage drop will add to the operating voltage. Sensing voltage must be accurate or the operating voltage will not be accurate. If you are testing a vehicle and find a higher than normal operating voltage, test the ground of the regulator with a sensitive voltmeter between the case of the regulator and the ground of the battery. With the vehicle running, there should not be any difference between the two ground points and the meter should read zero. Figure 11-17A shows this procedure. If the ground side is functioning correctly, use your voltmeter and test the sensing voltage input to the regulator as in Figure

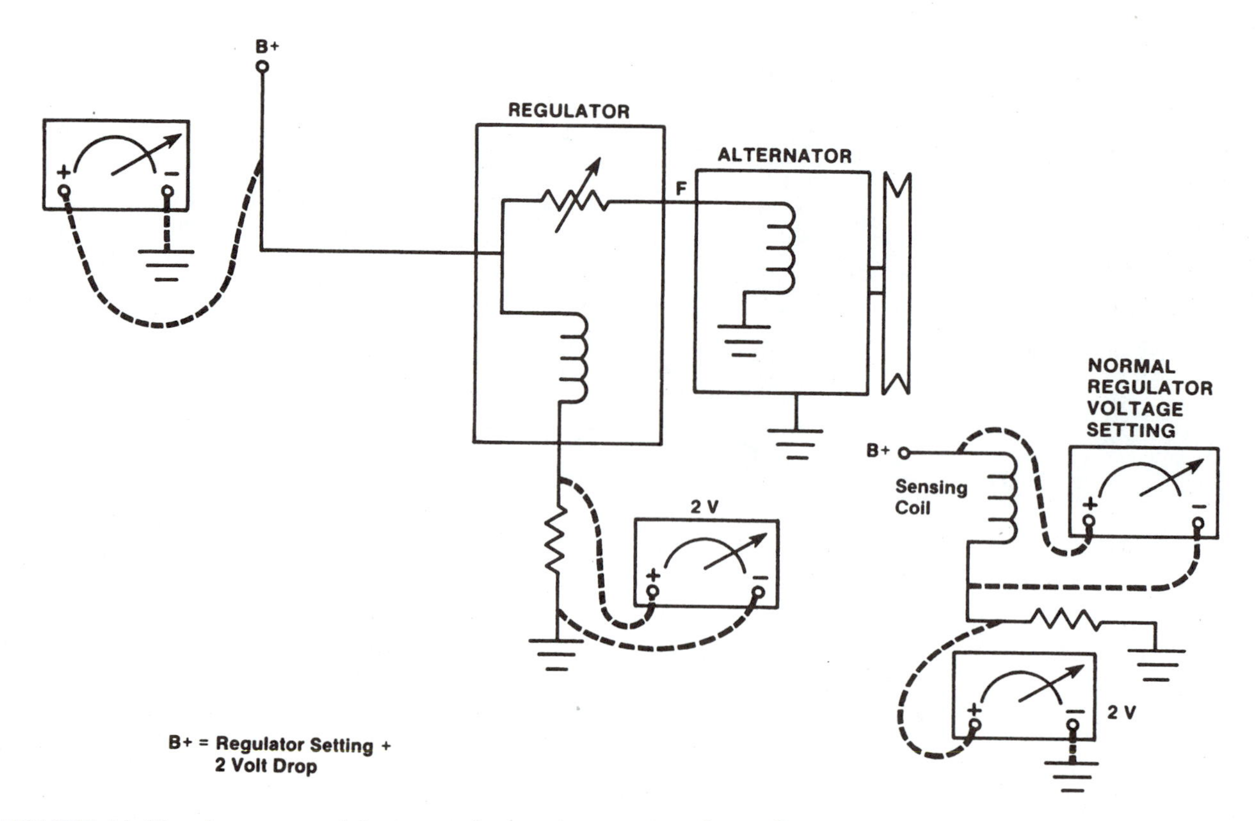

FIGURE 11-16 A poor regulator ground can raise up charging voltage.

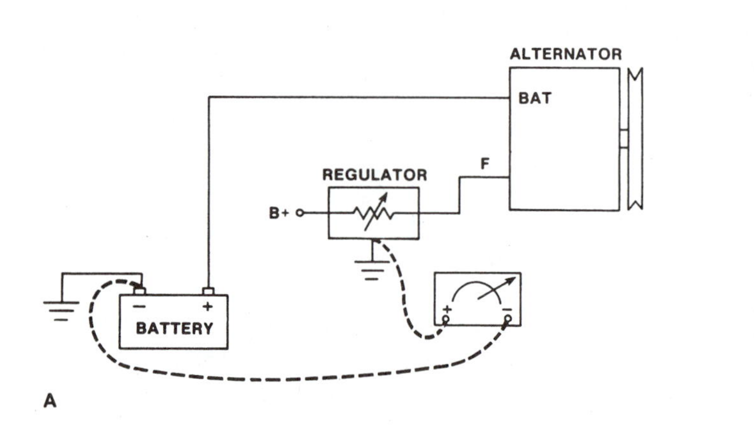

FIGURE 11-17 (A) ground circuit resistance check; (B) Sensing voltage check.

11-17B. The meter should read the same voltage as battery voltage. With a good ground and a good input voltage, the regulator should be capable of holding the voltage within specifications. If not, it should be replaced.

CHARGING VOLTAGE DROPS

The last on-car tests we will look at here are the voltage drop tests along the charging line. Testing for voltage drops is usually not routinely done on a vehicle but should be done any time a component is replaced or a stubborn charging problem cannot be identified. Simplified, it is the process to determine if the three major components, battery, regulator, and alternator, are all functioning at the same potential. We have seen why this is important for the regulator-sensing circuit, because any resistance will raise the charging voltage above the specification. Refer to Figure 11-18, which

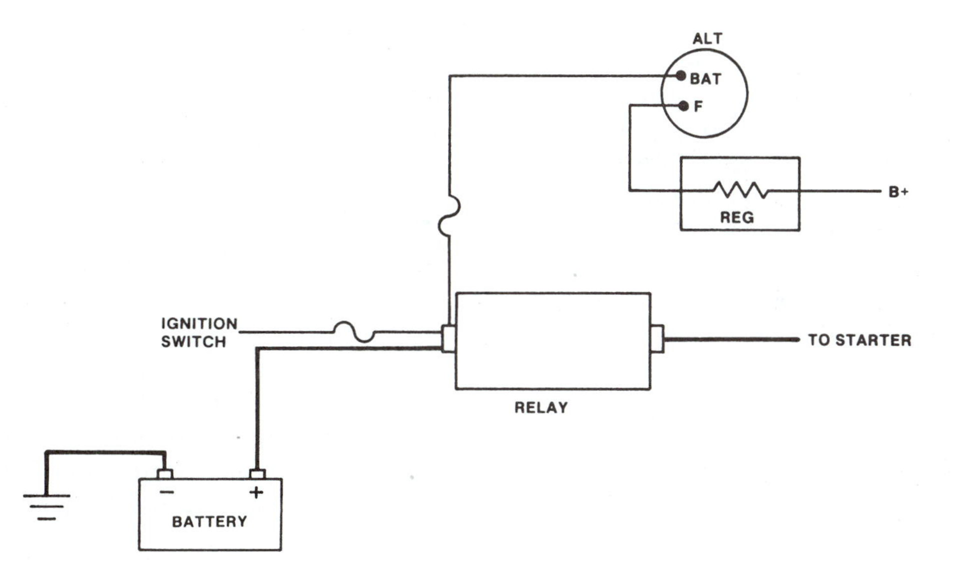

FIGURE 11-18 Typical external regulator.

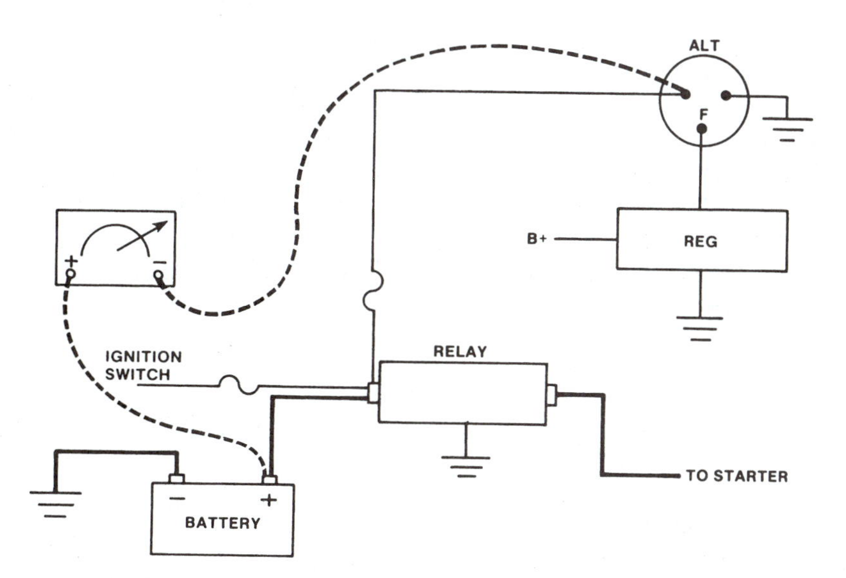

FIGURE 11-19 Voltmeter checking B+ path resistance.

is a block diagram of a typical charging system with an external regulator. Notice the use of a fuse link at the starting relay. Charging current to both run the vehicle and charge the battery must come through this link. It is a primary protection device that will have a slight voltage drop across with normal amperage. With increased amperage comes increased voltage drop and heat generation. Remember that a fuse link is two number sizes smaller than the wire it protects. This is so that it will burn before the wire or system it protects. By measuring the difference in potential at the alternator output terminal and at the battery positive post, we can look at the fuse link and assess its condition. Figure 11-19 shows the voltmeter connections for this test. It is necessary to have a small amount of amperage actually flowing through the circuit. Usually around 10 amps is specified for a maximum voltage drop of 0.3 volt with an indicator lamp or 0.7 volt with an in-dash ammeter. The circuit necessary for an ammeter is longer and will automatically have a greater voltage drop.

We must look at the ground path as well as the B+ path. With the 10 amps flowing as in the previous test, the voltmeter is moved over to the ground path. One voltmeter lead on the case of the alternator and the other on the negative post of the battery. There are not any fuse links on the ground return path and so we expect to have negligible voltage drops. Less than 0.1 volt will be acceptable. Figure 11-20 shows a drawing of the ground return path voltmeter connections. Once the alternator ground is verified, we will move the voltmeter lead from the alternator to the case of the regulator if it is externally mounted. With about 10 amps flowing, there should be no voltmeter reading. On internally regulated alternators, we skip this test as we cannot easily get to the regulator ground connection inside the alternator.

The B+ path and ground path voltage drop tests are all necessary because the charging system will be capable of functioning correctly only if all the major components are at the same potential. You will remember that voltage drops are present whenever current flows through resistance. Usually, this resistance does not appear overnight. It has taken years to corrode or burn. Testing for the resistance before it becomes an open circuit and the car dies is why this test is so important. It is preventive maintenance oriented and when done correctly, will usually identify a problem before it becomes a no-start and tow.

REVIEW OF CHARGING SYSTEM TESTING

Let us look at where we have been so far in this chapter. We have identified a series of tests to identify charging system difficulties. It began with a full series of battery tests because battery condition is fundamental to a properly functioning charging system. Next, we moved to the alternator and looked at V-belt tension, maximum output with the regulator in command, and then with the alternator full-fielded. This was done to pick out a bad alternator or regulator. We then checked out the actual voltage setting of the regulator. Rounding out the series was a check of the battery, alternator and regulator, B+ voltage, and ground potential, looking for excessive voltage drops that could have an effect on the systems performance. This series of tests is outlined for you in Figure 11-21.

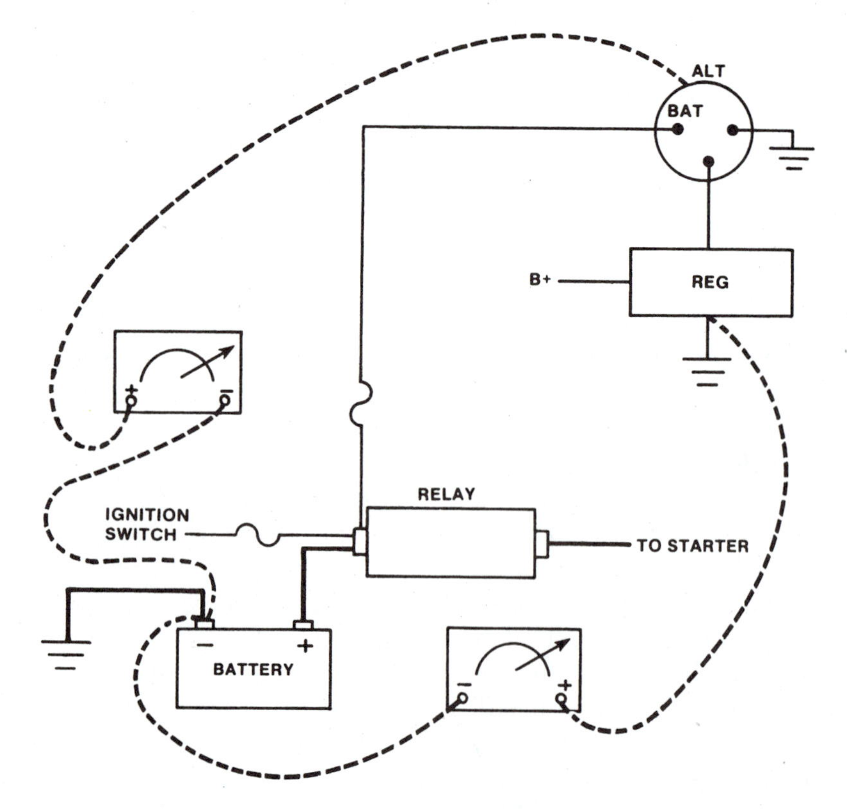

FIGURE 11-20 Voltmeter checking ground path for resistance.

Charging System-Alternator Output

1. With tester attached to vehicles battery as in starter draw, reconnect the ignition system and start the vehicle. Place fast idle cam to raise idle to 2,000 rpm approximately.
2. Change selector switch over to charging (or an amperage setting around 100 amps.)
3. Rotate the load control until the highest amp meter reading is obtained. Record here: ____________________ amps @ ____________________ voltage. Turn load off.
4. Turn vehicle off and disconnect the field lead from the regulator or the alternator. Note if A or B circuit.
5. Attach jumper wire or the SUN field lead to the F terminal.
6. Re-start the vehicle, raise idle to 2000 rpm and full field the alternator.
7. Rotate the load control to obtain the same voltage as in Step 3. Record here: ____________________ amps @ ____________________ voltage. Turn load off.
8. Compare reading obtained in Steps 3 and 7. They should both be the same and to specs. If not
   a. the regulator is faulty, or
   b. voltage drop exists between the alternator, regulator, and/or battery.

Voltage Regulator Setting

1. With the regulator connected, start the vehicle; idle speed.
2. Raise the engine speed to a fast idle. The voltmeter should start to climb. Maintain fast idle with *no* load until voltage peaks.
3. Record the highest voltage: ____________________ volts. The highest voltage read should match the specs. Any difference indicates
   a. the regulator is faulty, or
   b. a voltage drop exists between the alternator, regulator, and/or battery.

   Note: Position #3 on a VAT or a 1/4 ohm position can be used to simulate a fully charged battery for this test.

Charging System Voltage Drop

1. With the charging system hooked up. Start the vehicle and raise the speed up to a fast idle.
2. Vary the load until the alternators output is around 20 amps.
3. Maintain the load and use a voltmeter to measure the B+ drop between the output terminal of the alternator (positve voltmeter lead) and the battery positive post (negative voltmeter lead). Record below:

   ____________________ volts

   Note: Charging systems with warning lights should have less than 0.3 volt drop.
   Charging systems with ammeters should have less than 0.7 volt drop.
4. Place the negative lead of the voltmeter on the alternator case and the positive meter lead on the regulator case. Record below:

   ____________________ volts

   Note: Any reading more than 0.3 volt indicates excessive resistance. Most resistance problems are traced to improper, loose, or corroded terminals and connections.
5. Place the negative lead of the voltmeter on the alternator case and the positive meter leads on the battery negative post. Record below:

   ____________________ volts

   Note: Any reading more than 0.3 volt indicates excessive resistance. Most resistance problems are traced to improper, loose, or corroded terminals and connections.

Accessory Amp Draw

1. With the vehicle off and doors closed, record ammeter reading below:

   ____________________ amps

   _____ Good—0 reading

   _____ Any reading at all. Pull each fuse until meter registers 0. Short circuited section can be found from wiring diagram that shows fuses.

CHARGING SYSTEM DIODE CHECK

1. With the engine at a fast idle switch the VAT over to position #4 (Diode/Stator). The needle should stay within the O.K. band.

   or:

1. Use an AC voltmeter and measure the amount of AC available at the alternator output terminal. Record below:

   ____________________ Volts AC

   Note: Any AC voltage in excess of 0.2 Volt indicates diode problems.

FIGURE 11-21 Charging worksheet.

ALTERNATOR DISASSEMBLY AND REPAIR

In this section, we will look at some typical alternator repairs. It is important to note that in many sections of the country, alternators are not repaired. They are replaced with a new or rebuilt unit. The difference in consumer costs is what usually dictates whether you will rebuild the unit yourself or purchase it from a supplier. With the labor rate rising constantly, it is conceivable that in your part of the country it is cheaper and just as reliable to purchase a new or rebuilt unit than it is to purchase the internal components and pay the labor rate. Removing and replacing costs will be the same in either case. We will identify the repairs in this chapter, realizing that you might never attempt them out in the field.

The "when" of repair is just as important as the "why". We should realize that alternators are repaired usually because of noise, reduced output, or reduce rectification. We will also be capable of replacing internal regulators and diode trios. Let us begin our discussion of these repairs with a generic disassembly and testing procedure that will work on most alternators.

FIGURE 11-22 Scribe case.

FIGURE 11-23 Unbolt and separate case halves.

II. Procedure for Alternator Disassembly and Testing

1. Scribe the case to help in identification of correct position during reassembly, Figure 11-22.
2. Separate the case halves, Figure 11-23.
3. Remove the stator from the diodes, Figure 11-24.
4. Remove the brush assembly, Figure 11-25.

Many times a visual inspection at this point will reveal just what the difficulty is. Broken wires, missing insulation, burned conductors, short brushes, or corroded slip rings are all easily spotted and do not require specialized equipment.

FIGURE 11-24 Disconnect and remove stator.

FIGURE 11-25 Remove screws holding brush assembly, regulator, and diode trio.

Using an ohmmeter, measure the resistance of the rotor: slip ring to slip ring, Figure 11-26A, and slip ring to shaft, Figure 11-26B. Here we are looking at the actual resistance of the rotor and ensuring that it is still insulated from ground (shaft). You should have low resistance (specification) between the slip rings and infinity between the rings and the shaft.

NOTE: Do not forget to use the lowest scale for low expected readings and the highest scale if you are looking for or expect to find infinity.

Using an ohmmeter on the diode test position, test the individual diodes, Figure 11-27. They should exhibit low resistance in one direction and very high (infinity) in the other. All diodes should have about the same readings. Reversing the ohmmeter leads gives you the ability to look at the forward and reverse conduction of the diode.

Using an ohmmeter, test the stator windings. You should have zero resistance between the stator leads, Figure 11-28A, and infinity between the leads and the stator frame, Figure 11-28B.

Using an ohmmeter, test the diode trio (if the alternator uses one), Figure 11-29. Your readings should be the same as for any diode. High resistance in one direction and low resistance in the other with the leads reversed.

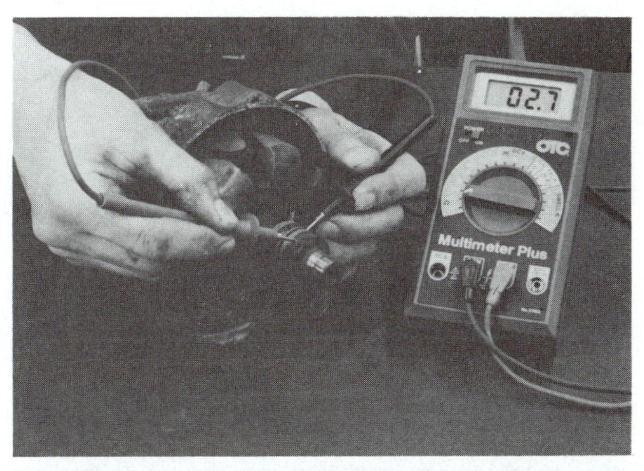

(A) Measure resistance of rotor, slip ring to slip ring.

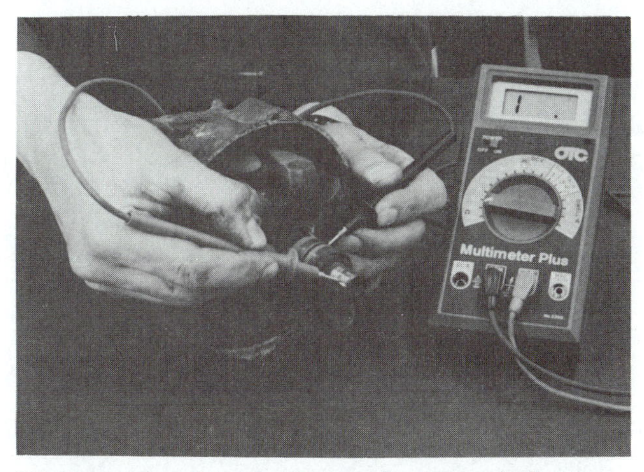

(B) Measure resistance of rotor, slip ring to shaft

FIGURE 11-26 Test the rotor for an open circuit.

FIGURE 11-27 Test each diode using an ohmmeter.

Use an ohmmeter and check the brush path. You should find 0 ohms of resistance through brushes, Figure 11-30.

By performing these tests, you will be able to identify which components will require replacement to restore the alternator's ability to produce maximum current again. The following series of figures will show how to replace some of these components. While each alternator differs slightly in design, you will find the procedures will work on the vast majority of both foreign and American alternators.

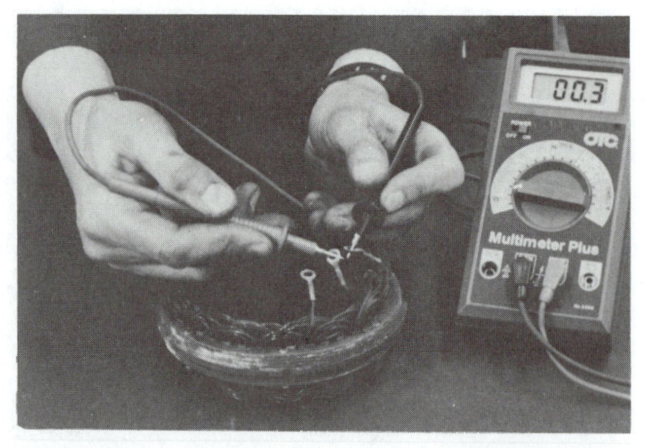

(A) Using an ohmmeter, test the stator windings. There should be zero resistance between the stator leads.

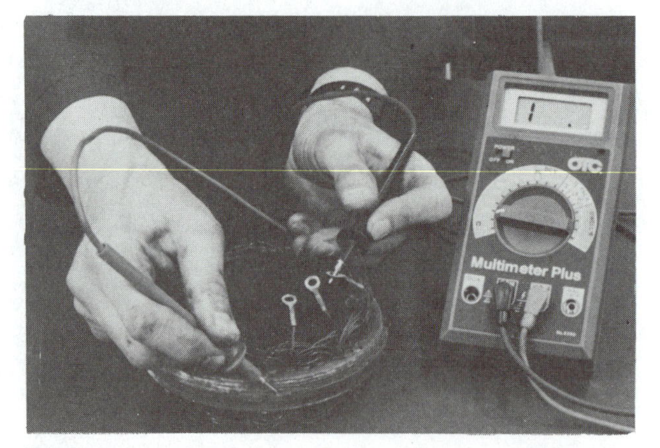

(B) Using an ohmmeter, test the resistance between the leads and the stator frame. It should measure infinity.

FIGURE 11-28 Testing each stator circuit.

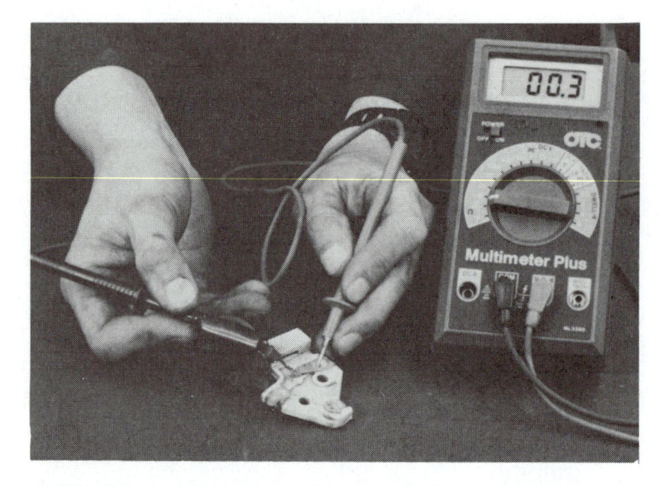

FIGURE 11-30 Test the brush field current path with an ohmmeter.

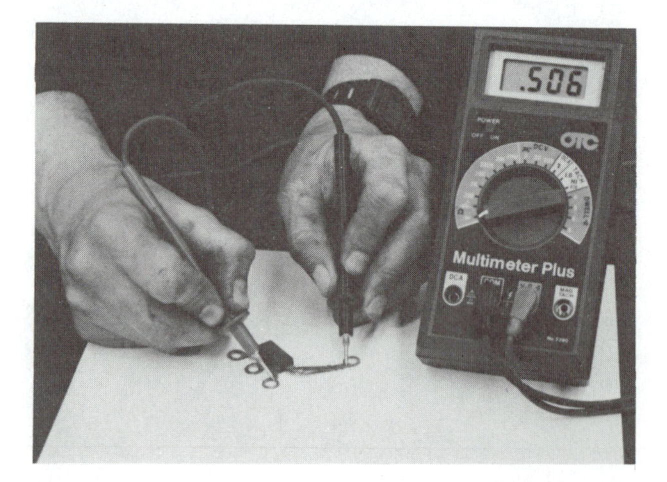

FIGURE 11-29 Test the diode trio with an ohmmeter.

III. Procedure for Alternator Repairs

Bearing Replacement

1. Hold the rotor shaft with an allen wrench and remove the shaft nut and washer, Figure 11-31.
2. Remove the rotor from the case half, Figure 11-32.
3. Remove the bearing retainer, Figure 11-33.
4. Remove the bearing, Figure 11-34.

FIGURE 11-31 Remove rotor nut.

FIGURE 11-32 Remove pulley, fan spacer, and locknut.

FIGURE 11-33 Remove bearing retainer.

FIGURE 11-34 Drive the bearing out of the end frame.

Replace in reverse order, paying particular attention to the torque of the shaft nut. Many alternators use a wave washer to hold the pulley to the rotor shaft. If the torque is too low, the pulley might slip on the shaft, reducing the output and quite possibly destroying the rotor shaft.

Figures 11-31 through 11-34 is also the procedure followed to replace a rotor.

Bearing Replacement—Brush End Frame

1. Drive the bearing out of the case being careful not to bend the aluminum case, Figure 11-35. Note that the case half is being supported on wooden blocks.
2. Inspect the bearing hole for burrs or cracks, Figure 11-36.
3. Drive the new bearing into position, Figure 11-37.

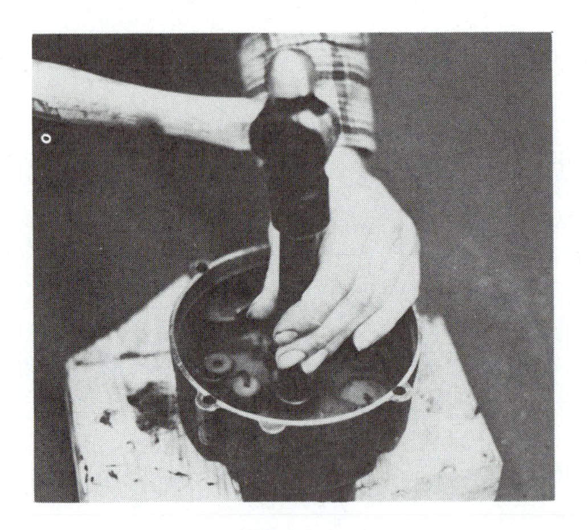

FIGURE 11-35 Drive brush end frame bearing out.

FIGURE 11-36 Inspect bearing hole for cracks and so on.

Diode Replacement—Pressed-In Design

1. Remove the stator connection from the diode, Figure 11-38.
2. Set up the diode press to push the diode out of the case or heat sink, Figure 11-39.

NOTE: It is possible to use two sockets and a vise if the press is not available.

Position the sockets with a large one over the diode and a smaller one pushing on the diode case. In this manner, you will push the diode into the larger socket.

3. Inspect the diode hole for cracks or burrs, Figure 11-40.
4. Replace the diode with the reversed press or sockets until its shoulder is flush with the case, Figure 11-41.

NOTE: Many diodes are currently in replaceable bridge assembly. They are replaced as a set rather than individually.

FIGURE 11-37 Drive new bearing in.

FIGURE 11-38 Remove stator connector.

FIGURE 11-39 Press diodes out.

FIGURE 11-40 Inspect diode hole.

Diode Replacement—Soldered-In Design

1. Use a soldering iron or gun capable of about 25 to 40 watts of heat, Figure 11-42. Do not use much more heat or you might damage the replacement diode.
2. Use a heat sink between the soldered connection and the diode, Figure 11-43.

NOTE: A pair of needle nose pliers held tightly onto the diode lead will act as a heat sink.

---

3. Pull on the connection lightly as you heat the solder, Figure 11-44. Do not overheat the connection. Use the minimum amount of heat necessary to melt the solder.
4. Resolder your new diode using the heat sink and a minimum amount of rosin core solder, Figure 11-45.
5. After cooling, recheck the diode with an ohmmeter to verify its condition, Figure 11-46.

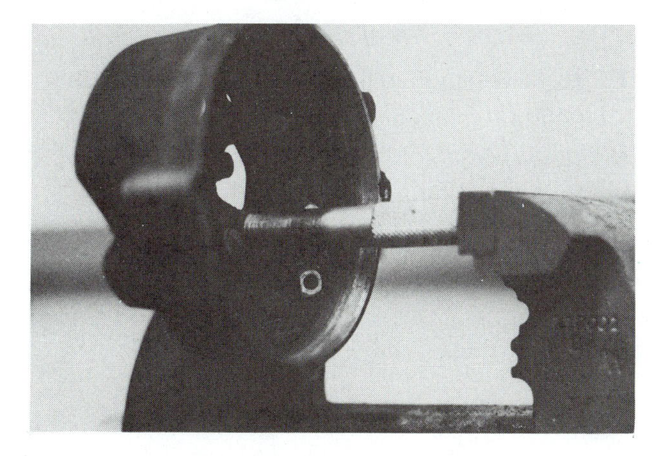

FIGURE 11-41 Diode press replacing new diode.

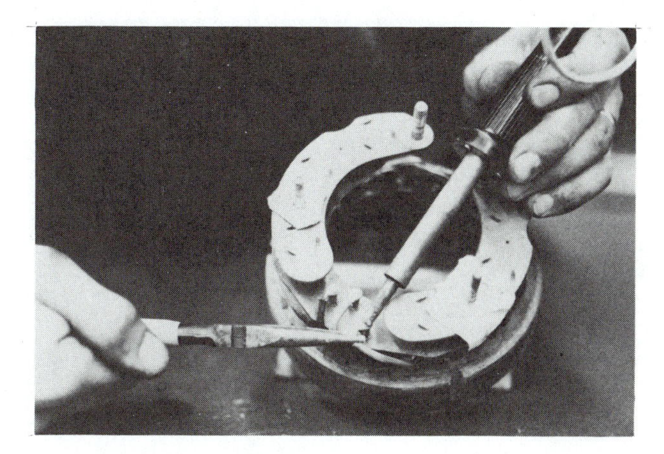

FIGURE 11-42 Soldering iron.

FIGURE 11-43 Needle nose pliers can act as a heat sink.

FIGURE 11-44 Unsolder old diode.

FIGURE 11-45 Resolder new diode.

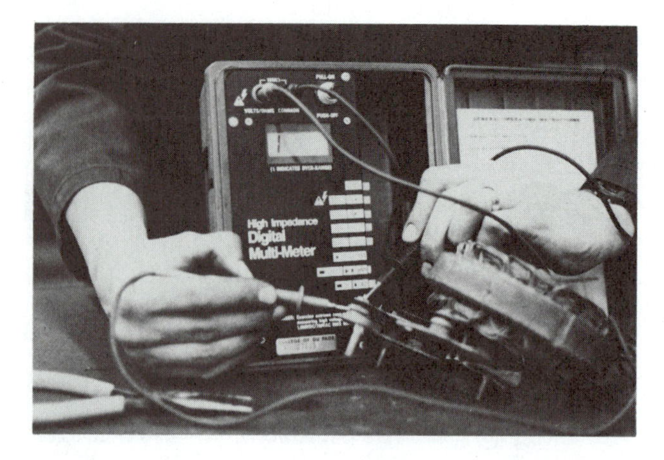

FIGURE 11-46 Use ohmmeter to check the new diode.

Brush Assembly

Most brushes are available as an assembly in their insulating holders.

1. Attach the brush assembly to the case, Figure 11-47. Do not over tighten the screws.
2. Use a paper clip through the case half to hold the brushes against spring tension, Figure 11-48.
3. Replace the rotor half with the slip rings into the bearing, Figure 11-49. Do not force it together. It should easily slip together. Make sure the stator wires are not touching the rotor.
4. Spin the rotor, Figure 11-50. It should spin quietly and easily.
5. Line up the scribe marks and replace the case half bolts or screws, Figure 11-51.
6. Remove the paper clip, Figure 11-52. The brushes will now be forced against the slip rings by their springs.

FIGURE 11-47 Install new brush assembly.

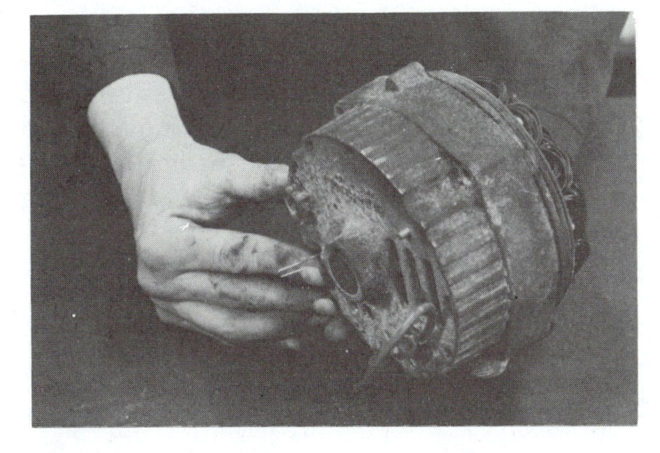

FIGURE 11-48 Paper clip holds the brushes against spring tension.

FIGURE 11-49 Replace rotor half.

FIGURE 11-50 Spin the rotor.

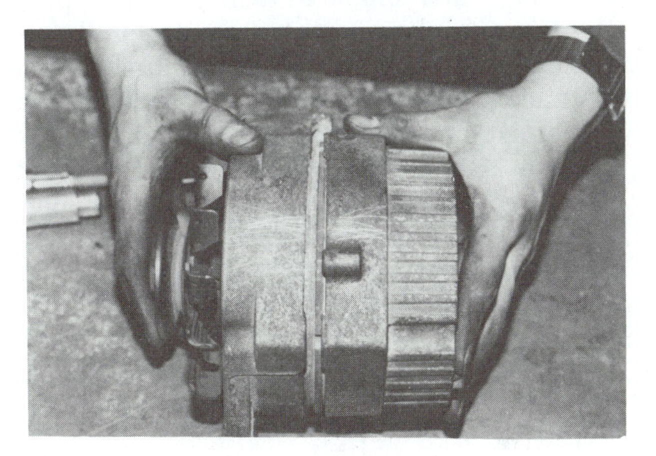

FIGURE 11-51 Line up the scribe marks.

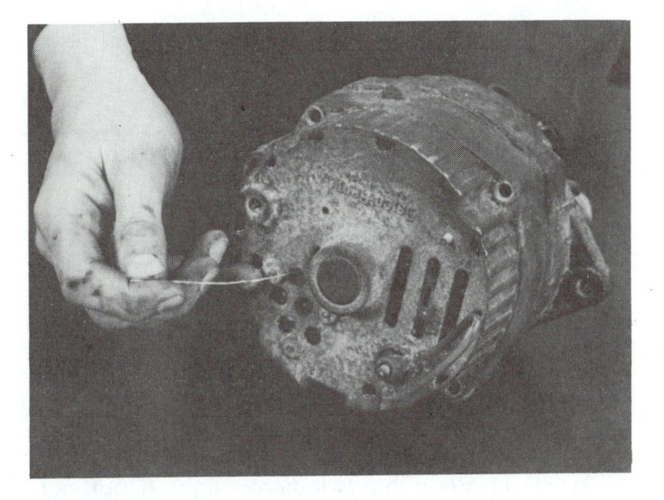

FIGURE 11-52 Pull out the paper clip.

CHARGING LIGHT DIAGNOSIS

The use of the in-dash charging light is rather limited from the technician's standpoint, but this is a good time to note its use. Let me emphasize that it is intended to be a warning light for the consumer rather than the technician, however, certain basic information can be interpreted off the light. It is especially important to discuss its operation with your customer and note if it has not been functioning in the normal manner. Let us define normal operation. With the key on and the engine off, the light should be on. Most vehicle charging systems run field current through the bulb with the key on. The very high resistance of the bulb, when compared to the field coil, puts the largest voltage drop across the bulb, resulting in it being on brightly. This high resistance also allows very little field current to flow. Just enough to begin operation once the vehicle is started. Once started, the light should go out and remain off during all operating conditions. Failure of the bulb to light with the key on and the engine off will indicate a problem in the field circuit. Failure of the bulb to go out once the engine is running indicates a lack of charging current. In either case, a complete performance test of the entire system beginning with the battery and ending with voltage drop testing is indicated.

The dimly lighted indicator lamp is harder to diagnose. Start with a complete charging system test and determine if there are any components at fault. If output is correct and regulator voltage is correct, it usually involves excessive resistance, causing a voltage drop somewhere in the circuit. Look at Figure 11-53. It shows the typical indicator lamp circuit for an externally regulated alternator. If with this vehicle we experience a glowing alternator light, we will have to assume that the voltage on the two sides of the bulb is different. One side of the bulb is fed off the ignition switch, while the other comes from the alternator. Getting out the voltmeter and testing

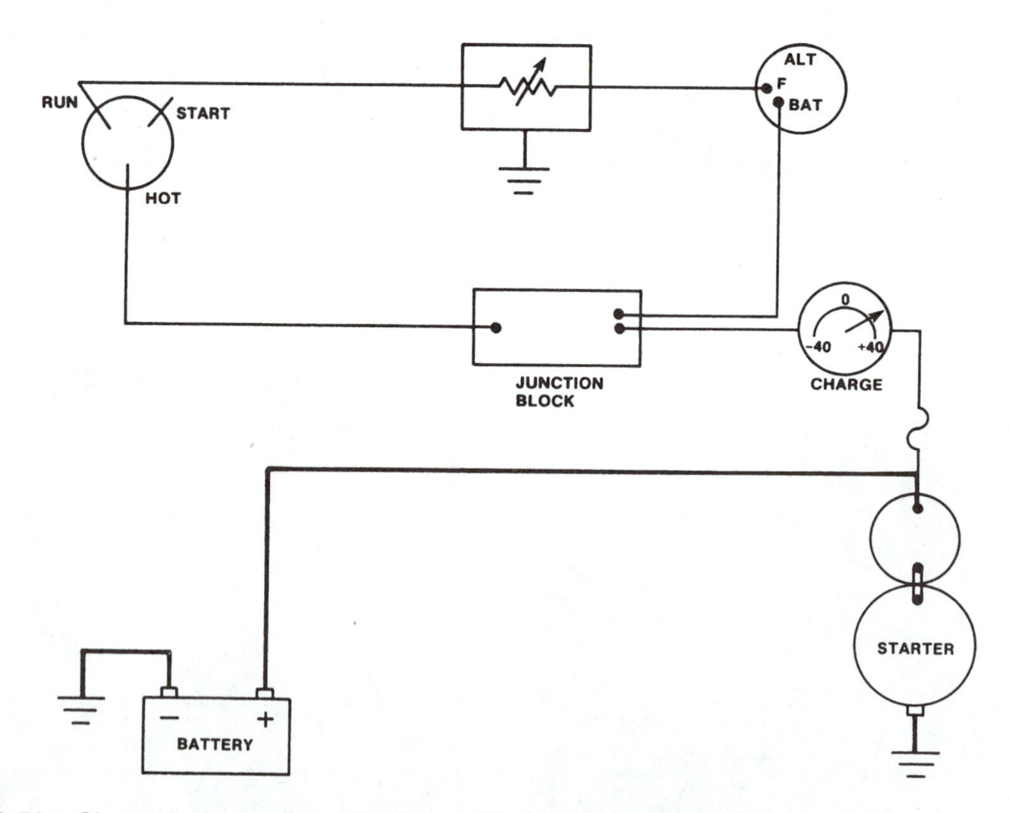

FIGURE 11-53 Charge indicator light. (Courtesy of Delco Remy Division)

the circuit voltage at both points and then working your way closer and closer to the bulb until the difference is found is usually the recommended procedure. When the voltage drops down, you have found the spot where the resistance is located.

DIODE PATTERN CHECKING

One last item for this chapter. The diodes are responsible for transferring the AC from the stator into the DC needed by the vehicle. If a diode is breaking down, it might allow some AC to get out of the alternator. Most of the components on the vehicle will not be affected by this AC. There is some electronic circuitry that could be damaged by it, however. As a result, it is a good idea to run a diode pattern check whenever failure of an electronic component has taken place. Testing is simple to perform.

With the scope on the lowest scale available, place the primary leads on the alternator output and alternator ground, Figure 11-54. Run the vehicle with a moderate load (about 50% of rated output). Figures 11-55 through 11-57 show the patterns for a four-cylinder vehicle. It is important to note that further testing should be done after the alternator has been disassembled. The primary reason for doing the diode pattern test will be to prevent a repeat failure of electronics. As more and more electronics is added to the vehicle, the diode pattern will be even more important.

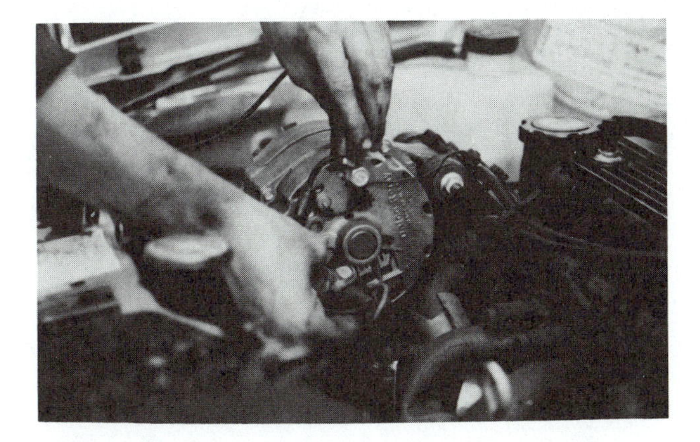

FIGURE 11-54 Setup for diode pattern check.

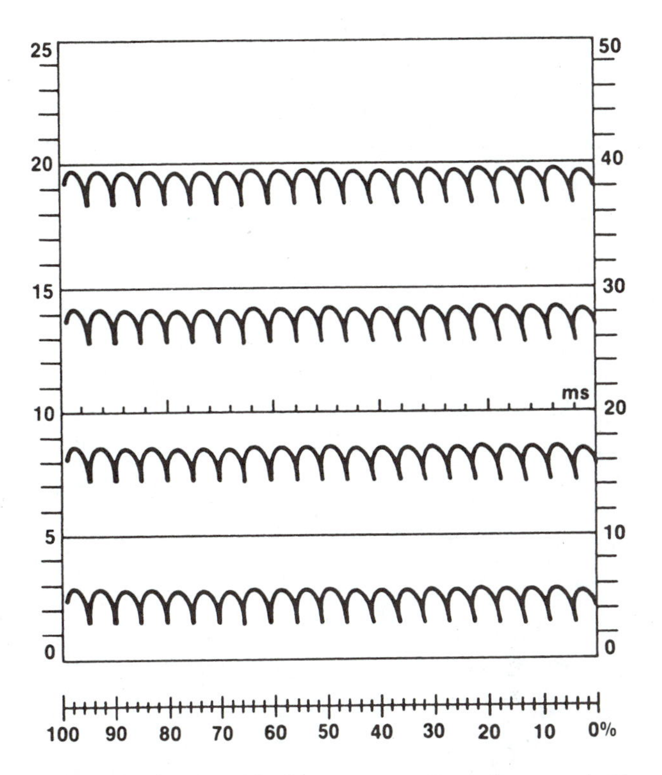

FIGURE 11-55 Diode scope pattern for good diodes. (Courtesy of Sun Electric Corporation)

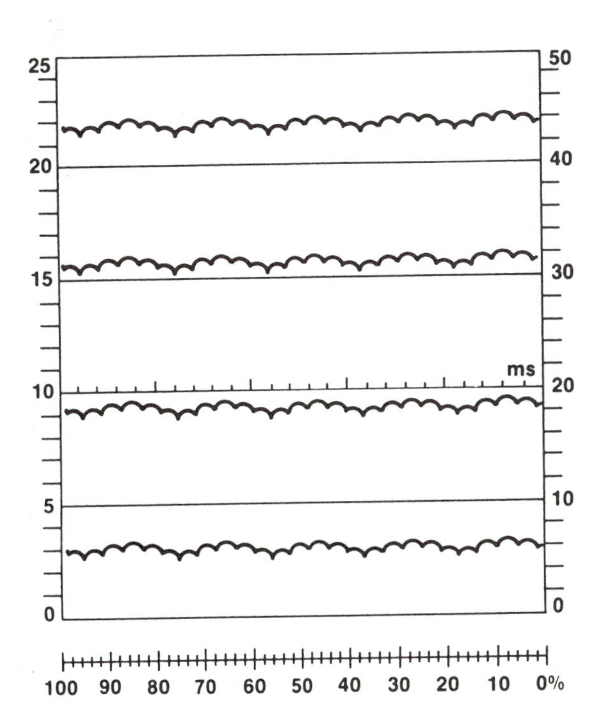

FIGURE 11-56 High resistance at the diodes. (Courtesy of Sun Electric Corporation)

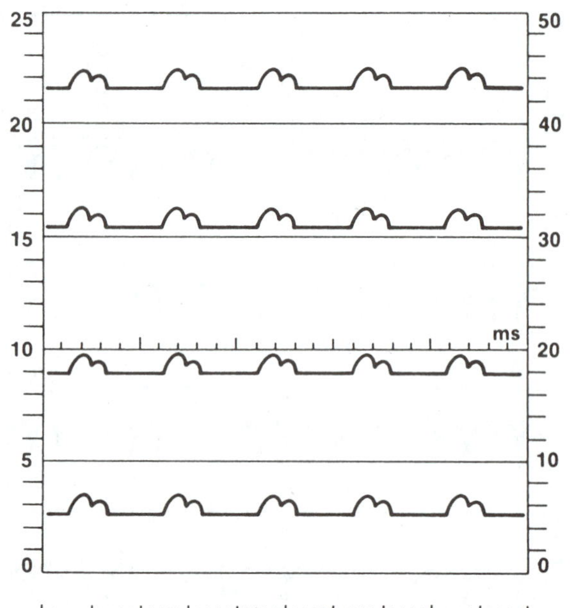

FIGURE 11-57 One open and one shorted diode. (Courtesy of Sun Electric Corporation)

In addition, there are some general rules to be observed when working on charging systems. They are:

1. Never run the vehicle with the battery disconnected. It helps to stabilize the voltage and absorbs any voltage spikes that could damage electronic components.
2. Do not allow the system voltage to go above 16 volts during full-fielding.
3. Jumper the battery off another vehicle positive post to positive post and negative to negative.
4. Charge the battery with the cables disconnected.
5. Never remove components on the vehicle with the battery connected.
6. Disconnecting and reconnecting electrical connectors should be done with the vehicle off.

SUMMARY

In this chapter, we have looked at on-vehicle charging system diagnosis. We stressed the importance of battery testing before any charging system diagnosis is attempted. Use of the VAT type tester to full-field the alternator was demonstrated. In addition, the most common alternator repairs were shown: replacing diodes, brushes, and bearings. Charging light diagnosis was shown. The routine voltage drop testing, sensing voltage testing, and fan belt tension were also demonstrated.

KEY TERM

Exciter

CHAPTER CHECK-UP

Multiple Choice

1. A battery that is overcharged can be due to
   a. loose drive belt.
   b. poor regulator ground.
   c. burned out diode.
   d. resistance in the field circuit.
2. An alternator should never be operated open-circuit (battery disconnected) because this could
   a. overcharge the battery.
   b. burn out electronic components.
   c. ruin the alternator stator.
   d. high-speed driving.
3. A battery that is undercharged might be due to
   a. poor regulator ground.
   b. open-sensing wire to the regulator.
   c. resistance between the output terminal and the battery.
   d. high-speed driving.
4. To full-field an alternator, Technician A says run a jumper wire from the F terminal of the regulator connector to ground on an A circuit. Technician B says run a jumper wire from the F terminal of the regulator connector to B+ on a B circuit. Who is correct?
   a. Technician A only
   b. Technician B only
   c. Both Technician A and B
   d. Neither Technician A nor B
5. A self-powered test light lights when hooked across a diode. Reversing the leads causes the light to go off. Technician A says this indicates the diode is faulty and should be replaced. Technician B says this indicates that the diode is functioning correctly. Who is correct?
   a. Technician A only
   b. Technician B only
   c. Both Technician A and B
   d. Neither Technician A nor B
6. Alternator output test shows 45 amps at 13 volts. Full-fielding produces 55 amps at 13 volts. This indicates
   a. nothing.
   b. alternator is faulty.
   c. regulator circuit problems.
   d. both the alternator and the regulator should be replaced.

7. Battery voltage is 19.6 volts while the vehicle is running. Alternator voltage is 19.9 volts and regulator voltage is 14.3 volts. Technician A says that this indicates a voltage drop between the regulator and battery. Technician B says that this indicates resistance that is causing the output voltage to rise above specification. Who is correct?
   a. Technician A only
   b. Technician B only
   c. Both Technician A and B
   d. Neither Technician A nor B
8. An alternator produces 55 amps all the time. Full-fielding it produces 55 amps also. This indicates
   a. the regulator is running the alternator full-fielded all the time.
   b. the alternator is faulty.
   c. the output wire has excessive resistance.
   d. not enough information is given.
9. Grounding the D hole on a GM alternator is done to
   a. isolate the diodes.
   b. full-field the alternator.
   c. test the system operating voltage.
   d. bypass the alternator.
10. Computer-controlled charging systems are full-fielded by
    a. special equipment.
    b. jumping a special terminal to ground at the computer.
    c. grounding or jumping to B+ the F terminal at the computer connection.
    d. none of the above.

True or False

11. Brushes and slip rings carry the full alternator output.
12. Grounding the D hole on a GM internally regulated alternator should force it into full-field operation.
13. The battery cable should be disconnected before removing any electrical unit.
14. Excessive output can be caused by a faulty battery.
15. During full-fielding testing it is okay to allow the system voltage to go as high as it can.
16. Less than rated output during full-fielding indicates need for a new regulator.
17. Sulfated batteries can be caused by a voltage regulator whose voltage is too low.
18. Diode pattern checks should be done anytime electronic components are being replaced.
19. An equal voltage on both sides of an indicator lamp will cause it to glow dimly.
20. Any sensing wire voltage drops will increase the system voltage by the amount of the drop.

CHAPTER 

12

Ignition Systems

OBJECTIVES

At the completion of this chapter, you should be able to:

- describe the different types of spark plugs currently in use.
- "read" a used plug for combustion chamber conditions.
- describe the major differences between solid-core, magnetic suppression, and resistive suppression ignition wire.
- describe the individual components and their operation of the typical ignition system.
- describe the operation of the advancing systems.
- identify the major differences between two-, three-, and four-function electronic ignition systems.

INTRODUCTION

Before we begin to look at the individual components that make up the ignition system of today, a brief overview might prove helpful. The basic purpose of the system will be to supply a high-voltage spark at the correct time to the compressed charge of fuel and air. It is important that you realize that these two purposes (high-voltage spark and correct time) are both equally important. The spark must have a voltage that is high enough and lasts long enough to actually ignite the mixture with the least likelihood of a misfire. This ignition must occur at the correct time relative to the piston position or the amount of power might be reduced. In addition, the level of pollutants out the tailpipe is usually increased if the ignition timing (when the spark is delivered) is not correct.

OVERVIEW OF IGNITION SYSTEMS

The ignition system has seen tremendous changes in the last 20 years. We have seen the system move from mechanical control to almost totally electronic and, in most cases today, computer controlled. In our discussion, we will separate the system into two halves: primary and secondary. We will then look at the different components that can be included. It is important to note that there are many different styles of ignition systems on the market today. You might not find all of the components present in each system. In the next chapter, we will discuss how each component is tested and verified.

The job of the ignition system should be kept in mind as we begin to look at individual components. A high voltage sparks to ignite the mixture at the correct time. Let us look at an item that all ignition systems have—the spark plugs.

Look at Figure 12-1. It is a drawing of a typical spark plug. The top terminal will be attached to the ignition system high voltage and the steel shell will be threaded into the combustion changer. The center electrode is insulated from the steel shell and the side or ground electrode is attached to the shell. This places the side electrode at ground potential and the center electrode at ignition potential. The air gap between the two electrodes is where the high voltage will arc to ground and, hopefully, ignite the mixture. This is the business end of the plug. Notice that the center electrode and the shell are insulated from one another by an insulator, usually porcelain. This insulator must be capable of withstanding thousands of volts. Remember the object of the system. We must deliver the voltage to the air gap. Any insulation breakdown before the air gap will result in a misfire.

Terminal
Insulator
Steel Shell
Side Electrodes
Center Electrode

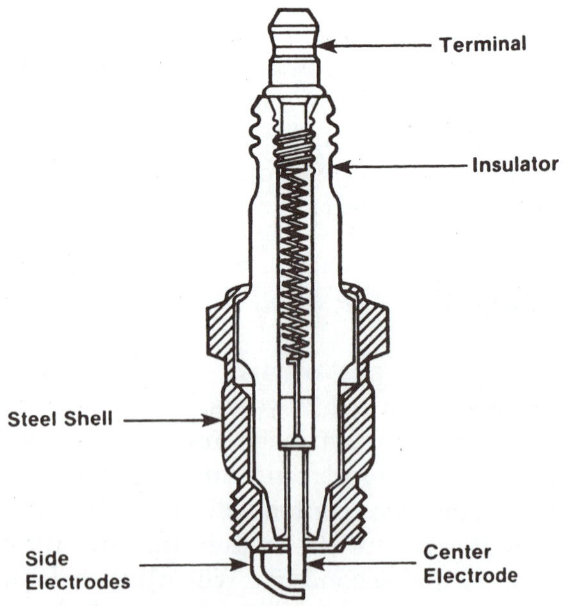

FIGURE 12-1 Typical spark plug.

There are a couple of design features that make the plug the correct one for a particular vehicle.

NOTE: A look through a parts catalog will show you that there are thousands of different types of plugs with only one correct one for the vehicle you are repairing.

Do not assume that the current plug in the engine is correct. Look it up and verify.

Reach

The reach of the plug is the length of the threaded portion of the plug steel shell. The object of it is to ensure that the air gap is correctly positioned inside the combustion chamber. Figure 12-2 illustrates the correct reach. In certain instances, the incorrect reach (too long) could result in extensive damage to the engine when the piston hits the plug. The reach must match the cylinder head design. Do not be surprised if the entire reach is not threaded.

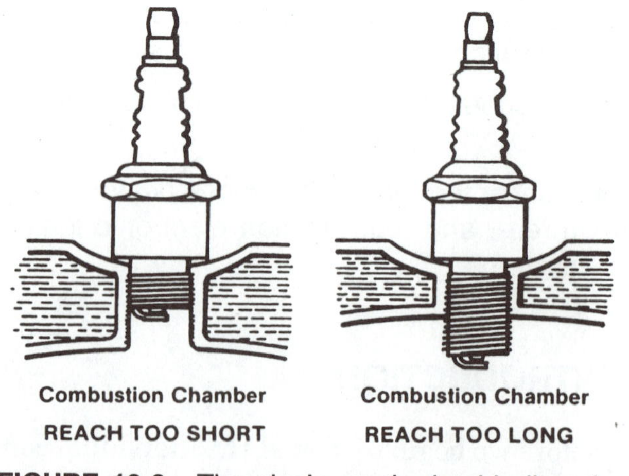

FIGURE 12-2 The plug's reach should allow the plug's tip to just enter the combustion chamber.

Heat Range

Heat range is another design feature. Simply, it is the temperature that the plug will operate at under normal driving conditions. If a plug tip is operated too cool, it will tend to foul. Fouling is the process where mixture residue, gas, or oil sticks to the surface of the plug, eventually filling in the air gap and eliminating the arcing. This shuts down the cylinder, reducing the engine's power, and pumps raw fuel out the exhaust system into the air. If a plug tip is operated too hot, the electrode life will be greatly reduced. The electrodes actually melt or burn away. This increases the air gap and forces the ignition system to produce a higher voltage. The larger air gap can be the cause of a misfire, also. Ideally, the temperature of the tip will be between the fouling level and the overheating level. This is the heat range of the plug. Figure 12-3 shows graphically that the internal distance from the tip of the plug to cooling water is what will determine the heat range of the plug. The range that the plug manufacturer recommends for a particular engine will ensure that the tip temperature will stay within the safe zone during driving conditions. Tip temperature will rise as one travels at highway speeds and fall as the vehicle slows down. The number of sparks per second and the mixture temperature are what change the plug's temperature.

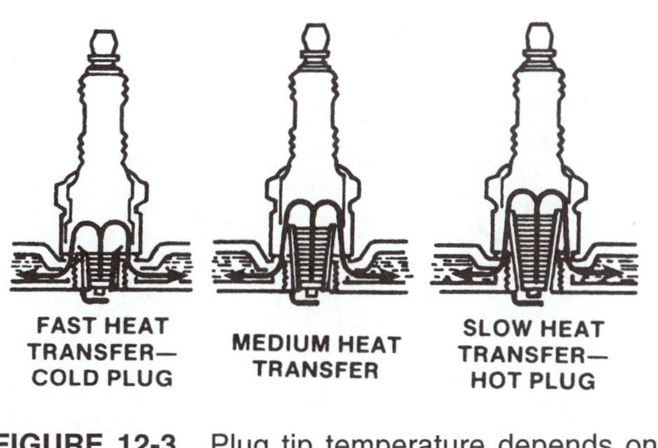

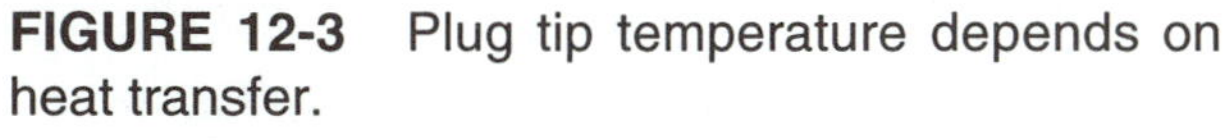

FIGURE 12-3 Plug tip temperature depends on heat transfer.

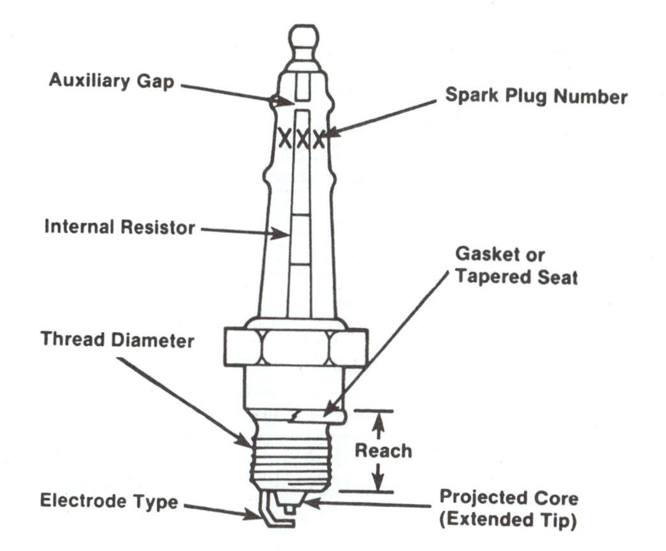

FIGURE 12-4 Resistor plugs are found on most modern engines.

RESISTOR AND NONRESISTOR PLUGS

The plug we have looked at so far has had a straight conductor between the terminal and the center electrode. With the increased use of on-board electronics, these plugs have been replaced with the resistor style shown in Figure 12-4. The additional resistor between the terminal and the center electrode will have two effects. First, it will raise up the firing voltage slightly, and second, it will suppress voltage spikes or AC while the plug is firing. This reduced AC produces less radio frequency interference. Always use a resistor plug if one is specified. Never replace a resistor with a non resistor plug because many of the components currently in use today operate at radio frequency. The additional radio frequency produced by the plug could interfere with the operation of a computer sensor, radio, or clock.

GAP

The gap of a spark plug is the distance in thousandths of an inch(millimeters) between the ground or side electrode and the center electrode. The importance of this gap cannot be overstated. It will have a direct effect on the

FIGURE 12-5 New plugs will have a flat, straight surface.

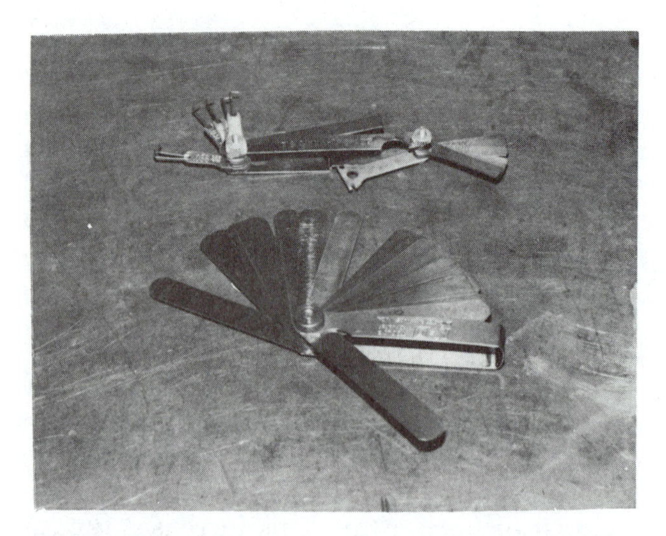

FIGURE 12-7 (Top) plug feeler gauges; (bottom) flat feeler gauges.

operation of the ignition system and on the firing voltage. When the plug is new, the electrode surfaces are flat, clean, and squared off, Figure 12-5. Once the miles begin to build on the plug, the electrodes take on a rounded, rough texture, Figure 12-6. The air gap will also get larger as the plug wears. Sometimes an old plug might have its gap changed by as much as 0.020" (0.5 mm). This increased gap is harder to fire efficiently, thereby forcing the rest of the ignition system to work harder. It is important that the air gap be set correctly when installing new plugs. Plugs are generally not found correctly gapped right out of the box. A plug feeler gauge, Figure 12-7, has rounded wires of the correct diameter. The ground electrode is bent down until the wire gauge passes through the gap, touching both the center and the side electrode. It is important for you to realize that the many different styles of plugs will require many different gap settings. Plugs should always be set to the correct gap as determined by the manufacturer. Once set, a correct plug gap will retain the squareness, Figure 12-8. If, however, the wrong plug or wrong gap is used, the plug might look like those in Figure 12-9. In either case, the plug will not have the life expectancy required. Make sure that the plug side electrode is perpendicular to the center electrode after gapping. If it is not, recheck the part

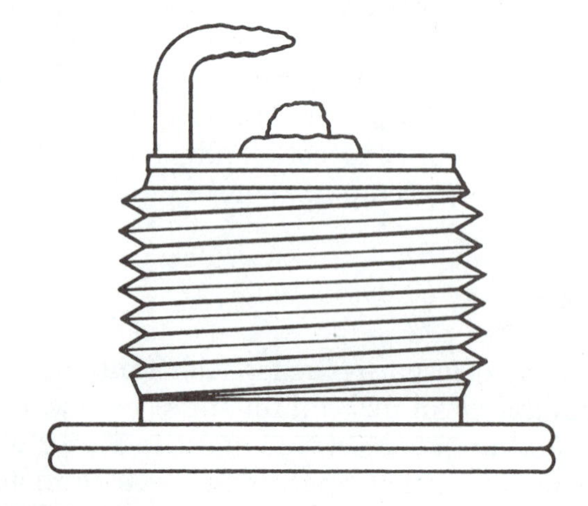

FIGURE 12-6 Plug gap will erode through use.

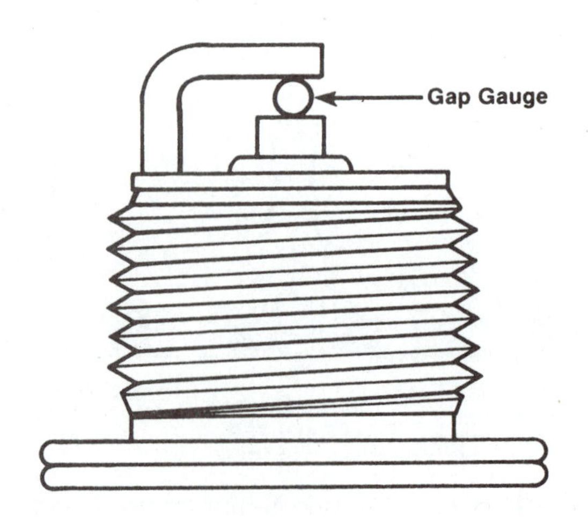

FIGURE 12-8 A squared off gap is correctly set.

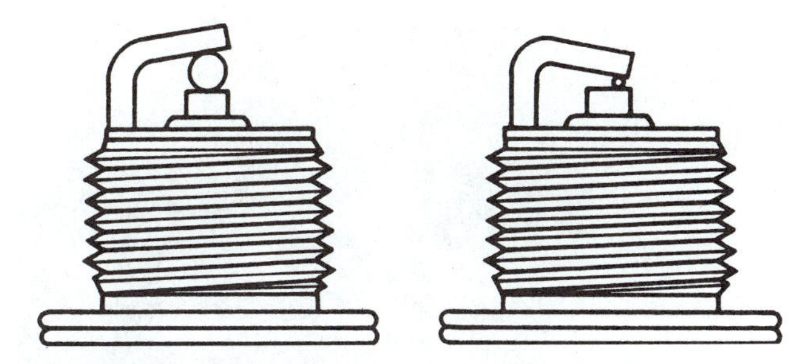

FIGURE 12-9 Incorrect plug gaps.

number of the plug and the manufacturer's gap. One or both of the two are wrong.

THREAD AND SEALING

There are a couple of different thread size plugs available and in use today. The same principle as in heat range should guide you. Use what the manufacturer specifies and you cannot go wrong. If the threads on the replacement plug are different from the plugs removed, recheck the part number to ensure that you are correct. In addition, you should realize that the plug and cylinder head might seal with a gasket or with a tapered seat. The torque (twisting effort) required to seal either plug is very specific. Over-torquing can result in damage to the plug or cylinder head (especially if it is aluminum). Over tightening will also reduce the heat range of the plug and tend to foul the plug easier. Under-torquing the plug has the opposite effect. The heat has a harder time leaving the plug and the tip has a greater tendency to burn. The plug threads might also allow some blow-by of hot exhaust gases on the power stroke. These gases could destroy the cylinder head threads.

The majority of plugs today use the tapered seat for sealing. However, you will find a substantial number of plugs that use the gasket. This gasket is usually hollow and seals by being crushed down between the plug and the cylinder head. New gaskets are required each time a plug is installed in the head. The increased use of aluminum cylinder heads has greatly reduced the use of gaskets for sealing.

ALUMINUM HEADS AND HELI-COILING

With the increased use of aluminum cylinder heads in vehicles today, the technician must pay attention to detail when installing or removing spark plugs. The steel shell of the plug can destroy the aluminum threads of the cylinder head if it is over tightened or removed with the engine still hot. For this reason, never remove a spark plug from an aluminum cylinder head if the engine is still warm. Allowing it to cool down completely will ensure that heat expansion has not locked the plug into the head. The increased head damage seen frequently today can also be lessened with the use of powdered graphite on the plug threads. A small amount of the dry compound will not affect the heat range and will help to prevent thread damage.

If you encounter a damaged spark plug thread hole in the cylinder head, it could possibly be repaired by a process of installing an insert thread. Heli-coiling is just such a process. The old threads of the head are drilled away, and the hole is rethreaded to a larger size. A special steel insert is then installed into the larger threads of the hole with a special tool called a **mandrel**. When the mandrel is removed, the steel insert remains in the head. Figure 12-10A to 10E shows this process. It is important to note that the cylinder head usually is removed for this machining process. The repair is effective and does save the cylinder head. The need for the repair can be traced to the removing of the previous plug while the engine was hot or the over tightening of the plug. In addition, if the customer has not had the vehicle tuned and the plugs replaced at reasonable mileage intervals, you might experience the steel threads of the plug destroying the cylinder head threads when you remove the plugs. Spark plug mileage intervals must not be exceeded or a costly repair might be required.

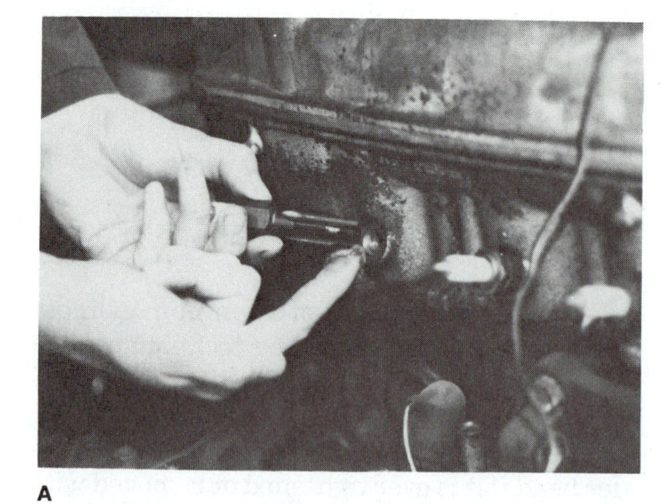
A

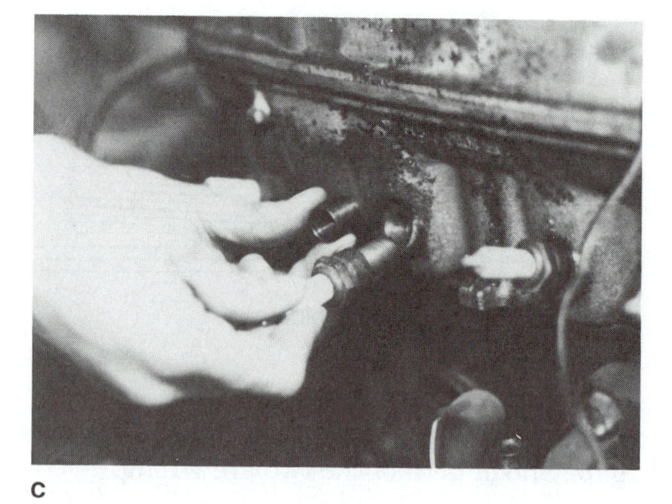
C

B

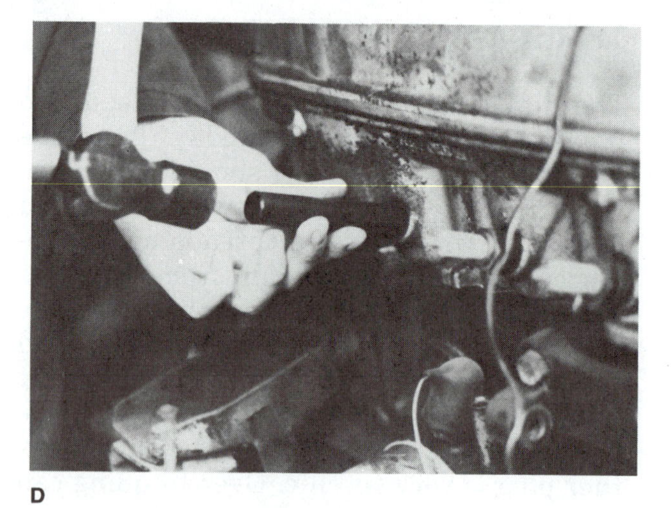
D

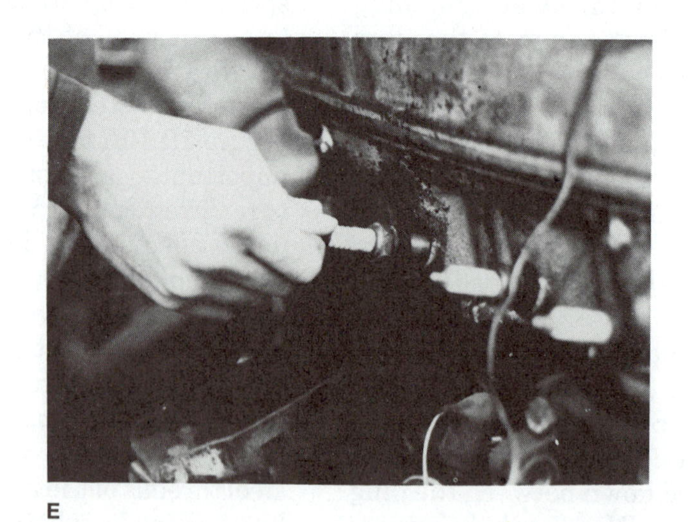
E

FIGURE 12-10 Helical sequence.

READING A USED PLUG

Mechanics must have the ability to look at an old set of spark plugs and determine the basic conditions that the plug has been operating under. Plug manufacturers publish wall charts that list the more common looking plugs. Figure 12-11 shows what some of the more common engine conditions will result in. The key here is to look for differences between cylinders first, because this might indicate that the engine cannot be tuned. For example, an oil-fouled plug on one cylinder indicates that oil is entering the combustion chamber past the rings or down past the valve guides. Major engine work might be necessary to restore the customer's driveablility. A tune-up and plug replacement might help but might not cure the entire problem. A careful explanation to your customer will help outline the options. Customers frequently are under the impression that a tune-up will cure anything! Reading the plugs as they leave the cylinder helps us determine the course of action we will recommend.

Another example involving all cylinders might be a set of plugs that are carbon fouled. The customer might bring the vehicle to you again for a driveability concern. Carbon forms on the plug if the amount of gas in the combustion chamber is greater than it should be. The system is running rich and fuel system maintenance is required. Installing another set of plugs will probably improve the driveability but only until the new set gets carbon fouled. By reading the carbon condition on the removed plugs, you can alert the customer to the need for fuel system maintenance in addition to the need for plugs. We will outline the removal, inspection, and replacement of spark plugs in the next chapter.

WIRES

Now that we have discussed just how we are going to use the high voltage developed by the ignition system at the spark plugs, we will begin to look at the production and transmission of this high voltage. Ignition cables or plug wires will deliver this high voltage from the ignition coil over to the plugs. Let us zero in on these wires and examine the most common types available, in addition to looking at just how the cable of today has evolved. Years ago ignition wire was just simply a conductor surrounded by lots of good quality insulation. The insulation was necessary because the ignition system was developing as high as 20,000 volts. The size or diameter of the actual conductor was very small, because the current was extremely low. Remember, the amount of current flowing determines the size of the conductor, while the voltage determines the amount of insulation necessary. For this reason, our typical solid-core plug wire was composed mostly of insulation to ensure that the ignition current would actually reach the spark plugs. This wire was popular from the first vehicles until radios began to appear as an option.

One of the difficulties with the solid-core wire was the fact that it produced strong magnetic fields around itself. These fields would find their way into the radio and produce a pop or click that could be heard each time the plug fired. It was not uncommon, at highway speeds, to be unable to hear the radio because the popping would become a buzz, which would drawn out the music. The noise that was finding its way into the radio was called **radio frequency interference**, or RFI for short. The tendency of the plug voltage to rise and fall rapidly once the plug fired was the cause of the noise. These alternating currents would have to be suppressed or reduced if radio was to become popular in vehicles. The earliest attempts at suppression were very costly as they involved placing grounded metal shields around the entire ignition system. This was very effective and can still be found on plastic or fiberglass vehicles like the Chevrolet Corvette, Figure 12-12. Setting up a grounded field around the high voltage prevents the RFI from getting into the radio circuits.

Eventually, the high cost of RFI shielding gave way to the most popular wire of the sixties and seventies, TVRS wire. TVRS stand for **television and radio suppression**. Generally, it is composed of a nylon strand with small

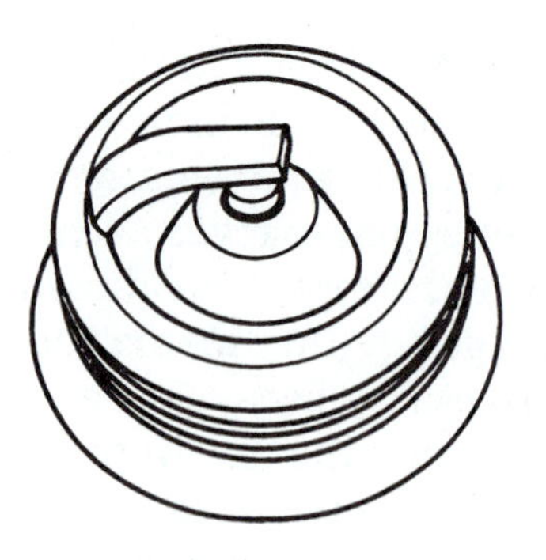

NORMAL

GAP BRIDGED

OIL FOULED

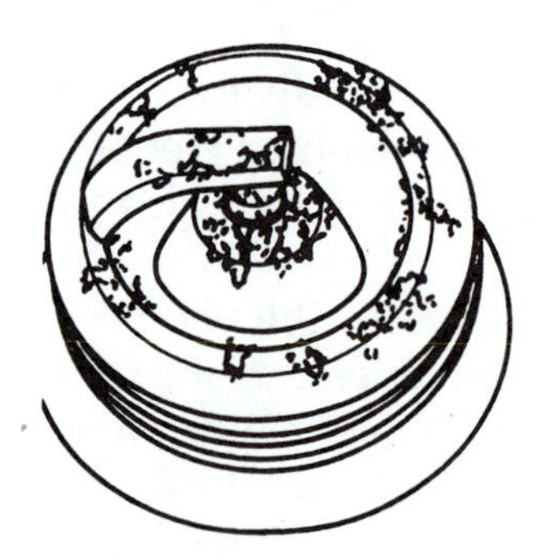

CARBON FOULED

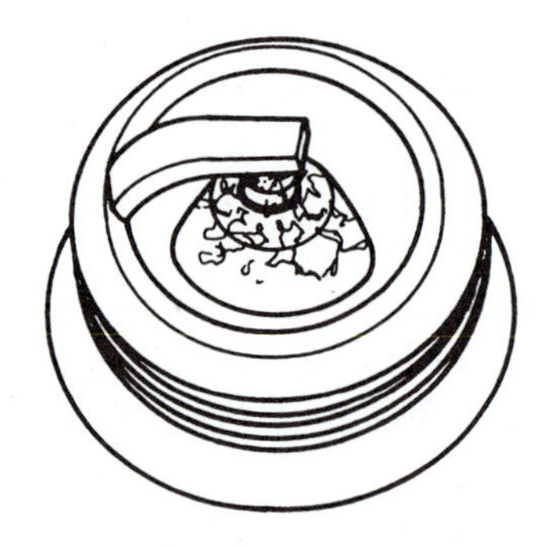

FUSED SPOT DEPOSIT

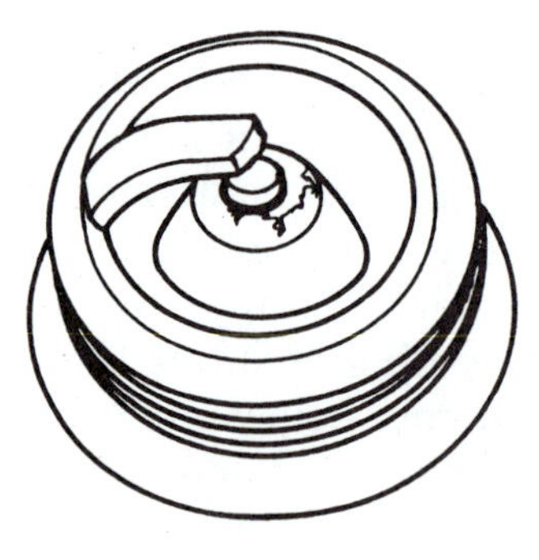

WORN

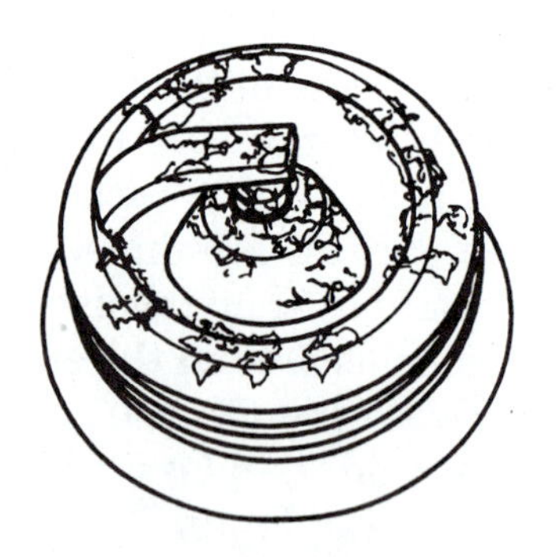

LEAD FOULED

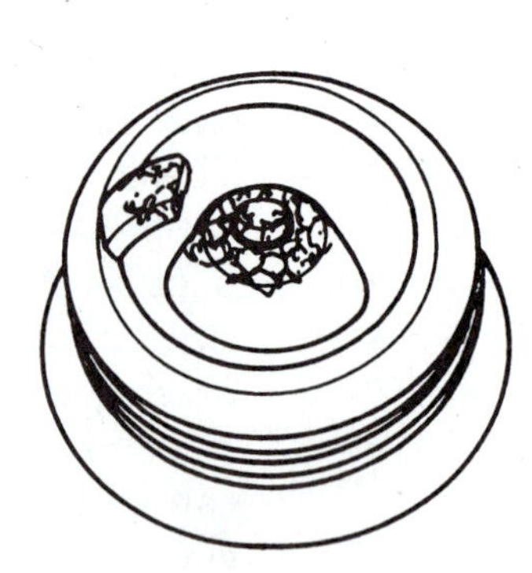

PRE-IGNITION

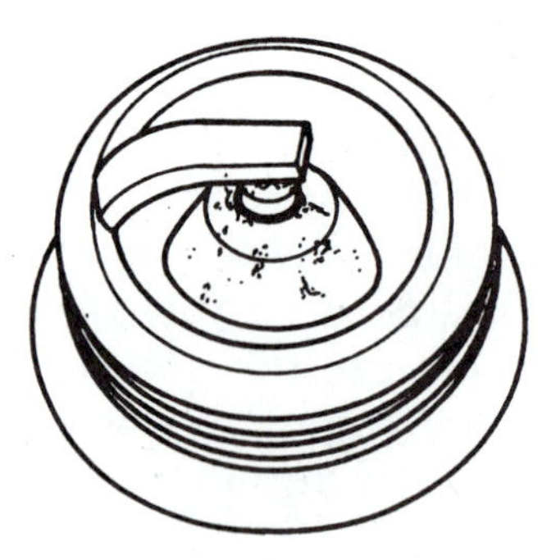

OVERHEATING

FIGURE 12-11 Common plug problems. (Courtesy of Ford Motor Company)

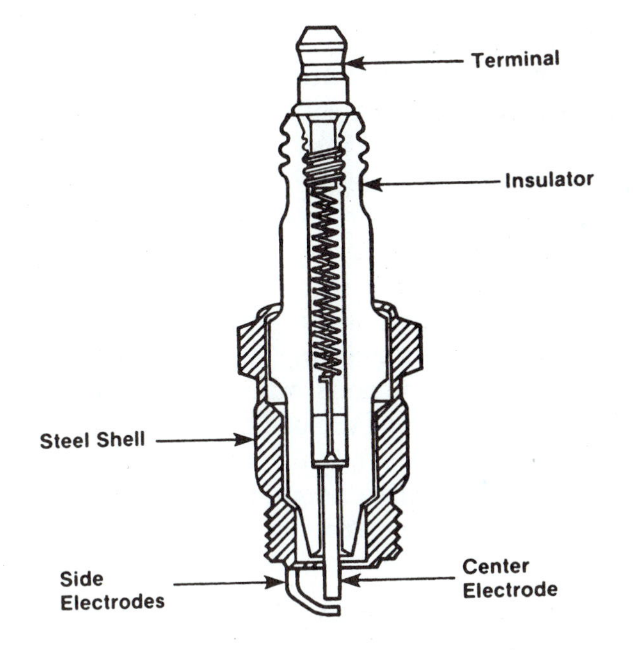

FIGURE 12-12 Corvette radio suppression cover.

pieces of conductive carbon impregnated into it. Figure 12-13 shows a simple drawing of the wire. The carbon offers resistance to the spark and drops the voltage as it travels to the plug. This voltage drop along the wire typically can be as high as 2,000 volts and has the effect of suppressing the RFI because the alternating currents usually present are prevented. In addition, the level of voltage along the entire length of the cable is different and will have a lower field strength than with solid-core wire where the voltage is the same and stronger. For example, if a plug is firing at a level of 5,000 volts with solid-core wire, the entire length of the wire will have a force field equal and, therefore, strong. If, however, a plug is firing with TVRS wire at 5,000 volts, the coil end of the wire could have as much as 7,000 volts. The 2,000 volt drop along the length of the wire puts each section at a different level, thus reducing the total field strength. Resistor wire, as TVRS is sometimes referred to, will have resistance per inch of a specified amount. The amount will differ but could be around 4,000 ohms per foot. A 2' wire will, therefore, have 8,000 ohms of resistance. A 3' wire equals 12,000 ohms, and so on. For the exact specification, consult a manual for the vehicle. TVRS cables should not be replaced with solid-core wire as the radio frequency interference might be objectionable to the customer.

MAGNETIC SUPPRESSION

Over the years, there have been many methods of trying to both suppress RF (radio frequency) interference and to make components less likely to be affected by the interference. One of the more recent has been the introduction of a solid-core wire that is wrapped in a conductive outer shell. Figure 12-14 shows the principle behind this type of **magnetic suppression**. The wire wound around the spark-carrying conductor shields the magnetic force field that is produced. No effort is made to alter the spark, as is the case with resistive suppression. Instead, the wire itself prevents the RFI from leaving the surface of the wire. An ohmmeter placed at the two ends of this type of wire will typically read 0 ohms of resistance. If the ohmmeter were to read anything other than zero, the wire would be faulty.

This is a good time to point out the need for knowing and using manufacturers' specifications. Ignition cables might look exactly alike and yet possess thousands of ohms per inch, or it could have a spec of 0 ohms for its entire

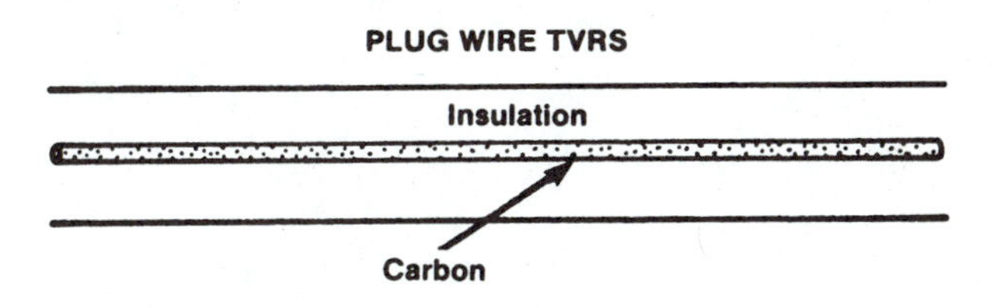

FIGURE 12-13 Carbon strand ignition cable.

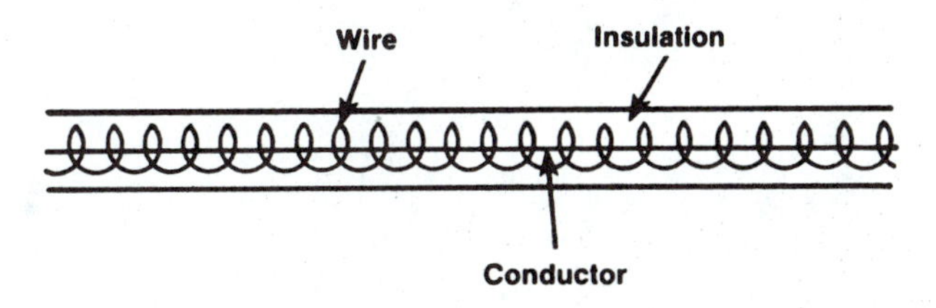

FIGURE 12-14 Magnetic suppression ignition cable.

length. Always refer to the manufacturer's specification if you are using an ohmmeter to diagnose a faulty cable. The exception to this is the use of an oscilloscope. A bad wire will usually result in a high firing line as you will see when we learn to use a scope in ignition diagnosis.

The insulation surrounding the ignition wire is just as important as the wire itself. The ability of the wire to deliver to the plug the high voltage is dependent on the insulating ability of both the jacket and the boots or caps at the two ends of the cable. Modern ignition systems have the capability of in excess of 50,000 volts. Electricity is always looking for the path of least resistance, which with poor insulation, might be through the side of the wire over to the engine block. Delivering the spark to the engine block results in a dead cylinder, rough running, and reduced mileage. In addition, with smaller engines, the power is so greatly affected by the loss of one cylinder that the safety of the driver and passengers could be at risk. Delivery of the spark to the plug is the job of the insulation, and it obviously must be capable of insulating up to the limit of the ignition system. In addition, it should be capable of doing it under adverse conditions, such as 20° below zero, 110° above, or extreme dampness. When we look at the use of a scope, we will discuss testing of insulation in depth.

The connections at the ends of the ignition cables are usually made with insulated crimped-on metal terminals, Figure 12-15. Removing a plug wire should only be done by grasping this metal terminal. Never pull on the wire or the end might break off. In addition, many manufacturers recommend the use of silicone dielectric compound in the boots to prevent their deterioration. Silicone **dielectric** compound does not conduct electricity and is used to lubricate, transfer heat, and suppress RFI. The compound also helps in preventing the boot from sticking and burning on the insulation of the plug. Plug wires are best removed by twisting slightly and pulling on the terminal end. Additional compound should be added to the boots to aid in their future removal. Your toolbox should contain a tube of dielectric compound because it is used frequently on modern vehicles.

CAP AND ROTOR

On many of the ignition systems in use today, a single ignition coil is used for a multiple number of spark plugs. The spark produced by this single coil is transferred over to the correct cylinder through the use of the cap and rotor. Figure 12-16 shows some of the different types of rotors utilized. For the most part, their principles are the same. A single coil wire brings the high voltage over to the center of the distributor cap where it is transferred to the center of the rotor. The rotor is attached to the distributor shaft and turns at camshaft speed. Because it is turning, it will act like a rotating bridge and bring the high voltage over to the outer distributor cap ter-

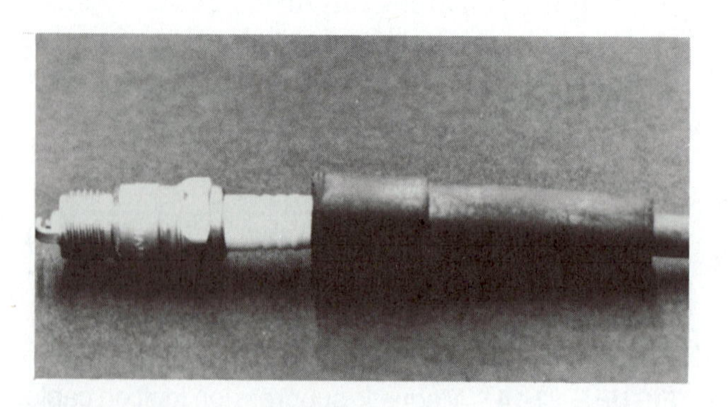

FIGURE 12-15 Plug boot.

FIGURE 12-16 Different types of rotors.

minals, which are in turn connected to the individual plug wires. In this manner, a single ignition coil can supply all the spark plugs with high voltage in a sequence called the **firing order**. This very specific order of spark firings is determined by the manufacturer. As each one of the cylinders is following its own intake, compression, power, and then exhaust sequence, the spark will be delivered at the correct time when the combustion chamber is full of compressed fuel and air. This firing order is set by the placement of the ignition wires and the direction the distributor shaft turns. For example, let us assume we are replacing the wires on a V-6 engine with a 165432 firing order. Simplified, this means that the cylinder the manufacturer has numbered as number 1 will fire first. Then, 60° later of camshaft rotation (a total of 120° of crankshaft rotation), the next cylinder (number 6) will be ready for the spark plug to fire and ignite the mixture. Again 60 degrees later cylinder number 5 will be fired and so on until all cylinders have fired. The entire sequence of all plugs firing will take two complete crank shaft rotations with the individual power impulses spread out 120° apart, Figure 12-17. The wire positions on the outer edge of the cap will determine the spark plug firings. They must always be correctly positioned or the spark might be delivered to the incorrect plug. By incorrect, we mean that the cylinder is not ready for the spark. Perhaps it is still on the exhaust stroke or on the intake stroke. It is important to note that there are many different styles of ignition systems and many different firing orders. Always make sure that you are rewiring the cap in the correct order, and verify by looking up the specifications for the vehicle by year, make, and model.

The end of the rotor is a spot that does not touch the distributor cap terminal. There is a small air gap of a couple of thousandths of an inch (0.2 mm). This air gap is necessary because the rotor would destroy itself if it came into contact with the cap inserts. Generally, the arcing of high voltage that takes place at this air gap eventually causes sufficient corrosion that the cap and rotor require replacement. Scope testing allows us to pinpoint excessive corrosion caused by arcing. Some manufacturers recommend coating the end of the rotor with silicone dielectric compound to cut down on the arcing. This helps in the prevention of radio frequency interference and extends the life of the cap and rotor.

In addition, some caps and rotors are utilized for other functions. Figure 12-18 illustrates this with a rotor that has metal shutters attached. We will study this in the future. Also, some rotors have multiple ends. See Figure 12-19, wisker, and dual plane rotors.

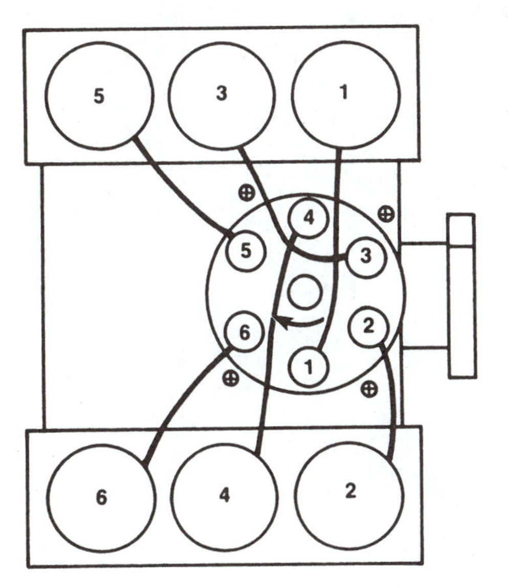

FIGURE 12-17 Firing order for a GM V-6.

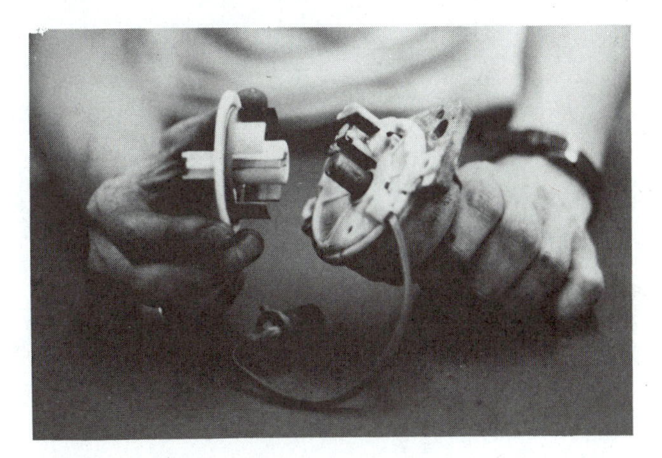

FIGURE 12-18 Hall effect shutter combined with a rotor.

FIGURE 12-19 Dual plane and wisker rotor.

Distributor caps on some vehicles house the ignition coil as in Figure 12-20. This GM high-energy ignition coil is housed inside the top of the distributor cap. The bottom of the coil feeds the rotor directly, thus eliminating the need for a coil wire. A small carbon button is placed between the ignition coil output and the center of the rotor.

The job of the cap and rotor is extremely important. It must deliver the high voltage to the correct plug without allowing the spark to find an alternate or short path to ground. High-quality caps and rotors have the ability to insulate 50,000 volts and last many thousands of miles. As mentioned previously, they are usually diagnosed with an oscilloscope and replaced as needed.

IGNITION COILS

The ignition coil is the next component we will discuss. It is found on all spark ignition systems. There will be at least one per vehicle or as many as one per cylinder. Their function and operating principles are essentially the same on all systems, however. The igni-

FIGURE 12-20 HEI coil mounted in the distributor cap.

tion coil is necessary because the 12.6 volts of the battery will not be sufficient to jump the air gap of the spark plugs. The ignition coil will transform the low voltage, high amperage from the battery into high voltage, low amperage. Just how high the "high" voltage will be, depends on many factors, such as plug gap, air/fuel ratio, plug wire resistance, and so on. In addition, the type of ignition system in use will have a large bearing on the level of high voltage that is available. Keep in mind that we need the high voltage because the extreme resistance of the air gap will require quite a bit of pressure to actually push electrons across. The air gap, the mixture, and everything from the ignition coil to the plug ground will be viewed by the coil as resistance. As this resistance increases, the voltage will have to increase if a spark is to take place. You will remember that Ohm's law stated that resistance will reduce voltage and yet I have just said that additional resistance after the coil output will increase voltage. These two statements are not as opposite as you might think, because the ignition coil output in watts will remain essentially the same. The resistance will have the effect of driving up the voltage, while the amperage is reduced. This transformer effect will take place within the ignition coil. Let us build a simple transformer or coil and analyze how it will produce the high voltage required.

We will start by wrapping some heavy wire around a soft iron core, Figure 12-21. This will be called the primary winding and it will have the job of giving the coil a strong magnetic field. Hopefully, you remember that soft iron is used because it is a good conductor of magnetic lines of force. When we put B+ to one end of the primary winding and ground to the other end, we will have a strong electromagnet. The strength of the field will be dependent on two things: the amount and size of the conductor. The longer and thicker the primary wire is, the stronger the field will be. The overall resistance of the primary winding will determine the amount of current that will flow through it. Keep this in mind later as we look at the different types of systems in use

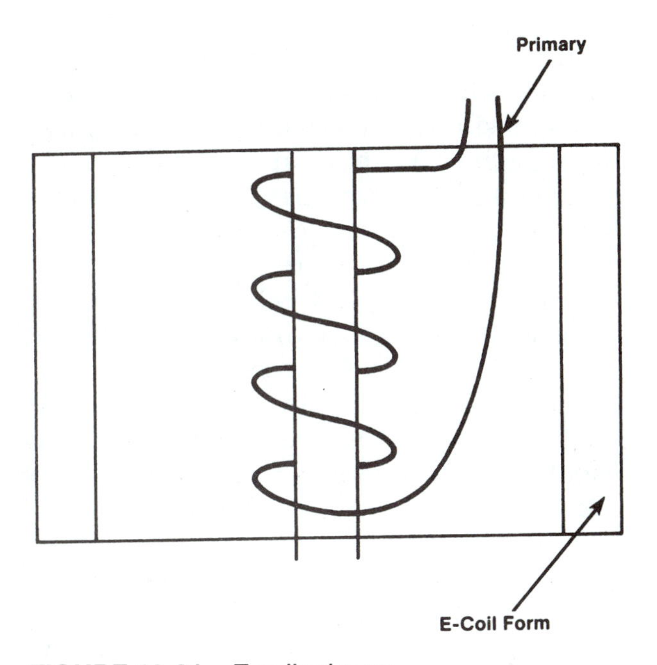

FIGURE 12-21 E coil primary.

today. One of their major differences is in the amount of current they will draw. This is called **primary current.**

Now let us take some additional wire and wrap it around this same iron core as in Figure 12-22. Notice that we have used many more turns of wire than we used in the primary winding. Also, this winding will ordinarily use smaller diameter wire than the

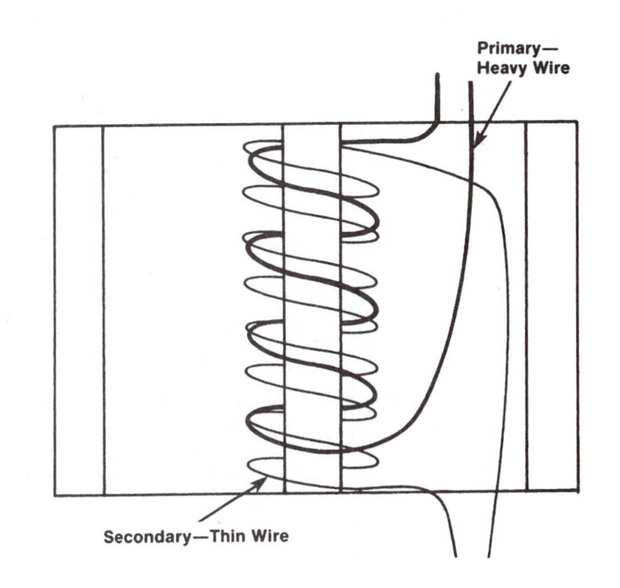

FIGURE 12-22 E coil primary and secondary.

primary. This winding is called the **secondary winding**. One end of the secondary winding will be connected to the spark plugs through the cap, rotor, and plug wires. The other end of the secondary will be connected to either ground or the primary winding. Typically, the ratio of primary to secondary windings is in the order of a few hundred to several thousand. The primary winding typically has less than 10 ohms of resistance and, remember, is composed of a few hundred turns of relatively thick wire. The secondary winding usually has in excess of 5 thousand ohms of resistance and is composed of thousands of turns of thin wire.

The action between the primary and secondary windings will result in the high-voltage spark we will need at the plugs. Let us follow the action through a typical ignition coil and see how we can increase our battery voltage. In sequence, the action looks like this:

1. Primary current is turned on by a primary control device.
2. Current flowing builds the magnetic field around the soft iron core.
3. The coil's magnetic field becomes fully developed (saturated).
4. The primary control device opens the primary circuit.
5. The magnetic field collapses.
6. The moving magnetic lines of force cut across the secondary windings.
7. The moving magnetic lines of force induce high voltage into the secondary.
8. This high voltage is delivered to the plugs.

This sequence will become second nature to you, especially when you begin to use the automotive oscilloscope. Taking each one in sequence and analyzing it will increase your understanding of the coil's operation. Primary current is turned on by a primary control device. This first action is important because we want the magnetic field to build. The electricity flowing through the primary winding will do just this. Our primary winding will, in effect, become an electromagnet. The use of the term primary control device might be new to you, but simply we will want to turn the current on and off with either a switch called **ignition points** or an electronic device such as a switching transistor. There are many different styles of primary control devices that we will look at in the very near future. But let us get back to our sequence. As current flows through the primary winding, the coil's magnetic field builds. This is called **saturation time**. Once the magnetic field is fully saturated as in Figure 12-23, the primary control device will open the circuit. This interruption of the primary current eliminates the electromagnetic field and the lines of force go back to their point of origination—the soft iron core. To get to the core, the lines of force will have to cut across both the primary and secondary windings. These moving lines of force cutting across conductors should sound familiar to you because it is the same condition that

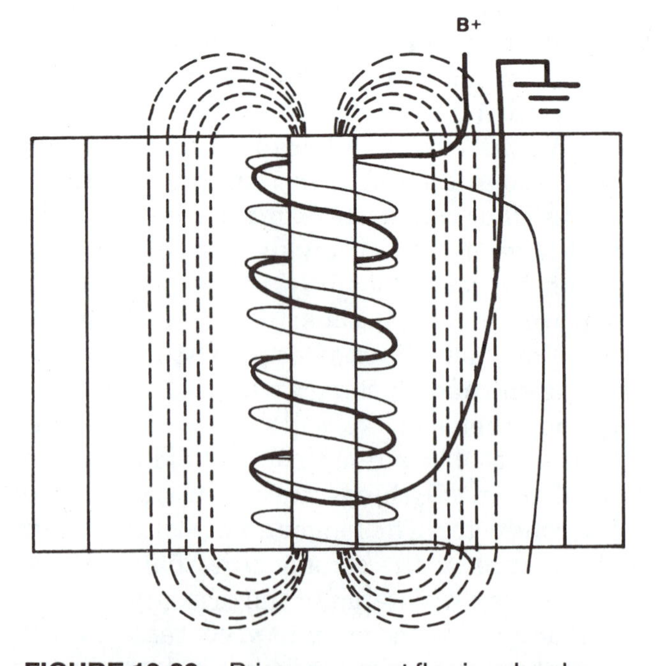

FIGURE 12-23 Primary current flowing develops a magnetic field.

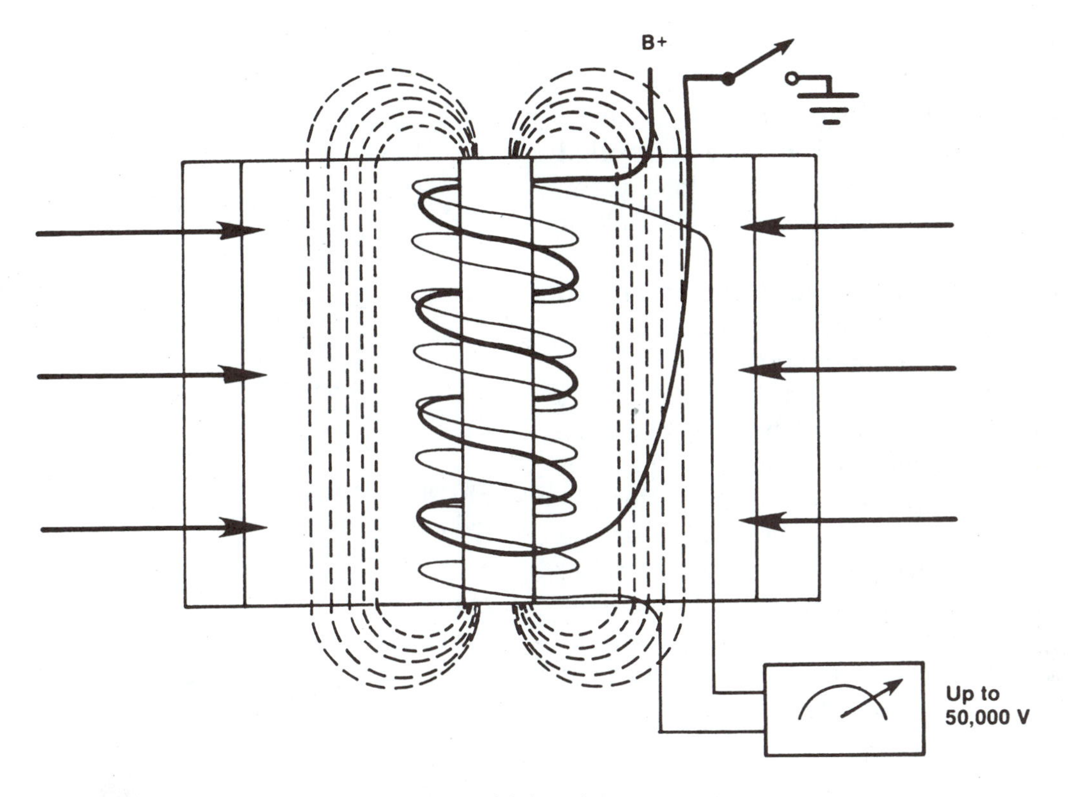

FIGURE 12-24 Moving magnetic field generates high voltage.

generated electricity in our study of the alternator. Here is where the additional turns of wire in the secondary will come into play. The amount of induced voltage in the primary and the secondary is relative to the number of turns of wire in each. Generally, a couple of hundred volts will be induced in the primary because of the couple of hundred turns of wire, Figure 12-24. However, in the secondary, the thousands of turns of wire are affected to a greater degree by the moving lines of force. Here is where the transformation really takes place. The motion between the fast moving lines of force and the thousands of turns of wire can boost the voltage up to as much as 50,000 volts or more. This high-voltage potential will pull the needed current from either the primary circuit or from ground, depending on where the secondary circuit is connected.

A couple of points should be emphasized here. Saturation time or the time that the magnetic field takes to completely build is dependent on the amount of current flowing through the primary circuit. Demands on the ignition system over the years have seen the amount of primary current steadily increase from as low as 2 amps years ago to as high as 20 amps on some of the modern electronic systems currently in use. In addition, the term **dwell** should become part of your vocabulary. Dwell is the amount of time, in camshaft degrees, that primary current flows through the ignition coil. On some systems, it is adjustable; on others it is just checkable. In either case, the time in degrees between primary current on and primary current off is called the dwell and it can be a fixed value such as 30° or it can vary with speed. For example, 15°at idle and 35° at 4500 rpm. We will cover more on this later. Remember also that this sequence will repeat over and over for each spark plug.

PRIMARY RESISTANCE

The use of some type of primary resistance is found in the vast majority of ignition systems in use today. It is important to note that this primary resistance can be external, as in Figure 12-25; internal (inside the coil), as in Figure 12-26; or the most common style, a variable resistance electronically controlled within the ignition module. Let us look at the why of primary resistance before we look at the how.

Primary resistance had its beginning in breaker point style ignition systems. The ignition points, or breaker points as they are sometimes called, have a limit to the amount of current they can handle. The points become the primary control device as illustrated in Figure 12-27. The amount of current flowing through the primary must be limited or the points will burn up from the constant arcing that occurs as they open and close. Remember that our previous discussion of the primary circuit discussed the necessity of turning on (closing) and turning off (opening) the primary circuit to allow the rapid saturation and induction. Most point type ignition systems limit the primary current to less than 3 amps and design the ignition coil with a maximum

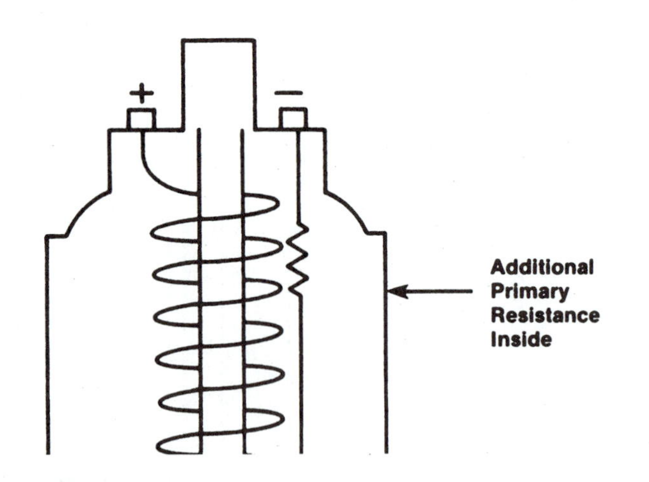

FIGURE 12-26 Primary resistance internally mounted in the coil.

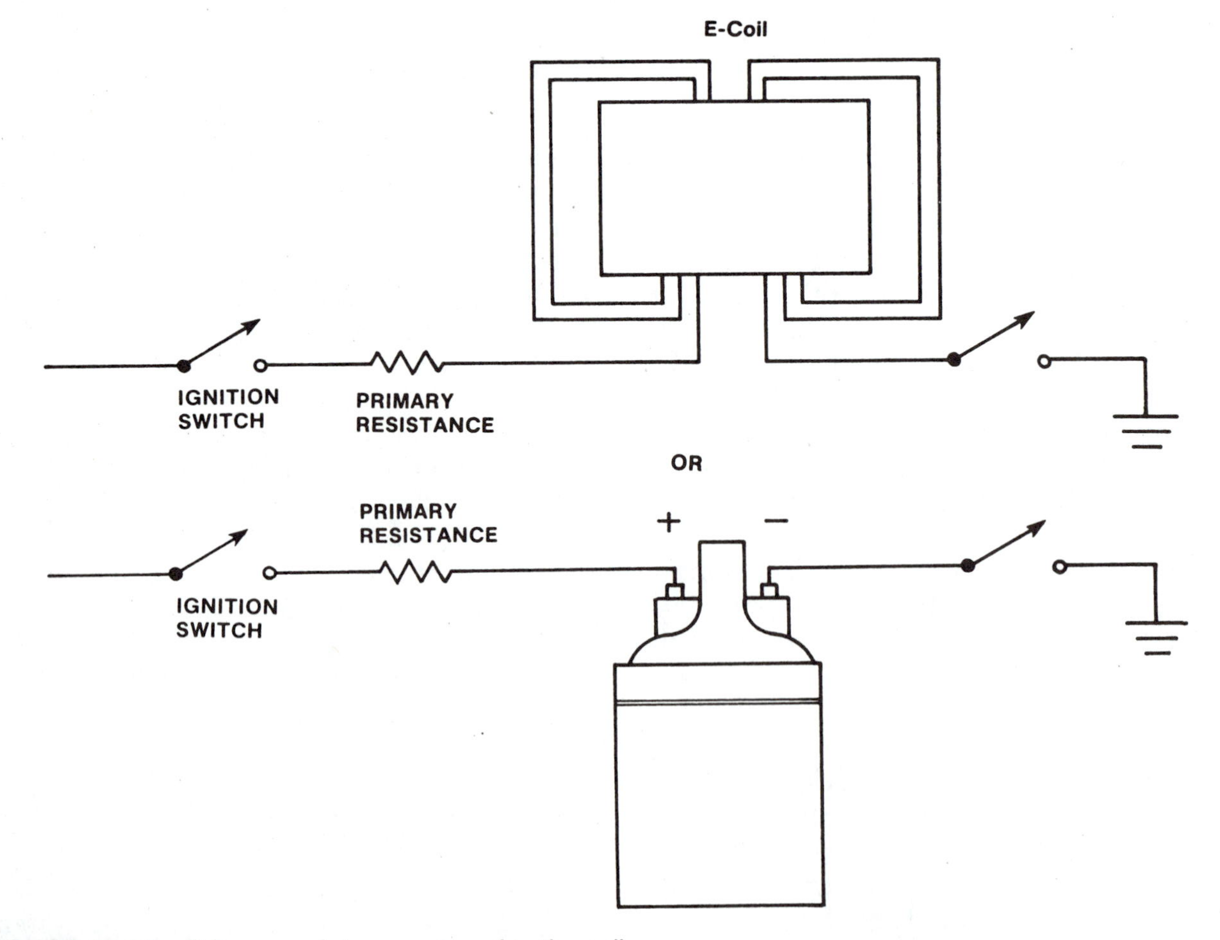

FIGURE 12-25 Primary resistance external to the coil.

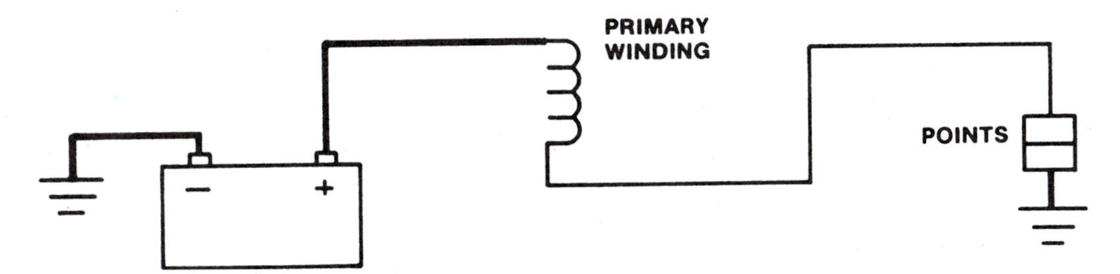

FIGURE 12-27 Primary control can be the ignition points.

output voltage of around 20,000 volts. The ignition resistor will limit the current because it is series resistance. Ohm's law has shown us that this series resistance will cause a voltage division and a reduction in current flowing. In reality, this voltage division usually will drop the coil input voltage to around 10 volts. The ignition coil will be capable of producing the 20,000+ volts necessary if it has around 10 volts across it.

This is a good time to discuss the bypass circuit. Primary resistance, especially in point style systems, has the job of limiting primary current to a safe level. "Safe" for the ignition points, that is. During actual running, the resistance is wired in series, as we have seen. However during cranking, this resistance would further reduce the coil's input voltage even lower than it will be normally. The normal cranking voltage drop of system or battery voltage has been shown in the battery and starting sections of this text, and hopefully you have observed the same during the on-car testing. Let us look at an example, and see what effect this will have on the coil's output. Keep in mind that one of the hardest things the vehicle's engine will do is start. Cranking action pulls the B+ of the battery down. If this reduced B+ is applied to the primary resistor, it will be reduced even further before it is applied to the coil, Figure 12-28. In this example, our coil input voltage has been reduced down to 6.6 volts and the current flowing reduced down to 1.3 amps. This will reduce the coil output down by 33%, making the starting of the vehicle even harder than it is normally. With this in mind, the bypass circuit was designed. Figure 12-29 illus-

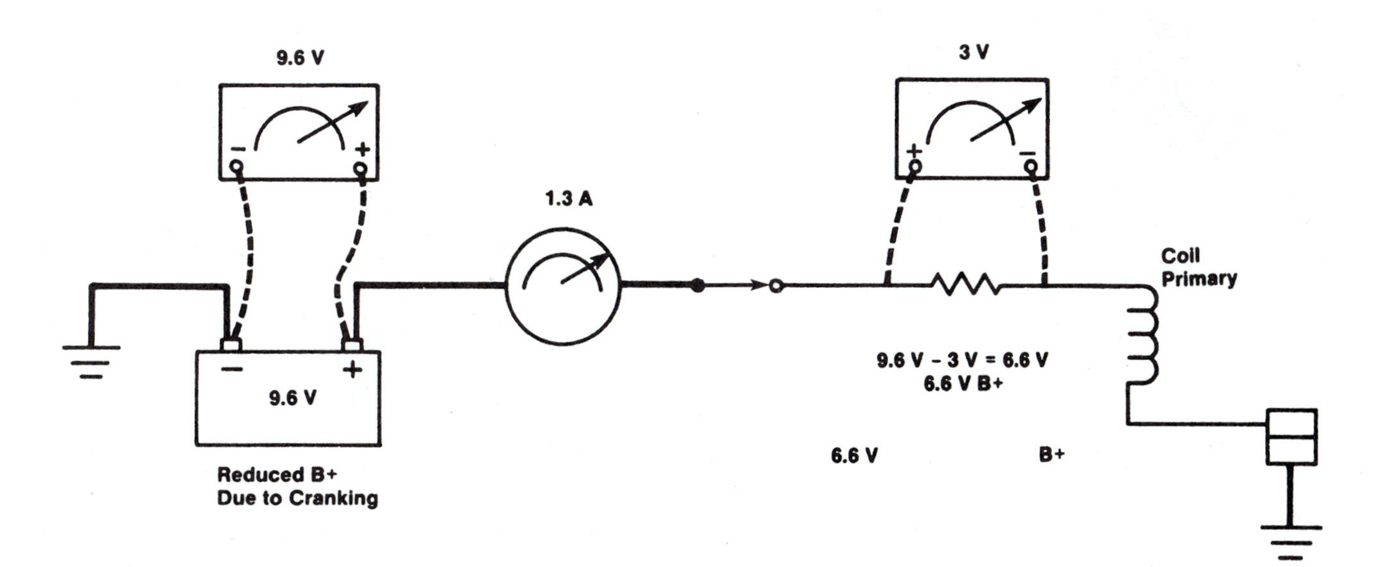

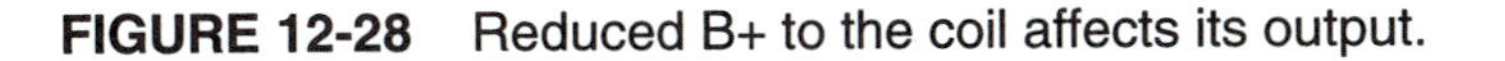
FIGURE 12-28 Reduced B+ to the coil affects its output.

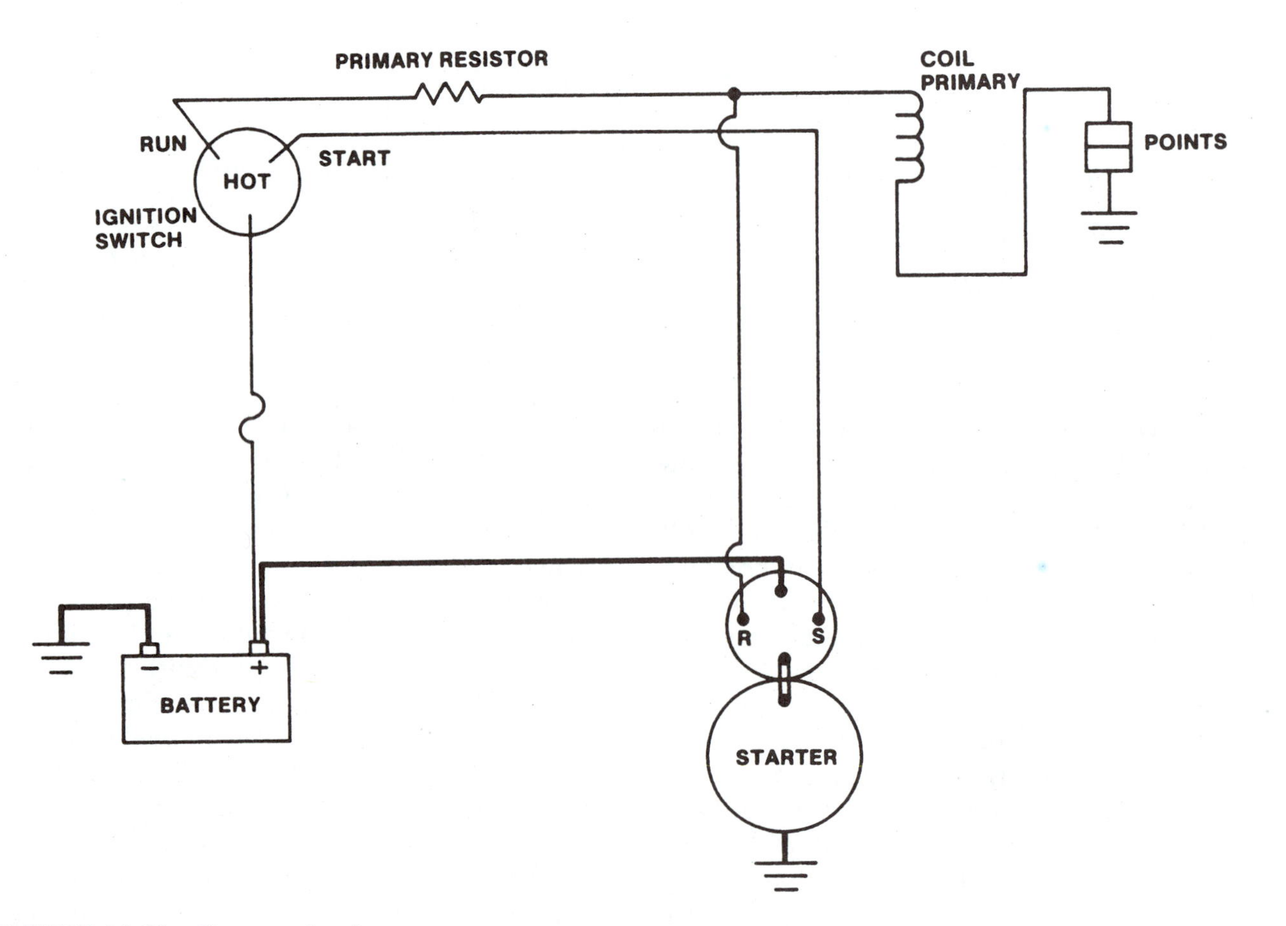

FIGURE 12-29 Bypass circuit.

trates the principle behind it. During cranking, full B+ is allowed to bypass or go around the ignition resistor. This increases the coil available output voltage, making starting easier. The bypass circuit can be powered off the ignition switch, the starting relay, or the starting solenoid. Any of these points will have full B+ available during cranking. Once the vehicle has started, the ignition switch becomes the source of B+ to the primary resistance and coil. This reduces the output to a reasonable level and again protects the ignition points. This protection should allow the points to operate with reasonable efficiency for the length of the tune-up, typically around 10,000 miles.

Breaker point systems almost always use external primary resistance, as we have shown. There are cases of internal (inside the ignition coil) resistance, however, found in older foreign vehicles. These ignition coils were usually labeled with "not for use with external resistance". Ignition available output is greatly reduced if, by accident, one were to install an internal resistance coil into a system that also had external resistance. When in doubt check the specification for the resistance and then check the new coil with an ohmmeter.

The use of external resistance has been reduced greatly. Very few manufacturers utilize it on all their models. Instead, a current-limiting type of electronically controlled resistance is used. With modern electronics, coil saturation is monitored. Once the coil if fully saturated, the electronic ignition module cuts back on the current with a process called **current limit**. This process greatly increases the life of the coil and yet keeps the available voltage up high in case it is needed. We will see how this circuit is utilized and tested in a later chapter.

PRIMARY CONTROL DEVICES

Up to this point, we have discussed the components that are common to most ignition systems. Secondary circuits are, for the most part, the same from manufacturer to manufacturer, including the foreign market. The real difference in ignition systems is realized when one begins to look at the different methods of turning the primary current on and off. We will subdivide the various types into two large categories, those using points and those using electronics. We will further divide the electronics group. We will not make an attempt to cover different designs. You should realize that there is tremendous variation even within a single manufacturer, not to mention the entire industry. If you become familiar with the types of systems, you will not need to know who made them to effectively repair them.

Let us start with the original points and capacitor system, which was the mainstay for many years. Even though there are few manufacturers who currently utilize ignition points, there are enough systems still on the road that you should become familiar with their operation and setting. As we go over them, keep in mind that the object of any primary control device will be to turn on the primary current, allow the magnetic field to build to full saturation, and then turn off the current to allow the field to collapse and energize the secondary. Ignition points do just that. They close and allow primary current to flow, and then open and allow the magnetic field to collapse. They are used most frequently with a capacitor wired in parallel. The capacitor does two important things. First, it allows the magnetic field to collapse quickly because it stops the flow of current through the coil primary quickly, and second it helps to prevent excess arcing at the contact surfaces. Stopping the flow of current through the primary quickly actually accomplishes both functions together. It prevents excess arcing because it allows a temporary place for electrons to flow just as the ignition points begin to open. Figures 12-30 through 12-31 illustrate this principle. It is important to note that this component, the capacitor, has been misnamed by the majority of the trade. It is most frequently referred to as a condenser; however, with either name goes the same function.

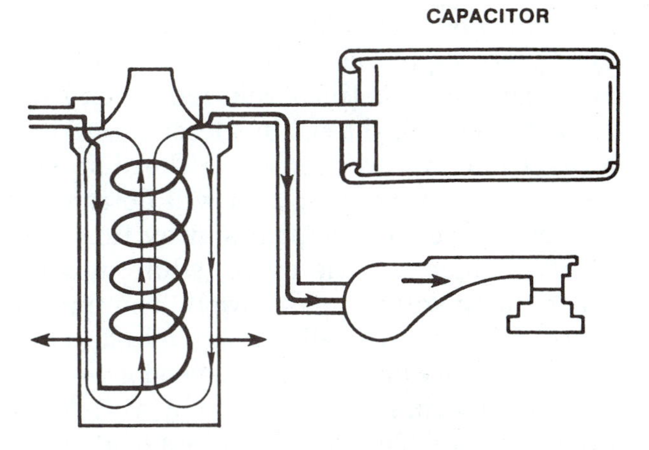

FIGURE 12-30 Current flow with points closed.

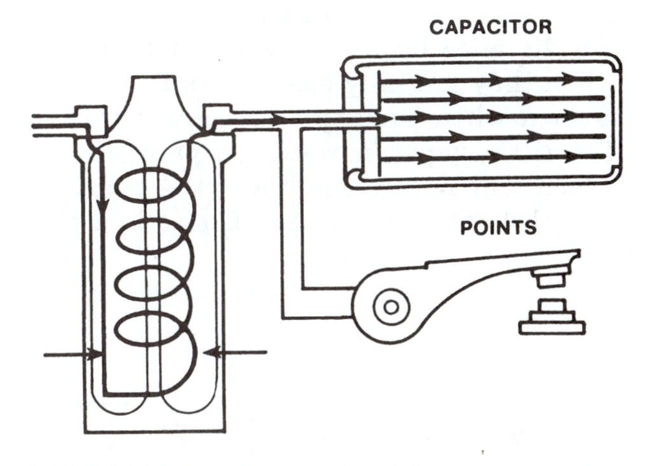

FIGURE 12-31 Current flow into capacitor when points open.

Let us turn our attention to the ignition points. They must make and break the path to ground as they open and close. Two settings are important here: dwell and gap. We have already defined dwell as the amount of time the points are closed, as measured in degrees of camshaft rotation. The dwell is the amount of time that the ignition coil primary will have to become saturated. Obviously, it must be high enough to allow for full saturation. Manufacturers supply us with the dwell, or cam angle as it is sometimes referred to, and we set it by moving the points closer or farther away from the breaker cam. The distance away from the breaker cam will determine the dwell. Moving the points away from the cam will increase the dwell, while moving them closer will decrease dwell. The importance of dwell, especially at high speeds, cannot be overemphasized. High speed gives the ignition less time (as measured by a clock) to fully saturate. Opening the points before full saturation results in reduced maximum output out of the secondary. As we change dwell, you should realize that we are also changing something called **point gap**. Point gap is the distance between the contact surfaces when they are fully opened. Its importance is seen when the gap is not great enough. Too little gap can result in point arcing, which will reduce the life of the points. It can also result in misfire. There is a relationship between point dwell and gap. It is illustrated in Figure 12-32.

Any discussion of ignition points has to ask the question: why are most manufacturers not using them in current production? There are many answers to this question. However, there are actually four principle reasons. First, the amount of current that can flow through them must be limited. Additional secondary output much above 20,000 volts just is not practical because the life of the points would be greatly reduced. Second, points require installation and adjustment, which is a maintenance cost absorbed by the consumer. Eliminating points reduces the consumer cost. Third, points gradually develop resistance as they age. This is due to the slight arcing that occurs each time they open or close. This increased resistance in the primary circuit reduces the current flow through the primary, thus reducing the secondary available output. Vehicles in need of a tune-up frequently misfire. As the spark plugs age, they require in-

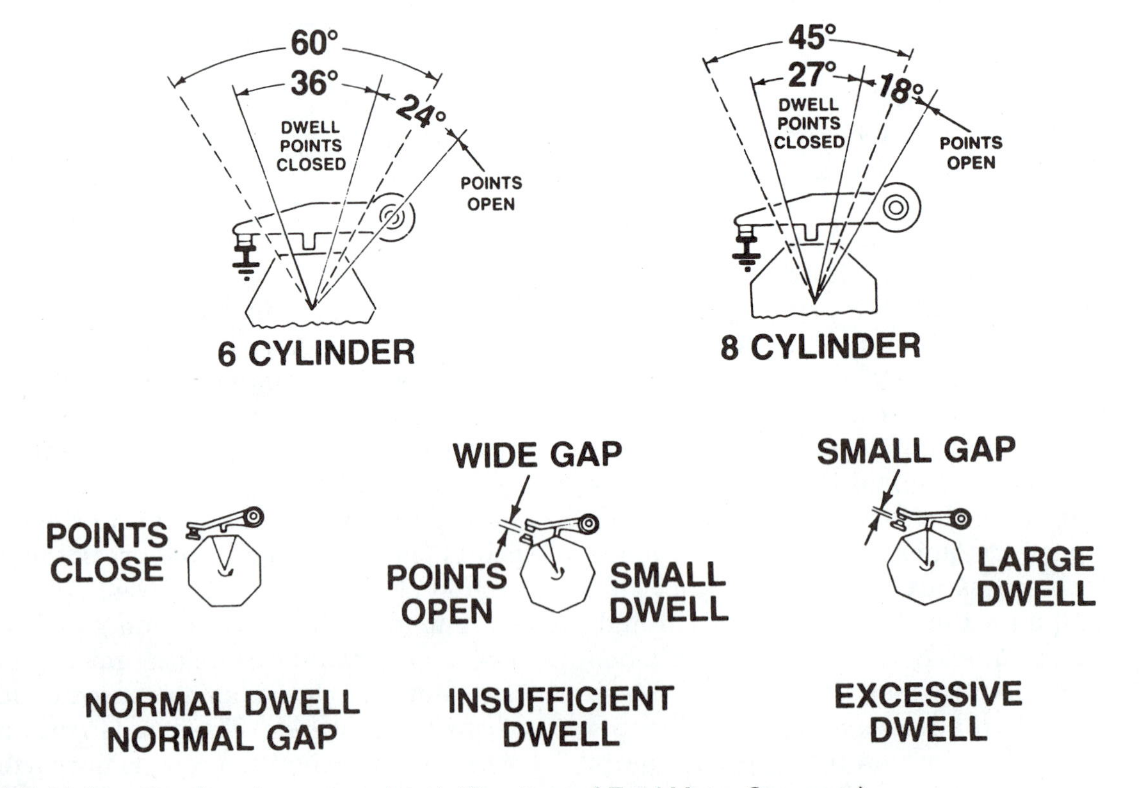

FIGURE 12-32 Dwell and gap are related. (Courtesy of Ford Motor Company)

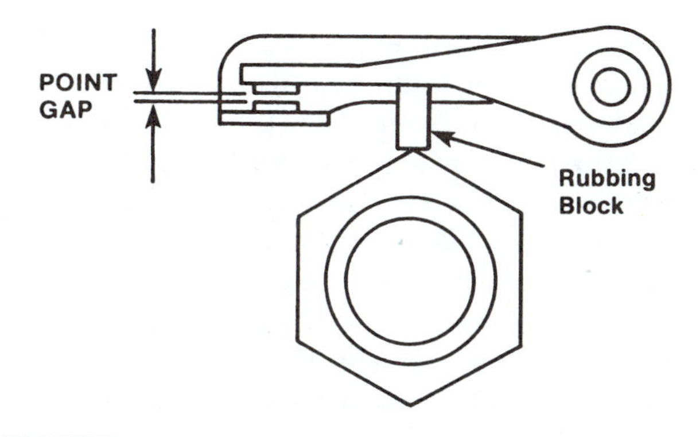

FIGURE 12-33 Rubbing block wear changes dwell and gap.

creased voltage at the same time that ignition output is being reduced because of the additional point resistance. Fourth is the timing change that is evident during the points life. Ignition points are opened because of the action of the breaker cam against a part of the points called the **rubbing block.** This fiber block shown in Figure 12-33 wears down and changes the dwell and gap of the points. By changing the length of time that the points remain closed, the timing of the spark is also being changed. Remember that delivery of the spark at the correct time is extremely important. Any change during the life of the tune-up could upset a balance, resulting in driveability problems or emission increases. Without a doubt, the advancement of electronics into the automobile signaled the end of the breaker point era.

ELECTRONIC IGNITION

Up to this point, we have discussed breaker point switched primary circuits and ended with a discussion of their disadvantages. Let us take a moment now and discuss the advantages to the use of electronics as the primary control device. Transistors or integrated circuits can easily handle two or three times the amount of primary current that a set of contact points can. We currently have systems on the road that are drawing close to 10 amps rather than the 3 amps, which is the approximate maximum current through points. Life expectancy of the points decreases as the current flow through them increases, whereas the current through the transistor or integrated circuit will have virtually no effect on its life.

A block diagram of the typical transistor ignition circuit, Figure 12-34, shows that the actual primary current flows through the emitter/collector, while some type of a piston position sensor feeds into a control circuit that will turn on and off the base of the switching transistor. In principle, most systems operate in this manner. The only distinguishing feature is the piston position sensor. In common use are four different types: contact points, magnetic AC generator, Hall effect, or an optical sensor. We also find systems on the road that have more than one of the sensors. They are usually limited to computer sys-

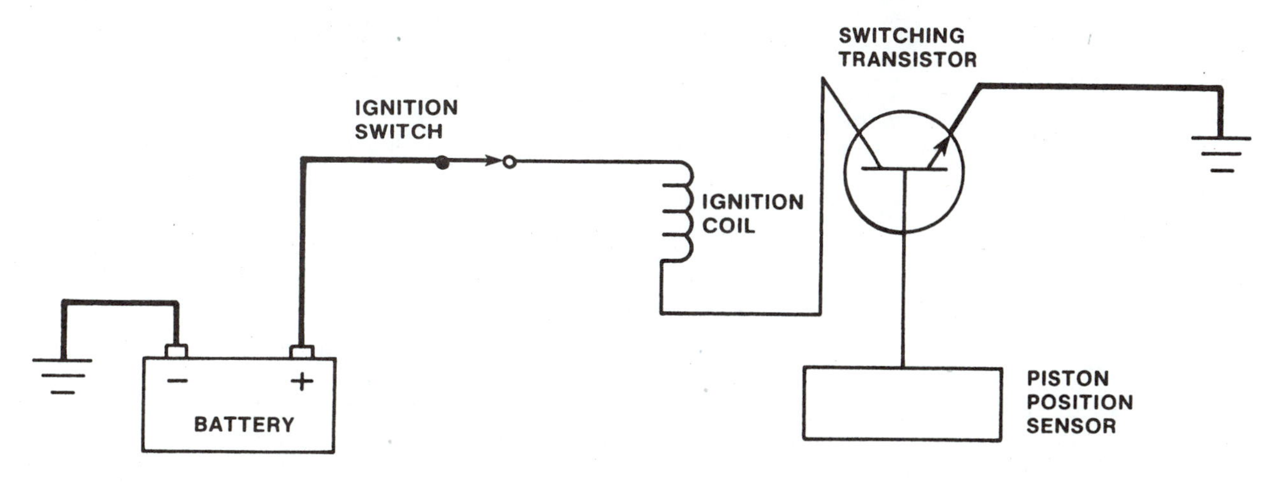

FIGURE 12-34 Electronic ignition simplified.

tems, which we will discuss later. For the most part, the majority of the electronic ignition systems on the road use one sensor to indicate piston position. Let us look at each of the sensors separately, and analyze what its signal looks like to the electronic ignition module. Keep in mind that the object of any of the sensors is to tell the ignition module the position of the pistons. This information will either directly turn the primary switching transistor on and off, or it will feed its position information to the engine computer. You will see in future sections how computer-controlled ignition systems use this information to determine just when to fire the plugs.

Let us take a look at a very simple transistor circuit that involves the use of a set of contact points as the piston position sensor. Remember the breaker point ignition that we previously studied? When the points opened, the plug fired. This is called **timing**. In our simple electronic ignition system, Figure 12-35, the contact points close and apply a ground to the base of the switching transistor. This ground turns on the emitter/collector circuit, and the primary current flows through the ignition coil. After the coil is saturated, the contact points open, removing the ground from the base of the transistor. Primary current stops flowing through the emitter/collector and the coil primary winding. The field collapses and induces the secondary winding. Timing occurs at the same time as the points open. The earliest form of electronic ignition was very similar to this simple adaptation or refinement on breaker points. Notice that many of the disadvantages to contact points are still present. Rubbing block wear still occurs as does contact point burning. The amount of current through the points is greatly reduced, however, as they are no longer carrying primary current. The small sensing current flowing through the points greatly increases their life. The main advantage to this style of EI (electronic ignition) is the ability to have greatly increased primary current over that which can normally flow through contact points. You will not find too many systems on the road that use contact points to trigger a switching transistor. Their use is almost totally in the aftermarket, add-on EI. Notice that this style does pick up some advantages but does not totally eliminate the disadvantages when compared to straight breaker point systems.

MAGNETIC AC SIGNALS

A further refinement to breaker point EI is a style that is very popular worldwide. The magnetic AC generator sensor replaces the contact points along with a more complicated ignition module. The principle behind this style is very similar to an alternator. Within our vehicle's alternator, a moving magnetic field generates an alternating current, or AC. In our typical EI

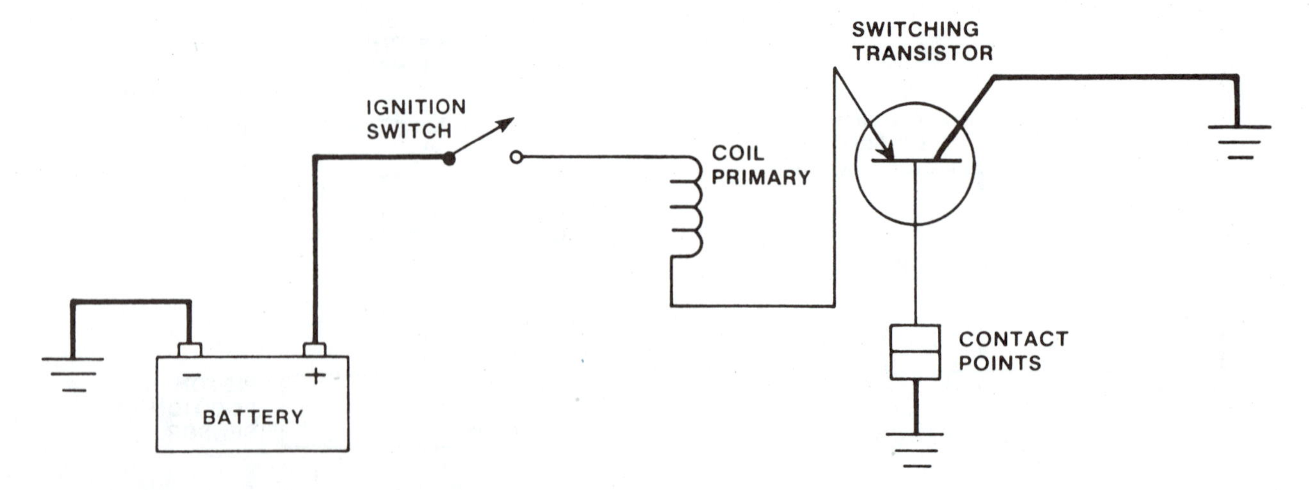

FIGURE 12-35 Points as the piston position sensor.

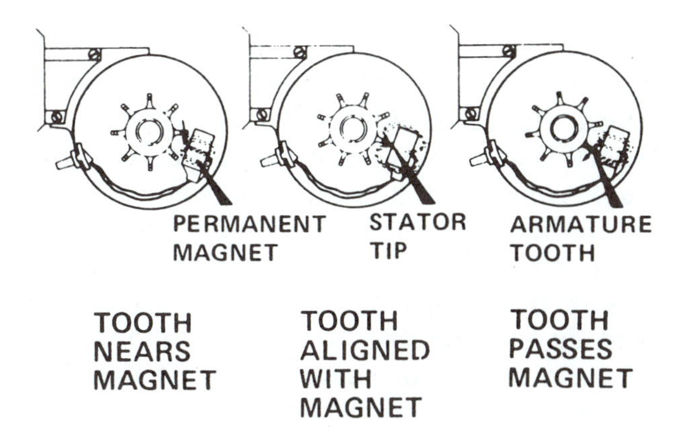

FIGURE 12-36 Moving distributor changes magnetic lines of force. (Courtesy of Ford Motor Company)

FIGURE 12-37 Armature and stator in line.

sensor, the generation of AC will be done inside the distributor. As the distributor is driven off the engine, piston positions can be sensed and the timing determined. Look at Figure 12-36. It shows how the moving distributor will concentrate the magnetic line of force until the armature lines up with stator tooth. As the armature passes the stator, the lines of force will unconcentrate, until the next armature tooth approaches. Approaching armature teeth concentrate the lines of force, while passing armature teeth unconcentrate the lines of force. What we have here are moving lines of force, just like our alternator has. The assembly also consists of a coil of wire wound around the end of the stator tooth. The magnetic lines of force are moving across this coil of wire. Moving magnetic lines of force cutting across a coil of wire will generate a voltage, and it is this voltage that will be sent over to the ignition module. The signal will be a type of sine wave with 0 voltage and a change in polarity evident when the armature is directly in line with the stator tooth, as in Figure 12-37. Most electronic ignition systems fire the plugs just as this polarity changes from positive to negative. The AC signal is amplified by the module and then applied to the base of the transistor switching circuit to turn on and off primary current and time the spark.

It is important to note that the operation of the stator and armature are found in virtually every manufacturer's design. The terminology stator and armature might not appear, however. For instance, General Motors calls the components that generate the AC signal the pick-up coil and pole piece, while Chrysler calls them the reluctor and pick-up coil. Their operation is essentially the same, as you will see in the diagnosis section of the next chapter. Magnetic generators usually have two wires leading from the unit to the ignition module. An exception to this is when computer control of timing is being used: that is, when the computer circuits are between the pick-up coil and the module as seen in Figure 12-38.

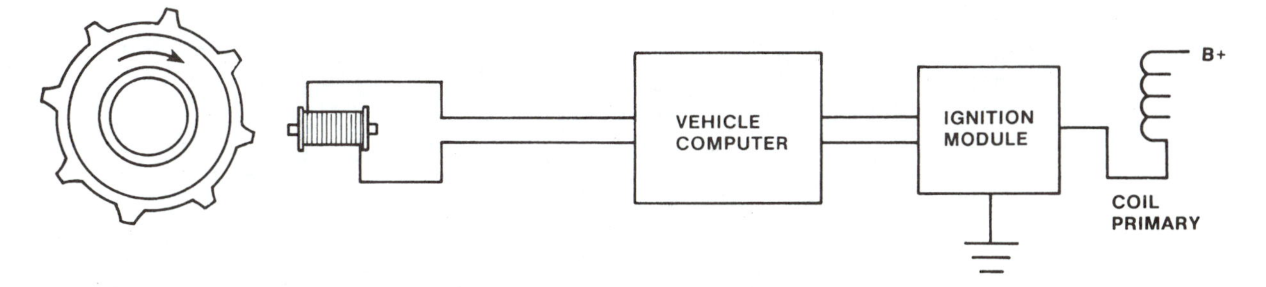

FIGURE 12-38 Vehicle computers are usually in charge of timing.

CRANKSHAFT POSITION SENSORS

A further refinement in this type of AC-generated crankshaft position sensing is the style of sensor shown in Figure 12-39. It is mounted on a bracket very near to the crankshaft, usually at the front of the engine. The vibration dampener, which is bolted onto the crankshaft, will have some magnets permanently attached or formed into its outer ring. The sensor is a simple coil of wire that will act like an alternator and generate a signal that the computer will use to turn the coil primary current through the electronic module on and off. The process here is exactly the same as that of the sensor mounted in the distributor. The magnet generates a positive voltage as it approaches the sensor's coil of wire. As the magnet just passes the coil, the polarity of the generated signal changes from positive to negative. This polarity change signals the crankshaft position to the computer. Keep in mind that the simplest electronic ignition or the most complicated computer-controlled ignition requires some basic input information. Crankshaft position is one of these basics. We would not want to fire the spark plug during the exhaust stroke, would we? Timing the spark is extremely important for driveability, fuel economy, and tailpipe emissions. The ignition module will not be able to turn primary current on and off without an input of crankshaft position. The crankshaft position sensor, used on some vehicles, is an input device. It gives basic information that will be used to determine timing. Again, timing is when the spark is delivered to the cylinder relative to the position and stroke of the piston. Its importance cannot be over stressed. It is basic and fundamental to the efficient operation of all spark-ignited engines that timing be correct under a variety of operating conditions.

The use of the crankshaft position sensor has grown over the last few years with computerization. An increase will no doubt occur as more and more manufacturers switch over to a style of distributorless, multi-coil ignition system. Without a distributor to signal piston position to the ignition system,

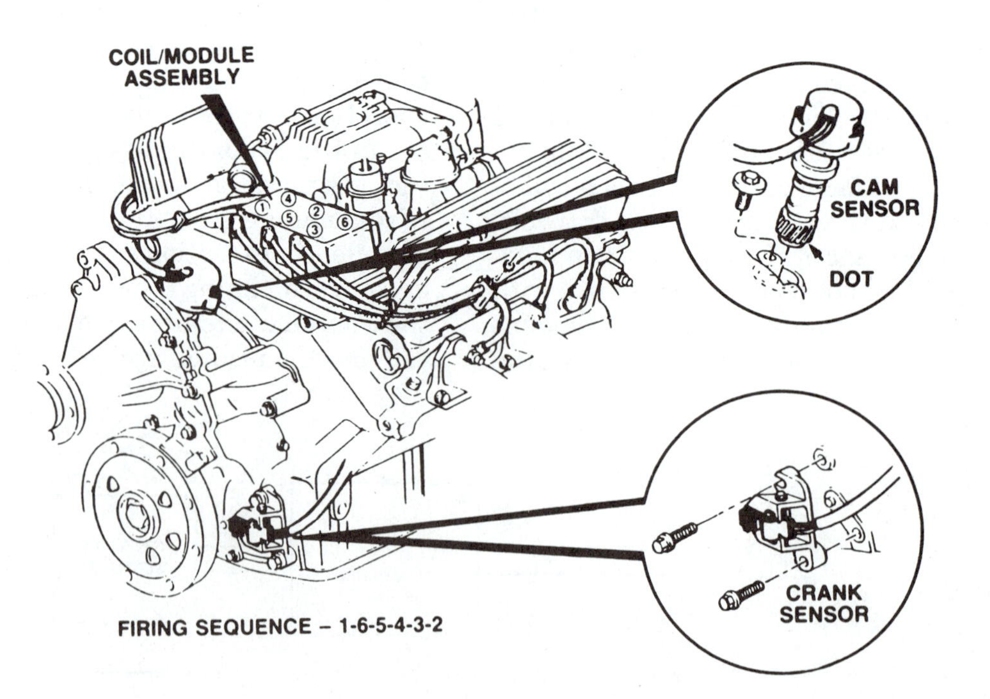

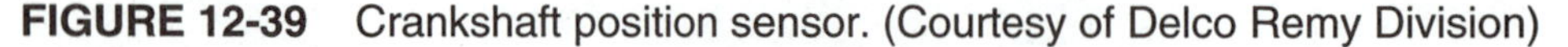

FIGURE 12-39 Crankshaft position sensor. (Courtesy of Delco Remy Division)

the crankshaft position sensor, mounted near the revolving magnets, supplies the needed signal. Chapter 14 will discuss DIS in detail.

HALL EFFECT SENSORS

Another type of crankshaft or piston position sensor in use is the Hall effect sensor. It is different from those sensors previously studied because it does not generate an AC signal from a moving magnetic field. Instead, its signal is rise in voltage followed by a drop in voltage. This changing up, then down, then up, and so on, signal is generated by a sensor such as that shown in Figure 12-40. The shutters connected to the bottom of the rotor are used as shields between the two sides of the sensor. One side of the sensor has a permanent magnet. When the magnet lines of force are allowed to reach the other side of the sensor, a relatively low voltage is passed out (near 0 volts on some models). As the shutter passes into the magnetic field, it blocks the sensor and the voltage rises sharply coming out of the sensor element. If the element is not exposed to a magnetic field, it produces a voltage; if it is exposed to a magnetic field, it does not produce a voltage. The shutters, in this example, are used to block the magnetic field from the sensor. Hall effect sensors are used on many vehicles. They are different from magnetic AC generators in two ways, generally. First, they require an input voltage source and second, their output is not dependent on engine speed. A Hall effect sensor will generate the voltage change discussed at any speed, even stopped. If an input voltage is applied to a Hall effect sensor, the output signal will depend on the shutter position not on its movement as in the AC generator style.

FIGURE 12-40 Hall effect distributor.

OPTICAL CRANKSHAFT POSITION SENSORS

The final type of crankshaft position sensor that we will look at is the optical style. It has had limited use but remains an option for the manufacturer. It consists of a light-emitting diode that will produce a small sharp beam of light when B+ is applied to it. The beam of light is focused through holes onto a photoelectric cell that generates a voltage when light hits it. Let us follow the sequence of events using the distributor shown in Figure 12-41. The light-emitting diode is on top of the rotor plate and the photo diode cell is below the rotor plate. The small slits in the rotor plate are placed at one degree spacing—360 slits around the edge of the plate. As the

FIGURE 12-41 Datsun distributor.

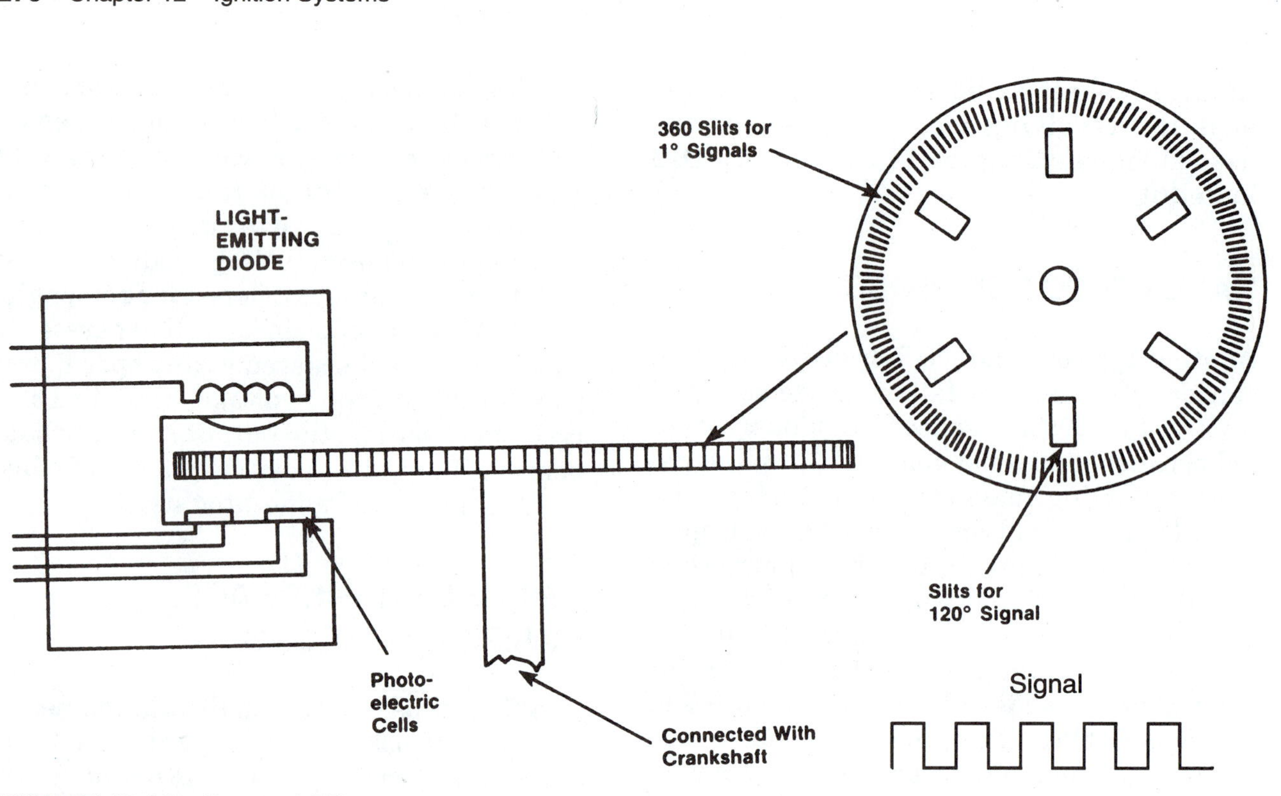

FIGURE 12-42 Datsun LED.

plate rotates with the engine's distributor, the photo diode cell will alternately see light and then not see light. The cell will generate a small voltage and then no voltage. This on-off pulsing voltage is sent to the computer to determine both engine speed and piston position. Notice that Figure 12-42 shows the signal that will be sent to the computer off the sensor. Also notice that our example has two photo diode cells. One for the 360° signal and one for the 60° signal. The 60° signal corresponds to the number of cylinders and crankshaft degrees, whereas the 360° signal is camshaft degrees. This unit would be for a six-cylinder engine with a cylinder firing every 120° of crankshaft rotation.

In addition, this type of sensor is frequently used up at the speedometer to determine vehicle speed for the computer as shown in Figure 12-43. The output from a light-emitting diode (LED) style sensor is a pulse voltage similar to a Hall effect sensor. Its level is also not dependent on speed. The number of pulses will change with speed but not the output level.

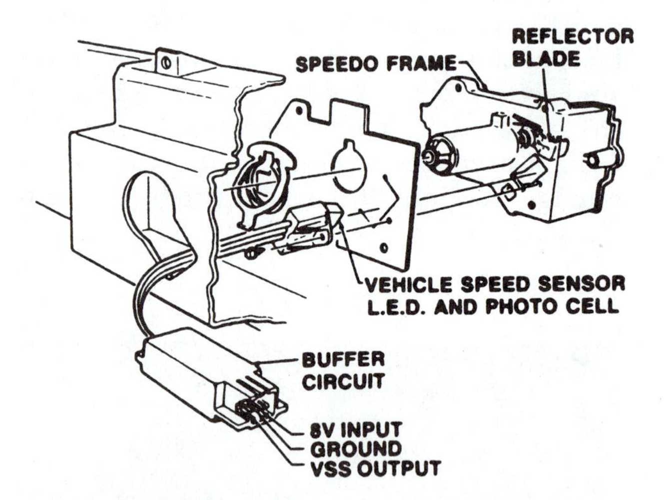

FIGURE 12-43 GM vehicle speed sensor. (Courtesy of GM Corporation)

ELECTRONIC IGNITION EXAMPLES

Now that we have a working knowledge of the different components that might make up a system, let us look at some actual examples. Our goal here is to show you how a manufacturer might put together certain components and how they would work. It does not make a lot of sense to try and memorize what a certain style is composed of because the following year might be totally different. Your skills will truly be adaptable if you can open the hood and recognize a sensor in a situation in which you were not aware that it was being used. When electronics began to find its way into the automotive field, many mechanics began to try to memorize certain styles of electronics. They were very used to working with breaker point ignition, and it did not vary from year to year or manufacturer to manufacturer. Not so with electronics! For example, the manufacturers are free to choose from five different methods of sensing crankshaft position. In addition, they might use more than one of the five in different applications. Your time will be more productive if you try to figure out how a sensor is used, and tested, rather than memorizing the applications where you will see it. In this way, you will not be too surprised when you open the hood and realize that the system you expected to find is not in use anymore. Transferable skills are absolutely necessary for the automotive technician of today. Let us look at the options available and see how they might be put together in a system.

The most common ignition system on the road today consists of an ignition coil (standard or E type), a module as the primary control device, turning primary current on and off, and a magnetic AC generator to signal crankshaft and piston position. Figure 12-44 shows Ford's TFI (thick film ignition) currently in use. Thick film is a term applied to the manufacturing process of the IC (integrated circuit) style of module that is mounted to the base of the distributor. There is no primary resistance in series with the coil primary. This would indicate that the module must be responsible for the limiting of the current. The AC signal from the sensor is used to directly turn the primary on and off with the module acting as the switch. Notice that Figure 12-45 is the same system with one difference. Rather than an AC generator in the distributor, a Hall effect sensor has been installed. The rest of the system appears to be the same at a quick glance. There is another difference, however; the module is not the same. This is a good example of why your ability to detect just what is in use is important. A call to the parts store for a new TFI module will result in the question being asked, "Is it with Hall effect or magnetic generator?" Your ability must enable you to determine which is most important. How do you tell the difference? Hall effect has shutters; magnetic generators do not.

FIGURE 12-44 TFI with magnetic pickup.

FIGURE 12-45 TFI with Hall effect switch.

Let us look at another style that is common. It is Chrysler's early electronic ignition with an externally mounted ignition module.

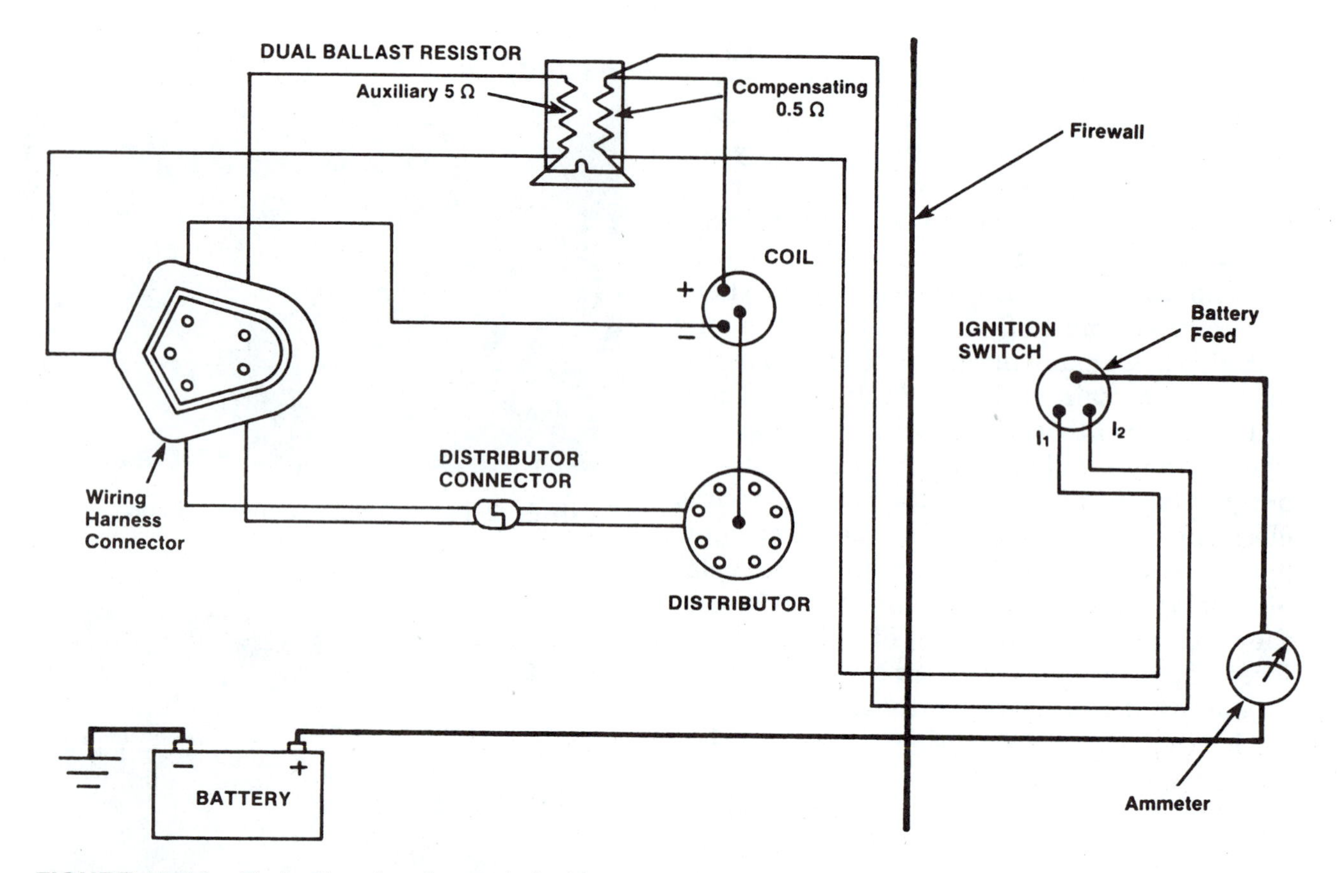

FIGURE 12-46 Early Chrysler electronic ignition.

It uses a standard type coil with a primary resistor wired in series as in the diagram, Figure 12-46. How is crankshaft position sensed? Notice inside the distributor the use of an AC generator with two wires leading directly to the ignition module. This system is different in that it uses a primary resistor to limit current just like the breaker point system we studied. In addition, it must have a bypass circuit to bring current directly to the coil during cranking. A later version of this system is shown in Figure 12-47. Notice the changes: no primary resistor and a Hall effect sensor. It is important to look at the manufacturer's literature and note differences from year to year and model to model. In addition, a visual inspection will reveal the different options picked by the manufacturer. Do not assume that you know what is under the hood until you open it!

CURRENT LIMIT FUNCTION

Electronic ignition modules have the obvious function of turning primary current on and off as we have discussed. In addition, many modules perform other functions. We have mentioned one, current limiting. Current limit function is found on ignition systems that do not have a primary resistor. Our unit on oscilloscopes will show what the pattern will look like with current limit, but let us analyze its function. When primary current is turned on, the magnetic field builds until the coil is fully saturated. With the current limit function, once full saturation has taken place, the module will cut back on primary current. This function increases the life of the coil by keeping it cooler. In addition, a bypass circuit is not needed. You can recognize a system that has current limit by looking at the wiring diagram.

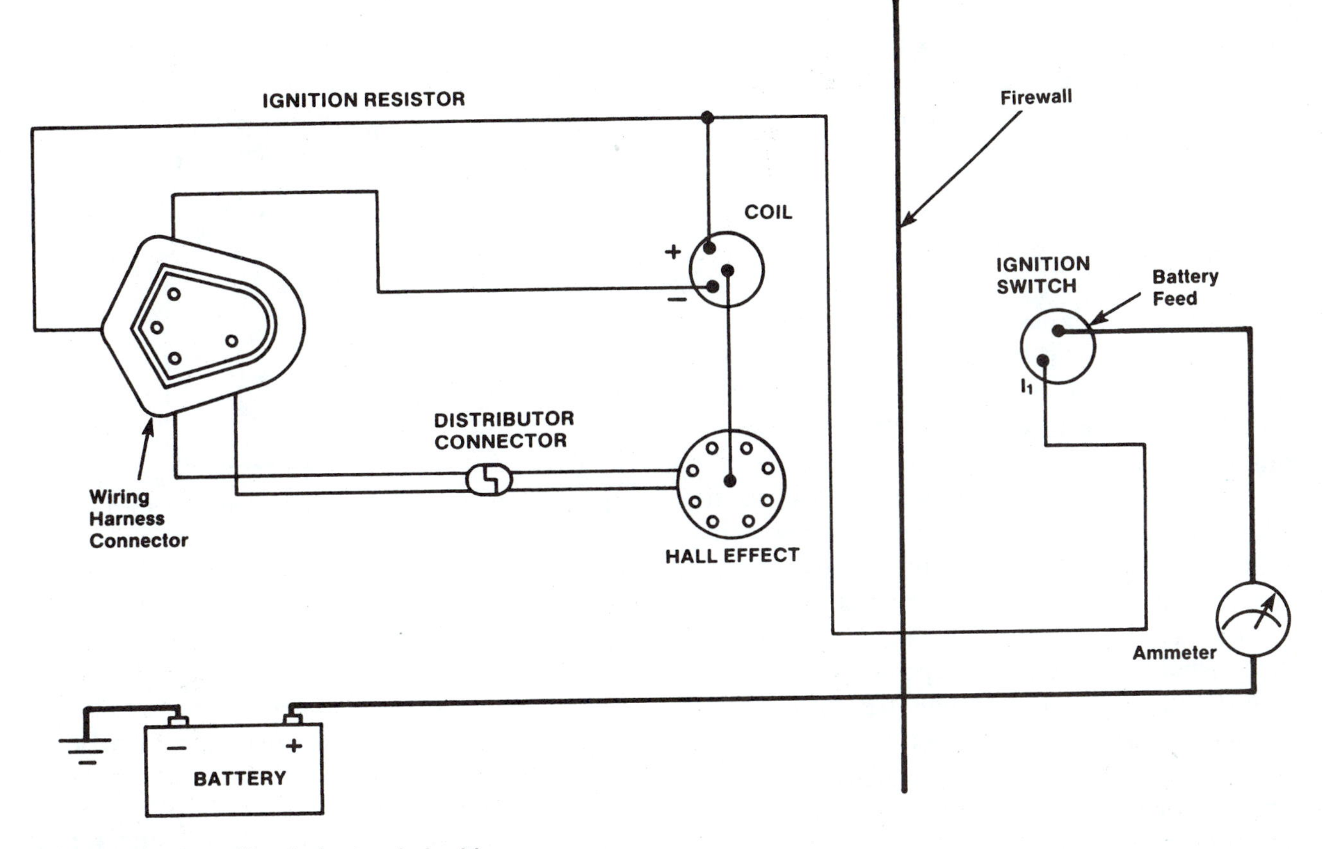

FIGURE 12-47 Omni electronic ignition.

If no bypass circuit or primary resistor is seen, the system must have the current limit function.

VARIABLE DWELL

Another function of many modern systems is variable dwell. On modules with this function, the dwell will be short at low engine speeds and higher as the engine speeds up. Dwell, you will remember, is the length of time in degrees of camshaft that primary current flows through the coil primary. A long enough dwell will result in sufficient coil saturation to attain full secondary output. Keeping the current turned on longer than is necessary accomplishes little at the plug end but can greatly decrease the life of the coil. As engine speed is increased, coil saturation time is automatically decreased if the dwell remains the same. For example, a dwell of 15° allows for a certain amount of coil saturation at a speed of 1,000 rpm. At 2,000 rpm, if the dwell were to remain the same, the amount of available coil saturation time would have decreased by 50%. This level of saturation might not be sufficient for full coil output. If, however, the dwell were to increase to 30° at 2,000 rpm, the amount of clock time would remain the same. For this reason, many of the manufacturers build their modules to have a variable dwell function.

To summarize, all ignition modules will have at a minimum two functions: primary current on and primary current off. In addition, they might have the current limit function to replace the primary resistor and bypass circuit or they might contain the variable dwell function. These function options, like the crankshaft position sensor's options, are found on a variety of different applications. They are easy to identify with an oscilloscope. The majority of the manufacturers currently use all four functions in their modules.

TIMING

Up to this point, we have mentioned timing only briefly and defined it as the moment that we delivered the spark to the cylinder. It is now time to discuss just how this is accomplished. We will begin with a discussion of a mechanical system that has been around for a long time and is still found on modern vehicles and end with the computer-controlled timing systems found on many vehicles.

Before we get into the specifics, let us discuss why we need any type of a timing system at all. If you remember, the four-stroke theory—intake, compression, power, and exhaust—assumed that between compression and power was a spark that would ignite the mixture and produce the power. In theory, the peak combustion power of the burning mixture was to be delivered to the piston just as it passed TDC (top dead center) at the beginning of the power stroke. This still remains the goal of the most modern timing systems: deliver peak combustive power to the piston just after TDC on the power stroke. The problem, however, is that the mixture will not develop peak combustive power at the same time as the spark occurs, Figure 12-48. The mixture will have to be ignited and burn for a period of time before sufficient fuel will be delivering power. This time between ignition and peak combustive power is the reason for the sometimes complicated components in the advance mechanisms that are considered to be part of the ignition system. Let us define some common terms that will be part of our discussion. You already know that **TDC** is **top dead center**, the point where the piston is as high in the cylinder as it will go and stops to reverse its direction. During the four stroke, there are two TDCs. One between compression and power, and one between exhaust and intake. Anything that occurs before TDC (for instance, during the compression stroke) is said to occur **BTDC (before top dead center)** and is usually expressed in crankshaft degrees, Figure 12-49. For example, a spark that is delivered to the cylinder 10° BTDC refers to the position of the piston during the last part of the compression stroke. As each stroke is equal to 180° of crankshaft rotation, you can tell that the compression stroke is just about over. The piston is on its way up and just about to begin the power stroke. Any-

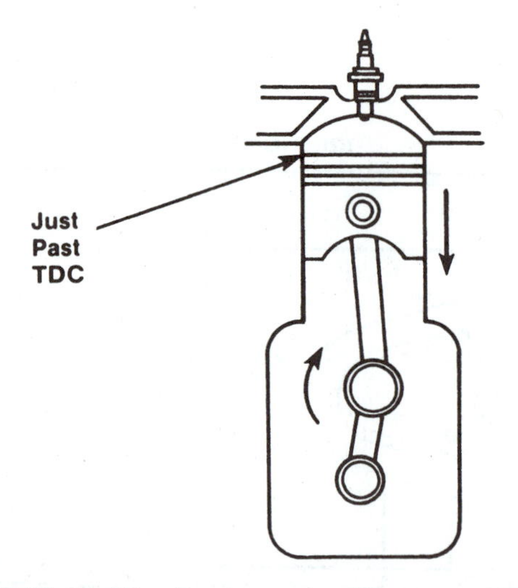

FIGURE 12-48 Peak combustion pressure should occur just after TDC.

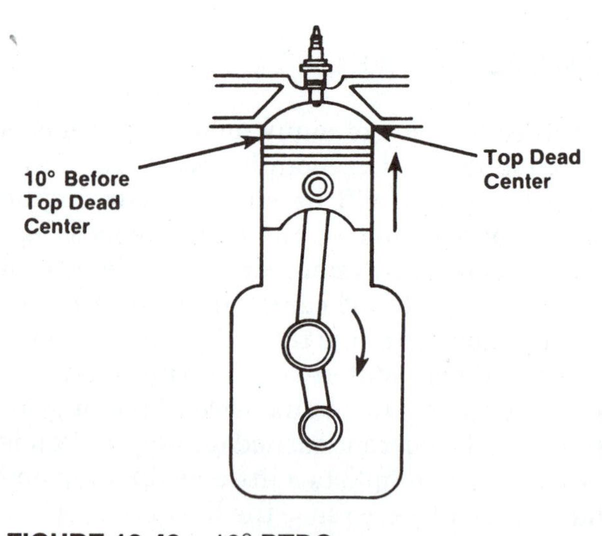

FIGURE 12-49 10° BTDC.

thing that occurs after top dead center is referred to as ATDC and again is usually expressed in degrees of crankshaft. Getting these two abbreviations mixed up while working on a vehicle can really cause some terrible driveability complaints. Another term to define is **advance**. To advance the spark is to deliver it sooner. A change in timing from 10° BTDC to 20° BTDC is an advance of 10°. The opposite of advance is **retard** and it means that the spark is being delivered later. A change in timing from 20 BTDC to 10 BTDC is a retard of 10°. Notice that the spark is still being delivered before top dead center. It is possible under certain circumstances to have the spark delivered ATDC. If you think about ATDC timing, it really does not make a lot of sense because obviously peak combustive pressure will also be delivered after top dead center, wasting power. Generally speaking, the reasons for ATDC timing are for tailpipe emissions. Certain vehicles produce less pollution at idle if operated with ATDC timing. The use of this has greatly diminished, especially with computerization of timing. As we look at different advancing and retarding systems, keep in mind the basic idea of timing. The spark will generally arrive before TDC so that peak combustive pressure will push the piston down on the power stroke. Also, keep in mind anything that is done to change around the amount of time it will take between ignition and peak combustive power will have to be compensated for in the advancing systems of the vehicle. The vehicle's ignition system will be constantly compensating for changes in the time between ignition and the power stroke.

The conditions that will need to be compensated for are generally engine speed and engine load. These two have the greatest effect on the burning time (time between ignition and peak combustive power). In addition, many modern systems compensate for altitude and engine temperature. We will start with the mechanical systems, compensating for speed and load only, and move into the computerized systems.

MECHANICAL ADVANCE

The earliest form of speed-compensated advancing system was the mechanical advance. It was designed during the early years of the automobile and is still in use today on certain vehicles. It operates on the simple principle of centrifugal force. Centrifugal force is directly related to speed and so can be calibrated to give an increased advance with speed. Figure 12-50 shows a mechanical advance unit in conjunction with an AC generator style position sensor. As the distributor speed increases, the centrifugal weights fly out and put the stator ahead of the shaft. The number of degrees advance is equal to how far the stator is ahead of the distributor shaft. The faster the distributor turns, the farther ahead the stator gets and the more advanced the spark is. As the engine slows down, the amount of centrifugal force decreases and the timing retards because the advance springs pull the weights back.

Typically, the mechanical advance system is relatively smooth operating, giving an equal advance with speed. The graph in Figure 12-51 shows that at 500 distributor rpm the distributor should give an additional 2° of distributor advance. Both these numbers need to be converted over to crankshaft degrees by multiplying them by 2. The distributor is traveling at half engine speed, so 1° of distributor is equal to 2° crankshaft. Based on this information, the timing will change (advance) by 4°

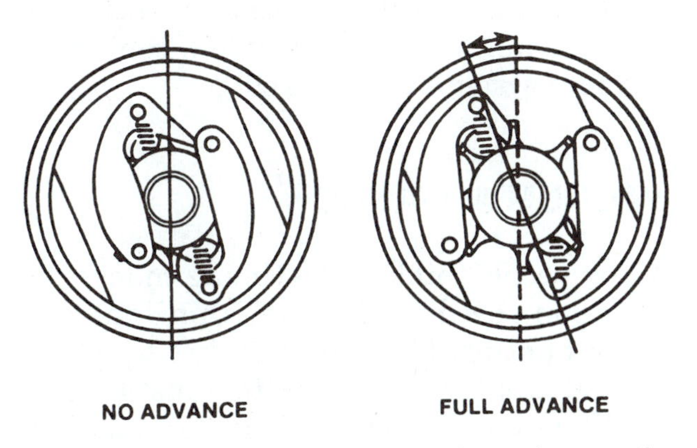

FIGURE 12-50 Distributor weights advance the timing.

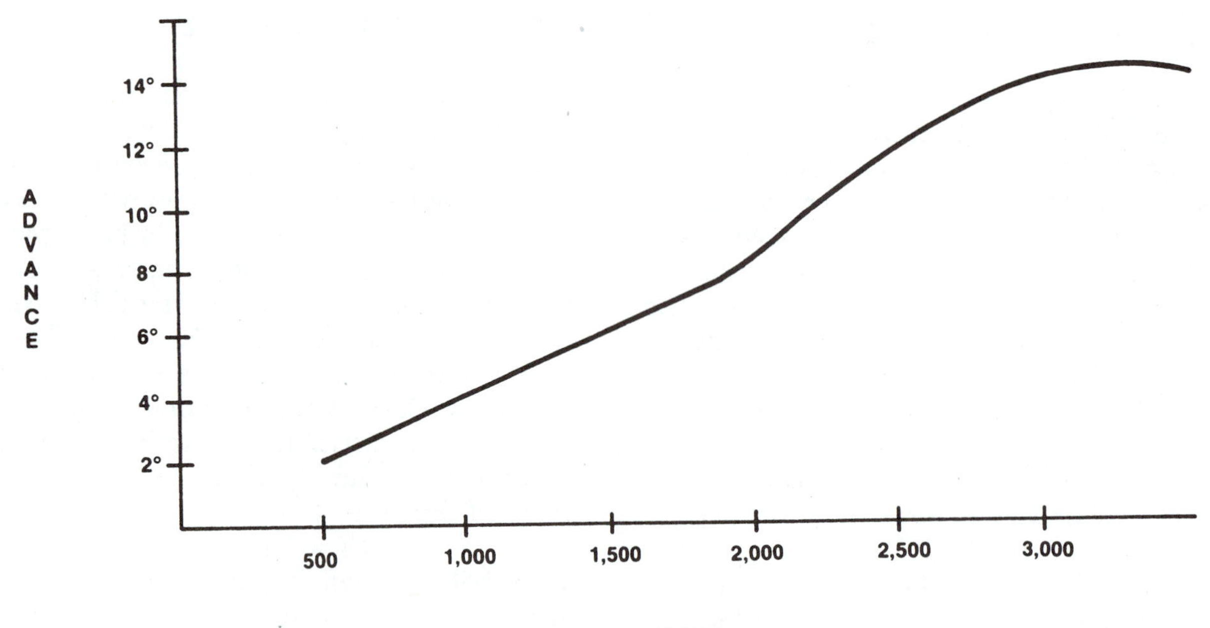

FIGURE 12-51 Speed affects timing.

at 1,000 engine rpm. What will happen at 2,000 engine rpm? The chart shows that at 1,000 distributor rpm, an additional 6° of advance will take place. This will convert over to 12° crankshaft at 2,000 rpm. Notice that we have said additional degrees of advance. Additional degrees over something called **base timing**. Base timing is the minimum advance required by the engine at the rpm that it is usually tuned at. Normally, this is the minimum speed that the engine operates at, and it is the speed that the engine base or initial timing is set at. We will cover more of this in the next chapter when we look at how timing and advance systems are checked.

VACUUM ADVANCE

Many people think of the timing change due to piston speed as the same thing one does when hunting. If a hunter wishes to shoot a duck that is flying, the hunter must lead the duck so that the shot reaches the spot where the bird is. The hunter actually aims ahead of the flying bird. The piston moving through the cylinder is fundamentally the same idea. The spark must lead the piston, compensating for its speed. Double the speed and you need to advance the timing twice as far. The principle is the same as our flying duck who suddenly starts flying faster. To have duck dinner, we will have to lead the bird even more to compensate for the increased speed. The increased advance supplied by the centrifugal weights will compensate for increased piston speed and does it very well if the same volume of air and fuel is constantly in the cylinder. The time between ignition and peak combustive force is dependent on the amount of fuel in the cylinder. The greater the amount of gas and air, the faster the burning will be. The greatest volume of combustible mixture will be in the cylinder if the driver of the vehicle has his or her foot to the floor. This opens up the throttle and allows the engine to breathe in the most air and fuel. For this reason, the mechanical advance is usually calibrated for the timing necessary with the throttle wide open. Anything less than wide open throttle will reduce the volume of fuel and air and slow down the burning. Remember, the goal of the advancing systems: to deliver peak combustive pressure to the pis-

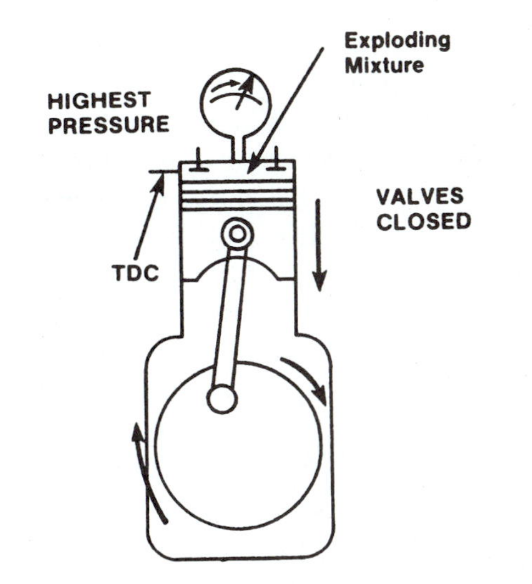

FIGURE 12-52 Peak combustion pressure should arrive.

ton just as it passes TDC, Figure 12-52. When do you think we will deliver peak combustive pressure if the throttle is partially open with only a mechanical advance? Let us analyze this question and see if we can figure it out together. If the mechanical advance is calibrated for WOT (wide open throttle) and the throttle is partially open, the burning will be slower, so the peak combustive pressure will be delivered later. The pressure will strike the piston after it has passed TDC and is on its way down. This will reduce both the power and the economy. Do not hesitate to review the last few sentences if the conclusion is not quite clear at this point.

So, we have found out that a mechanical advancing system will not be calibrated correctly except for WOT conditions. Obviously, most people cannot drive with their foot to the floor all the time, so the manufacturer generally supplies an additional advancing system that will compensate for less than WOT conditions. This system is called the **vacuum advance**, and it will use engine vacuum to pull on a diaphragm. Figure 12-53 shows that the diaphragm is attached to the pick-up coil plate. The movement of the plate will pull the pick-up coil in the opposite direction to that in which the distributor shaft is rotating. This will advance the spark proportionately to engine vacuum. The greater the vacuum, the more the advance; the lower the vacuum, the less the advance. Vacuum and quantity of air/fuel in the cylinder are one and the same. As the amount of air/fuel is increased (larger throttle opening), the vacuum is decreased, and as the amount of air/fuel is decreased (smaller throttle opening), the vacuum is increased, Figure 12-54. The vacuum advance can be connected to a port that has vacuum anytime the engine in on. This port is called a source of manifold vacuum. The other choice for advance vacuum is usually a port or hole above the throttle plates. It will be a source of vacuum when the engine is above

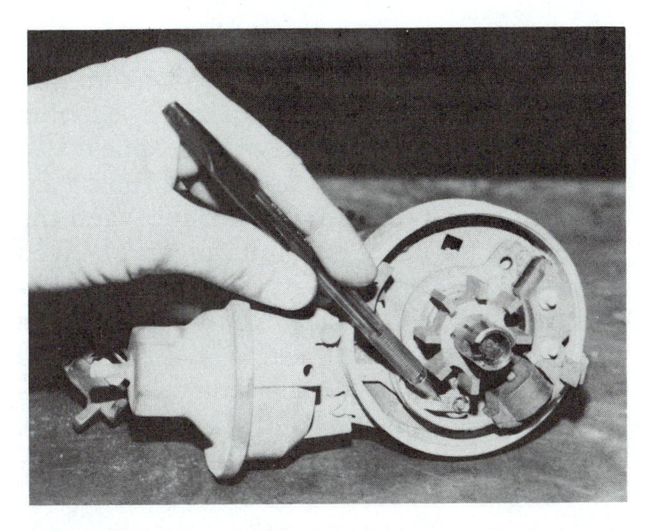

FIGURE 12-53 The vacuum advance diaphragm is attached to the stator.

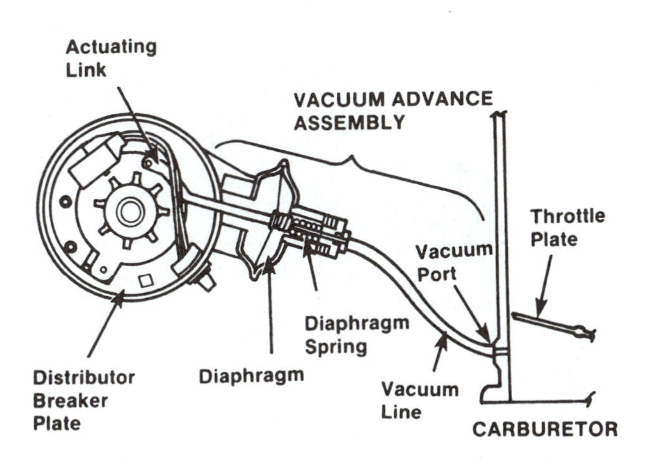

FIGURE 12-54 Manifold vacuum is sometimes used for the vacuum advance.

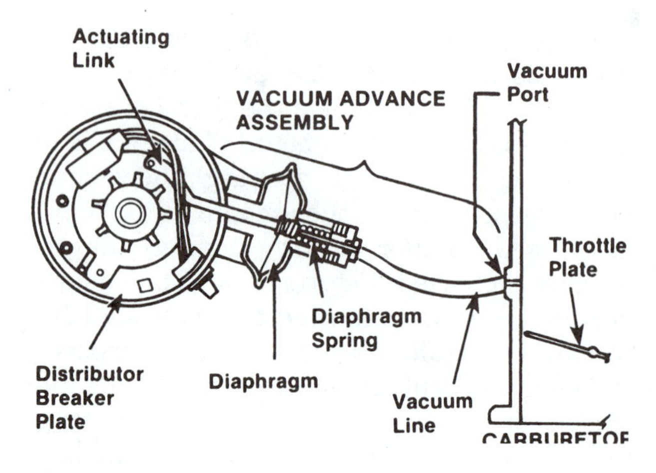

FIGURE 12-55 Ported vacuum is sometimes used for the vacuum advance.

idle, Figure 12-55. This type of vacuum is called **ported vacuum** and has vacuum present only when the port is exposed by opening the throttle. In this manner, the vehicle systems will always give the correct timing for both the speed (mechanical advance) and the quantity of air/fuel (vacuum advance). Keep in mind that these two systems work in addition to the initial timing. You will see in the next chapter that these systems can be checked against specifications with an advance timing light or a digital timing indicator.

In addition, there are some timing systems with another diaphragm that will pull the timing into a retarded condition. This type of system is shown in Figure 12-56. Usually, manifold vacuum is applied to the retard diaphragm, which is small, and ported vacuum is applied to the advance diaphragm, which is large. In this way, the system will go into a retard mode at idle when the small diaphragm only has vacuum applied to it, and go into an advance mode when vacuum is applied to the larger diaphragm. Remember, ported vacuum is vacuum above the throttle plates. At idle, there will be no ported vacuum, because the throttle plates are closed, blocking the port. Once the throttle is opened up and the engine is above idle, there will be vacuum present at this port. In this manner, the timing is retarded at idle and advanced above idle. Emission control is usually the reason for this dual diaphragm advancing retarding system.

Another type of retarding system has been built into certain electronic ignition systems. Ford's dual mode is an example of this. It will retard the timing during cranking to reduce the cylinder pressure. This reduced pressure allows the starting system to spin the engine over faster, greatly increasing the chances of the engine starting. Figure 12-57 shows a dual mode electronic ignition module.

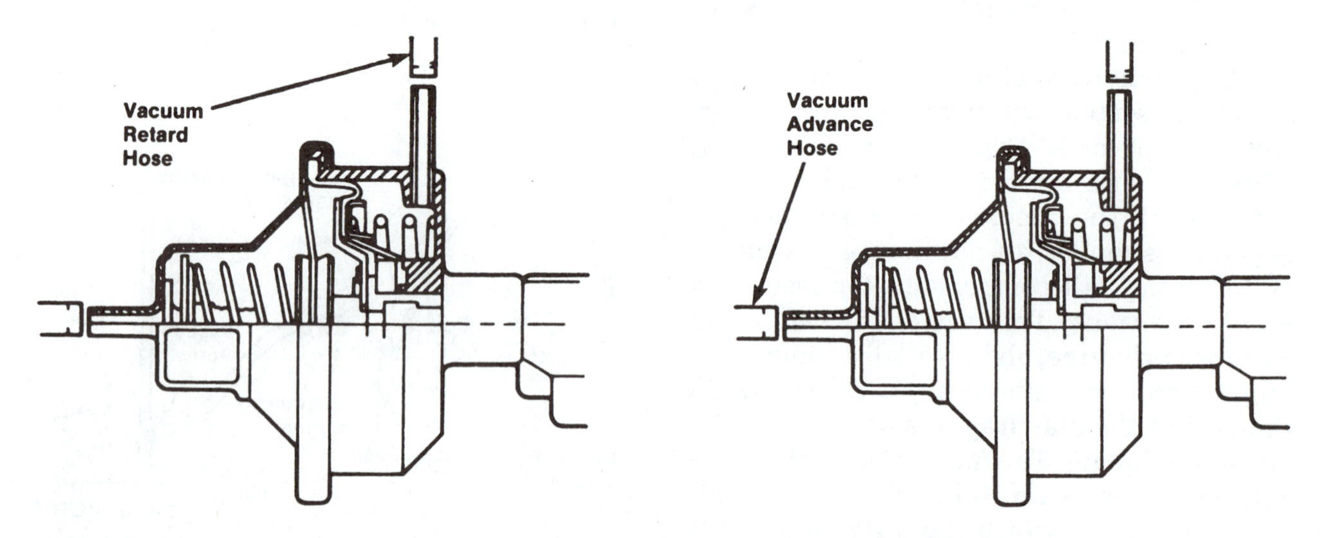

FIGURE 12-56 Vacuum advance/retard unit. (Courtesy of Ford Motor Company)

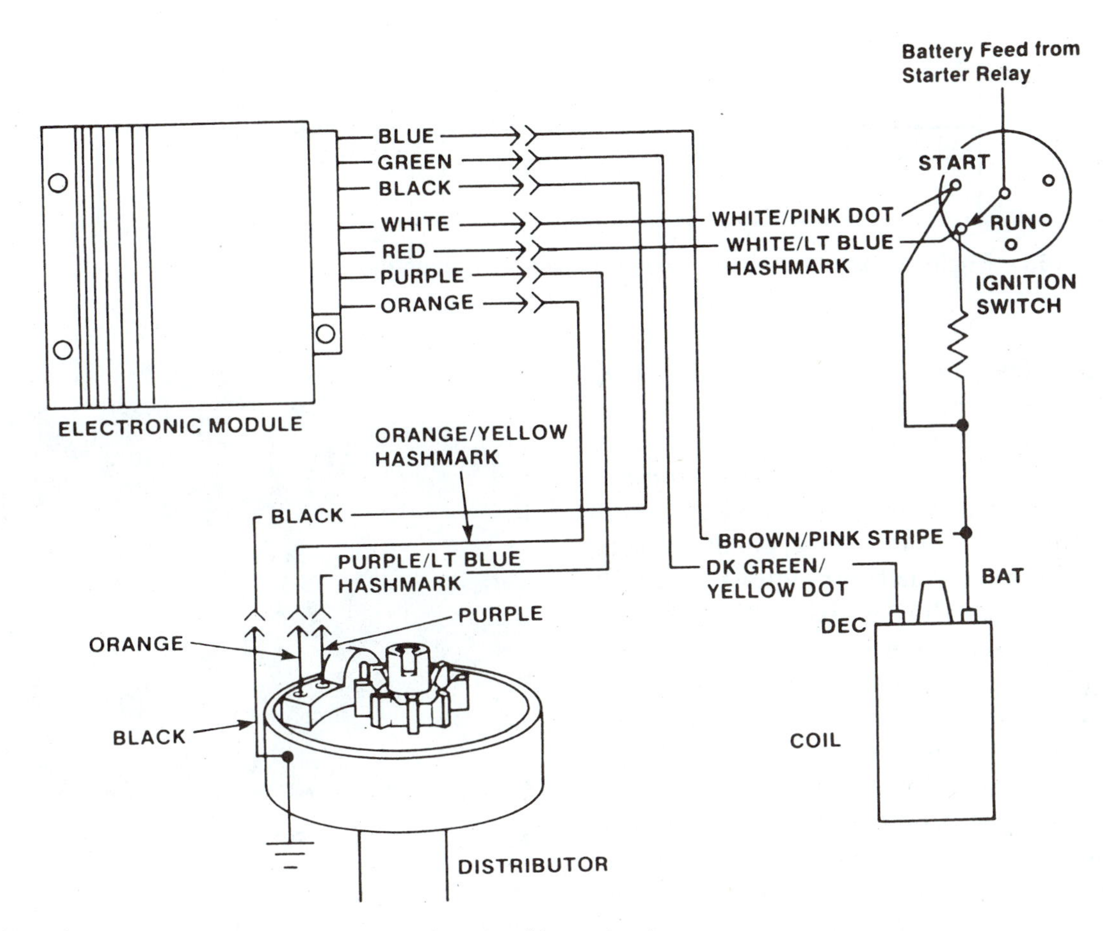

FIGURE 12-57 Dual mode module provides cranking retard.

COMPUTER-CONTROLLED TIMING

The mechanical and vacuum advance do a good job but have their limitations. One of these limitations is the inability of the system to generally recognize that different cylinders might require different timing. In addition, engine temperature changes, air temperature changes, and altitude generally cause the timing requirements to be different. With its inability to adapt to these changes, the traditional advancing systems have generally been replaced with computer-controlled systems, especially on American-produced vehicles. Ordinarily, the diagnosis of these systems is not difficult, especially if you have the basics of the mechanical and vacuum advance in mind. In general, the computer systems perform the same functions, compensating for piston speed and quantity of air and fuel in the cylinders. They just do it with more precision to increase both economy and driveability. Let us look at the different sensors that an engine computer might use to determine the cylinder timing that is best. Try to keep the mechanical and vacuum advance in mind, as this usually makes the computer systems easier to understand.

The block diagram in Figure 12-58 shows the system most frequently used in computer-controlled timing systems. The signal from the pick-up unit is sent to the computer where it is modified or compensated before it is allowed to control the ignition module. Do not forget the module is turning on and off primary current based on the signal it gets from the computer. This signal is supplied to the module from the computer after timing alter-

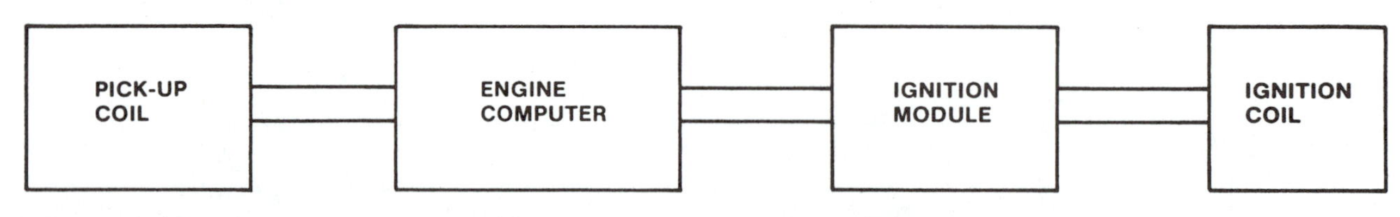

FIGURE 12-58 Engine computer between pick-up coil and ignition module.

ations have been added. The distributor does not have any mechanical or vacuum advance units. The signal coming of the pick-up unit is always the same, relative to the piston positions. The input sensors are the key to understanding just how this type of system operates, because it is their signals that will tell the computer the conditions that the engine is operating under.

The first sensor is the throttle position sensor (TPS) and it is just what the name implies. Its job will be to tell the computer the position that the throttle is in. It usually is a voltage-dividing circuit like we discussed in a previous chapter. It will send a voltage back to the computer, which is read as throttle position. Many systems use 0.5 volt as idle, 2.5 volts as half throttle opening, and 4.5 volts as WOT. The computer uses this information in its calculation of timing. Figure 12-59 shows a TPS on an injection unit. Figure 12-60 shows how the voltage divides through a TPS at half throttle.

The second sensor that might be used is the manifold absolute pressure sensor (MAP). Its job will be to tell the computer what the pressure is inside the intake manifold. Do you remember the vacuum advance? The MAP sensor is the replacement for it. In addition, pressure changes caused by altitude can be recognized by the computer. As we mentioned earlier in this text, some MAP sensors are voltage dividers like TPS and others recognize pressure changes and change around the frequency of the signal delivered to the computer, Figure 12-61 shows a photo of a typical MAP sensor.

FIGURE 12-59 TPS mounted on injection unit.

The third sensor is the pick-up unit inside

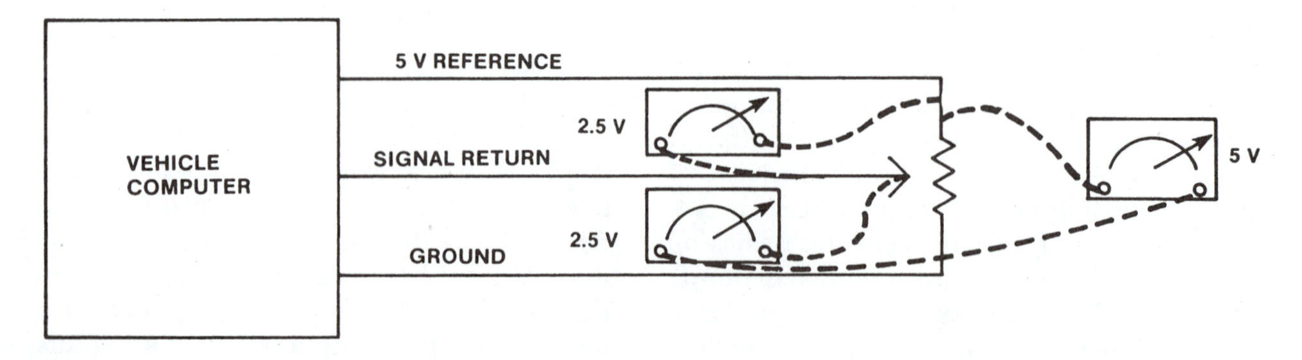

FIGURE 12-60 Voltage divides through a TPS at half throttle.

FIGURE 12-61 MAP sensor.

the distributor. We have seen how the signal from the pick-up unit is used to trigger the electronic ignition. It becomes a simple procedure for the computer to count the number of pulses per minute and convert it over to engine speed. Do not forget engine speed is a necessary computer input before the output of correct timing can be calculated.

Controlling Timing Changes with the Computer

If you look at a typical computer controlled ignition system such as that shown in Figure 12-62, you will see that this common system uses four wires to connect the computer to the module. The object of the four wires is to allow the computer to: 1) recognize engine speed; 2) send a timing pulse to turn off the primary current; 3) tell the module that it (the computer) is in charge of timing; and 4) maintain a connection between the computer and the ignition module used to keep both at the same ground voltage. Most systems in use are the same as this EST circuit for a General Motors vehicle. Only slight differences exist. Let us look at each of these wires or functions, and use the EST as our example by following along on the wiring diagram.

The first function is to recognize engine speed. This is done through the HEI reference wire (purple/white circuit #430), which runs between the module and the ECM (the computer). This signal comes off the pick-up coil through the signal converter. The converter changes the AC signal from the pick-up coil

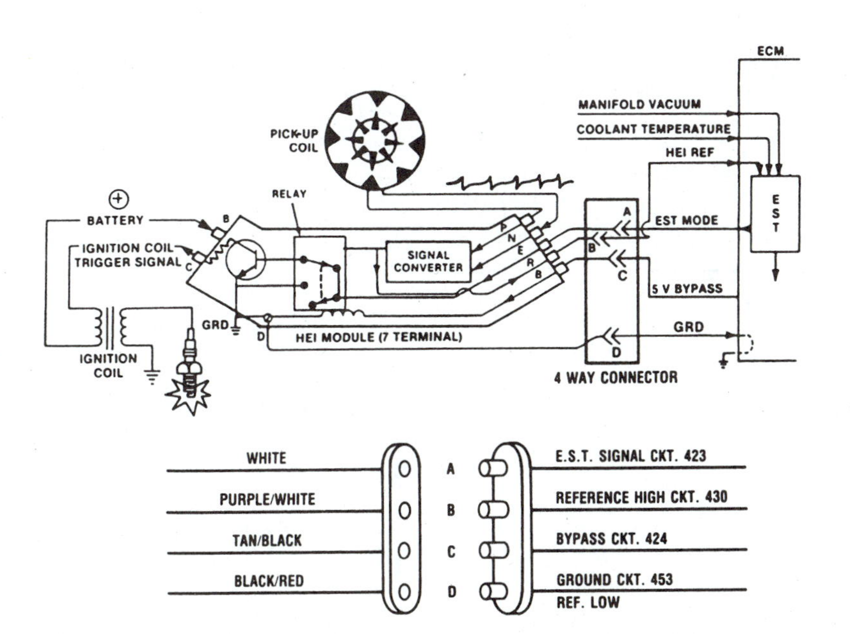

FIGURE 12-62 Standard EST interface. (Courtesy of GM Corporation)

into a digital square wave pulsing DC. Reference gives the ECM the engine speed information that it will need to recognize that the engine is in fact running and at what speed.

Once the engine is running, the ECM will begin to generate timing pulses, which are also digital square wave pulsing DC. These pulses will be generated off the timing information that the various sensors have supplied to the computer and will be sent to the module along the EST signal wire (white wire, circuit #423). The module will begin using this signal coming from the ECM to control timing once the third wire, 5-volt bypass (tan/black wire, circuit #424) is energized.

The bypass wire is the control circuit for the relay function of the module. Simplified, this bypass tells the module to switch from the pick-up coil signal to the ECM's EST signal for timing. The ECM will now generate the pulse which will turn off the coil primary and fire the plug based on other inputs such as coolant temperature, engine load, and throttle opening. Computer controlled timing has been around since 1980, and requires the setting of initial or base timing just like any other vehicle we have looked at.

Figure 12-63 is another example of computer controlled ignition, but with slightly different terminology. The computer is the EEC-IV module. Profile ignition pick-up (PIP) is the signal coming from the Hall effect device which is sent to both the ignition module (TFI-IV module) and the computer. This square wave digital signal does not have to be converted because the Hall effect device produces just what the computer and module require. PIP is the equal of reference in our previous example, and it gives the computer the speed information it needs. Spark output (SPOUT) is the timing signal generated in the computer which the module will use to determine timing. IGND is the ignition ground.

Have you noticed the only difference between these two systems? It is that TFI-IV does not use a bypass signal. When the module sees a SPOUT signal it switches from using PIP for timing to SPOUT without the

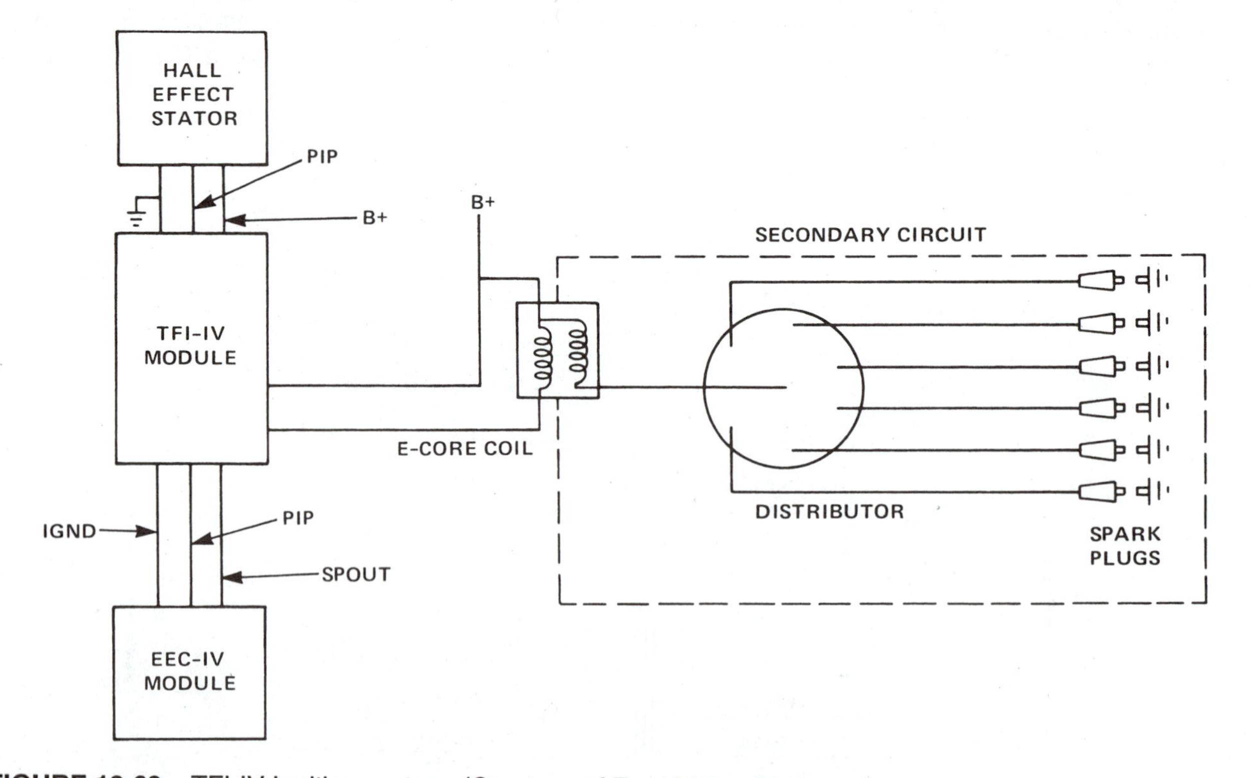

FIGURE 12-63 TFI-IV Ignition system. (Courtesy of Ford Motor Company)

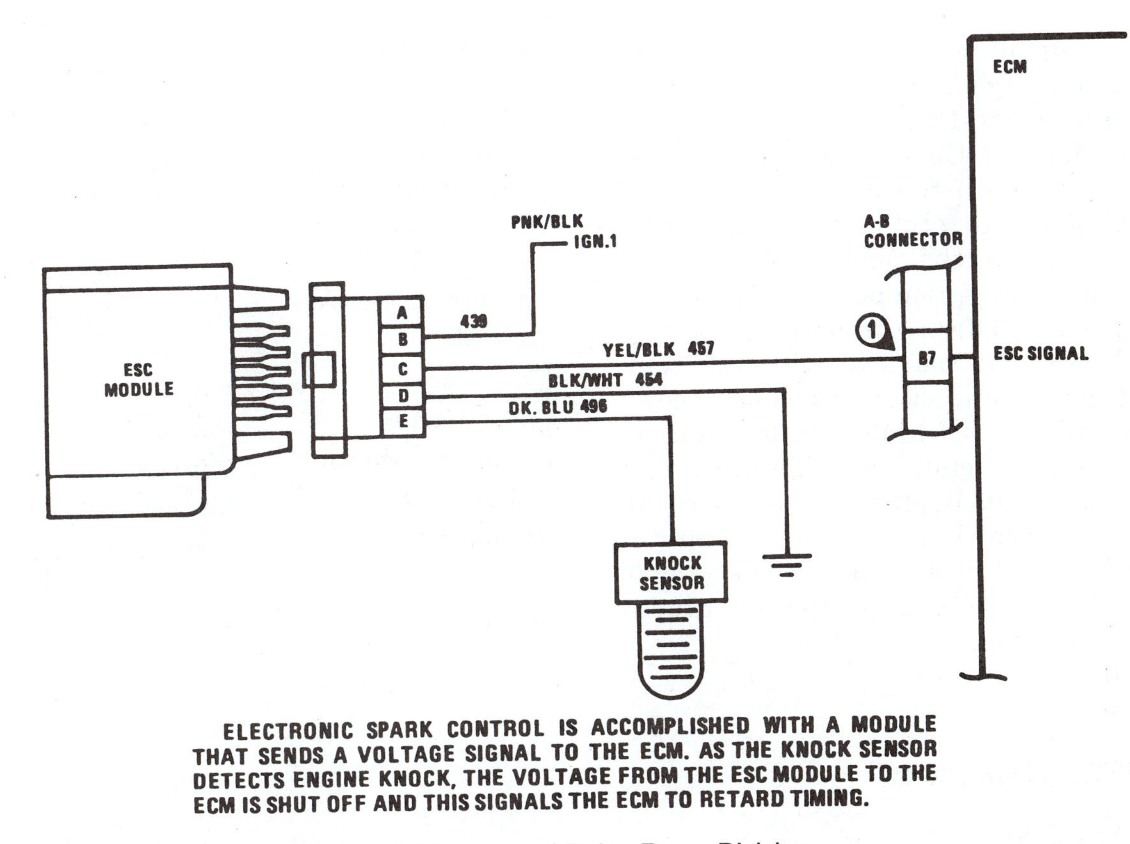

FIGURE 12-64 Knock sensor circuit. (Courtesy of Delco Remy Division

need for an additional signal from the computer. If there is no SPOUT, the TFI-IV module will use PIP for timing, the vehicle will not have any timing change, and will run at base or initial timing.

In both of these examples, if the technician wishes to check or adjust timing, he would pull either the four wire connector or the SPOUT wire. Once the module no longer sees the timing pulse, it will use the signal coming from the PIP or the pick-up coil as its timing pulse and the engine will run at base timing.

It is very important that you understand basic computer control of timing at this point. Take the time NOW to review not only the preceding information but also the following information in this chapter until it makes sense. Do not go on until you are ready.

Recent years have seen the increased use of knock sensors by the manufacturers of computer-controlled systems. The theory behind them is that they are generally piezoelectric crystals tuned to recognize the frequency of an engine knocking. This pulse that is generated by the crystal is sent over to the computer, which retards the timing until the knocking is eliminated. The circuit for a typical knock sensor is shown in Figure 12-64. This circuit allows the maximum advance just short of knocking. This is the timing that will give the greatest economy and the best driveability.

SUMMARY

In this chapter, we have looked at the ignition system and stressed the many changes that have occurred within it in the past few years. Starting with the spark plug, we discussed the different types, heat ranges, reaches, and styles of sealing used. We also looked at some repairing procedures used frequently on aluminum cylinder heads. Ignition secondary wires were looked at next with the three most common styles currently in use outlined. Caps and rotors were discussed along with the basic principles of ignition coils. Control of primary current for the generation of the spark was discussed beginning with ignition points and running through the modern computer-controlled primary current. Included with the computer discussion was a review of the necessary sensors used to determine piston position, engine speed, and engine load. Timing of the spark ended the chapter with mechanical, vacuum advance, dual mode, and computer control. In addition, a section was devoted to the use of knock sensors as a computer input for timing control.

KEY TERMS

Mandrel
RFI (radio frequency interference)
TVRS (televised and radio suppression)
Magnetic suppression
Dielectric
Firing order
Primary current
Secondary winding
Ignition points
Saturation time
Dwell
Current limit
Point gap
Rubbing block
Timing
TDC (top dead center)
BTDC (before top dead center)
Advance
Retard
Base timing
Vacuum advance
Ported vacuum

CHAPTER CHECK-UP

Multiple Choice

1. When primary current is turned off, the coil's magnetic field
   a. builds.
   b. collapses.
   c. impedes.
   d. relucts.
2. Which winding directly supplies the spark plug with high voltage?
   a. primary
   b. secondary
   c. neither primary or secondary
   d. both the primary and the secondary
3. Which of the following is not part of the primary circuit?
   a. rotor
   b. control module
   c. condenser
   d. points
4. Vacuum advance compensates for varying
   a. engine speeds.
   b. weather conditions.
   c. engine size differences.
   d. loads.

5. Which of the following components is not part of the secondary circuit?
   a. spark plug wires
   b. distributor cap
   c. rotor
   d. condenser
6. Normally, mechanical advance will
   a. advance the spark with higher vacuum.
   b. advance the spark with additional speed.
   c. cause little change in timing.
   d. compensate for load and weather conditions.
7. Current limit is found on electronic ignition systems that are missing the
   a. coil secondary.
   b. coil primary.
   c. Hall effect sensor.
   d. primary ignition resistor.
8. The ignition module
   a. turns the magnetic pick-up on and off.
   b. turns primary current on and off.
   c. changes the ignition timing.
   d. raises and lowers the secondary voltage.
9. The ignition primary resistor is
   a. bypassed during the normal running of the engine.
   b. important in increasing coil life.
   c. used to increase plug life.
   d. used on all ignition systems.
10. Too high a heat range spark plug will
    a. increase performance.
    b. decrease plug life.
    c. reduce ignition firing voltage.
    d. increase the chances of fouling.

True or False

11. Vacuum advance units advance the spark through the distributor weights.
12. Spark plugs should never be removed from an aluminum cylinder head that is still hot.
13. Manifold absolute pressure sensors will advance the timing if higher engine vacuum is sensed.
14. Variable dwell will result in a shorter primary current on time at high speeds than low speeds.
15. A knock sensor will retard the ignition timing until knock is eliminated.

16. A colder than specification plug will usually result in fouling.
17. The spark is delivered to the plug just as primary current is turned on.
18. Primary resistors are in use on all ignition systems.
19. Computer-controlled ignition systems adjust ignition timing through the use of sensors rather than by a vacuum and mechanical advance.
20. Current limit is a function found frequently on modern electronic ignition systems.

CHAPTER 

Vehicle Ignition Testing and Service

OBJECTIVES

At the completion of this chapter, you should be able to:

- use a feeler gauge to set air gaps in spark plugs and pick-up coils.
- use a tach dwellmeter to measure engine speed and ignition dwell.
- use a timing light to set initial timing.
- use an advance timing light or a digital timing meter to measure advance.
- use an oscilloscope to test ignition system operation.
- use appropriate diagnostic equipment to determine which electronic ignition component is responsible for a no-spark condition.

INTRODUCTION

In the previous chapter, we looked at the individual components that make up the modern ignition system. With these in mind, we will begin our discussion of vehicle testing. It is important that you realize that the tools of the trade have changed just as much as the systems have. The phrase "correct tool for the job" cannot be overemphasized because the repair or diagnosis of certain items is impossible without both a knowledge of the component and the correct piece of test equipment. At minimum, the mechanic of today needs a digital volt-ohm meter, a method of measuring ignition advance, and an oscilloscope with its additional gauges to accurately set or diagnose electronic ignition. In addition, the term "tune-up" has become very misleading. Years ago it meant basically the same to all mechanics—points, plugs, filters, and setting the idle speed, timing and point dwell. Obviously, many of these items are no longer in use or require less setting, while many new items are found that require a different level of expertise and diagnostic skills. We will emphasize the tools necessary as we go along.

FEELER GAUGES

One of the earliest ignition tools was the feeler gauge. Whenever we wish to measure the distance between two items, we use a feeler gauge. It is still in use today for pretty much the same function and comes in three basic designs: flat steel, flat brass, and round. The thickness of the two flat feeler gauges is normally printed on them. One is made of steel, while the other one is made of brass. You will see the use of the brass one when we set air gaps on pick-up coils. Its nonmagnetic properties make it the ideal choice when a magnetic field is present. The round feeler gauge is the type we use to set spark plugs. The diameter of the wire is printed on the holder and should match the specification of the air gap between the center electrode and the side electrode. Figure 13-1 shows the air gap of a plug being measured with the wire gauge, The wire should pass through the gap with some effort. This indicates that the plugs are gapped to the diameter of the gauge. Most wire gauges have a bend bar attached that is useful if the air gap is not correct. By bending the side electrode slightly, we should be able to get the air gap within specification. Remember from the previous chapter that the finished air gap should put the center electrode and the side electrode at right angles to one another. Figure 13-2 shows what a plug's air gap looks like if it is gapped either too small or too large. Usually, an incorrect looking plug is the result of either the wrong gap specification or the wrong plug for the vehicle. In either case, verify both the plug and the gap before proceeding. The wrong plug can cause some strange driveability problems. Do not assume that the correct plug is in the vehicle either, because this assumption can cause you lots of lost time in diagnosis. Always verify the component if there is any doubt in your mind. Also remember that new plugs should always be gapped before they are installed in an engine. Their gap will be close, right out of the box, but exactness is important on today's vehicles. The right plug correctly gapped will function as it was designed for the recommended tune-up interval.

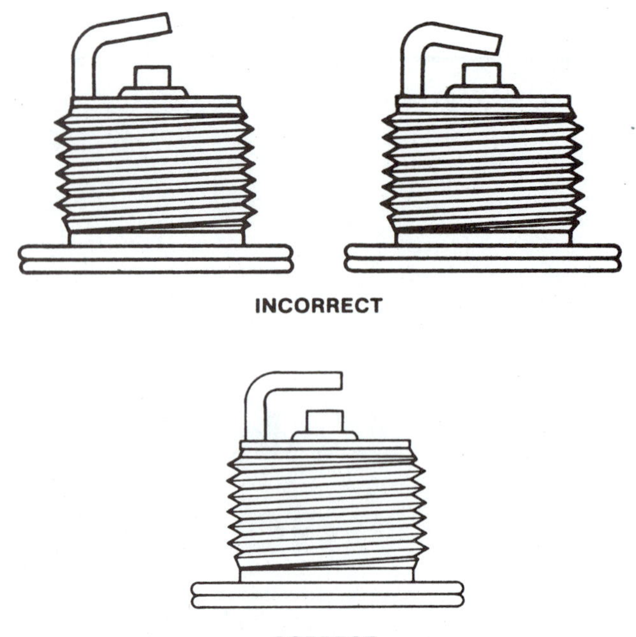

FIGURE 13-2 Incorrect gap versus correct gap.

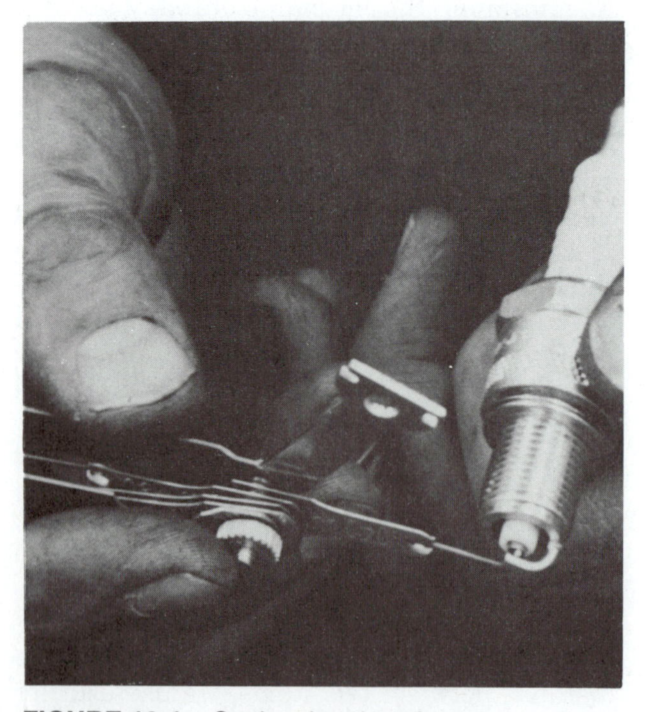

FIGURE 13-1 Setting the plug air gap.

Removing the plugs from the engine should be done only with a plug socket, and do not forget that aluminum engines should be cooled down before the removing is attempted. Reading what the old plugs look like will give much more information regarding the conditions inside the cylinder such as rich running or oil fouling. For this reason, keep the plugs in order as you remove them and note on the customer ticket any problems found. Over tightening of plugs is a major reason behind cylinder head failure. Use a torque wrench to tighten the plug in and you will be assured that the threads will not be damaged.

Feeler gauges are sometimes used to set ignition point gap. Point gap is the distance between the points when they are open all the way. The preferred tool for setting points is the dwellmeter, but because of the relationship between dwell and gap, a feeler gauge can be used. This relationship is illustrated in Figure 13-3. Notice that in theory, if the gap is correct so is the dwell. The opposite is also true, set the dwell correctly and the gap will be correct. Feeler gauge setting of the point gap is done with the rubbing block of the points on the highest point of the cam as shown in Figure 13-4. The point base is then moved either toward the breaker cam if the gap is too small or away from the cam if the gap is too large. Once set, the points are locked down with a screw. The exception to this procedure is the GM points that are adjusted with an allen wrench through a door in the distributor cap, Figure 13-5. GM is the only manufacturer that could have the points

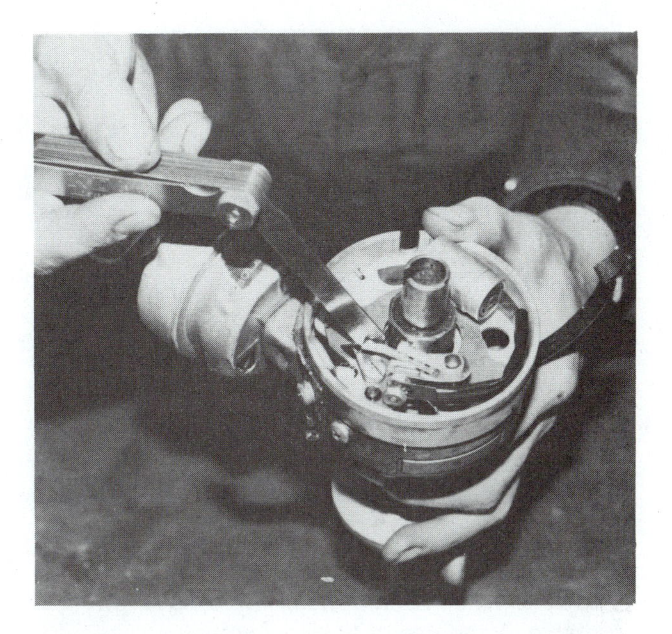

FIGURE 13-4 Point gap is measured with the points fully opened.

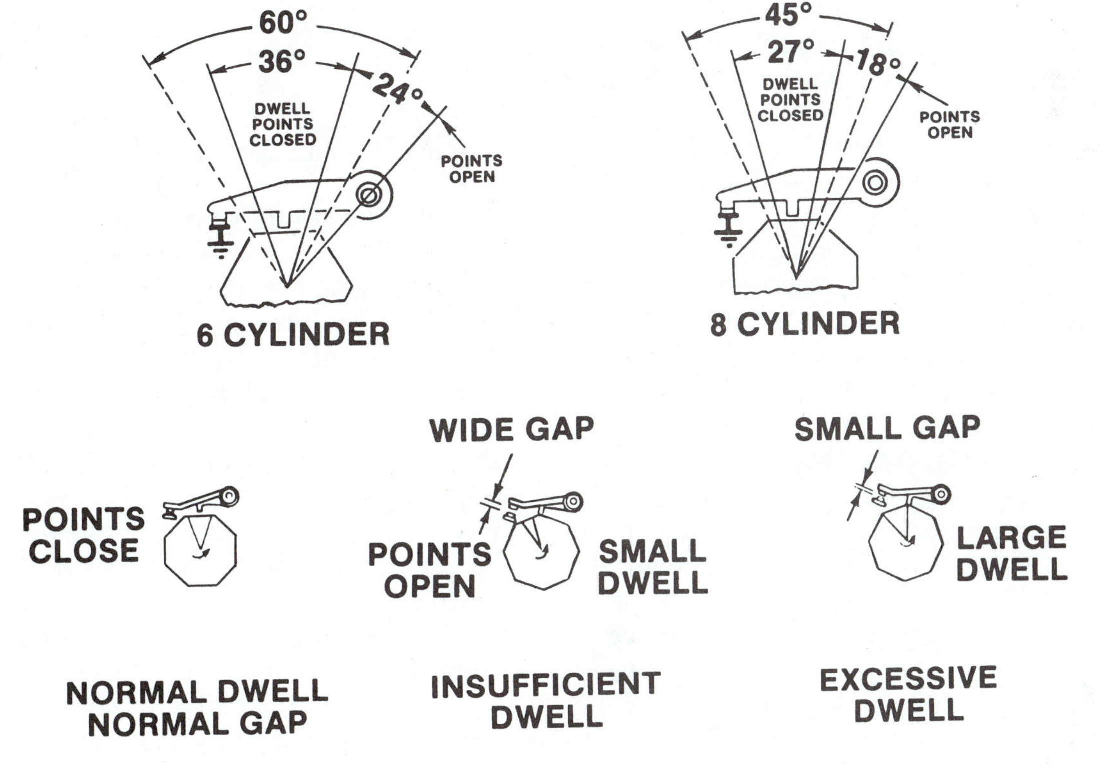

FIGURE 13-3 Dwell versus gap.

FIGURE 13-5 Some distributors have dwell adjustable by the use of an allen wrench.

adjusted with the engine running. Gap was still set with the engine off, however. Point gap is always the first to be set because it will have an effect on the other two tune-up specs, timing and idle speed. Changing dwell or point gap will change timing and idle speed, whereas timing changes or speed changes will have no effect on point gap. For this reason, the points are always set first. The limited use of breaker point systems has reduced the number of vehicles you will be required to set dwell or gap on, but you should still be familiar with their operation.

The nonmagnetic feeler gauge is ordinarily used to set distance between the reluctor and the magnetic pick-up coil on an electronic ignition system. Many systems have adjustable air gaps like the Chrysler, Figure 13-6. The distance is set much the same as it would be with a set of points. Once set, it is locked down in place by means of a hold-down screw. Too little gap might result in the reluctor and pick-up coil coming into contact. This could destroy both. If the gap is too great, the signal to the control module might be weaker, especially at slow engine speeds. This could make starting very difficult, because the module would not be able to determine piston position accurately. Remember that the signal from the

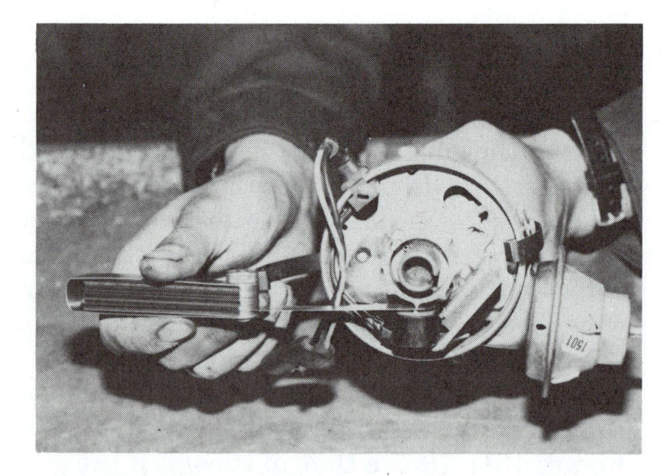

FIGURE 13-6 A brass feeler gauge is used to set electronic ignition pick-up air gap.

pick-up coil gives the system information that it will need to determine just when to turn off the primary current and fire the plug. If this information is weak, the module will not turn off primary current and the plug will not fire. Air gap setting and/or testing is important, especially if hard starting is the customer complaint. We will see later how the signal from the pick-up coil can be tested with an oscilloscope. Also, keep in mind that many pick-up coils are nonadjustable, making their replacement even simpler.

TACH DWELLMETERS

Another old standby type tool is the tachometer, which usually also has a method of measuring ignition dwell in degrees of camshaft rotation. The tachometer part of the meter is fairly straightforward with usually two leads and a switch for four, six, or eight cylinders. The leads normally attach to the negative of the ignition coil and ground, Figure 13-7. Some tachometers are inductive pick-ups and require a spark plug wire connection. Engine speed can now be measured directly and adjusted if necessary.

Dwell connections usually are the same as they are for the tachometer—negative of the coil and ground. Some ignition systems identify the coil negative by a – symbol or "tach" or "neg". If there is a doubt in your mind, put a

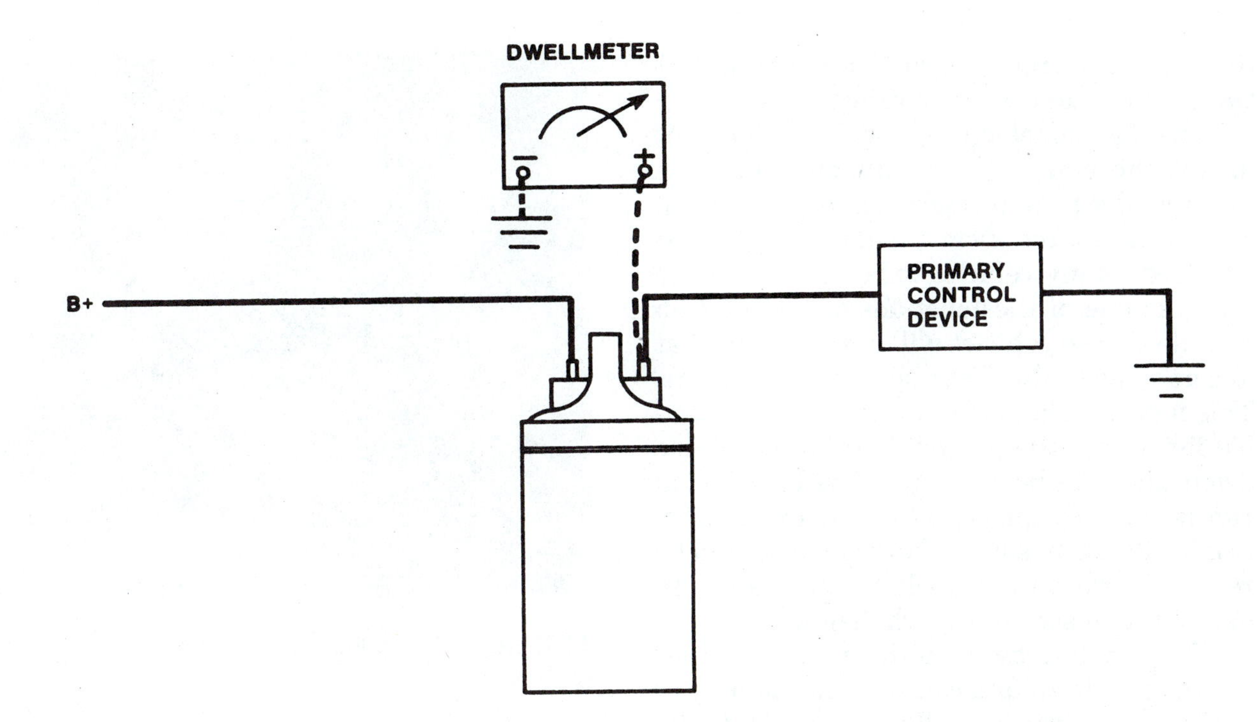

FIGURE 13-7 Dwellmeter connections.

12-volt test light on the terminal in question and crank the engine. The positive of the coil will have the test light on continuously, while the negative will blink on and off as the engine cranks. Dwell reading should be taken from only the negative of the coil. Reading and interpreting dwell varies with the type of system. We have briefly discussed dwell on a point type system and defined it as the amount of time that primary current is flowing through the coil's primary winding. This is the time that the points are closed and measured in degrees of camshaft rotation. Camshaft degrees are used because the points are opened and closed by distributor action, which is spinning at camshaft speed. For example, if a four-cylinder engine has a dwell specification of 50°, the points will remain closed for 50°, then open for 40°, then close for 50°, then open for 40°, and so on, Figure 13-8. The 50° and the 40° add up to the 90° camshaft rotation between plug firings. How many degrees of crankshaft rotation will occur between plug firings on a four-cylinder engine? The crankshaft is traveling at twice the speed, so the relationship between cam and crank will be 1 to 2. For every camshaft degree,

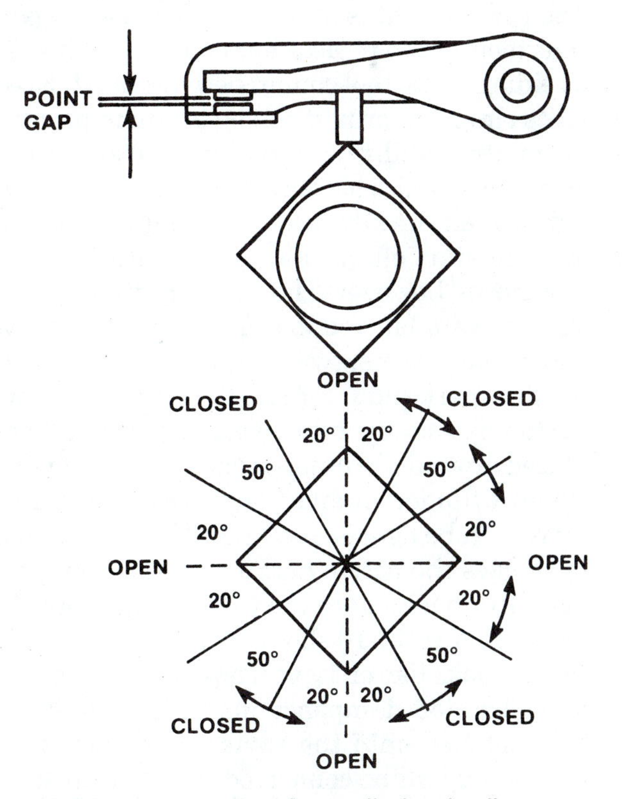

FIGURE 13-8 Typical four-cylinder dwell.

the crank will travel 2°, so 180° of crankshaft travel separates each plug firing.

The age of electronic ignition has not meant the end of the dwellmeter. Ignition dwell or the time primary current flows can still be measured even though it cannot be adjusted. Many vehicles today have variable dwell as one of the functions of the ignition module. These vehicles will have low dwell at idle and higher dwell at higher engine speeds. This function should be checked as it is possible for an electronic module to lose variable dwell and still be running. How long it will run is the only question here. Once an electronic module has lost a function, it should be replaced before the vehicle becomes a candidate for a no-start/no-spark diagnosis.

Dwellmeters average the actual cylinder to cylinder dwell and can be either analog or digital. In addition, dwell can be individually measured with the automotive oscilloscope.

THE TIMING LIGHT

The timing light is still one of the most important tools of the professional mechanic. Its use as a diagnostic tool cannot be overstated. Even in its simplest form, it can be useful in pinning down driveability complaints. Most professional mechanics use an adaptation of the basic light called the advance timing light or meter. We will start off our discussion with the simplest form, however. The standard timing light that is available and in common use today has three connections, two for the battery and one for the spark plug, Figure 13-9. The plug connection is usually inductive and designed to be placed around the number one spark plug wire. When cylinder number one fires, the timing light's strobe light is also fired. The strobe stops or freezes the action of the vibration dampener or flywheel where the timing marks are located. Figure 13-10 shows what a set of timing marks looks like on typical engines. Notice that the vibration dampener might have a single line notched onto the surface. The vibration dampener will be connected to the crankshaft and will be spinning at engine rpm. The small

FIGURE 13-9 Inductive timing_light.

plate with numbers is attached to the timing cover or the front of the engine. Notice the 0 on the plate. When the notch on the vibration dampener lines up with this 0, cylinder number one will be in top dead center position. The strobe of the timing light will freeze the position of the vibration dampener and allow us to note what position cylinder number one is in relative to top dead center. This is called **timing the engine** and it is basic and fundamental to just about all driveability complaints. Timing must be correct. Let us use an actual example and see how and what we might do to time a non computerized vehicle. The specs for this engine read:

9° BTDC @ 650 rpm dr. Vacuum advance disconnected and plugged.

Do you remember the relationship between speed and timing that we discussed in the previous chapter? As the engine speed increased, the timing was usually advanced. This is the reason for the timing check being done at the very specific speed of 650 rpm. The manufacturer is specifying a speed below the point where the mechanical advance will begin to operate. The manufacturer is also saying that the point of ignition must be 9° before the piston reaches top dead center. This will allow the

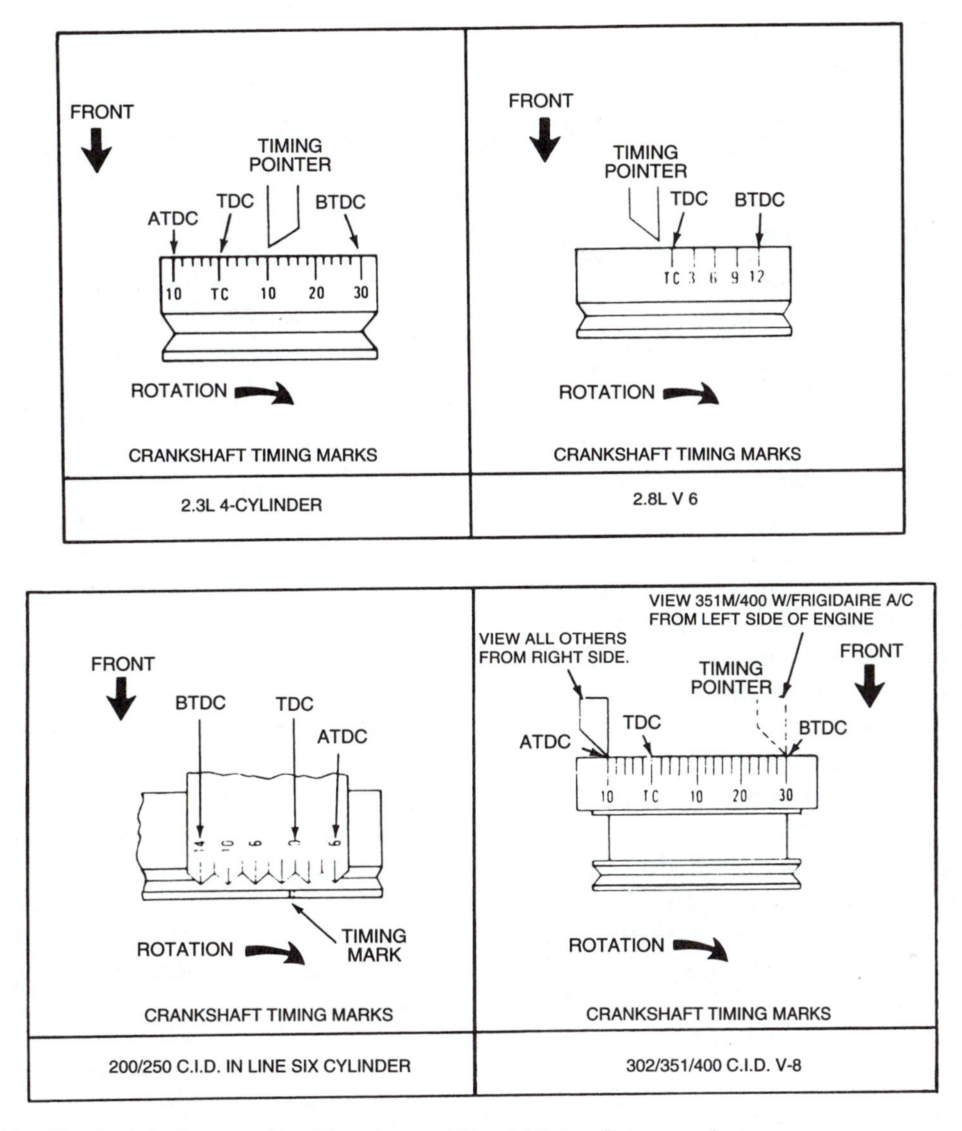

FIGURE 13-10 Typical timing marks. (Courtesy of Ford Motor Company)

burning process to reach its peak combustive pressure by the time the piston reaches the end of the compression stroke and is on its way down on the power stroke. In addition, the speed of 650 rpm should be set with the automatic transmission in drive (dr stands for drive). The last piece of information that is important is the fact that the vacuum advance is disconnected and plugged. Why do you think this is necessary? Remember that manufacturers have two choices for their supply of vacuum for the advance. One choice is above the throttle plates. This is called ported vacuum and will only be present with the engine off idle. At idle, no vacuum is present at a ported source. The other choice is to have **manifold vacuum** as the source for the vacuum advance. Manifold vacuum will be present anytime the engine is running at something less than full throttle. From the specs, it appears that manifold vacuum is being used as the source for the vacuum advance and the manufacturer wants the initial timing to be done with no vacuum advance present. This is why manufacturers have specified that the advance be disconnected and plugged.

If we have met all of the conditions that the manufacturer has specified, we should expect the actual timing to be 9° before top dead center. Shining the timing light down on the

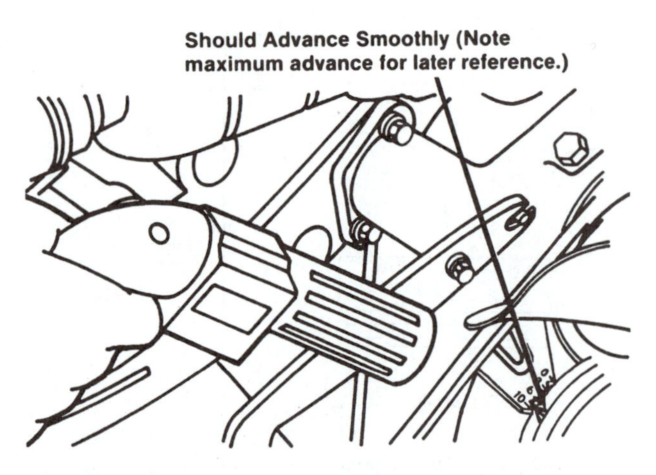

FIGURE 13-11 Timing marks illuminated by a timing light at 9° BTDC.

marks and vibration dampener should show the marks as illustrated in Figure 13-11. If this were the case, we would say that the timing is correct. What do you think we would do if the timing marks lined up as in Figure 13-12? Notice that the vibration mark is lined up with the number 3 on the timing plate. We are off by 6°. The timing is retarded by 6°. Remember, retarded timing is timing that is behind what it should be. The opposite is advanced timing, which means timing that is ahead of where it should be. In this example, the specification is 9 BTDC and the actual is 3 BTDC. We will have to change the timing and advance it by 6° to make it correct. This will be done by loosening the distributor hold-down bolt and turning the distributor housing until the number 9 is in line with the vibration dampener mark. Once the engine is timed, we can retighten the hold-down bolt and recheck. It is important to note that changing timing might have an effect on engine speed. This will mean that you will have to readjust the engine idle speed back to the specification. When you are completely finished, all conditions of the specifications should be met. The timing marks should line up at the 9° BTDC position with the engine speed at 650 rpm and the distributor vacuum line disconnected and plugged. If the conditions are met, the vacuum line can be reconnected. You have just set or checked timing. Remember, every engine has different conditions that must be met. Do not assume that all engines are timed exactly the same. In addition, realize that the timing is set ordinarily on the number one cylinder with the assumption that all other cylinders will have the same timing. Some vehicles use a different cylinder for the timing check. Do not assume that all engines are the same.

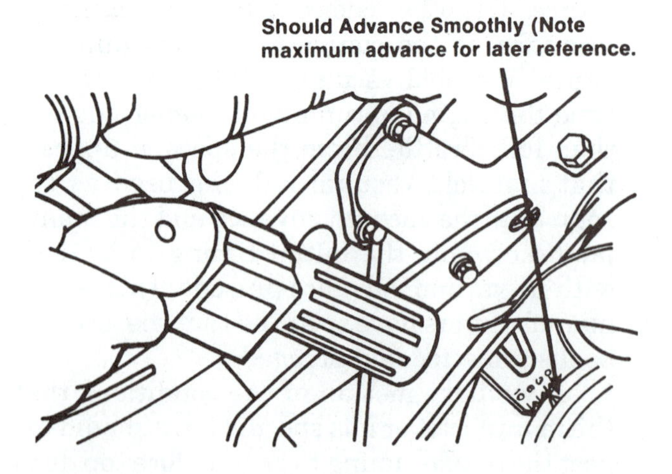

FIGURE 13-12 Timing marks at 3° BTDC.

In recent years, the computer has taken on the responsibility of timing along with fuel control. This will allow the computer to compute the required timing and actually change it to match the conditions. This does not mean that initial timing is not set or checked. The computer makes the assumption that initial timing is correct and computes running timing changes off initial timing. Timing a computer-controlled engine is just as easy as timing a non computer-controlled engine. The only difference is that the computer must be told to not control timing. Figure 13-13 shows the spout wire located at the distributor of a computer-controlled vehicle. SPOUT stands for **spark out**. By disconnecting the connector, we will eliminate the computer from the timing circuit. The engine will now be running in a non advanced base or initial timing mode. We can now turn the distributor until the timing marks line up just as we did on a mechanical advance system. Once the initial timing is correct the spout is reconnected and the computer takes over the timing function. Notice that the procedure is actually easier, because we do not have to worry about the

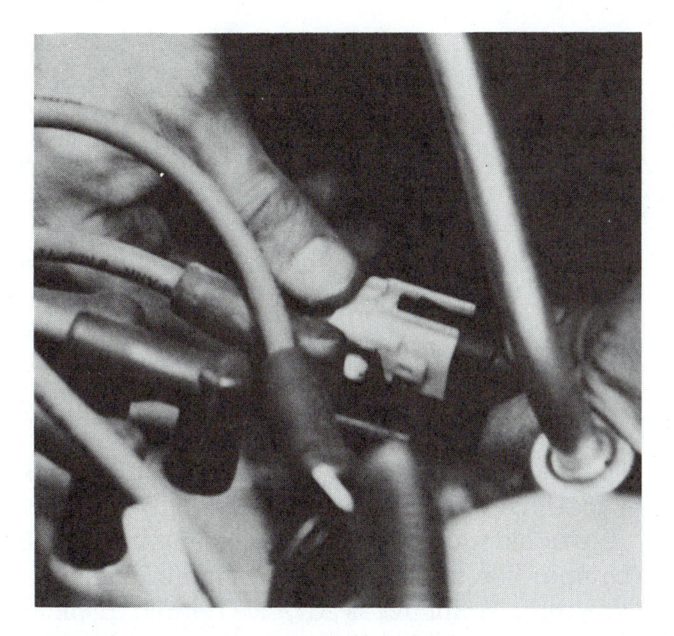

FIGURE 13-13 Opening the spout connector to check timing.

speed the engine is running at. With no centrifugal advance system changing the timing with speed changes, the engine can be timed at most any speeds.

Not all vehicles have a spout as in the Ford example. Toyota's procedure is to disconnect the labeled connector next to the distributor, while General Motors' carbureted vehicles are timed with the four-wire connector disconnected, Figure 13-14. Hopefully you are beginning to realize that each manufacturer will have different procedures that must be followed exactly. Most computerized systems will change timing from a base or initial timing position. If the initial timing is incorrect, virtually all calculations will also be incorrect and a driveability complaint might be present. Generally, over advanced timing can cause a reduction in fuel economy and a ping or knock, while retarded timing can cause hesitations and reduced fuel economy. Greatly over advanced timing can also do engine damage as the ping or knock is indicative of uncontrolled combustion and greatly raised cylinder temperatures. Extreme temperatures can even melt pistons. Make sure that initial timing is correct. It is basic and fundamental to a properly running engine.

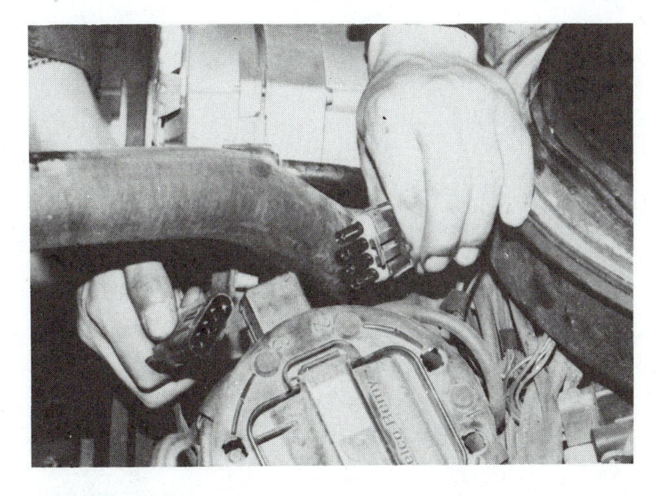

FIGURE 13-14 Opening the four-wire connector to check timing.

THE ADVANCE TIMING LIGHT

The basic timing light is necessary and should be used frequently to either set initial timing or determine if it is correct. It does, however, have its limitations if the mechanic wishes to check something other than initial timing. If, for example, the customer complaint is poor economy and high-speed performance, the mechanic might decide that a check of the advancing systems might be helpful. A normal inductive pick-up timing light would be limited to an idle only timing check. This is where the advance timing light is useful. Figure 13-15 shows a common advance timing

FIGURE 13-15 Advance timing light.

light. Notice that the back of the light has a meter calibrated from 0° to 45°. The knob that is part of the handle controls this meter and the amount of delay between the time the spark plug fires and the time that the light flashes. By turning the knob, the mechanic will be able to delay the flash of the light by the number of engine degrees shown on the back of the meter. The delay of the light allows the use of the engine's timing marks as a reference and enables the mechanic to note the timing change. Let us take an example and try to explain just how the light allows us to read at how many degrees advance the engine is running. The procedure is quite simple and starts with a check of the initial timing and a reset if necessary. Let us assume that the engine is timed to 9° BTDC @ 650 rpm and we wish to determine if the mechanical advance will advance the timing by an additional 10° by the time the engine has reached 1500 rpm. With the vacuum advance disconnected and plugged, we would raise the engine speed up to 1500 rpm and turn the knob on the timing light handle until the timing marks again line up at the 9° spot. The meter on the back will now tell us how many degrees we have delayed the flash of the light. This will then be the amount of mechanical advance at 1500 rpm. Note that the timing light just delays the flash, it does not change the actual ignition timing. With an advance timing light, we can advance curve the distributor against the manufacturer's specifications. Advance curving is just a fancy way of saying we have checked the mechanical advance at various speeds and the vacuum advance at various vacuum levels, and compared both to the specifications supplied by the manufacturer. The comparison allows us to determine if the driveability complaint is advance related. Look at Table 13-1 and try to figure out just what the complaint might in fact be.

The following graph, Figure 13-16, is for the same vehicle. Notice that the dashed line is the specification, while the heavy line is the actual. What kind of driving condition might be experienced by the customer with this me-

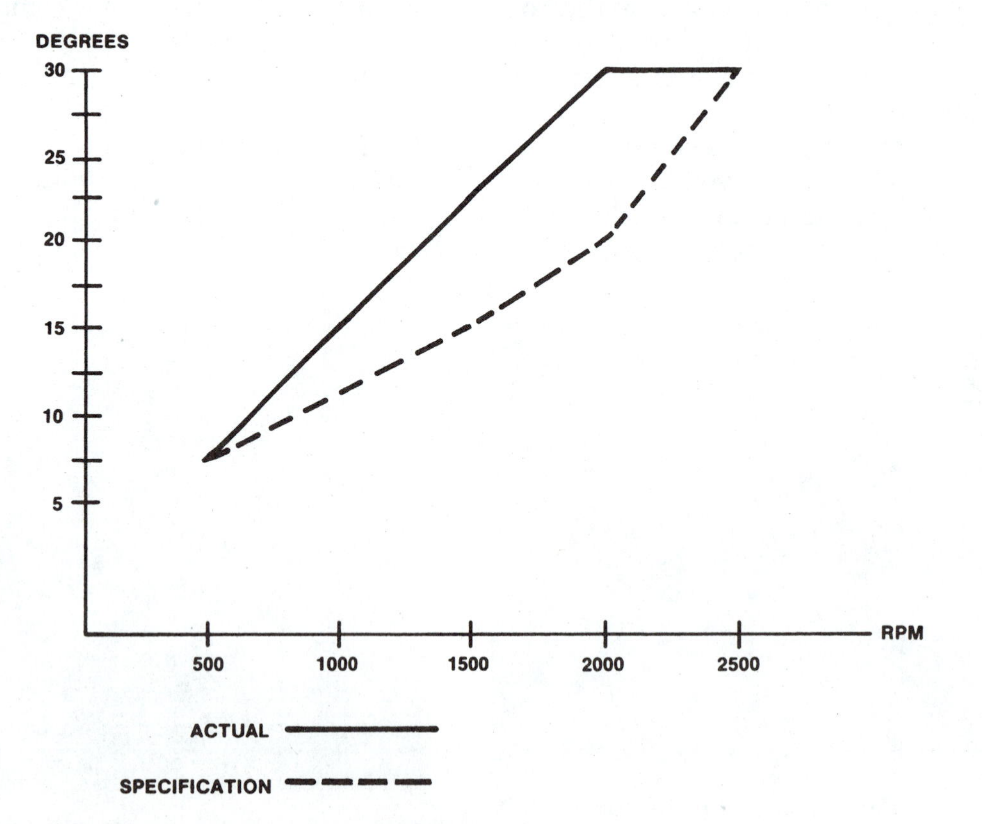

FIGURE 13-16 Actual versus specification timing.

TABLE 13-1: COMPARISON OF SPECIFICATIONS AND SPEED

Speed	Actual	Specification
650	9 BTDC	9 BTDC
1000	16 BTDC	12 BTDC
1500	25 BTDC	15 BTDC
2000	30 BTDC	20 BTDC
2500	30 BTDC	28 BTDC

chanical advance? Remember what we have repeatedly mentioned: Retarded timing can result in hesitations and reduced economy, while advanced timing can result in pinging. Well, what do we have here? In the mid range of 1,000 to 2,000 rpm, the actual timing is advanced by as much as 10° with the low end (650 rpm) correct and the upper end (2500 rpm) just about correct (off by 2°). We would expect this vehicle to exhibit a ping in the mid range, which gradually goes away as the customer drives faster and faster.

We can check the vacuum advance the same way. Use a vacuum gun to draw on the diaphragm, Figure 13-17, until the vacuum specification is achieved, and turn the timing light knob until the timing marks line up again. The amount of flash delay gives us the amount of vacuum advance. Let us look at

FIGURE 13-17 Using a vacuum gun to check vacuum advance.

TABLE 13-2: COMPARISON OF SPECIFICATIONS AND VACUUM ADVANCE

Vacuum	Actual	Specification
0 in. hg	0 BTDC	0 BTDC
5 in. hg	2 BTDC	4 BTDC
10 in. hg	3 BTDC	9 BTDC
15 in. hg	3 BTDC	18 BTDC
20 in. hg	4 BTDC	28 BTDC

Table 13-2, like the table we examined for mechanical advance.

What driving condition might the customer experience? Keep in mind that the only purpose of the vacuum advance is to improve economy under conditions of high vacuum. High vacuum is achieved during cruising or light loads and is measured in **inches of mercury (in.hg)**. Our actual timing advance is quite a bit less than the specification. It is retarded from where it belongs and might result in hesitations or reduced economy.

The repair of the two vehicles just tested is simple once the diagnosis is complete. The first mechanical advance problem might require a new set of weights and springs, while the second vehicle will require a new vacuum advance unit and a check of the movement of the pick-up. In some cases, a replacement distributor might be installed in the vehicle if the cost of repairing exceeds the cost of a rebuilt unit. Distributor rebuilding procedure will be covered in detail in another section of this chapter.

DIGITAL TIMING INDICATORS

Recently, the automotive industry has seen increased use of digital timing indicators like those shown in Figure 13-18. They do essentially the same thing as an advance timing light with the exception that they give a direct readout of total timing (initial + advance) without having to turn a knob or watch timing marks. Vehicles produced since the late seventies generally are fitted with probe holders or holes, and magnets mounted in the vibration

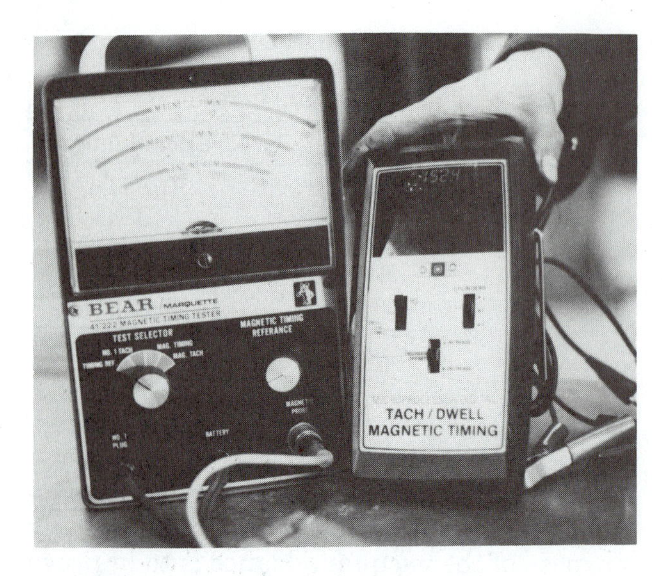

FIGURE 13-18 Digital timing indicators.

FIGURE 13-19 Offset probe hole.

dampener. The digital timing probe is placed in the hole with its end just touching the vibration dampener. As the dampener magnet spins close to the probe, it generates an AC signal that the timing indicator compares to the spark plug pulse. This comparison allows the unit to determine total timing and displays it digitally. On vehicles where the timing marks are hard to see, such as some front wheel drive engines, the digital timing indicators have become popular.

One other item must be understood before digital indicators will make sense: offset. When manufacturers design and build an engine, they frequently cannot place the timing marks and the probe hole in the same place, Figure 13-19. The probe hole and bracket sometimes used would interfere with the view of the conventional timing marks, making their use difficult. For this reason, the probe hole is separated from the conventional marks. This distance between the probe hole and the timing marks in degrees is the amount of offset. Offset must be dialed into the digital timing indicator before the vehicle is running. Most manuals will indicate the amount of timing offset in the section where timing is discussed. By dialing in offset, you are telling the digital timing indicator where the probe hole is located. Note that offset and initial timing are not the same. Offset is preset into the unit with the vehicle off and must be correct or all timing checks will be off by the amount of the error. For example, if offset is –135° and the mechanic dials –125° into the digital timing indicator and then sets timing to 10° BTDC, the actual timing of the vehicle is off by 10°. The 10° offset error will give the wrong information to the mechanic and quite possibly cause a diagnosis error. Accuracy with digital timing indicators is very high, assuming that the offset has been dialed into the unit correctly. Some of the popular digital units can measure timing to a tenth of a degree.

OSCILLOSCOPES

No discussion of ignition system testing would be complete without a comprehensive discussion of an oscilloscope. The scope, as it is referred to, gives us the opportunity to look inside the ignition system, including the cylinder. The scope is one of the most used testers in modern repair centers. The results are accepted by all of the major manufacturers worldwide for warranty claims. Do not be misled, however. The scope will not make you a better mechanic. It will not normally give you information that you could not have found by other methods. The difference and one of the main advantages to its use, comes from the speed with which a diagnosis can be made. A good mechanic can, for instance, find a faulty spark plug wire with an ohmmeter or a substitute wire. The same mechanic will also find the faulty wire with a scope but will probably find it much faster. Auto mechanics is a speed and quality business. The best mechanic will starve if he or she is too slow. The scope can be the edge that the good mechanic needs to become a great technician. In addition, scope testing is a form of quality control that allows the technicians to check their tune-up for accuracy. Preventive maintenance should also utilize the scope. The same procedure with which we verify the accuracy of the tune-up can be applied to a preventive maintenance check-up to determine exactly what the customer needs.

There are just as many reasons to not use a scope as there are to use it. We have mentioned one already. It will not make you a better mechanic. You must have the basics and a good working knowledge of ignition systems or the patterns will probably not make a great deal of sense. The scope is also not a replacement for common sense. You should not need an expensive scope to tell you that the ignition wires are oil soaked and require replacement. Visual inspection and common sense is still one of the best diagnostic tools to use.

Before we zero in on ignition system test procedures, let us review the information presented in an earlier chapter about scopes. You will remember that an oscilloscope is nothing more than a voltmeter with a time frame built

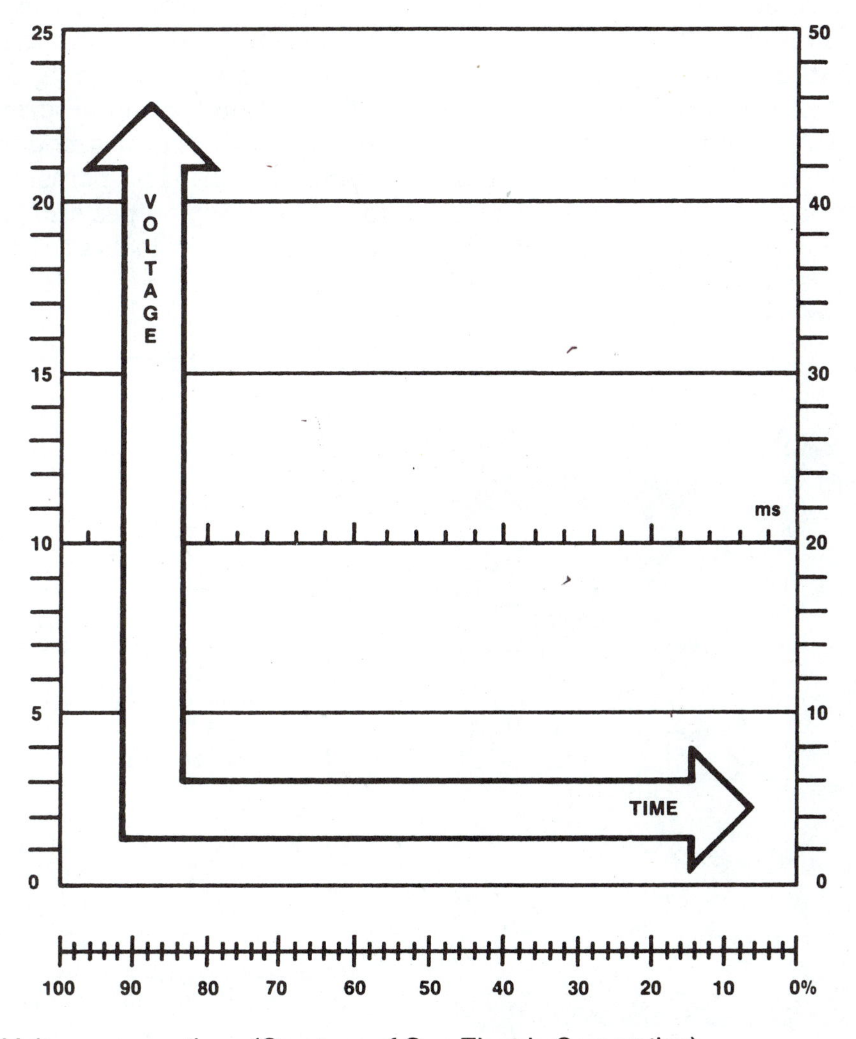

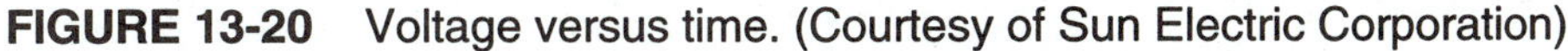

FIGURE 13-20 Voltage versus time. (Courtesy of Sun Electric Corporation)

into it, Figure 13-20. A voltmeter with a clock attached is how some mechanics refer to a scope. The time frame or clock function is important because of the tremendous speed with which the ignition system produces and delivers sparks. This speed makes the traditional voltmeter almost worthless. For example, a four-cylinder vehicle running at 2500 rpm, or about 55 mph, is producing 83 sparks per second. Even if our meter could follow the voltage changes, our eyes and brain probably could not. This makes the scope very valuable because we can display the changing voltages on a screen that will stay lit long enough for us to observe and mentally record. It is somewhat like an electronic piece of graph paper with both positive and negative voltages displayed. Figure 13-21 shows the scope radicule or screen. Notice the zero line down toward the bottom of the screen. Voltages or traces that are above the zero line are positive, while voltages that are below the zero are negative. As is the case with any voltmeter, the scale or range of readable voltages must be determined and set. In most cases, the highest scale is used when dealing with ignition systems. Usually, this scale is in the 50,000 (50 K) range. Notice the lines on the radicule and the numbers on the two sides of the screen. These are the scale numbers that you will use to determine the actual voltage levels. In addition, remember that the screen is read from left to right with readings on the left of the screen happening before those that are on the right side. Many scopes also have the ability to read degrees or percent of dwell and milliseconds as in Figure 13-22. Notice that the vertical scale on the left is divided into spaces equal to 1 kV up to a maximum of 25 kV. The right side of the screen has the same divisions equal to 2 kV up to the maximum of 50 kV. In addition, the screen can measure 0 to 25 volts on the left and 0 to 50 volts on the right. These two lower scales are used during testing of the primary circuit. The millisecond scale located in the center of the screen is used to measure the actual time that the spark will be jumping the air gap in the cylinder. Some manufacturers specify a spark that lasts a certain amount of time in milliseconds (mS). The percent of dwell scale is found on many scopes and gives the mechanic the ability to look at the amount of time

FIGURE 13-21 Scope radicule.

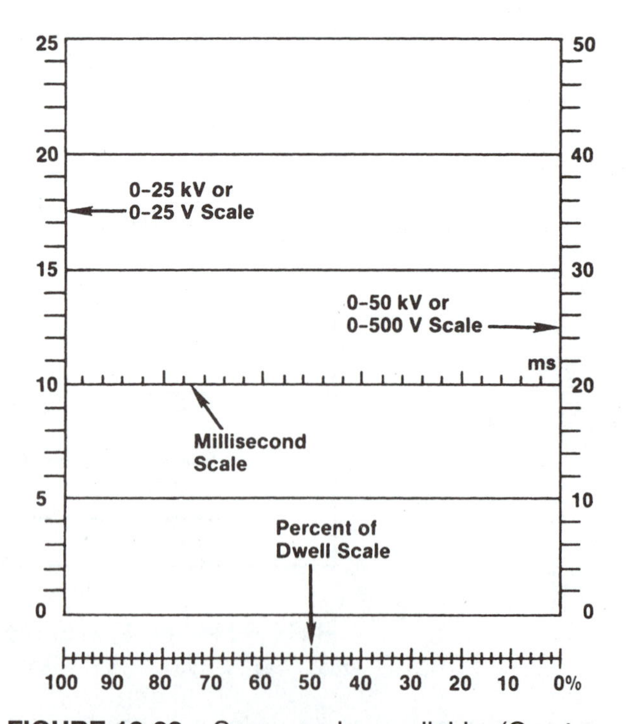

FIGURE 13-22 Scope scales available. (Courtesy of Sun Electric Corporation)

that current is flowing through the coil's primary circuit. Keep in mind that not all scopes will have the same capabilities or the same scales. Different manufacturer's scope screens might look different from that shown. Do not hesitate to review the literature supplied with the unit to really familiarize yourself with its functions.

Scope Pattern Selection

Most scopes currently in use in the automotive field have some available patterns that one can use to look in detail at different sections of the ignition system. Both primary and secondary patterns can usually be displayed in one of four different patterns. Let us look at each one separately, realizing that not every scope has the same patterns or calls them by the same name. **Display** or **parade** is shown in Figure 13-23. All of the ignition firings are in a row in the firing order. Remember, firing order is the order in which each cylinder will have its spark plug fired. Our illustration shows an eight-cylinder engine with a firing order of 18436572. Notice that the upward lines are all labeled for you and that cylinder number one's upward line occurs at the end of the screen. Some scopes will have the first cylinder's upward line at the beginning, while others will have it at the end as in our illustration. This upward line represents the initial firing of the spark plug as you will see in a later section where we will learn what each line is. Do not forget that in display, the scope is read from left to right. The next pattern usually available is called **raster** or **stacked**. Notice Figure 13-24. The cylinders are again labeled with the same firing order and the patterns are read from the bottom up with the first cylinder in the firing order being on the bottom of the screen. You can see that this pattern will be useful in comparing all the cylinders against one another. The third pattern usually available is called **superimposed** and is shown in Figure 13-25. In this pattern, all of the cylinder patterns are on top of one another, making cylinder identification impossible. Some scopes, in this position, have the ability to put individual cylinders above the main trace in a split screen. Buttons on the scope allow any combination of cylinders to be shown on top or on the bottom. Not all scopes have this ability, nor the ability to display firing times in milliseconds. A millisecond is a thousandth of a second. Figure 13-26 shows what a pattern looks like with the scope in the millisecond sweep (mS) position. If the scope is in the 5 mS position only that portion of the firing pattern that occurs during 5 milliseconds will be shown. Some manufacturers will specify just how long the actual spark is supposed to last in milliseconds. The ms sweep pattern gives the mechanic the ability to measure it directly, usually in either a 5 or 25 milliseconds scale.

Each of the scales or patterns available have a use or function in diagnostics. We will see the advantages and the disadvantages to each one after we look at what a typical cylinder looks like on a scope.

Single Cylinder Patterns

Let us develop a single cylinder pattern so that you will be able to see the ignition firing through the eyes of the scope. Let us start with the firing line shown in Figure 13-27. An upward line, you will remember, signifies voltage. This upward line is indicative of the voltage that is necessary to start the spark. The ignition coil's pressure rises up to the point where it can overcome all of the resistance of the secondary circuit. The conditions inside the cylinder will also look like resistance to the coil. The greater this resistance, the higher the voltage; the lower this resistance, the lower the voltage. You will see when you begin testing with the scope that typically it takes around 10,000 volts to overcome this resistance and get our spark to begin. Keep in mind that cylinder conditions will have an effect on this resistance. Leaner air/fuel ratios will increase the resistance and raise up the firing voltage. This line is called the **firing line**. Do not forget that it represents the voltage necessary to overcome all of the secondary resistance.

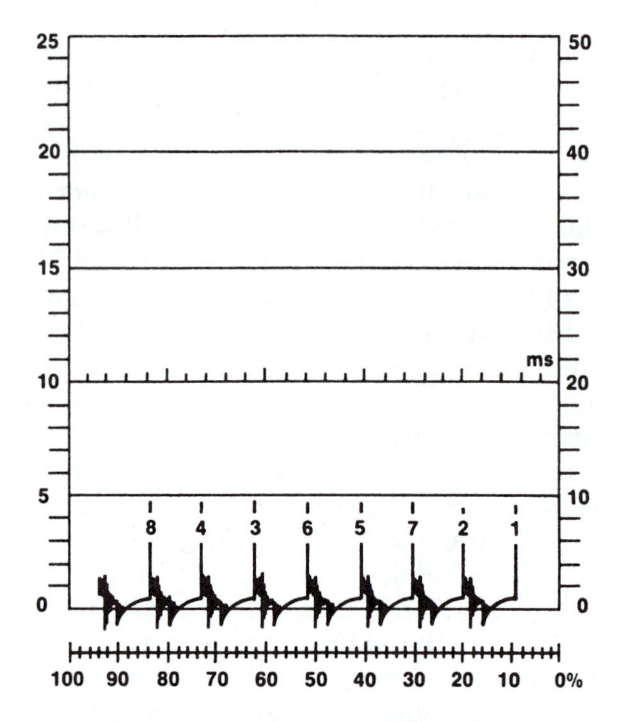

FIGURE 13-23 Scope showing display or parade pattern. (Courtesy of Sun Electric Corporation)

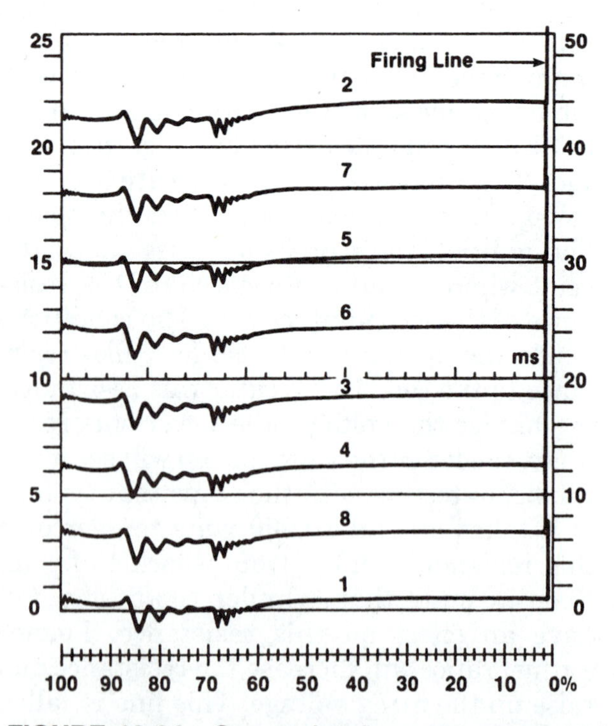

FIGURE 13-24 Scope showing raster or stacked pattern. (Courtesy of Sun Electric Corporation)

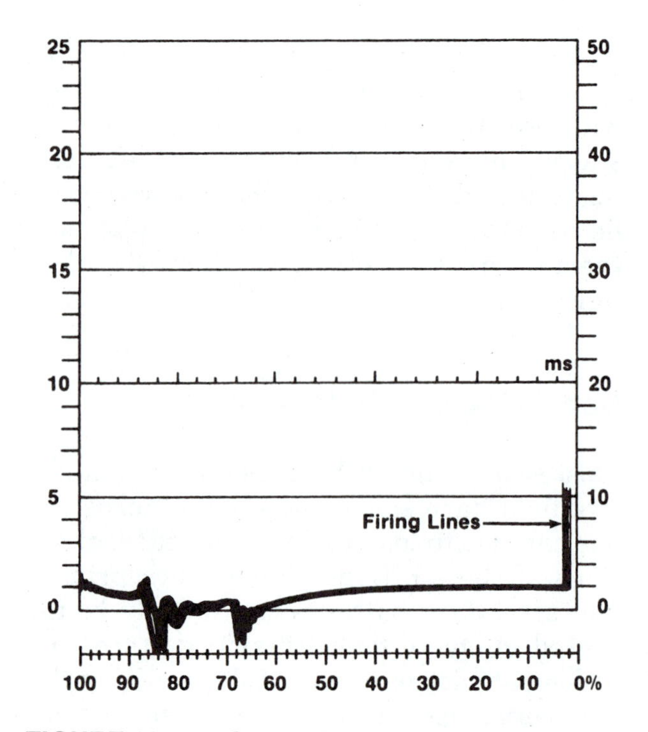

FIGURE 13-25 Scope showing superimposed pattern. (Courtesy of Sun Electric Corporation)

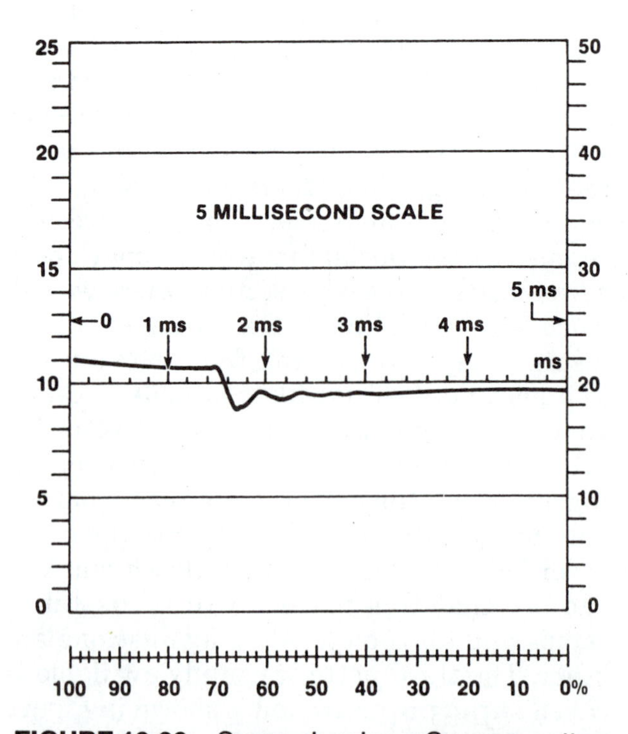

FIGURE 13-26 Scope showing mS sweep pattern. (Courtesy of Sun Electric Corporation)

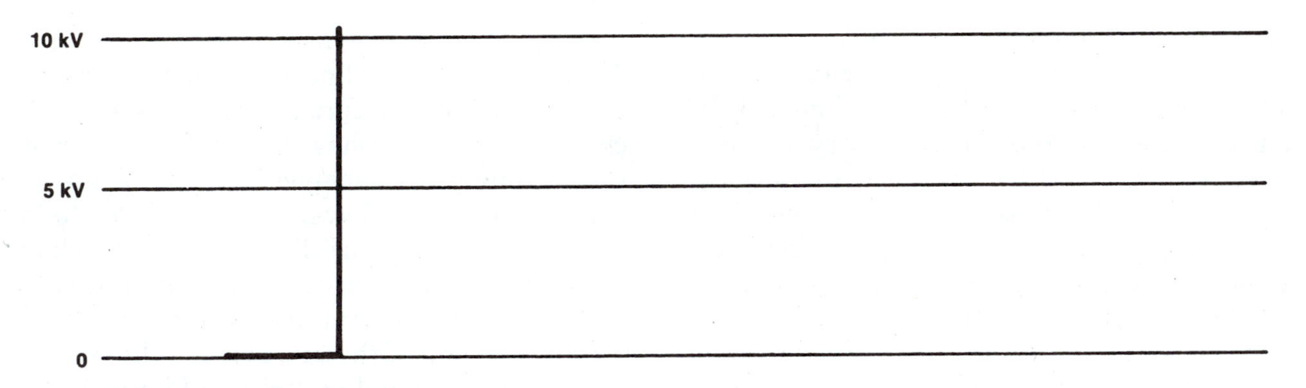

FIGURE 13-27 Firing line scope pattern.

Once this secondary resistance is overcome, electrons begin flowing across the air gap of the plug, hopefully igniting the mixture in the cylinder. This next section of the pattern is called the **spark line**, Figure 13-28. It represents the length of time that the spark actually lasts. The spark line begins at the firing line about one fourth to one third of the way up from the zero line and will continue until the coil's pressure drops below the level necessary to keep the electrons flowing. Simply stated, the spark is draining the pressure off the coil's secondary. As long as the coil has sufficient pressure, the spark continues. Once this pressure is reduced below the minimum needed, the spark ends, and the next section of the scope trace begins. This section is called the **intermediate zone** and is shown in Figure 13-29. This section represents the remaining coil voltage being dissipated through

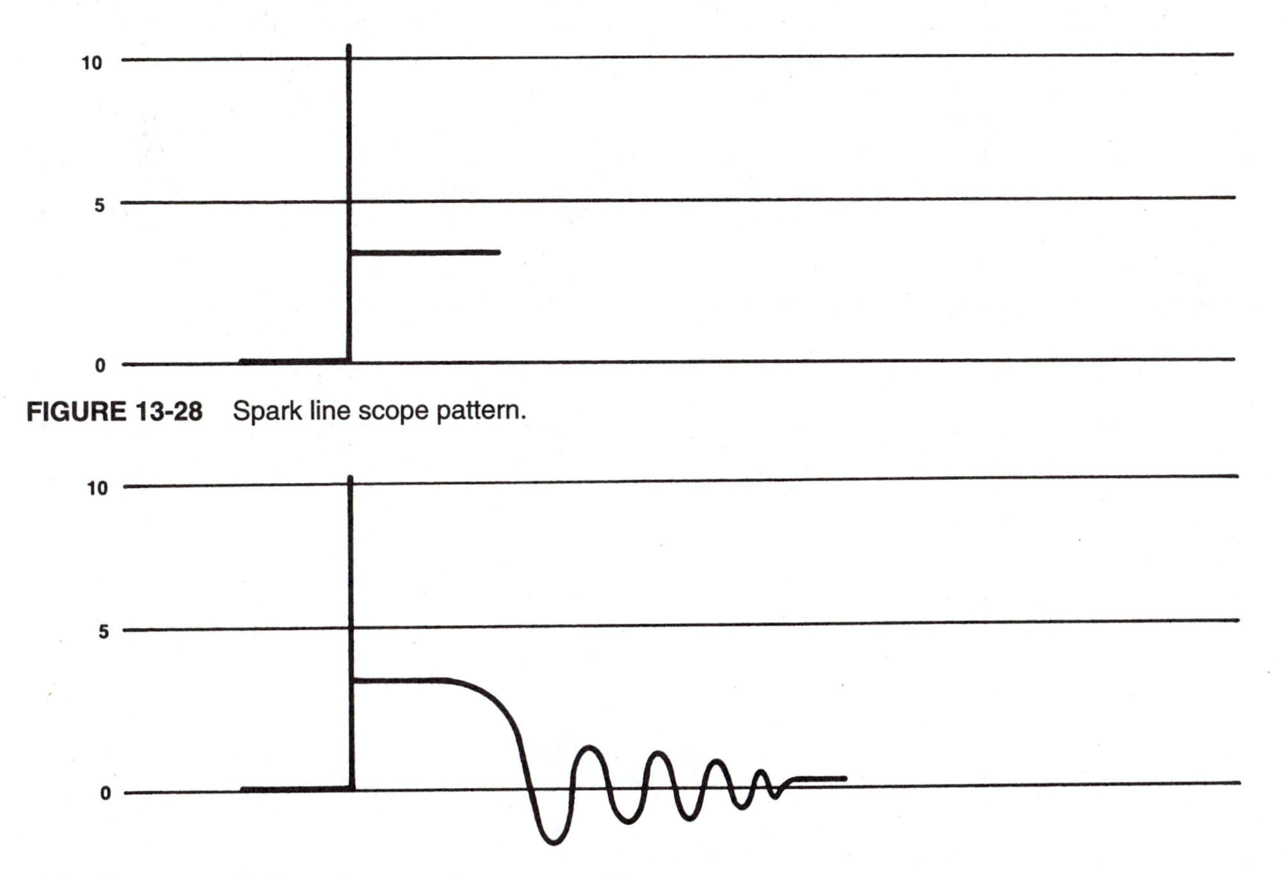

FIGURE 13-28 Spark line scope pattern.

FIGURE 13-29 Coil condenser action scope pattern.

action involving both the coil and a capacitor. Once the spark has ended, there will be quite a bit of voltage still remaining at the coil. If, for example, 2,000 volts is necessary to maintain the spark, it will continue until the coil's voltage drops below the 2,000 volt level. The remaining 1,999 volts will alternate between the coil primary and the capacitor as alternating or oscillating current until the pressure is down to zero. Notice that the lines steadily diminish from the high level they were at as the spark ended.

The next part of the trace is the primary current on signal. Figure 13-30 shows this to be a downward turning of the trace followed by some small oscillations. Current going into the coil is represented by lines below the zero line. The short downward line occurs just as the current begins flowing through the coil's primary circuit. The oscillations that follow represent the beginning of the magnetic field build up. As primary current flows, the magnetic field builds until the coil is fully saturated. The line after the primary current on signal gradually curves toward the zero line as this saturation occurs. The end of the line is when the primary current off signal occurs like Figure 13-31 shows. The upward turning of the trace is at the end of the dwell zone. Turning the primary current off will collapse the magnetic field and generate another high voltage spark for the next cylinder in the firing order.

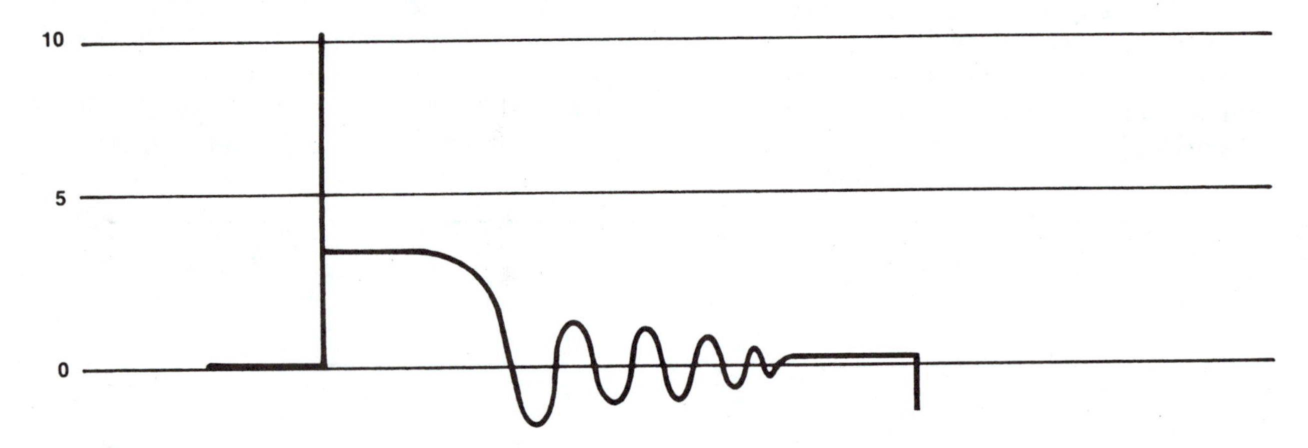

FIGURE 13-30 Scope pattern showing primary current on.

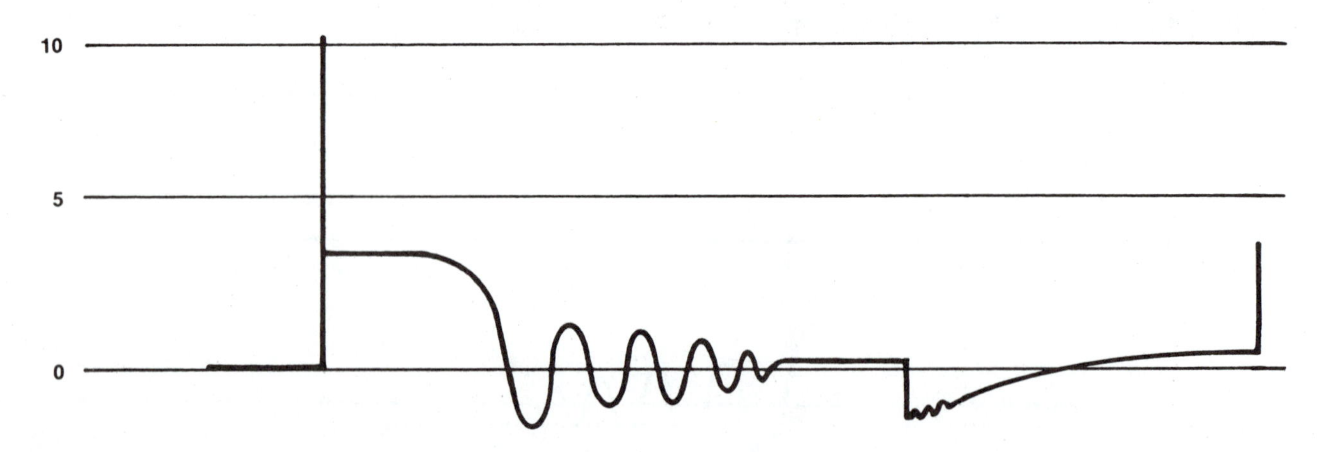

FIGURE 13-31 Scope pattern showing primary current off.

NOTE: The primary current off signal is the same as the firing line for the next cylinder.

Putting the whole sequence together, Figure 13-32, gives a trace starting with the firing line that runs into the spark line and ends with the coil condenser action. This is followed by a primary current turn on signal that begins the dwell zone and the coil's saturation period. The dwell zone ends with the primary current turn off signal that is also the firing line for the next cylinder. Keep this basic pattern in mind as you use the scope. There will be some differences form one vehicle to another, but basically all will look about the same.

Dwell Zone Variations

Generally speaking the majority of the scope patterns will look just about the same from one vehicle to another. The firing line, for instance, will be between 5 kV and 12 kV at 1200 to 1500 rpm on virtually all vehicles recently tuned. It will not matter whether the vehicle is breaker point controlled or any one of the many forms of electronic ignition. Secondary resistance will have the same effect on all systems: raising up the firing voltage. The exception to this is the dwell zone. The primary current on to the primary current off distance might have some additional variables and characteristics significant to a particular ignition system. These differences or variables fall into four categories.

1. fixed dwell short
2. fixed dwell long
3. variable dwell
4. current limit

Let us look at each one separately and see how the scope can identify problems based on the type of ignition system in use on the vehicle. The early breaker point systems had a fixed dwell as do some of the current electronic ignition systems. By fixed dwell we mean that the number of degrees of dwell remains the same during engine speed changes. If the vehicle has 30° of dwell at idle, it should have 30° at 2,000 rpm. This is not to say that the amount of time of dwell has remained the same, however. Remember that a dwell of 30° at 2,000 rpm will give the ignition coil one fourth of the actual time to fully saturate that it has at 500 rpm. The scope will show that the number of degrees or the percent of dwell remains the same as the engine speed is

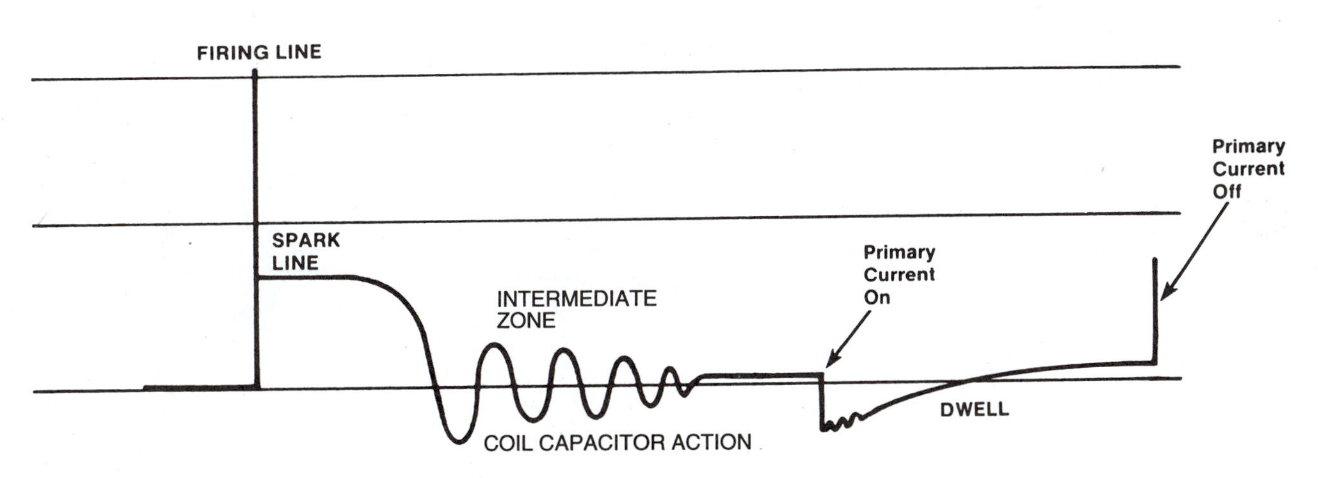

FIGURE 13-32 One cylinder scope pattern.

changed, as in Figure 13-33. Many of the early electronic ignition systems had a dwell that was fixed at a value quite similar to their breaker point predecessors. As a matter of fact, many early systems were impossible to identify as electronic strictly from their scope patterns. The exception to this was the early chrysler electronic ignition that has a very distinctive long dwell as in Figure 13-34. Notice that the dwell zone begins just as the spark ends and is much longer than any breaker point ignition system. This dwell will actually change slightly as the engine speed is increased. As the spark line slightly increases, which it will generally do at higher speeds, the dwell zone will actually begin later. When looking at a scope pattern, it will appear that the amount of dwell decreases as the engine speed is increased. A dwellmeter will confirm this. The change is slight, however, usually less than 5°, so many mechanics put this ignition system into a category of long fixed dwell. Keep in mind that this change in dwell has no effect on the timing of the engine,

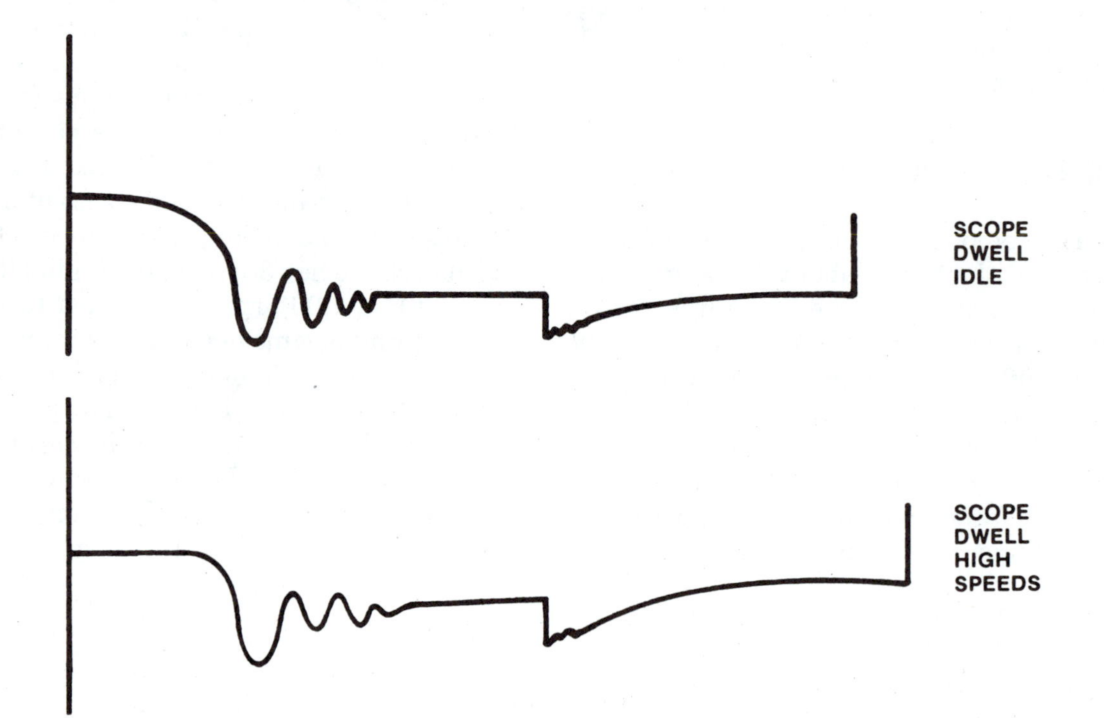

FIGURE 13-33 Scope pattern showing fixed dwell—short.

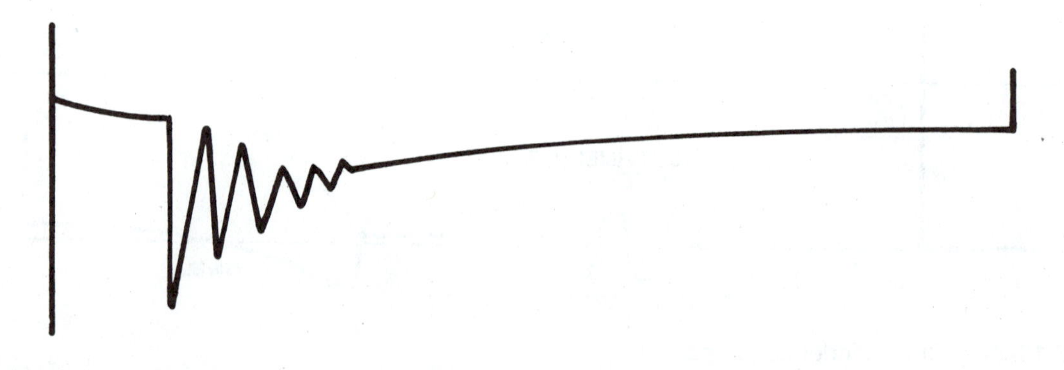

FIGURE 13-34 Scope pattern showing long dwell.

because timing is when the primary current off signal occurs. The change takes place from a shortening of the dwell from the primary current on signal rather than from the primary current off signal. Figure 13-35 illustrates this change.

The third dwell zone difference you might encounter has become very popular in recent years: variable dwell. Variable dwell is just what the name implies, the dwell will actually change with engine speed. We previously discussed the reasons why it is desirable. The engine's ignition coil does not require much dwell at low engine speeds because saturation time is extremely fast. However, as the engine speed is increased, the amount of time for coil saturation is reduced. Remember our example of one fourth the actual time at 2,000 rpm over that at 500 rpm? The variable dwell partially compensates for this condition. You will notice on the scope that the primary current on signal will be very close to the primary current off signal at idle (less than 20°, typically) as in Figure 13-36. As the engine speed is increased, the dwell will increase by moving the primary current on signal farther to the left as in Figure 13-37. By accelerating the engine, you should observe the dwell getting larger with

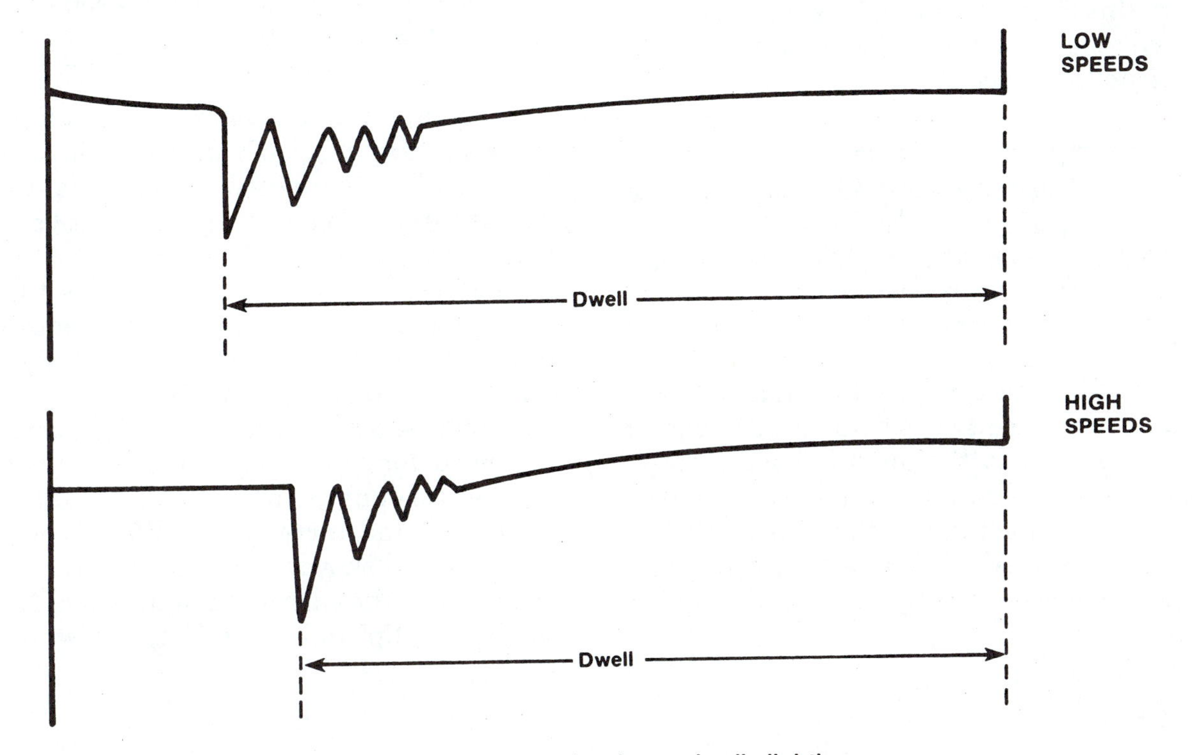

FIGURE 13-35 Scope pattern showing how speed reduces dwell slightly.

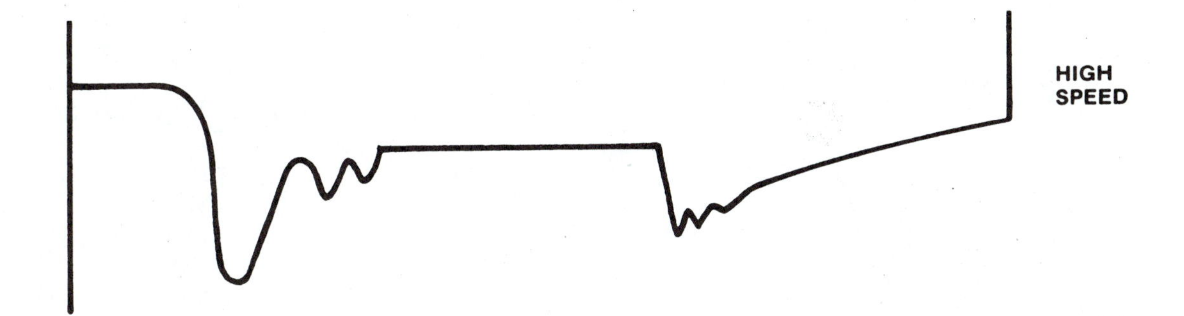

FIGURE 13-36 Scope pattern showing short low-speed dwell.

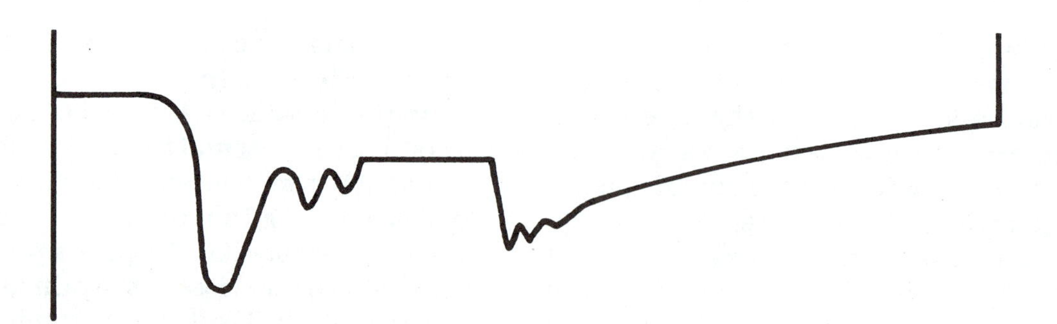

FIGURE 13-37 Scope pattern showing how high-speed increases dwell.

increased speed and getting shorter with decreasing engine speed. A dwellmeter will also record this changing dwell. Most modern ignition systems have the variable dwell function built into their modules.

NOTE: The method of determining timing or piston position, whether it be Hall effect, magnetic generation, or light-emitting diodes, has no effect on the dwell zone.

As a matter of fact, you will be unable to determine which method is being used to signal piston position to the module by the scope pattern. Generally, you will have to research the manuals or actually look at the system to pinpoint the method. The scope, especially the dwell zone, will not give you this information.

NOTE: It is possible for the variable dwell function to burn out of the module and still have the vehicle run.

This condition is not common but is a possibility. Finding a lack of variable dwell on a vehicle that is supposed to have it would necessitate the module replacement because variable dwell is a module function.

The last dwell zone variation is that of current limit. Many ignition systems currently in use saturate the coil quickly by running very high primary current through them for a split second. Once the coil is saturated, the need for this high current is gone. A smaller amount of current will keep the coil saturated and increase the life of the coil. Figure 13-38 shows an ignition pattern for a vehicle with the current limit function. Notice the little "blip" in the dwell zone toward the

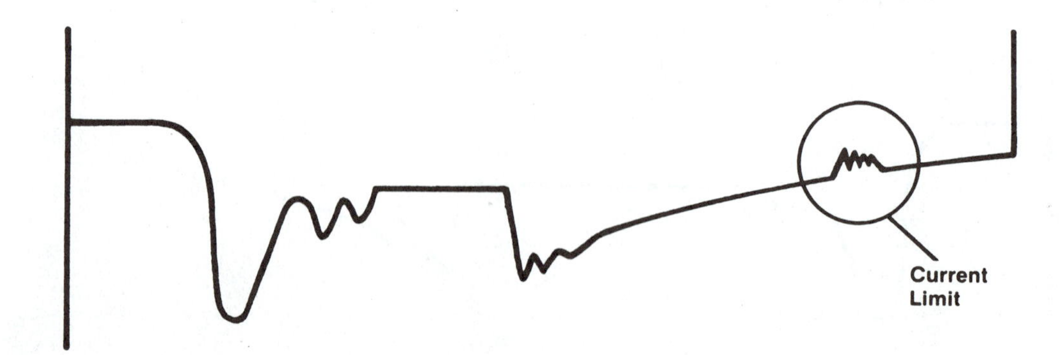

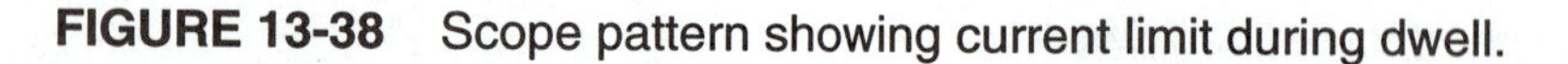

FIGURE 13-38 Scope pattern showing current limit during dwell.

primary current off signal. This "blip" is the module recognizing that the coil is fully saturated and cutting back on the primary current. You will again find, as was the case with variable dwell, that this current limit function is becoming increasingly common on modern electronic ignition systems. It is also a module function. Very high speeds will sometimes find the "blip" missing. At high speeds, the coil might not become fully saturated and the "blip" will not be present because the module will keep the higher current flowing in an effort to get the coil saturated. In addition, if the primary of the ignition coil has developed excessive resistance or is faulty, the "blip" might be missing because the coil's primary never saturates. As was the case with the variable dwell, it is possible for the current limit function to not be present and still have the vehicle running. Further testing of the coil primary and the primary B+ feed circuit would be necessary to determine why the "blip" is missing. Current limit is a module function that will also limit the number of kilovolts available at the secondary output. Do not touch the secondary wires, cap, or coil if the current limit "blip" is missing on a vehicle that should have it. The secondary available voltage could easily exceed 100,000 volts with full primary current flowing.

Scope Check-Out

Now that we know what a basic scope pattern should look like and also can identify the variables that might be present in the dwell zone, it is time to put it all together and come up with a basic ignition check-out. There are many things that you will be able to do with the scope once you are experienced in its operation. The basic check-out, which we will detail here, is a good preventive maintenance or after tune-up quality control check. Before we begin, let us discuss electronic ignition and open circuits. When ignition systems were all breaker point, a common test was to see just how much available voltage the coil could produce into an open circuit. Removing the coil wire and cranking the engine over or pulling a plug wire with the vehicle running would force the ignition coil to produce its maximum output and try to overcome the resistance. Typically, the output would exceed 20,000 volts and we would know that the system was able to work under adverse conditions. Open circuiting a modern ignition system is not advised, however. Most electronic ignition systems are four function, with variable dwell and current limiting circuits built into their modules. These systems will pull out all primary resistance if for some reason a spark does not occur. This will allow tremendous primary current to flow through the coil, producing an extremely high-voltage spark looking for a path to ground. A path that is frequently used by the spark is through the side of the coil into the metal laminated form. This would result in an insulation breakdown at the coil and a future site for arcing to ground. For this reason, whenever we refer to a maximum coil output test, we will have a special plug, Figure 13-39, installed in place of the spark plug. This plug has a very large gap. When it is grounded by means of the special clip, the ignition coil will produce a higher voltage to overcome the additional resistance and we will be able to read this voltage on the scope. Typically, about 35,000 volts is necessary to fire this test plug. This level is sufficient to stress test the system and yet will not result

FIGURE 13-39 Test plug.

in insulation breakdowns at the coil. Never run any modern electronic ignition open circuited as damage might occur.

Scope Check-Out Procedures

The first test we will do is a very thorough visual inspection under the hood. As we mentioned earlier, you do not need a scope to find oil-soaked wires or an insulation breakdown. Look over the entire system carefully and note on the check-out sheet, Figure 13-40, any irregularities you find. As you look, keep in mind that the vehicle might not be back to a mechanic for another year or 25,000 miles. Try to anticipate if the component will, in fact, be likely to last the duration of the tune-up. After the inspection, we can move onto a cranking output test.

I. Procedure for Cranking Output Test

Install the test plug in the coil wire, or a plug wire if there is not a coil wire, Figure 13-41. This controlled air gap of the test plug will

Scope Check Out

Name ____

Vehicle type ______________________ **Year** ____________

Ignition type ______________________

1. Visual inspection	Good ___	Bad ___
2. Cranking coil output	Good ___	Bad ___
3. Coil polarity	Good ___	Bad ___
4. Firing voltage	Good ___	Bad ___
5. Available voltage	Good ___	Bad ___
6. Secondary insulation	Good ___	Bad ___
7. Plugs under load	Good ___	Bad ___
8. Secondary circuit resistance	Good ___	Bad ___
If breaker point ignition:		
9. Coil and condenser	Good ___	Bad ___
10. Point action	Good ___	Bad ___
11. Cam lobe accuracy	Good ___	Bad ___
If electronic ignition, check one.		
12. 2 function ___		
Primary current on	Good ___	Bad ___
Primary current off	Good ___	Bad ___
13. 4 function ___		
Primary current on	Good ___	Bad ___
Primary current off	Good ___	Bad ___
Variable dwell	Good ___	Bad ___
Current limit	Good ___	Bad ___

Based on above information, what is your recommendation?

FIGURE 13-40 Scope tests.

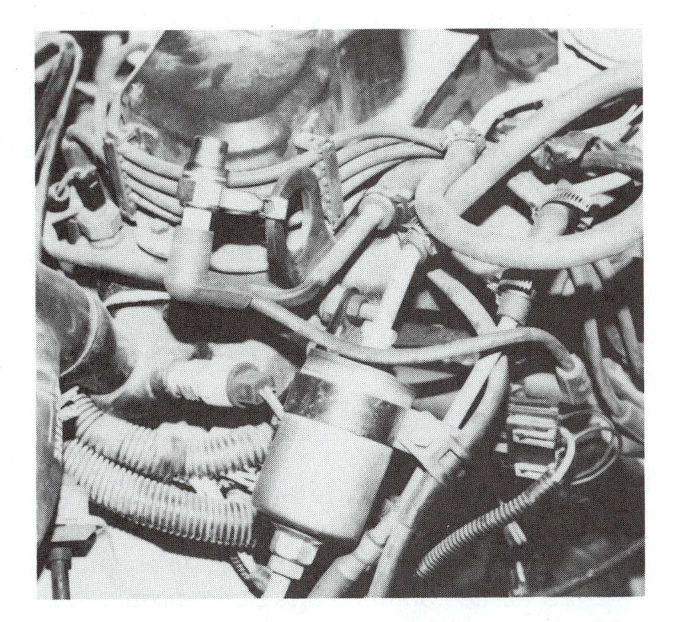

FIGURE 13-41 Test plug installed.

stress the system sufficiently to reveal any weaknesses. Crank the engine over with the scope on display and a voltage range around 50 kV. The firing line should exceed 35 kV and be consistent. Less than specified or irregular should be noted on the check-out form as it is indicative of problems. Let us look at some of the possibilities. If a lower than specified voltage is present, it is likely that the spark plug tester did not arc. A lower than normal available voltage can be the result of lower than normal B+ primary voltage, or the module might have developed resistance, or the ignition coil could be faulty. At this point, the exact cause of the failed cranking test cannot be pinpointed, and we will have to do additional testing to determine what is wrong. Do not make the very common mistake of trying to pinpoint the difficultly with one test only. Write the results down and move on.

Replace the coil wire or the plug wire and start the engine. Allow it a moment or two to stabilize and then raise the engine speed up to 1200 to 1500 rpm and if possible block the throttle open at this speed. The majority of the testing will be done at this speed. All cylinders should be showing on the scope in a parade pattern. Look at the firing lines. They should all be within 3 kV of each other and between 6 kV and 12 kV high. The range of 6 kV needs to be explained in detail. New plugs with a correct secondary resistance, a correct air/fuel ratio, and properly functioning cylinder will generally fall at the lower end of the 6–12 kV range. As the plugs wear, their air gap increases and the firing voltages rise up. The goal, if the vehicle has been recently tuned, will be to have the firing voltages as low as possible. Low firing voltages with high available voltages give the greatest margin of unused voltage for the future. This margin of voltage is necessary for two reasons: acceleration and mileage between tune-ups. Acceleration might require an additional 50% firing voltage if the engine is to accelerate smoothly. If the firing voltage were abnormally high under cruising conditions, the acceleration voltage might exceed the insulation capability of the coil or the wires and an insulation breakdown would occur. If you have recently tuned the engine, the firing voltage should be low. If it is not, secondary resistance must not be to spec. The length of time between tune-ups is also a reason for wanting the greatest margin between firing voltages and available voltage. New spark plugs require the lowest voltage to fire. As they wear, the voltage required increases. As this voltage increases, the likelihood of misfires will also increase if we approach the available voltage level. Keep the firing voltage low and the available voltage high and you eliminate the majority of ignition-caused driveability complaints. Now is when we begin to look at the firing lines and compare them. If they are all high as in Figure 13-42, it is likely that one item is responsible for their height. Common components include rotors, coil wires, coil secondaries, and air/fuel ratios. If on the other hand, only one or two firing lines are high as in Figure 13-43, the cause must be a component common to only the one cylinder, like the plug or the plug wire. At the end of this section, we will see how the scope and a ground probe can be used to determine the cause of high firing lines. At this point, we would just make a note of the results on the check-out sheet and move onto the next test.

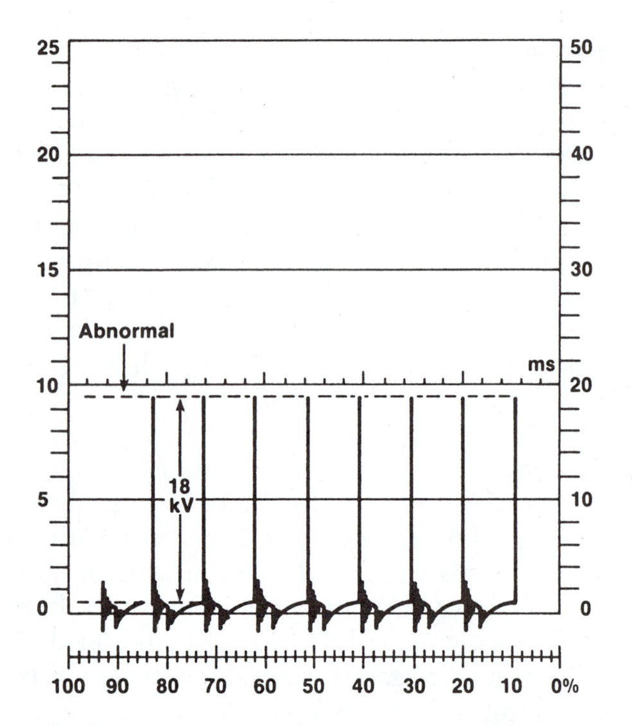

FIGURE 13-42 Scope showing all firing lines higher than normal. (Courtesy of Sun Electric Corporation)

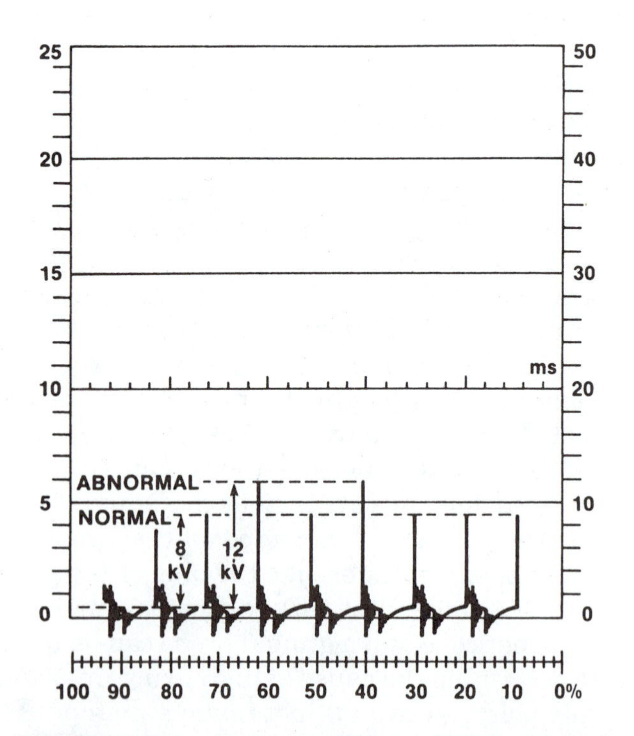

FIGURE 13-43 Scope showing two firing lines higher than normal. (Courtesy of Sun Electric Corporation)

While we have the ability to look at the firing lines on a display type of pattern, we should do two additional tests. The first is a coil output test and the second is a plugs under load test. The coil output test is similar to the cranking coil output test and is done with a test spark plug installed in the end of a spark plug wire.

II. Procedure for Coil Output Test

Turn the engine off while you insert the test plug and ground it. Once the engine is running, a firing voltage for the test cylinder should exceed 35,000 volts and be consistent. With the engine off, the test plug can be removed and the spark plug reconnected.

III. Procedure for Spark Plugs Under Load Test

The next test we will do is the spark plugs under load test. With the engine at idle, rapidly open the throttle and then close it a few times. Do not allow the engine speed to rise above 2,000 rpm when you open the throttle. The faster you can open and close it, the better the test results will be. The open throttle will increase the pressure on the plugs and increase their likelihood of misfire. As the pressure increases, the firing lines should all rise up equally. There should be no more than a 3 kV difference between the highest and the lowest. Anymore of a voltage difference indicates that the conditions inside the cylinder, including the spark plug, must be different. For example, a spark plug with a core nose fracture might tend to arc over to the ground shell at a lower voltage than that required for acceleration. This cylinder would then have a lower spark plug under load firing voltage than the rest of the cylinders.

The spark section of the pattern is the next section to be viewed. It can best be seen on a different scale, usually raster or stacked. This pattern selection allows the greatest opportunity to compare the lengths of all the spark lines and the angle at which they come off the firing line. They should be approximately the same length and should have a slight slope downward as in Figure 13-44. If the scope you are using is equipped to read millisecond sweep, a properly functioning ignition system will usually have a 1.5 ms spark line. Most manufacturers specify a range of from 0.8 ms to 2 ms. Lower than specification might result in incomplete combustion or misfire, while longer than the 2 ms might be indicative of a fouled plug, lower compression, or a plug with a narrow gap. In addition, a four-function ignition system with the current limit function burned out will produce a spark that lasts longer than normal. Figure 13-45 shows a normal spark line that lasts about 1.3 ms. If your scope does not have the ability to measure milliseconds directly, you can use the percent of dwell scale and convert it over to milliseconds using Table 13-3. When using the percent of dwell scale, it is important that the speed of the engine be at 1200 rpm or the readings will not be accurate.

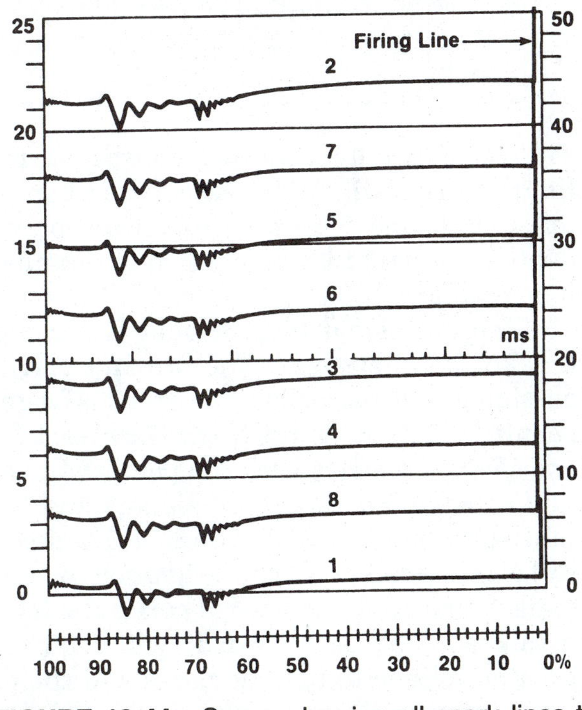

FIGURE 13-44 Scope showing all spark lines to be equal. (Courtesy of Sun Electric Corporation)

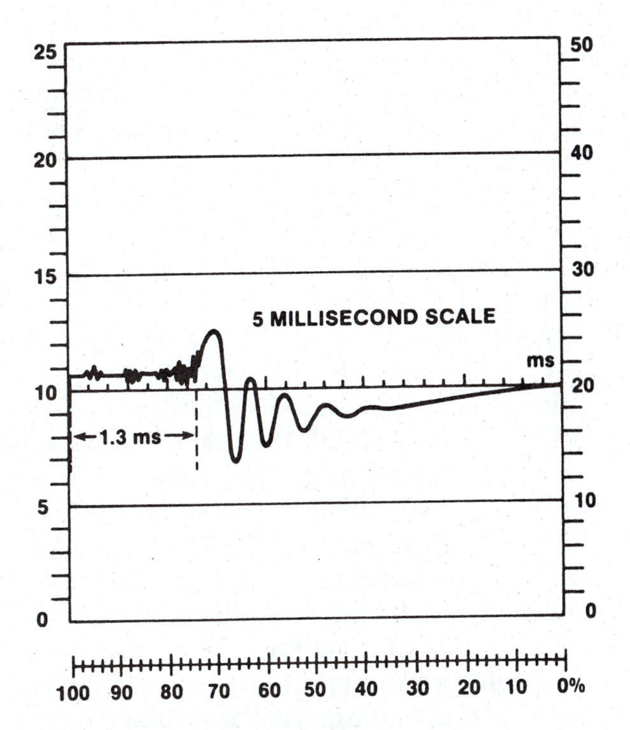

FIGURE 13-45 Scope showing 1.3 ms spark. (Courtesy of Sun Electric Corporation)

The length of the spark line, assuming the firing line was normal, gives us an indication of the amount of voltage in reserve. Do not forget this reserve voltage becomes more important as we begin to accelerate. As much as 50% more voltage might be required if the vehicle

TABLE 13-3: CONVERTING THE DWELL SCALE TO MILLISECONDS @ 1200 RPM

Milliseconds	Percent of Dwell	Number of Cylinders
.5	4%	8
1.0	8%	8
1.5	13%	8
2.0	17%	8
.5	3%	6
1.0	6%	6
1.5	9%	6
2.0	12%	6
.5	2%	4
1.0	4%	4
1.5	6%	4
2.0	8%	4

is to accelerate smoothly. In addition, this reserve voltage will be needed as the customer drives the vehicle and the plugs get worn and their gaps get larger.

The next section of the trace that we look at is the coil condenser zone. A steadily diminishing series of oscillations, beginning at the spark line and ending at the zero line, indicates that all is well after the spark and before primary current is turned on. There will usually be five to seven oscillations in this section, Figure 13-46. Problems in this area are usually found in shorted turns of the coil, especially with modern electronic ignition. Further component testing as we will outline later will be necessary to identify if the coil primary or secondary is faulty.

The remaining section of the trace is the dwell zone beginning with the primary current on signal and ending with the primary current off signal or firing line for the next cylinder in order. Again, we keep the scope on the raster or stacked pattern and observe the dwell zone. Keep in mind that we have three variables to work with. All patterns will show primary current on and primary current off. In addition, some systems will show variable dwell and/or current limit. By accelerating the engine from idle to about 2,000 rpm, the change in dwell will be apparent. It should be smooth and even as it increases with speed and then decreases as the engine slows down. This change can also be observed with a standard dwellmeter, Figure 13-47. Remember that a lack of dwell change on a system that is designed with variable dwell usually indicates the need for a new module. The current limit function is also easily checked. The presence near the end of the dwell zone of the blip as in Figure 13-48 indicates that the coil has reached full saturation. If you are doubtful as to whether the system was designed with the current limit function, check the wiring diagram. The majority of the systems currently in use will have either a primary resistor or the current limit function in the module. A system with no primary resistor on the wiring diagram should show the current limit blip on the dwell zone.

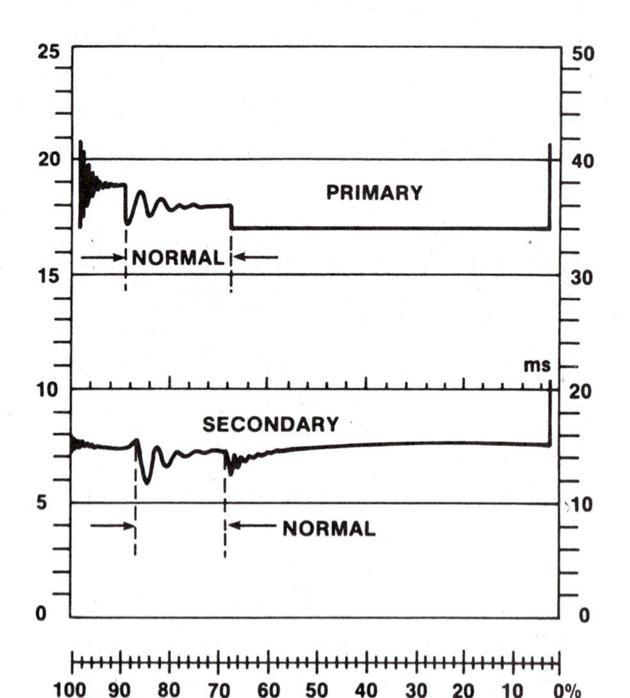

FIGURE 13-46 Scope showing coil condenser trace. (Courtesy of Sun Electric Corporation)

NOTE: The lack of a blip might be indicative of a bad coil or reduced primary current flow. It is not automatically indicative of a bad module.

The coil must fully saturate before the current limit function will "blip". Any increase in the amount of time necessary for saturation will delay the current limit function or eliminate it completely.

Two additional tests should be done on breaker point systems. The primary current on signal and the primary current off signal should both be 90° turns of the trace as in Figure 13-49. Problems with the points will usually show up on these two signals and look much like Figures 13-50 and 13-51. The second test shows whether or not the ignition coil is installed correctly and is called **coil polarity**. If a coil is wired backward, it will require a greater voltage to fire the plugs and the entire pattern will appear upside down as in

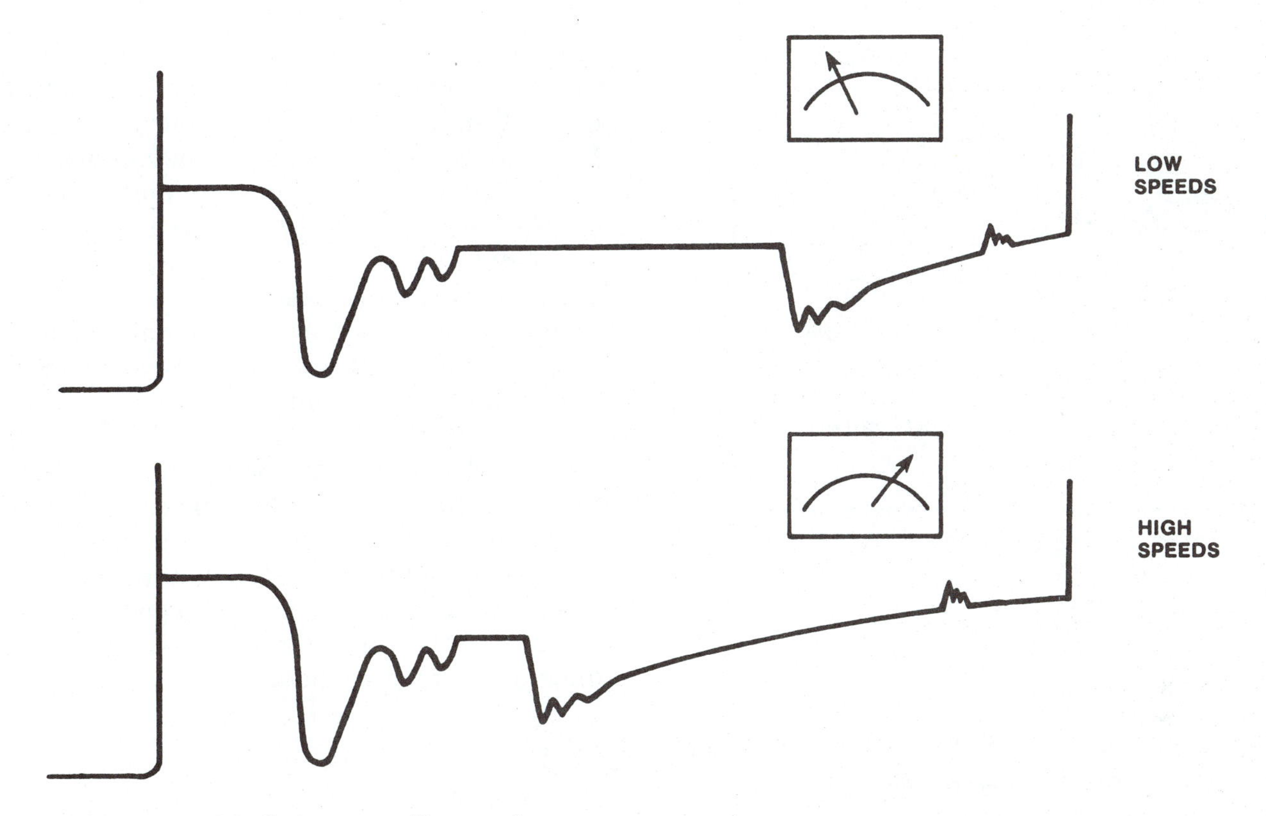

FIGURE 13-47 Dwell changes with speed.

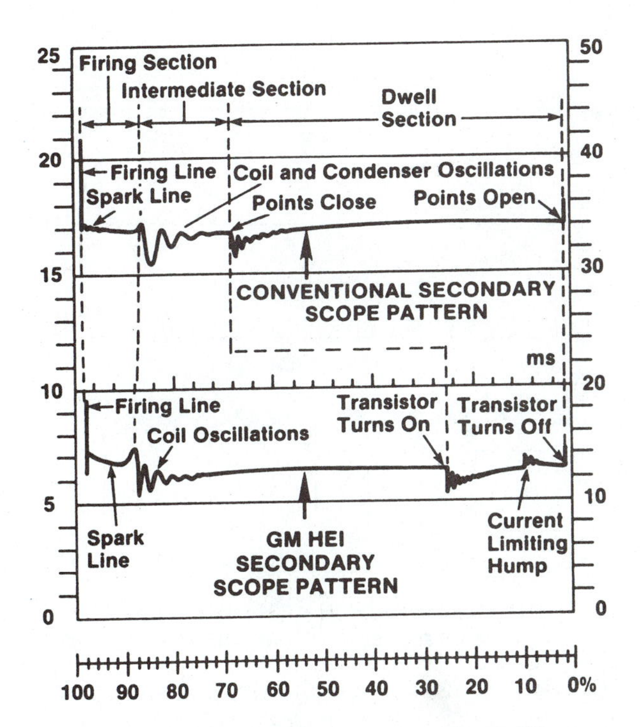

FIGURE 13-48 Scope showing current limit. (Courtesy of Sun Electric Corporation)

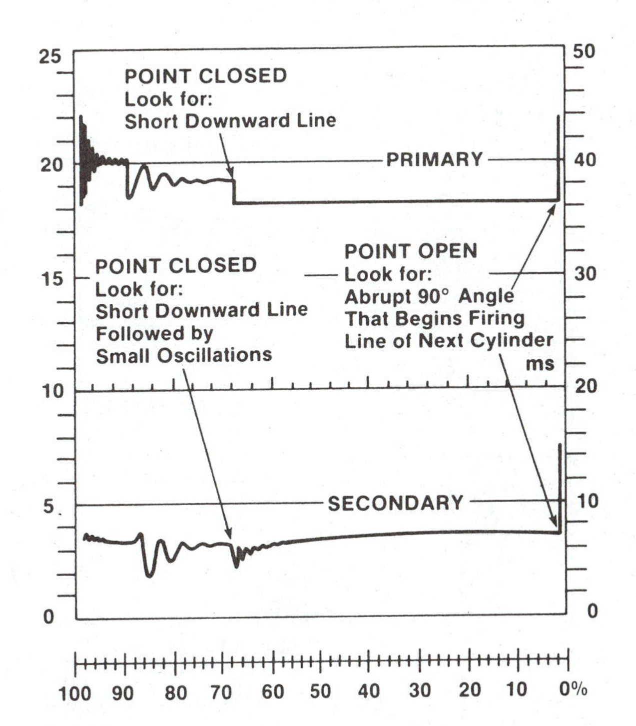

FIGURE 13-49 Scope showing primary current on and off signals. (Courtesy of Sun Electric Corporation)

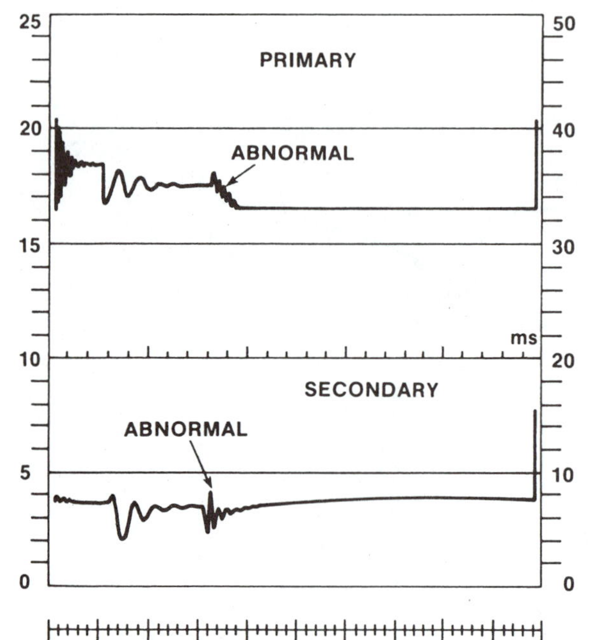

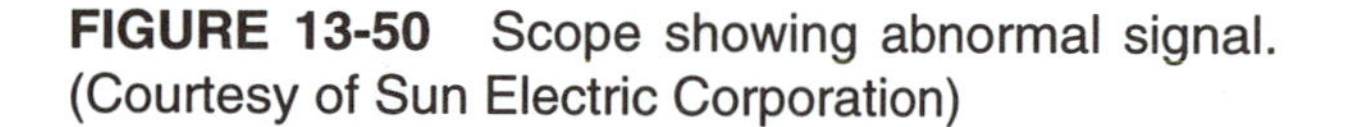

FIGURE 13-50 Scope showing abnormal signal. (Courtesy of Sun Electric Corporation)

Figure 13-52. With the use of electronic ignition, reversed coil polarity has been eliminated. Connectors like that shown in Figure 13-53 are almost impossible to install reversed.

This ends the usual testing done with an ignition oscilloscope. At this point, upon the completion of all the tests, you should look back at the results and try to identify the faulty components. Save your final diagnosis until you have a minimum of the above tests completed. Do not jump to conclusions based on a single test or your diagnosis might be incorrect. The interrelationship between all of the components in the system can give you false impressions if you rely on a single test. If a vehicle's ignition system passes the basic tests outlined, it is essentially a good system. Additional testing using the scope will be outlined in the next section.

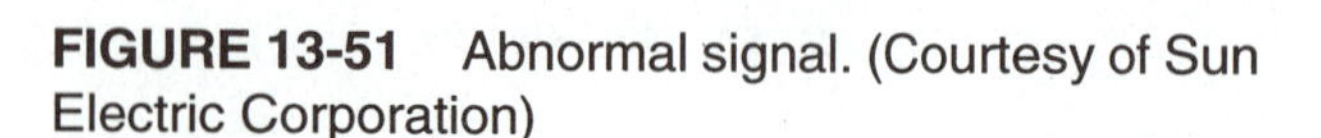

FIGURE 13-51 Abnormal signal. (Courtesy of Sun Electric Corporation)

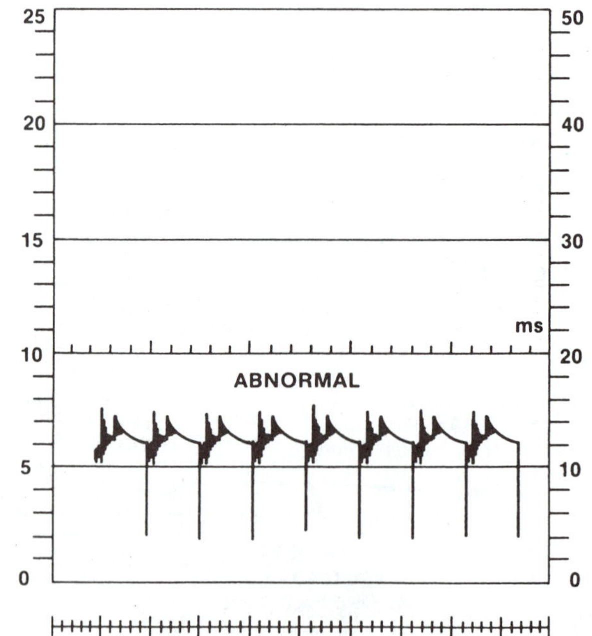

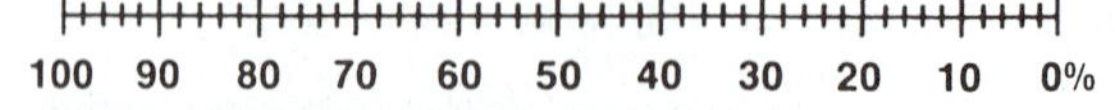

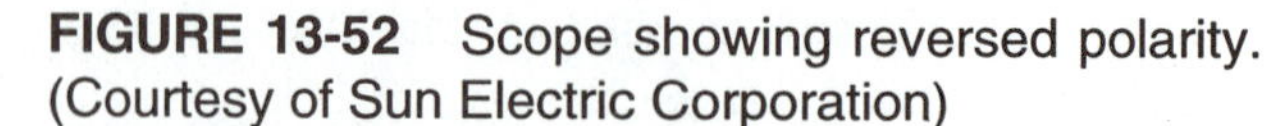

FIGURE 13-52 Scope showing reversed polarity. (Courtesy of Sun Electric Corporation)

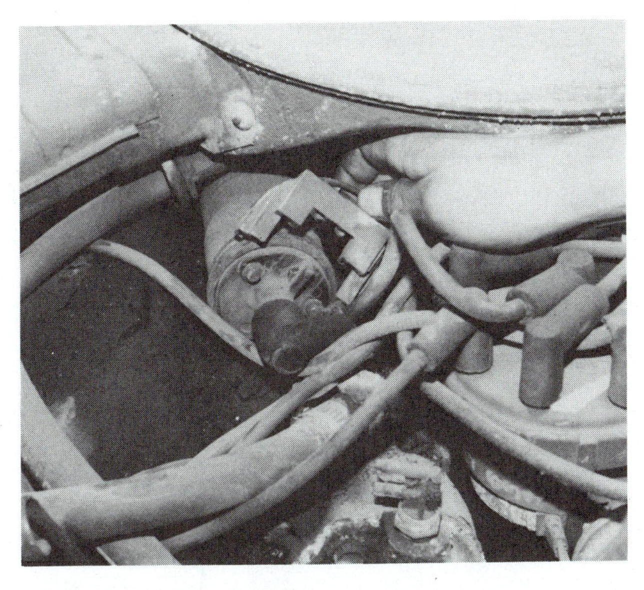

FIGURE 13-53 Polarity protected connector.

Additional Scope Testing

There are many things that can be done with an oscilloscope to further test the ignition system. We will look at a few of these now. The first was mentioned during the discussion of the testing of the firing lines. We noted that the cause of excessive firing voltage should be eliminated to increase the level of reserve voltage. The simplest method for finding the cause is to use a ground probe and eliminate resistance while observing what the firing line does. Excessive resistance causes excessive voltage. The two go hand in hand. When we have eliminated the resistance, the firing line will go down to an acceptable level. Ground probes are commercially available or they can be made from an insulated screwdriver and a clip lead. Let us take a simple example and go through the procedure keeping a few things in mind. The first is that the typical ignition system has the greatest amount of resistance across the plug tip. It should therefore have the greatest voltage drop when its resistance is eliminated. All of the rest of the ignition components: coil secondary, coil wire, distributor cap rotor, and plug wire will usually have just enough resistance to cause about a 2,000 to 3,000 volt drop, with the majority of this drop across the air gap of the cap and rotor. With this in mind, let us look at a typical procedure for isolating the cause of a single cylinder with an excessively high firing line.

IV. Procedure for Isolating the Cause of a Single Cylinder with an Excessively High Firing Line

With the vehicle off, place one end of the ground probe into the spark plug boot until it makes good contact with the plug connector. The other end of the ground probe should be connected to a good clean engine ground, Figure 13-54. Start the engine and observe the firing line for the grounded cylinder. The change in the firing voltage is the amount of voltage that the plug and cylinder took. The firing line should now be at the 2 to 3 kV level, Figure 13-55. If we started with 25 kV, for example, with the cylinder firing and now had 2.5 kV with the plug wire grounded, we would say that the cylinder was responsible for 22.5 kV (25 kV – 2.5 kV = 22.5 kV). The ground probe has told the story. The spark plug and the conditions inside the cylinder have excessive resistance. Remember that the spec is 6 to 12 kV with everything but the plug and cylinder taking 2 to 3 kV. This then means that a properly functioning plug and cylinder

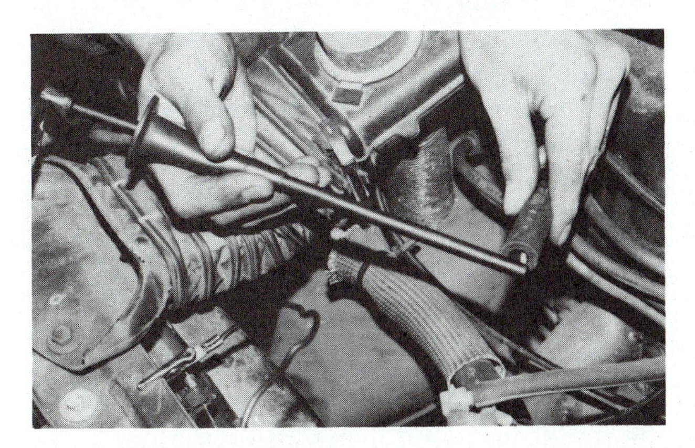

FIGURE 13-54 Ground probe in spark plug boot.

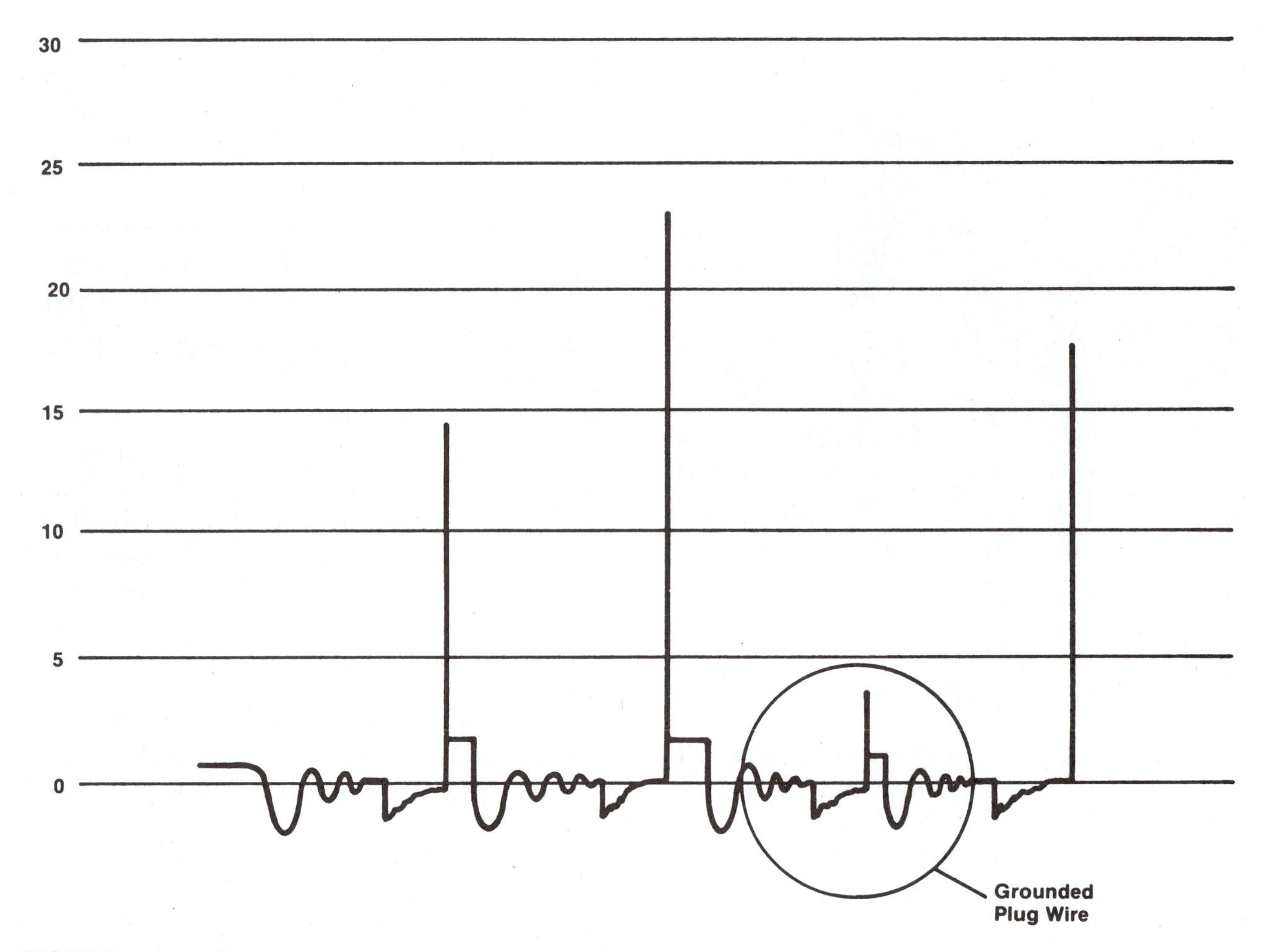

FIGURE 13-55 Scope pattern showing third cylinder grounded.

should require no more than 4 to 10 kV. The remedy here is simple, replacement or swapping the plug for a known good one will pinpoint the problem to either the plug or the cylinder.

If on the other hand the firing voltage remains high with the plug wire grounded, excessive resistance must still be a part of the circuit. With the vehicle off, pull the plug wire out of the distributor cap and put the ground probe in its place, Figure 13-56. Start the engine and again observe the firing line. If the firing line drops to normal (2 to 3 kV), the plug wire has excessive resistance and should be replaced. If the firing line remained high, the resistance still has not been pinpointed. We know that it is in the cap, rotor, or coil

FIGURE 13-56 Grounded distributor cap terminal.

wire but will have trouble running the engine with any of these components grounded. Do you remember one of the first scope tests that we did? Cranking coil output with the coil or coil wire grounded will indirectly tell us the resistance of the coil wire, Figure 13-57. If we know the voltage that the coil wire requires and we know the voltage that the plug wire, spark plug, and cylinder require, it is simple math to determine just how much voltage the cap and rotor require.

This procedure is outlined in worksheet form in Figure 13-58. As you do it, do not allow the ignition system to operate open cir-

FIGURE 13-57 Grounded coil wire.

Firing Voltage Worksheet

Firing Voltage 1200 to 1500 rpm	Grounded Plug Wire	Grounded Dist Cap
1. ______ V	______ V	______ V
2. ______ V	______ V	______ V
3. ______ V	______ V	______ V
4. ______ V	______ V	______ V
5. ______ V	______ V	______ V
6. ______ V	______ V	______ V
7. ______ V	______ V	______ V
8. ______ V	______ V	______ V

Grounded Coil Wire Voltage ______ V

Grounded Coil Voltage ______ V

COMPONENT VOLTAGE SECTION

Cyl & Plug	Plug Wire	Cap/Rotor	Coil Wire
1. ______ V	______ V	______ V	______ V
2. ______ V	______ V		
3. ______ V	______ V		
4. ______ V	______ V		
5. ______ V	______ V		
6. ______ V	______ V		
7. ______ V	______ V		
8. ______ V	______ V		

FIGURE 13-58 Firing voltage worksheets.

cuit with no electrical path. The high voltage produced and the lack of module current limit may damage the ignition coil, cap, or rotor.

IGNITION STRESS TESTING

Periodically, an intermittent problem might appear at the front door of your shop. The customer describes a driveability problem that occurred "yesterday" but appears to be gone right now. The most viable approach available to you is to start with a full scope check-out as we have outlined, then do a series of individual tests that you think might reveal the intermittent defect. Sometimes this approach forces us into replacing items because we "think" they are the most likely causes of the complaint. The scope allows us to view the inside of the ignition system in detail and observe just what is happening under various conditions. In addition, we can stress the ignition system with three items: heat, cold, and moisture. Many of the typical intermittent driveability problems occur under conditions of heat, cold, moisture, or a combination of all three. The first part of the procedure is perhaps the most important—a discussion with the customer! Ask him or her exactly what conditions are present when the driveability complaint begins. What temperature is the engine? What temperature is the outside air? Is it raining or humid out? The more information you have, the easier it will be to try to duplicate the conditions that generally produce the complaint. Once the information is compiled, you can begin to stress the system with three basic pieces of equipment. A heat gun (hair dryer) of about 500 watts, a spray bottle of water, and a choke checker that supplies cold air.

V. Procedure for Ignition System Cold Stress Test

With the scope on raster, begin to cool the major electronic ignition components with cold air: the pick-up unit, the module, and the major connectors. Directing the cold air onto these components will cool them down to below freezing. Move the air around the components, directing the spray about 3" away from them, Figure 13-59. Cool the system down as you observe the scope pattern, looking for changes especially in the dwell zone.

CAUTION: Do not use liquid freon (R-12) to cool components. Freon has been shown to be destroying the earth's ozone layer.

Once the component is cold, allow the system to reach normal temperature as you observe the scope and run the engine speed up and down. Any change in the dwell zone will indicate that the component you have frozen is likely to be defective and should be replaced. Make sure that you do not freeze more than one component at a time or your results will be inconclusive. Cooling the pick-up unit or a module that is located in the base of the distributor is done by directing the cold air up at the base, Figure 13-60.

FIGURE 13-59 Cooling down a module.

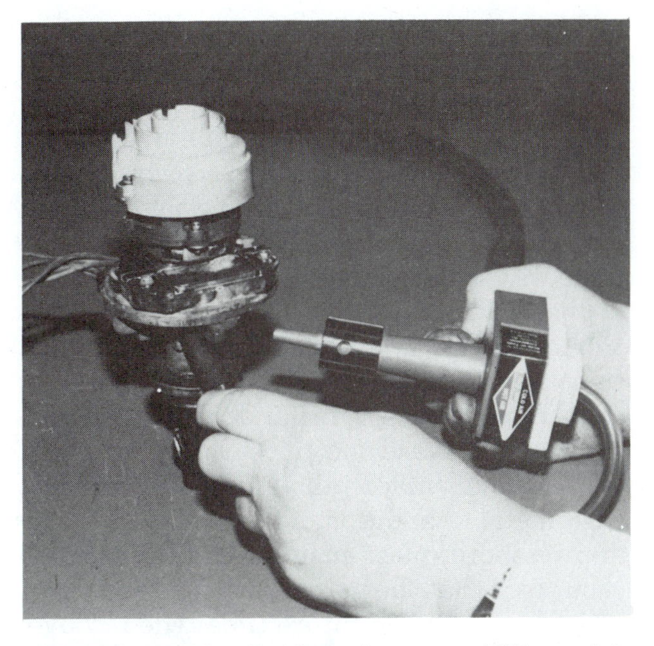

FIGURE 13-60 Cooling down an HEI module mounted inside the distributor.

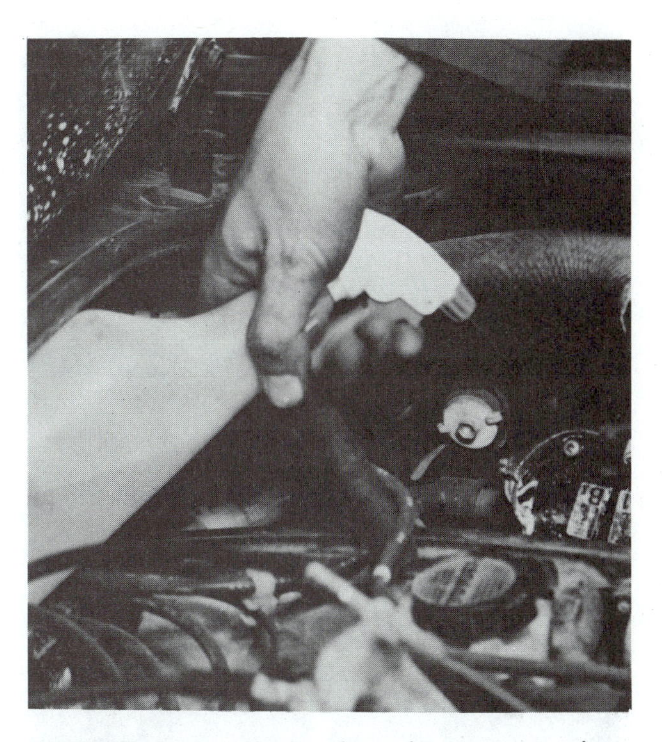

FIGURE 13-61 Wetting down the system to simulate damp weather.

VI. Procedure for Ignition System Moisture Stress Test

The next stress that we will throw at the ignition system is moisture. As in the cold test, look at the scope on a raster pattern and pay particular attention to the dwell zone as you spray the primary ignition components: the ignition module, the pick-up unit, and all the interconnecting wires. If no problems show up, switch the scope over to display and, while looking at the firing and spark line, spray down the distributor cap, rotor, and spark plug wires, especially where they run next to metal (ground), Figure 13-61. If the secondary insulation will break down and arc to ground, it is most likely to occur under engine load and moisture. Load the engine down slightly in drive with your foot on the brake. Any arcing or scope changes indicate insulation breakdowns. Do not forget the coil area.

VII. Procedure for Ignition System Heat Stress Test

The last stress test utilizes heat, Figure 13-62. With the scope on a raster type pattern, use a hair dryer or a heat gun of about 500 watts and heat the module, then the pick-up unit, and finally the electronic ignition connectors. If heat is the cause of the driveability complaint, the dwell zone will usually show changes, especially in the variable dwell or current limit sections. If the connectors appear to be the source of the difficulties, pull them apart, clean them, and then coat them with dielectric compound, which will seal out dirt and moisture, Figure 13-63.

FIGURE 13-62 Heating the system to simulate hot weather.

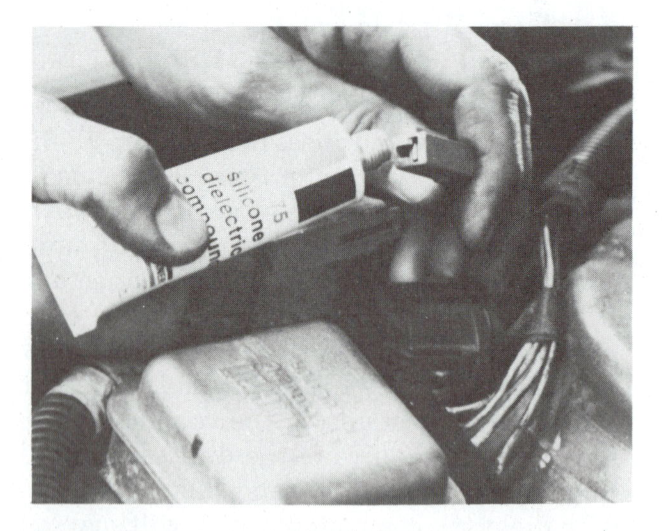

FIGURE 13-63 Dielectric compound should be used on all connections.

NO START, NO SPARK

A good diagnosis opportunity will frequently appear in shops today. The engine cranks well, has fuel, but lacks spark. The technician is faced with the problem of picking out the failed component, replacing it, and trying to determine why it failed so that a repeat repair will not be necessary in a short period of time. Electronic ignition is extremely reliable and has some expensive components. The high cost of each item makes a hit-and-miss approach, where we replace the "most likely" cause of the no spark, unacceptable, especially to most customers. In addition, if the module is faulty and replaced, most customers will like some assurance that the module will not require replacement at some future time. The inconvenience of having to face the day without the use of the vehicle makes customers seek out a repair that will last. Fix it right, the first time!

There are many different methods of diagnosing a no-spark condition in addition to many different types of equipment that claim to make the technician's job easier and faster. Generally speaking, the most readily available test tools work out the best. They might not necessarily be the fastest, but they frequently are the most reliable. The test procedure outlined here makes use of two pieces of equipment: a 12-volt test light, a multimeter (preferably digital)—and visual inspection. It also ignores the direct testing of the module and instead tests all of the other components in the system. If all other components are found to be within specification, the untested module is replaced. This procedure allows you to test the items that can be tested with the test light and the multimeter. Most modules require expensive testers that usually work on only selected systems. By ignoring the module until we have tested the rest of the system, we will be indirectly testing it. If you do the procedure exactly the same each time, you will get very good and fast. Your accuracy will also come naturally. Do not jump around during the procedure and keep in mind that we do not want to connect or disconnect the module with the ignition key on. The spike that might occur could destroy an otherwise good module leaving you with the original cause of the no spark in addition to the burned module.

VIII. Procedure for No Start, No Spark

1. Locate a wiring diagram that shows the ignition system in detail. An example of a current system is shown in Figure 13-64. Notice that the diagram shows the module, the ignition coil, the pick-up coil, the ignition switch, and all the connectors. The color of the wires is also noted for location assistance.
2. Use your 12-volt test light to determine if the system has power both during run and crank positions. Use the wiring diagram and find the correct wires. In our example, the run terminal of the ignition switch has two wires connected to it. Both are white with a light blue dashed stripe (hashmark) and run to the electronic module and the ignition primary resistor and coil. With the ignition key off, disconnect the two connectors at the module. The top connector has three wires and the bottom has four wires. Once the connectors are disconnected, turn the key on and probe the white/light blue cavity testing for power. The test light should light brightly, indicating power. Do the same with the Bat terminal of the coil. It should also light the test light with the key in the run position, Figure 13-65. Once we have determined that we have B+ available with the key in the run position, we need to determine that B+ is also available with the key in the start position. Notice that the start position has a white wire with pink colored dots. Probe the cavity in the disconnected module connector and put the key in the start position. If you disconnect the S terminal of the solenoid, the engine will not crank. Power should also be

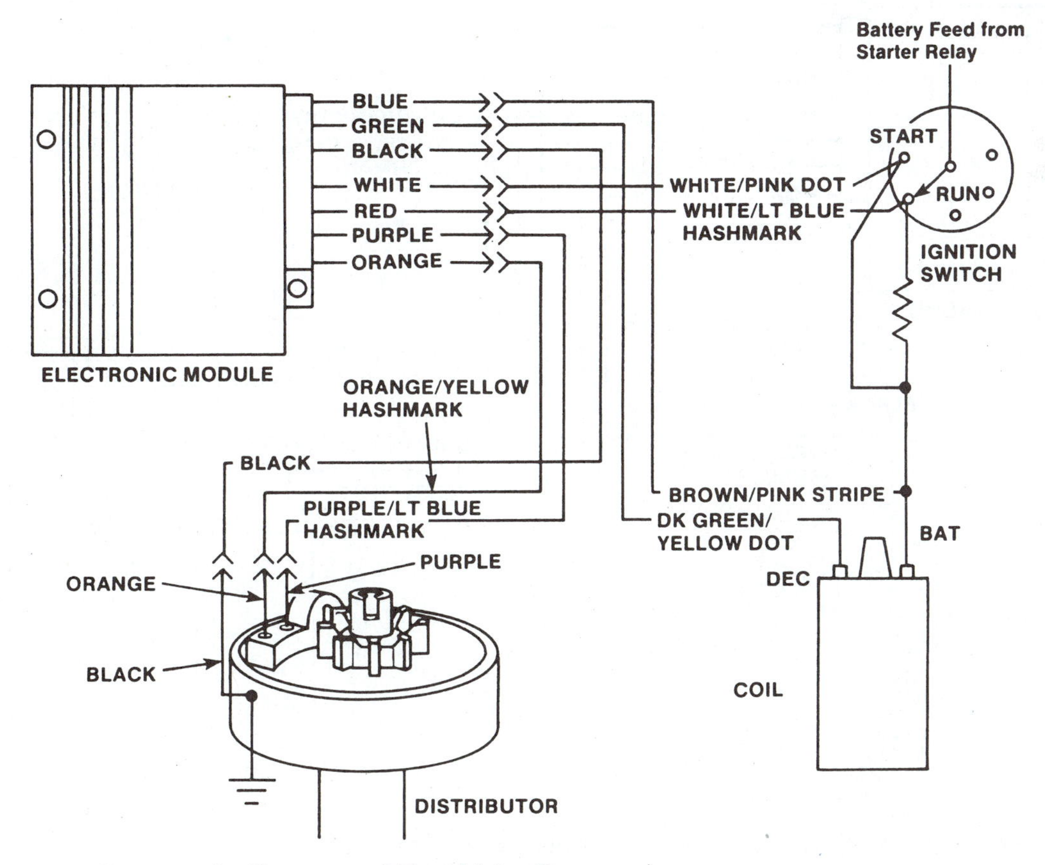

FIGURE 13-64 Dura spark. (Courtesy of Ford Motor Company)

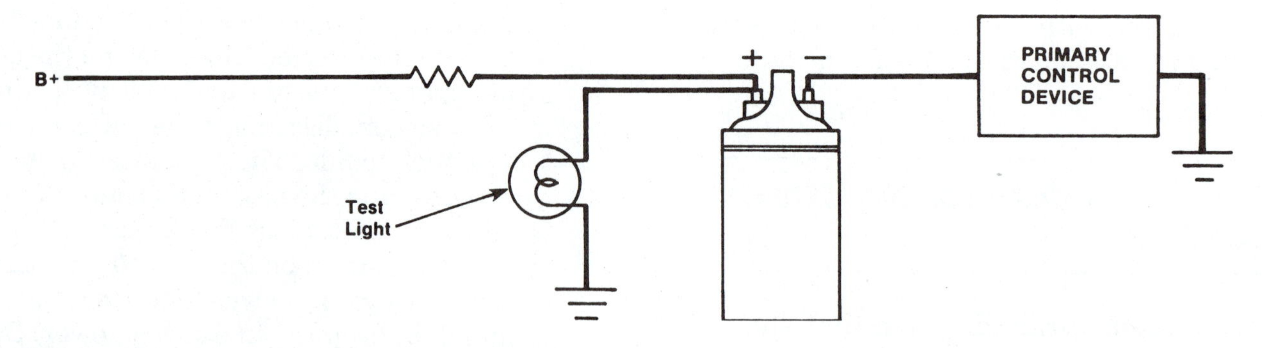

FIGURE 13-65 Dura spark testing for power.

present at the Bat terminal of the coil. If B+ is not present at either test point, the ignition system cannot function. A repair must be completed after the exact spot of the open circuit is located. B+ must be present before you go on.

3. With the key off, use an ohmmeter and measure for the resistance of the primary resistor, Figure 13-66. The easiest method on this example will be to locate two convenient terminals that will be connected directly to the resistor. The Bat terminal of the coil is one and the white/light blue hashmark is the other. Using these terminals will save you the trouble of locating the primary resistor in the wiring harness.

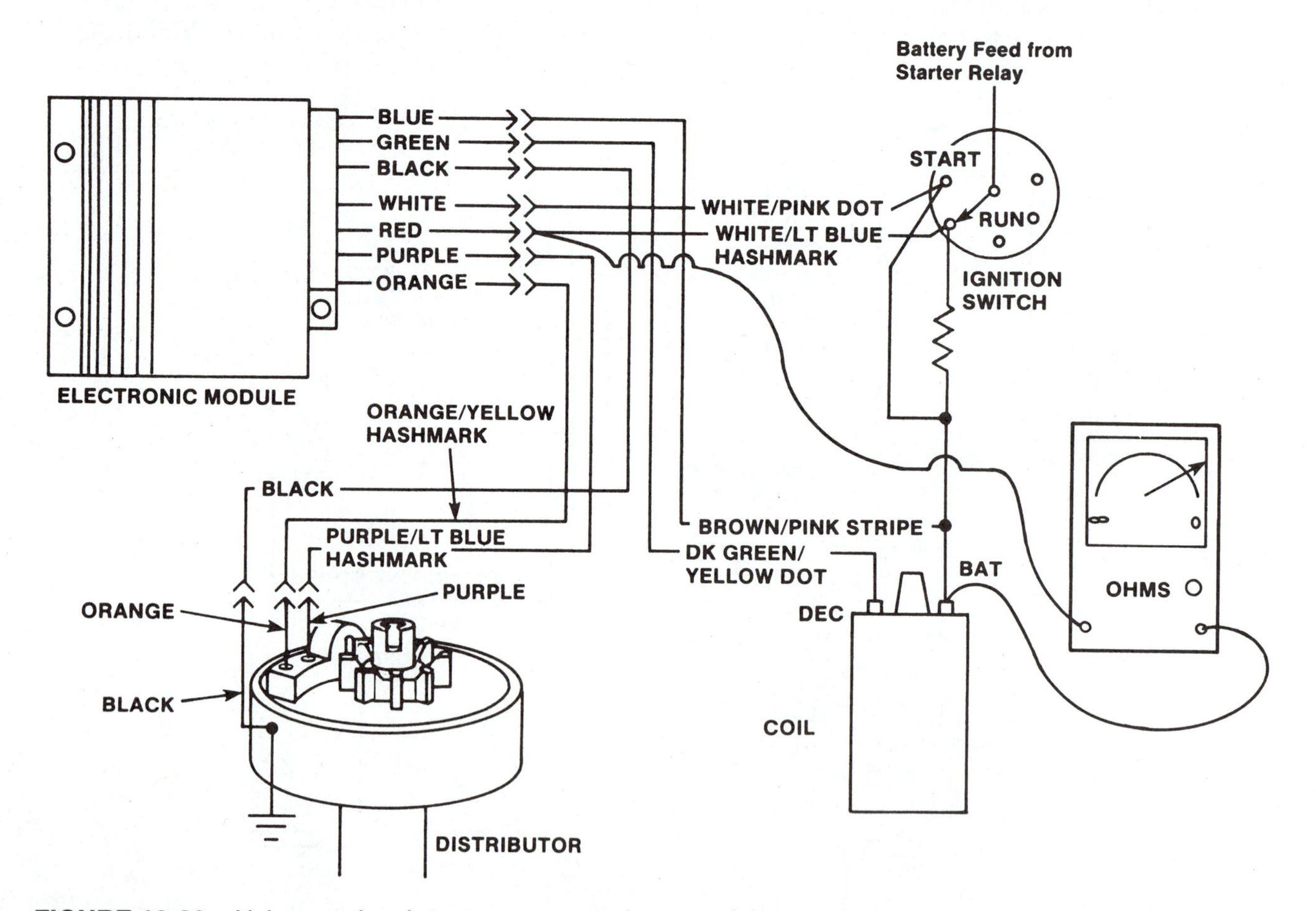

FIGURE 13-66 Using an ohmmeter to measure primary resistance.

FIGURE 13-67 Measuring coil primary resistance.

FIGURE 13-68 Measuring coil secondary resistance.

NOTE: Not all systems use a resistor and require this test.

4. With the key off, use an ohmmeter and measure the primary resistance of the coil, Figure 13-67. The specification listed is 3 to 4 ohms. After the primary resistance is verified as being correct, move one of the leads over to the coil tower, Figure 13-68. Notice that the coil secondary winding is connected to the primary winding inside the coil case. Secondary resistance will be measured with the ohmmeter leads on one of the primary terminals (it does not matter which one) and the other in the coil tower as in Figure 13-69. The specification for this system is 7,000 to 13,000 ohms. In addition, look

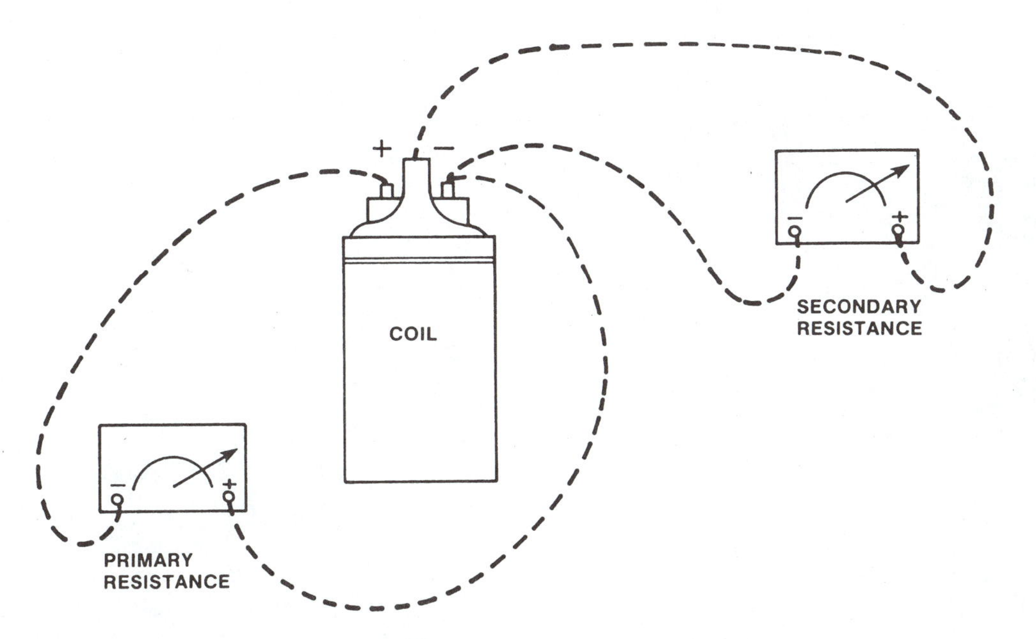

FIGURE 13-69 Ohmmeter connections for primary and secondary resistance.

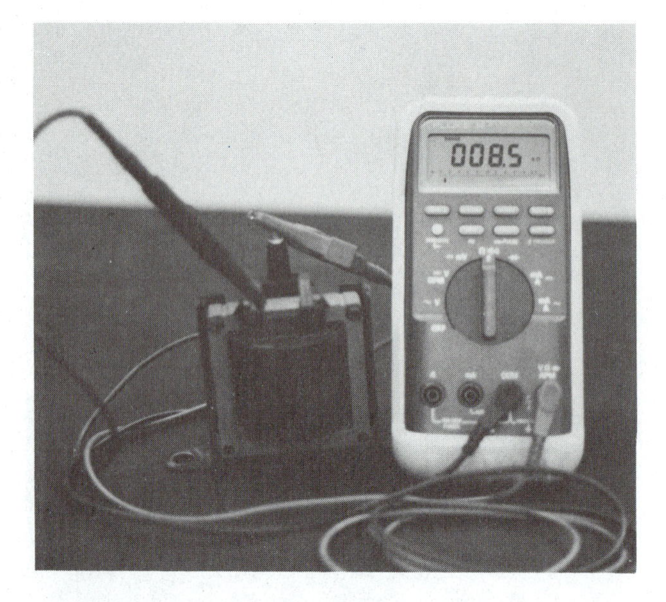

FIGURE 13-70 E-coil.

over the coil closely for cracks in the tower or indications of insulation breakdowns. Many of the newer style coils are of the open E-coil design shown in Figure 13-70. Look over the insulation near the metal of the form. Frequently, breakdowns will occur in this area as Figure 13-71 shows. Coils that have insulation breakdowns will frequently be capable of producing a spark, but usually the level of maximum output will be reduced. This can cause hesitations, especially

FIGURE 13-71 Burned E-coil.

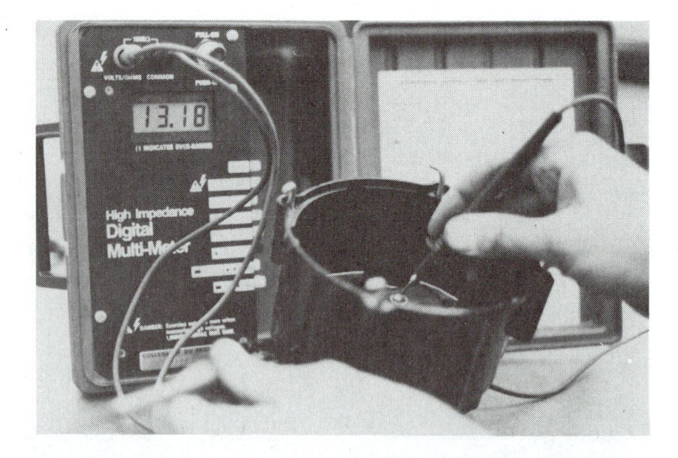

FIGURE 13-72 Coils installed inside distributor caps can be tested without removing them.

during humid weather. If the ignition coil's primary and secondary resistance is within specification and no cracks or insulation breakdowns are present, the coil is capable of producing the required kV of output. Figure 13-72 shows the testing of a coil installed inside the cap.

5. The next component to test is the pick-up coil. This style is the magnetic AC generation style. The purple and orange/yellow wires at the connector carry the signal to the module. Use your ohmmeter and measure the resistance of the pick-up coil. The specification is 400 to 800 ohms. Figure 13-73 shows this procedure. If the resistance is correct, verify that the distributor does turn when the engine cranks. In a later section,

FIGURE 13-73 Ohmmeter checking stator.

we will look at an additional method of testing pick-up coils. Hall effect sensors can generally be tested in the same manner with an ohmmeter.

6. The last check necessary will be to ensure that the module ground is good. Notice that our example has a black wire from the module to the distributor where it is attached to ground. Again, we can use the ohmmeter with one lead connected to engine ground and the other connected to the black wire in the three wire connector. Zero ohms indicates continuity to ground. Any resistance is unacceptable and should be eliminated, Figure 13-74.

With the above six tests, you will be able to determine the validity of the major components with the exception of the module. If all of the tests were good, we would know that the module is not doing its job and that it should be replaced. By systematically testing around the module, we verify that all inputs are functional and that the coil is also functional. Use any manual that will give both a wiring diagram and the necessary specifications. There are additional methods of testing the various components that will be covered in a later section of this chapter.

COMMON SCOPE PATTERNS

What we will try to cover here are some of the more common scope patterns that occur. The first one we will look at is an open circuit, such as a broken spark plug wire or terminal. The open circuit causes the ignition coil to raise up the voltage in an effort to overcome the resistance of the open circuit. This causes the firing line to generally go off the screen, Figure 13-75. Keep in mind that you can use the ground probe to find the exact cause of the high firing line. Remember also that this open circuit might cause the ignition system to fail. The lack of any actual spark will cause most systems to eliminate the current limit function and run full primary current through the ignition coil. This full ignition primary cur-

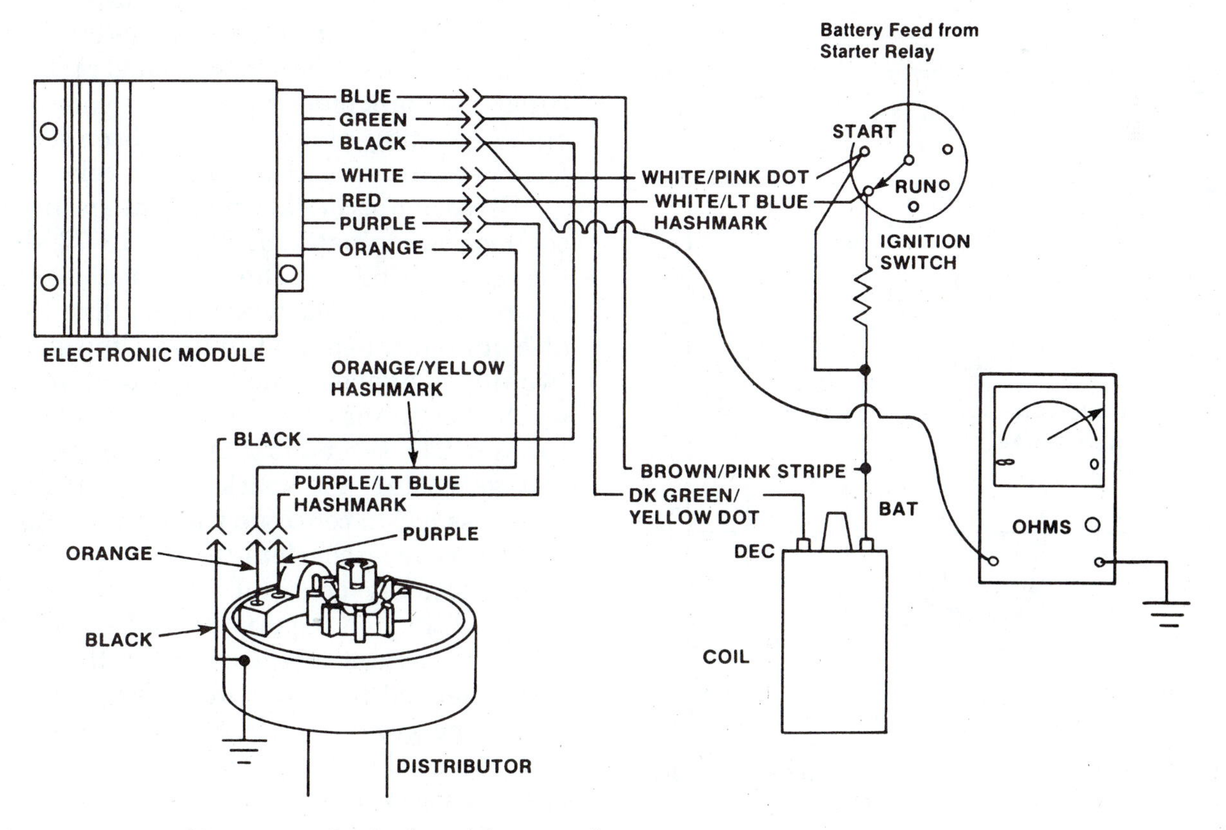

FIGURE 13-74 Ohmmeter check of module ground.

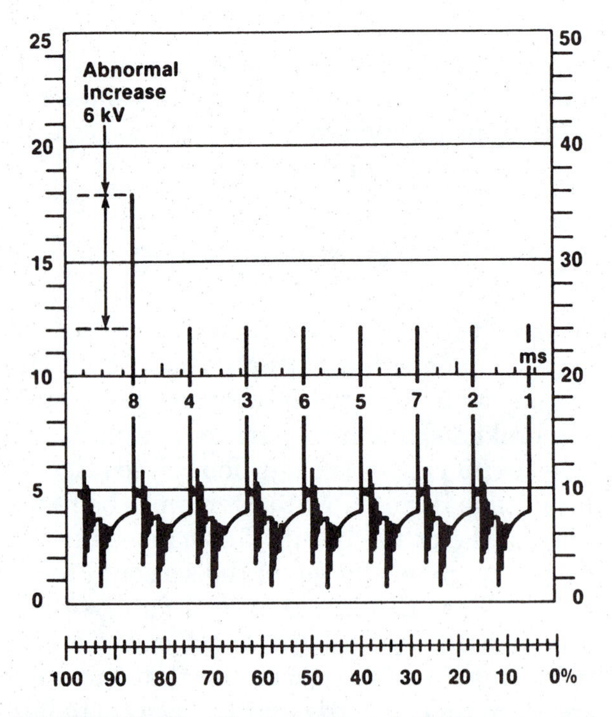

FIGURE 13-75 Scope pattern indicating open circuit. (Courtesy of Sun Electric Corporation)

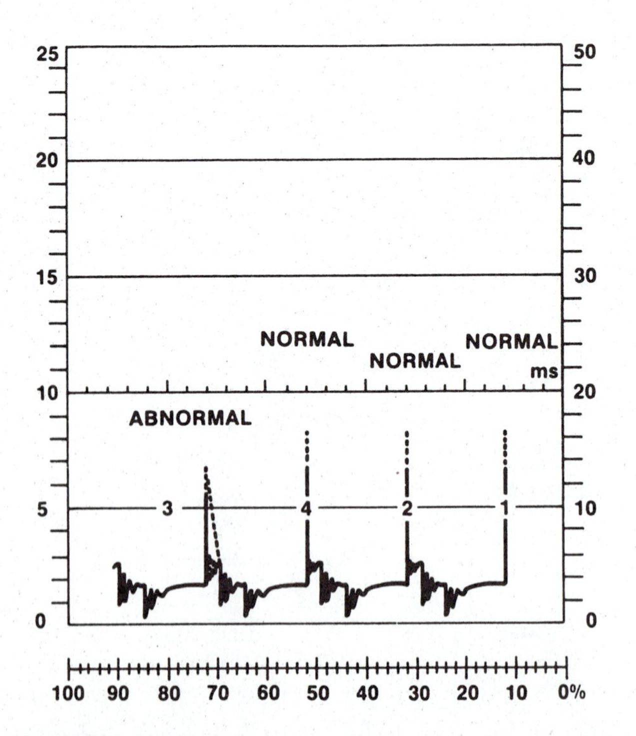

FIGURE 13-76 Scope pattern indicating fouled plug. (Courtesy of Sun Electric Corporation)

rent produces extremely high secondary voltages that might destroy secondary insulation. The coil and rotor are the most frequent locations for this insulation breakdown.

The next reasonably common scope patter is that from a fouled plug. Notice the pattern in Figure 13-76. The firing line is short and the spark line starts up at the top of the firing line. In addition, the spark line is quite long because very little voltage is necessary. This leaves a large amount of coil power to have a very long spark line. The plug gap is clogged with something that conducts electricity. Fuel, carbon, or other deposits have fouled the air gap. With no gap there will be no spark and the cylinder will be dead. The greatly inclined spark line can sometimes be misdiagnosed as excessive secondary resistance. If you take note that the spark line starts at the top of the firing line, the pattern for a fouled plug is easy to pick out. Keep in mind that a fouled plug is usually indicative of some other problem that eventually resulted in the fouling condition. By removing the plug and examining it closely, you should be able to tell what caused the fouling condition. Fuel and oil are the two most common causes. Winter starting with enriched fuel systems can lead to fuel fouling. Once the excessive fuel problem is cured, the spark plugs should be replaced. Generally speaking, plugs that have been severely fouled can never be cleaned effectively.

Excessive secondary resistance is another common scope pattern, Figure 13-77 shows the raised firing line, the shortened, less than the required 0.8 ms sweep spark line that is also quite inclined. As the firing line goes up, two things happen. The first is that you pull up the beginning of the spark line, which inclines it. The second is that you will use more voltage to start the spark and will therefore have less left to continue the spark. This will greatly reduce the length of time that the spark actually continues and reduces the length of the line. Again, the millisecond that the spark occurs might be below the 0.8 ms recommended by many manufacturers. The use of the ground probe as we outlined can again be used to determine the exact location of the resistance.

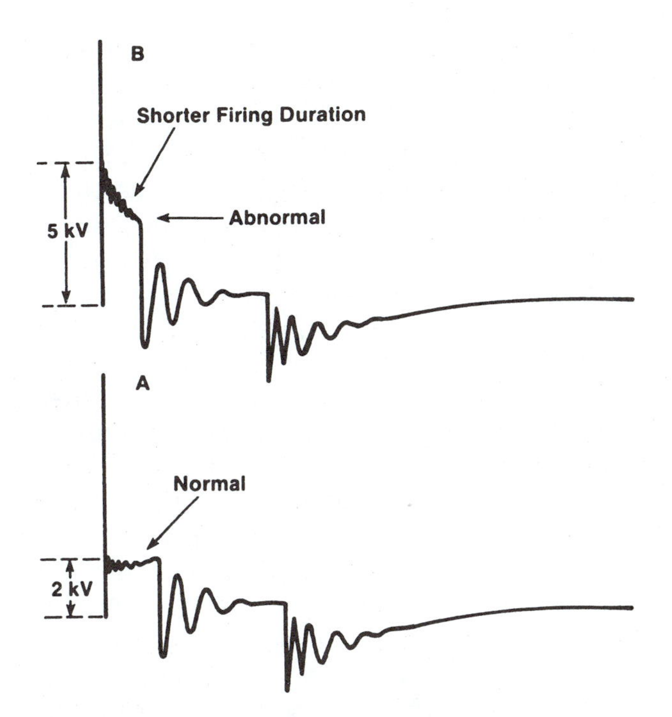

FIGURE 13-77 Scope pattern indicating excessive secondary resistance. (Courtesy of Sun Electric Corporation)

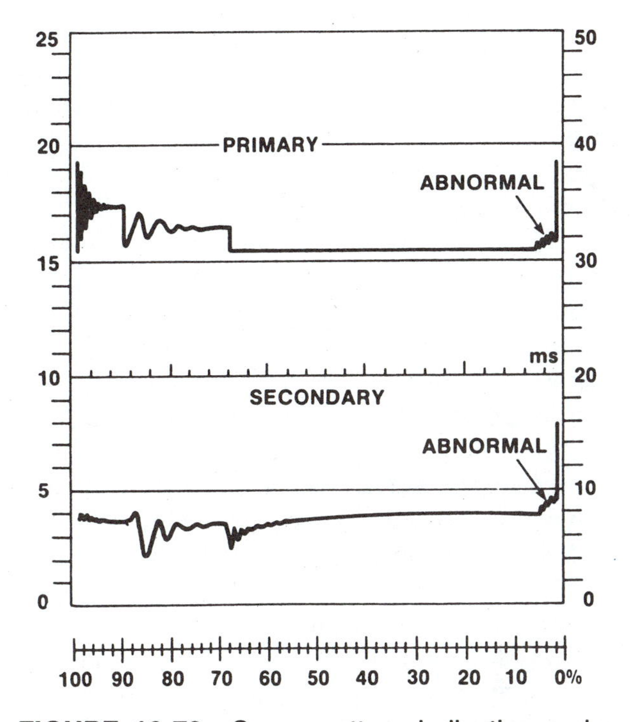

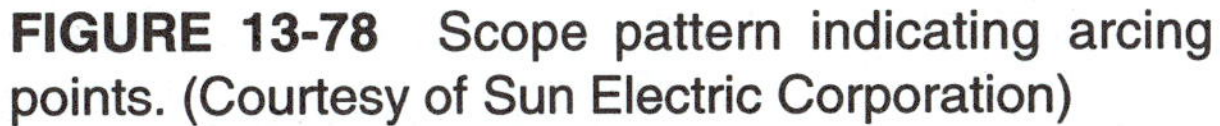

FIGURE 13-78 Scope pattern indicating arcing points. (Courtesy of Sun Electric Corporation)

Contact point problems are becoming a thing of the past as the number of vehicles with breaker point ignitions is reduced. We will, however, still find them on the streets for many more years, and you should be familiar with some of the patterns that indicate problems. Figure 13-78 shows what an arcing set of points looks like. Notice that the point's open signal is not a clear 90° upward turning of the trace as we have seen before. The stair step trace at the end of the dwell is caused from the points arcing. Primary current is not turned off quickly. Instead, it slowly ends with some arcing across the partially opened points. The scope might not tell you what caused the arcing to occur. Further testing will be necessary to tell whether the condenser (capacitor) is faulty; the points are burned or not set correctly; or the distributor is at fault. Looking at other sections of the pattern might give additional information about the cause of the arcing.

A shorted ignition coil or condenser (on a point type system) will generally look like the pattern shown in Figure 13-79. Notice the al-

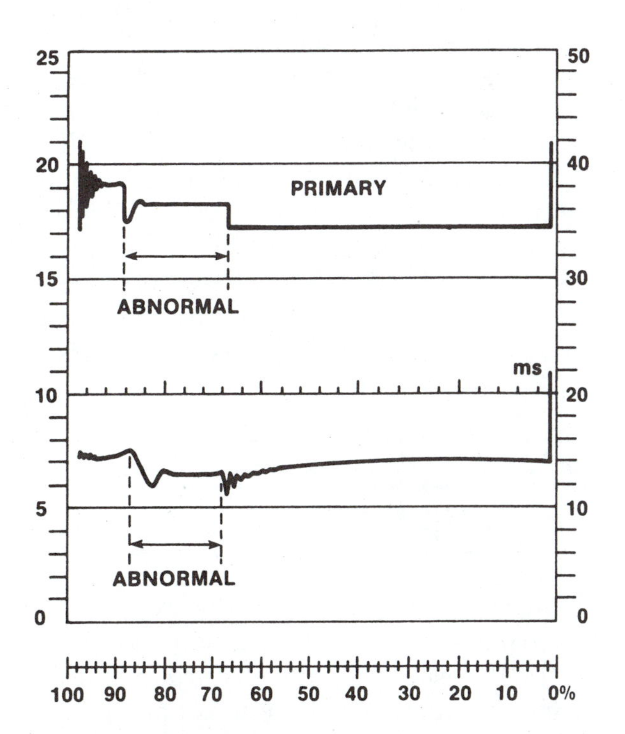

FIGURE 13-79 Scope pattern indicating shorted coil windings. (Courtesy of Sun Electric Corporation)

most total lack of oscillations after the spark line. In addition, you will probably find that this will be accompanied by a reduced maximum coil output of less than 20 kV on a point type system or less than 35 kV on most electronic systems.

These are by far the most common scope pattern defects that you will run into. Analyzing the defect first from the standpoint of what section it occurs in and then noting whether it is common to all or just one cylinder is the best method for curing ignition problems. As you use the scope, remember to do an entire series of tests before coming to any conclusions. Jumping to a snap decision before all the testing is accomplished is all too common and leads to misdiagnosis. With the high cost of some of the replacement components today, to diagnose and fix it right the first time is more important than ever.

ADDITIONAL COMPONENT TESTING

When dealing with electronic ignition, testing of the individual components is by far the easiest and most popular method of diagnosis. We have covered extensively the testing of the timing device with an ohmmeter. You should realize that the majority of the pick-up coils can be tested with an ohmmeter. We will now look at additional methods of testing these devices, starting with the oscilloscope. The scope can show voltage changes quite easily. Both the pick-up coil and the Hall effect sensor will change the voltage if they are functioning correctly. Let us look at the pick-up coil first.

IX. Procedure for Testing a Pick-Up Coil

Using the primary leads of the scope, connect to the pick-up coil leads. The scope needs to be on the lowest scale, typically around 20 volts. When the distributor shaft is spun, the coil will have an AC voltage generated in it, which will show on the screen. Figure 13-80 shows a true AC sine wave. The wave off the typical pick-up unit will not be exactly a sine wave, but it will consist of a positive and a negative pulse and be apparent even at cranking speeds. Another method of measuring this AC signal is with a simple AC voltmeter. Again the pick-up unit is connected to the voltmeter leads and the meter is placed on a very low scale in the 20-volt range. The meter will register the AC voltage during cranking. Measure this voltage as close to the ignition module as is possible, because it is the module that must see this voltage if it is to turn off primary current and fire the plugs.

To accurately test a Hall effect sensor, we can again look at the voltage that is changing from a high level to a lower level as the magnetic field is blocked and then allowed to pass into the sensor. This is not a sine wave signal, because the voltage never changes polarity, but is pulled down lower than the source. A square wave like that shown in Figure 13-81 will be produced by a properly functioning Hall effect sensor. Remember that an input voltage, usually the same as the electrical system, must be present at the sensor before it can produce the square wave that will give piston position to the computer. To independently

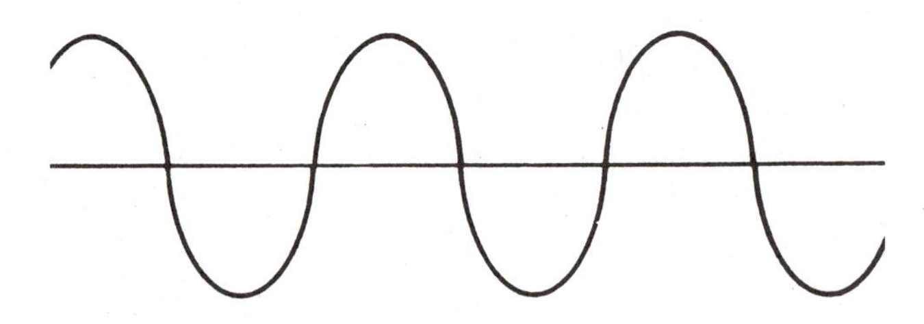

FIGURE 13-80 AC sine wave.

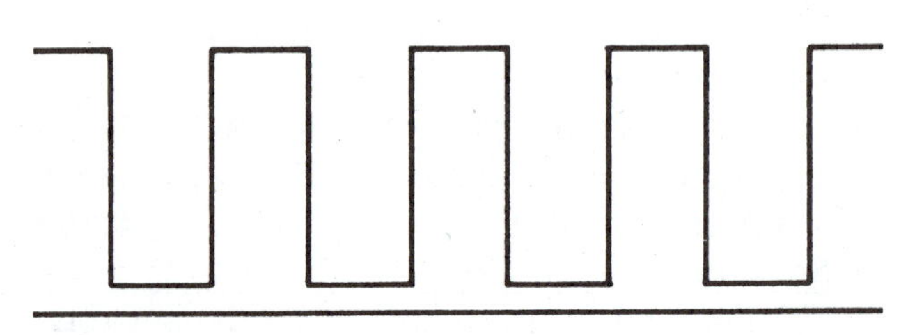

FIGURE 13-81 Square wave.

test the sensor, you will have to supply input voltage and ground and then turn the distributor of the engine (for crankshaft sensors). The output can be measured on a sensitive scope or a voltmeter. When using voltmeter, an analog meter is preferred because the swinging needle is easier to observe than the changing readout of the typical digital meter. Figure 13-82 shows the set up for testing Hall effect sensors.

Electronic control of timing requires some additional information if you are to understand it well enough to diagnose problems. Refer to Figure 13-83. It is a common circuit that General Motors uses in its electronic spark timing (EST). It requires a special module with additional connections. Notice the first extra connection, a ground that puts the module and the computer at the same ground potential. The other additional connections

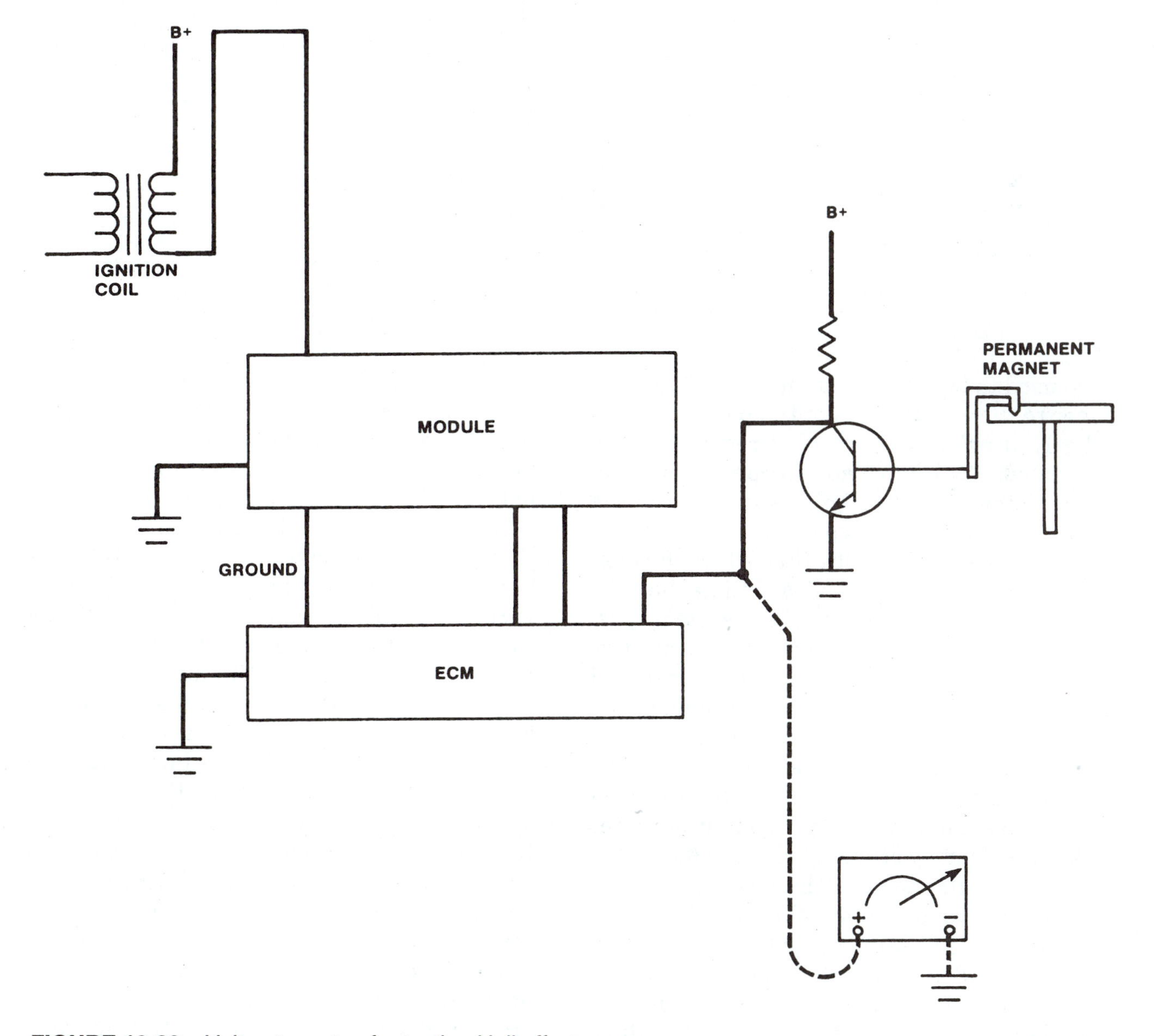

FIGURE 13-82 Voltmeter setup for testing Hall effect sensors.

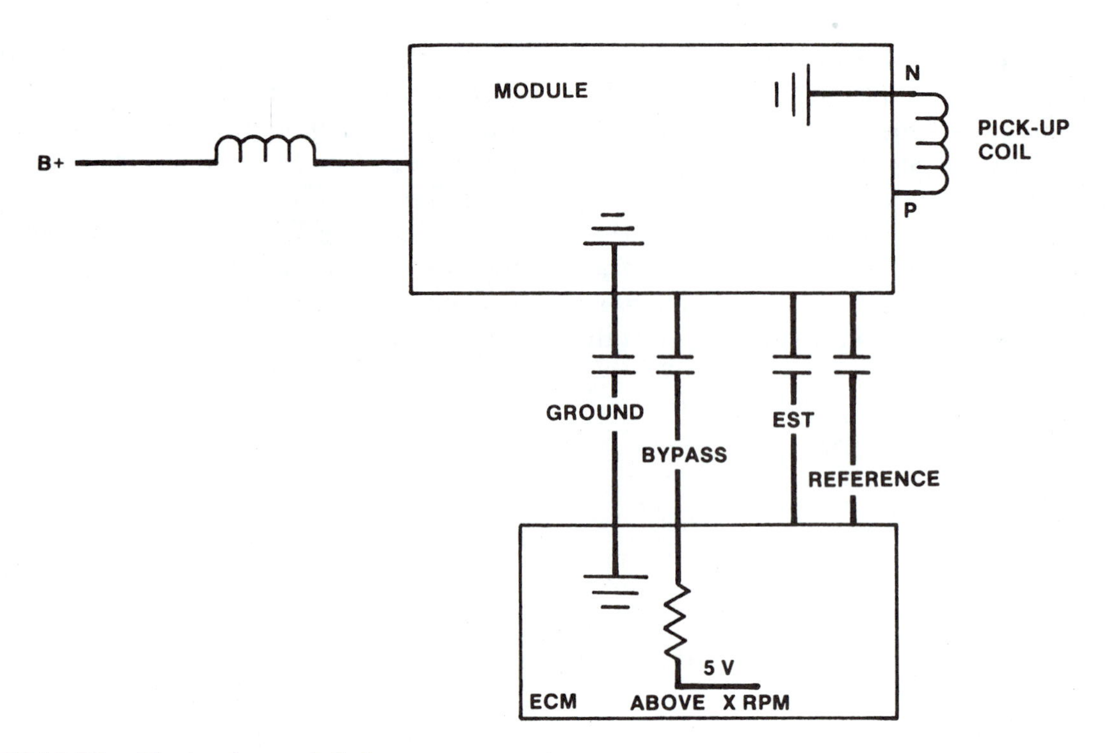

FIGURE 13-83 Electronic spark timing.

are used to bring a speed and position signal over to the computer from the pick-up coil or the Hall effect sensor called reference. The distributor will have no mechanical advancing systems so the signal will remain the same number of degrees before top dead center at all speeds. Usually, this will be the base timing of the vehicle and a timing that the system will revert back to if the system fails. The bypass is a signal from the computer back to the ignition module, which will tell the module that ignition timing should be determined by the computer rather than using reference. The EST is the electronic spark timing signal. It is the advance that the computer has calculated. The system operates like this. During cranking there is no bypass voltage to the module. The module uses the reference signal coming off of the Hall effect sensor or the pick-up coil to turn off primary current and fire the plugs at base timing. Once the vehicle is running, the computer sends a 5-volt signal to the module, which switches the timing signal from the reference over to the EST coming from the computer. A lack of timing advance on a vehicle with electronic spark timing could be caused from a lack of EST signal, a bypass signal that does not come from the computer, or a module that will not switch over from one to the other. Generally, the diagnosis charts for EST will have you sample the bypass once the vehicle is running. In this way, you are seeing if the computer is telling the module to switch over to computer-controlled timing. In addition, many vehicles will have you disconnect the bypass wire to time the engine. Just because the computer is in control of timing with the vehicle running does not mean that the initial timing is not set. Initial or base timing is set in the same manner as with mechanical/vacuum advance. The only difference is that the bypass line is disconnected. Not all manufacturers refer to this signal as bypass. Ford Motor Corporation refers to it as **spout**, which stands for spark out. The spout is disconnected when initial timing is checked or set.

DISTRIBUTORLESS IGNITIONS

Most manufacturer's ignition systems are currently distributorless systems. The system use multiple coils, one for every two cylinders. The two sides of the secondary are attached to two different spark plugs. Both plugs fire at the same time. One on the compression stroke, while the other is on the exhaust stroke. The system still must receive information about piston position and this is generally done through a crankshaft position sensor. This sensor supplies the equal to reference. In addition, a cam sensor, which can be a pick-up coil or a Hall effect sensor, sends speed information to the computer, Figure 13-84. This information is used by the computer to determine engine timing for each pair of cylinders. Normal test procedures involving a scope, ohmmeter, and timing light can be used on this style of ignition, keeping in mind that problems involving one of the cylinders might also show up on the companion cylinder firing off the same coil. In addition, most scopes will need their pick-ups placed on each pair of cylinders to view all patterns. Many scope manufacturers now make a single adapter that will allow the viewing of all six patterns at the same time.

Most modern systems today utilize knock sensors to retard the timing under ping conditions. The procedure for diagnosing a bad one will vary from one manufacturer to another and you should reference the correct literature before attempting to test them. Many of the more popular ones are piezoelectric devices, which generate a signal at the specific frequency that has been identified as ping. Rapping lightly on the manifold near the sensor can be the method of testing it. A timing check while rapping should show the timing retarding.

R AND R PICK-UP UNITS

The following procedure, Figures 13-85 to 13-104, will allow you to remove and replace many of the pick-up coils currently in use.

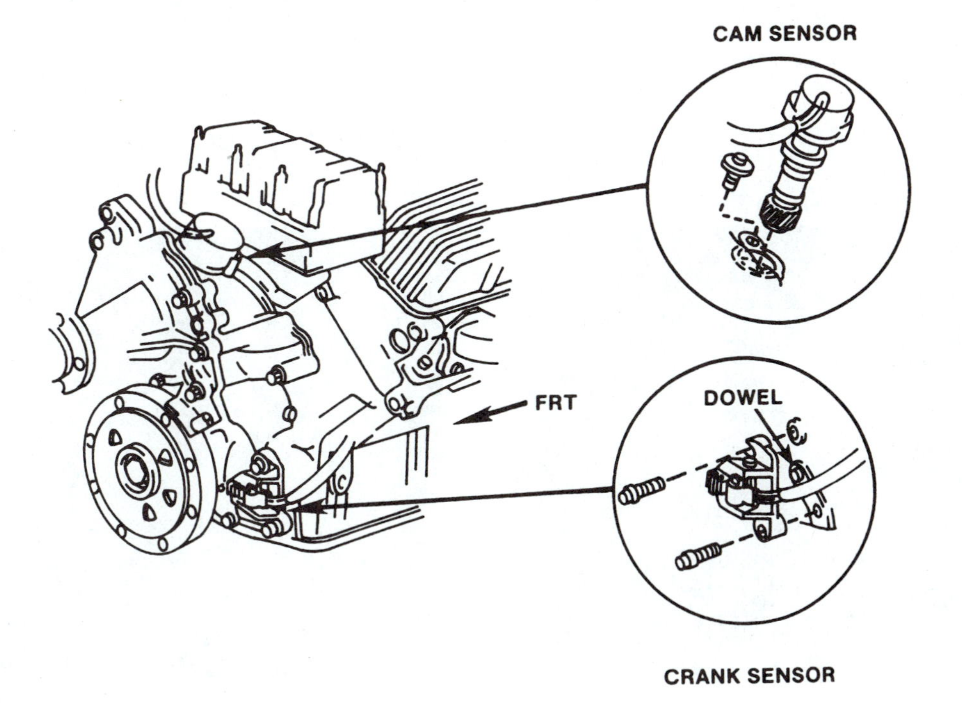

FIGURE 13-84 Distributorless ignition.

NOTE: It is necessary to remove the distributor from the vehicle if the coil is surrounding the distributor shaft.

XX. Procedure for Removing and Replacing Pick-Up Coil

1. Release hold-down clamps, Figure 13-85.
2. Remove distributor cap, Figure 13-86.
3. Drive pin out of gear, Figure 13-87.
4. Remove the distributor shaft, Figure 13-88.
5. Remove the pick-up coil hold-down screws, Figure 13-89.
6. Remove the pick-up coil, Figure 13-90.
7. Install a new pick-up coil, Figure 13-91.
8. Replace the distributor shaft and drive gear, Figure 13-92.
9. Drive the pin in place, Figure 13-93.
10. Wind the spring around the pin, Figure 13-94.

FIGURE 13-86 Remove the distributor cap and rotor.

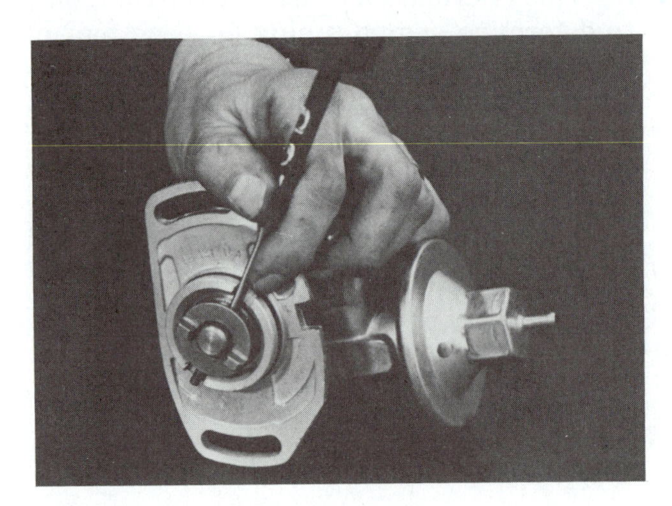

FIGURE 13-87 Drive pin out of gear.

FIGURE 13-85 Release hold-down clamps.

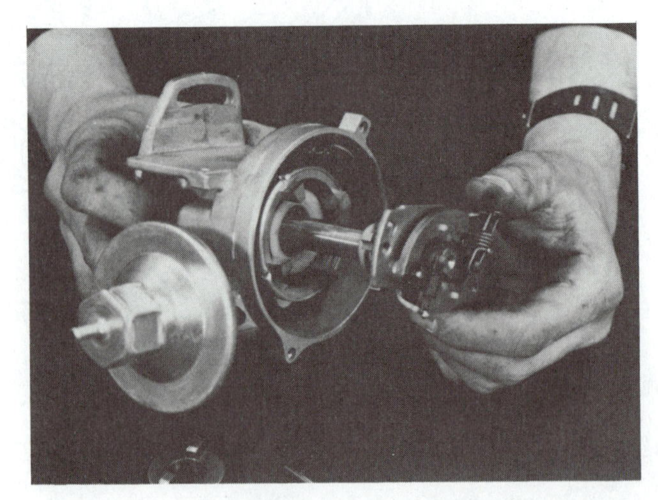

FIGURE 13-88 Remove the distributor shaft.

FIGURE 13-89 Remove the pick-up coil hold-down screws.

FIGURE 13-90 Remove the pick-up coil.

FIGURE 13-91 Install a new pick-up coil.

FIGURE 13-92 Replace the distributor shaft and drive gear.

FIGURE 13-93 Drive the pin in place.

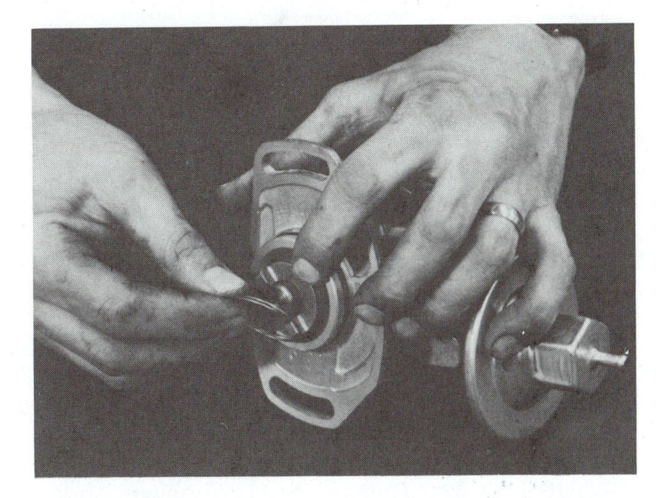

FIGURE 13-94 Wind the spring around the pin.

XXI. Procedure for Removing and Replacing GM Pick-Up Coil

1. Remove the three-wire connector, Figure 13-95.
2. Release the fold-downs and remove the cap, Figure 13-96.
3. Remove the distributor drive pin and gear, Figure 13-97.
4. Remove the hold-down screws, Figure 13-98.
5. Release the spring clip and remove the coil, Figure 13-99.

FIGURE 13-97 Remove the distributor drive pin and gear.

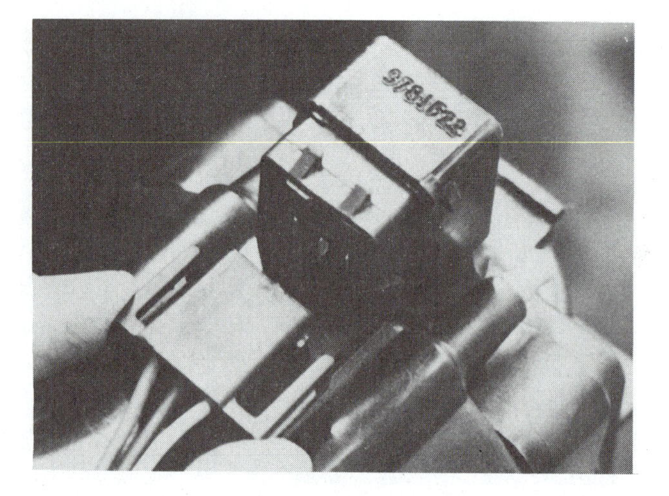

FIGURE 13-95 Remove the three-wire connector.

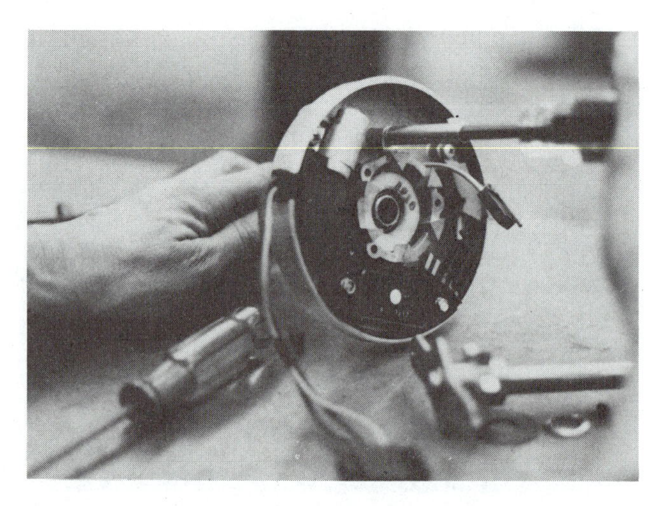

FIGURE 13-98 Remove the hold-down screws.

FIGURE 13-96 Release the hold-downs and remove the cap.

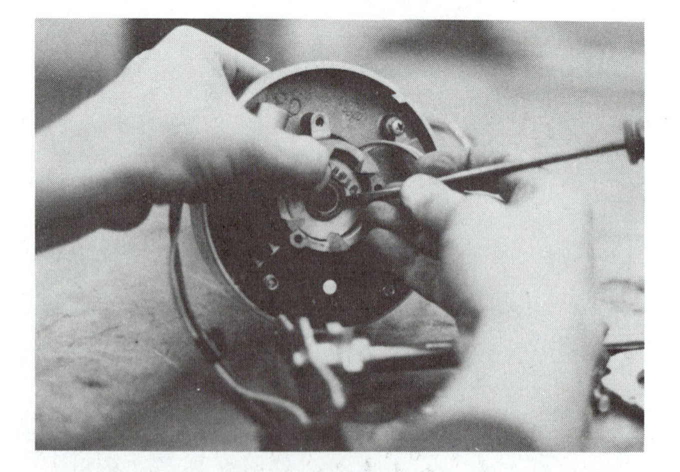

FIGURE 13-99 Release the spring clip and remove the coil.

XXII. Procedure for Removing and Replacing Distributor Assemblies

1. Chalk distributor vacuum advance or wires for position of base in engine, Figure 13-100.
2. Chalk rotor position, Figure 13-101.
3. Remove hold-down bolt, Figure 13-102.
4. Pull distributor out, Figure 13-103.
5. Time engine after reversing removal for installation, Figure 13-104.

FIGURE 13-100 Chalk distributor vacuum advance or wires for position of base in engine.

FIGURE 13-101 Chalk rotor position.

FIGURE 13-102 Remove hold-down bolt.

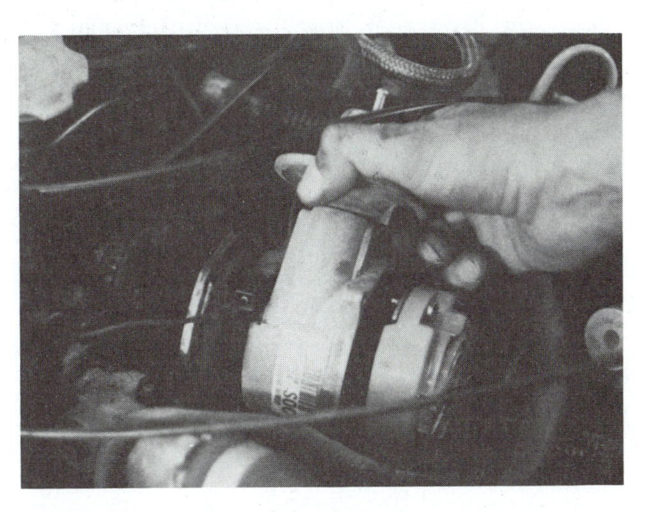

FIGURE 13-103 Pull distributor out.

FIGURE 13-104 Time engine after reversing removal for installation.

This procedure will work on many of the current systems in use today. It is important to time the engine when you are finished.

SUMMARY

In this chapter, we have looked at the most common ignition service, starting with the tools of the trade, including the timing light, tach dwellmeter, feeler gauge, digital timing indicator, and oscilloscope. Extensive time was spent on each, so that you would be able to use them efficiently. In addition, we have looked at a common scope check out procedure, including causes of high firing lines. We have examined heat-caused failures, plus moisture and cold, and have seen how the technician can induce them for testing purposes. A no-start resulting from a no-spark test procedure was examined, which utilized simple diagnostic tools. Routine R and R functions on pick-up coils were also demonstrated.

KEY TERMS

Timing the engine
Manifold vacuum
(SPOUT) spark out
Inches of mercury (In.hg)
Display
Parade
Raster
Stacked
Superimposed
Firing line
Spark line
Intermediate zone
Coil polarity

CHAPTER CHECK-UP

Multiple Choice

1. Brass feeler gauges are generally used to set
   a. spark plug gap.
   b. Hall effect sensor gap.
   c. magnetic pick-up air gap.
   d. none of the above.
2. Tach dwellmeters are generally attached to the
   a. positive of the coil and ground.
   b. positive and the negative of the coil.
   c. negative of the coil and ground.
   d. battery positive and coil positive.
3. Initial timing on a non computerized vehicle is generally set at a specified speed with the
   a. transmission in reverse.
   b. vacuum advance connected.
   c. engine at 1000 rpm.
   d. vacuum advance disconnected and plugged.
4. Mechanical advance is checked with a(n):
   a. advance timing light and a tach.
   b. oscilloscope.
   c. inductive timing light and a dwellmeter.
   d. feeler gauge and a spring tension tester.

5. A firing line that is excessively high on only one cylinder can be indicative of a(n)
   a. shorted rotor.
   b. open coil wire.
   c. lean carburetor.
   d. open plug wire.
6. A four-function ignition system is being tested on the scope. Technician A says that variable dwell will be observed between the primary current on and off signals. Technician B says that it can be observed on a conventional dwellmeter. Who is correct?
   a. Technician A only
   b. Technician B only
   c. Both Technician A and B
   d. Neither Technician A nor B
7. A ground probe placed in a distributor cap terminal for a spark plug reduces the firing voltage down to 2500 volts. Technician A states that the cap and rotor have excessive resistance. Technician B states that the cap and rotor are within specs. Who is correct?
   a. Technician A only
   b. Technician B only
   c. Both Technician A and B
   d. Neither Technician A nor B
8. The current limit blip will appear
   a. once the coil is fully saturated.
   b. before the coil is fully saturated.
   c. after the cylinder plug fires.
   d. just after the primary current on signal.
9. A primary ignition winding is being tested. Its resistance is three times the specification. This will cause
   a. longer spark lines.
   b. longer dwell.
   c. reduced secondary output.
   d. nothing.
10. Less than 0.5 ms of spark is observed. Technician A says that this will increase the changes of misfire. Technician B says that this might be caused by the firing voltages being excessive. Who is correct?
    a. Technician A only
    b. Technician B only
    c. Both Technician A and B
    d. Neither Technician A nor B

True or False

11. Hall effect voltages will increase with speed.
12. Magnetic pick-up voltages will increase with speed.

13. Initial timing must be correct on computer-controlled vehicles or the entire curve will be off.
14. To initially time a computer-controlled vehicle, the engine must be at a specific speed and the vacuum advance must be disconnected.
15. Excessive secondary resistance will increase the slope of the spark line.
16. Current limit blips are only found on systems without primary resistors.
17. Excessive coil primary resistance will decrease the available voltage.
18. The point dwell must be set before the timing is set.
19. Reversed coil polarity will show up as firing lines going up on the scope.
20. Insulation breakdown will reduce the available voltage.

CHAPTER 

Distributorless Ignition Theory

OBJECTIVES

At the completion of this chapter, you should be able to:

- explain the major differences between distributed spark and distributorless spark.
- identify companion cylinders on common engines.
- explain the waste spark theory.
- figure total voltage for one coil.
- explain the function of and how the signal is generated in a magnetic sensor.
- explain the function of and how the signal is generated in a Hall effect sensor.
- trace the operating principles of DIS using both a crank and a cam sensor.
- trace the operating principles of DIS using a combination sensor.
- trace the operating principles of a dual-plug-per-cylinder DIS.

INTRODUCTION

Up to this point we have looked at a steadily evolving ignition system. Starting with breaker points, moving into electronic ignition, E-coils, and computer control of timing advance curves, we have seen electronics taking over mechanical devices. In addition, we have seen increased service intervals, from 10,000 miles for the old fashioned tune-up to 30,000 miles or more between spark plug changes. Obviously, the majority of the consumers are pleased with increased service intervals, even if they are not aware of the evolution of the ignition system. However, even with the many changes studied so far, the major automotive manufacturers are still faced with minor problems centered around the distributor. Being a mechanical device it is subject to wear, and can be responsible for unwanted timing change. In addition, consumers or uninformed technicians could easily set timing different from the specifications and increase tailpipe emissions. Longer emission warranties and more concern for "total" control of timing made **distributorless ignition systems (DIS)** a natural evolution for the ignition engineers. In addition, DIS allows for greatly increased available dwell. A six cylinder vehicle with DIS has 360° of available dwell (measured at the crankshaft) while the same vehicle with a distributor would have a maximum of 120°. One less mechanical part to wear out, one less place where high ignition secondary voltage might find a path to ground, and one less component to have an effect on emissions.

In this chapter we will look at some of the common DISs on the road. You will see that their introduction does not mean that you will have to start over and learn volumes of new information. Actually, you will find that the technology used is basically the same as in computer controlled ignition systems which have been around since the early 1980s. The manufacturers took existing technology, gave it a new twist and adapted it to the DIS of today. We will look at the most popular systems on the road and divide them up by manufacturer and by the number of sensors used. This approach should greatly simplify your learning the systems. As usual, do not go on in your reading until the section just finished makes sense. The systems are not complicated if taken in small, bite size pieces, which is exactly how we are going to approach them.

WASTE SPARK THEORY

In the ignition systems that we have studied and hopefully repaired in past chapters, one end of the ignition coil's secondary winding was connected to either ground or the primary winding. The other end was connected to the spark plug through a cap, rotor, and wires combination. Figures 14-1 and 14-2 illustrate these different methods of wiring an ignition coil's secondary winding. The early years of ignition systems saw the majority of manufacturers using secondary connected to primary wiring as in 14-1. In the mid 1970s, some of the manufacturers began using a

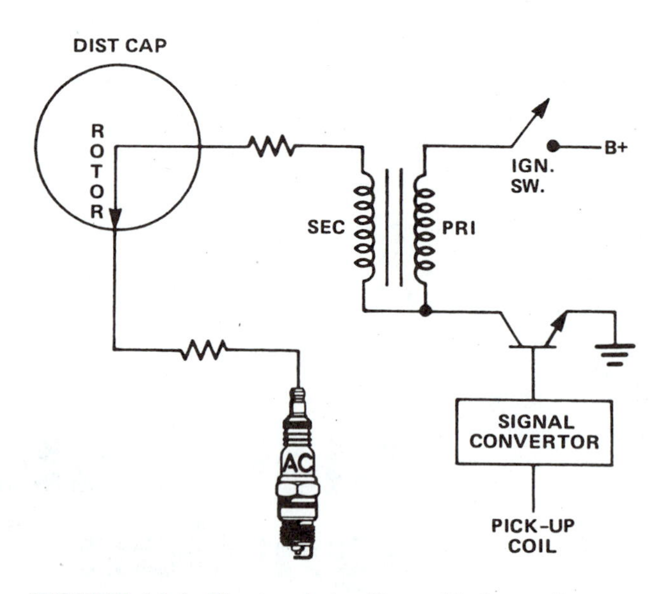

FIGURE 14-1 Electronic ignition with the coil's secondary winding connected to the primary.

grounded secondary as in 14-2. Both of these methods result in the same high voltage spark being delivered to the plug; however, grounding the secondary appeared to have less radio frequency interference (RFI) on some applications.

Look at Figure 14-3. It shows a typical DIS where both ends of the secondary winding are attached to spark plugs. The two plugs are in companion cylinders (cylinders whose pistons are in the same position, moving in the same

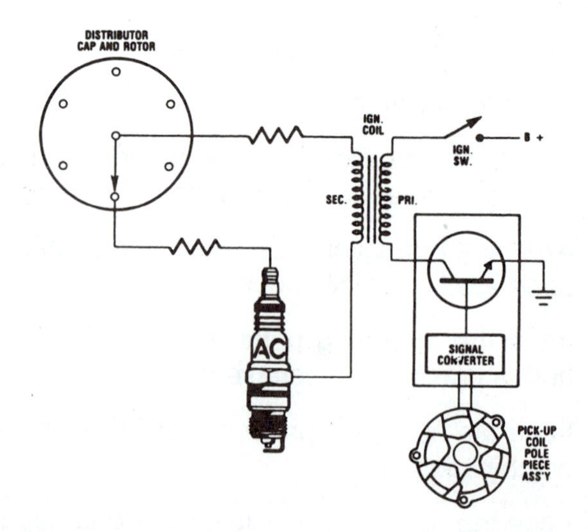

FIGURE 14-2 Conventional electronic ignition system. (Courtesy of GM Corporation)

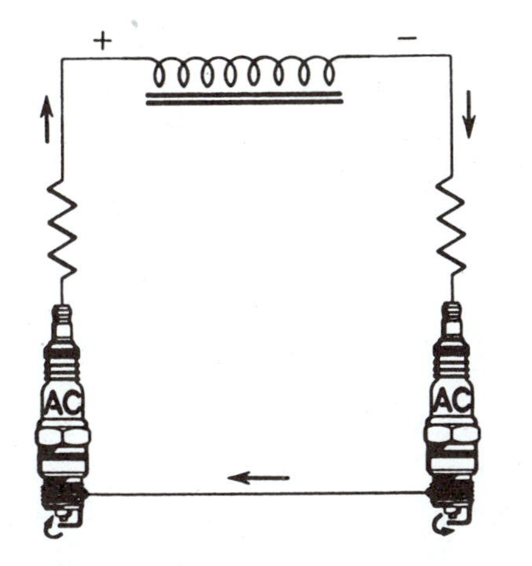

FIGURE 14-3 DIS current flow. (Courtesy of GM Corporation)

direction, but on different strokes). Cylinders on intake and power or compression and exhaust are examples of companion cylinders. Notice that the primary winding is not connected electrically to the secondary as in early ignition systems. This is a series circuit where both plugs will fire at the same time when the coil has reached sufficient voltage to overcome all the resistance of the circuit including the two plugs. In this case, the cylinder head is the wire or conductor between the two plugs. In this way, we are firing both plugs at the same time in companion cylinders. The high compression and combustible mixture in the cylinder which is on the compression stroke will require a high voltage. We are able to see this on an ignition scope when we accelerate. The added pressure of additional mixture in the cylinder increases the firing voltage. We called this test "plugs under load", and hopefully, you have done a few on some live vehicles.

The opposite is also true. If less pressure is placed on the plug, it will require less voltage to fire because its effective resistance has been reduced. This means that the plug firing on exhaust will fire at a lower voltage. The scope pattern reproduced here from a Bear ACE (Figure 14-4) shows that cylinders on compression (igniting plug) have greatly increased firing lines while those firing on exhaust (**wasted spark**) have greatly reduced voltages. Putting the pattern into chart form shows that each plug will fire at the voltage required of its cylinder. Each plug will fire high while on compression and low while on exhaust, Figure 14-5.

Hopefully, you have figured out that one plug will have current flow from the ground electrode to the center electrode while the other plug will have current flow from the center electrode to the ground electrode. In ignition systems of the past, coil polarity was extreme-

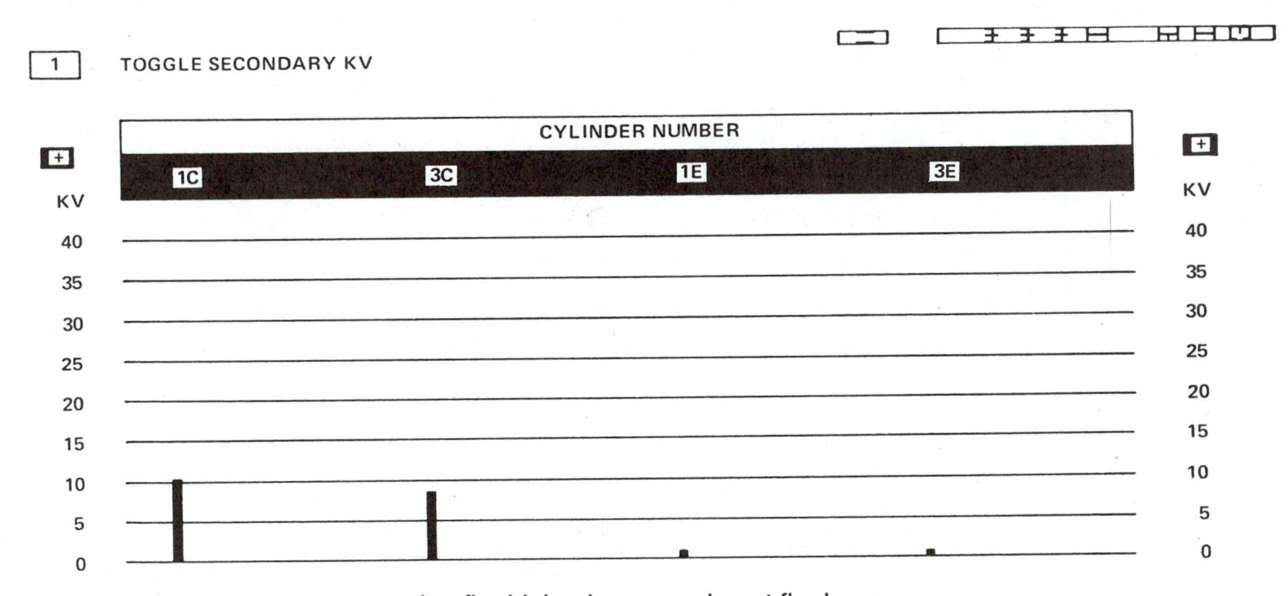

FIGURE 14-4 Plugs on compression fire high; plugs on exhaust fire low.

CYL #	COMPRESSION VOLTAGE	EXHAUST VOLTAGE
1	10,000 V	600 V
2	9,500 V	500 V
3	9,000 V	400 V
4	10,200	550 V

FIGURE 14-5 Compression voltage high; exhaust voltage low.

ly important. Plugs fired with opposite polarity required greater voltage, which was not available. For years technicians checked to make sure that the firing line went straight up when using a scope. Firing lines going down indicated reverse coil polarity and a problem. With only 20kV available, misfires occurred with reverse coil polarity. However, with DIS, available voltages are in excess of 50kV, which is enough coil power to fire just about anything, including a plug with reverse polarity.

With this style of DIS, a single coil is required for each two cylinders. Two coils for a 4-cylinder engine, three coils for a 6-cylinder, etc. There are exceptions to this however. We will look at the exceptions of dual plugs per cylinder, or one coil per plug later on in this chapter and in the DIS service chapter.

Keep in mind that the function and control of the primary will be exactly the same as it was in past ignition systems. Rapidly build and collapse a strong magnetic field across the secondary winding to produce the high voltage necessary to fire the plug, or in this case both plugs. In addition, realize that the total voltage required of the coil secondary will be the total of both spark plugs. If the igniting plug requires 12kV and the waste plug requires 2kV then the ignition coil secondary will produce 12kV + 2kV = 14kV. Most scopes on the market allow you to read the individual voltages at each plug. You must, however, add them together if you wish to know the total voltage produced by the coil.

DIS PRIMARY CIRCUIT

Let us look at the typical primary circuit for DIS. Figure 14-6 shows a simplified diagram of a typical GM DIS circuit. Notice the ignition coils drawn on the right side. Two coil secondaries connected to four spark plugs. This must be for a four cylinder engine, right? Follow along in the primary circuit starting at the battery and switch. Full B+ is applied to the ignition module. No resistance in series, full battery voltage available. This B+ is also applied to the primaries of both coils wired in parallel. Notice that the primary coils are not electrically connected to the secondaries.

The primary circuit continues out the primary windings and into two switching transistors to ground. Just like other electronic ignition systems, it is these two transistors that will turn on primary current and allow time for full coil saturation (dwell) and then turn off to fire the plugs. This is no different from electronic systems that we have looked at in previous chapters. Four-function electronic ignition used variable dwell and current limiting to control primary current flow with two goals or objectives. First, to allow for the quick full saturation of the coil primary giving the greatest spark potential and, second, limiting the current after saturation to a safe level.

A "safe level" is one which the system can handle without damaging itself. DIS coils typically have less than 1 ohm of resistance in the primary. Put 14 volts across this 1 ohm and you have 14 amps minimum current! This could stress the system way beyond its capabilities. It is the object of the internal control circuit within the module to limit the current flow through the switching transis-

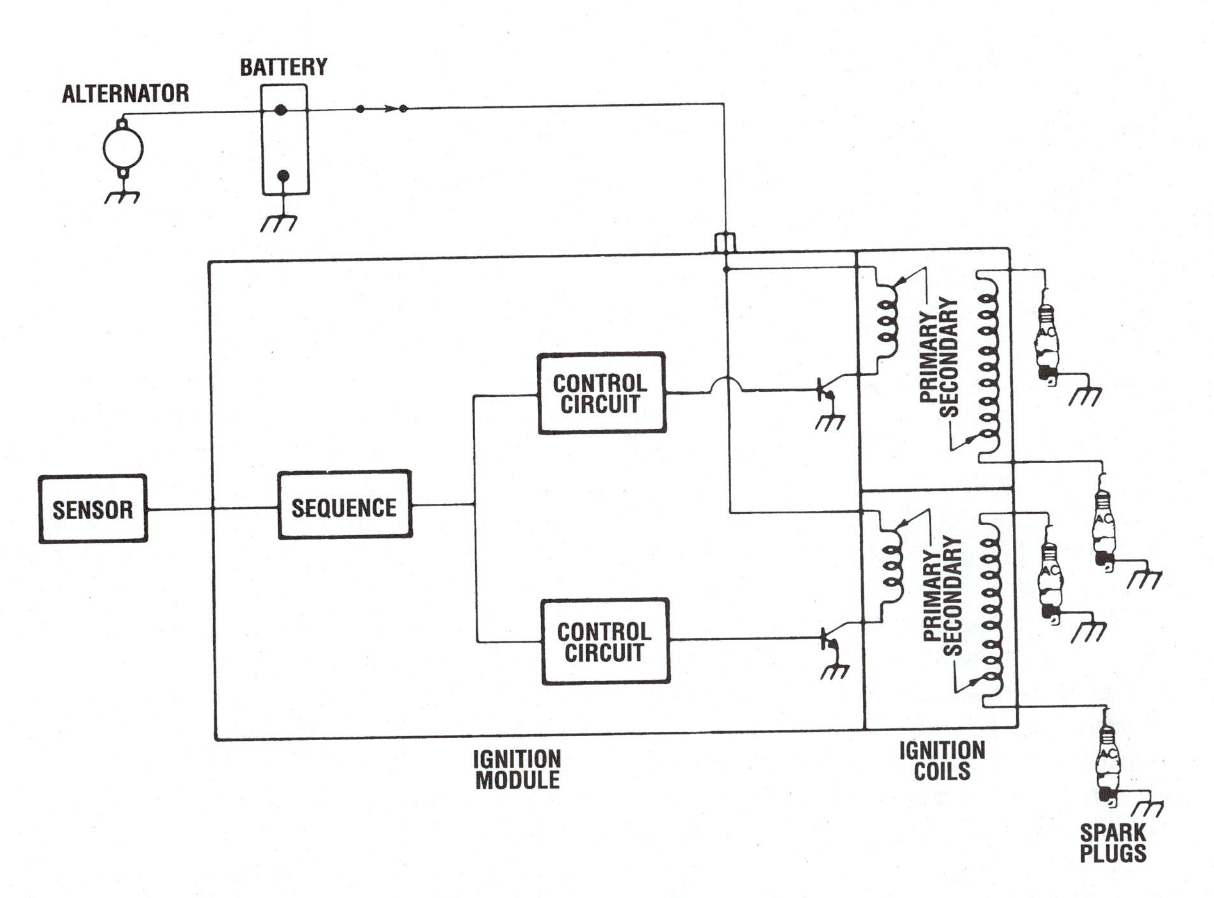

FIGURE 14-6 Direct ignition system. (Courtesy of GM Corporation)

tors to a level low enough to prevent damage to system parts. This current rating or limiting is different for each manufacturer. For example, GM limits their primary current to 8.5–10 amps, while Ford limits theirs to 5.5 amps. Same theory, slightly different circuitry.

This limiting is done by the ignition module monitoring primary current and modifying base current flow of the switching transistor. By limiting the base current flow, the switching transistor limits the collector emitter current to the required level. In addition, the module will control or vary the dwell as it has done in the past by looking at the previous spark. If maximum current flow was achieved, the dwell is shortened. If it was not achieved, it will lengthen the dwell to allow for full coil saturation. Sounds very familiar, doesn't it? DIS is basically a four-function ignition system which:

1. Turns on primary current.
2. Turns off primary current.
3. Limits primary current flow.
4. Varies the dwell.

You now have the basic theory of how DIS will function. Feel free to review or completely reread this section until it is second nature to you. Once you are comfortable with the information, we will begin looking at the sensors in common use with DIS.

DIS SENSORS

Currently, DIS is no different from any other ignition system. It must know where the pistons are, how fast they are moving, and which spark plugs need to be fired. Think about the differences between DIS and a typical electronic ignition system. The distributor played a very important part. It was responsible for delivering the spark to the right plug at the right time. Picking the correct plug was easy. The rotor position determined it. With a single coil delivering spark to the center of the rotor, the outer blade coming close to the distributor spark terminal gives us the correct plug, assuming the distributor is in correctly and plugs are wired correctly. We call this firing order, right? Timing was accomplished either directly off a position sensor or through the vehicle computer. Base or initial timing was a determination of the distributor position. Sounds like this distributor will be hard to replace, doesn't it? To replace the distributor will require that the module have information about which coil to fire. In other words, a number one cylinder position sensor. The module will be told which coil to fire first during cranking. The signal from #1 is located on either the crankshaft or the camshaft and will perform two functions. It will identify which coil to fire and is usually used to identify which fuel injector needs to be fired. Firing the injectors and the ignition coil at the correct time is a dual or shared function of this sensor's signal.

The second sensor's signal is required to identify small speed changes which occur and will be used to determine crankshaft speed, piston positions relative to TDC, and whether the engine is speeding up or slowing down. This sensor is usually off the crankshaft. Do not forget that both of these signals, piston speed and position, are necessary for any ignition system to function properly. As you look at DIS keep in mind the basic function of any system—to fire the right plug at the right time.

MAGNETIC SENSOR

As mentioned before, there are different types of sensors in use and their signals are used differently by the module and computer. Let us start with a familiar sensor, the magnetic type. We discussed this type in other applications. They have been around for quite a while. They were first used on Chrysler electronic ignition during the early 1970s. Let us review their principles of operation. Remember that moving magnetic lines of force near coils of wire produce electrical impulses (voltage changes). These voltage changes can determine position.

Look at Figure 14-7. It shows a GM magnetic sensor and a reluctor with seven notches machined into it. The **reluctor** is actually part of the crankshaft. The sensor is composed of a permanent magnetic and a coil of wire. The sensor is stationary and placed about .050" away from the reluctor so that they will not touch. Having the small space between the magnet and the reluctor allows the magnetic field to flow easily through the metal. The magnetic field will remain relatively constant and not move until a notch arrives. The notch causes a change in the intensity of the field because the magnetic lines of force are now traveling through air. This changing of the intensity of the field induces a very small AC voltage in the coil of wire, which is part of the sensor.

Figure 14-8 shows the relationship between the notch and the AC signal that is generated. This voltage is very small, especially during cranking, when the crankshaft is moving slowly. Some vehicle sensors will generate only 250 millivolts. That is only 1/4 of a volt. By this time you probably have figured out that the synch notch is how the vehicle computer and ignition module identify where cylinder number one is. That is, the computer will look for two pulses 10 degrees apart to identify where number one is. It will then charge the correct

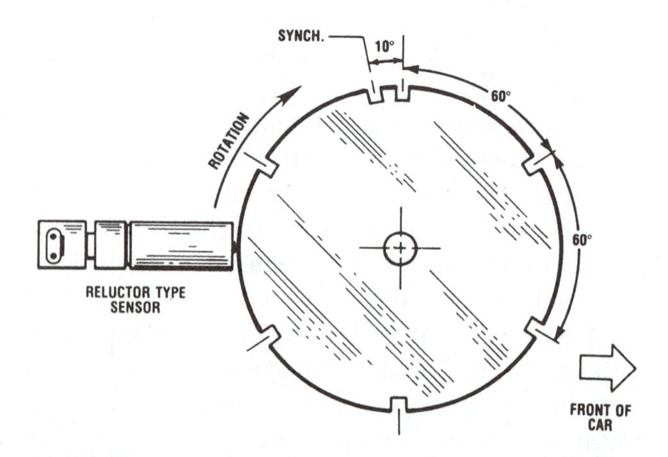

FIGURE 14-7 GM magnetic sensor and reluctor. (Courtesy of GM Corporation)

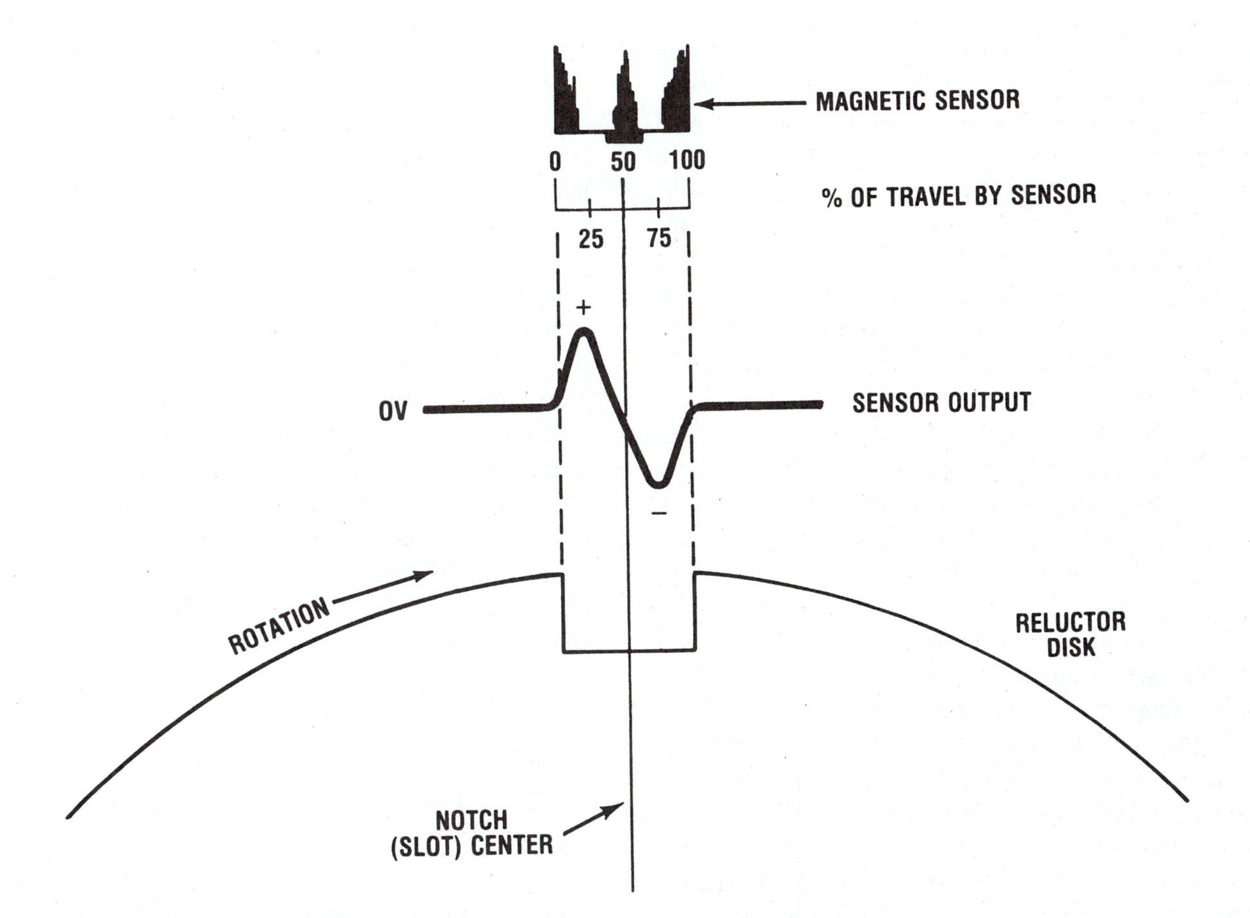

FIGURE 14-8 Notch effect on output signal. (Courtesy of GM Corporation)

coil and fire the correct injector. This combination sensor will identify cylinder #1 and determine engine speed. Both pieces of essential information, piston position, and engine speed will come off this sensor.

Figure 14-9 shows how specific computer programming is set up to look at specific pulses and ignore others. Let us look a these pulses for a typical 6-cylinder engine. As notches six and seven pass the sensor, the

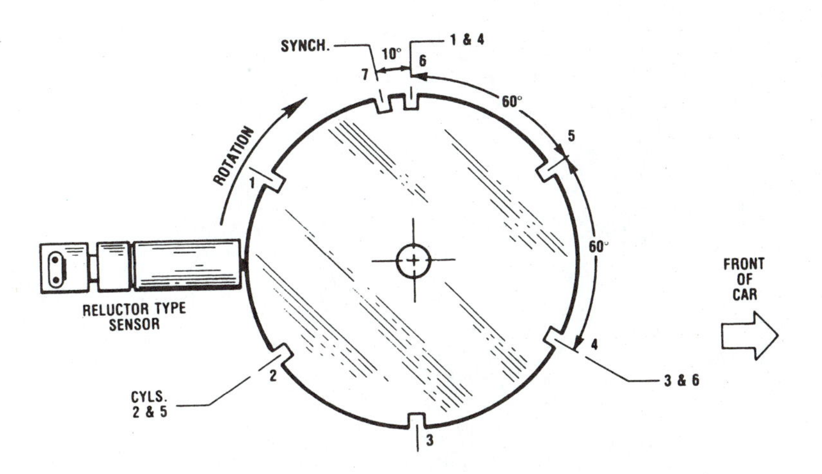

FIGURE 14-9 2.8L V-6 engine showing how specific computer programming is set up to look at specific pulses and ignore others. (Courtesy of GM Corporation)

synch pulse is detected. The module skips notch #1 but accepts notch #2 and fires cylinders two and five. Then it skips notch #3 and accepts notch #4 to fire the coil for cylinders three and six. Skipping notch #5 it accepts notch #6 to fire the coil for cylinders one and four. Notice that notch #6 was ignored the first time it went past, but used the second time to fire an ignition coil. The firing order for this V-6 is 1-2-3 during the first crankshaft rotation and 4-5-6 during the second.

Figure 14-10 shows the same sensor and reluctor in use on a four-cylinder engine. Specific programming within the computer and module now looks at notches two and five, which are opposite, to fire the coils. A different system of which notch to skip and which to use allows the same components to be used on many different engines. These distributorless engines are using one sensor to give the two required bits of information, cylinder position and engine speed. Again, specific programming within the module and computer looks at different pulses and interprets them. General Motors uses this single magnetic style of DIS sensor on many of its engines.

For Motor Company has some of their DIS-equipped engines also with one magnetic sensor supplying both engine speed and piston position. Figure 14-11 shows the crankshaft sensor for a 4.0 liter engine using **EDIS (Electronic Distributorless Ignition System)**. Notice the trigger wheel with many notches and a VRS (variable reluctance sensor) mounted alongside it. In function, it is exactly the same as the GM magnetic sensor that we have looked at. The difference is in the number of teeth or notches in the "trigger wheel". The trigger wheel has 35 teeth spaced 10 degrees apart with one empty space for a missing tooth. By monitoring the signal coming off the **variable reluctance sensor (VRS)**, the module and computer will again know where the pistons are, and how fast they are moving. The signal that will be generated by the VRS will look like Figure 14-12. The missing tooth directly corresponds to piston #1 position. The use of this one sensor is not different from the previously discussed GM sensor. Both are magnetic AC voltage generators just like the pick-up coil of distributed ignition systems. In Chapter 15 we will go over the correct diagnosis procedure for this type of sensor.

MULTIPLE SENSOR DIS/HALL EFFECT SENSORS

Some of the vehicles currently on the road use more than one sensor to get the required information to the computer. Do not forget that this required information is piston position and speed. It is important to note that some vehicle manufacturers will place both sensors into one housing. So actually this becomes one sensor with two functions based on two outputs. We will look at these sensors later.

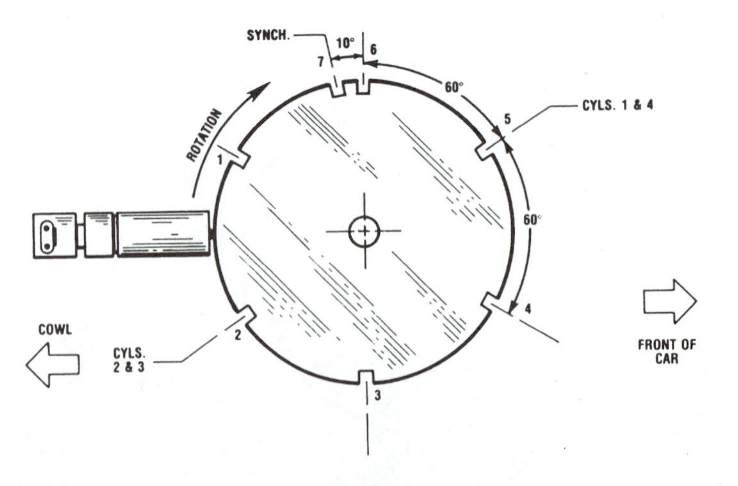

FIGURE 14-10 Same sensor and reluctor on 2.0L, 2.3, 2.5L 4-cylinder engine showing how specific programming looks at different pulses and interpets them. (Courtesy of GM Corporation)

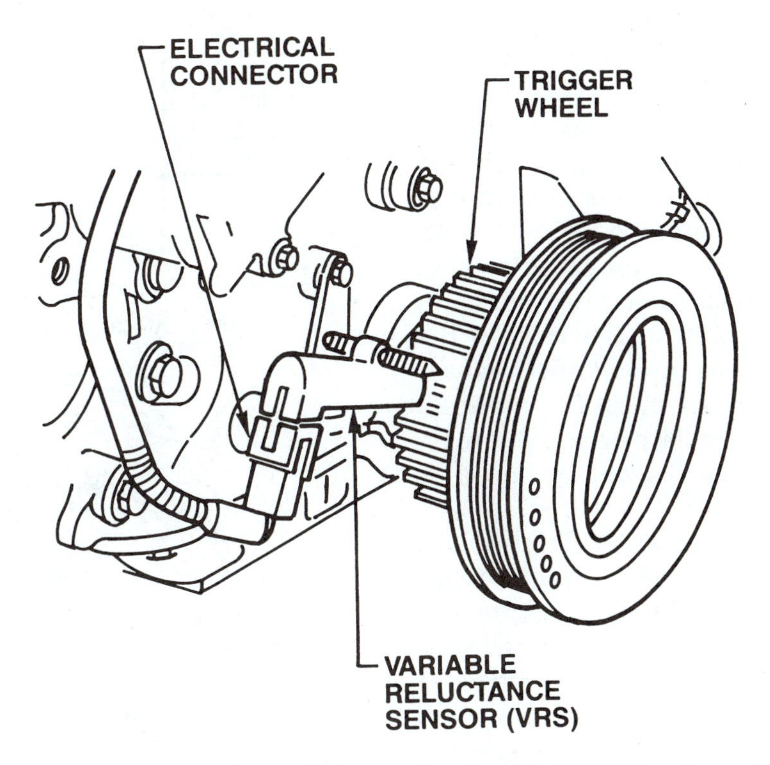

FIGURE 14-11 Variable reluctance sensor and trigger wheel—4.0L V-6. (Courtesy of Ford Motor Company)

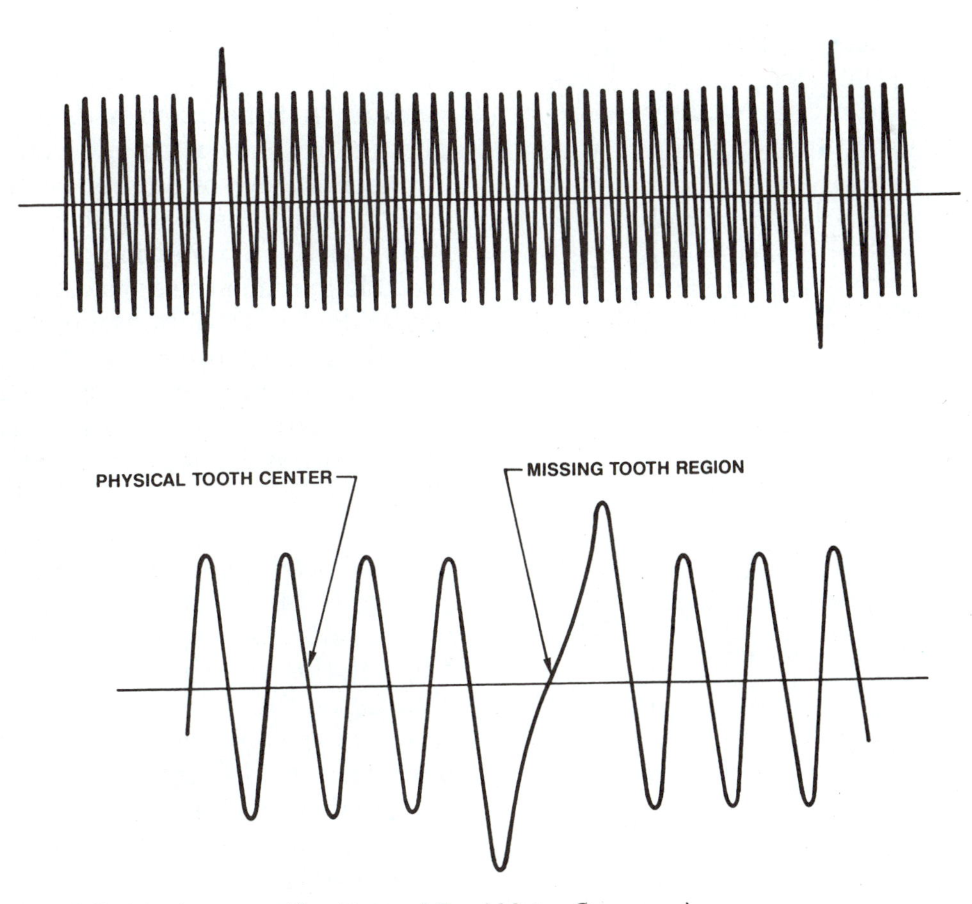

FIGURE 14-12 VRS signal waves. (Courtesy of Ford Motor Company)

However, let us first look at a very popular style of sensor, the Hall effect sensor. Hopefully, this sounds familiar to you since we have looked at their use in past ignition systems. Remember that their output is a pulsing DC voltage as opposed to an AC voltage signal that comes off a magnetic sensor. Simplified, a Hall effect sensor has the ability to sense magnetic fields. Generally, if the magnetic field is sensed, a Hall effect sensor will produce a low voltage output (sometimes even zero volt). The opposite occurs if a magnetic field is not sensed; the voltage output is high (sometimes full B+).

Figure 14-13 illustrates a PIP sensor. **PIP** is Ford's identification of a **profile ignition pick-up**, which is a Hall effect device. The digital signal that this PIP will produce is illustrated in Figure 14-14. If you look at both of these illustrations together, they will make sense. Virtually all Hall effect sensors operate in this manner. The three leads coming into the bottom of the device, which is another way of saying sensor, are B+, ground, and sensor output. When the magnetic flux lines from the permanent magnet are able to reach the device, the voltage output is low (off). When the metal vane is placed in the gap between the magnet and the device, the flux lines are shunted or diverted. Without the flux lines reaching the device, its output goes up to battery voltage. This digital on, then off, DC signal will be sent to the module or computer (or both) to indicate position and/or speed.

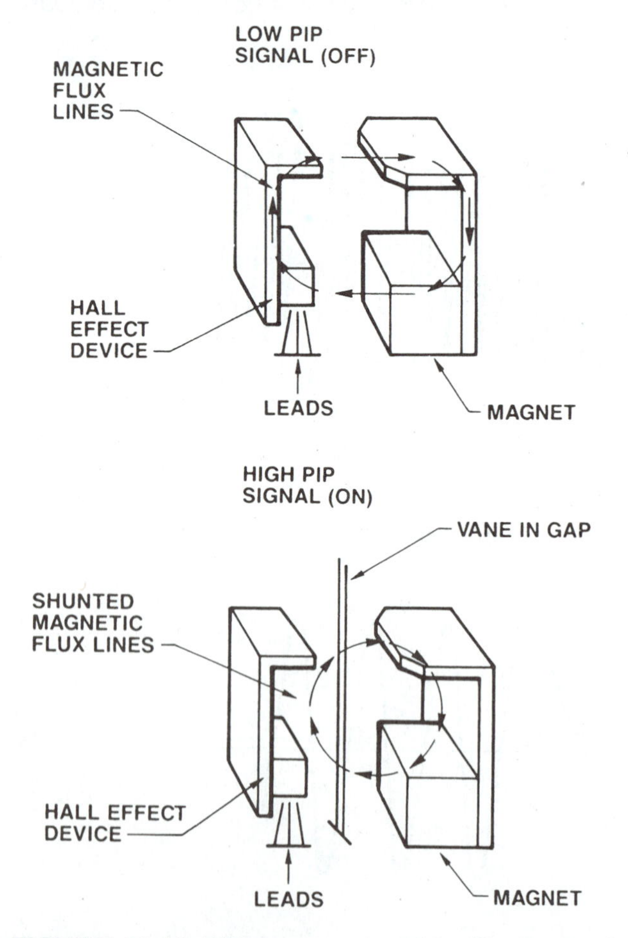

FIGURE 14-13 PIP sensor operation. (Courtesy of Ford Motor Company)

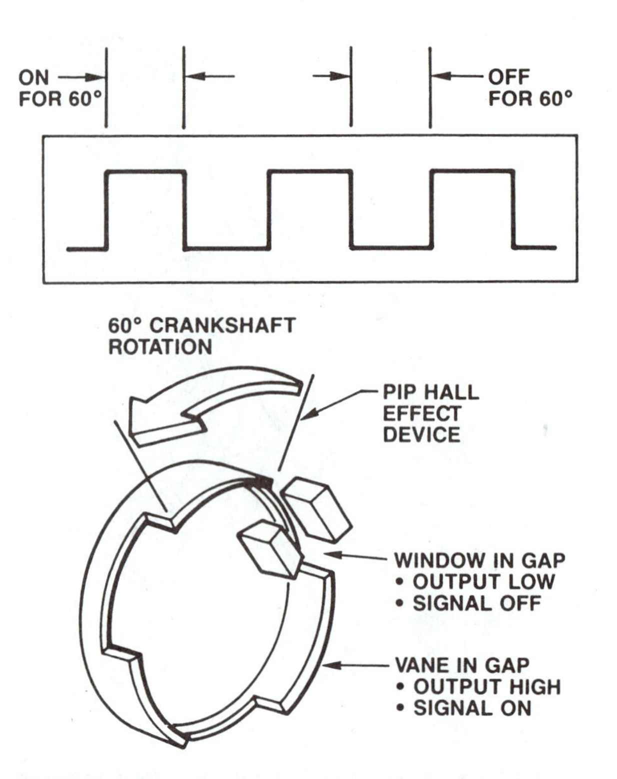

FIGURE 14-14 Digital signal produced by PIP sensor (top) and PIP sensor (bottom). (Courtesy of Ford Motor Company)

Figure 14-15 illustrates the use of the PIP to determine piston position. As the leading edge of the vane enters the opening and shunts the flux lines the voltage rises. This occurs in the illustrated engine at 10 degrees before top dead center (BTDC). By counting the number of signals, the vehicle computer also knows the speed of the engine.

The terminology of PIP has changed slightly over the evolution of DIS. Prior to DIS

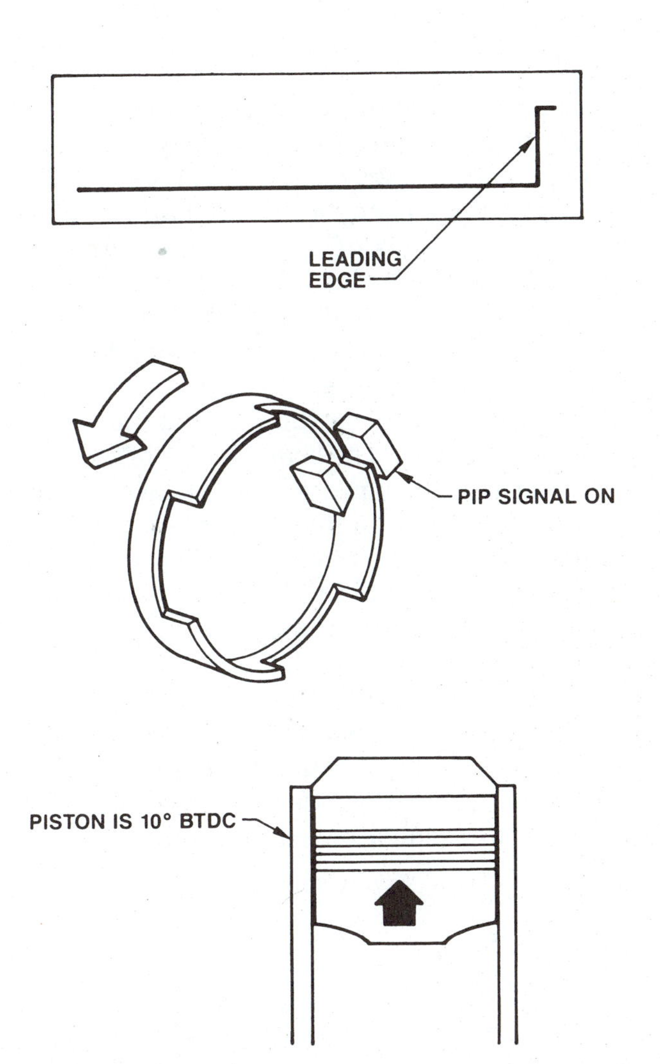

FIGURE 14-15 PIP leading edge 10° BTDC—3.8L and 3.0L SHO. (Courtesy of Ford Motor Company)

with TFI (thick film ignition) and a vehicle computer a Hall effect device was used in the distributor and was referred to as the PIP. However, in DIS the PIP is now the sensor mounted alongside the crankshaft as Figure 14-16 shows. The function of either PIP is the same, only the location has changed.

CAMSHAFT POSITION SENSORS

Notice that Figure 14-16 shows a CID (Cylinder Identification) sensor mounted where a distributor used to be. If you look closely at it you will notice that it resembles a distributor (Figure 14-17). some of the manufacturers removed the distributor and replaced it with

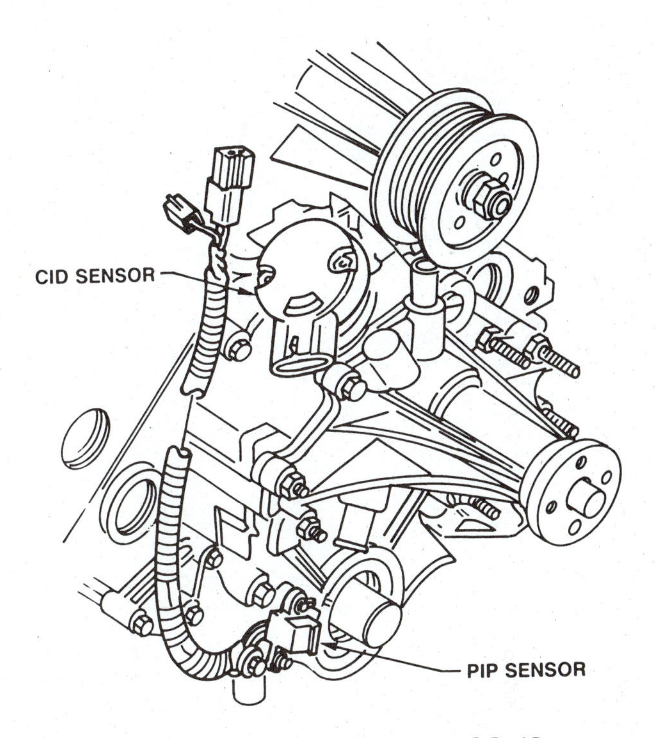

FIGURE 14-16 DIS components—3.8SC. (Courtesy of Ford Motor Company)

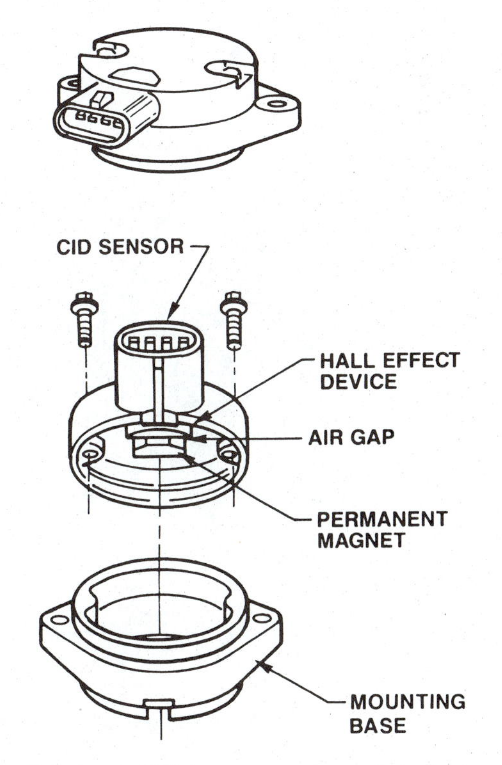

FIGURE 14-17 CID sensor and synchronizer—3.8L SC. (Courtesy of Ford Motor Company)

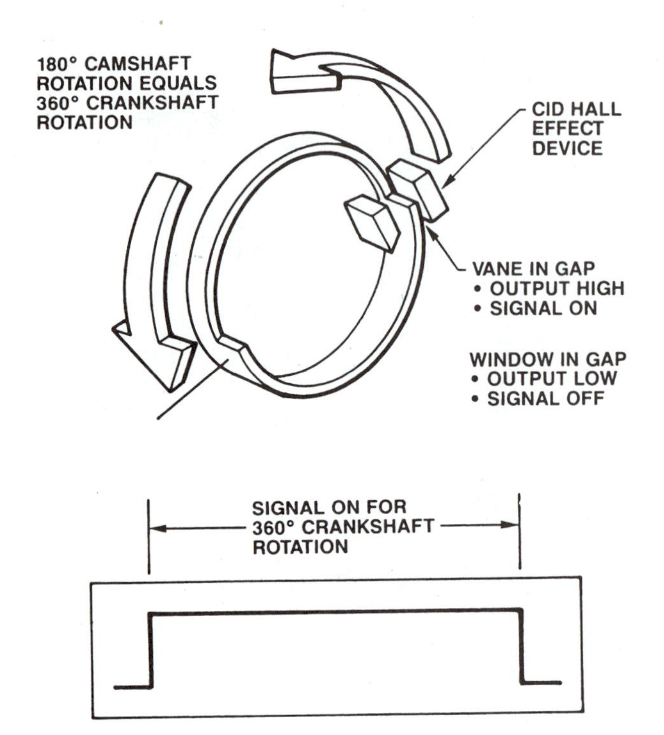

FIGURE 14-18 CID signal—3.8L and 3.0L SHO. (Courtesy of Ford Motor Company)

some type of a camshaft sensor on early DIS. Just like the crankshaft sensor this is a Hall effect device (Figure 14-18) which will produce a digital square wave signal. Notice that the device is housed in a distributor base. Because this signal comes off the camshaft and the camshaft is moving at 1/2 engine speed it will be slightly different from PIP. Generally, it will only have one vane and one window and will produce the signal illustrated.

In operation, it is the same as any other Hall effect device. The leading edge of the vane will cause a rise in the voltage, while the leading edge of the window will cause a drop in voltage. On many of the Ford produced engines, this signal is used to fire the sequentially fired fuel injectors. Its signal identifies the compression stroke of piston #1 during cranking.

Figure 14-19 shows a very similar multiple sensor system using a cam sensor and a crank sensor. This illustration is for a Buick using an early DIS system referred to as C31 which stands for three-coil ignition.

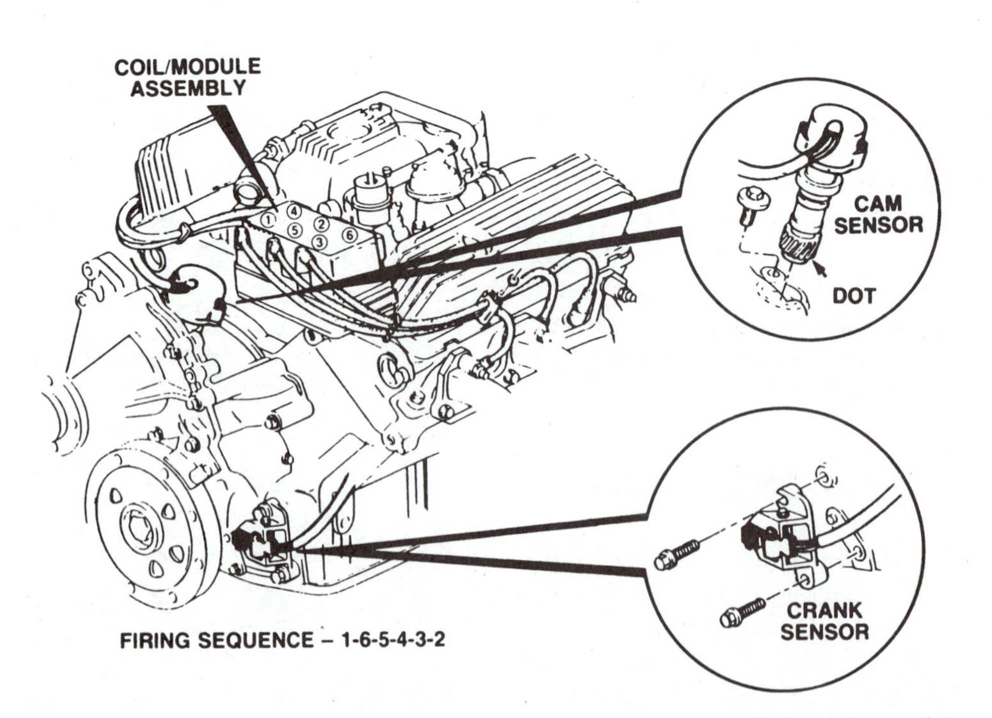

FIGURE 14-19 Crankshaft position sensor. (Courtesy of Delco Remy Division)

MULTIPLE OUTPUT SENSORS

Some of the vehicle manufacturers use a single sensor with two outputs called a combination sensor. Figure 14-20 shows a popular Buick sensor which is mounted near the crankshaft. A single magnet supplies flux lines to two separate Hall effect devices. Two separate sets of windows/vanes, which are part of the vibration dampener, produce the two outputs shown on the bottom of the illustration, Figure 14-21. The outside Hall effect device has 18 equally spaced vanes and will produce 18 pulses per crank rotation. This signal is known as the 18X signal. The inside set of windows/vanes is different from anything we

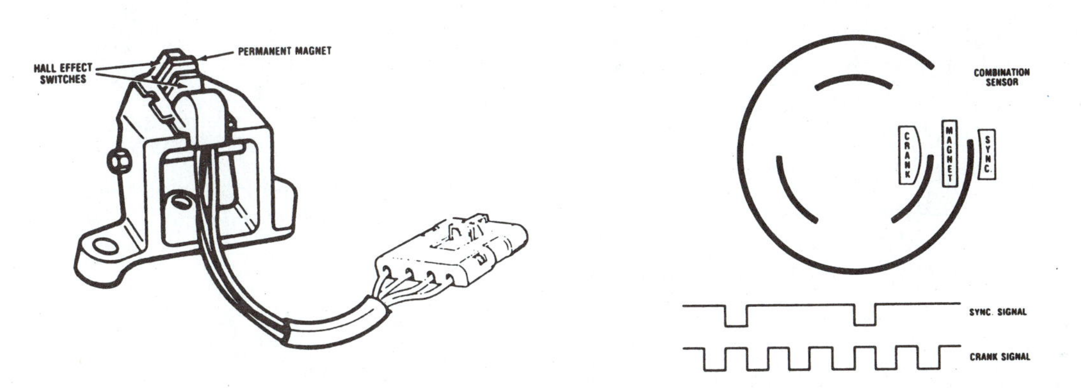

FIGURE 14-20 (Left) 3.0L V-6 crankshaft sensor; (right) 3.0L MFI V-6 engine. (Courtesy of GM Corporation)

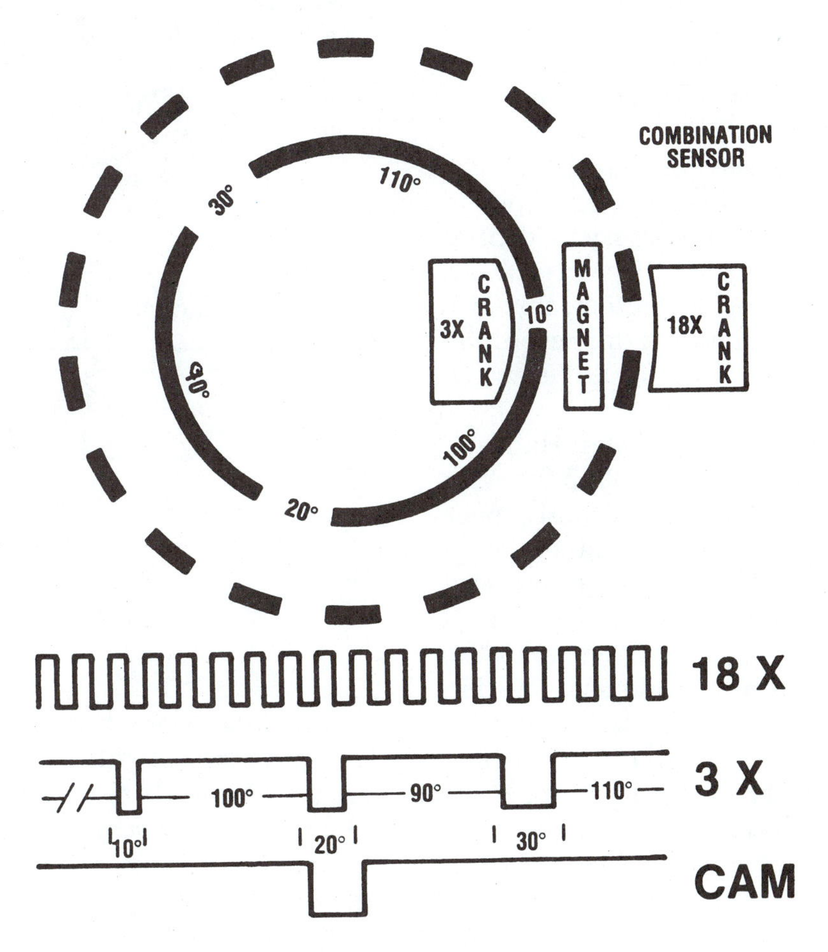

FIGURE 14-21 3800 Buick V-6 engine. (Courtesy of GM Corporation)

have looked at so far. Notice that each window is a different size. There is a 10 degree window, a 20 degree window, and a 30 degree window, producing the signal shown. This is the 3X signal. General Motors uses this system on some of their engines and refers to it as the "fast start". The different spacing of the windows of the 3X sensor allow the computer and module to identify which coil needs to be charged in as little as 120° of crankshaft rotation. Hence the term "**fast start**". The module will use the 18X pulse to measure the duration of each of the 3X pulses. Because the 18X pulse will change state once during the 10 degree window, twice during the 20 degree window, and three times during the 30 degree window, the module can pick out cylinder position as soon as any window of the 3X pulse passes.

Not all GM vehicles use this "fast start" system. Some use a dual Hall effect sensor with equally spaced vanes such as that shown back in Figure 14-20. Notice that the output from this sensor will have three pulses from the inside Hall effect. When used in this manner, the outside Hall effect is called the synchronizer or sync pulse while the inside pulse relates to the number of cylinders and piston speed. The sync pulse is necessary to identify which coil and injector needs to be fired. With only one sync pulse per revolution, this system could turn over almost a full revolution before the module would have identified which coil to fire.

A slightly different style of dual Hall sensor is illustrated in Figure 14-22. In function and design it is the same as any dual sensor we have looked at: a shared permanent magnet and a separate PIP and CID sensor with rotating cups (windows/vanes). The signal that comes off this sensor (Figure 14-23) gives the two pulses per crankshaft rotation needed for a 4-cylinder PIP and one pulse per revolution needed for CID. The only difference in this sensor is that it will produce a signal that will have longer time between leading and/or trailing edge. The length of time that the signal is on or off is equal. This is usually referred to as 50% duty cycle. Simplified, this means that during any period of time, the signal will be on or off for 1/2 of the total time. **Duty cycle** is frequently used when describing computer controlled circuits where something must be on or off for a specific period of time. Fuel injectors are good examples of duty cycle. A 10% duty

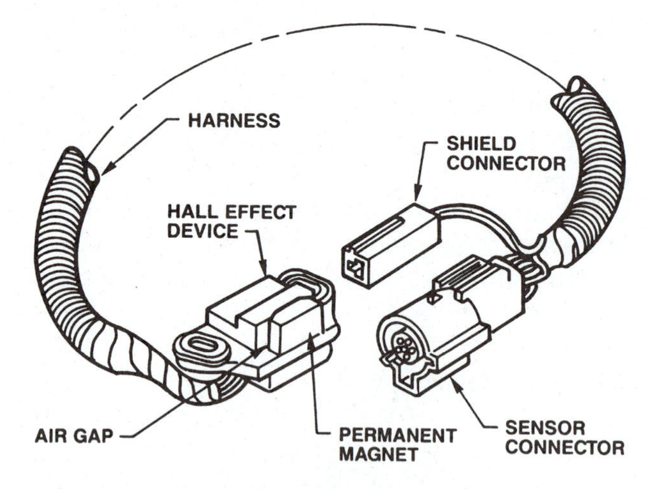

FIGURE 14-22 Dual Hall sensor—2.3L DP. (Courtesy of Ford Motor Company)

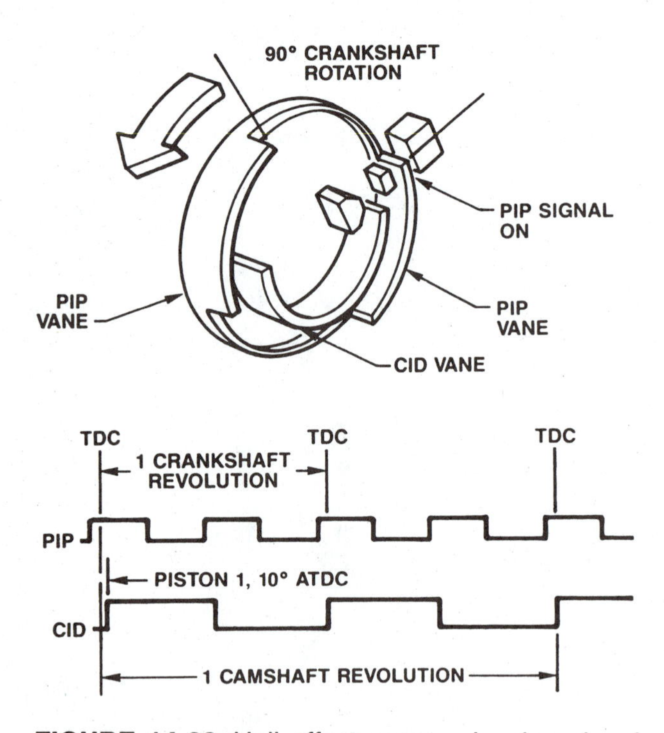

FIGURE 14-23 Hall effect sensor showing signal trace. (Courtesy of Ford Motor Company)

cycle means that the injector will deliver 10% of its maximum fuel, while a 75% duty cycle will have much more fuel injected.

DIS TIMING CONTROL

The timing of the spark on a DIS vehicle is very complicated, and involves the computer even during cranking. However, setting timing is not necessary in most cases. This makes the system actually seem much simpler than the other systems we have looked at. On distributed vehicles, the position of the distributor determined base or initial timing. With DIS, the computer will be in total control of timing using only the signals from the speed and position sensor to determine initial timing. Once the vehicle is running, the information from the MAP (manifold absolute pressure), the TPS (throttle position sensor), and perhaps the coolant temperature sensor will be used to help determine the maximum advance required.

On distributor vehicles, you set timing by opening a connector or shorting a terminal. What you were actually doing was telling the computer that it should not try to change timing so that you could check the position of the distributor. With the computer out of the timing picture, you could adjust and correctly place the distributor in position. The correct position was the one which would allow for the rotor to change position during timing advance/retard while always keeping the tip of the rotor alongside the correct distributor cap terminal. When maximum advance occurred, the rotor still had to be able to deliver the spark to the correct terminal. This back and forth moving of the rotor required that initial timing be correct.

Now think about DIS. No rotor or distributor terminal to worry about keeping aligned; so most systems do not require setting initial timing. The computer has been programmed to deliver to the ignition module correct timing for whatever conditions are present. The information from the TPS, the MAP sensor, and on some vehicles the coolant temperature sensor will be analyzed to determine the necessary timing required. The chapter on electronic fundamentals covered these important sensors.

In addition, many vehicle manufacturers are turning to circuitry which allows the computer to advance the timing along a curve which is determined off a signal from the knock or detonation sensor. This sensor is a device which will generate a signal when the engine is knocking or pinging. These sensors have been in use on many vehicles with distributed systems. Figure 14-24 shows a knock sensor circuit, with an ESC (electronic spark control) module.

This module sends a signal to the vehicle computer if the engine is knocking. When this signal is received by the computer, it will retard the timing until the knocking is eliminated. In this way, the computer advances the timing to the maximum, where the greatest fuel economy will be achieved, just short of knocking. The sensor is either a piezo crystal or a coil of wire.

Either of these styles will generate a voltage if knocking is occurring. They are generally threaded into the block or cylinder head and may be connected directly to the computer or to a separate module which is, in turn, wired to the computer as the example shows. Some of the systems look at the signal when the vehicle is cranking and look for knocking as the computer advances the timing. When knocking occurs, the computer calculates the octane of the fuel and determines what the advance curve will be for operation until the vehicle is turned off.

Compensating for varying octanes of fuel is the greatest reason for a knock sensor. If the octane is low, the advance curve will not be great. There will be less advance for the same speed/load combination than there would be if the octane was high. Again, the goal is to have the maximum advance, just short of knocking, because this is the point where maximum fuel economy will be achieved.

DIS has simplified the tune-up procedure because the setting of timing has been eliminated. In fact, some of the manufacturers are

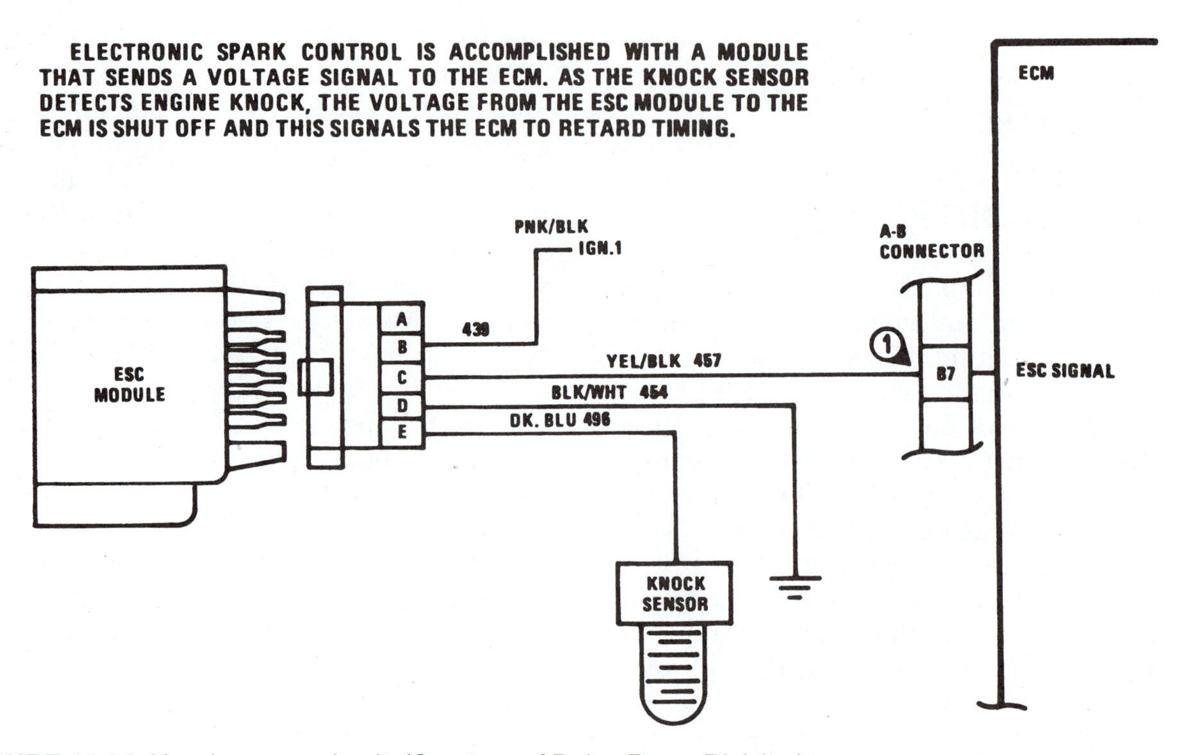

FIGURE 14-24 Knock sensor circuit. (Courtesy of Delco Remy Division)

eliminating the timing marks or the probe holes which allowed the technician check timing. On those vehicles which still have timing marks, check their advance curves and compare them to a mechanical/vacuum advance distributor. You will be surprised at how much more advance the computerized vehicles have. The new systems take advantage of the technology available and advance to the maximum. They can do this because the computer is in total control of both fuel and ignition. Correct timing involves the use of the same information necessary to determine correct air/fuel ratio.

DIS VARIATIONS

Two Plugs per Cylinder

We have looked at the most common types of DIS that use one plug per cylinder and one coil for every two spark plugs. A variation to this system is the Ford 2.3 liter four-cylinder with four coils in two packs and two plugs per cylinder. Figure 14-25 shows this system. The right coil packs and plugs are in operation all the time and are similar in function to the other DIS we have looked at. The left bank is where the major differences occur. These coils and plugs can be switched on or off by the vehicle computer called the EEC-IV (electronic engine control) module as necessary.

A circuit within the EEC-IV module is called the dual plug inhibit (DPI) and allows the use of the left bank plugs as needed. Single plug mode is used during cranking and dual plug mode is used once the engine is running. Both plugs within the one cylinder are fired at the same time but are fired with opposite polarity. This is done through the plug wiring and coil pack wiring. For example, the center electrode of the #1 plug on the right side is negative while the center electrode of the left plug cylinder #1 is positive as both plugs fire together.

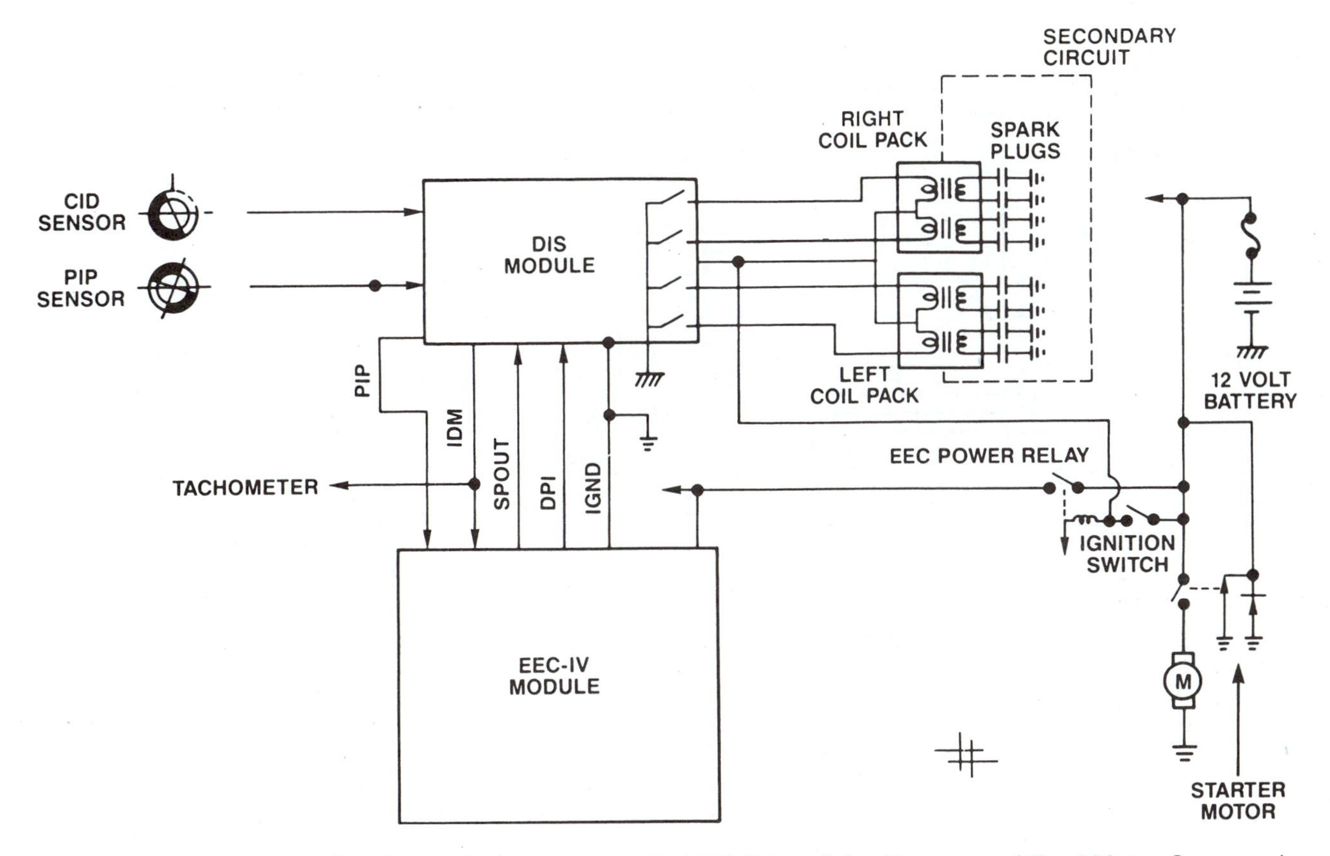

FIGURE 14-25 Ford 2.3 liter four-cylinder system with EEC-IV module. (Courtesy of Ford Motor Company)

Let us follow this ignition system during start-up and running modes. During cranking, the DIS module looks at CID (cylinder identification) sensor and SPOUT (spark output) as displayed in Figure 14-26. The EEC-IV module knows that the vehicle is cranking so it inhibits the left coil pack by sending a 12-volt DPI signal to the DIS module. The injectors fire and the vehicle starts. Once running, the EEC-IV module removes the 12-volt DPI signal. When the DIS module sees zero volt on the DPI circuit, it will fire both coils as Figure 14-27 shows. Once the vehicle is running, timing changes are sent to the DIS module along the SPOUT circuit just like any other DIS. The only difference between having the DIS module fire one or both plugs is the voltage on the DPI circuit. If it is high, only the right bank plugs will fire. If it is low, both sets of plugs will fire.

FAILURE MODES

Most of the systems on the road today have some type of failure mode available that allows the vehicle to be driven in for service. Usually timing changes are eliminated, so that the vehicle does not run as well, but it will start and allow the owner to get the vehicle into your shop.

Ford vehicles experiencing a failure in the CID circuit will have the DIS module randomly select a coil to fire. CID is used to determine which coil should fire and start the sequence based on cylinder position. Without CID, the module picks a coil and fires it. If it is the wrong coil pack, the engine will not start and cranking will be very erratic. The next time the ignition key is turned to the start position, the DIS module will select a different coil pack to fire. Usually after three or four

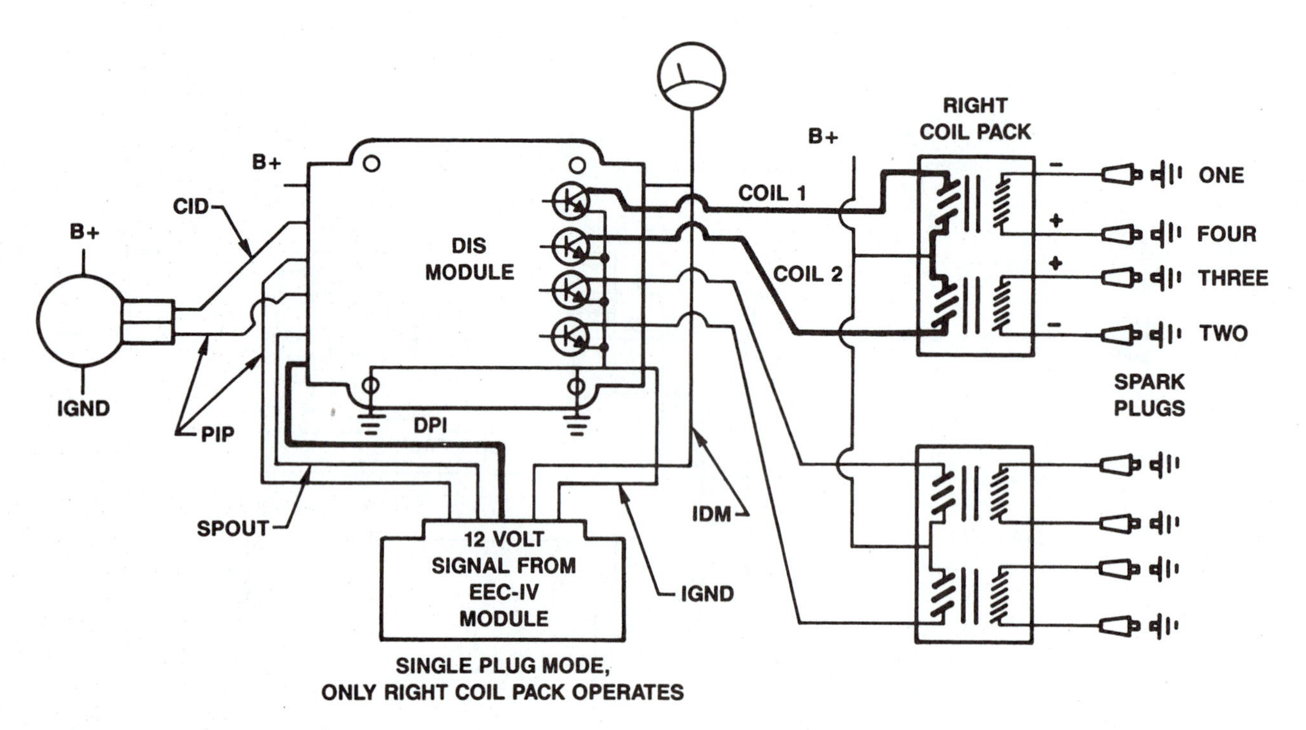

FIGURE14-26 Single plug mode at startup—2.3L DP. (Courtesy of Ford Motor Company)

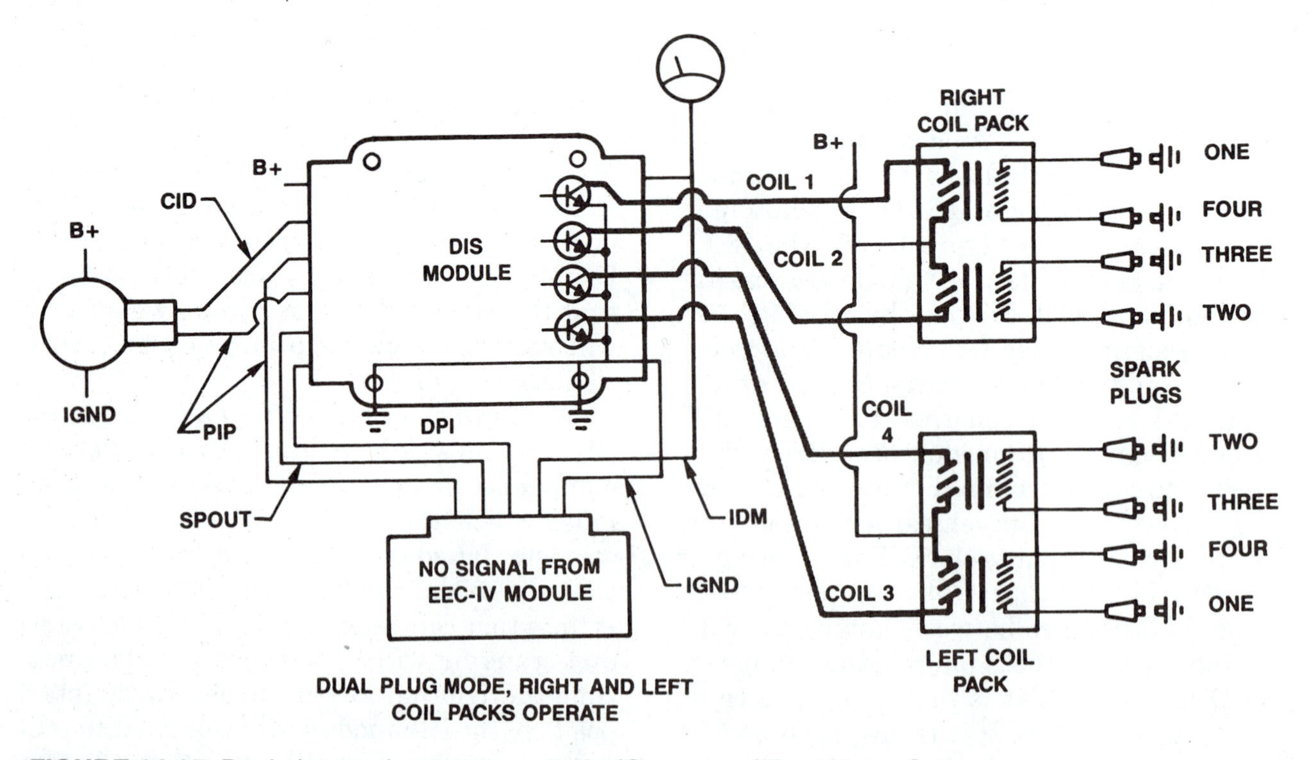

FIGURE 14-27 Dual plug mode at startup—2.3L. (Courtesy of Ford Motor Company

attempts the module will select the correct coil pack and the engine will start. Without all the information it is used to receiving, the EEC-IV module will open the SPOUT and fix the timing at 10° before top dead center. In this way, the engine runs, but not well, because it is operating with no advance. Most of the modern systems today have some type of failure mode system, making them able to function even with a sensor or circuit out.

SUMMARY

In this chapter, we have looked at the common styles of distributorless ignition. We have seen that most systems use one plug per cylinder and one coil for every two companion cylinders. We defined companion cylinders as cylinders whose pistons are moving in the same direction on opposite strokes, and in the same position. The ignition coil fires the two plugs in companion cylinders at the same time. The two ends of the coil's secondary winding are connected to different plugs. One plug will be in a cylinder that is on the exhaust stroke while the other plug will be in a cylinder that is on compression.

The plug polarity on companion cylinders will be opposite, with one firing with positive polarity and the other firing with negative polarity. The total voltage produced by the coil will be the combined voltage of the two plugs. We examined the use of the magnetic sensor or the Hall effect sensor to identify both cylinder position and speed. In addition, the signals generated by both of these sensors were examined. The sine wave from the magnetic sensor was compared to the square wave, on/off signal from the Hall effect sensor. The use of different spaced windows or missing tooth patterns to provide a synch pulse for quick start was discussed. Finally, we looked at the use of dual plugs per cylinder with one set being used for starting and both sets for running.

KEY TERMS

DIS (distributorless ignition system)
Wasted spark
Reluctor
EDIS (electronic distributorless ignition system)
VRS (variable reluctance sensor)
PIP (profile ignition pickup)
Fast start
Duty cycle

CHAPTER CHECK-UP

Multiple Choice

1. Technician A states that most DIS have no method of setting ignition timing. Technician B states that advance curves are determined by the vehicle computer. Who is correct?
   a. Technician A only
   b. Technician B only
   c. Both Technician A and B
   d. Neither Technician A nor B
2. DIS requires inputs from crank and/or cam sensors to know
   a. the speed of the vehicle and load on the engine.
   b. piston position and the speed of the engine.
   c. which fuel injector is firing.
   d. throttle position and engine speed.

3. Waste spark refers to the
   a. plug that is firing while the cylinder is on the compression stroke.
   b. timing change that the computer will produce.
   c. reference pulse that the computer will use to determine engine speed.
   d. plug that will fire while the cylinder is on the exhaust stroke.
4. Companion cylinders are cylinders that are
   a. in the same position.
   b. going in the same direction.
   c. on opposite strokes: i.e., intake/power or compression/exhaust.
   d. all of the above.
5. Technician A states that the signal from a Hall effect sensor will be a sine wave. Technician B states that the signal from a magnetic sensor will be a square wave pulse (high voltage/low voltage). Who is correct?
   a. Technician A only
   b. Technician B only
   c. Both Technician A and B
   d. Neither Technician A nor B
6. The object of the synch pulse is to
   a. identify companion cylinders.
   b. identify which coil pack must be fired.
   c. give engine speed information to the computer.
   d. produce the reference pulse used for engine timing.
7. On a two-plug-per-cylinder DIS
   a. all plugs are firing during normal driving conditions.
   b. all plugs are fired during starting.
   c. only one set of plugs is used during acceleration.
   d. all of the above.
8. Fast start DIS uses
   a. the signal from the injector to fire the correct coil.
   b. spacing changes in the Hall effect windows to identify cylinder position.
   c. information stored in the module's memory to fire the correct coil.
   d. all the coils at the same time until the engine starts.
9. Firing voltage for the plug on the compression stroke are
   a. lower than for the plug on the exhaust stroke.
   b. higher than for the plug on the exhaust stroke.
   c. the same for both plugs.
   d. always 6kV.
10. Technician A states that combination sensors usually eliminate the cam sensor. Technician B states that combination sensors are usually mounted alongside the crankshaft. Who is correct?
    a. Technician A only
    b. Technician B only
    c. Both Technician A and B
    d. Neither Technician A nor B

True or False

11. Companion cylinders are always on the same stroke.
12. Waste spark is the spark for the cylinder which is on the exhaust stroke.
13. DIS uses a single coil for each pair of cylinders.
14. A Hall effect sensor produces a square wave signal.
15. A magnetic sensor produces a sine wave.
16. A magnetic sensor usually has three wires connected to it.
17. A combination sensor will usually produce two outputs.
18. The synch pulse will identify cylinder number one's position.
19. The module used with DIS is usually a two function (primary current on—primary current off).
20. DIS usually has no method of setting timing.

CHAPTER 15

Distributorless Ignition Diagnosis and Repair

OBJECTIVES

At the completion of this chapter, you should be able to:

- follow recommended procedures to safely diagnose and repair distributorless ignition systems.
- use an automotive oscilloscope to diagnose an intermittent DIS problem.
- use an ohmmeter to check an ignition coil.
- use an oscilloscope to check a magnetic sensor.
- use a recording voltmeter (DVOM) to check a Hall effect sensor.
- use an oscilloscope to check a Hall effect sensor.
- use a scanner to monitor cranking rpm.

INTRODUCTION

Up to this point, we have been dealing with the theory and operation of distributorless ignition systems. It is now time to take a closer look at the diagnosis and repair of some of the common systems on the road today. If you do not feel that you know how the systems operate at this point, reread the DIS theory chapter until the information presented is clear in your mind. Just like the other systems covered in this text, you should have a working knowledge of the theory of operation before you attempt a repair. Let us cover some general repair considerations before we get down to the specifics.

1. Never disconnect or reconnect any connector with the ignition key on, unless repair procedures specifically call for you to do it. You run the risk of voltage spiking a component. With the extremely high cost of electronics on the vehicle of today, spiking a device is the last thing you, or your employer wants.
2. Do not use a battery charger or a starting unit unless you disconnect the battery. The unit might be producing either too high a voltage or a high level of unfiltered alternating current. Either condition can cause expensive damage to the onboard electronics.
3. Use only a good quality 10-megaohm input digital volt ohmmeter. A lower input

impedance meter might load down the circuit and give you wrong answers.

4. When diagnosing problems, make sure you are following the manufacturer's recommended procedures. The generic ones presented in this text will work on most models; however, the manufacturer's literature should be your first choice.
5. If there is a self-diagnostic procedure utilizing a scanner and the vehicle's memory—access it first. Frequently, the stored fault codes give clues as to the problem. In addition, it is very common for the manufacturer to use the signal from one sensor to do more than one thing. This overlap may help guide us to the failed component.
6. Be very aware that static electricity may destroy a component before you have the ability to install it. Keep all components in the original packages until you are ready to use them. Never slide across the seat while holding a component. Always ground yourself to the vehicle chassis before you touch the terminals of a component. These simple precautions, in addition to others that the vehicle manufacturers may supply in the form of service bulletins, will help prevent unnecessary component failure caused by our service procedures. You must fix the vehicles of today with the tools and techniques of today—they will not be repaired easily and totally with 1970 technology!

We will divide the diagnosis of DIS into three sections; secondary, primary, and sensors. This is also the most sensible approach to actually testing the vehicle. The procedure that we will examine is really not that much different from the distributed spark procedures. There are some slight differences, though.

Let us first look at the secondary side of the ignition system. With the very high voltages available, it is always a good idea to look for additional paths that might develop over the years. Moisture can become a real problem and can give the high voltage a shorter path than the plugs offer. In a previous chapter, we discussed using a spray bottle of water to simulate very high moisture levels that might become a path to ground. This technique works equally well on DIS vehicles. Do not forget that DIS uses a spark plug on both sides of the secondary winding. Any problems on either side of the winding will usually have an effect on both plugs and cylinders.

SCOPE TESTING

The automotive oscilloscope is still a very valuable tool when preventive maintenance checking a DIS, when looking for intermittent problems, or when checking up on the tune-up you have just performed. Even some older scopes can be adapted for use on DIS vehicles. If your scope does not have DIS capability, use the pattern pickup around each plug wire and read the firing voltages individually. Although this is not the easiest method, it does work and will give you the information you require. Most new scopes have the built in capability of looking at DIS. Look at Figure 15-1. It is a reprint of the scope form we used on a distributed spark vehicle. You will find that it is very applicable to DIS. Let us take each item and discuss it.

Visual Inspection

Do not let the obvious escape you. Look over the coils, wires, plug boots, and other connections which might be loose, corroded or disconnected. Pull off the plug wires at the coil and at the plugs. Look for signs of failure. If the wire falls apart in your hands, you have found a problem True, it may not be the one which brought the vehicle into your shop, but it is a problem nonetheless. You should not need some fancy piece of diagnostic equipment to detect simple problems. Just make

Scope Check Out

Name ______________________________

Vehicle type ______________________ Year __________

Ignition type ______________________

1. Visual inspection Good ___ Bad ___
2. Cranking coil output Good ___ Bad ___
3. Coil polarity Good ___ Bad ___
4. Firing voltage Good ___ Bad ___
5. Available voltage Good ___ Bad ___
6. Secondary insulation Good ___ Bad ___
7. Plugs under load Good ___ Bad ___
8. Secondary circuit resistance Good ___ Bad ___

If breaker point ignition:

9. Coil and condenser Good ___ Bad ___
10. Point action Good ___ Bad ___
11. Cam lobe accuracy Good ___ Bad ___

If electronic ignition, check one.

12. 2 function ___
    Primary current on Good ___ Bad ___
    Primary current off Good ___ Bad ___
13. 4 function ___
    Primary current on Good ___ Bad ___
    Primary current off Good ___ Bad ___
    Variable dwell Good ___ Bad ___
    Current limit Good ___ Bad ___

Based on above information, what is your recommendation?

FIGURE 15-1 Scope tests.

sure you open your eyes and look for the problem. Do not forget to check the connection at the ignition module and at any sensors used.

Cranking Coil Output

For this test, we must make sure that we understand that most modern DIS can produce more voltage than their own insulation can stand. Do not, under any circumstances, open circuit one of these plug wires. Some of the systems are capable of producing close to 100,000 volts! You will destroy something with voltages this high. Instead, rely on a spark plug tester, as we have done in the past. If it has an adjustable air gap, set it for around 40,000 volts. If the system is capable of 40kV, it is good. There is no need to stress it any more than this.

Keep in mind that you should test each coil pack. They are individual coils, and it is possible to have failure on one, while the other one(s) are okay, Figure 15-2. Do not crank the

FIGURE 15-2 Spark tester in plug wire.

engine over for excessive periods of time during this test. Do it just long enough to get a scope reading and see the quality of the spark at the tester. While the engine is cranking, the fuel system will be pumping fuel into the manifold. If extensive cranking is done, you will flood the engine—and now you have two problems. The one the vehicle came in with and a flooded engine. If you are dealing with a throttle body fuel injection system, disconnect the injectors to prevent flooding during the cranking test. Do not forget to test each coil pack. If you find a weak coil, place your test plug in the other plug wire from the suspected coil and retest. This will isolate the coil from a possibly bad plug wire.

Coil Polarity

Coil polarity is no longer important, as we fire one plug positive and the other negative. This test should be eliminated for DIS. The extremely high voltages available will fire the plugs with either positive or negative polarity.

Firing Voltage

Your scope will probably show plug firing with the compression and exhaust plug identified as Figure 15-3 shows. Some of the newer digital scopes will give you the actual voltages printed as in Figure 15-4. Usually the exhaust plug will be barely noticeable on the pattern and usually be around 1kV or less. For the plug firing on compression, use the 6 to 12kV specification for a recently tuned engine that we used on other electronic ignition systems. Old plugs or out-of-specification conditions will raise up the firing voltage, especially on the compression plug, just like on other systems. Worn plug electrode, lean mixtures, vacuum leaks, open or high resistance windings or wires will have the same effect that they have had on past ignition systems.

Available Voltage

Many times this test is not necessary as a cranking coil output test has sufficiently stressed the coil. If you do it, remember that

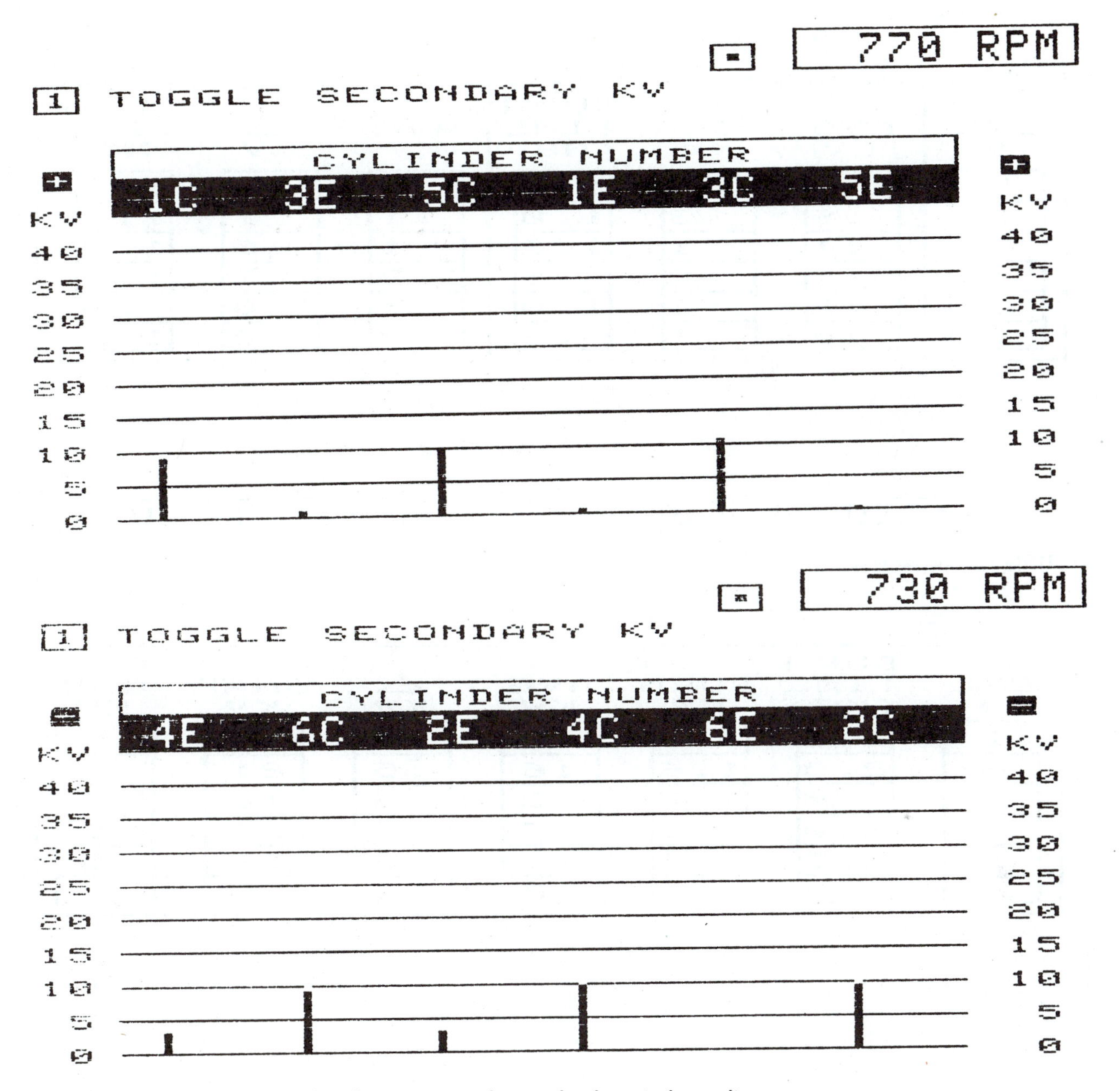

FIGURE 15-3 Bear pattern showing compression and exhaust plug voltages.

you cannot run the secondary open circuited. Install a test plug, start the engine, read the highest voltage and shut the engine off as quickly as possible. This prevents the flooding of the cylinder with the fuel. Fuel injection has greatly increased the amount of fuel available at each cylinder. Without the plug firing, the fuel runs through the cylinder and into the catalytic convertor. Too much fuel will overheat the convertor and quite possibly permanently damage it. Run this test for the smallest amount of time that will give you results.

Secondary Insulation

Lightly mist down the secondary wires and coil with water while you watch the pattern on the scope. Listen for any arcing, watch for any

D.I.S. SECONDARY TEST

CYL #	COMPRESSION AVG KV	COMPRESSION Δ KV	COMPRESSION BURN TIME (MS)	EXHAUST AVG KV	EXHAUST Δ KV	EXHAUST BURN TIME (MS)
1	9	3	1.1	1	0	1.0
6	9	3	1.0	1	1	1.1
5	11	1	1.2	1	0	1.0
4	11	2	.9	2	3	1.1
3	12	3	1.0	1	0	.8
2	10	1	.8	2	2	1.2

CONTINUE REPEAT PRINT -

FIGURE 15-4 Bear voltage chart.

CYL #	BURN TIME (MS)	BURN KV	AVG KV	Δ KV	SNAP KV	CKT GAP KV
1	2.0	1.7	10	5	14	5
6	2.0	1.6	16	5	21	6
5	2.0	1.7	13	3	18	5
4	2.0	1.6	17	3	21	6
3	2.0	1.7	13	3	19	5
2	1.8	2.1	17	2	20	6

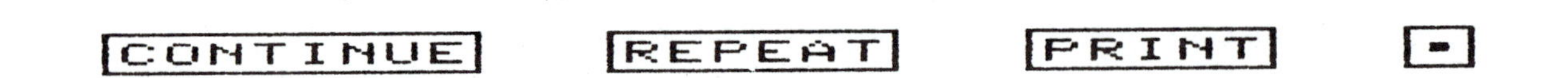

FIGURE 15-5 Bear plugs under load chart.

missing, and/or changes in the pattern. Plugs under load: rapidly accelerate the engine and look at the firing lines. Just as in other electronic ignition systems, they should go up. Usually a 50% rise in voltage is normal with good plugs. Figure 15-5 shows this plugs under load chart for a 4 cylinder engine. Notice that the firing voltage went up by volts from what it was previously. This will be especially seen on the plugs firing on compression.

Milliseconds of Spark

Look at the chart in Figure 15-5 again. This digital scope prints the actual length of the spark in milliseconds. We use the same specification of .8 to 2.0 mS of spark, while the plug is on compression. The length of spark for the exhaust firing is of no real importance. Do not forget, the length of time that the spark lasts is very important and one of the major reasons for DIS. Increased spark length at high speeds. Less than .8 mS and we get frequent misfires. Over 2.0 mS and we erode the plugs away quickly, greatly reducing their life. Hopefully, you remember that a millisecond is actually one thousandth (.001) of a second. Skip past the breaker point ignition section and go to the electronic ignition section.

Four Function

Most distributorless ignition systems on the road are 4 function. You should be able to see this on scope. Let us review and use Figure 15-6. The first two functions (primary current on and primary current off) will be seen on the scope at the beginning and end of the dwell section. It is impossible to have the module lose these functions and still have the engine run. As the engine speeds up the module allows more time for full coil saturation. This is done by starting the dwell section earlier. This is the third function—variable dwell. Longer dwell allows more time for primary current to flow, thus giving the primary more time to fully saturate. Full saturation equals full secondary output when it is needed. Once

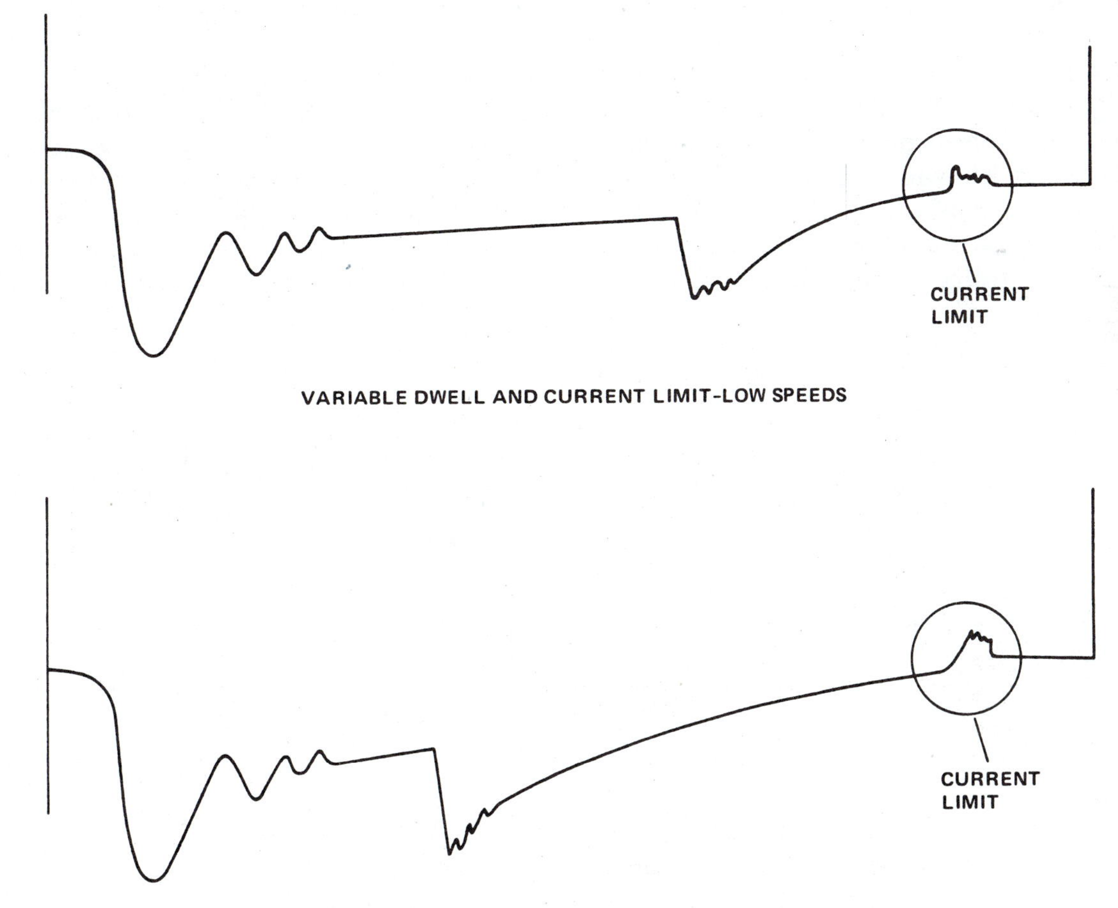

FIGURE 15-6 Comparison of variable dwell and current limit.

full saturation occurs, the fourth function—current limit—comes into play and is usually visible toward the end of the dwell section as a blip or rise. Figure 15-7 shows an actual pattern from a GM 6-cylinder. Notice the circled area in the dwell section. This is current limit occurring.

The scope test is now complete. If you have any questions on it, please reread this section until it makes sense. Do not go on until you understand how to proceed with a scope check-out. Try to test a few vehicles as soon as possible and practice. Discuss your results with your instructor.

TESTING IGNITION COILS

The distributorless ignition systems currently on the road use basically the same style of ignition coil that has been in use for years. Testing it really does not involve any new technology or any new equipment. The most effective method of diagnosing a faulty coil is to substitute a known good one. This has really been the manufacturer's preferred diagnostic method for years. However, DIS improved our ability to substitute or swap coils. Having an extra coil or coils around the typical repair facility that fits a particular application was not very common in the past. With DIS though, the extra coil is part of the system. If you are faced with the situation of no spark out of one coil and spark out of the other (on a 2-coil 4-cylinder vehicle), change the position of the coils on the module. Crank the engine over. Did the spark follow the good coil, and is the previous no spark coil still not producing a spark? If so, you definitely have found a faulty coil If however, the previous good coil will not produce a spark in its new position, you have a module problem. If this sounds simple, great! This process is really not difficult. Just keep in mind a few things.

During testing do not allow the coil to function open circuit; without plug wires on. The resulting high voltage could destroy a good coil Make sure the module to coil connection is good and tight. Tighten all screws, bolts, etc. so that the coils are functioning in their normal manner. Keep the plug wires with the spark plugs and cylinders that they belong on. Move the wires from one coil to the other so that their position on the module remains the same. **DO NOT** keep the wires with the coil they were originally on. The engine could be

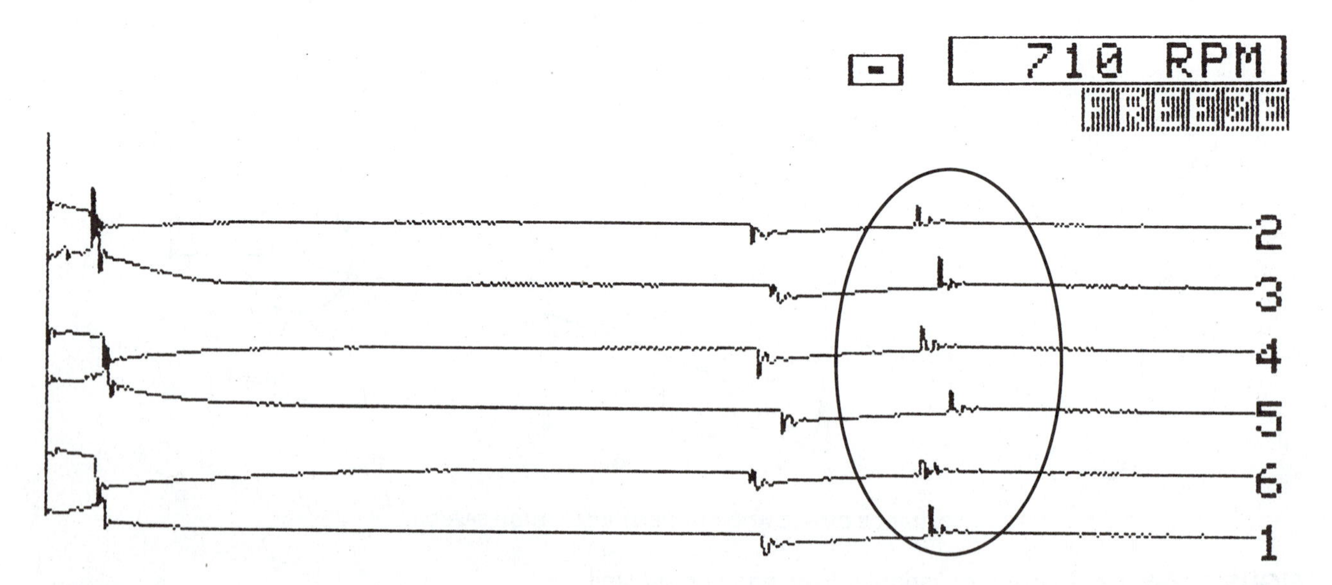

FIGURE 15-7 Bear printed raster pattern with current limit.

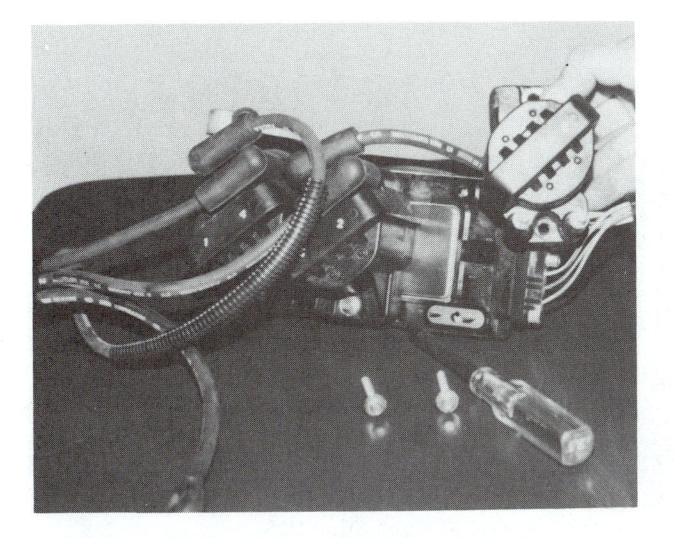

FIGURE 15-8 Coil pack with one coil removed.

damaged because the timing to each cylinder if off by 180° (on a 4-cylinder vehicle)!

If you are faced with the situation of a no spark out of one coil and the coils are not individual pieces like Figure 15-8 shows, you will have to rely on ohmmeter readings to steer you in the correct direction. Measuring primary and secondary resistance usually will identify the bad coil Figure 15-9 shows this process. Notice that ohmmeter #1 is connected across the primary winding of one coil The resistance will be very low; typically less than one ohm. Refer to the manufacturer's specification for the exact resistance range. Be sure to compare the resistance readings between all coils in the pack. They should all be within .1 ohm. Once you have verified that the primary is okay, test the secondary. Typically your reading will be in the 5,000 to 50,000 ohm range. Make sure you are using the correct range on your meter. Again, compare the readings to each other in addition to the manufacturer's specification. The last thing to do is to make sure that the primary is insulated from the secondary, and neither winding is grounded, Figure 15-10. Keep in mind that these ohmmeter readings will locate open or shorted windings. They will not find poor insulation between windings of either the primary or the secondary.

For example, if a vehicle has been driven with an open spark plug wire, the resulting high voltage might have destroyed the insulation on the secondary winding. It takes the high voltage output of the coil to break down the insulation and produce the no spark condition. The ohmmeter will not typically find these insulation breakdown conditions. Make sure you visually inspect the coil for evidence of insulation breakdown. There are cases where the insulation is destroyed, you are not able to find any burns, the ohmmeter readings are correct, and the coil will not produce spark. There is not much else you can do in

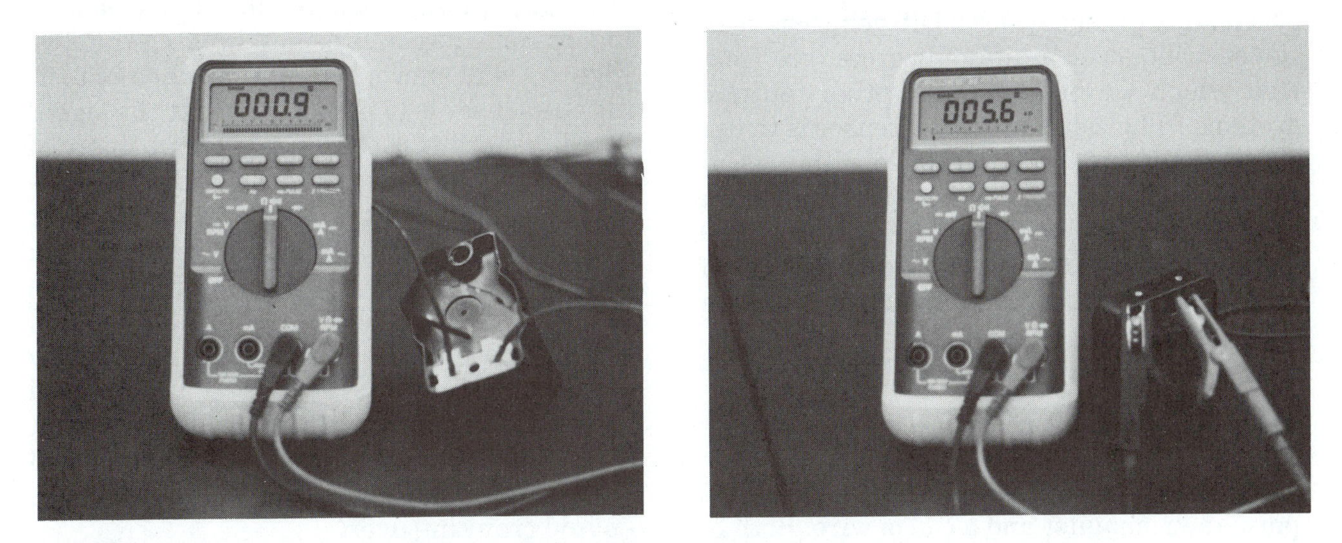

FIGURE 15-9 Two ohmmeters: (left) primary; (right) secondary.

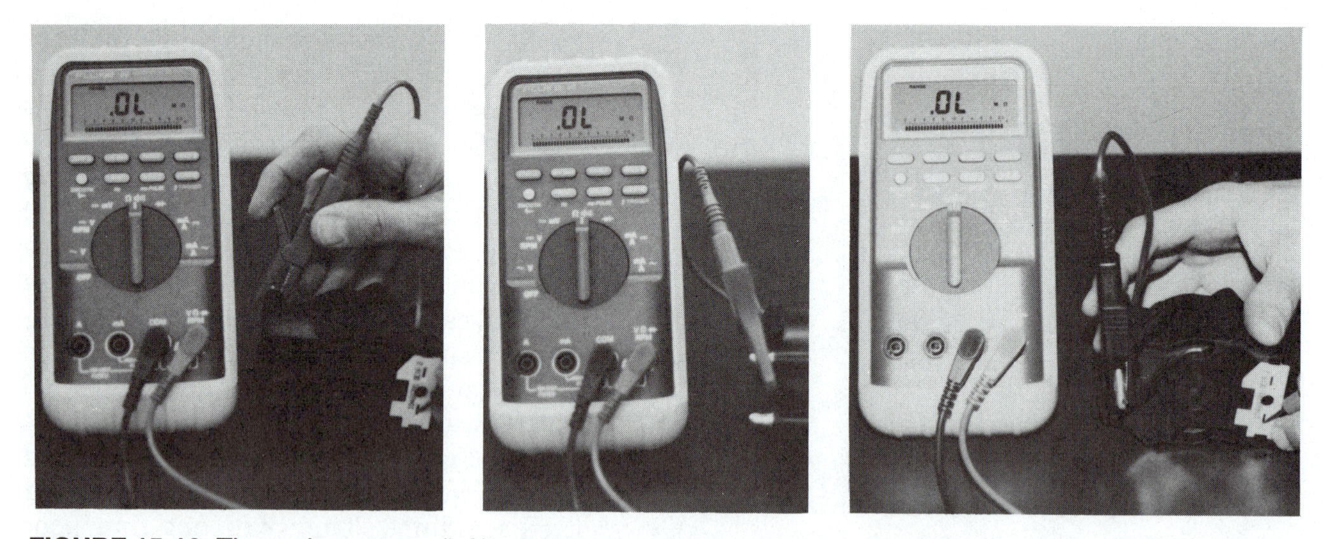

FIGURE 15-10 Three ohmmeters: (left) primary to secondary; (middle) ground to secondary; (right) ground to primary.

this example except substitute a known good coil Just make sure your customer understands the process of testing that you have to use. If they understand what you are doing, they will not think that you are trying to find the burned out component by replacing parts until it runs! Customers are generally understanding of the technician if they are informed along with way.

TESTING SENSORS

The testing of sensors for a DIS vehicle is really not difficult or a different procedure from that which we have done on other ignition systems. Whether a magnetic sensor is in use on an electronic ignition vehicle with a vacuum and mechanical advance or on a modern state of the art DIS vehicle, the testing will be the same. We will however stress the use of additional tools such as scanners later on in this chapter.

MAGNETIC SENSORS

Magnetic sensors are designed around a permanent magnet and a coil of wire. Both of these can be tested easily. If resistance specifications are known, we can make the two required ohmmeter checks. Figure 15-11 shows the ohmmeter setup for these readings. Do not forget that you are trying to see if the coil can generate a signal (meter #1) and deliver it to the module (meter #2). Meter #1 should be to the manufacturer's specification while meter #2 should be infinity (∞). Infinity ensures that the signal will not be delivered to ground. In addition to the resistance checks, make sure that the sensor is still magnetized. It should be capable of picking up some paper clips and holding them as Figure 15-12 shows. Another method of testing the sensor is to actually look at the signal it will generate. Keep in mind that many of the sensors in use generate very small voltages in the .250 volt range. Set your meter or scope up accordingly. Use the record function of your meter, if it has one, and crank the engine over. Just about any voltage should trigger the module and indicate a correctly functioning sensor. A scope can also be used to look at the sine wave coming off the sensor. Figure 15-13, page 380, shows what a good sensor will look like on a scope with the engine cranking over.

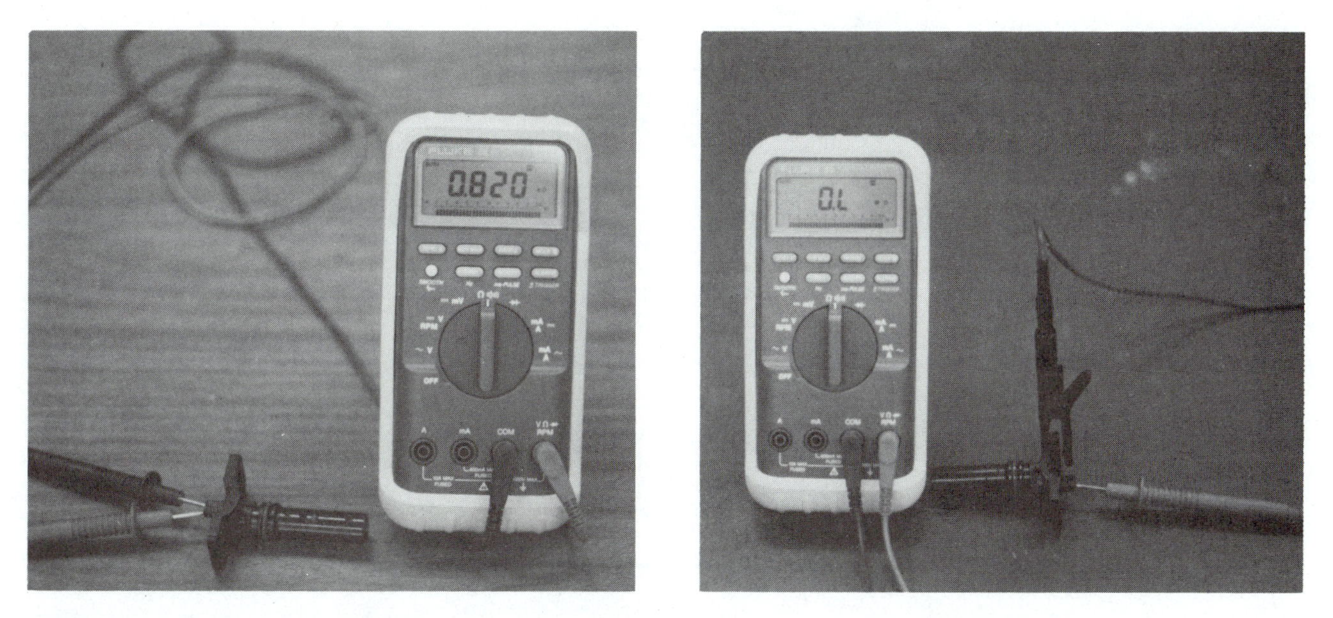

FIGURE 15-11 Pick-up coil with two ohmmeters: (left) across leads; (right) lead to ground.

FIGURE 15-12 Magnetic sensor with paper clips.

HALL EFFECT SENSORS

Hall effect sensors require a different test procedure. Keep in mind that they require voltage and ground applied before they will generate a signal. Most Hall effect sensors cannot be easily tested with an ohmmeter. Instead they should be tested using a digital voltmeter, preferably one that will record the voltage changes while you crank the engine over. The procedure is as follows:

1. Identify which leads are B+, ground, and sensor output.
2. Probe the sensor output wire and crank the engine over.
   a. Try to avoid piercing the wire if possible.
3. Test at connectors by backprobing.
   a. If you cannot test at a connector, gently pierce the insulation of the wires with a sharp pin.
4. Do not use your test light as its tip is too dull and will leave behind a large hole in the insulation.
5. When you are through testing, make sure you cover this small hole with insulating material, especially if it is exposed to road splash. Hardening gasket cement works well for this purpose.

The recording VOM should show you that a good sensor generated a high (on) then low (off) signal. A good sensor will consistently generate this signal. When the Hall effect is in the window, the voltage should go low. When the shutter is blocking the Hall effect, the voltage should go high. How high and how low the voltage swings will depend on the manufacturer. Generally, DIS sensors deliver full B+ for the high and zero volt for the low. Figure 15-14 shows the set-up for the output test.

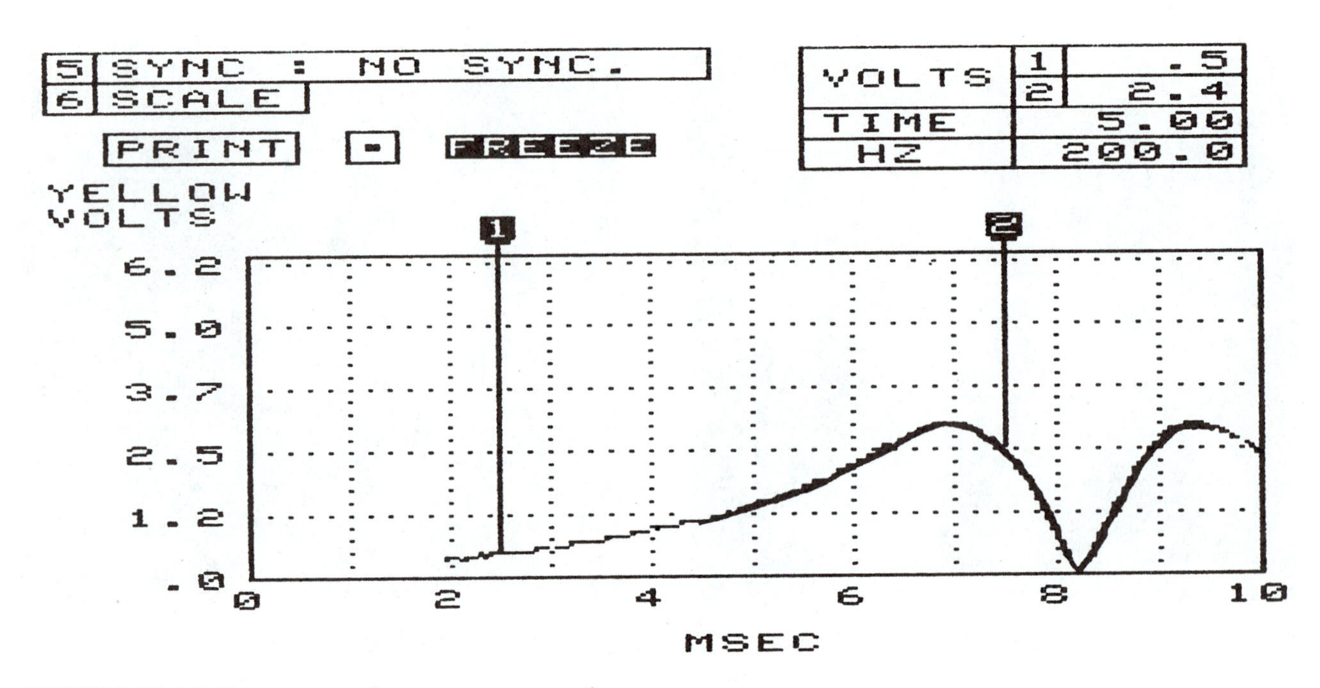

FIGURE 15-13 Bear scope sine wave—magnetic sensor.

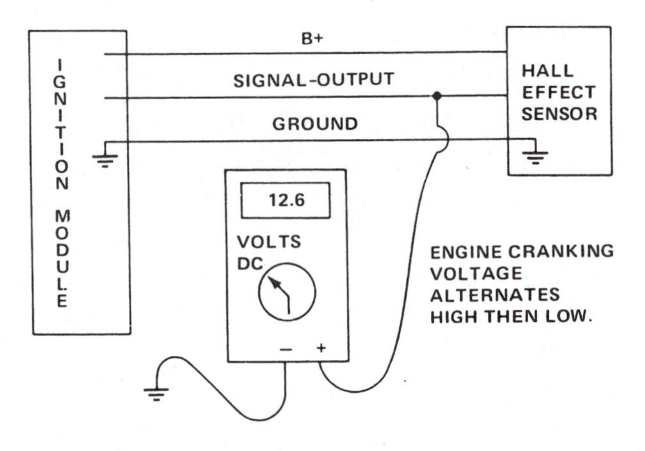

FIGURE 15-14 Hall effect—testing for output.

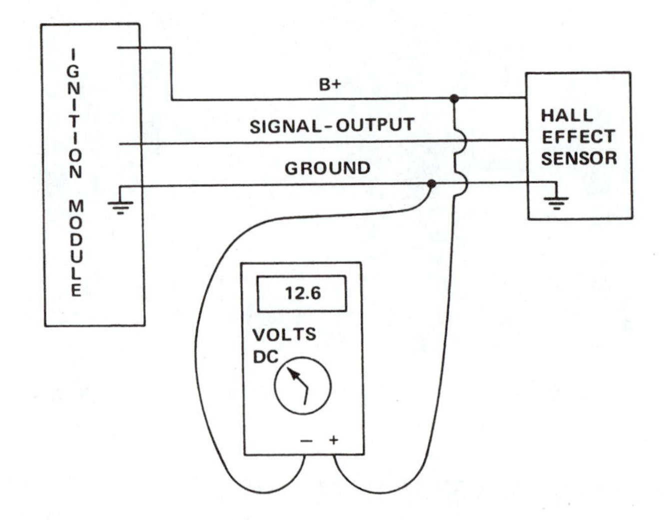

FIGURE 15-15 Hall effect with VOM testing for power and ground.

If you do not have sensor output, you must verify that B+ and ground are available at the sensor. You can do this by measuring the applied voltage and ground as close to the sensor as you can get. Figure 15-15 shows this procedure. Do not forget, the sensor needs B+ and ground or it cannot function. When you connect your DVOM to the B+ and the ground that the sensor will be using, and see a voltage, you have verified that the sensor has everything it needs to function. No voltmeter reading indicates a problem in either the B+ or ground circuits. You can also check out a Hall effect sensor with a scope. Probe the sensor output wire and crank the engine over. The square wave seen in Figure 15-16 indicates a correctly functioning sensor.

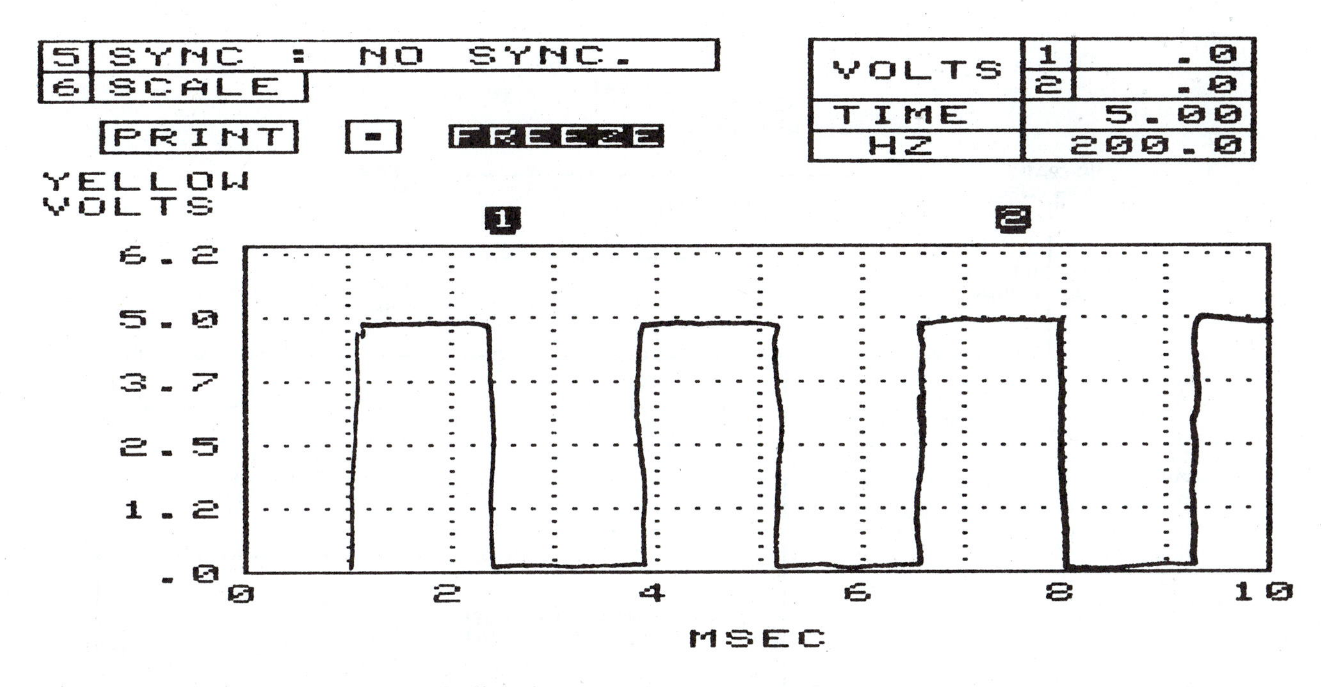

FIGURE 15-16 Bear printout—Hall effect sensor output square wave.

NO SPARK—USING SCANNER

Let us turn our attention to a no spark condition and discuss some general practices that can be used on most DIS vehicles. A no spark condition can be the result of many different problems, so it is important to follow some logical method of testing. In addition, the computer has many interrelated systems. The control of both fuel and ignition by the computer is a good example of this interrelationship. If we understand that both systems rely on the same information, our diagnosis job will be easier and more effective. Refer to Figure 15-17, which is a generic DIS test sequence.

We will use a **scanner** if the vehicle has a test connection and the ability to deliver computer information while cranking. This information is referred to as serial data. It is the sensor information that the computer is reading (inputs), in addition to the jobs or functions (outputs) that it is performing. For example, the signal from a crank sensor that goes to the computer is an input, while the timing signal (such as EST for a GM vehicle) is the output. The ability to read the inputs and outputs is not available on all vehicles. Some systems do not allow us to look at serial data. However, if you can monitor the inputs and outputs of the computer, you will save yourself a lot of diagnosis time.

Think about the process that the ignition module and computer will go through as they generate and deliver both fuel and ignition and the diagnosis will be simpler. First of all, a signal must come from the crank sensor indicating that the engine is cranking. Next some signal must be generated that indicates cylinder number one's position. Both the vehicle's computer or processor and ignition module must have both bits of information so that they can turn on the correct injector and the correct coil With both spark and fuel being delivered a the correct time, the engine should start. This system of inputs and outputs is block diagrammed for you in Figure 15-18. Notice that serial data is considered an output. With this in mind, let us look at the use of a scanner to

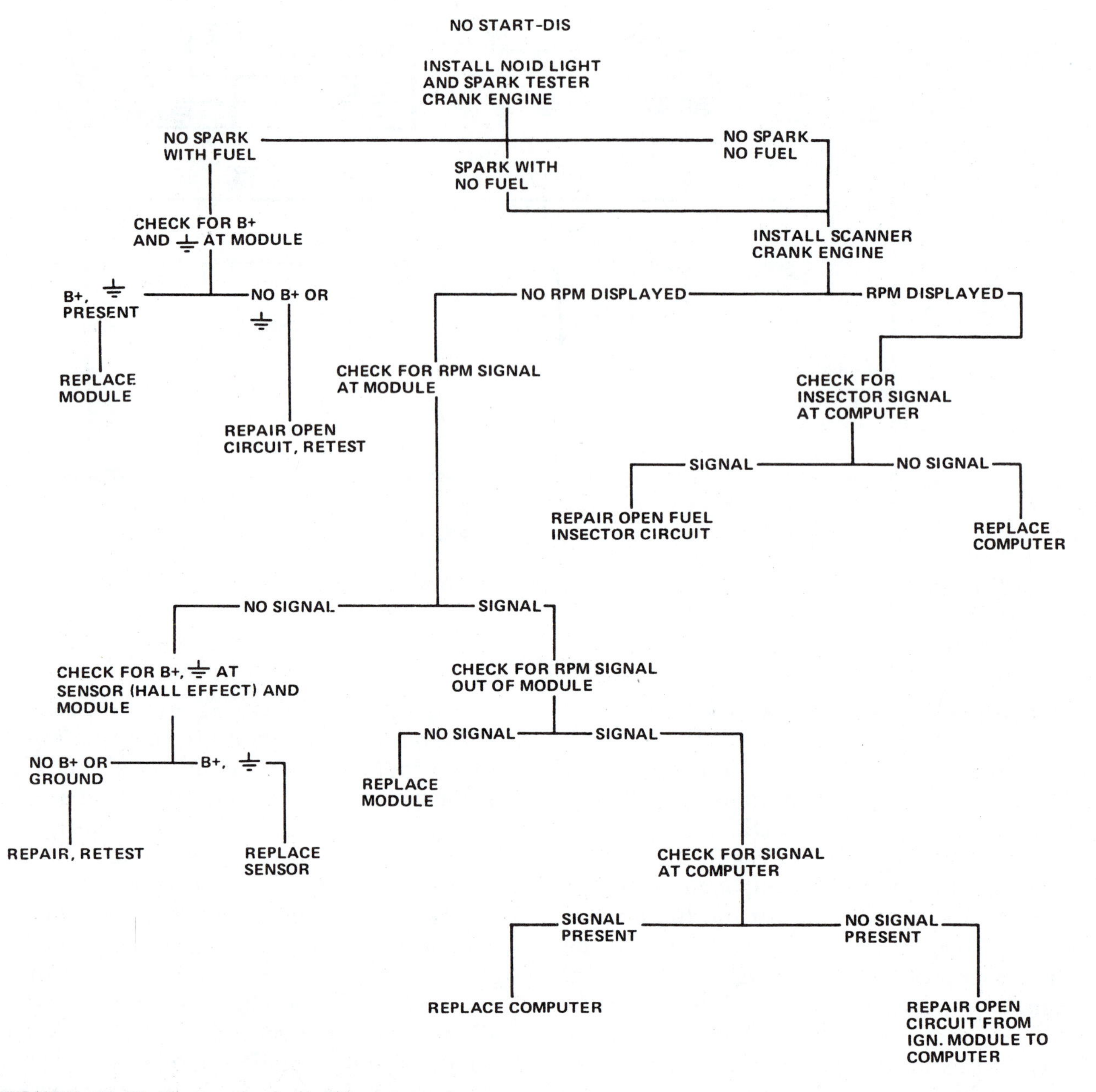

FIGURE 15-17 Diagnosis chart with scanner.

guide our diagnosis. Figure 15-19 shows a popular scanner which can be plugged into the serial data connection. It will read various inputs and outputs for us.

It is important that the scanner be programmed for the correct vehicle application, year, model, etc. Make sure you have set the scanner up correctly or the data you receive may not be correct. Read the scanner manual! Once the scanner is installed in the serial data line, we program it to read cranking rpm. If cranking rpm is available off the serial data line, we know that the crankshaft sensor must be functioning. For the computer to be able to read rpm and deliver it on the serial data line, the crankshaft sensor must have been generating a signal. This signal went through the ignition mod-

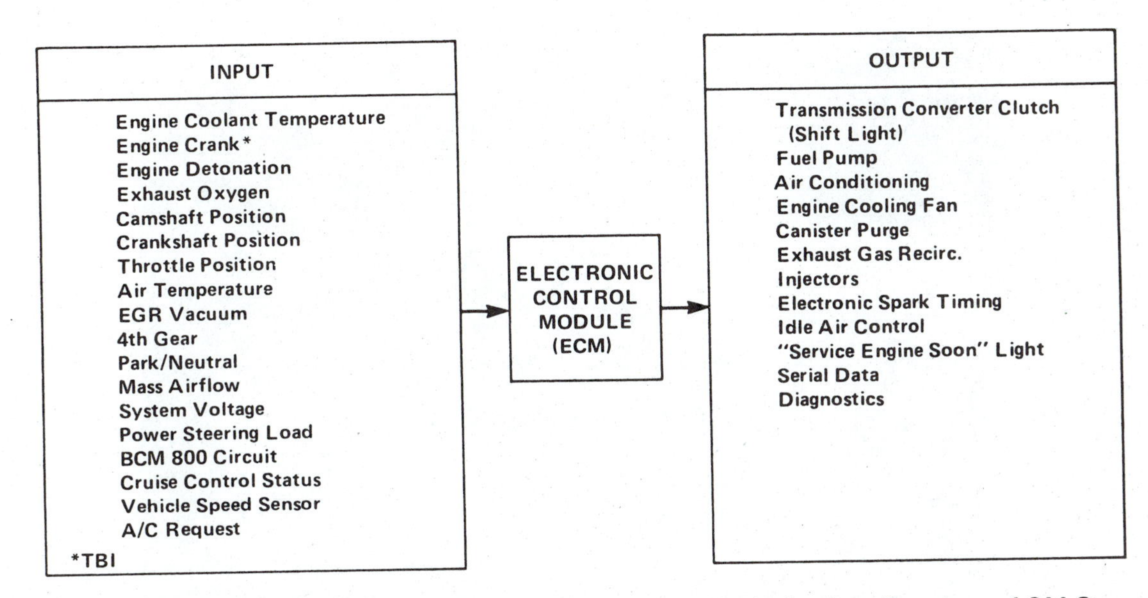

FIGURE 15-18 Inputs and outputs (serial data) for GM advanced fuel injection. (Courtesy of GM Corporation)

FIGURE 15-19 OTC Scanner. (Courtesy of OTC Division)

ule and on to the computer. Now, do we have fuel delivery?

Let us introduce another real simple and effective tool called a noid light. A noid light will flash each time a signal is sent to an injector to fire. Figure 15-20 shows a noid light installed in a fuel injector connector. If we crank over the vehicle, and have a fuel signal, the computer must have recognized cylinder number one's position. Note that a scope can be used to look at a fuel injector signal. Figure 15-21 shows a correctly functioning injector pattern. What do you think could be the problem if we have cranking rpm and no fuel delivery signal? Analyze this situation as follows.

Cranking rpm indicates a signal from the crankshaft sensor. No fuel means either no cylinder identification or no signal generated by the computer. The scanner and noid light have quickly reduced the components requiring testing. How about the condition of no cranking rpm and no fuel delivery signal? This condition means that the computer is not turning on the fuel system because it does not know that the engine is cranking. More testing will be necessary. Now let us go through the last variable—spark. Knowing whether we have spark will be the last bit of information we should need to get us on the way to a repair. With a test plug installed, or better yet, one test plug per ignition coil, crank the engine over and test for spark. No spark out of one coil, spark out of the other coil or coils would indicate that the ignition system is definitely in need of attention.

Now comes the advantage to many DIS. With multiple coils available right on the vehicle, we can change their position around. Moving the no spark coil to another position on the module and cranking the vehicle over again should tell the story. Note: make sure that you keep the spark plug wires on the correct plugs and coil The wires must stay in

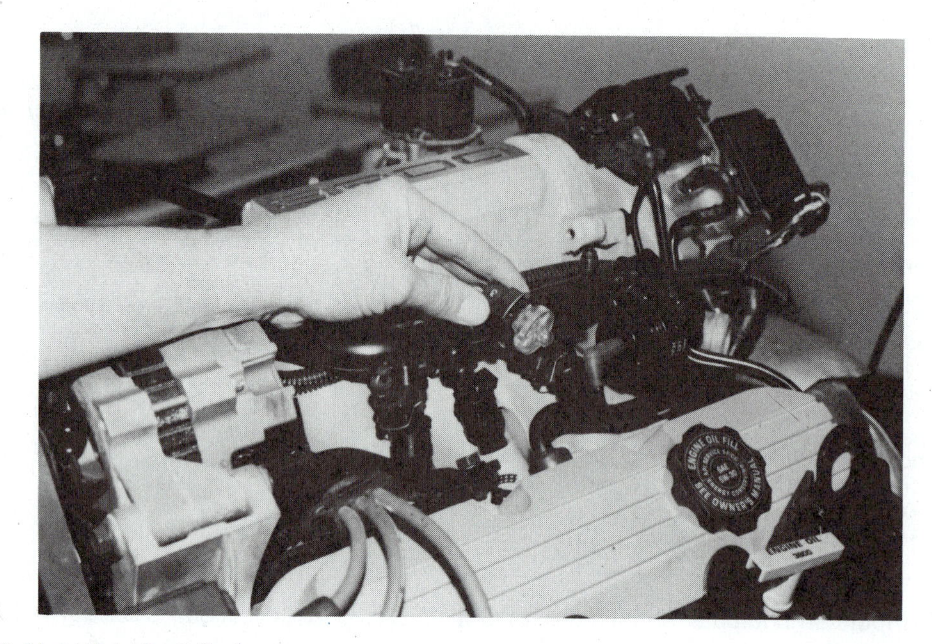

FIGURE 15-20 Noid light installed.

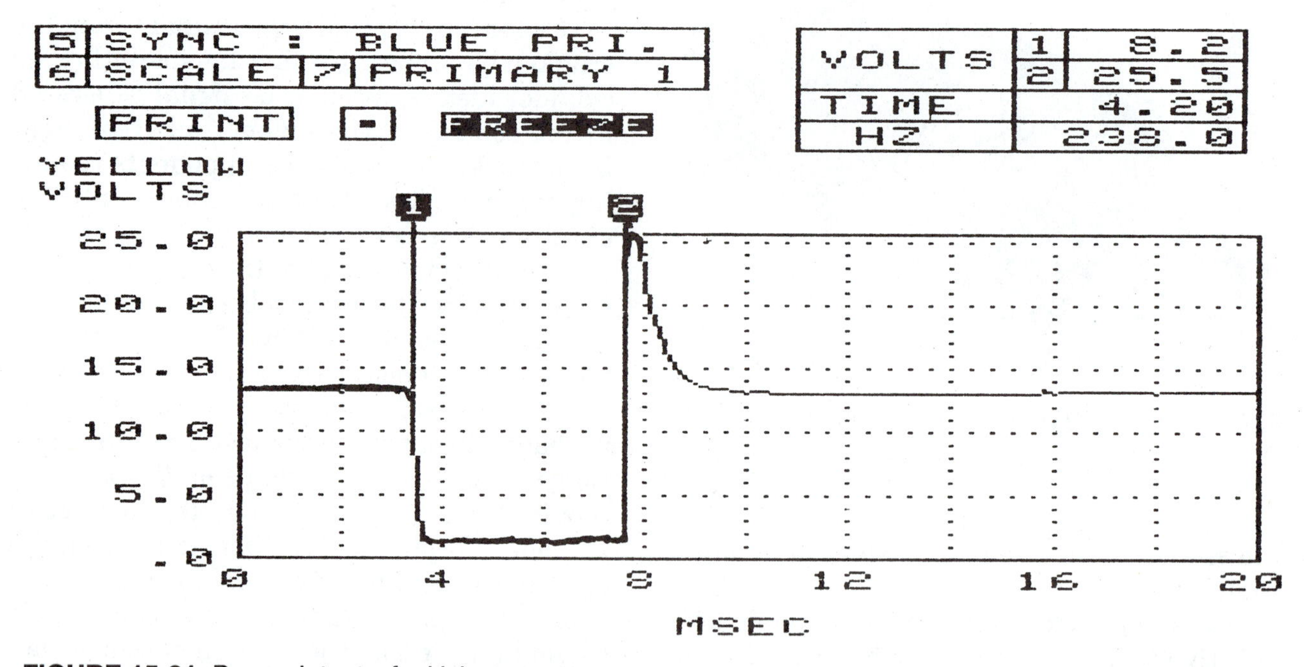

FIGURE 15-21 Bear printout—fuel injector pattern

the original coil position. They **SHOULD NOT** be moved with the coil as this would fire the plugs at the wrong time, possibly damaging the engine. While cranking the engine, did the no spark move with the coil to two different plugs? If so, the coil is bad. Or, did the no spark remain at the same two cylinders even with the new coil? If so, the module is bad.

Do not forget that inside the module there are actually 2, 3, or 4 individual modules each responsible for the firing of one coil. It is possible for one coil firing section to be burned out and have the rest of the module be okay. What if we crank the engine over and have no spark out of any of the coils? Think your way through this one. If no spark is being generated, either we have no power to the coil packs, or no rpm/cylinder identification input, or a bad module. Combining the spark information with cranking rpm and fuel injector signal (noid light), the pieces of the puzzle should come together.

Cranking rpm indicates a good crankshaft sensor and fuel indicates a good cylinder identification. Knowing what you have or do not have should lead you to the correct component test. This process sounds complicated and probably requires that you read it over a few times for it to make sense. Once you understand it, it will save you diagnosis time though. Noid lights are available at most professional parts stores, and scanners are in just about all repair facilities. Scanners are another example of using the right tool for the technology.

SUMMARY

In this chapter, we have taken a look at some of the common diagnostic repair procedures for the modern distributorless ignition system.

We started by reviewing the precautions necessary when dealing with onboard electronic systems, emphasizing never to disconnect or reconnect with power on, always observing polarity, and guarding against high voltage by checking battery chargers, starting units, etc. The oscilloscope was used to diagnose either an intermittent problem or to quality check a tune-up or repair.

We then tested ignition coils for opens, shorts, and faulty insulation using an ohmmeter. The next task was the testing of both magnetic and Hall effect sensors using a new type of digital volt ohmmeter, with recording capabilities. We used this recording DVOM to look at both the on/off signal from the Hall effect sensor and the AC sine wave from the magnetic sensor.

We finally looked at a procedure to test a vehicle that displays serial data using a scanner. We learned how to use this data to shorten our diagnosis time and guide us to failed components quickly. We used the same principle of testing around the module that we used with other electronic ignition systems. The process of testing the sensors, the coils, wires, and plugs, in addition to verifying that both B+ and ground are available to the module, indirectly tests the module. We saw that if all other components are functioning correctly, and we still do not have spark, the module must be at fault.

KEY TERMS

Impedance
Scanner
Magnetic sensor
Hall effect sensor

CHAPTER CHECK-UP

Multiple Choice

1. Technician A states that the battery should be disconnected before it is charged on a DIS vehicle. Technician B states that systems are protected against reverse polarity. Who is correct?

    a. Technician A only
    b. Technician B only
    c. Both Technician A and B
    d. Neither Technician A nor B

2. When doing a cranking coil output test with a scope
   a. hold a plug wire away from ground.
   b. disconnect the injectors and crank the engine.
   c. use a test plug set for about 40kV.
   d. you cannot do a cranking coil output test on a DIS vehicle.
3. Technician A states that mS of spark should exceed .8 or the vehicle may misfire. Technician B states that ms of spark should be less than 2.0 or the plug life will be decreased. Who is correct?
   a. Technician A only
   b. Technician B only
   c. Both Technician A and B
   d. Neither Technician A nor B
4. An ohmmeter reads .2 ohm with the leads placed across the primary connections of a DIS coil. This indicates that the primary
   a. has continuity.
   b. is open.
   c. is shorted.
   d. is grounded.
5. A recording voltmeter is placed across a magnetic sensor and registers .248 volt AC with the engine cranking. This indicates that the sensor
   a. is shorted.
   b. is open.
   c. is generating a signal.
   d. is grounded.
6. A Hall effect sensor has no output, when tested with a voltmeter and the engine cranking. Technician A states that the sensor may be faulty. Technician B states that the sensor may not have B+ or ground applied. Who is correct?
   a. Technician A only
   b. Technician B only
   c. Both Technician A and B
   d. Neither Technician A nor B
7. On a vehicle with no spark, a scanner indicates no cranking rpm with the engine cranking. This indicates that
   a. the crank sensor may not be generating a signal.
   b. the module may be faulty.
   c. an open exists between the sensor and the vehicle computer.
   d. all of the above.
8. Testing verifies that coil #1 is not producing a spark, and coil #2 is. To isolate the problem to the coil or the module, Technician A states to swap the coil positions on the module. Technician B states to substitute a known good coil pack. Who is correct?
   a. Technician A only
   b. Technician B only
   c. Both Technician A and B
   d. Neither Technician A nor B

9. A scanner shows cranking rpm, a noid light placed in the injector connector blinks, and neither coil on a four cylinder vehicle will produce cranking spark. The most likely cause is a faulty
   a. crankshaft sensor.
   b. coil pack.
   c. module.
   d. computer.
10. An ohmmeter reads 0 ohm with the leads across the connector to a magnetic sensor. This indicates that the
    a. sensor is shorted.
    b. signal will be alternately high then low.
    c. sensor is open.
    d. none of the above.

True or False

11. An infinity reading across a DIS ignition coil primary indicates a good coil.
12. A zero-volt reading with the window in the Hall effect sensor is a good reading.
13. A scanner indicates cranking rpm. This indicates that the crankshaft sensor is functioning.
14. It is okay to leave a battery connected while charging it.
15. 2.8 mS of spark will give extended spark plug life.
16. The firing voltage of the waste spark should be between 6kV and 12kV.
17. A Hall effect sensor should produce an AC sine wave.
18. An ohmmeter reads infinity with one lead on a magnetic sensor wire and the other lead to ground. This is a good reading.
19. An ohmmeter reads zero ohm between the DIS coil primary and the secondary winding. This indicates that the coil is shorted.
20. Cranking coil output should only be done with a spark plug tester inserted in one of the plug wire ends.

CHAPTER

Accessories

OBJECTIVES

At the conclusion of this chapter, you should be able to:

- diagnose and repair lighting circuits, including headlight systems, taillight systems, and turn signal systems.
- diagnose and repair relay and non relay horn systems.
- diagnose and repair standard windshield wiper systems, including the washer systems.
- diagnose and repair interval wiper systems.
- diagnose and repair front and rear electrical defoggers.
- diagnose and repair power window systems, including station wagon rear tailgate windows.
- diagnose and repair engine cooling fan circuits for standard cooling and air-conditioning equipped engines.

INTRODUCTION

The age of electrical accessories on the modern vehicle began years ago. Today the technician is faced with the opportunity, and sometimes the challenge, to diagnose and repair many different types of accessories. Some are considered standard equipment, such as headlights, turn signals, and horns, while others are considered optional, like front and rear defoggers. In this chapter, we will look at some of the more common standard and optional circuits that you will be expected to diagnose and repair once you are a working technician. You will not find them difficult to comprehend if you keep in mind the basics of series and parallel circuits. In addition, if your understanding of relays is weak, review the appropriate section of this text. Many of the common vehicle accessory circuits utilize relays. A complete and thorough understanding of relays is a must if accessories and their control circuits are to make much sense to you.

LIGHTING CIRCUITS

The lighting circuits on today's vehicles are just as important as they have always been. Their contribution to safety and carefree driving is taken for granted by most of us. This circuit is an example of one which has not seen the need for change and remains the same as it was in the fifties. The exception to this is the automatic dimming circuits that we will look at later in this section. For now, let us look at the typical headlight circuit that is manually operated, starting with the headlight switch. Figure 16-1 shows a typical schematic diagram of a headlight switch. As the light switch knob is pulled, it will pull the wipers inside to the different positions shown in Figures 16-2 and 16-3. The wipers will allow the switch to direct the B+ to the parking lights and the headlights. Let us follow the current through the switch and see how it works. With the switch off, B+ flows through the right circuit breaker and is made available to the horn and stop light circuits. Notice that the circuit from the B+ common point to the left is open through the internal circuit breaker and the switch wiper if the switch is in the off position. As the switch is moved to the first position, power through the right circuit breaker flows through the wiper to the parking lights. Notice that the left wiper is now connected to a vacant contact on the switch. The only light circuits powered are the park lights and the dashboard lights. As the switch is pulled to the full on position, notice that power becomes available to the headlights while remaining on to the dash lights and park lights.

As the knob is turned back and forth, the variable resistance is added or subtracted off the right circuit breaker to dim the dash lights. If the knob is turned completely to the counterclockwise position, the additional B+ will be directed through the additional contacts to the dome lights, turning them on. Notice that the headlight switch is far from simple because it contains six contacts, two wipers, two circuit breakers, and a variable resistance.

If, in the process of a repair, you have to remove and replace the headlight switch, you will have to remove the knob before you remove it from the dash. Most switches have a spring-loaded release button that is pushed in to remove the knob, Figure 16-4. An ohmmeter or self-powered test light can be used to de-

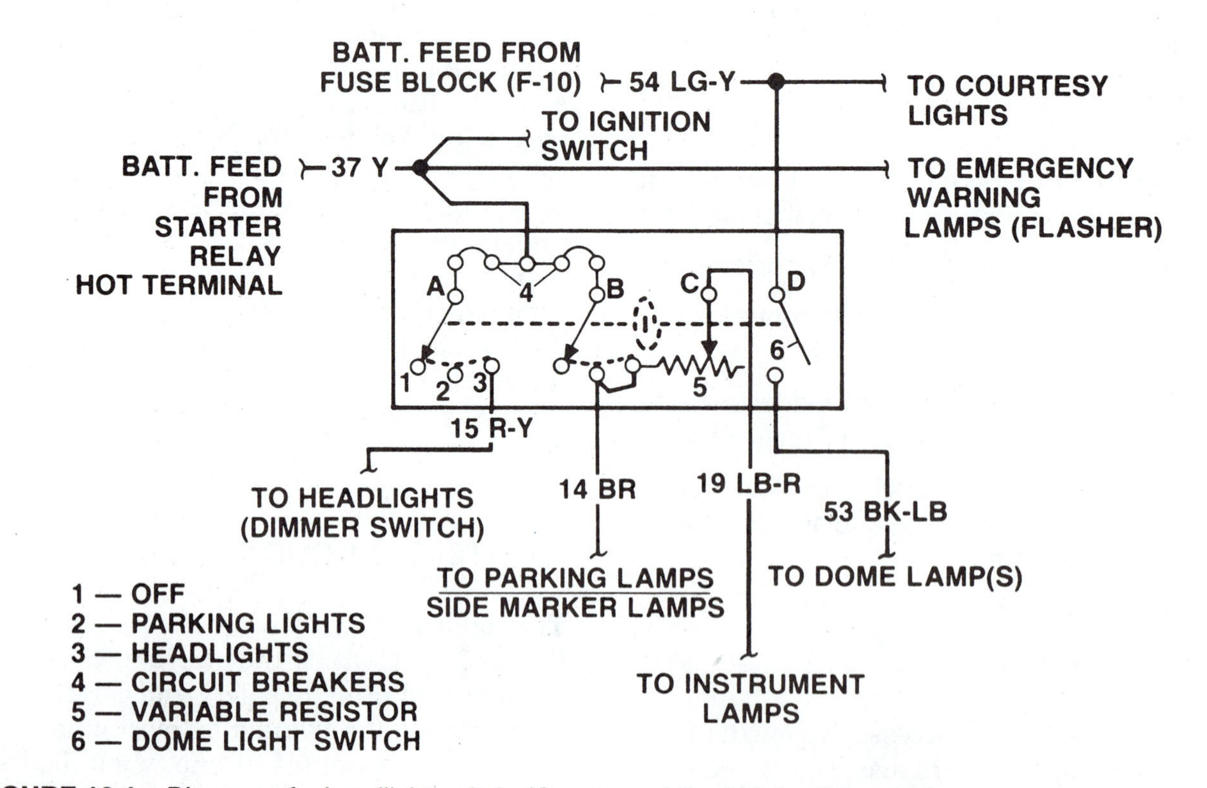

FIGURE 16-1 Diagram of a headlight switch. (Courtesy of Ford Motor Company)

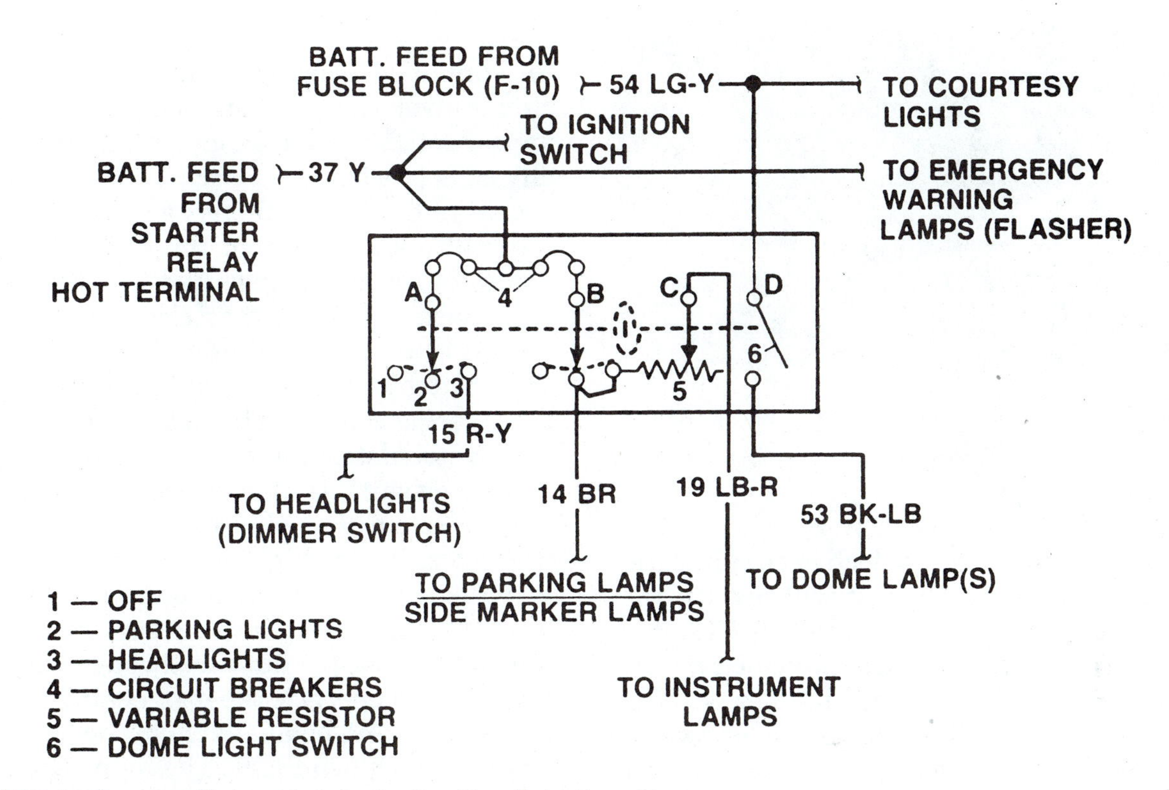

FIGURE 16-2 Headlight switch in the "parking lights" position.

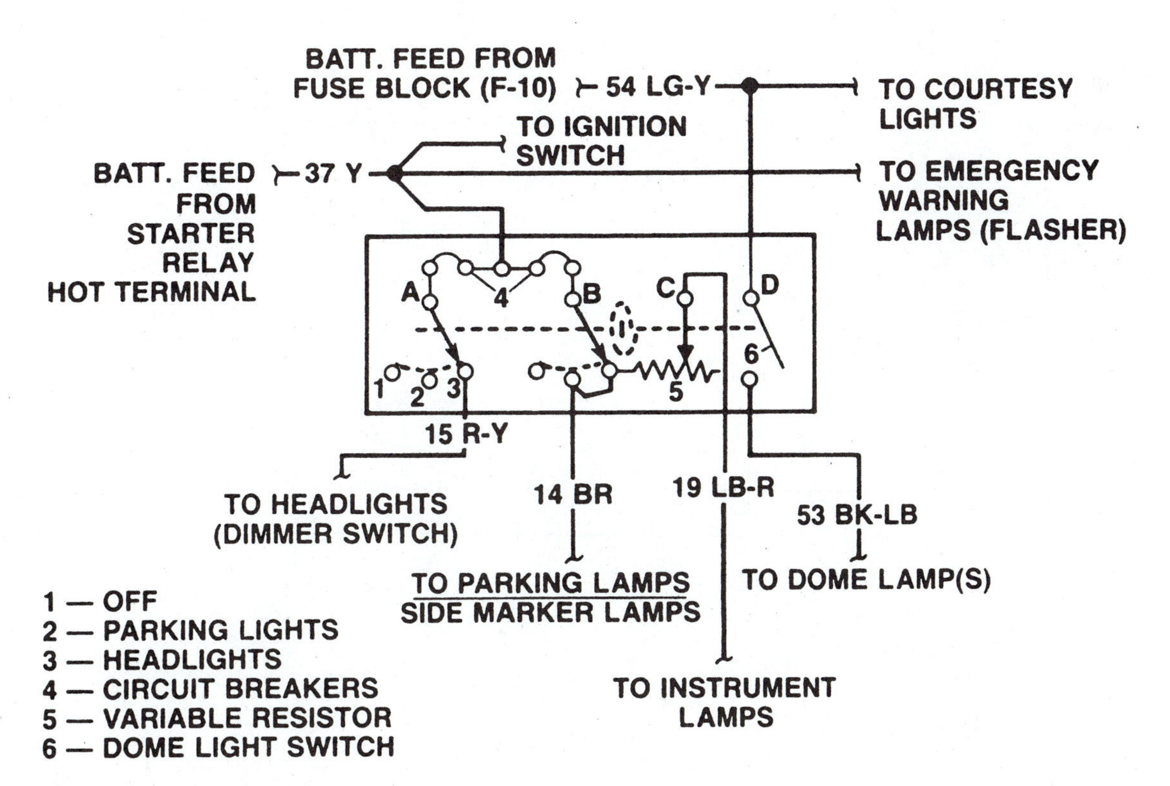

FIGURE 16-3 Headlight switch in the "headlights on" position.

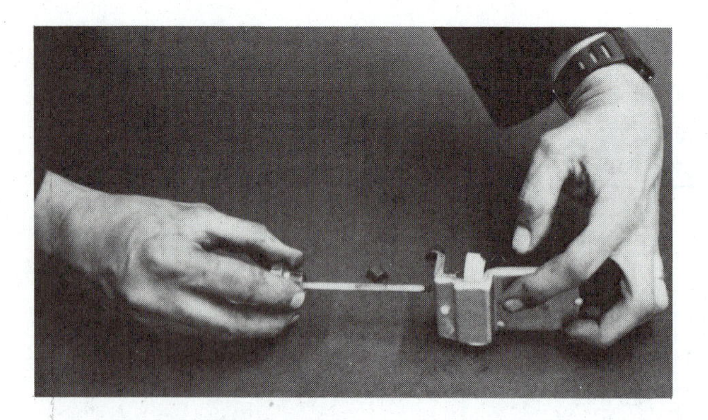

FIGURE 16-4 Pushing a release button in allows the removal of the knob.

termine and verify the defective circuit. A repair of the defective circuit will involve the complete replacement of the switch. The control of the headlights is accomplished by the headlight switch and the dimmer switch. Whether they are square, round, or two or four lights, the circuit is basically the same for the majority of vehicles. B+ flows through the closed contacts of the switch, through a low/high beam selector switch, the filaments of the bulbs, and the ground circuit. Notice from the simplified drawing of Figure 16-5 that the circuit involves many common points and can appear much more complicated than it actually is. Let us trace the circuit from the power source to the bulbs. Do not forget that there are multiple contacts in the switch. At this point, we are only concerned with the headlight contacts. When they close, B+ becomes available to the headlight dimmer switch. This switch can be located on the floor or on the steering column. The position of the dimmer switch determines the path that current will have available. It is drawn in the high position. Notice that this two position switch will select either high or low, but not both. We will discuss a "flash to pass" style switch later that can select both sets of filaments at the same time. B+ flowing through the high contacts will energize the high beam indicator on the dash and be available to the larger filament of each headlight. The high beams are brighter than the low beams. They shine brighter and illuminate the road farther away from the vehicle than the low beams do. The

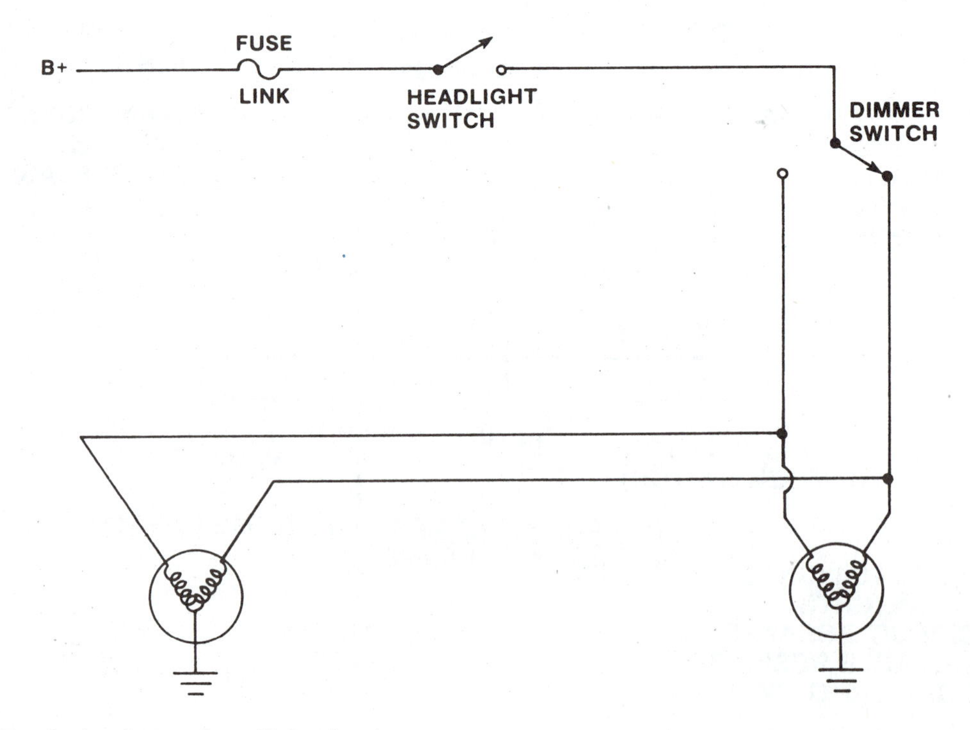

FIGURE 16-5 A simple two-headlight circuit.

ground islet grounds the filament away from the bulb. A case ground is not used because most headlights are glass, which requires a separate ground wire. We identified this type of ground as a remote ground. One filament in each headlight has B+ and ground available and should be lit, assuming it is not burned out. The brightness of each filament will depend on the voltage applied across it and its resistance.

Figure 16-6 shows basically the same circuit with an additional set of headlights. Many vehicles have four headlights. Two of the headlights are double filaments, while two are single filaments. Generally, the outside lights are double filaments and are in use for both low and high beam operation, while the inside lights are for high beam operation only. Electrically, they function the same. Notice that one ground has been used for each side of the vehicle for cost reduction. On some vehicles one ground is used for all four lights.

Diagnosing this circuit is very simple, if you remember that B+, the headlight switch, and the dimmer switch are wired in series, while the actual bulb filaments are wired in parallel. If the entire circuit is dead, the open must be in the series section. If just one bulb is off, then the open must be in the parallel

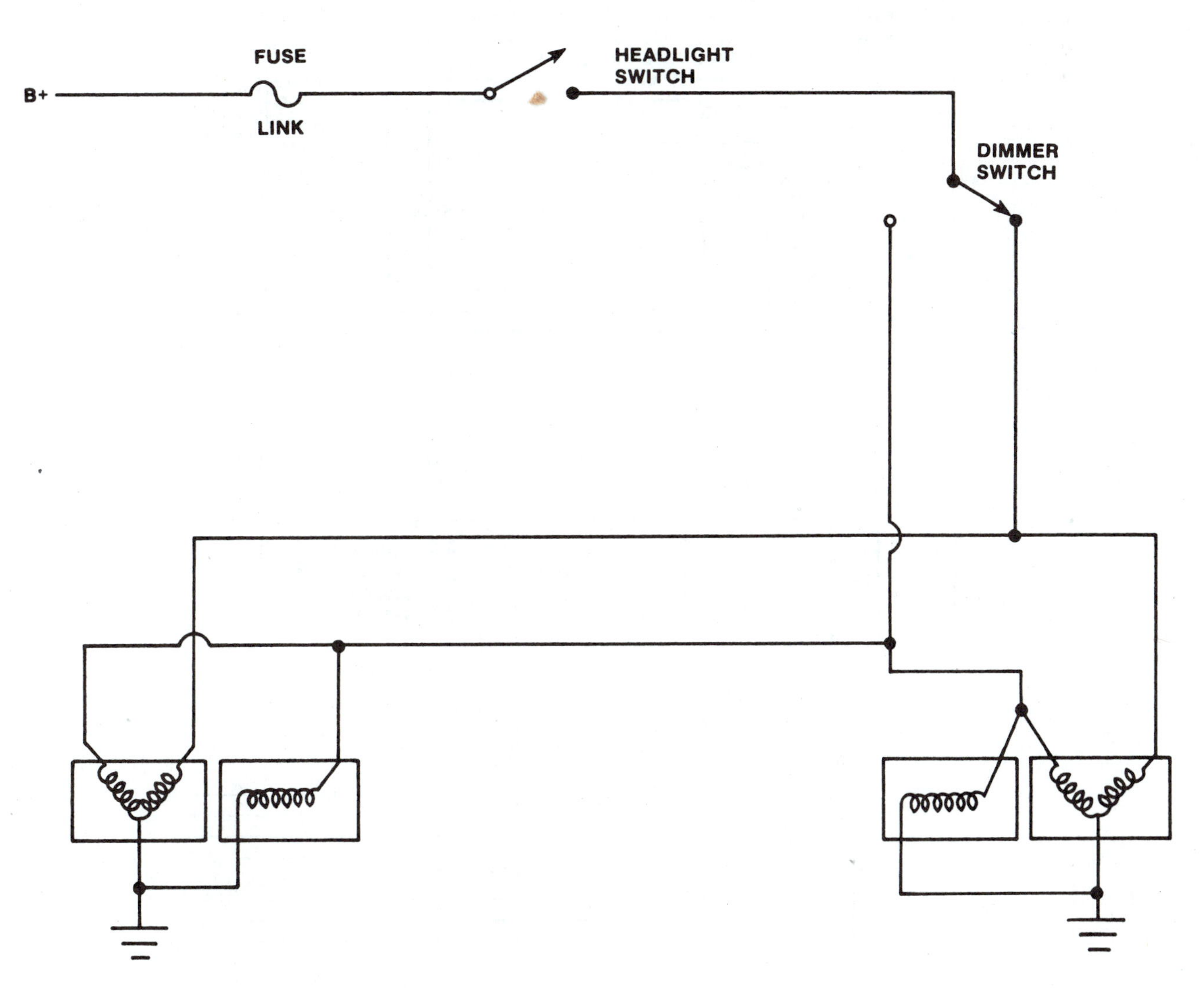

FIGURE 16-6 A four-headlight system.

section of the circuit. The use of the 12-volt test light is usually all that is needed for the testing and repair of the circuit. The majority of the difficulties involves excessive resistance in either the controls or the connections. In your repair and diagnosis, do not forget the remote grounds because they can be a source of resistance just like the B+ path.

"FLASH-TO-PASS" CIRCUITS

Some dimmer switches have the additional feature of the ability to energize both the high beams and the low beams even if the headlight switch is off. These circuits are usually referred to as "**flash to pass**". Their circuit is shown in Figure 16-7. By pulling back on the dimmer stalk, located as part of the turn sig-

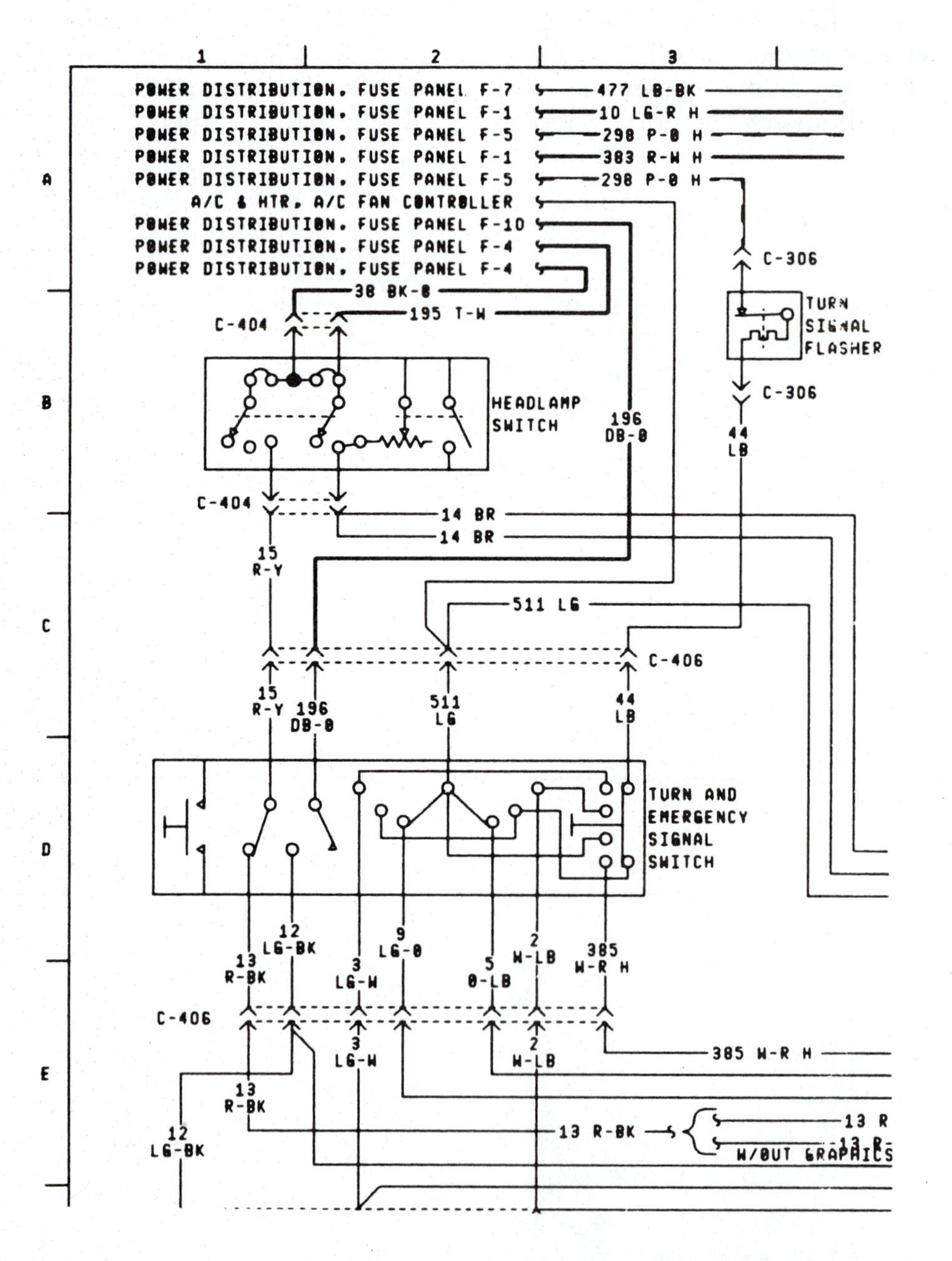

FIGURE 16-7 A "flash to pass" headlight dimmer switch. (Courtesy of Ford Motor Company)

nal switch, the entire headlight circuit has B+ applied to it and both sets of filaments light. In addition, the high beam indicator on the dash will light.

A further refinement of the dimmer circuit is found on certain vehicles. Automatic dimming using a photoelectric cell is available as an option on many luxury vehicles. The photoelectric cell is located either on the dashboard or by the grill where oncoming headlights will shine on it. As the vehicle gets closer, the voltage generated by the cell rises. Eventually, it reaches a present level and triggers the dimmer module, automatically dimming the headlights. Once the car passes, the headlights automatically return to the high beam position. This circuit is shown in Figure 16-8. It is generally diagnosed by shining a light at the cell and measuring the voltage generated. The dimming module is diagnosed by bypassing it with a jumper wire to manually energize the different filaments. No repair of either the module or the photoelectric cell is attempted in the field. They are replaced as units.

FOG LIGHT CIRCUITS

Many vehicles have fog lights, either from the factory or as add-ons. Generally, their circuits are the same. They usually involve a relay, switch, and the lights themselves. A relay is used because the amount of current that fog lights require, especially halogen ones, can be quite high; 25 amps is not unusual. Figure 16-9 is a typical circuit. Notice that an ignition switched hot (drawn as a dash) is supplied to the switch from the fuse box's auxiliary fuse. Fog lights are generally not wired directly from an unswitched B+, because this would allow them to be left on with the vehicle off. Their high current draw would quickly drain the battery. The dash switch

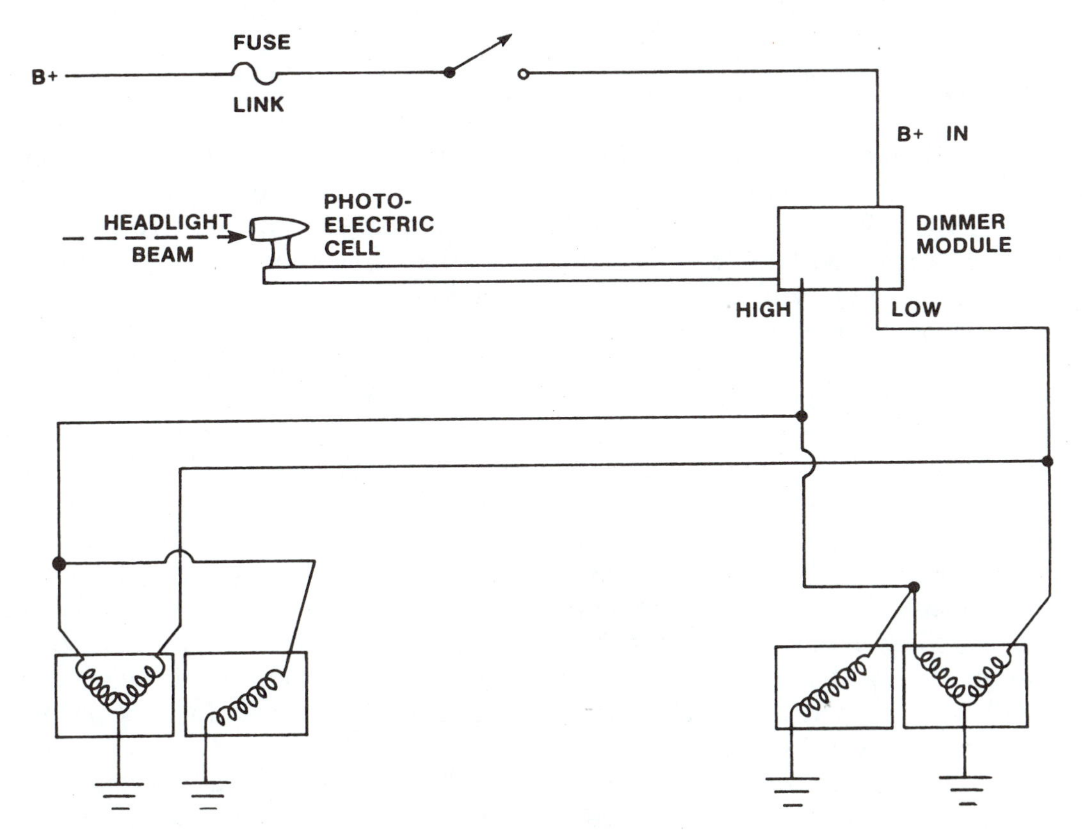

FIGURE 16-8 Automatic headlight dimming circuit.

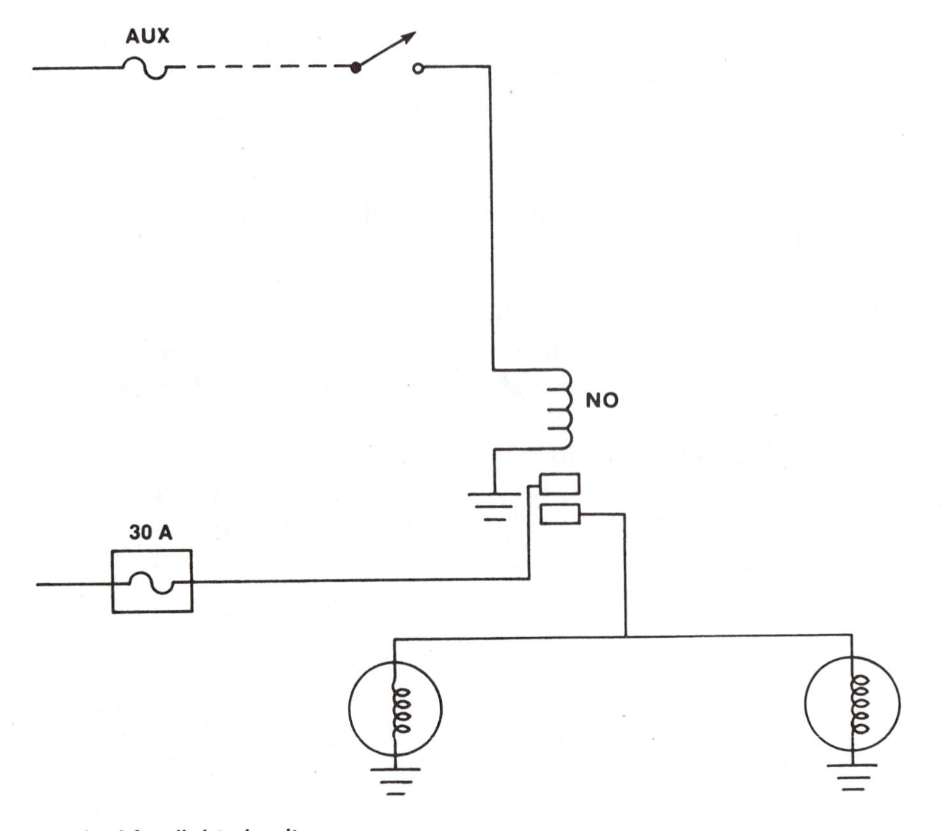

FIGURE 16-9 A typical fog light circuit.

controls relay coil current to one side of the coil. A direct ground is supplied to the other side of the relay coil. With both B+ and ground applied, current will flow through the coil and a magnetic field will develop. The field will close the contacts in the relay. Notice that the one side of the contacts are connected to a 30 amp fused source of B+, while the other side of the contacts runs to the fog lights, wired in parallel. Each filament has its own remote ground connection. Again, the circuit is diagnosed with the 12-volt test light as two separate circuits. First, the relay control circuit and then the high current circuit that powers the filaments. Do not forget that the fuse might be located under the hood where it will be subjected to the elements. Corrosion from the weather is a common problem. In-line fuses should be protected with dielectric compound to reduce the corrosive effects. If you add fog lights, or for that matter, any accessory, fuse it as close to the source as possible. Picking up B+ under the hood is easy, because there are many choices available: solenoids, horn relays, alternators, and even directly to the battery are some of the common choices. An in-line fuse should be placed as close as is practical to this source. An overload or a short circuit would safely open the circuit protector and eliminate B+ for the entire circuit. If the circuit protector is, for example, placed two feet from the battery , the two feet of wire is basically unprotected against overload. A short in this section could start a fire. In addition, do not forget that you should fuse for the size of the wire, and your choice of the wire size should be based on the load it will be powering. Overfusing and underwiring are two common problems that should be avoided. Always determine the wire size necessary for the load and then fuse for the wire size.

TAILLIGHTS

The rear of vehicles have many lighting circuits. Most cars have brake lights, run lights, turn signals, and back-up lights. In addition, all vehicles produced since 1986, have collision avoidance lights. Let us look at these circuits and see that their diagnosis is very simple once you are aware of how they are wired. We will start with the brake lights. The easiest circuit to look at first will be the three-bulb circuit found on many vehicles. Figure 16-10 is a drawing of a typical three-bulb brake system. The term "three-bulb" indicates that there are three separate filaments for each side of the rear of the vehicle. A separate filament for each function, brake, turn, and run. A constant source of fused B+ is made available to the brake switch. The brake switch is usually located on the brake pedal and is closed by pushing down on the brake, Figure 16-11. B+ is now available to the bulbs, wired in parallel, at the rear of the vehicle. Releasing the brake pedal allows the spring-loaded NO (normally open) switch to open and turn the brake lights off. A simple circuit that only requires a 12-volt test light or a voltmeter for diagnosis. The most common cause of failure are bulbs that burn out. Testing for B+ and ground at the bulb socket should verify the circuit. If B+ is not available at the socket, test for power at each connector, mov-

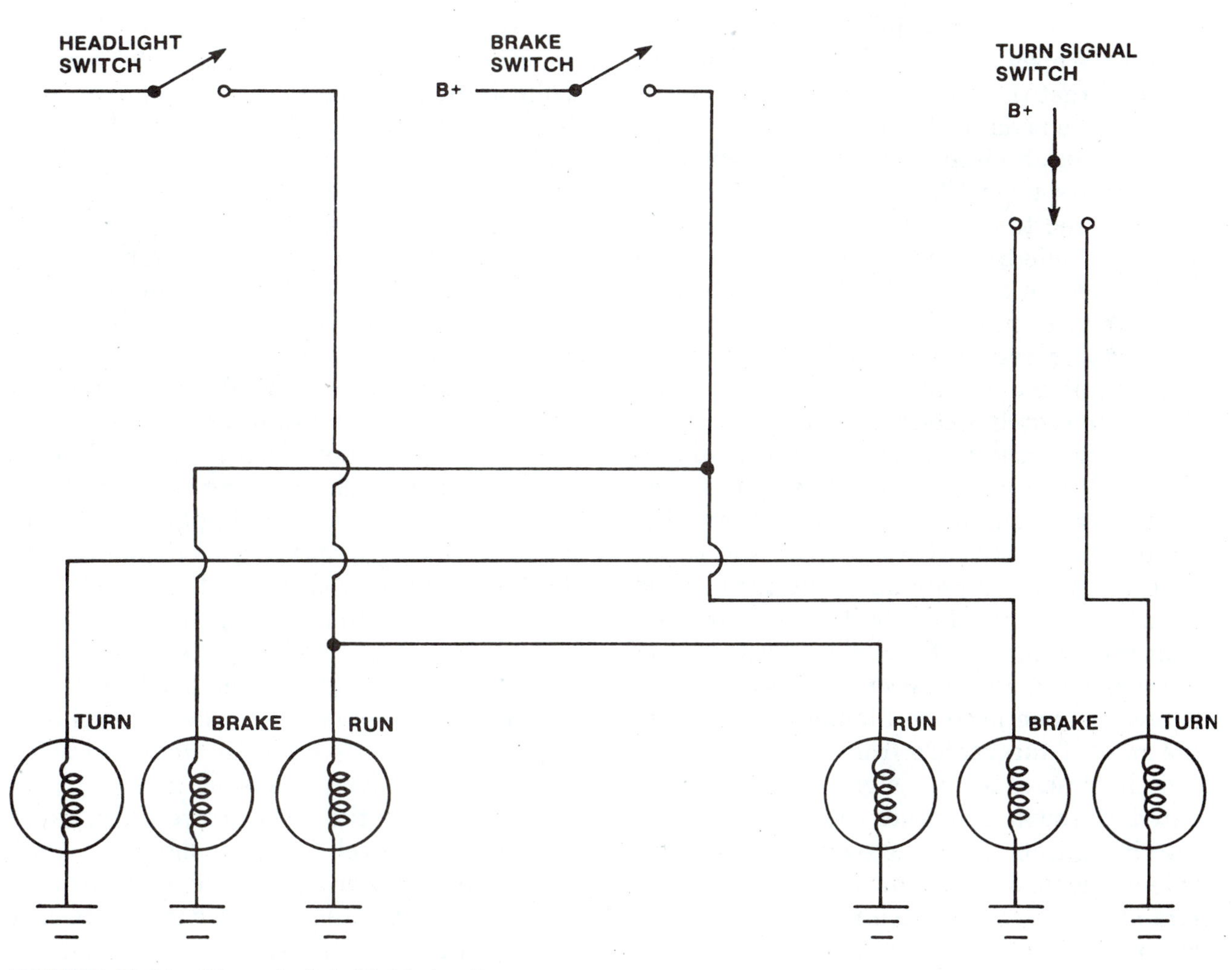

FIGURE 16-10 Three-bulb taillight circuit.

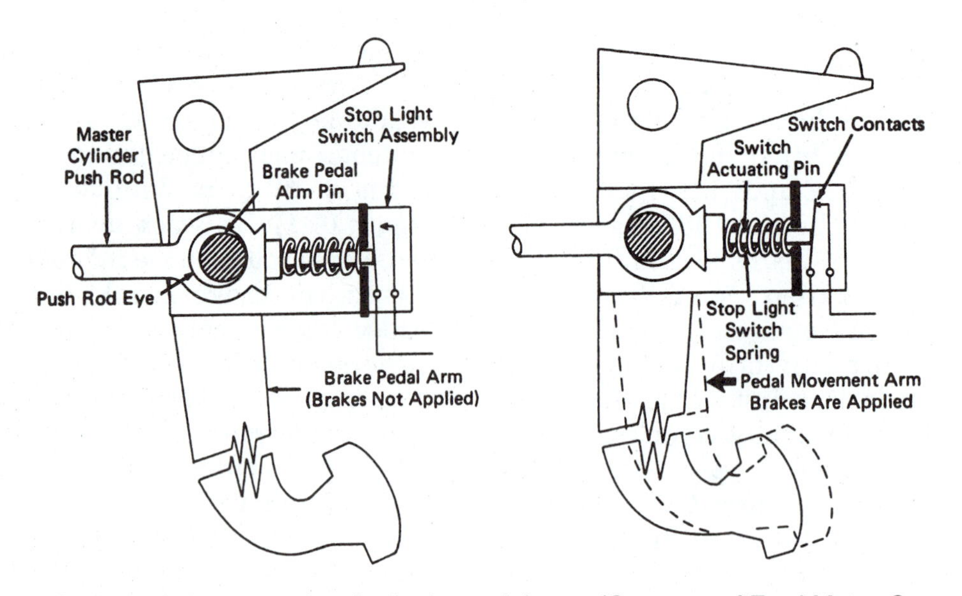

FIGURE 16-11 Brake switch mounted to the brake pedal arm. (Courtesy of Ford Motor Company)

ing back toward the switch until you find it. Repair the open and you are back in business. Do not forget that the circuit will only be hot if the brake pedal is depressed.

This simple circuit is modified slightly for a two-bulb system. This style of system is usually referred to as a combination system because a single filament will have two jobs to do, brake light or turn signal. We will be covering turn signals in depth a little later in this chapter. Figure 16-12 shows a simplified drawing of the circuit. Notice that B+ from the brake switch is connected to the turn signal switch. If the turn signals are not in use, B+ will flow through the turn signal switch and become available at both combination lamps. Notice that this style has the series part of the circuit only up to the turn signal switch, while the three-bulb system did not split the circuit into parallel until the rear of the vehicle. A combination lamp system has two wires, one for each filament, running to the rear of the vehicle from the turn signal switch. When the turn signal switch is activated, the switch removes the turning side of the circuit from the brake switch B+. Now one side of the vehicle will have the combination lamp being used as a turning light, while the other side will have it being used as a brake light. The key to understanding how this system functions is the turn signal switch. It will function just as a series to parallel common point if the turn signals are not in use, allowing the brake switched B+ to flow through it to the rear of the vehicle. If the turn signals are activated, it will remove one of the parallel paths from the brake circuit and use it for a turning light, while the remaining parallel path is supplying brake light power. Do not forget that the turn signal switch is an integral part of the brake circuit on a two-bulb system. In addition, B+ flows through the steering column and some additional connectors. This does not make the circuit more difficult, just longer. Testing is again accomplished with a 12-volt test light or a voltmeter after observing both the turn signal and the brake circuit operation. With the additional connectors and the turn signal switch in series, voltage drops could add up sufficiently to reduce the bulb wattage and brightness.

Beginning in 1986, collision avoidance lights have been part of the required safety equipment on vehicles sold in the United States. Having the light at eye level, rather than taillight level, has been shown to significantly reduce rear end collisions. In a three-bulb system, the filament is wired in parallel off the brake switch B+ as Figure 16-13 shows. The three-bulb system greatly simplifies the

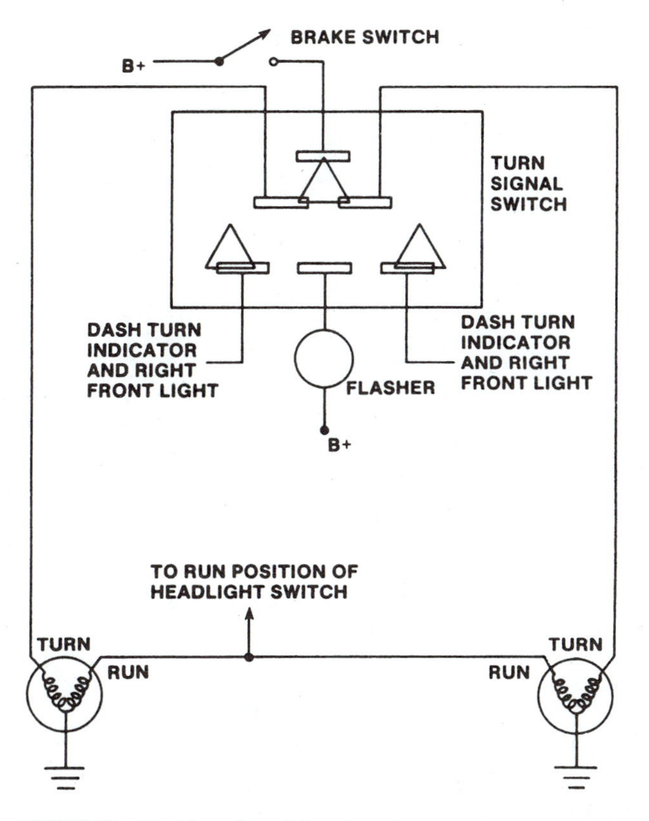

FIGURE 16-12 Combination lamp taillight circuit.

circuit for this light. The combination systems are a bit more difficult because of the dual use of the filaments. Generally, the manufacturers use a set of diodes in the conductors from the brake/turn signal filaments. Figure 16-14 shows this. If the turn signals are not in use, B+ from both brake lights is applied to the avoidance light through the diodes. When the turn signal is activated for a left turn, the left filament has the pulsing B+, while the right filament has the constant B+ required for brake light. Power for the avoidance light will come from the right circuit and the right diode will conduct. This power will be blocked by the left diode, preventing it from reaching the left filament. If B+ were to flow through the left diode, the filament would be powered off the right path and the bulb would not flash. The action of the diodes is critical to the function of the three filaments. They allow the separation necessary to the two sides of the vehicle. There are some add-on collision avoidance systems available that require the technician installing them to run a conductor from the output side of the

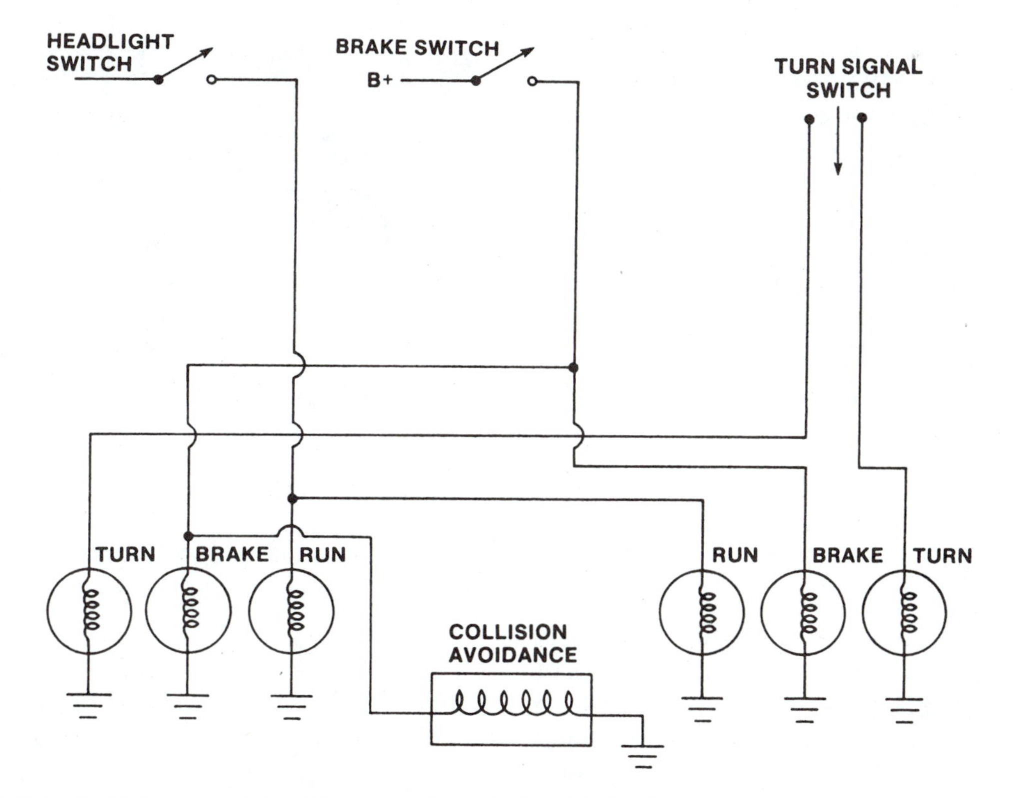

FIGURE 16-13 Collision avoidance light in a three-bulb taillight circuit.

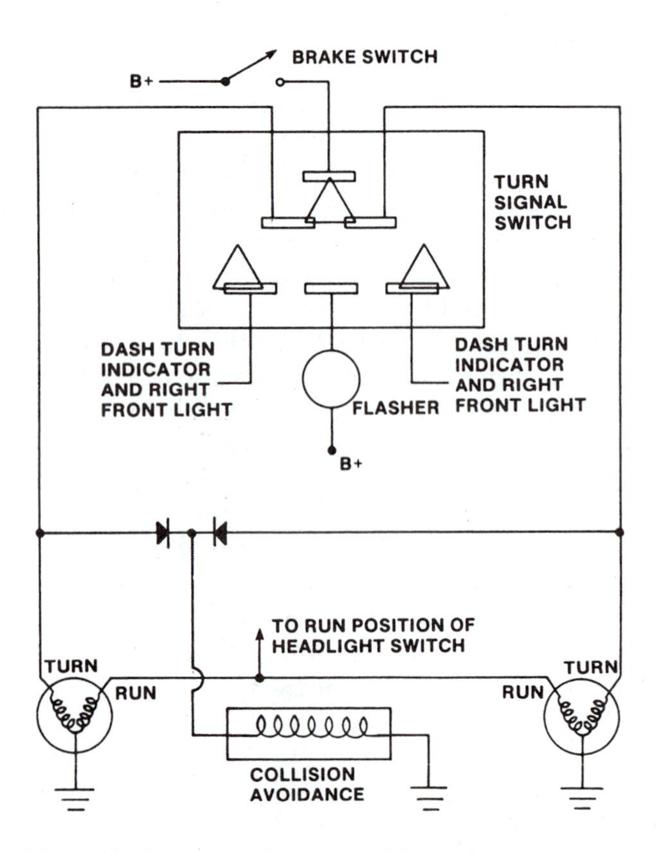

FIGURE 16-14 Collision avoidance light in a combination lamp-taillight circuit.

brake switch to the filament. The two-bulb system is simply a parallel connection from one of the brake lights to the extra light as Figure 16-15 shows.

Before we leave this section, we should mention that trailer lights can be wired in the same two choices, two- or three-bulb circuits. There are special convertors available that allow a two bulb trailer to be wired to a three-bulb vehicle. The convertor installation is extremely simple. Figure 16-16 shows a typical installation inside the trunk of a vehicle.

TURN SIGNALS

Turn signal circuits are frequent sources of difficulties. Their diagnosis, however, is not difficult and can usually be accomplished with just a 12-volt test light or a voltmeter. We will look at the common circuits, starting first with the flasher. The flasher is actually a type of cir-

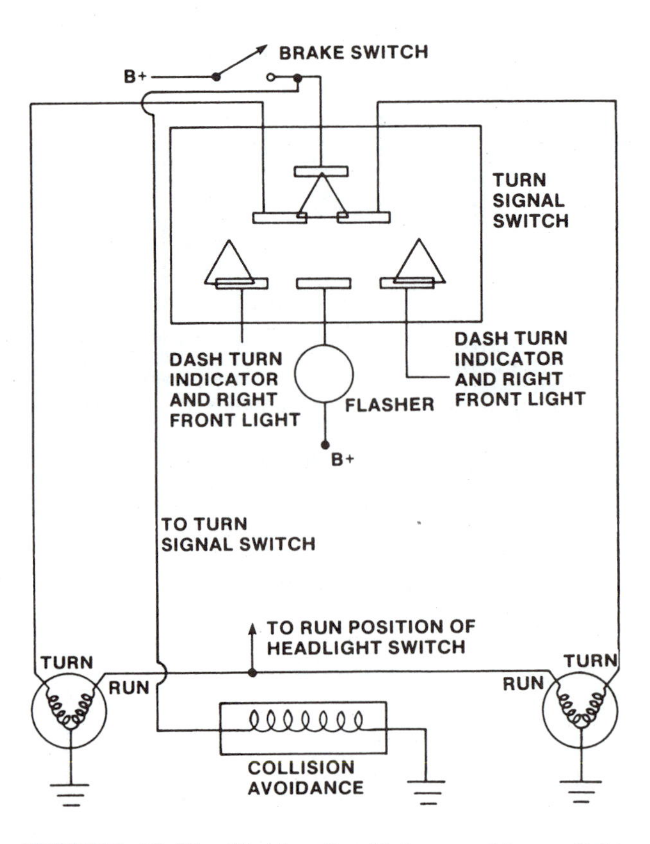

FIGURE 16-15 "Add-on" collision avoidance light.

FIGURE 16-16 Three-bulb combination converter installed in the trunk of a vehicle.

cuit breaker. You probably remember that a circuit breaker is an overload protection device, designed to open the circuit because of the heat developed from excessive current. Flashers are usually mounted in the fuse box and made up of a fixed contact and a moveable bi-metallic contact. The bi-metallic contact will heat and cool, making and breaking the circuit. As current flows through the contacts, they will heat. The heating will expand the bi-metallic side until it breaks open the circuit. With no current flow, the trip will cool and close the contacts. The alternate heating to open, cooling to close the circuit, continues as long as B+ is applied to one side of the flasher, and a grounded load is applied to the other side. Original automotive manufacturers frequently supply flashers that are rated for the number of bulbs the vehicle has. These flashers must be used only with the number of loads that they are rated for. The size of the bulbs (in terms of resistance) must be kept within the current capacity of the flasher or the flash rate will be affected. If the resistance of the bulbs is reduced or more bulbs are added, such as with the addition of a trailer, the flash rate will increase. If the resistance of the circuit is increased, or the number of bulbs decreased such as when a bulb is burned out, the flash rate will decrease. Some flashers will not flash at all if a bulb is burned out or missing.

Most replacement flashers are of the heavy-duty type and will flash with one or more bulbs as loads. These flashers generally have a heating circuit designed into them, Figure 16-17. If B+ is applied to one side and a load of at least one bulb to the other, current will flow through the flasher contacts and the heater coil at the same time. The heater heats the bi-metallic strip and the contacts open. With the extra circuit doing the heating, the flash rate is independent of current flow through contacts. Most shops stock only the heavy-duty flashers because they can be used in virtually all vehicles.

Let us look at the turn signal circuit in detail now. Figure 16-18 shows a common style of drawing the inside of the turn signal switch for a two-bulb system. Remember that a two-bulb system requires that a single bulb be used for either brake or turn signal. The turn signal switch must determine whether the bulb will be used for turning or brake lighting.

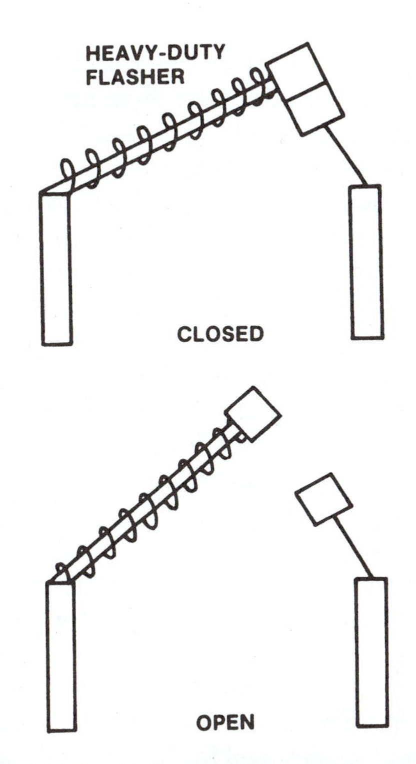

FIGURE 16-17 Coil around moveable arm will heat the bimetallic strip to open the flasher circuit.

The rectangular bars on the diagram are stationary contacts that the circuit wires connect to. Each contact has one wire connected to it. The top connection is from the brake switch and will be B+ if the brakes are applied. The middle row connections are for rear combination lamps (combination brake/turn signal).

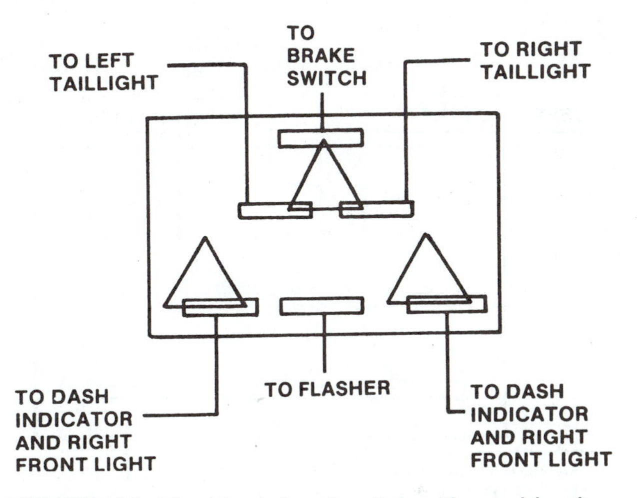

FIGURE 16-18 Turn signal switch with combination lamp taillights.

The bottom row of connections are for the front lights, including the dash indicators and the B+ coming from the flasher. The triangles drawn over the bars are a set of three moveable conductive pads which will connect the different bars together depending on the position of the switch. They are drawn in the no turn or neutral position. This will allow B+ from the brake switch to activate both rear lights at the same time. Figure 16-19 shows the same switch in a left turn. Notice that the conductive pads and triangles have moved to the right. This allows the brake switch to only power the right taillight, while the flasher connection is now in contact with the left rear taillight and the left front/indicator lights. The right taillight is being operated as a brake light, while the left one is in a turn signal operation. Figure 16-20 shows the same switch in a right turn mode. Notice that the conductive pads have moved to the left and have connected the brake switch to the left taillight, while the right is now powered off the turn signal flasher. This style of switch is very popular and normally is very durable. The most common problem encountered with the switch is usually mechanical rather than electrical. As the vehicle is driven around the corner, the cancelling mechanism must put the switch back into a neutral position so that both taillights can be used for brake warning. When this cancelling does not take place, the turn signal switch will usually need to be replaced.

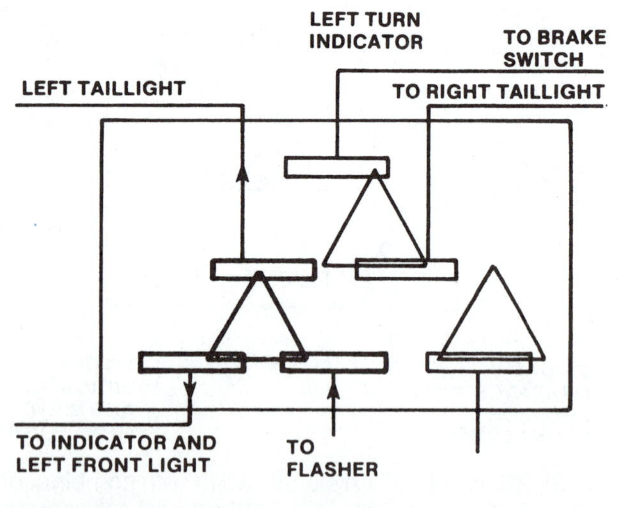

FIGURE 16-19 Left turn indicated.

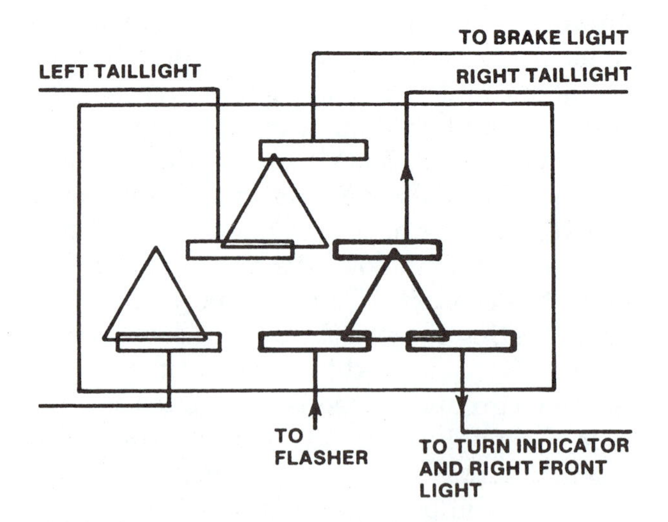

FIGURE 16-20 Right turn indicated.

The three-bulb switch is far simpler to understand because it is just a three-position switch. Because there is a separate circuit for the brake lights, there is no need to switch between the turn and brake B+. Figure 16-21 shows this greatly simplified switch. Moving the switch powers the front lights, the dash indicator, and the rear turn signal lights off the turn signal flasher. All the lights for one side of the vehicle are wired in parallel off the common point. Notice that there is no brake circuitry running through the turn signal switch. This simplifies the steering column wiring, even though it increases the number of wires running to the back of the vehicle. Most of the manufacturers are currently using this style of circuit, because it lends itself to the eye level brake light easier.

Either the two-bulb or the three-bulb system will power all bulbs off another source of B+ through a hazard flasher. All bulbs will flash together off this separate heavy-duty type flasher. Some hazard flashers have three terminals: an input, an output, and a ground. The flasher heater coil is grounded so that the rate of flash will not be dependent on the number of bulbs. The remainder of the hazard flashers are just two connection, heavy-duty flashers, that are capable of handling the current draw of all bulbs. The hazard circuit will

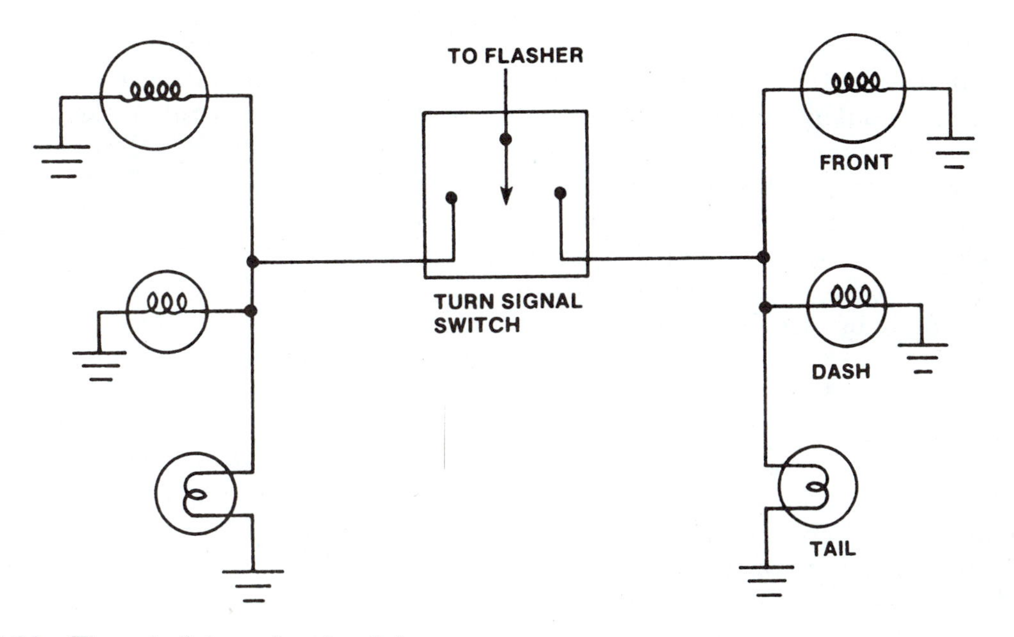

FIGURE 16-21 Three-bulb turn signal switch.

be powered off a B+ source, which is hot all the time, rather than ignition switched, as the turn signals are.

SIDE MARKERS

Side markers that also blink on turns are in common use today. Their circuitry is very simple and relies on a special side marker bulb to function correctly. Figure 16-22 shows the wiring for this style of bulb. Notice that it is between the B+ sources for the turn signals and the run lights. If the run lights are on, B+ will be applied to the side marker. Current will flow through its high resistance and use the turn signal light as its ground. The turn signal light will not light because it will have a much lower resistance. The majority of the voltage drop will be across the side marker. It will be on full brightness. If the turn signals are on and the run lights are off, Figure 16-23, the side markers will blink with the turn sig-

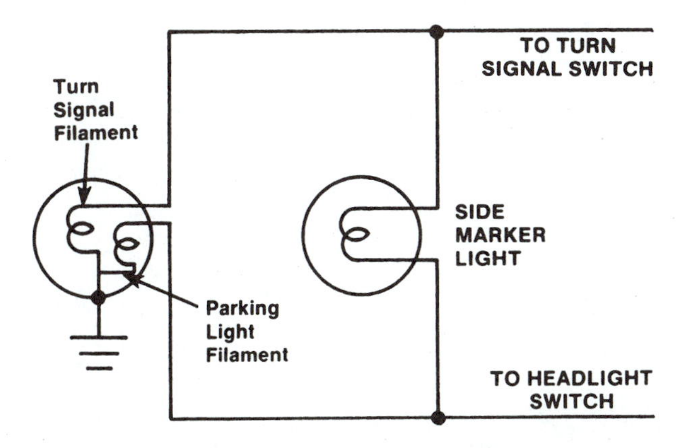

FIGURE 16-22 Wiring for side marker light.

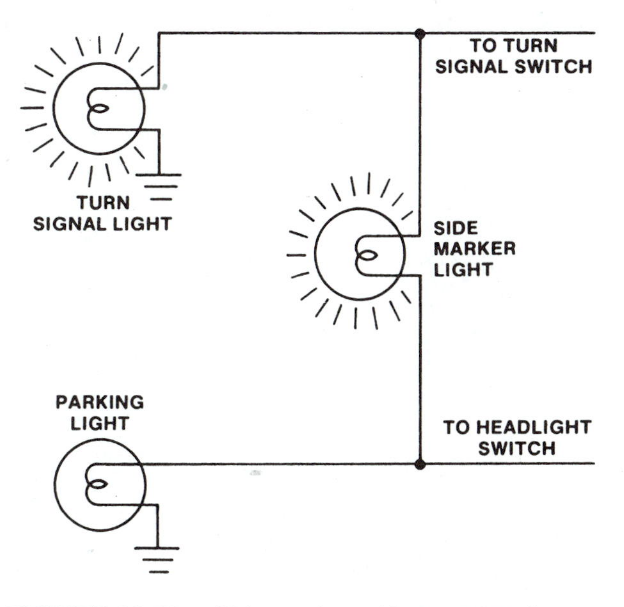

FIGURE 16-23 Side marker with turn signal on.

nals. This side marker will use the run light as its ground and the turn signal light as its B+. As the turn signal light cycles off, the side marker comes on. As the turn signal comes on, the side marker goes off because B+ is applied to both sides of the bulb. This circuit allows the side marker to "wink" on and off or be on steady, depending on whether the turning and running lights are on or off.

HORN CIRCUITS

Common horn circuits in use today fall into two groups: those with relays and those without. Both are reasonably simple, quite reliable, and very easy to repair. Figure 16-24 shows a two-horn circuit without a relay. A fused B+ goes up the steering column to the horn button, which is normally open (NO). Closing the switch closes the circuit, bringing B+ down the column through the bulkhead connector to a common point where the two horns are connected in parallel. On the other side of the horn is a ground and the circuit is complete, a good example of a simple series-parallel circuit with one control. The circuit is easily diagnosed with a 12-volt test light, following the usual procedure of looking for B+ with the horn button pushed in. Correct diagnosis is important, especially because many horn switches are part of a multiple function switch, which is very costly. The same stalk on the steering column might be used for the horn, the turn signals, the windshield wipers, and the cruise control. Make sure your diagnosis is correctly arrived at before you spend your customer's money. Power in at the switch and no power out with the button depressed would be the only reason to replace the unit.

Figure 16-25 shows an adaptation of the simple horn circuit. Notice that a fused B+ is run to the horn relay rather than to the switch. A NO relay is used. Notice that internally the relay coil receives B+ from the same fused source as the contacts do. The other end of the relay coil runs through the bulkhead connector and up into the steering column where a NO switch is located. Closing the switch grounds the relay coil, which develops a magnetic field, closing the contacts. The closed contacts bring B+ over to the common point for the two horns, which are grounded. You can see that the circuit is slightly more complicated, but an excellent example of the use of a relay and ground side switching, which we have discussed before. Remember, any relay circuit must be diagnosed as two separate circuits wired in parallel. The relay coil is a separate circuit that needs a path for it to develop the magnetic field necessary to close the contacts. Closing the contacts allows current flow through the other circuit with the horns as the load. There are actually three loads in the previous example, the horns (2) and the relay coil Diagnose them separately,

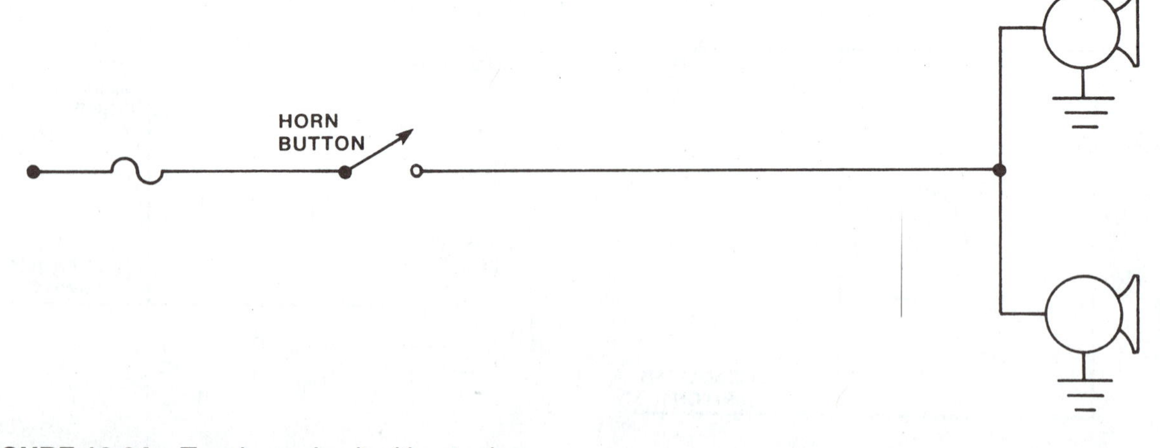

FIGURE 16-24 Two-horn circuit without relay.

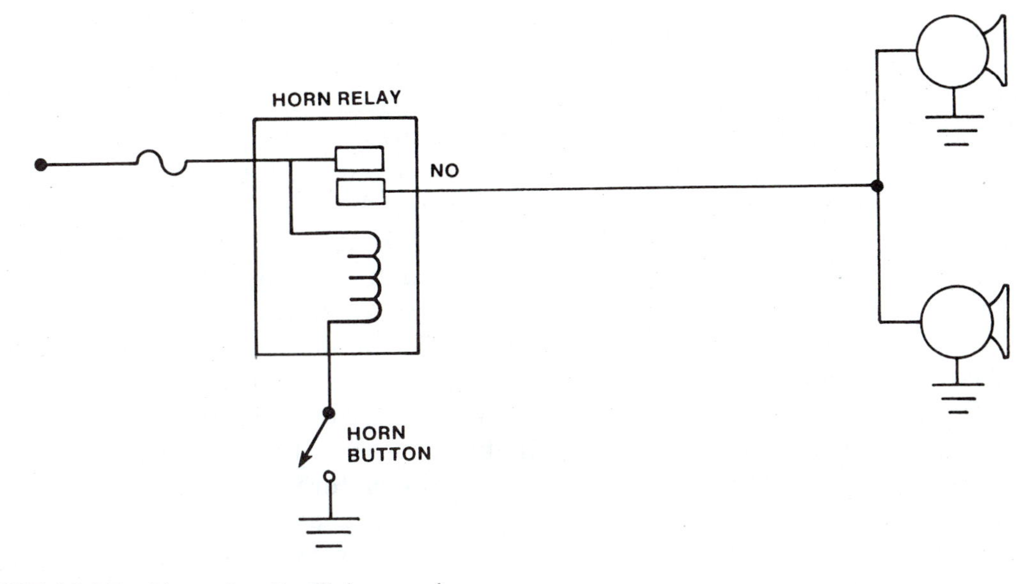

FIGURE 16-25 Horn circuit utilizing a relay.

as you would any relay circuit, and you are less likely to make mistakes. Again, the 12-volt test light is your tool along with a wiring diagram to pinpoint the connectors where you will test for power.

WINDSHIELD WIPERS AND WASHERS

Windshield wipers were one of the first extras added to vehicles. Their contribution to safety over the years cannot be underestimated. Basically, they fall into three categories: those that wipe and park at the end of their stroke; those that park in a different position (usually lower) than they wipe; and those that are **intermittent**. Most manufacturers' systems are similar in design, but differ enough that a diagnostic procedure cannot be given that would work for all. Be sure to follow the diagnostics that apply to the specific vehicle you are working on. Also realize that a single manufacturer might use more than one design in a car line. Figure 16-26 is an example of a nondepressed park, two-speed wiper system. Let us follow the circuit. Two sources of B+ are available to the switch. Circuit 296 is the power for the windshield washer motor. Pushing in on the button will pass B+ through to the pump

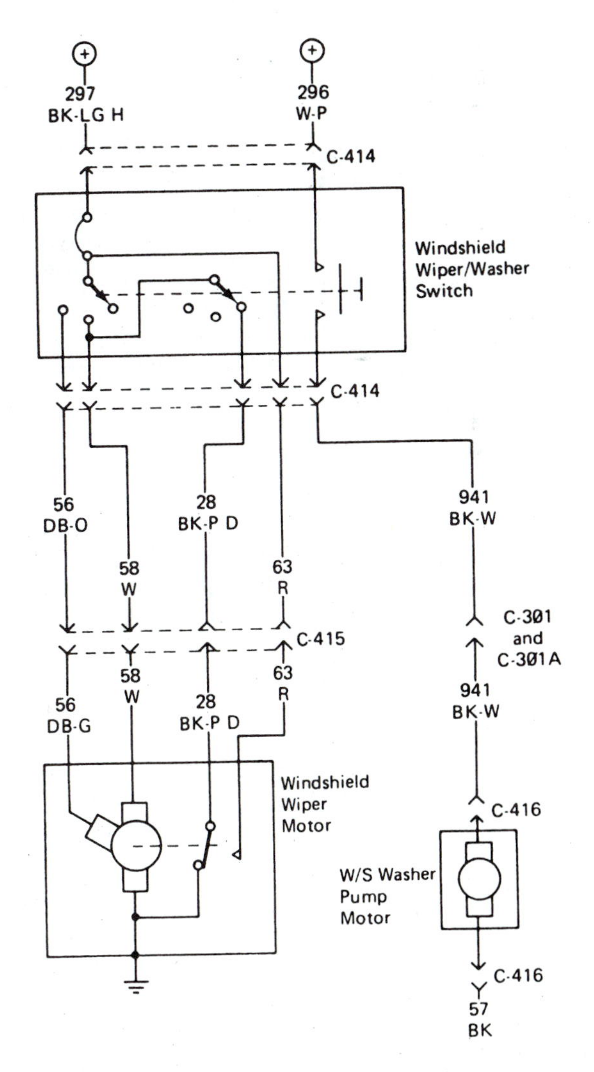

FIGURE 16-26 Nondepressed park, two-speed wipers. (Courtesy of Ford Motor Company)

motor, which is remote grounded. As long as the button is depressed, the motor will run. The button is mechanically locked out and is not functional if the wiper switch is in the off position. Moving the switch to one of the two speeds will allow B+ from circuit 297 through the internal circuit breaker and over to the motor. The switch contacts that are drawn to the right of the armature will also close at this time but will have no function until the motor is turned off. These contacts are the park switch contacts. The motor will function in a normal manner with two speeds being available through different sets of armature brushes. When the switch is turned off by the driver, power will still be available to the armature through the closed park switch. As the wiper arm moves through its travel, a cam will open the park switch stopping the wiper arm in the park position.

Depressed park or parking off the glass is accomplished usually by reversing the motor. When the switch is turned off, power flow through the motor is reversed through the park contacts in the motor housing, Figure 16-27. The motor reverses itself until the park cam opens the contacts. Reversing the motor allows the crank mechanism to travel farther and park the blades off the glass. Reversing the motor is a very common method of placing the blades off the usual wiping surface. The crank mechanism is the key to this

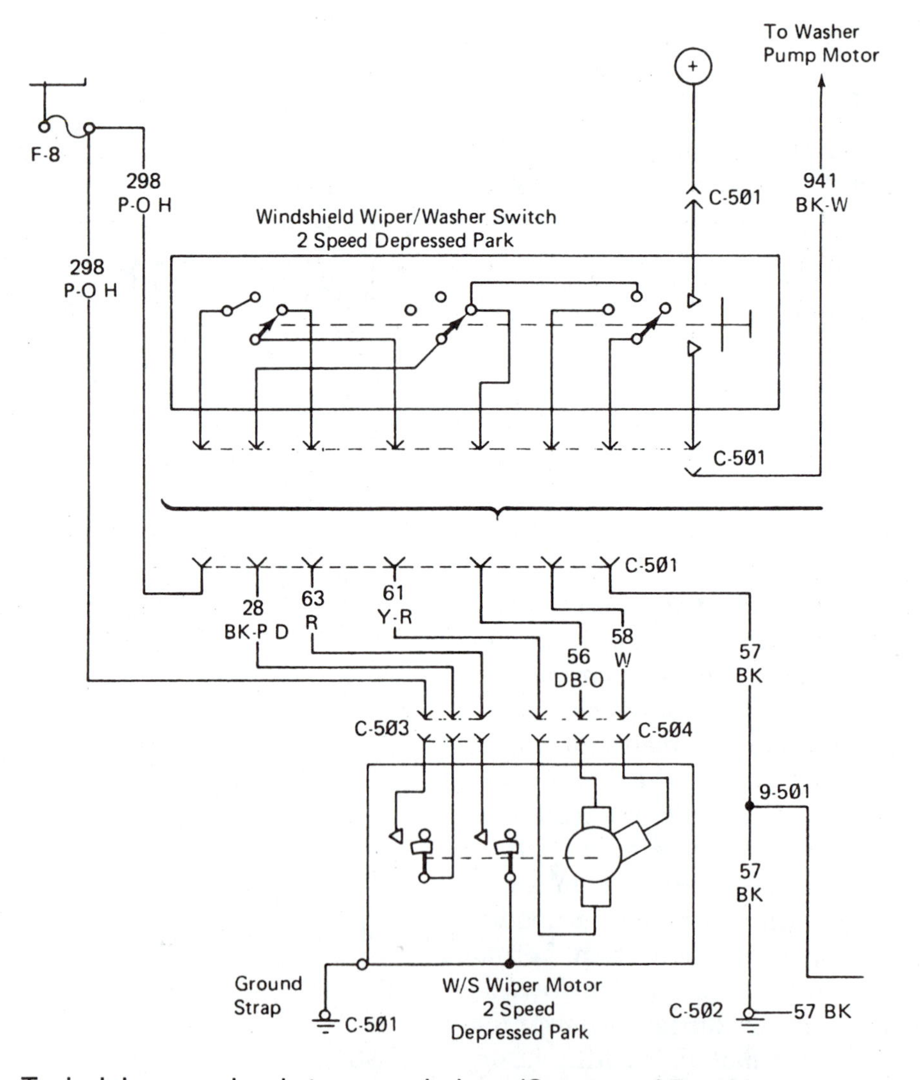

FIGURE 16-27 Typical depressed park, two-speed wiper. (Courtesy of Ford Motor Company)

function. Typically, an additional 1/4" of crank travel is achieved at the pivot. This 1/4" will move the blades an additional few inches in reverse. When the wiper motor is again turned on in the normal direction, the linkage will change back to its original length. In this manner, the beginning of the wiping stroke can be higher than the park position.

An additional feature of either style of windshield wipers is the intermittent function. A variable resistor controls current to a pause control module. The pause control module powers the wiper motor directly just like a switch would. When it opens the circuit, the motor parks. When it closes the circuit the motor runs. The variable resistance controls how long the motor is off. The greater the current flow through the resistance, the less the control unit will pause the wiper motor in the park position. The diagnosis of intermittent wiping systems involves testing the voltage to the pause unit. The electronic circuitry in the pause control usually is not repairable and is replaced as a unit if it is defective.

Again, follow the diagnosis that is specific for the vehicle being repaired. Keep in mind that the pause control unit is separate from the switch and the motor. The three components are generally diagnosed and replaced separately.

DEFOGGERS

The use of electric defogging or defrosting systems has increased over the last few years. The grid work of thin wires in the rear window is a very common sight, especially in the North where winter nights can leave frost behind. Rear defoggers have become one of the most common options available to the consumer. Their working principle is that current flowing through resistance will generate heat. It is not uncommon for these rear window heaters to draw 25 amps or more. Because of the very high current draw, most of the major manufacturers use a relay with a timer. We have studied the use of relays in other automotive circuits, and their operation remains the same: to control high current with low current. The simple drawing of a rear defogger circuit in Figure 16-28A shows a slight difference, however. Notice that the relay current ground flows through a timer-circuit with a NO spring-loaded switch. The switch will energize the relay coil and the timer. The relay will open up if either B+ is removed from the coil by turning the ignition switch off or the timer runs out. Typical timers have about 10 minutes programmed in. This allows the rear window sufficient time to defrost. If additional time is needed, the driver can reactivate the switch and start the timer over again. The timer circuit prevents the high-current draw from being left on by accident over an extended period of time. The high current could, in some cases, prevent the battery from receiving a full charge. The manufacturers assume that if there is a need for the rear defogger, there might also be a need for the heater, windshield wipers, and quite possibly the lights. This total draw might be more than the alternator can produce at idle. The timer prevents the additional draw for extended periods of time. Normally, the timer is built into the switch/relay and is not serviceable. Diagnosis is again accomplished with a 12-volt test light. The use of the light on the rear window grid can indicate if any of the grids are open. By moving the test light along the width of the window and gently piercing into the strips, the light will dim as you get closer to ground and brighten as you get closer to B+. A broken conductor strip will not have the change in brightness. It will remain bright until you just pass the open spot. At this point, it will go off completely, Figure 16-28B. Open grid conductors can be repaired with special conductor paste. Figures 16-29 through 16-32 show the procedure.

It is important to realize that most factory systems will come with a larger alternator to compensate for the additional amperage draw. Replacement alternators should always be sized with this in mind. It is also possible to add a rear defogger to an automobile that did not have one installed at the factory. Alternator output should be tested to make sure

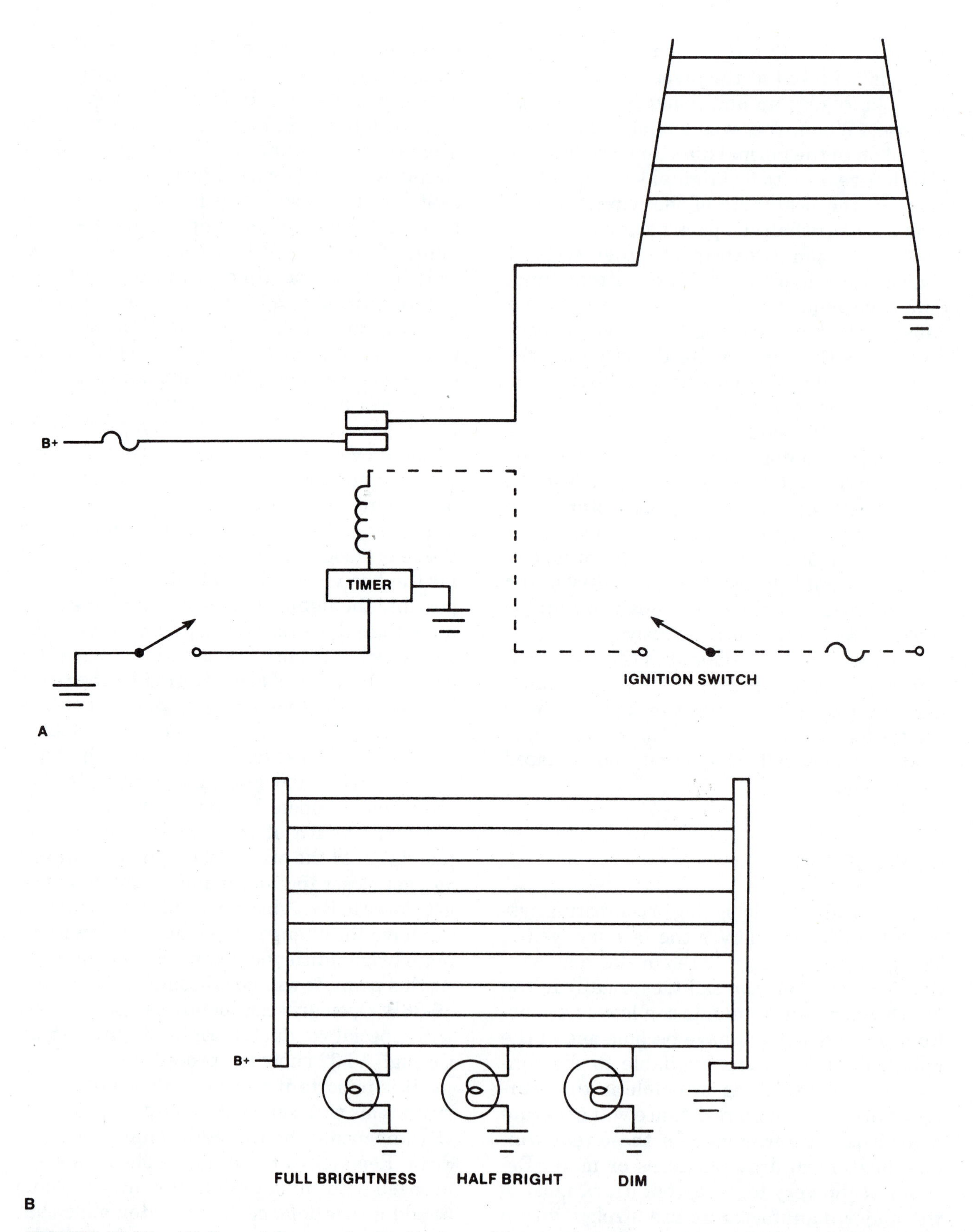

FIGURE 16-28 (A) Rear defogger circuit with timer; (B) using a test light on a heating grid indicates open circuits.

FIGURE 16-29 Mask repair area.

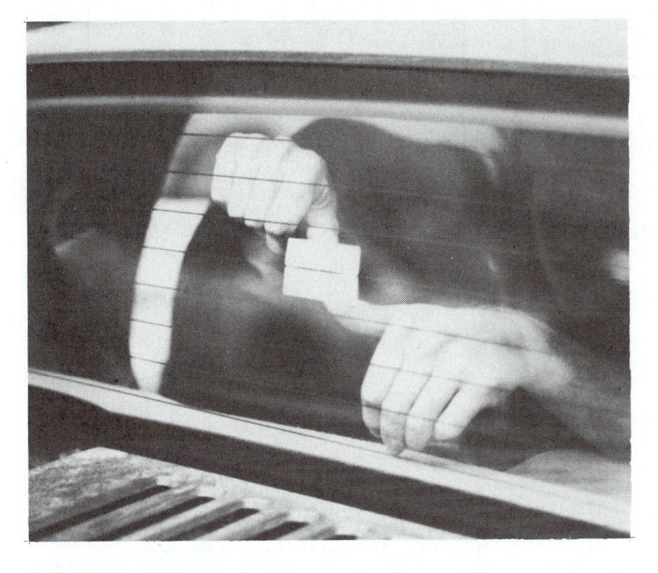

FIGURE 16-30 Apply conductive paste.

that it can handle the additional demands of the system.

FRONT ELECTRICAL DEICING

Recent Ford products have seen a front windshield deicing option that is electrically powered. In principle, it is similar to the rear defogger that we have just looked at. It differs because it uses very fine wires that cannot be seen by the driver. The finer wires require that the voltage applied be increased to reduce the current to a level that will heat but not damage the windshield grid. Figure 16-32 shows the circuit. When the dash switch is closed, the output control relay diverts full alternator output to the field circuit inside the alternator and the windshield grid. In addition, it removes the battery from the alternator circuit. The battery is now running the vehicle, while the alternator is powering the front windshield deicer. With full-field current flowing and no battery to hold down the voltage, the alternator's voltage will increase to as high as 80 volts. This high voltage will heat and defrost the window in usually less than 3 minutes. During deicing, the control module is looking at the time the circuit has been on, and the battery/system voltage. As the battery is running the vehicle, it is imperative that system voltage remain high enough for the ignition and fuel systems or the vehicle might die and be very difficult to start. The module makes sure that the voltage does not drop down too low. Usually, 11 volts is required for the module to allow front deicing. It will switch the system over to normal operation and allow battery charging if the voltage is lower than the spec of if the timer has run out. Dual voltage systems such as this might become more popular in the future.

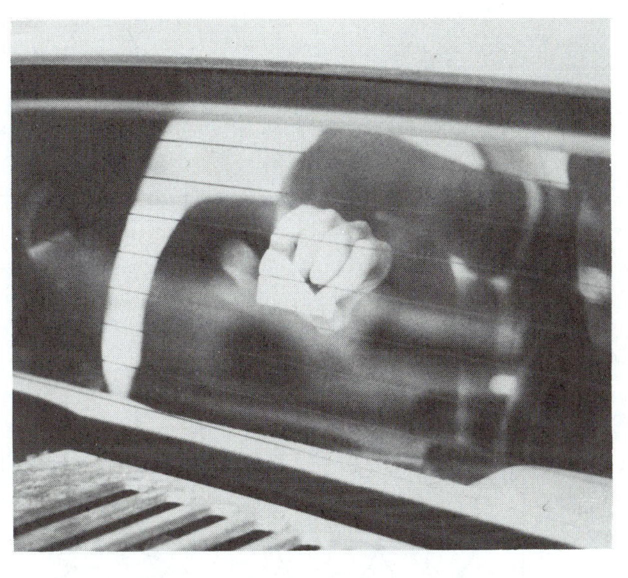

FIGURE 16-31 Remove mask.

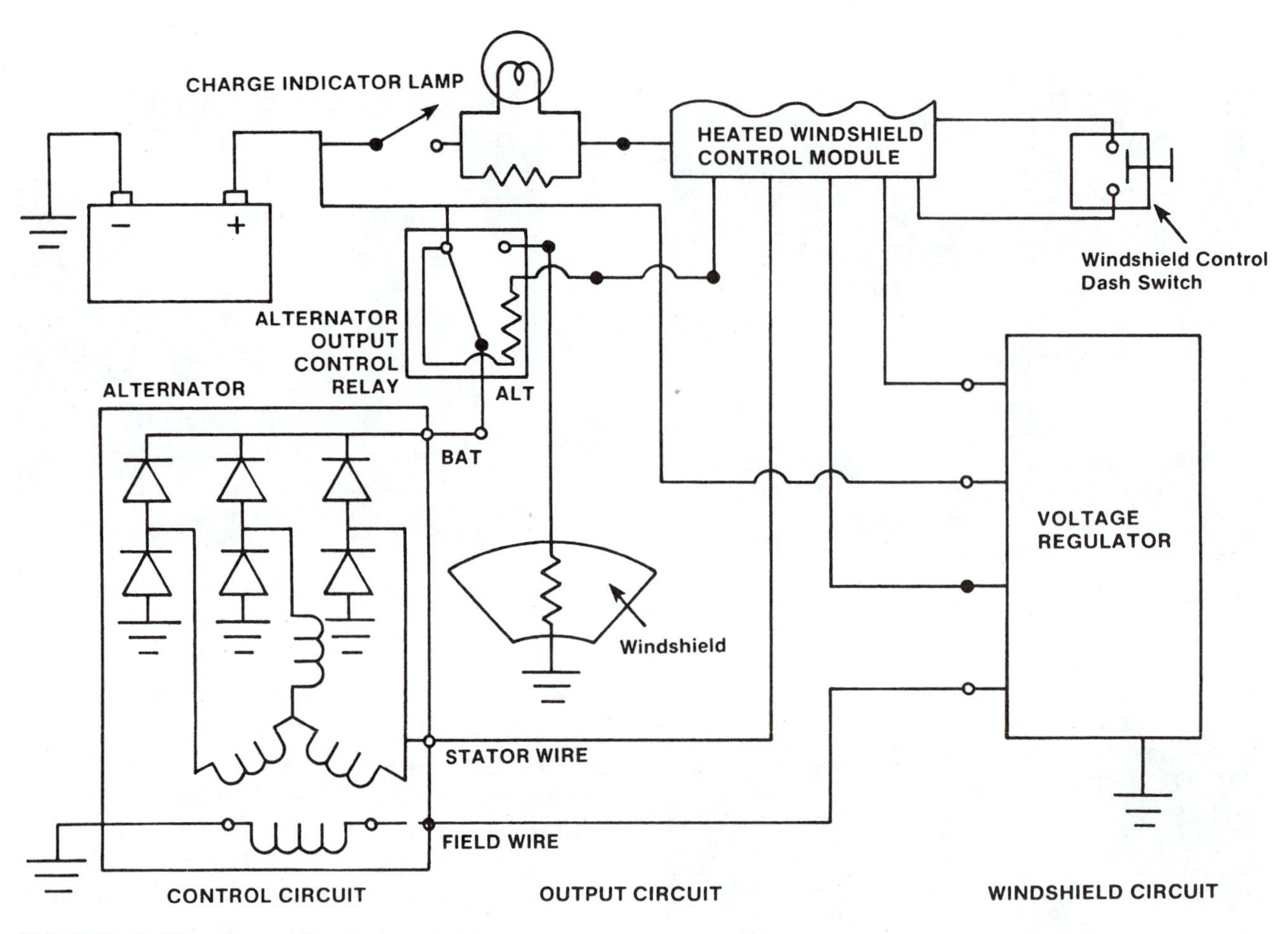

FIGURE 16-32 Taurus/Sable ice-melting circuit.

Testing of this system should be done only by following the manufacturers' procedures.

CAUTION: Voltage this high is dangerous. Be very careful during testing.

POWER WINDOWS

Power windows, whether they are for the side windows or the rear of a station wagon, are generally wired and controlled in the same manner. Figure 16-33 shows a typical circuit. An ignition switched B+ is run through a circuit breaker to the master switch. A lock switch is paralleled off the breaker and becomes the source of B+ for the window switch. Conductors are run to the insulated **reversible DC motor**. Let us examine each

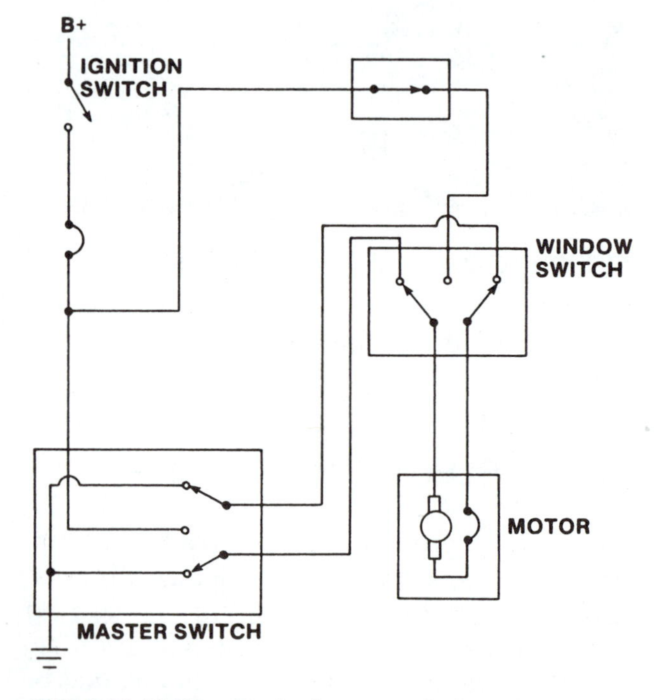

FIGURE 16-33 Typical power window circuit.

component separately. Circuit breakers are generally used on power windows because they will open if an overload occurs. Winter weather up North might freeze the window closed. Without a circuit breaker to open, the motor might be damaged trying to move the window against the ice. As the ice is removed, the breaker will cool, close, and allow future window operation. The lock switch will prevent operation of the window switch if it is open, because it is the source of B+ for the switch. Generally, the lock switch is placed on the driver's door or on the dash to allow the driver control over the doors. This is especially important with small children in the back seat. The motor is usually an insulated directional style that will run up or down, depending on the direction of the applied B+ and ground. The master switch has its own B+ source and is the ground for the entire system. Notice that the style of switch used at the window and the master control is basically the same. The switch will allow B+ to be directed through the motor from the center terminal and then flow back through the other side of the switch. The window switch and master switch are actually wired in series. When they are in a neutral position, as they are drawn, they are actually series connections. A problem with either switch will affect motor operation. A dead motor can be diagnosed with a 12-volt test light or voltmeter by first ensuring that B+ is available to both switches. Moving the switches should switch the B+ from one side to the other. Wire the 12-volt test light across the switch, Figure 16-34, because the switch is a path for both B+ and ground. The motor is checked easily by jumping the terminals to B+ and ground. Reversing the connections should reverse the direction of the motor, raising and lowering the window. Corroded switch contacts will generally reduce B+, slowing or stopping the motor. Measuring the voltage applied across the motor will generally indicate the presence of series resistance.

Tailgate window motors are usually wired in the same manner with the exception that they require an inner lock switch that must be closed for the motor to function. The inner lock switch opens and closes with the door opening and closing. In this manner, Figure 16-35, the window cannot be raised or low-

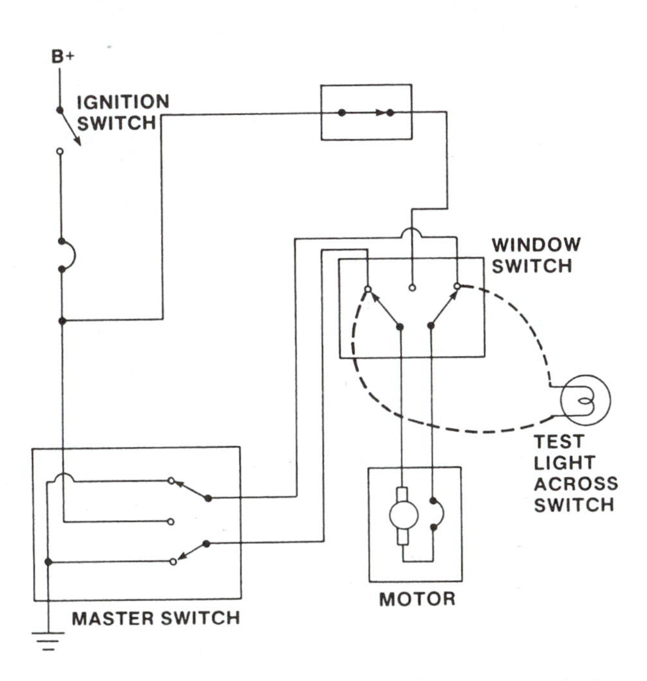

FIGURE 16-34 Testing a switch in a power window circuit.

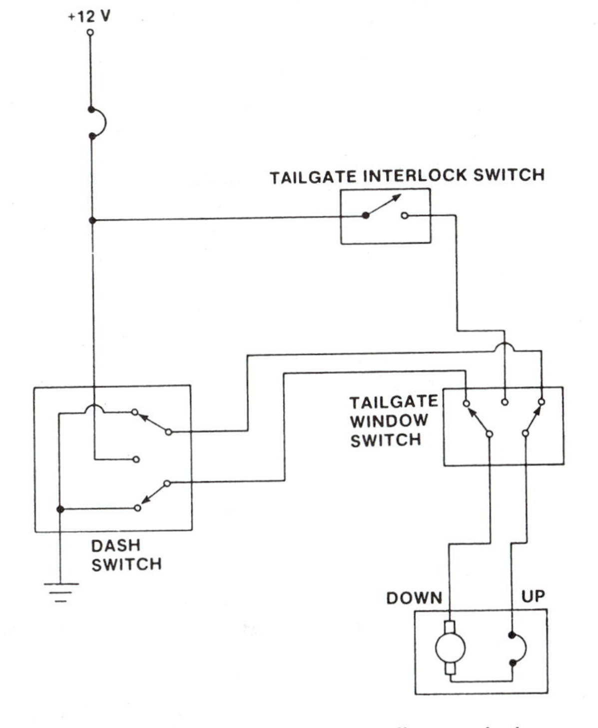

FIGURE 16-35 Typical power tailgate window on a station wagon.

ered if the tailgate is open. The same style of switch and motor are generally used for both side windows and tailgate windows.

ENGINE COOLING FANS

With the advent of transverse-mounted engines, electrical cooling fans found their way under the hood of the modern automobile. They offer advantages over the mechanical cooling fans because of their ability to move large amounts of air independent of engine speed. For this reason, some of the rear drive vehicles are currently coming factory equipped with electrical cooling fans. Their circuitry is very simple, especially on a non air-conditioned vehicle. Figure 16-36 shows an electric cooling fan circuit. Notice that the relay used is powered through a fuse link off the starting relay. It has B+ available all of the time. This will allow the fan to continue to cool the engine compartment and the radiator coolant even if the ignition switch is off. Power for the relay contacts and the relay coil come off this fused B+ lead. Notice that the other end of the relay coil runs to an engine coolant switch (NO). This switch will be located in the engine block or cylinder head coolant passage so that it will be able to sense coolant temperature. Notice the 226°F printed next to the switch? This tells you that when the coolant reaches 226°F, this switch will close and apply a ground to the relay coil. This should close the relay contacts and B+ will be applied to the cooling fan. With the remote ground on the other side of the motor, it will run at full speed and cool the radiator coolant down. Typically, the coolant has to drop about 20° before the switch will open and turn the fan off. This cycling of the fan, on at high temperature and off about 20° cooler, continues to keep coolant temperature hot enough to be efficient and cool enough to be safe. The system is easily diagnosed with a 12-volt test light and a thermometer. Remember that the relay actually has two circuits that should be independently tested. Keep in mind that the speed of the motor will be dependent on the voltage applied across it. Any resistance will drop the voltage and the current to the motor. This will slow it down

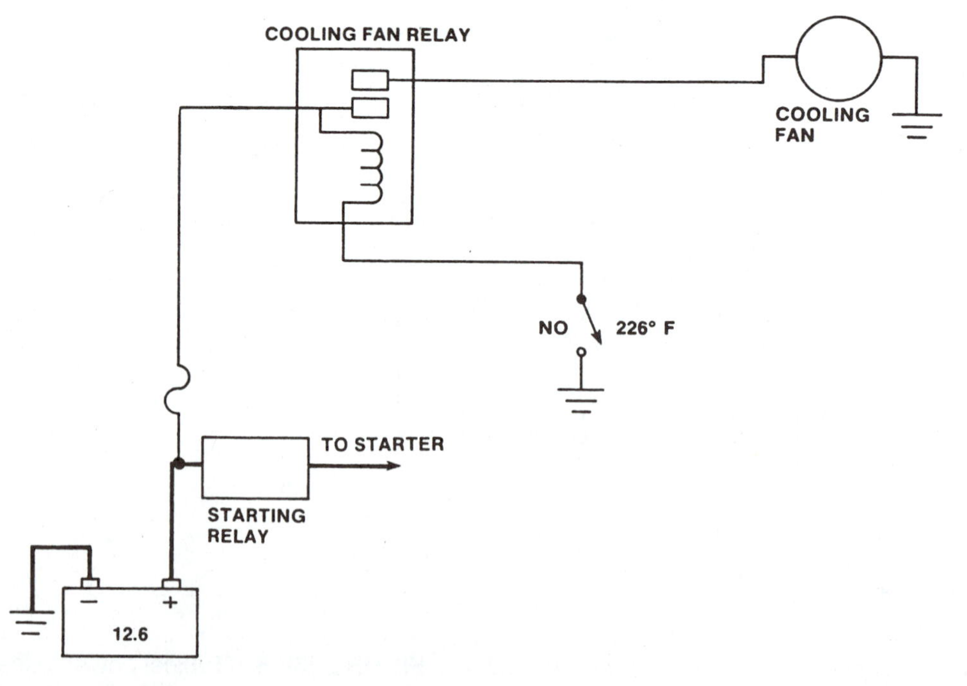

FIGURE 16-36 Cooling fan circuit, non-AC equipped.

and might possibly cause the engine to overheat. This is an extremely important circuit. Overheating an aluminum engine can cause thousands of dollars worth of damage. Periodic testing of the circuit, especially before summer, should be done to ensure that all is well. Make sure to test the circuit for a couple of cycles to guarantee that it will consistently operate and keep the engine within the safe temperature range. Test for three conditions:

1. relay coil and switch circuit
2. relay's high-current circuit
3. voltage applied across the fan motor

Air-conditioned vehicles, especially those with small engines, usually have additional circuitry to ensure that the cooling fan will come on when the compressor cycles on. Airflow through the condenser must be present for the AC to function correctly. With the condenser mounted in front of the radiator, the logical method of ensuring airflow is to turn the cooling fan on as the AC compressor cycles on. Most of the manufacturers follow one of two methods. Either they cycle the compressor and the fan at the same time as Figure 16-37 shows, or they cycle the fan on when AC high side pressure reaches a predetermined level, Figure 16-38. In either case, this ensures that airflow through the condenser will keep both the AC and the engine cool. Diagnosis of either system is easily accomplished again with a 12-volt test light. Keep in mind that the cooling fan motor might be different for the AC equipped vehicle. Some of the manufacturers use a higher speed motor to pull additional air

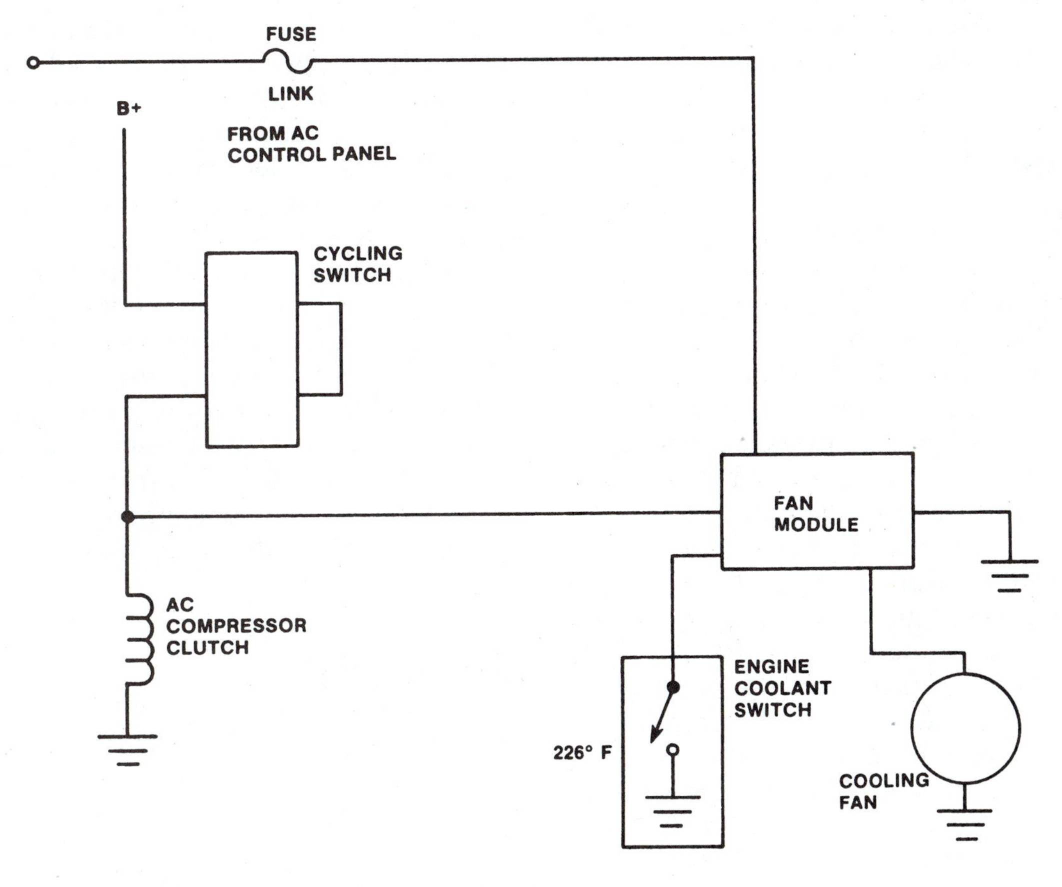

FIGURE 16-37 Cooling fan circuit, AC equipped.

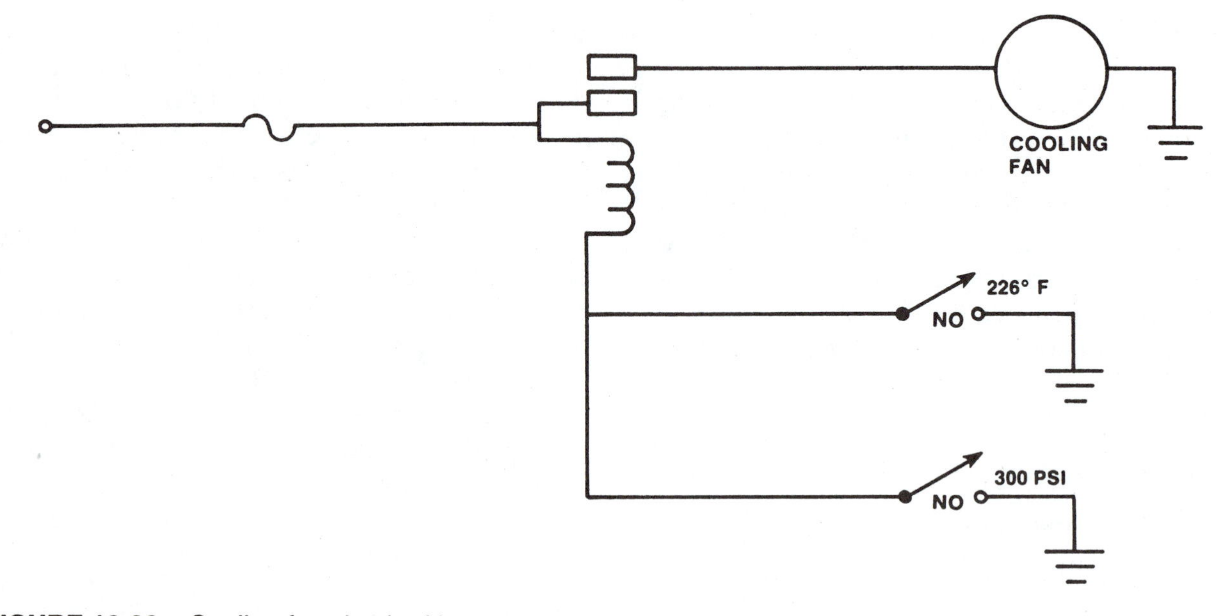

FIGURE 16-38 Cooling fan circuit with pressure switching.

through the radiator and the condenser to improve efficiency. Replacing the motor with the wrong one could have drastic effects on AC performance and/or engine cooling.

SUMMARY

In this chapter, we have attempted to look at some of the more common accessory circuits. We began with headlights and examined switches, high and low beams, and fog lights. In addition, we looked at common "flash to pass" and automatic dimming circuits. We then moved to the rear of the vehicle where we looked at the common two- and three-bulb systems currently in use. We also examined the circuitry used for the eye level collision avoidance light. Turn signals, flashers, and hazard warning light systems for both the two- and the three-bulb systems were traced. The circuit for front side markers, which flash during turning, was analyzed. This concluded the lighting circuit section of the chapter. Horn circuits with and without relays were diagrammed before we looked at the most common types of windshield wiping systems. The standard non depressed park, the depressed park, and the intermittent wiper systems were included. Reversing the motor to park was discussed and diagramed. Front and rear electric defoggers were discussed, including the "insta-clear" system, which uses full-fielded alternator output to power the front deicer. Reversible DC motors used for power windows were discussed. These included both side windows and station wagon tailgate windows. Chapter 16 ended with the circuitry used for front-wheel drive electric cooling fans with and without factory installed air-conditioning. During the chapter, emphasis was placed on following a diagnostic procedure that was specifically written for the accessory. The small differences from one manufacturer to the next are enough to make a generic procedure impossible or inaccurate.

KEY TERMS

Flash to pass
Intermittent
Depressed park
Reversible DC motor

CHAPTER CHECK-UP

Multiple Choice

1. The right headlight does not function on either high or low beam. Technician A states that this could be caused by an open ground on the right side. Technician B states that the B+ to the dimmer switch might be open. Who is correct?
   a. Technician A only
   b. Technician B only
   c. Both Technician A and B
   d. Neither Technician A nor B
2. A three-bulb taillight system is wired with
   a. a common brake turn signal bulb.
   b. separate bulbs for turn signals and brake lights.
   c. a collision avoidance light.
   d. side markers that flash in the rear.
3. Side markers that flash have
   a. no ground wire connected.
   b. B+ applied to them from both turn signal and run lights.
   c. require special bulbs with high resistance.
   d. all of the above.
4. Technician A states horn buttons in circuits with relays are B+ switches. Technician B states horn buttons in circuits without relays are actually switches to ground. Who is correct?
   a. Technician A only
   b. Technician B only
   c. Both Technician A and B
   d. Neither Technician A nor B
5. Depressed park windshield wiper systems park the blades by generally
   a. turning off B+ to the motor.
   b. applying B+ to the motor.
   c. ungrounding the switch.
   d. reversing the motor direction.
6. Rear defoggers generally have a relay with a timer. This allows
   a. the defogger to operate for only a specific amount of time.
   b. the defogger to function just until the rear window is clear.
   c. the defogger to be independent of the ignition switch.
   d. none of the above.
7. The right front window will not function on a power window vehicle. Technician A states that the window motor might be insulated from ground causing the problem. Technician B states that either switch could be the problem. Who is correct?
   a. Technician A only
   b. Technician B only
   c. Both Technician A and B
   d. Neither Technician A nor B

8. On an air-conditioned vehicle the cooling fan will not turn off. A possible cause could be
   a. a grounded coolant temperature switch.
   b. an open relay.
   c. a grounded cooling fan motor.
   d. a burned out fuse link.

True or False

9. Automatic dimming circuits ordinarily use photoelectric cells.
10. On a two-bulb system, a burned out brake light will have no effect on the turn signal.
11. OEM flashers usually are rated for a specific number of bulbs.
12. The horn button on a relay equipped horn system will apply a ground to the relay coil to blow the horns.
13. Cooling fans are usually relay powered with the relay coil wired to a cooling temperature switch.
14. Power windows will work from the driver's switch even if the door switch is removed.
15. Interval or intermittent wipers use a variable resistor to directly control current flow to the monitor.
16. Front electric deicers are frequently run on the battery current, while the alternator runs the rest of the vehicle.

CHAPTER 17

Using a Digital Storage Oscilloscope (DSO) for Diagnostic Purposes

OBJECTIVES

At the conclusion of this chapter, you should be able to:

- set the time per division and the volts per division on a DSO.
- set the DSO correctly for single or recurrent patterns.
- analyze an AC or DC pattern for its voltage, frequency, pulse width, or time.
- capture and analyze the patterns for common sensors including, MAP, TP, O_2, MAF, and CKP (both Hall effect and variable reluctance)
- capture and analyze the patterns for common outputs including injection and idle speed.

INTRODUCTION

Every few years the automotive repair industry sees a "new" or "improved" tool come from the back of the tool salesman's truck. Generally, the tools are successful in making the technician's job easier or more accurate. However, few tools have had the impact of the digital storage oscilloscope **(DSO)** on the technician. As vehicle systems have become more and more complicated and sophisticated, diagnosis has become more technical. When faced with especially intermittent problems, more and more technicians are turning to the DSO as their tool of choice. Within this chapter we will look at how to set up a typical DSO and use it to analyze various computer inputs and outputs. Even though computer control is not the only scope of this text, the DSO is becoming more of a basic tool to technicians every year. It is the equal of the digital multimeter of a few years ago.

Every good technician has a DVOM in his tool box and the future will no doubt see a DSO right beside it.

VOLTAGE OVER TIME

DSOs will display a changing voltage over a period of time. We have seen this concept when we used an ignition scope. One of the major differences in the typical ignition scope and a DSO is the ability to change around the time and voltage setting so that we can view just about any voltage over just about any pe-

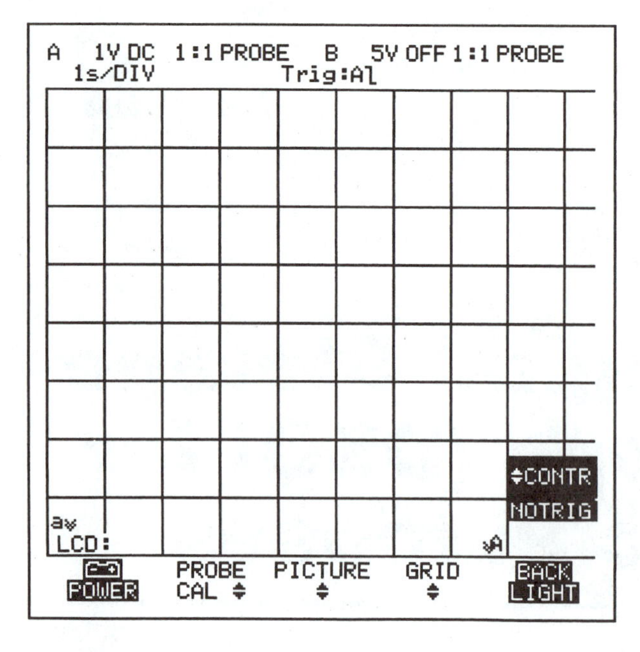

FIGURE 17-1 10 divisions across, 8 divisions up and down.

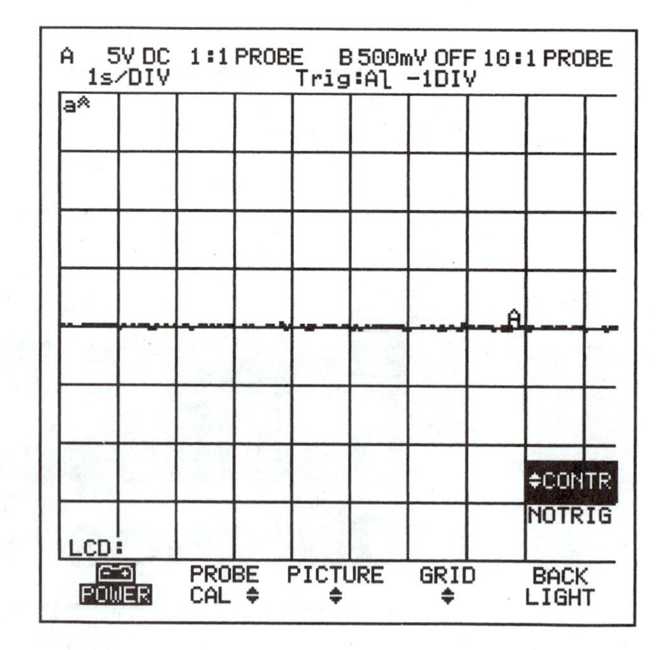

FIGURE 17-2 5 V per division. Zero in the middle of the screen.

riod of time. The scope screen is usually divided into a series of **divisions**, 10 across and 8 high like Figure 17-1 shows. The scope will display voltage vertically up and down. If the voltage rises, so does the waveform. If the voltage goes down, so does the waveform. Time is measured and displayed horizontally across from left to right. Notice that the scope screen has been divided in 80 squares or divisions. Each square will be a certain amount of time and a certain amount of voltage that you will set from the front panel controls. Let's set the scope to some very common voltages and times as an example. If we wish to measure DC charging voltage which we expect to be around 14 V, we would touch the voltage button until 5 V/Div appears in the upper left corner of the screen as Figure 17-2 shows. This means that each vertical division would be equal to a reading of 5 V. If zero volt is exactly in the center of the screen, then we would have 4 divisions above zero to register positive voltages and 4 divisions below zero to register negative voltages. The total available voltage would be 20 positive to 20 negative or 40 V total. However generally DSOs allow us to move the zero point up or down to suit the situation. Think about the charging system. Would we ever need to be able to read 20 V negative? Not likely. So let's move the zero down to one division from the bottom like Figure 17-3. Now what do we have? If the scope

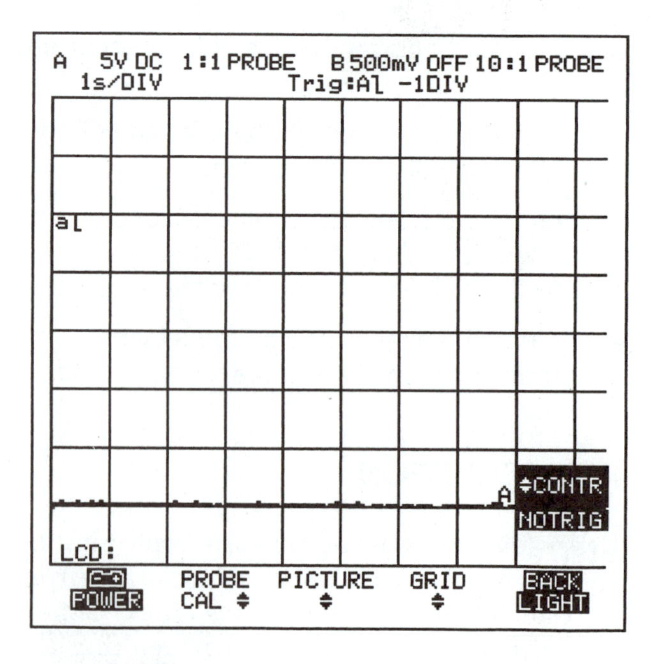

FIGURE 17-3 Zero moved down to one division from the bottom.

is still set on 5 V/Div, we will be able to see 35 V positive and 5 V negative. Again, using our example of charging system voltage, we have more volts available than we should need. If we touch the mV side of the volts button, we will reduce the scale. Our next lowest choice is 2 V/Div. Will this be an O.K. setting? With 7 divisions above the zero and one below, we will have a range of 14 V positive and 2 V negative. This should do it, right? Now, how about the time scales. How long do you want to view the charging pattern? How about 1 second total. With ten divisions worth of time across the screen, a 100 mSec/Div sounds right as Figure 17-4 shows. Notice that the amount of time per division shows in the same upper left corner of the screen along with the volts per division. Remember a millisecond is one thousandth of a second (.001) and carries the small "m" as its abbreviation, so 100 mSec is actually a tenth of a second (.100 sec). With ten divisions each equal to one tenth of a second, the total screen will equal 1 second. By touching the "S" side of the time button we can increase the time per division all the way up to 60 seconds. At 60 Sec/Div, we would be able to view ten (10) minutes worth of voltage. On the other hand, by reducing the time/division we will be able to capture time in as short as 10 nanoseconds (nS). A nanosecond is a billionth of a second. Within the automotive field, we generally find little application where we would need to measure down to a billionth of a second! Not all DSOs are capable of measuring down to a nanosecond. However most hand held units will go down to a microsecond per division. A microsecond is a millionth of a second. Still pretty fast! Generally speaking, a technician should set the time per division so that he can see about 2 to 4 complete cycles of whatever he wishes to look at.

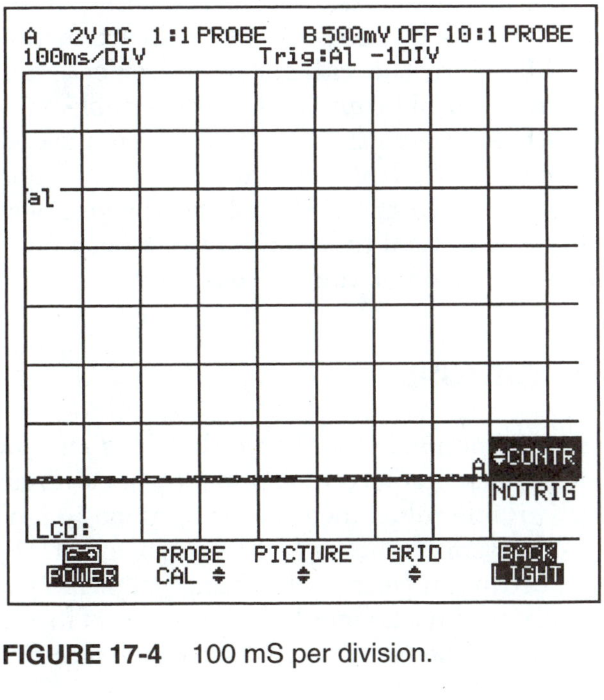

FIGURE 17-4 100 mS per division.

TRIGGER AND SLOPE

Once the time per division and volts per division have been set, the DSO is ready for work. Signals will be displayed and updated across the screen as they are measured. Sometimes we wish to capture a specific signal and "freeze" it on the screen for analysis. Here is where the settings of **trigger** and **slope** come into play. The trigger point is the point where the DSO will begin to draw the trace on the screen. For example, if the trigger point is 1 V, the DSO will not begin to trace the pattern on the screen until 1 V is achieved. If the voltage never rises above the 1-V trigger point, nothing will be displayed on the screen. As soon as 1 V is achieved, the screen will show the voltage trace for as long as the time frame allows. It will not re-trace over the existing pattern until the 1-V trigger point is again achieved. Slope is closely related to trigger. A positive slope represented by the ʃ symbol means that the trace will begin on the screen as the trigger point is passed by a rising voltage.

The ˥ symbol is the opposite. The trigger point must be passed by a falling voltage. Putting together the trigger and slope is what is typically done by the technician. For example, if the technician wishes to capture the throttle position sensor voltage change, he would set the slope for positive and the trigger point at a voltage slightly above the closed throttle voltage. If the closed throttle voltage is .5 V, a trigger point of .6 V would be reasonable.

The scope will now begin tracing the TP (throttle position) sensor voltage just as the throttle begins to open. The trigger level that the scope is set for usually shows on the screen along with the slope. The trigger level in Figure 17-5 is shown in the dark box on the lower right side of the screen, just below the title "LEVEL". The +SLOPE symbol is highlighted on the bottom of the screen. Using slope and trigger to capture only the patterns you want is helpful. It is especially useful if one wishes to capture a specific problem voltage that you as a technician believe might be occurring. For example, if you suspect that system voltage to the **PCM** (power control module) drops off periodically, causing the engine to die, you can set the trigger voltage slightly lower than system running voltage. Setting the slope negative will only turn on the trace if the voltage to the PCM drops below system voltage. Setting the DSO in this manner to "monitor" system voltage as the vehicle runs can quickly determine if your problem is oriented toward the source of power. When the vehicle dies, all you have to do is look at the screen. If the voltage did in fact drop off, even if only for a split second, there will be a saved pattern on the screen. If no pattern was captured your diagnosis will not include a PCM power problem. The tremendous speed of the DSO (as little as ten billionths of a second) insures that the slightest glitch in the power will be captured on the screen and saved for your observation. Trigger and slope are options that have some benefits, especially in intermittent diagnosis. If you want a continuous pattern, turn off trigger and slope. If you want to capture something very specific, turn it on. One additional comment on trigger; most of the DSOs on the market allow for either internal or external triggering. Internal is what we have discussed up to this point, where the voltage from the actual signal acts as the "turn on" trigger. External triggering is what you have been used to when using an ignition scope. The inductive pick-up around plug wire #1 started the scope pattern over each time it fired. The actual signal displayed was triggered by the #1 plug—externally. Most superior DSOs allow for either internal or external triggering.

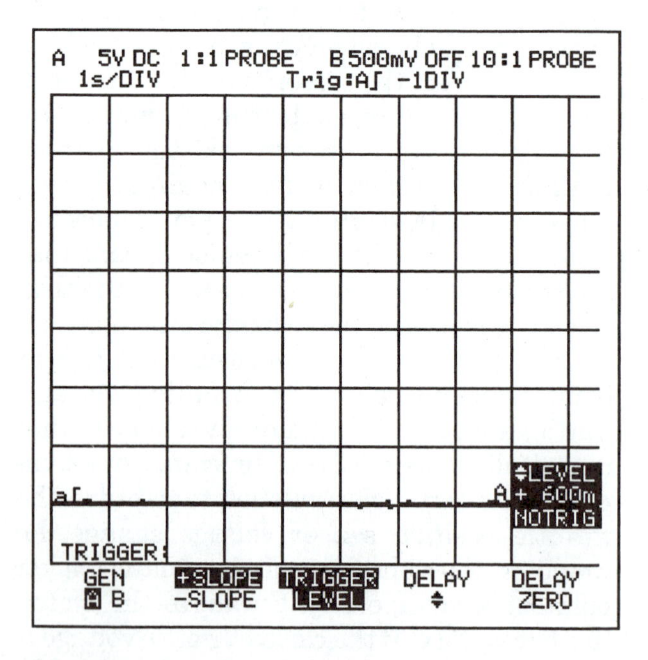

FIGURE 17-5 Setting trigger and slope.

READING AND INTERPRETING A DSO PATTERN

Let's look at the most common types of patterns usually encountered by the technician. Generally these fall into the categories of AC, DC, or pulsed. The pulsed signals are further divided into the categories of variable frequency, variable duty cycle, and variable pulse width. Although this may sound complicated, it really isn't. It is however, important information which the DSO will be showing you, and it is critical that you have a basic understanding of what you are looking at.

AC SIGNAL

You are probably familiar with AC signals and do not even know it. Remember the pick-up coil or variable reluctance sensor common to ignition systems? It actually is a form of an AC signal because it periodically changes it polarity or direction. An AC signal will have part of its signal above the zero line, in the positive direction and part of its signal below zero in the negative direction as Figure 17-6 shows. Follow along as we analyze this pattern. The voltage first begins in a negative direction and drops for slightly less than 2 divisions until it peaks. At this point, the voltage begins to rise until it reaches and crosses the zero volt line 1 division later. The exact reverse of the pattern now takes place. An increase in the positive voltage of about 2 divisions, over a period of time equal to about 1 division and then drop back to zero again over a time of 1 division. At this point the voltage again crosses zero in the negative direction and the pattern repeats itself over. This is an AC signal usually called a sine wave. What voltage was actually achieved? What is the scale or V/Div? The upper left of the screen shows 500mV AC/Div. Remember that this means that each division, either negative or positive equals 500mV. Our pattern reached a positive peak of 2 full divisions so 500mV × 2 divisions = +1.0 V. Because the negative pattern was the mirror image of the positive pattern, it must equal –1.0 V. How about time? A full positive and negative cycle actually lasted for about 3 divisions and again the upper left of the screen shows that each division is worth 10 mS/div (milliseconds) so 10 mS × 3 = 30 mSec. The entire event only took .030 seconds. The number of complete AC cycles that occur during 1 second is referred to as the frequency of the signal. If our example had occurred during one second it, would have a frequency of 1 cycle per second. Frequency measurements are usually in hertz, which is actually the number of cycles per second. Hertz is usually abbreviated as Hz. How fast is a 60 Hz signal? If you said that it is 60 times a second, you are on the right track. Remember that a complete cycle is necessary. Both the positive and negative voltage waveforms must be present. Generally a technician should set the time base to see a couple of complete cycles. Setting the time base higher will increase the number of cycles on the screen, but will make your interpretation difficult because of the closeness of the waveforms. Adjust the time/division so that 2 to 4 full cycles are on the screen and your analysis will be easier.

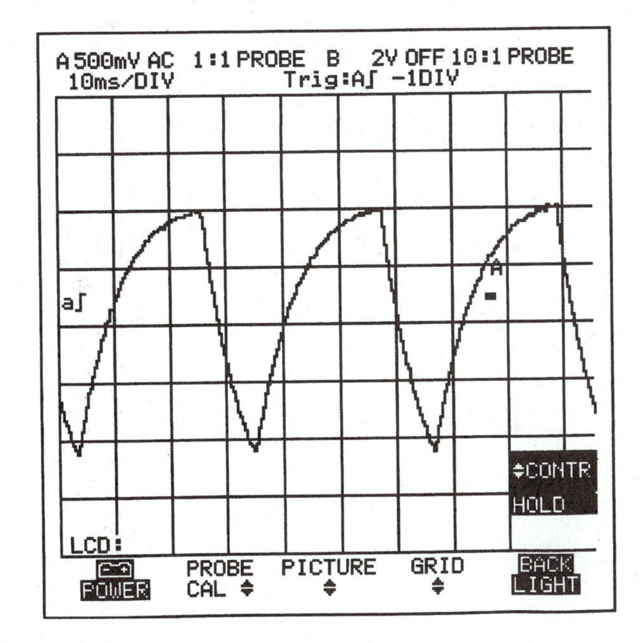

FIGURE 17-6 AC signal.

DC SIGNAL

A DC signal is a voltage signal that only contains a positive or negative polarity. Generally, positive voltages are encountered in the automotive electrical system. A useful feature of most DSOs is the ability to move the zero down so that a DC signal fills the screen. The move button will move the zero up or down as needed as Figure 17-7 shows.The zero line is where you see the A printed above a small black box in the lower right of the screen.

Make sure that you have one or two divisions below the zero as shown. If you place the zero at the bottom of the screen, you will not be able to see any negative pulses that might be present. Preset the scope with the ability to display some negative voltage, just in case. With the zero down on the 2nd division line, you will have 7 full divisions worth of voltage to display. Pick a V/Div that will allow for the greatest pattern height. This makes reading the signal easier and more accurate on your part. Let's set up the DSO for a battery voltage reading with the vehicle running. Typical charging voltage is around 14 V. If we move the zero down to the top of the first division,

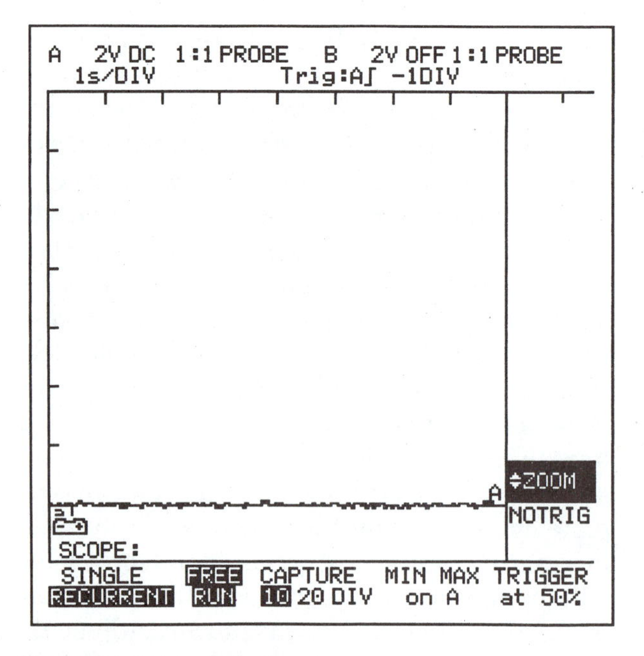

FIGURE 17-7 DC signal.

we will have 7 divisions worth of voltage available for measurement. 14 V with seven divisions sounds like 2 V/Div right? This would be the correct setting, but we must make sure that all of the pattern is displayed.

Every peak must be visible. If all are not, increase your V/Div until the entire pattern is visible. You do not want to miss any voltage surges that might have an effect on PCM circuits. How about the time division across the screen? Your setting will depend on what you are looking for. If for instance, an intermittent problem is suspected, increase the time base so that the pattern will be present for a longer period of time, up to the maximum of the DSO. Some DSOs allow for capturing 20 divisions worth of time, even though they are only displaying 10. This gives you the ability to capture up to 20 minutes worth of patterns and view them by moving the 20 divisions across the screen in a parade fashion. However, frequently the DC pattern will not be a constant flat line, but instead shows up in a pulsing format. The type of pulsing pattern can be characterized by its being frequency, duty cycle or pulse width modified. Let's look at the these different forms of pulse patterns.

VARIABLE FREQUENCY

If a DC signal is pulsing on and off like that shown in Figure 17-8, and the on and off time is equal, the signal is frequency oriented. Remember, we discussed the frequency of a signal when we looked at an AC signal. Frequency measured in hertz is the number of complete cycles per second. A frequency of 95 Hz means that the signal cycled 95 times every second. Keep in mind that this cycling might be an on then off signal or a voltage pulse that never reaches zero. The important component is the fact that it actually cycles. Let's look at some examples that might make this concept easier to comprehend. Look again at Figure 17-8. It is an example of a cycling signal that starts slightly above zero (where the black box is) rises over 5 divisions above zero, stays at this level for approximately 1 division and then drops back down the 5 divisions to zero. The signal stays at zero for 1 division and then repeats itself. Notice that the signal repeats 5 times across the screen. This is a frequency oriented signal. It repeats over and over and the rise time of the

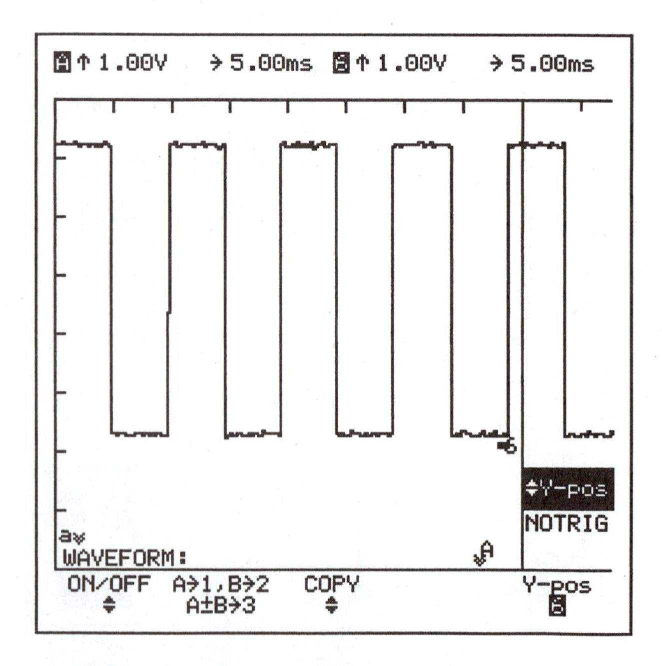

FIGURE 17-8 A frequency oriented signal.

voltage equals the fall time. The equal rising and falling voltage and the equal time high and low are the keys to this being a frequency oriented signal. Figure 17-9 is also an example of a signal which has frequency orientation. The voltage rises from zero straight up, stays high for less than a division, then falls back to zero where it stays for less than a division. This signal is faster than Figure 17-8 isn't it? How do you know? Hopefully, you have realized that the time base is the same, 5.0 mS, and the number of full cycles is greater, so the frequency must be higher. Again the characteristics of equal rise and fall voltages plus equal high and low time make this signal frequency oriented. There are many examples of frequency oriented signals on vehicles. Most of the Hall effect sensors, some **MAP** (manifold absolute pressure) sensors, and mass airflow sensors generate a frequency oriented signal. The number of pulses or cycles per second (Hertz) will be counted by a processor or module and be interpreted as speed, pressure, or mass. We will look at some very specific waveforms that are frequency oriented later.

DUTY CYCLE PATTERNS

A duty cycle oriented pattern is frequently encountered in automotive electronics. Specific examples include timing signals, mixture control solenoids, some charging systems, and some idle air systems. Keep a frequency oriented pattern in mind and a duty cycle pattern is a snap. Duty cycle refers to the amount of on time during the length of one complete pattern or cycle. As the on time increases, the off time decreases.

A mixture control solenoid is a great example to start with. Most solenoids cycle on and off 10 times a second, so their frequency remains relatively the same. However, the on time is varied based on required air fuel ratio. If more fuel is needed, the off time is increased and the on time is decreased. This is also refered to as **pulse width modulation**.

Figure 17-10 shows a signal with a 25% on time and a 75% off time. Notice that the complete on off cycle lasted about 45 mS. If the pulse width of the signal changes to 75% on and 25% off, the pattern would look like Figure 17-11. Notice that the complete on off cycle again lasted about 45 mS. The key to the understanding of a duty cycle or pulse width

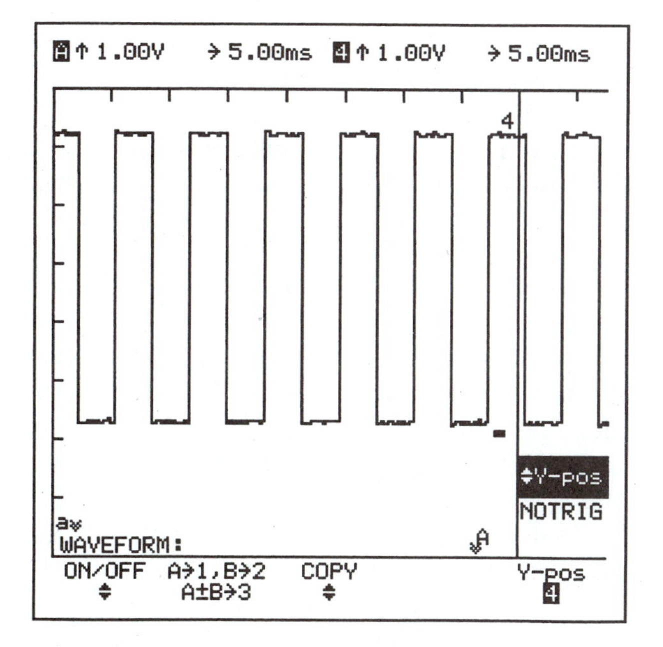

FIGURE 17-9 A repeating signal.

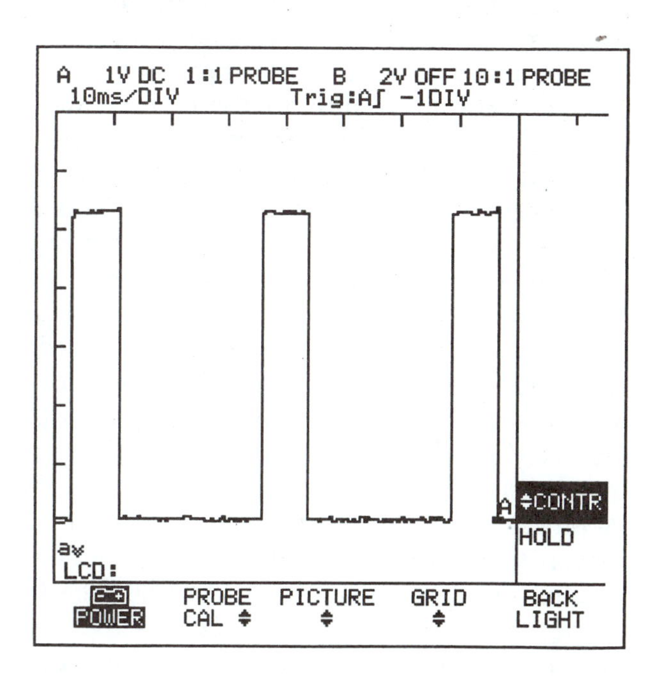

FIGURE 17-10 25% on—75% off signal.

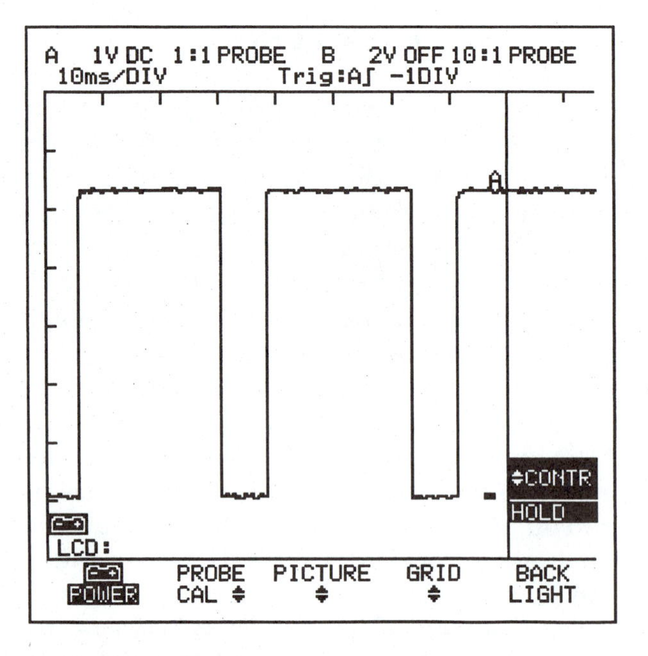

FIGURE 17-11 75% on—25% off signal.

modulated signal is in realizing that the number of pulses per second (the frequency) does not change. Only the relationship between the on and off time changes.

PULSE WIDTH

There is a specific example of a pulse width signal that will change frequency under various conditions. A fuel injector's on time is usually measured in mS and will vary with increasing and decreasing engine loads. Additional load will require additional fuel, so the injector's on time will be increased. This is another way of saying that the pulse width will be increased. There is a difference between pulse width and duty cycle. Remember that duty cycle signals will generally remain at a fixed frequency. If we continue with our fuel injector example, can the number of pulses per minute remain the same? Obviously not! As the engine speed changes, the number of pulses must also. Usually one pulse per rotation of the engine is used, so if the engine is running at 2000 rpm we will need 2000 pulses per minute. If the engine speed is reduced down to 900 rpm, the number of pulses must also be reduced.

TYPICAL SET-UPS AND PATTERNS

Now that we have the basics of the DSO covered, let's get down to actually using one on some typical signals. We will cover inputs and outputs that typical vehicles have. Keep in mind that it is not within the scope of this text to completely cover computer control. The DSO has become a part of the typical technician's tool box, however. It is becoming more of a basic tool each year. It will give you the opportunity to have additional methods of looking at varying signals. It is to the automotive repair industry what the digital multimeter was 10 years ago. We will start with a TP sensor and use the trigger and slope function to capture the signal.

THROTTLE POSITION SENSOR (TP)

The TP sensor is a good example of utilizing the capabilities of the DSO. We will set up the DSO for a voltage sweep. Sweep means that the entire voltage change cycle will show on the screen and be frozen for our interpretation. Set the scope up as follows: 1 volt per division and 1 second per division. Move your zeros down so that you have at least 5 full division above zero volt showing.

Probe the TP and read the closed throttle voltage. Note: Do not pierce the wires of any sensor if possible, as this will allow corrosion to begin. Back probing a connector is better. If you have absolutely no other option, use a piercing probe that makes a very small hole. Equipment manufacturers like Fluke offer the probes. Smear a little silicone over the hole to completely seal the wire once you are finished. Once you have the closed throttle TP voltage, set the trigger point slightly higher and the slope positive. Set the DSO for a single sweep and you are set to capture a TP. Let's analyze these setting so that you understand what you have done and why. Keep in mind that our goal is to capture the sweep and only the sweep. By setting the trigger voltage slightly higher than the closed throttle voltage and the slope positive, the screen will be blank until the TP begins to move. Once the trigger voltage is passed, the DSO will begin to trace the sweep. Move the throttle slowly as you count. You have a full ten seconds to move the throttle from closed to wide open and back to closed. A good TP sensor will have a sweep that looks like Figure 17-12. If the TP had an open section, its pattern would look like Figure 17-13.

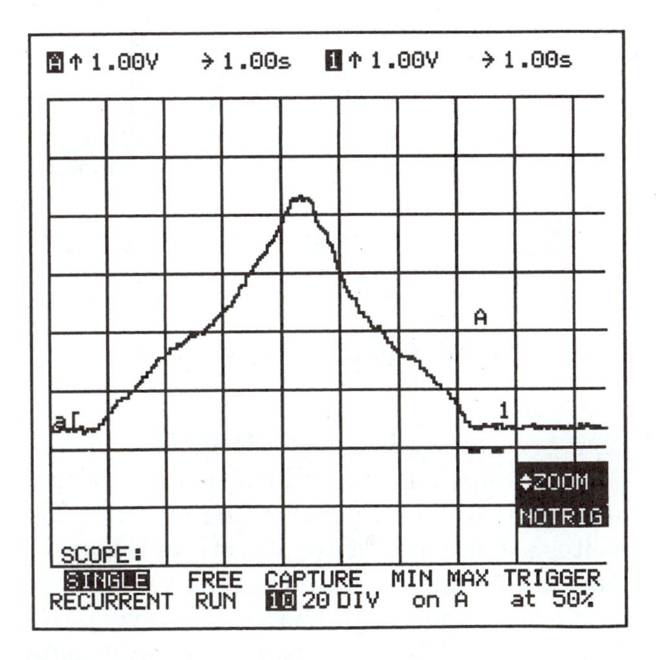

FIGURE 17-12 Functional TP.

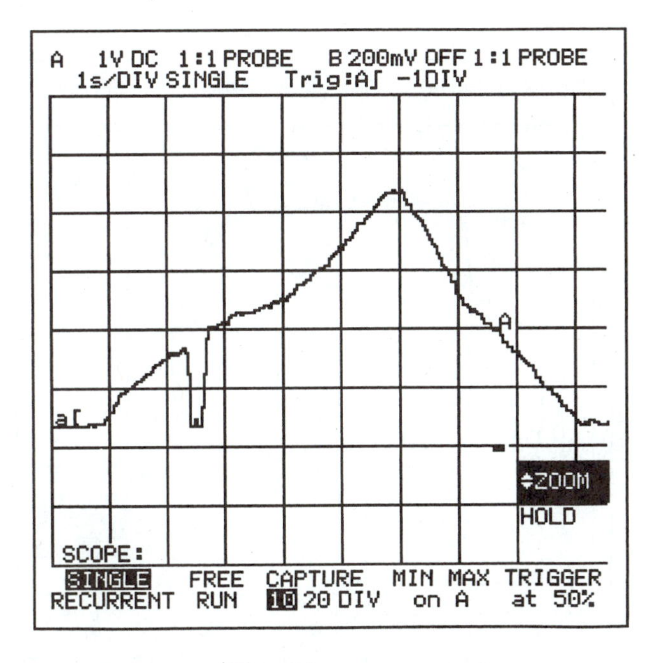

FIGURE 17-13 TP with an open section.

This is a common setup for many sensors on the vehicle. Voltage dividing MAP sensors, variable height sensors, and vane airflow meters are examples. Remember to determine what the range should be and set the scope with adequate volts per division. As these are DC sensors, move your zero volts position down to take advantage of most of the screen. Determine the closed voltage and set the trigger slightly higher. Now sweep the sensor slowly. Count to yourself and use the full amount of time that the DSO has been set for. Generally 10 seconds is a good number (1 sec/Div).

OXYGEN SENSOR (O_2S)

Setting up a DSO for an oxygen signal is slightly different because the signal repeats itself many times per minute. For this reason, we will not use the trigger or slope. Instead we will want a continuous pattern across the screen. Let's analyze what an **O_2S** does. There is additional information in the computer control chapter on how an oxygen sensor functions. For our purposes, realize that it will generate a voltage between about 100 mV and 800 mV. When fuel is burned, it leaves behind a level of oxygen that can be indicative of a rich or lean burn. If the exhaust has excessive oxygen (lean air fuel ratio), then the O_2S will generate a lower than 450 mV. If the exhaust has reduced oxygen (rich air fuel ratio), then the O_2S will generate a higher than 450 mV. The air fuel ratio should float between rich and lean as the vehicle power control module (PCM) controls the amount of fuel based on the signal from the O_2S.

A DSO is the perfect tool for looking at the oxygen sensor. Even the fastest digital multimeter has difficulty following a signal that can float from rich to lean and back again as many as 5 times a second. Even if the meter could follow it, your eyes would have difficulty.

Set the DSO with 500 mS per division and 200 mV per division. This will allow 5 full seconds of DSO operation if you capture 10 divisions or 10 seconds if you capture 20 divisions. Again, move the zero down to the 2nd or 3rd division from the bottom as Figure 17-14 shows. This will allow us to view negative voltage and yet have the capability of viewing a full volt. The 6 divisions up from zero will allow viewing of 1200 mV or 1.2 V. When testing an O_2S it is necessary to be able to view above 1 V in addition to reading below zero. As the sensor operates, it will float between rich and lean, crossing over the 450 mV mid-point. However, a correctly functioning one will never go above 1 V or below zero. Additional settings on the DSO should allow it to "free run" or trace continually. A correctly functioning oxygen sensor will switch back and forth across the 450 mV mid-point many times a second. Figure 17-15 shows a correctly functioning O_2S. Notice that there are about 10 full cycles during the entire trace. With each division lasting 500 mS, 5 seconds worth of trace are on the screen. 10 cycles during 5 seconds mean that the trace cycles 2.0 times a second. This is generally accepted as being an indication of a correctly functioning O_2S. The frequency of the trace indicates the activity level. Acceptable sensors will have a frequency of between .5 to 5 Hz. A frequency lower than .5 Hz can indicate a dead sensor while a frequency higher than 5 Hz is usually an indication of a misfire. There are additional tests that are usually done on O_2S that involve forcing the vehicle into a full rich condition, usually with propane, and watching the trace. The voltage should rise above 800 mV as Figure 17-16 shows. This indicates that the sensor is capable of measuring the decreased oxygen that is a result of the rich condition (additional fuel). An additional test tests the ability of the

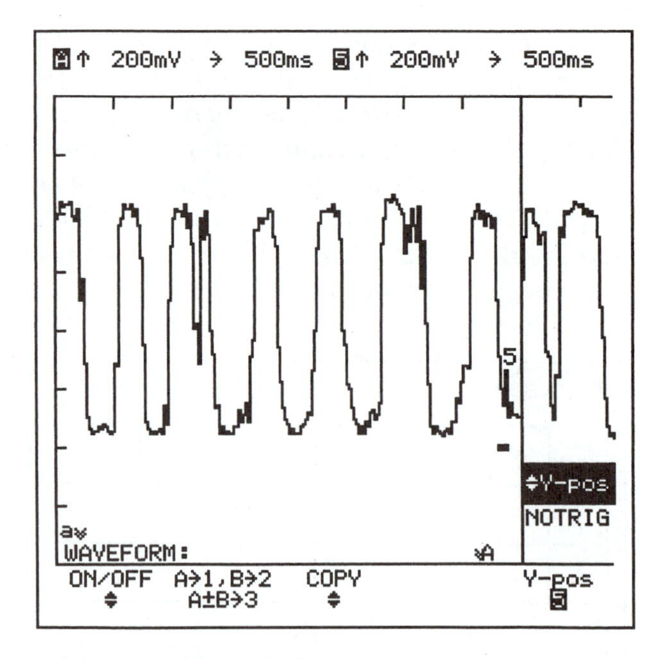

FIGURE 17-15 O_2S operating at 1.5 Hz.

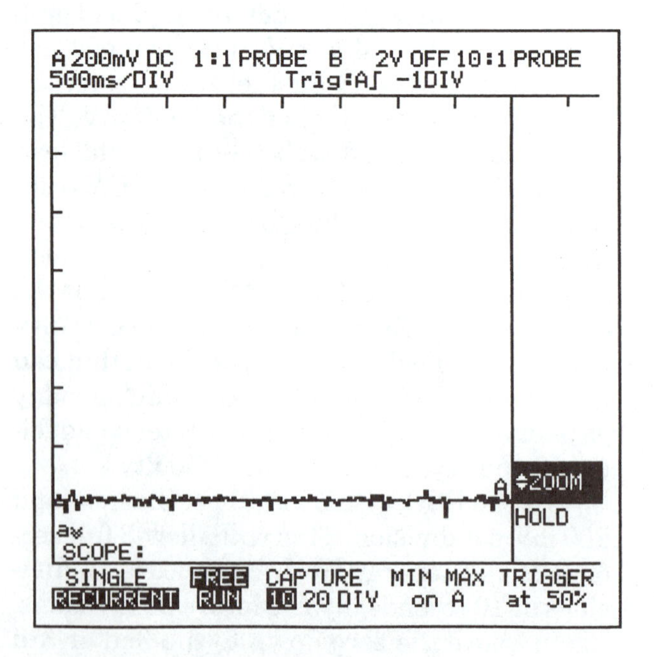

FIGURE 17-14 DSO set-up for O_2S testing.

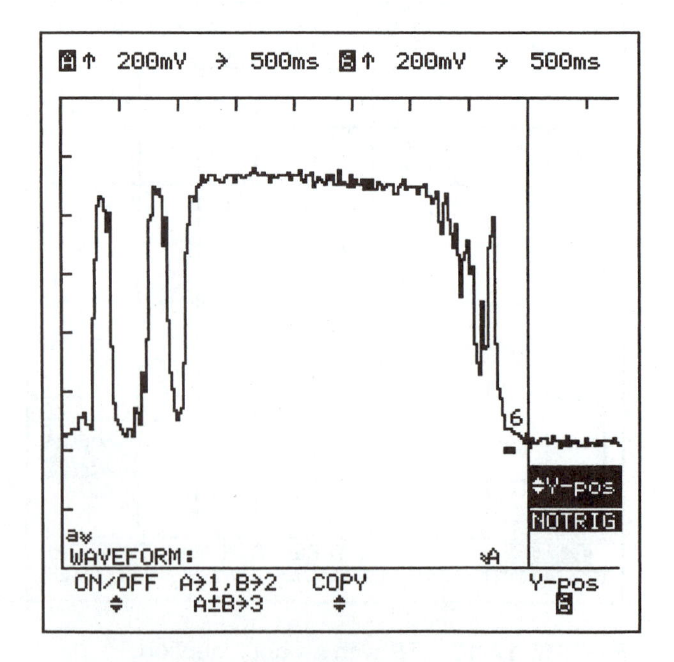

FIGURE 17-16 O_2S full rich signal.

sensor to indicate a full lean condition. The technician causes a lean condition (vacuum leak or turn off the flow of propane) and watches the trace. It should drop below 175 mV, indicating that the sensor can recognize a lean condition as Figure 17-17 shows. Notice that after the propane is on (high voltage), the trace drops down to about 50 mV. This is the sensor's lean voltage which occurred just as the propane was turned off causing the lean condition. It is extremely important that a O_2S be able to recognize a changing fuel condition quickly. By rapidly accelerating the engine, the amount of fuel is greatly increased. The trace should change from lean to rich very rapidly. The change from 300 to 600 mV should occur in less than 100 mS (Figure 17-18). Notice the extremely rapid, almost instantaneous rise after the lean condition in the middle of the screen. It is this fast reaction time that makes for a good clean running engine. This procedure of testing the sensor for frequency, a rich condition, a lean condition, and looking at the amount of time between rich and lean is courtesy of ASPIRE, Inc., and is based on the testing of hundreds of vehicles that passed enhanced emission testing at their Philadelphia Enhanced Emissions (IM-240) test lane.

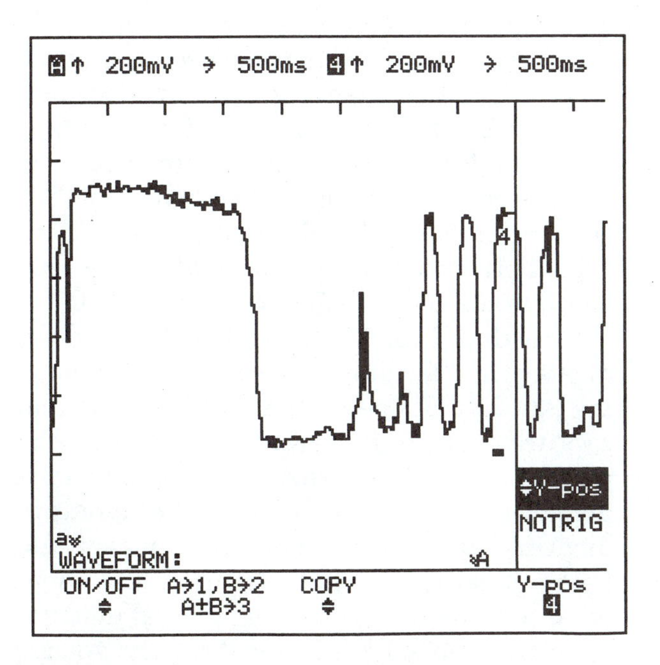

FIGURE 17-17 O_2S full lean signal.

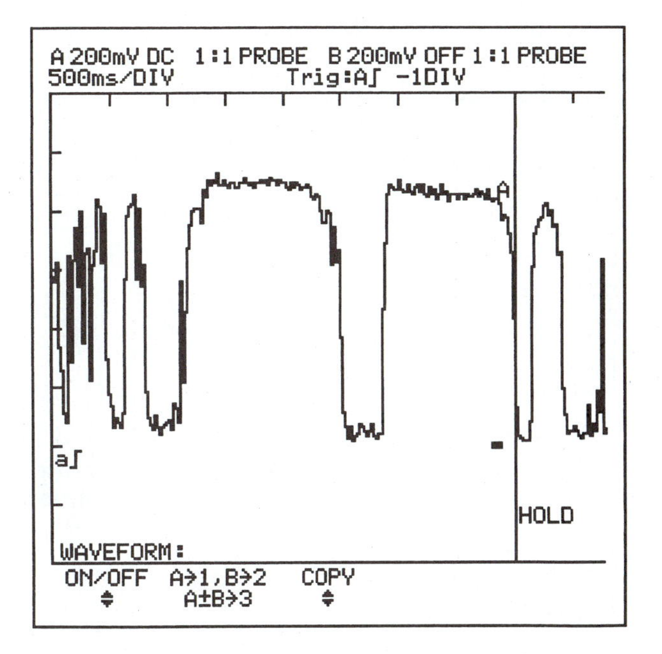

FIGURE 17-18 O_2S lean to rich signal.

AC SENSORS

As we have mentioned prior to this chapter, there are not a lot of examples of AC signals on vehicles. However, we find anti-lock braking systems and some ignition systems use a variable reluctance sensor that generates an AC signal. The signal will not look exactly like a sine wave but will have the characteristics of both a positive and negative signal. Let's set up the DSO to look at a crankshaft position sensor (**CKP**). This sensor will be used to indicate the crankshaft position for both the ignition and fuel systems. Figure 17-19 shows that the DSO is first of all set for AC, with 100mV/DIV and 50 mS/DIV. Notice that the zero indicator is exactly in the middle of the screen at the top of the fourth division. What does the trace indicate? Look at the voltage generated. The peak positive voltage is about 2 divisions or 2 V and the negative is about 2.5 V. Notice the short spacing between the last two cycles? Remember back in the DIS chapter, we looked at a GM engine with the synch pulse as illustrated in Figure 17-20. The two close pulses are the number 6 notch passing the sensor followed by synch pulse (#7). After the sync pulse, the longer space (over 50 mS) is the 60 degrees of crankshaft rotation prior

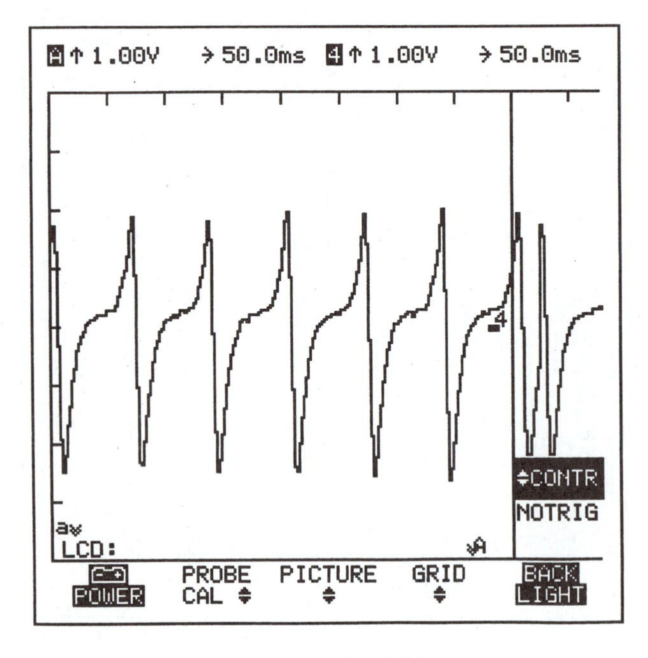

FIGURE 17-19 DSO set for AC input.

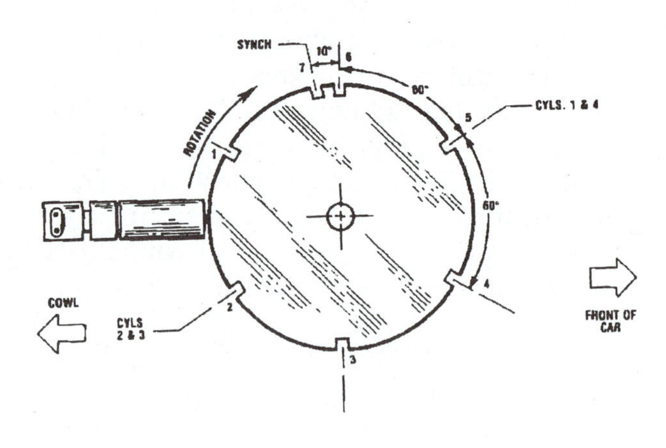

FIGURE 17-20 Synch pulse pattern. (Courtesy of GM Corporation)

to the number 1 notch which will be ignored by the ignition module. The next notch (#2) will follow 60 degrees of crankshaft rotation where cylinders 2 and 3 will be fired. Notice that the DSO gives us the ability to actually see the pulses that the ignition module and power control module will use to fire the ignition and fuel systems. In addition, most manufacturers require that a magnetic sensor generate at least 250 mV. Does our sensor generate the required voltage? It sure does and this tells us that the sensor is OK. The DSO has given us the ability to not only look at the voltage generated, but also to look at the timing of the pulses. In addition, the quality of the signal can be analyzed over any engine speed. There are some very fast digital multimeters on the market, but none of them are as fast as a DSO. At 3000 engine rpm, the sensor we have looked at will generate 21,000 pulses per minute or 350 pulses per second! The typical DVOM will update around 4 times a second. Realistically, the majority of the pulses will never be captured by the meter, but every one can be captured by the DSO. The technician's ability to find intermittent signals is greatly enhanced by the DSO. Learn to use it correctly and you will never want to be without it on those tough computer problems. It may very well be the tool of the 90s.

OPTICAL DISTRIBUTORS

Let's look at another example where a DVOM is not a realistic tool to use during diagnostics. The optical distributor shown in Figure 17-21 will generate 360 pulses per rotation of the distributor from the outside sensor and 6 pulses from the inside sensor. The inside pulses indicate crankshaft position while the outside pulses indicate engine speed. Think about the outside sensor at 3000 engine rpm. It will be generating 540,000 pulses per minute. Note: The distributor is spinning at 1/2 engine speed or 1500 rpm: $1500 \times 360 = 540{,}000$.

Dividing 540,000 pulses by 60 seconds indicates that the outer sensor will be generating 9,000 pulses per second! Only a scope can capture a signal this fast. By setting the DSO up for multiple patterns, a technician will be able to see the spacing between the signals indicating whether every signal is actually being generated or not. Figure 17-22 shows what a correctly functioning optical distributor signal looks like. Notice that the spacing

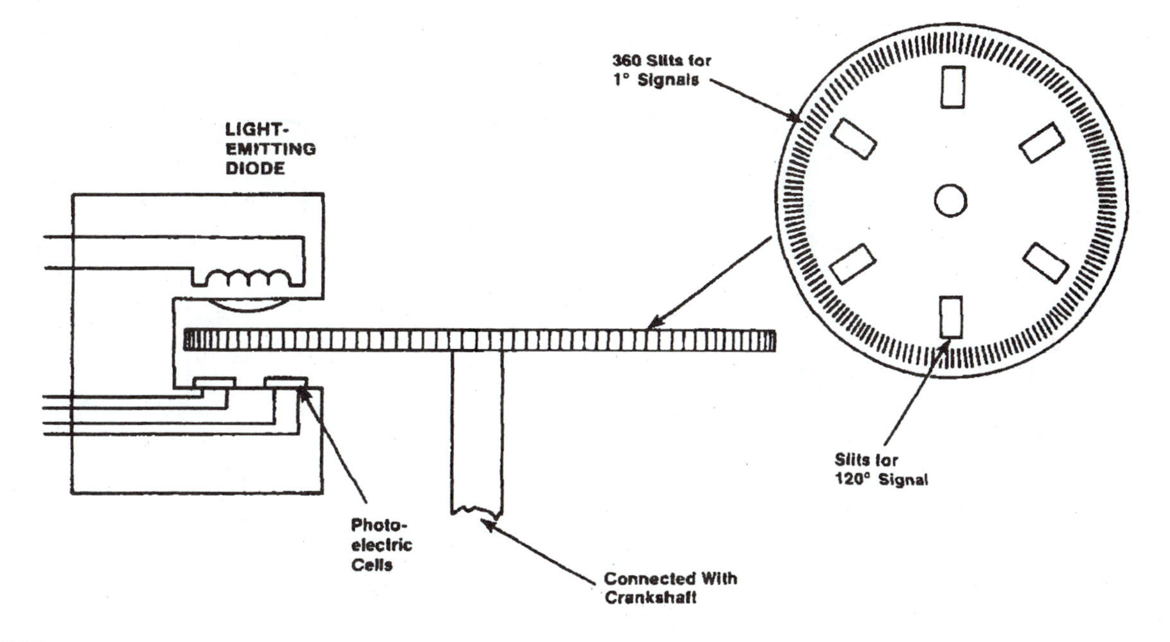

FIGURE 17-21 Optical distributor.

is relatively the same and both the high and low voltage is the same. This is an example where the technician will look for consistency of cycles and the generated voltage to determine the ability of the sensor to generate the signal required.

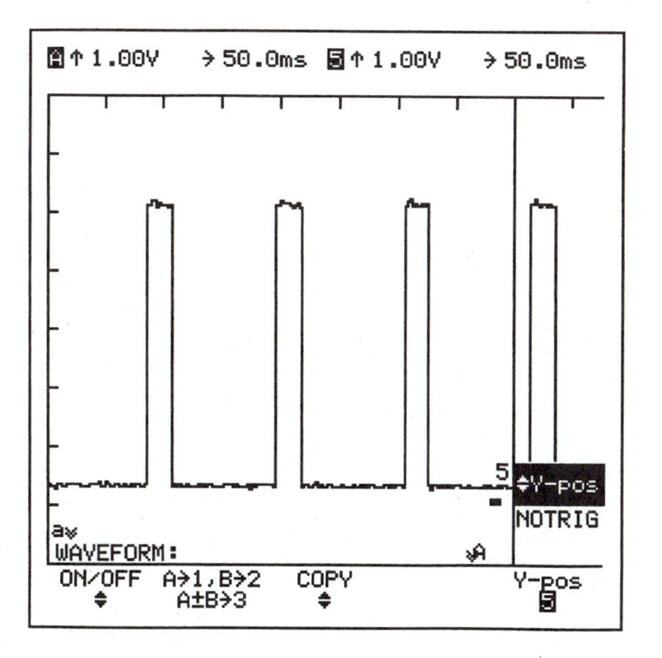

FIGURE 17-22 Distributor—optical signal.

MANIFOLD ABSOLUTE PRESSURE SENSOR (MAP)

MAP sensors come in two common varieties. The variable voltage style and the variable frequency. Testing the variable voltage style is very similar to testing the TP sensor, so we will not go into details here. The variable frequency MAP used by Ford is different however, so let's take the time to set up the scope to correctly capture the signal. The sensor is fed with 5 V from the PCM, so the DSO must be capable of displaying the full 5 V. Either 1 or 2 V per division will do. The MAP will vary the frequency of the pulse as the manifold pressure changes. The chart (17-1) reproduced here shows the relationship between the pressure (vacuum) and the output frequency in hertz. Remember that the frequency is the number of full cycles per second. Figure 17-23 shows a correctly functioning MAP with no vacuum applied. With a setting of 5 mS per division the trace shows 7 full cycles during the 50 mS. What frequency is this? There are actually two separate methods to determine the actual frequency. The easiest way is to set the scope up to actually read the frequency. The other way involves figuring out the frequency from the

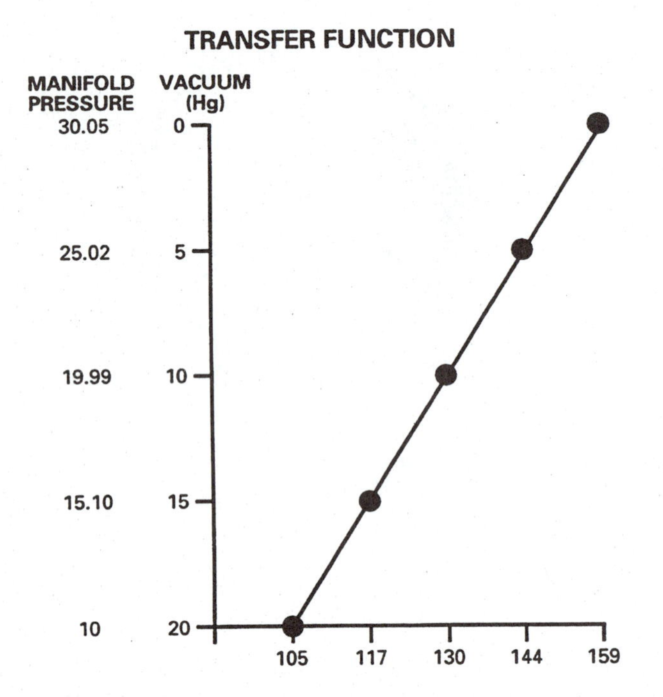

CHART 17-1 Relationship between pressure (vacuum) and the output frequency in hertz.

information on the screen. Let's do the math together. It would take 20 screens at 50 mS per screen to equal 1 second: .050 × 20 = 1.000 Sec. How many cycles would occur during the same 20 screens? If 7 cycles occurred during the one

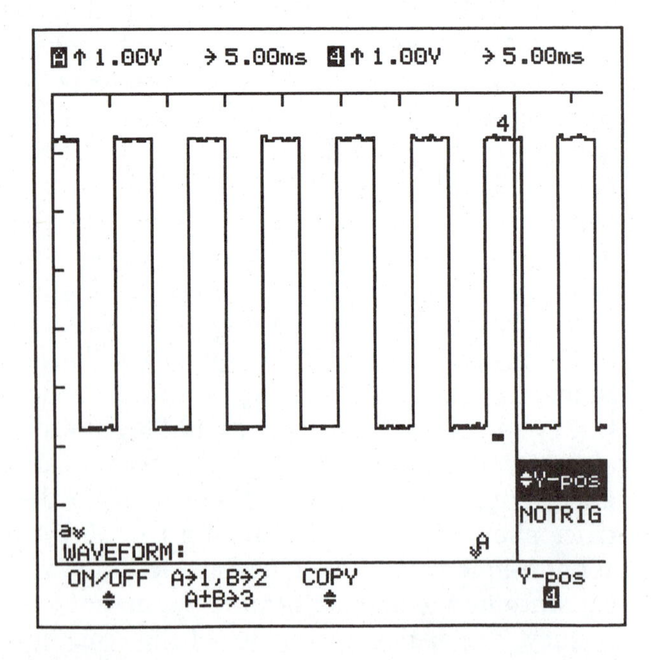

FIGURE 17-23 Ford MAP—KOEO (Key On Engine Off).

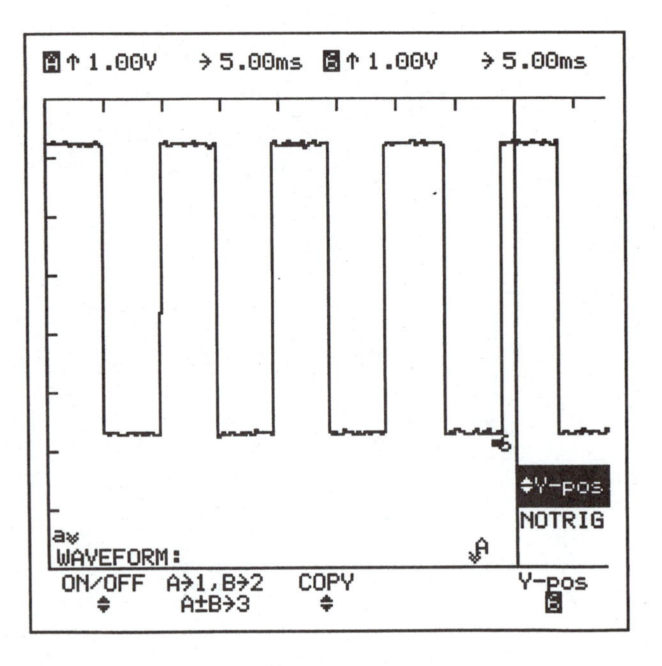

FIGURE 17-24 Ford MAP—20" vacuum.

screen, then the actual frequency is 140 hertz: 7 × 20 = 140 Hz. With no vacuum applied to this MAP sensor, is it outputting the correct frequency? The chart reproduced above indicates that this MAP is slightly off in frequency. Figure 17-24 shows the same MAP with 20 inches of vacuum applied to it. Can you figure out the actual frequency? The screen shows 5 cycles during 50 mS. What do you have to multiply 50 by to get to 1 second? 20, right? 50 mS × 20 = 1.000 Sec. Multiply the number of cycles on the screen by the same number, 20 and you get a frequency of 100 Hz. Comparing the frequency to the chart tells the accuracy of this very important sensor.

MASS AIRFLOW SENSORS (MAF)

We are seeing more and more applications of mass air flow sensors (**MAF**). Let's look at one type that is especially easy to test accurately with a DSO. Figure 17-25 shows both the setup and the resulting pattern from a MAF off a 3800 Buick engine running at about 2000 rpm. Notice that the pattern cycles very rapidly across the screen. With only 200 mS per division we see 9 complete cycles.

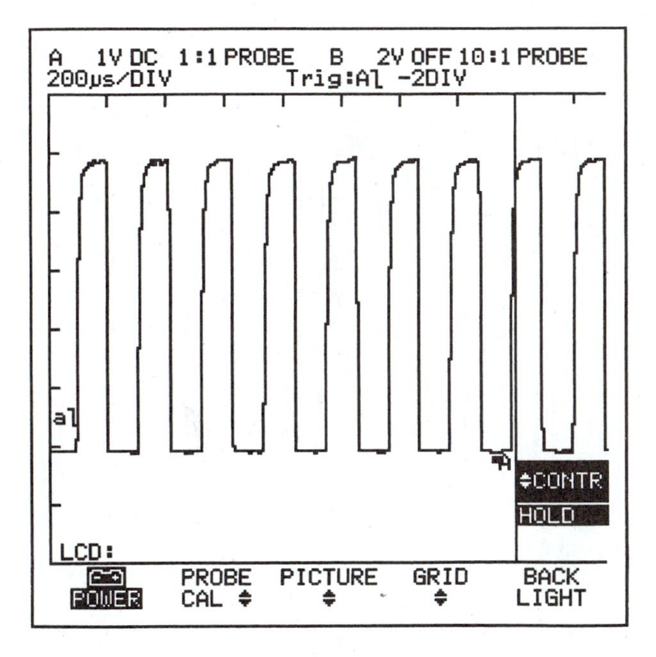

FIGURE 17-25 Buick 3800 MAF signal at 2000 engine rpm.

Notice that the relative spacing between cycles is about the same, and that most lines are squared off and clean. Compare these patterns to Figure 17-26. This is the same vehicle with the engine running at idle and minimum air passing through the MAF. Did the frequency change? There are about 5 full cycles showing during the same time period. Is this MAF able to recognize air flow? It sure seems like it can. The frequency almost doubles from idle to 2000 rpm. Notice also that the signals appear to be sharp and reasonably clean. The slight unevenness present on the horizontal lines is normal for a DSO. The relative evenness of the pattern and the squared off top and bottom voltage indicates a correctly functioning MAF. Keep in mind that different MAFs produce different signals. The output on some of them may be a DC voltage that varies with engine speed.

HALL EFFECT SENSORS

Let's look at a typical Hall effect signal from a Ford PIP sensor. Figure 17-27 is the DSO pattern for a correctly functioning Hall. Remember from our discussion of ignition systems, the Ford PIP sensor passes the applied voltage out when the shutter blocks the magnetic field. The DSO shows this voltage at about 13 V. (2 V/Div - 6.5 divisions). When the window unblocks the field, the Hall turns on and the output voltage drops down generally to near zero. A DSO is a logical tool to use to observe the signal coming

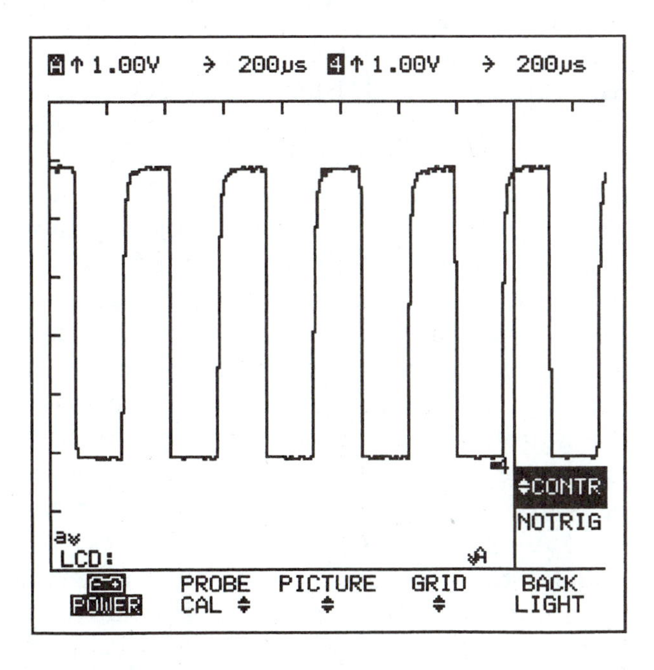

FIGURE17-26 Buick 3800 MAF signal at idle.

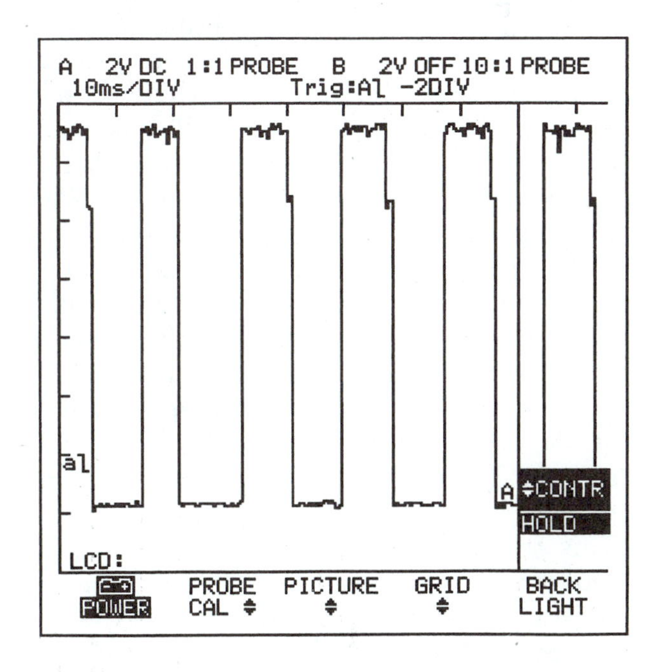

FIGURE 17-27 Ford PIP signal.

off the Hall sensor because it allows us the ability to look for clean crisp transitions between the high and low signals. In addition, a DSO allows a technician to check for the consistency of the signal. A misfire that is sensor oriented might not be detected by any other method. For example, a DVOM will average the high and low signals or will give the high voltage level and low voltage level if a min/max averaging feature is available, but will not be able to show each and every signal like a DSO can. What would a DVOM on DC volts show on the PIP? About 6.5 V average. The average reading which is about 50% is due to the relationship between the amount of time the sensor is on and off. In this example, the sensor is off (higher voltage) for about 8 mS and on (low voltage) for about the same 8 mS. Notice the wider window and narrower shutter on the left of the screen? This is the signal for the #1 piston in firing position necessary for sequential fuel injection. The PCM will look at this signal and see the different spacing only once per distributor rotation. This information is necessary because the injectors must be fired in the correct sequence starting with #1. This is another example of advantages to the DSO over the multimeter. The extreme speed of the signal frequently prevents even the fastest digital multimeter from capturing the changing voltage. A GM 18X Hall effect is another good example of this. Remember from the DIS chapter that many GM vehicles use a Hall sensor that has 18 windows and vanes and produces 18 voltage cycles per crankshaft rotation. At 2500 rpm that means that 45000 pulses per minute or 750 pulses per second will occur. Only a scope can capture a signal this fast and display it as shown in Figure 17-28. Notice that the DSO has been set for only 5 mS/Div.

That's only a total time across the screen of 50 mS and yet we have displayed about 16 signals. If we multiply the number of cycles present by 20 we get a frequency of 320 Hz. This equates over to almost 1100 rpm. 320 pulses per second × 60 seconds divided by 18 pulses = 1066 rpm. Again, the ability to, observe each signal may be important, especially in the case of an intermittent misfire.

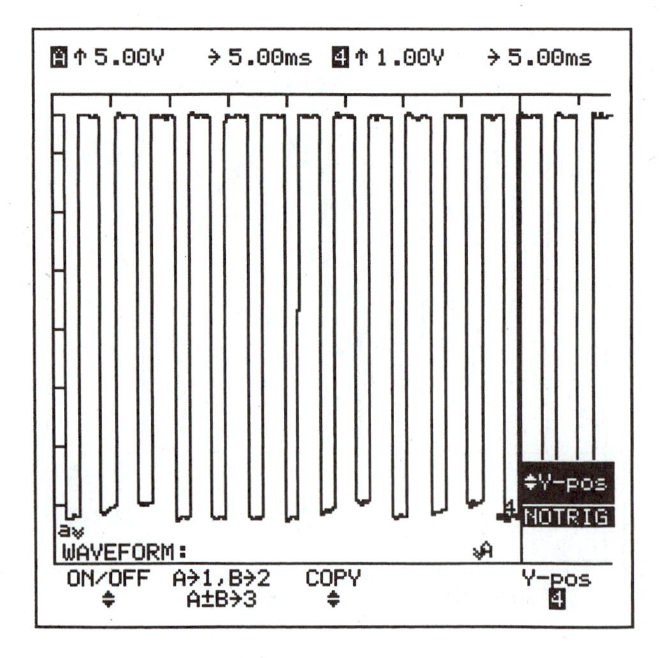

FIGURE 17-28 GM 18X signal.

FUEL INJECTOR PATTERNS

DSOs are especially useful when observing a fuel injector pattern. The DSO allows us to calculate how long the injector is open, in addition to verifying that the injector actually has the signal necessary to open. Fuel injectors are another example of using trigger and slope to make sure that the pattern actually can be captured easily. Figure 17-29 shows the screen for a GM port fuel injector. The trigger has been set for a negative slope of about 8 V. The fuel injector is fed with about 14 V (the beginning horizontal line) The PCM turns on the injector by applying a ground (the lower horizontal line at about 0 volt) for about 5 mS (a full division). When the injector is turned off, a voltage spike of about 30 V occurs just at the turn off moment. This spike is the result of the collapsing magnetic field of the injector. This illustration shows an additional DSO technique called **delay**. Notice in the upper right section of the screen the Trig:A–4DIV? This means that the DSO will display the pattern on the screen 4 divisions from the left edge, once the trigger and slope are achieved. This allows placement of the pattern in the center of the screen, if desired.

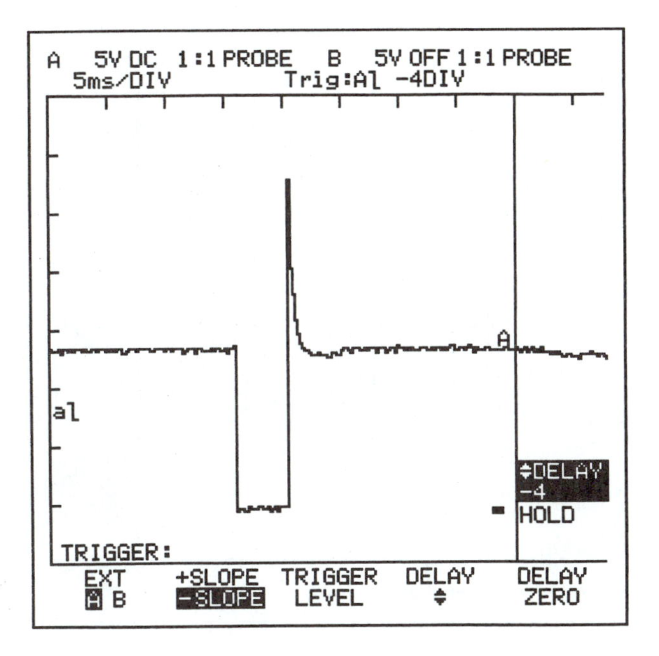

FIGURE 17-29 GM port fuel injector pattern.

This pattern shows a functional fuel injector signal. The technique of using trigger, slope and delay to display only the pattern you want, where you want it on the screen is especially helpful in intermittent problems.

IDLE SPEED OUTPUT

Lets look at another DSO example. That of an idle speed signal from a Ford vehicle. Figure 17-30 shows the signal with the vehicle at idle with a loaded engine. Perhaps the power steering is turning and the air conditioning compressor is on. This load shows up in the on time of the sensor. This on time indicates that the amount of time the air bypass is actually open. This time is from the beginning of the drop in voltage to the beginning of the rise in voltage. This illustration shows that period of on time to be about 4 mS, and the off time to be about 2+ mS. By decreasing the load on the engine, the air bypass should close down for longer periods of time. This is represented by Figure 17-31. Notice that the time from the beginning of the drop in voltage until the beginning of the rise in voltage is about 2 mS while the off time (time from rise in voltage

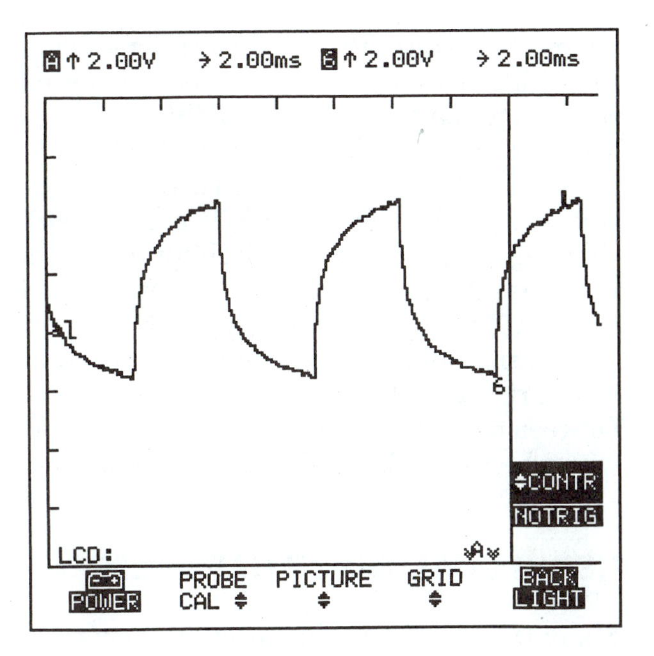

FIGURE 17-30 Ford idle air control signal with engine loaded down.

until drop in voltage) is now about 4 mS. Every time the idle air solenoid opens, the idle increases, and every time the solenoid closes, it decreases. This is actually a type of pulse width modulation as the number of traces on both illustrations is the same. The

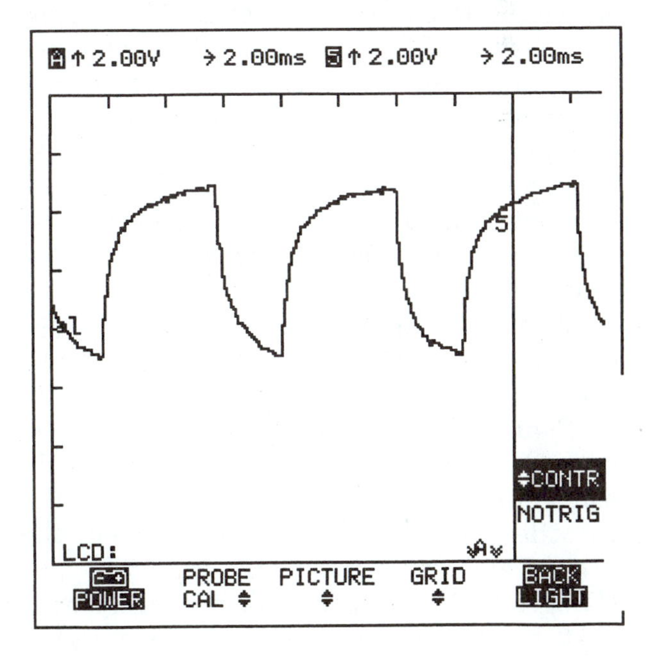

FIGURE 17-31 Ford idle air control signal with light load.

difference between the two illustrations shows that this idle air bypass system is functioning with longer on time during conditions of high load and lower on time for light load conditions. This shows that the PCM does have control over the idle speed. This is not a square wave signal because we are looking at a coil of wire that resists changes in current. When we turn it on, by applying a ground, the voltage doesn't fall instantly because the current flow is rising. It is the relative time difference between on and off time that we are concerned with when we look at this type of PCM output.

USING DUAL TRACE CAPABILITY

Many DSOs that are currently on the market allow the technician to display two or more patterns on the same screen. This is especially helpful for comparison purposes. We will look at two examples of the use of the **dual trace** option. The first is that of viewing both of the CKP sensors (crankshaft) on a fast start GM system. Remember the fast start from the DIS chapter: Two Hall effect sensors with 3 pulses per rotation from one and 18 pulses per rotation from the other. Figure 17-32 shows the 3X sensor on top of the 18X sensor. The relative size of the different 3X windows is apparent and it is this size difference that allows the PCM to know exactly where the crankshaft is in only 120 degrees of crankshaft rotation. The dual trace shows that all windows and vanes are present and that the signals from both sensors are clean, clear, and consistent. This is a good sensor.

Another example of using the dual trace option might be in the comparison of two changing signals to see if they change together. Figure 17-33 shows a TP (throttle position sensor) on the bottom and MAP (manifold absolute pressure sensor) on the top. As the throttle opens from idle to wide open, the pressure changes inside the manifold also changes as shown by the rising voltage. Notice that the pressure change occurred just as the throttle moved. Closing the throttle caused the voltage to drop (bottom pattern) at the same time that the pressure changed in the manifold (top pattern). Did the movement in the throttle cause a corresponding change in manifold pressure? Will the PCM get the in-

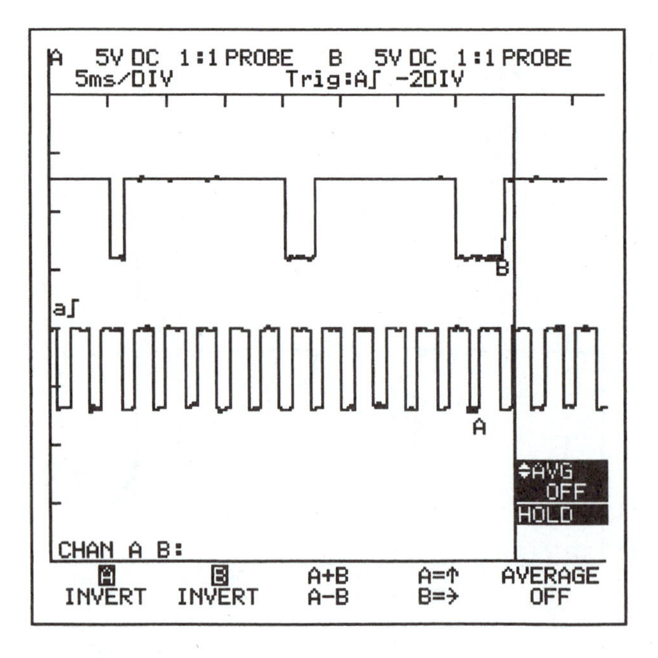

FIGURE 17-32 Dual trace showing 3X (top) and 18X (bottom) sensors.

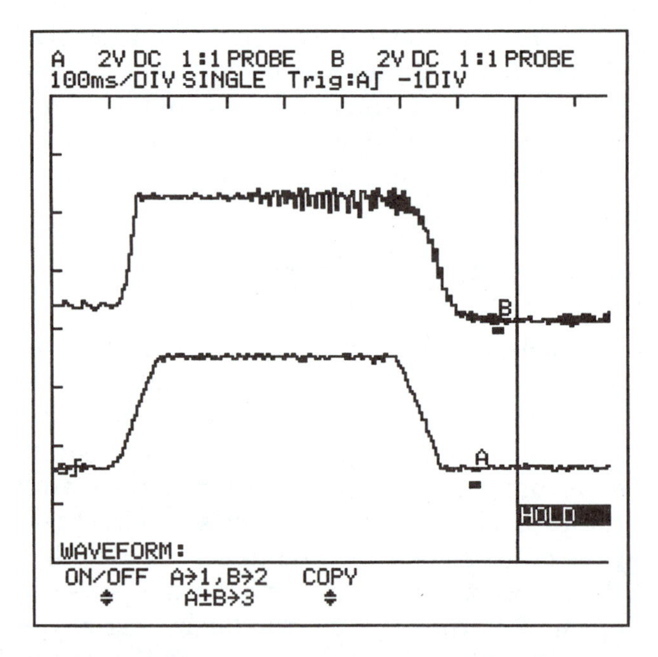

FIGURE 17-33 Dual trace showing MAP (top) and TP (bottom) sensors

formation it requires from these two sensors? The answer to both of these questions is yes!

SUMMARY

In this chapter we have looked at one of the newer diagnostic tools available to the modern technician—the digital storage scope. We have analyzed the basic screen adjustments involving time and voltage. We have seen common set-ups for looking at typical sensors and viewed patterns from both good and bad sensors. In addition we have looked at some common output wave forms on the DSO.

KEY TERMS

DSO (digital storage oscilliscope)
Divisions
Trigger
Slope
TP (throttle position)
PCM (power control module)
MAP (manifold absolute pressure)
Pulse width modulation
O_2S (oxygen sensor)
CKP (crankshaft position)
MAF (mass airflow)
Delay
Dual trace

CHAPTER CHECK-UP

Multiple Choice

1. Technician A states that the DSO is actually a voltmeter that is time based. Technician B states that the DSO can be used to view voltage changes over a period of time. Who is correct?
   a. Technician A only
   b. Technician B only
   c. Both Technician A and B
   d. Neither Technician A or B
2. The V/Div is actually the
   a. number of volts per division up and down the screen.
   b. number of volts per division across the screen.
   c. variable factor of time.
   d. none of the above.
3. The DSO will display volts
   a. up and down the screen.
   b. across the screen.
   c. from 0 to 25V only.
   d. that have a positive polarity only.
4. The purpose of using the trigger function is to
   a. turn on the cursors for screen evaluation.
   b. have the scope display everything that is happening in a live format.
   c. display only patterns that have certain characteristics.
   d. display the frequency of the sensor being tested.

5. Technician A states that a DSO set to 200mV/Div and 200 mS/Div will display a signal for a total of 2 seconds. Technician B states that a DSO set to 200mV/Div and 200 mS/Div will be able to display a 800 mV positive and a 800 mV negative signal. Who is correct?
   a. Technician A only
   b. Technician B only
   c. Both Technician A and B
   d. Neither Technician A or B
6. An O_2 signal is displayed on a DSO for a total of 10 seconds and has a frequency of 12 Hz at 2500 rpm. This indicates that the
   a. vehicle is in fuel control.
   b. fuel system is running rich.
   c. fuel system is running lean.
   d. engine is misfiring.
7. A Ford MAP sensor generates a signal of 200 Hz at idle. This indicates that the
   a. MAP is functioning correctly.
   b. engine is under heavy load.
   c. MAP is not functioning correctly.
   d. engine is under very light load.
8. A Hall effect sensor is being observed on a DSO. Technician A states that the high output voltage will generally be the input voltage. Technician B states that the low output voltage will usually be a negative signal. Who is correct?
   a. Technician A only
   b. Technician B only
   c. Both Technician A and B
   d. Neither Technician A or B
9. A pulse width modulated signal is one that will have
   a. a varying on/off time.
   b. the same number of voltage changes per second.
   c. either a DC or AC signal.
   d. all of the above.
10. A high voltage (45 V) is observed on a DSO just as a fuel injector is turned off. Technician A states that this is the spike generated by the turned off magnetic field. Technician B states that this is induced voltage and is normal. Who is correct?
    a. Technician A only
    b. Technician B only
    c. Both Technician A and B
    d. Neither Technician A or B

True or False

11. A 1 S/Div setting will allow 1 minute of signal changes to be displayed.
12. A negative slope setting is used if a technician wishes to capture a signal that is negative.
13. A rising voltage signal that crosses 1 V will be displayed on a DSO that has a trigger of 1 V with a positive slope setting.
14. A typical frequency modulated signal will have a variable on and off time.
15. A good TP sweep shows a gradual rising and then falling voltage.
16. A good Hall effect sensor will generally display a pulse width modulated signal as the engine speeds up and slows down.
17. O_2 sensors should never display voltages above 1 V or below zero V.
18. An mS is actually a thousandth of a second.
19. An mV is actually a millionth of a volt.
20. A DSO pattern cannot be frozen on the screen for analysis at a later time.

CHAPTER

Computerized Engine Controls

OBJECTIVES

At the completion of this chapter you should be able to:

- identify the various sensors that are typically used as inputs.
- identify the various actuators that are typically used as outputs.
- understand the relationship between inputs and outputs.
- have a working knowledge of how typical sensors and actuators function.

INTRODUCTION

The past years have seen increased reliance on the vehicle processor or processors to perform driveline operations. The typical vehicle of today has multiple processors that communicate to one another and are in direct control of fuel delivery and metering, spark output and timing, transmission pressures and shifting, air conditioning, cruise control, anti lock braking, digital dashboards, and emission control. The list gets longer each year as most automotive manufactures find some additional jobs for the processor to get involved in. Within this chapter we will take a look at some typical computerized systems that directly involve the control of the power train. We will look at some typical sensors and analyze how they function, in addition to looking at some typical outputs. We will end the chapter with a discussion of how the power control module (PCM) uses inputs to control the various functions assigned to it.

THE POWER CONTROL MODULE

The PCM on today's vehicle is a fast, efficient, over 100 pin processor with over 20 operating strategies, available from over 100k of ROM (Read Only Memory). It typically uses over 20 inputs and controls or outputs to over 25 devices. It can be reprogrammed in the dealership with the latest information downloaded from CDs. The earliest PCM's generally had less than 10 inputs and controlled 3 or 4 devices. The PCM on today's vehicles remains the heart of the modern system. To be able to make accurate decisions, it must have correct information that comes from many sensors.

COMMON SENSORS

Sensors are the input devices that we have previously commented on. Their job is to relay the important operating information that the PCM needs to be able to control the actuators. Some of the required information will be used to determine the operating strategy of major systems such as fuel control, timing control and transmission operation. Additional information will be used to modify or adapt major systems in an effort to "fine tune". In addition, some sensor inputs are used for non-essential systems such as cruise control or air conditioning. We will start with the most important or first sensors that the PCM will use to get the engine running.

ENGINE COOLANT TEMPERATURE (ECT)

The ECT is one of the most important sensors. It has the primary function of telling the PCM at what temperature the engine is operating. Before the PCM can determine how much fuel will be necessary to get the engine started, it will have to know what the engine temperature is.

ECTs are generally negative coefficient thermistors which are threaded into the coolant passage usually in the hottest section of the engine, Figure 18-1. As the coolant temperature goes down, the resistance of the ECT will go up. You can see from the chart in Figure 18-2 that a coolant temperature of 32° F (0°C) will result in a resistance of 95 K ohms, while a coolant temperature of 248°F (120°C) will result in a resistance of 1.18 K ohms (1180 ohms). The term negative, in negative coefficient, refers to the fact that temperature and resistance are inverse. As one goes up the other goes down. You probably realize that a computer cannot directly read resistance, so we will have to do something with the resistance that the PCM will be able to interpret as engine temperature.

Figure 18-3 shows that the typical ECT circuit has a 5-V feed through a pull-up resistor as the power to the thermistor. To complete the circuit, a conductor returns to the PCM ground. The PCM senses the voltage at point A, which is in between the pull-up resistor and the ECT's thermistor. Remember Ohm's law? It stated that the voltage would divide between two resistances in series if current was flowing. As the thermistor's resis-

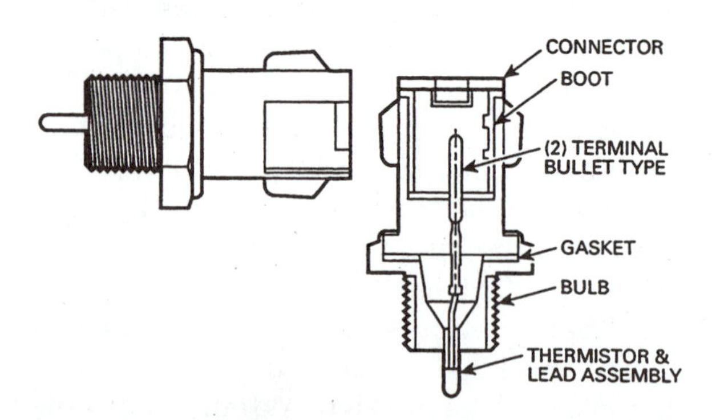

FIGURE 18-1 ECTs are threaded into the coolant passage usually in the hottest section of the engine. (Courtesy of Ford Motor Company)

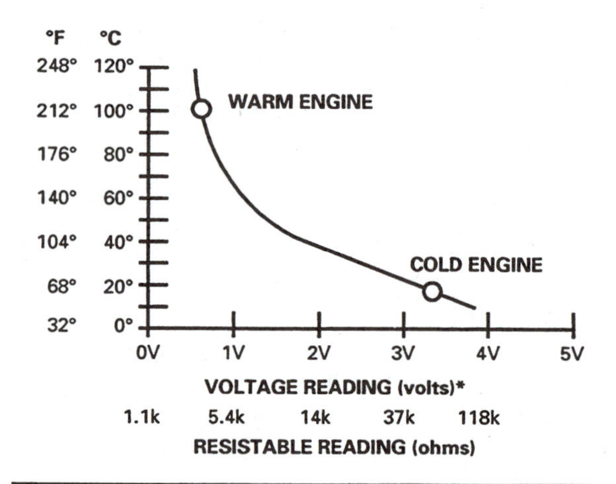

Temerature °F	Temerature °C	Voltage* Volts	Resistance K Ohms
WARM ENGINE 248	120	0.28	1.18
230	110	0.46	1.55
212	100	0.47	2.07
194	90	0.61	2.80
176	80	0.80	3.84
158	70	1.04	5.37
140	60	1.35	7.60
122	50	1.72	10.97
COLD ENGINE 104	40	2.16	16.15
86	30	2.62	24.27
68	20	3.06	37.30
50	10	3.52	58.75

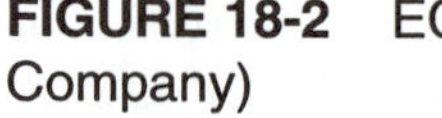

FIGURE 18-2 ECT chart. (Courtesy of Ford Motor Company)

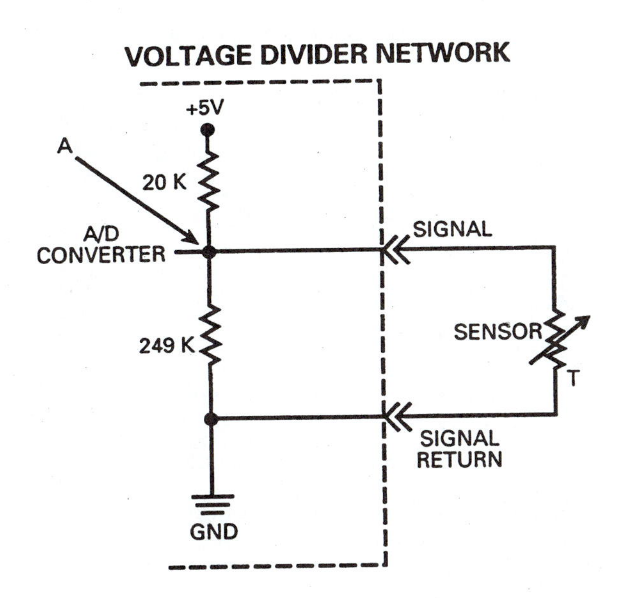

FIGURE 18-3 The typical ECT circuit has a 5-V feed through a pull-up resistor as the power to the thermistor. (Courtesy of Ford Motor Company)

tance goes down, because the engine is warming up, there will be less of a voltage drop. More of the 5 volts will be dropped across the pull up resistor, whose resistance is fixed. The voltage at point A will therefore go up. The graph shows the relationship between temperature and voltage at point A. As soon as the customer turns the ignition key on, the PCM will know what engine coolant temperature is. This input will be used to determine just how much fuel will be injected. The colder the engine is, the more fuel will be necessary. The warmer the engine is, the less fuel will be necessary. Obviously, the amount of fuel injected must be accurate if the engine is to start and run efficiently. Too little or too much fuel may result in reduced fuel economy, driveability problems, or excessive emissions. The ECT is extremely important because it determines, indirectly, the air fuel ratio. Additionally, the ECT will be used to turn on the cooling fan, and adjust timing during warm-up. Usually the timing is advanced slightly during warm-up because the additional fuel being delivered will be burned more completely. This helps to reduce emissions during warm-up in addition to lessening off idle hesitations.

INTAKE AIR TEMPERATURE SENSOR (IAT)

An equally important sensor is the intake air temperature sensor, the **IAT**. The air in the intake manifold will be used for combustion and its temperature is extremely important to performance and emissions. Cold air contains more oxygen and requires more fuel, while hot air contains less oxygen and requires less fuel. The IAT is exactly the same as the ECT as Figure 18-4 shows. It is a negative coefficient thermistor with the same resistance to temperature relationship.

The only real difference between the IAT and the ECT is the open tip which allows intake air to flow around the thermistor tip. The voltage to temperature relationship is also usually the same as for the ECT. The PCM will read the voltage at point A and know what the temperature of the air in the intake manifold is as soon as the key is turned on. This information is especially important in the winter up North when the engine is warmed up, requiring less fuel. If the PCM knows that the air is not as warm as the engine coolant, it will not lean out the air fuel ratio as much. In this way cold, weather driveability problems are prevented.

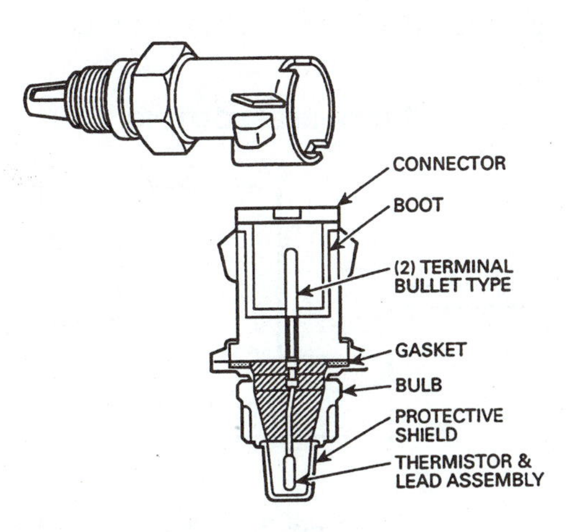

FIGURE 18-4 The IAT is exactly the same as the ECT. It is a negative coefficient thermistor with the same resistance to temperature relationship. (Courtesy of Ford Motor Company)

THROTTLE POSITION SENSOR (TP)

The position of the throttle is another very important input that is required for maximum driveability, fuel economy, easy starting, and lowest emissions. The information that this input will deliver will be used to determine air fuel ratio, timing changes, and transmission shift characteristics. As the driver depresses the gas pedal, the PCM changes the operating conditions of the engine so that it will begin to accelerate. The position of the throttle will be measured by a three wire sensor, as Figure 18-5 shows.

The TP has a 5-V feed and a ground return to the PCM. As the throttle is moved, a wiper will contact a different part of the sensor and deliver a divided voltage. For example, if the throttle is at 1/2 open position, the wiper will contact the middle of the fixed resistor. The voltage at this point will be 1/2 of the applied 5 V or 2.5 V. The wiper will contact the fixed resistor at the 2.5 V-position and deliver this voltage signal to the PCM. When the PCM sees this 2.5 V, it will know that the throttle is at the 1/2 or 50% position. How about 3/4 throttle? What will the voltage be? 3/4 of 5 V is 3.75 V. What will a 1/4 throttle voltage be? 1/4 of 5 V is 1.25 V. Hopefully you are getting the idea. The position of the throttle will be sent back to the PCM as a percentage of the 5 V which was applied to the sensor in the first place. Figure 18-6 shows the typical idle

TROUBLESHOOTING

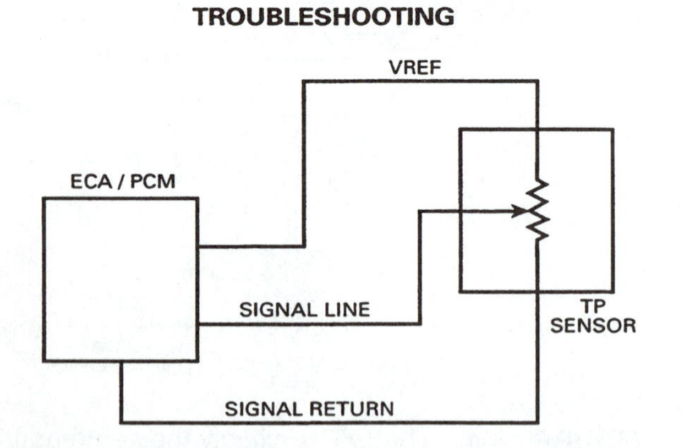

FIGURE 18-5 Throttle position will be measured by the TP, a three wire sensor. (Courtesy of Ford Motor Company)

TRANSFER FUNCTION

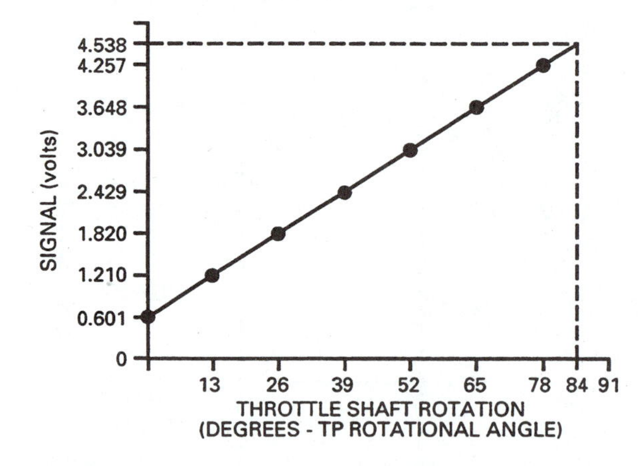

FIGURE 18-6 Typical idle through wide open throttle voltages. (Courtesy of Ford Motor Company)

through wide open throttle voltages. In the previous DSO chapter, we looked at a voltage sweep of a known good sensor between idle and wide open throttle. Figure 18-7 shows a good sweep of a TP.

What effect do you think this information will have on the operation of the engine and transmission? The engine will require more fuel and slight retarding of the timing when the consumer initially opens the throttle slightly. Additional fuel is required because the opened throttle allows the engine to breathe

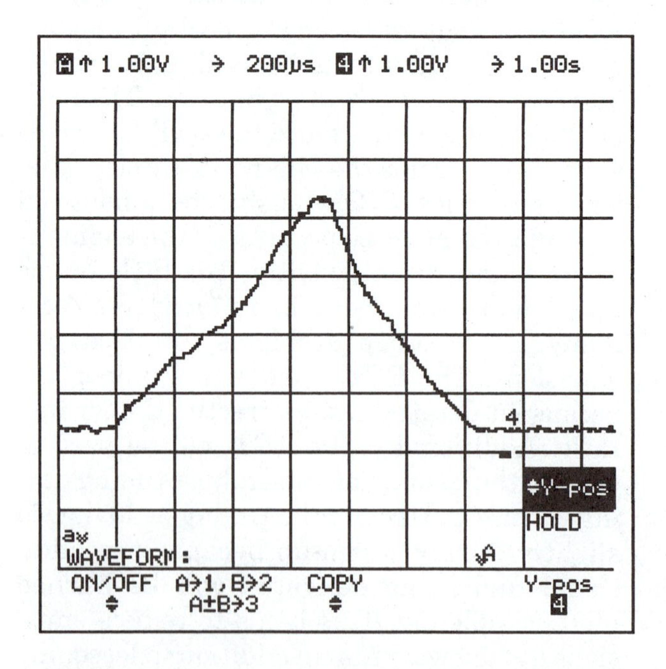

FIGURE 18-7 Scope pattern showing functional TP.

additional air. The increased amount of gas and air speed up the burn within the cylinder thus requiring a later spark (retarded). If the spark timing is not retarded during throttle opening, the peak combustion pressure of the burn might hit the piston before it reaches top dead center causing detonation and possibly piston damage. As the engine speeds up the timing will again be advanced. Remember the vacuum and centrifugal advancing systems? For the most part, the computer will follow the older mechanical advance/retard conditions of increased speed = increased advance and increased vacuum = increased advance. There are exceptions to these basic advance curves. Acceleration is one. If the throttle is opened quickly, additional fuel must be injected in accelerator pump fashion as Figure 18-8 shows. Air is lighter than fuel so as soon as the throttle is opened, the air volume increases. Fuel lags behind slightly during the acceleration. A slight amount of additional fuel and advanced timing will compensate for the additional air. The PCM will look at the position and rate of throttle opening to determine just how much fuel is necessary and whether to advance or retard the timing. Once the initial acceleration has been compensated for, the timing will be retarded. While all of this changing to the air fuel ratio and timing are taking place the actual load on the transmission will be increasing. Higher internal pressures are needed so that the transmission will not slip and destroy clutches. The PCM will boost up the operating pressure of the transmission to compensate for the additional load of additional gas and air burning. In addition, if the throttle is opened a lot, the transmission will downshift to a lower gear. You can see that the TP sensor is an important input to the PCM. One additional use of the TP sensor is when wide open throttle is achieved with the engine cranking. This combination of signals, WOT and cranking tells the PCM to greatly reduce fuel delivery because the engine is flooded. This is usually referred to as the "clear flood mode".

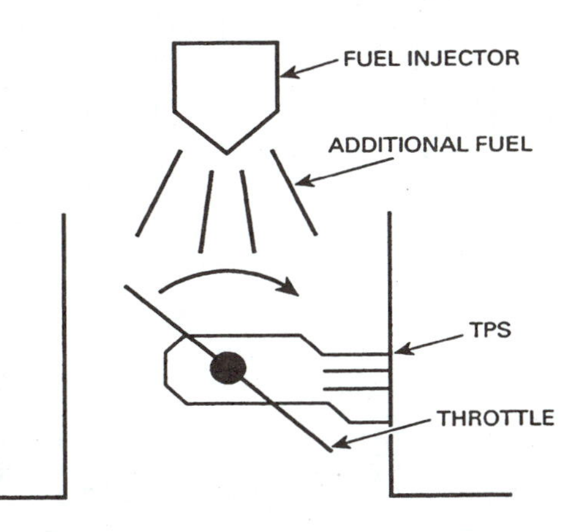

FIGURE 18-8 If the throttle is opened quickly, additional fuel must be injected in accelerator pump fashion. (Courtesy of Ford Motor Company)

MANIFOLD ABSOLUTE PRESSURE SENSOR (MAP)

The last "major" sensor is the manifold absolute pressure sensor, the MAP. The MAP has two functions. It will indicate engine load to the PCM and it will indicate altitude changes. The PCM needs to know the amount of load because it will require changes in the air fuel ratio, transmission pressure, and timing of the spark. In addition, in the mountains, where the air is thin, information from the MAP will allow the PCM to reduce the amount of fuel delivered. MAP sensors come in two varieties. One operates exactly like a TP in that it will convert manifold pressure into a voltage signal that varies. It is powered off 5 V from the PCM and will send back a signal that corresponds to engine vacuum or load. Its function is so similar to a TP that we will not look at it here. Instead we will look at a different type of MAP, one that will vary the frequency of its output to indicate manifold pressure. This style of MAP is common to Ford Motor Company vehicles.

Figure 18-9 shows that within the MAP is a silicon capacitive absolute pressure sensor and electronics to convert pressure/vacuum over to a frequency that the PCM can us.

Notice that we again use three wires feeding a 5-V source to the sensor and return a ground wire to the PCM. The signal return however is not a division of this 5-V source like the TP was. Instead Figure 18-10 shows

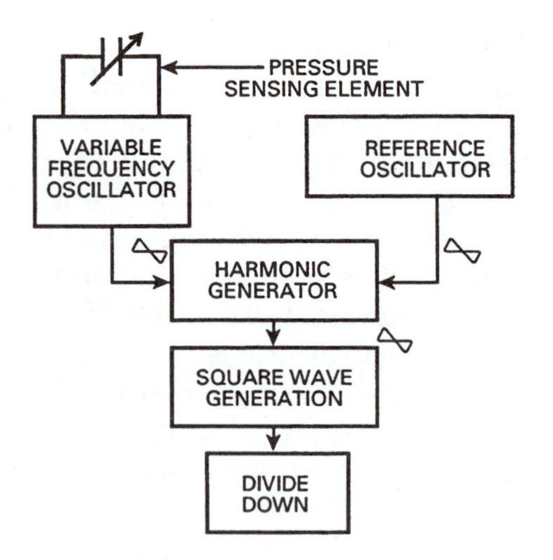

FIGURE 18-9 MAP sensor operation. (Courtesy of Ford Motor Company)

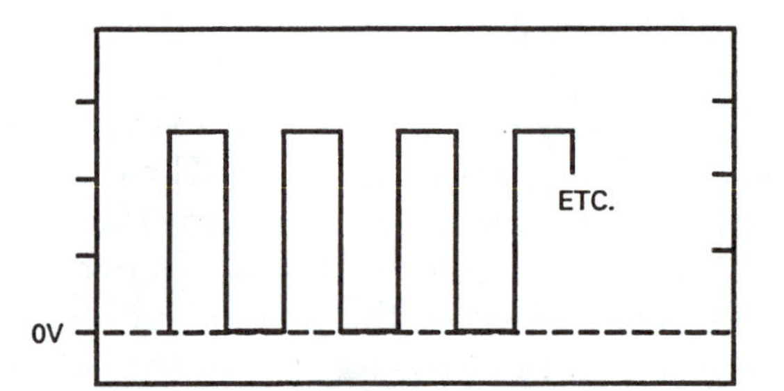

FIGURE 18-10 A full cycle starts at zero, rises to 5 V, stays on for a period of time, drops back down to zero, and remains off for the same period of time that it was on. This 50% on, 50% off signal is a complete cycle. (Courtesy of Ford Motor Company)

that the MAP generates a signal that varies its frequency with pressure as the chart shows. Frequency is the number of times per second that the signal is turned on and off. Counting these "on and off" signals is an easy thing for the PCM to accomplish. Don't forget, we have previously discussed the difference between frequency and pulse width modulation. Frequency is the number of times per second that a signal changes. It is usually a signal that is on for 50% of the time and off for 50% of the time. Figure 18-11 shows the DSO pattern off a functioning MAP during KOEO (Key On Engine Off) conditions. Notice that there are over 7 patterns on the screen. The scope has been set up to measure and display frequency also. On the right side of the screen

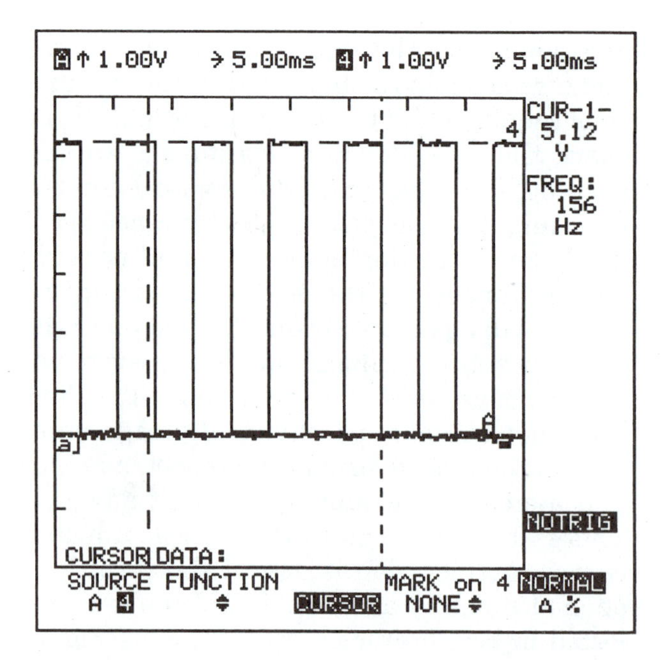

FIGURE 18-11 Scope pattern showing Ford MAP with KOEO (Key On Engine Off).

is printed FREQ: 156 Hz. This is the operating frequency of this MAP during KOEO conditions. With the engine running, the MAP must "see" engine vacuum and this frequency must change. As we start the engine, the DSO displays the patterns shown in Figure 18-12.

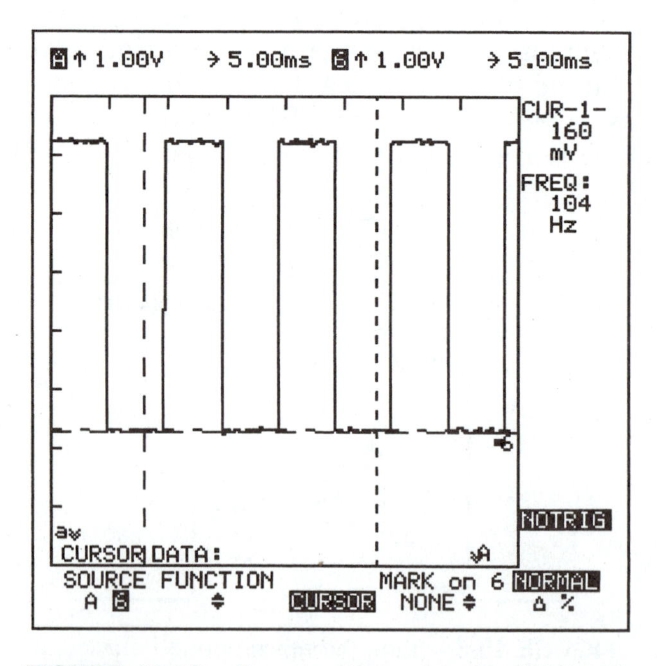

FIGURE 18-12 Scope pattern showing MAP with engine idling.

Did the frequency change? We now have only 4 patterns on the screen during the same time period, and printed on the right side is FREQ: 104 Hz

Figure 18-13 shows the conversion from manifold vacuum, atmospheric pressure to frequency.

Let's look at a few examples of how this works and look at what the PCM will do with the information. The customer turns the key on prior to cranking the engine over. The PCM sends out the 5-V power to the MAP. Without the engine running or even cranking, the pressure in the manifold will equal the atmospheric pressure. If we are at sea level, we will probably be close to 14.7 pounds or 30 in Hg. The MAP will generate a frequency of around 159 Hz. The PCM now knows what the pressure is in the manifold and will use this bit of information to help determine the amount of start up fuel required. Higher altitudes will require less fuel, while lower altitudes will require more. This necessary bit of information is captured within the short period of time that the ignition key is in the run position on its way to the start position. With the engine running, various outputs will require the information from the MAP. Opening the throttle will increase the pressure in the manifold and the frequency output will increase. The PCM will respond to this higher frequency by enriching the air fuel ratio, retarding the timing, increasing the transmission pressures, and possibly delaying upshifts. The more the throttle is opened, the greater the pressure inside the manifold. When the pressure increases, the engine load is increasing and load calculated outputs like fuel, timing, and transmission operation must be changed. In addition, certain emission related outputs like EGR (exhaust gas recirculation) need to be recalculated. Increased pressure within the manifold will generally require less EGR flow, while decreased pressure will require more EGR.

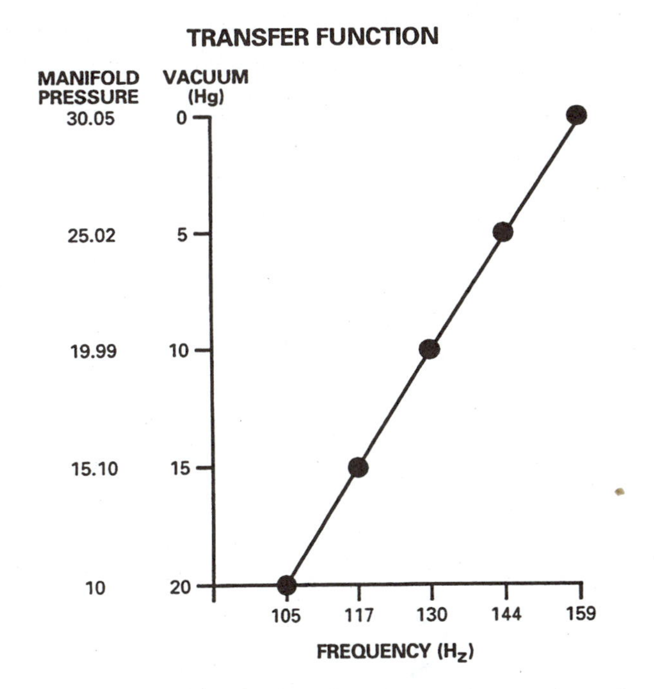

FIGURE 18-13 The PCM will count the number of complete cycles per second and convert it over to manifold vacuum and/or atmospheric pressure as the chart shows. (Courtesy of Ford Motor Company)

EXHAUST GAS RECIRCULATION (EGR)

EGR is a critical emission component responsible for reducing NO_x or oxides of nitrogen, Figure 18-14. EGR flow is regulated or changed partially based on information that the PCM receives from the MAP. As the vehicle accelerates, the pressure within the intake manifold

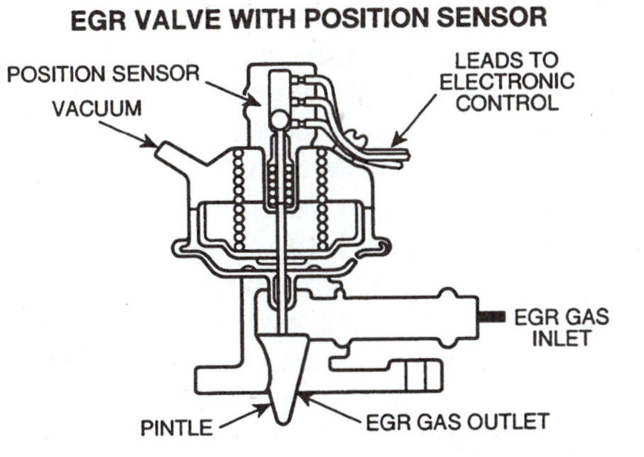

FIGURE 18-14 EGR is a critical emission component responsible for reducing NO_x or oxides of nitrogen. (Courtesy of Ford Motor Company)

will increase because the throttle is opened. More fuel and less EGR will be required. The frequency of the MAP will increase until the vehicle reaches cruising speed. As the driver cuts back on throttle position, the pressure within the manifold will decrease causing a drop in MAP output frequency. The information from the MAP will be used to cut back on fuel delivery, advance the spark, shift the transmission, reduce the operating pressure of the transmission, and increase EGR flow. You can see why the MAP is considered a major sensor. Its information is used to control power, mileage, and emissions. Good service technicians will frequently test the MAP because they recognize that proper operation is required on today's vehicles. An out of range MAP could easily be responsible for decreased performance, mileage, or a failed Enhanced Emission test. As Enhanced Emission testing comes on line in all states, the MAP, among other sensors, will become even more important, especially because it is a major player in determining EGR flow. Remember that EGR flow is the most frequent reason that a vehicle fails the NO_x (oxides of nitrogen) part of the emission test. NO_x is a major player in the air pollution game. EGR is not the only reason for high levels of NO_x, but it is a major one.

ENGINE SPEED SENSING

The PCM must have additional information if it is to do its job effectively. Another important input is engine speed and on some applications, additionally, the position of #1 cylinder. These two bits of important information are gathered by the PCM through the use of either Hall effect sensors or variable reluctance sensors. We looked at the application of these sensors within the ignition chapter, so we will not spend additional time covering their operation here. However, let's look at how the information will be used by the PCM.

Figure 18-15 shows that virtually all PCM functions require the input of engine speed to determine advance, transmission operation, and fuel delivery. An engine turning at 3000 rpm will require more fuel than one spinning at 1500 rpm. In addition, the identification of #1 is important for the correct coil charging on a DIS system and for correct injector firing on a specific type of fuel injection system called **SFI**, (sequential fuel injection). Figure 18-16 shows that SFI will fire an injector at a specific time related to cylinder piston position.

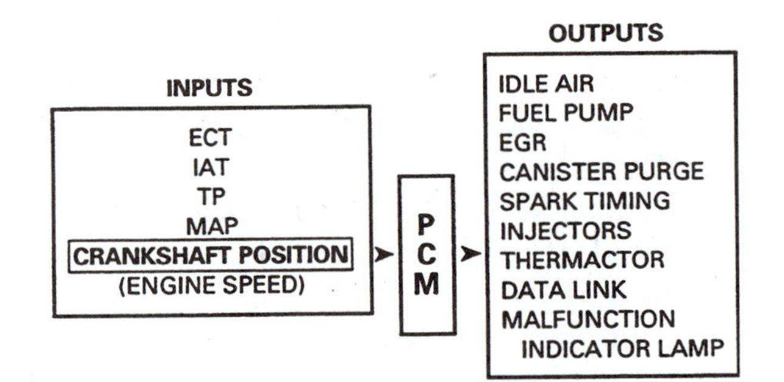

FIGURE 18-15 Virtually all PCM functions require the input of engine speed to determine advance, transmission operation and fuel delivery. (Courtesy of Ford Motor Company)

To be able to accurately determine when to open the injector, the PCM will look at the sensor. Hopefully, you remember from the ignition chapters that we looked at the CKP (crankshaft position sensor) as a Hall effect sensor that would produce a square wave signal whose spacing would be determined by the spacing of the windows and vanes. By counting the wave forms, the PCM or ignition

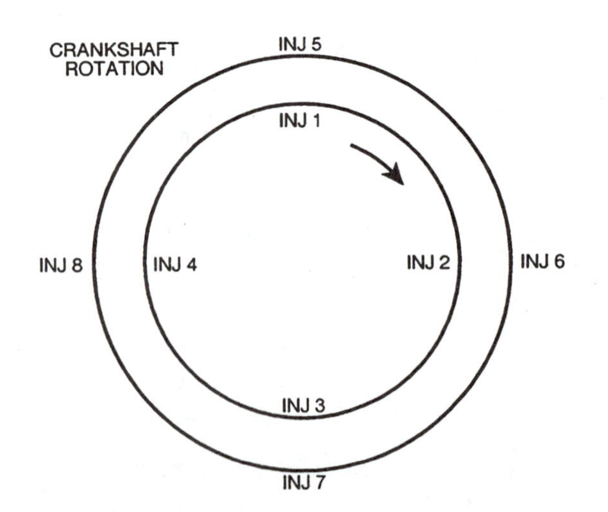

FIGURE 18-16 Sequential fuel injection (SFI) will fire an injector at a specific time related to cylinder piston position. (Courtesy of Ford Motor Company)

module is able to determine engine speed. For example, Figure 18-17 shows that a six cylinder CKP will generate 3 square waves per crankshaft rotation.

If the PCM sees 6,000 square wave pulses per minute, it will know that the engine speed is 2,000 rpm, 6,000 / 3 (per rotation) = 2,000. In addition, the number of pulses per second compared to the previous second will give the PCM the information it requires about changing engine speed. When we discuss **OBD-II**, you will see that this information is also used to diagnose a misfire. OBD-II is the on board diagnostic, 2nd generation that all manufactures are currently building into their vehicles.

In some applications, the manufacturer adds an additional sensor, the **CMP** (camshaft position) which gives the PCM cylinder identification information. Generally, this type of sensor produces one pulse per camshaft rotation or two rotations of the crankshaft. Figure 18-18 shows a CMP and the information it will supply about camshaft and crankshaft position required to control ignition coil operation and fuel injection.

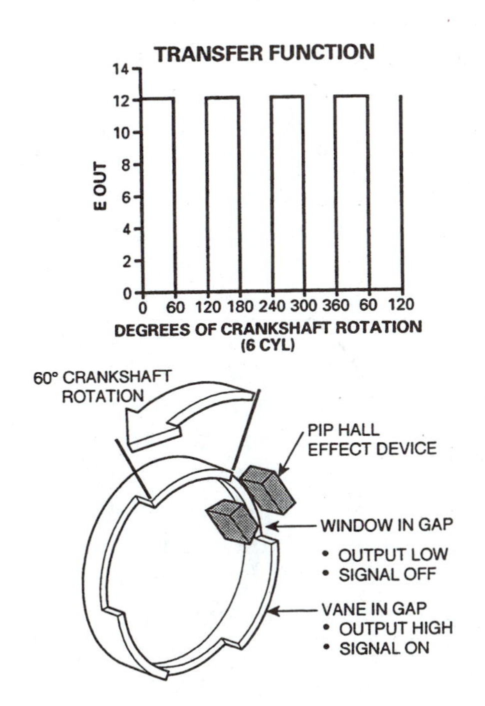

FIGURE 18-17 A six-cylinder PIP will generate 3 square waves per crankshaft rotation. (Courtesy of Ford Motor Company)

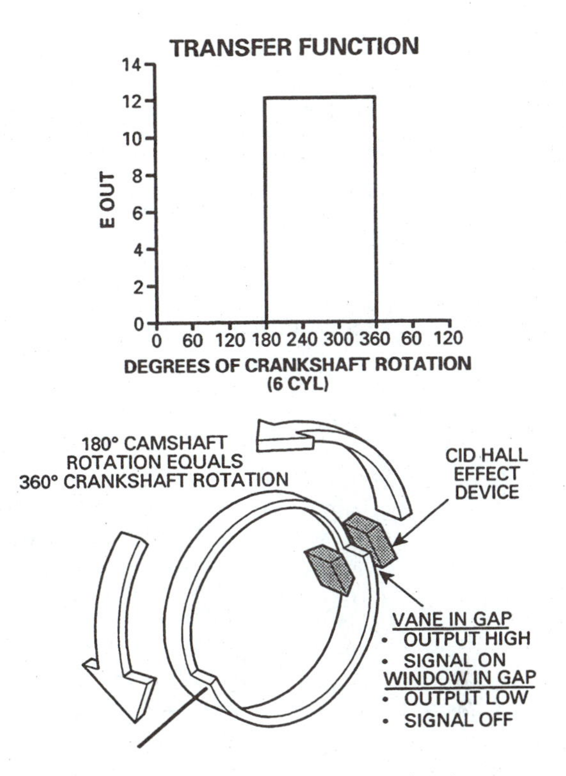

FIGURE 18-18 A CMP and the information it will supply about camshaft and crankshaft position required to control ignition coil operation and fuel injection. (Courtesy of Ford Motor Company)

Don't forget that many manufacturers may use a VRS or Variable Reluctance Sensor as the CKP to indicate engine rpm and cylinder position, in place of the Hall effect CKP or CMP.

Figure 18-19 shows a Ford 5.0 liter CKP with 35 teeth spaced 10° apart and one empty space for a missing tooth.

By monitoring the signal coming off the CKP, the module and PCM will again know where the pistons are, and how fast they are moving. The missing tooth directly corresponds to piston #1 position. The change in amplitude of the voltage signal will indicate #1. This is a good example of how fast the computer is at counting. How many pulses will the PCM see if the engine is at 3000 rpm? The VRS sensor generates 35 pulses per rotation, so the answer is 3000 × 35 or 105,000 pulses per minute. Another way to look at it is that 1,750 pulses per second indicate 3,000 rpm, which is a high speed cruise on many vehicles. Remember a DSO is the only practical tool that can display a sensor this fast.

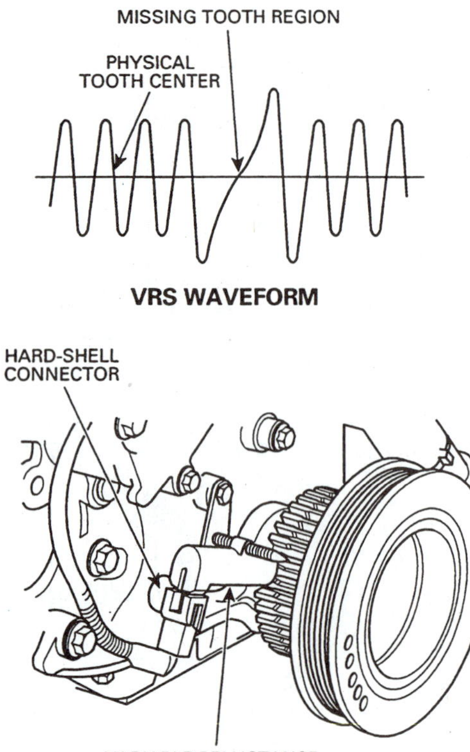

FIGURE 18-19 Ford 5.0 liter CKP with 35 teeth spaced 10° apart and one empty space for a missing tooth. (Courtesy of Ford Motor Company)

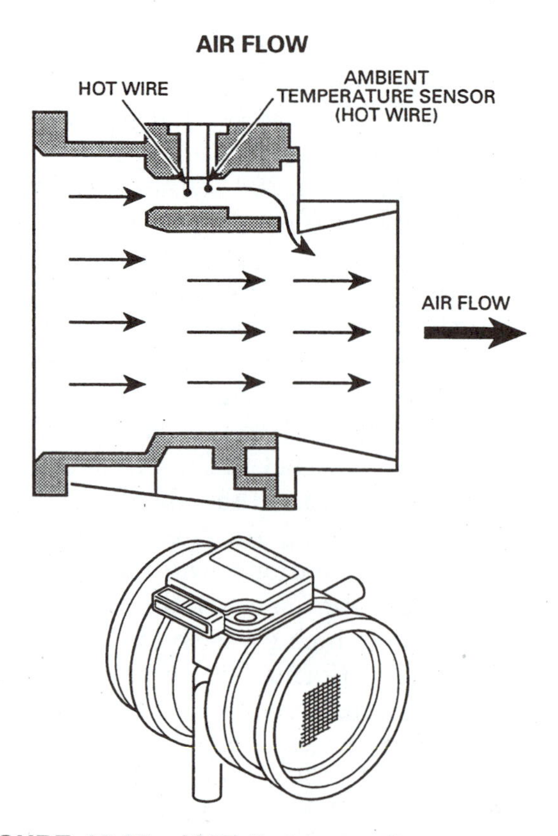

FIGURE 18-20 MAF that is a voltage generator. Ford MAFs are of a hot wire design with a percentage of the actual engine air flowing past an electrically heated wire. (Courtesy of Ford Motor Company)

MASS AIR FLOW (MAF)

Whenever we look at current systems, it is important to note that the all important air fuel ratio is calculated off the actual amount of air flowing into the engine. This amount will either be calculated through what is usually referred to as a "speed density" system with the inputs of engine speed and MAP being used or a direct reading of the mass of the air off the mass air flow meter. The mass of the air is the weight of the air, usually measured in pounds per minute or kilograms per hour. MAFs generally are of either a frequency style or a DC voltage generator.

Figure 18-20 shows a MAF that is a voltage generator. This type is a hot wire design with a percentage of the actual engine air flowing past an electrically heated wire. The amount of heat dissipated by the airflow is an indication of the amount or mass of the air. The more air flowing, the greater the cooling. An electronic circuit will convert the cooling of the wire into a voltage signal that the PCM can use to have a direct indication of the mass.

Figure 18-21 shows the chart with typical MAF values ranging from about .60V at idle to 2.10V at around 60 mph.

Generally the MAF is located between the air cleaner and the throttle body, or integrated into the air cleaner assembly on some applications. It is usually powered directly off vehicle power rather than from a 5-V PCM voltage like most sensors. Reading mass directly off a MAF appears to be slightly more accurate than calculating it based on engine speed and MAP. If a vehicle uses a MAF, it generally does not have a MAP.

Figure 18-22 shows MAF location. Note that MAFs require clean air and are prone to failure if the customer removes the air cleaner allowing dirt to come in contact with the sensing element.

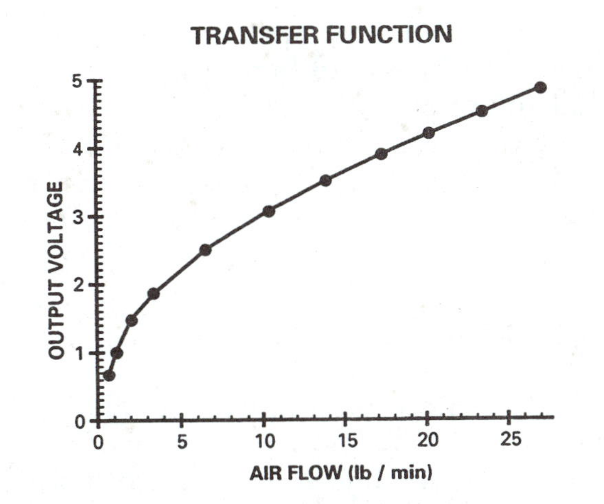

FIGURE 18-21 Chart showing typical MAF values ranging from about .60 V at idle to 2.10 V at around 60 mph.. (Courtesy of Ford Motor Company)

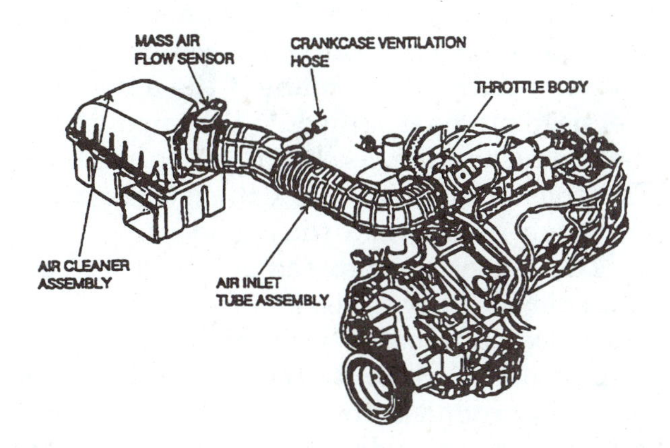

FIGURE 18-22 MAF location. MAFs require clean air and are prone to failure if the customer removes the air cleaner allowing dirt to come in contact with the hot wire sensing element. (Courtesy of Ford Motor Company)

In addition, it is important to note that all of the air being drawn into the engine must flow through the MAF. Any leakage after the MAF will be unmeasured air and will tend to produce leaner than desired air fuel ratios. Don't forget that all vehicles do not have a MAF. Some vehicles will use the combination of a MAP and an engine speed sensor to calculate the mass of the air. These systems you will remember are called speed density.

The other style of MAF that is common generates a signal that is frequency oriented. As the mass of the air changes, so does the frequency. We looked at the signal using a DSO in the previous chapter. Figure 18-23 is the signal from a Buick 3800 MAF at idle. Notice the number of pulses across the screen. There are about 6. Now look at the same MAF signal at 2000 engine rpm in Figure 18-24. Notice that the number of pulses across the screen has now gone up to 9. The PCM will

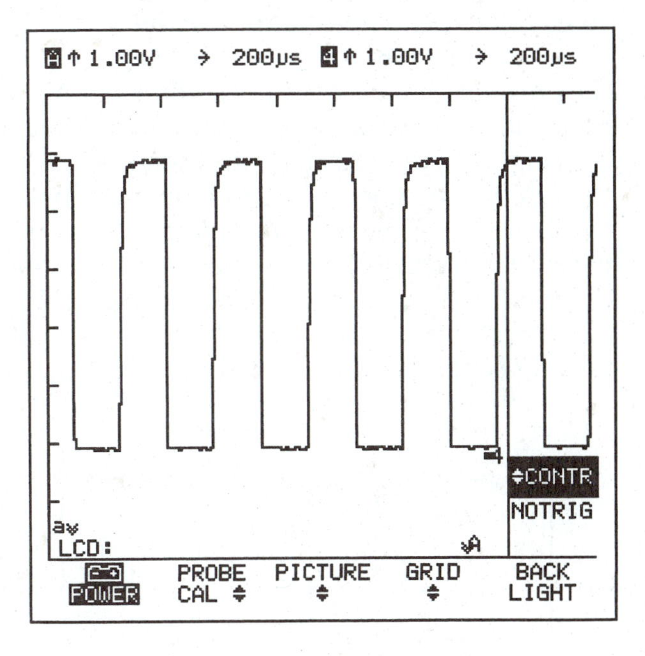

FIGURE 18-23 Scope pattern showing MAF signal at idle.

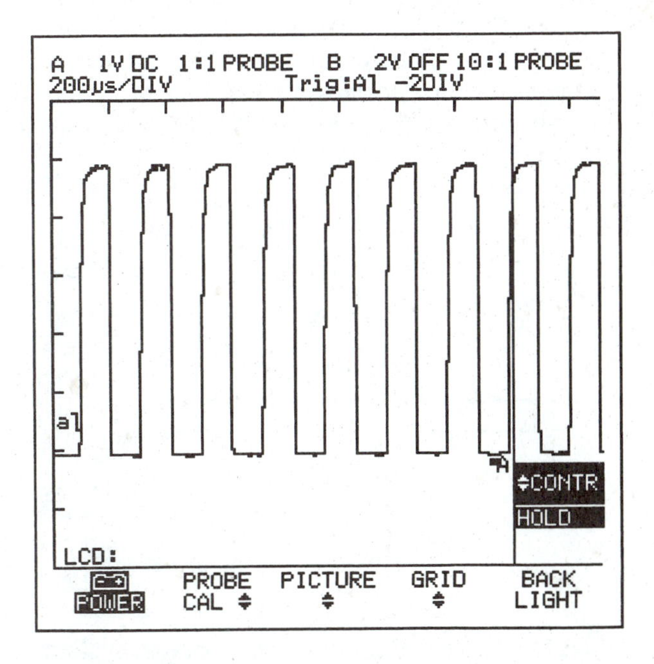

FIGURE 18-24 Scope pattern showing MAF signal at 2000 RPM.

definitely know that this engine is running faster, just by counting the number of pulses. It is this changing frequency that will be read by the PCM and used in the calculations for fuel and ignition.

The basics of engine operation are now present for the processor to calculate air fuel ratio and timing changes necessary. Some technicians refer to the sensors that we have looked at as the primary sensors. Primary in that they are used to determine the first and perhaps the most important operating conditions.

Engine temperature, air temperature, throttle position, engine speed, and the mass of the air either directly through a MAF or indirectly through a speed density system using a MAP are the primary sensors as Figure 18-25 shows Once these extremely important operating conditions are known by the PCM, it will be able to control spark and fuel. However, sometimes these conditions are not totally all of the information that is required. So some of the sensors are better called modifiers and come into play after the basic air fuel ratio and timing calculations have been made.

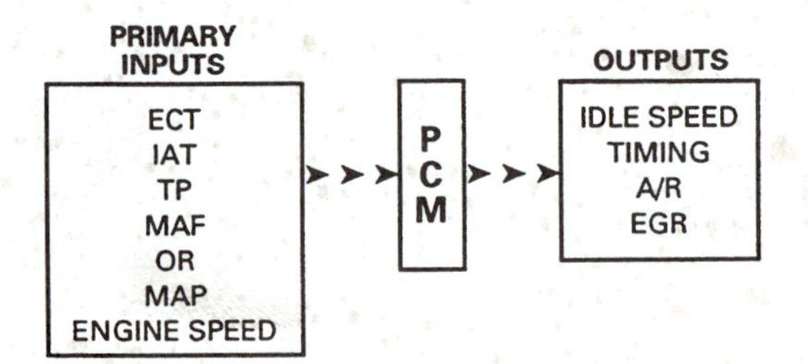

FIGURE 18-25 Engine temperature, air temperature, throttle position, engine speed and the mass of the air either directly through a MAF or indirectly through a speed density system using a MAP are the primary sensors. (Courtesy of Ford Motor Company)

OXYGEN SENSOR(S) O_2S

One of the most common fuel modifiers is the oxygen sensor. We refer to it as a modifier because the basic injector pulse width will be determined off the coolant temperature, manifold absolute pressure, air temperature, throttle position, speed/position sensors, and possibly a MAF. The initial air fuel ratio is calculated and injected. After the fuel burns, the PCM will look at the signal coming off the oxygen sensor to determine if the amount of fuel needs to be modified. O_2S is mounted in the exhaust system as shown in Figure 18-26. The sensor is exposed to exhaust gases immediately as they exit the cylinders.

The sensor is composed of a hollow zirconium dioxide body closed at one end. The faces of the body are coated with thin films of platinum which serve as electrodes. Exhaust oxygen flows around the outside of the sensor and is exposed to one side of the sensor while outside ambient air is allowed to flow to the inside. With ambient oxygen in contact with one side of the sensor and exhaust gas oxygen ex-

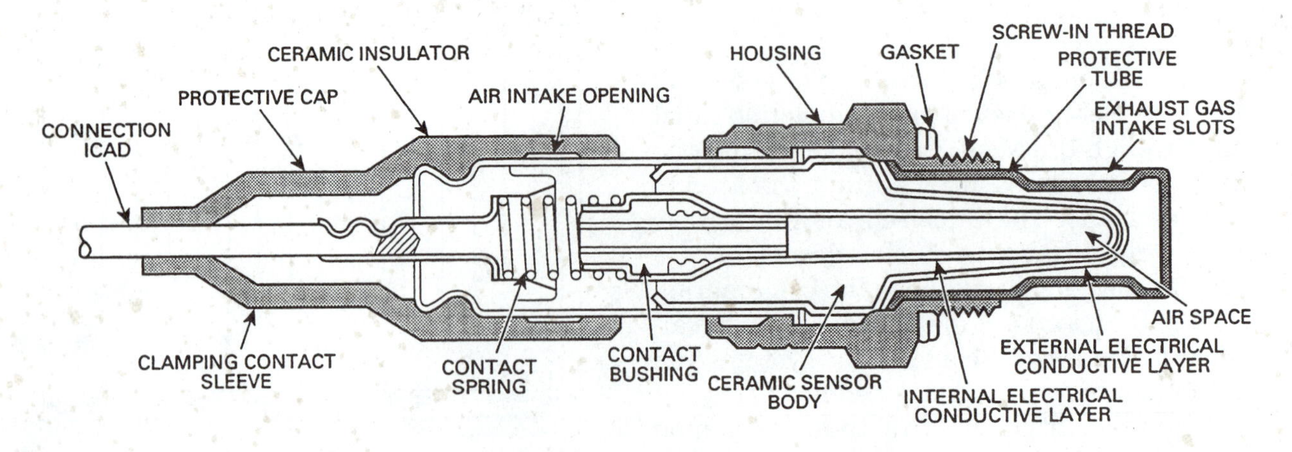

FIGURE 18-26 Mounted in the exhaust system as shown, the oxygen sensor is exposed to exhaust gases immediately as they exit the cylinders. (Courtesy of Ford Motor Company)

posed to the other, we have the making of a basic cell or battery. Remember that two dissimilar materials in an acid produce voltage. The dissimilarity in an oxygen sensor comes from the two different levels of oxygen exposed to the two sides of the zirconium dioxide. The greater the difference, the greater the voltage generated. Air has approximately 20% oxygen prior to combining with fuel in an engine. The hydrocarbon from the fuel combines with the oxygen reducing the 20% down to usually 1 to 2% for a good running engine. At this level of oxygen, the sensor will generate a voltage around 450 mV (.450V). Most vehicles today have the PCM producing a "set-point" voltage of 450 mV which is sent to the oxygen sensor. Anything higher seen at the PCM is considered a rich exhaust, while anything lower is considered a leaner exhaust. The PCM will take this information and convert it over to an opposite command.

A DSO can be used to monitor the action of the O_2S . If the system is running rich, less oxygen will be available to the sensor and the voltage will rise as Figure 18-27 shows in the middle of the screen. Notice that the voltage was cycling up and down until the third division when it shot up and stayed up until the 7th division. Each division of the scope has been set for 200 mV. Zero V has been moved down to the second division from the bottom. Is our high voltage above 450 mV? Counting up from the bottom we have 4.5 divisions, don't we? This equals about 900 mV which is an indication of a rich running fuel system.

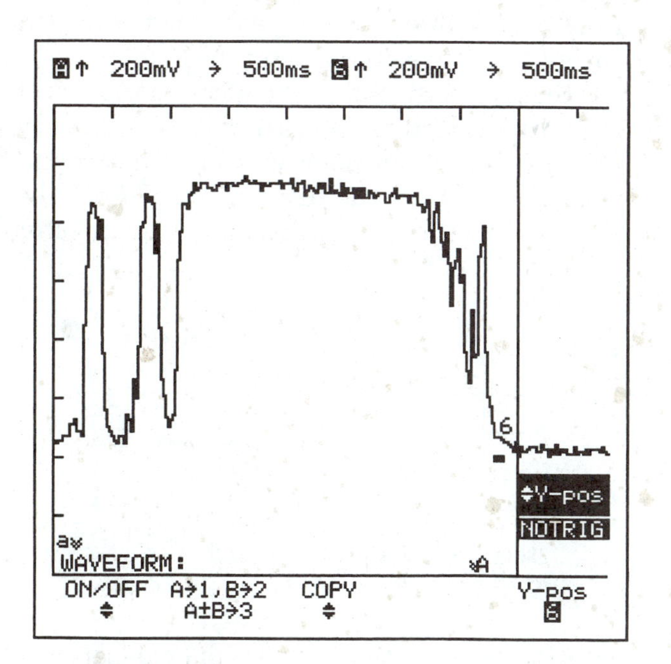

FIGURE 18-27 Scope pattern showing that O_2S voltage rises during rich conditions.

Analyzing just how the voltage went above .450 V for a rich exhaust goes back to the basics of the sensor. Remember that the exhaust stream contains the leftover oxygen from the burn. Whatever is left over flows past the sensor. If the vehicle is running rich, the burning fuel will have used up all or most of the 20% oxygen that came in during the intake stroke. More fuel takes more oxygen to burn. The lower exhaust oxygen at the sensor is seen as a greater difference, when compared to the ambient air available to the sensor. The greater the difference between exhaust O_2 and ambient O_2, the greater the potential voltage generated by the sensor. Remember that the PCM will interpret this higher than .450 V as a rich condition and cut back on the fuel injected into the engine. By reducing the fuel delivery, the opposite condition should occur at the sensor. Less fuel burned, leaves more oxygen remaining in the exhaust stream. Now the comparison between exhaust O_2 and ambient O_2 has less difference which in turn generates a lower than .450 V or lean signal.

Figure 18-28 shows the relationship between O_2S voltage and the amount of time the fuel injectors will stay open. If the exhaust is lean, the PCM will cause the injectors to stay

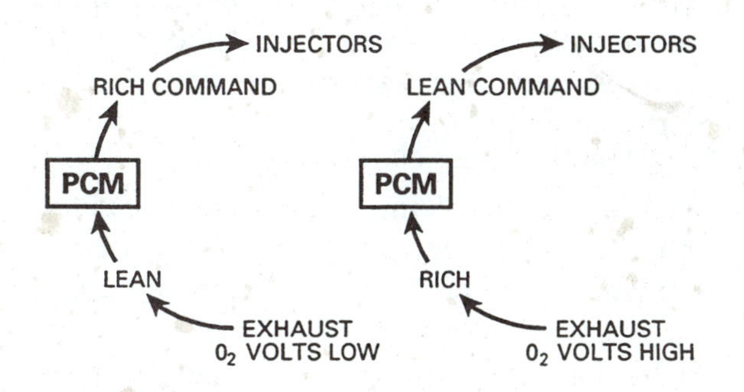

FIGURE 18-28 O2S vs. fuel injectors. (Courtesy of Ford Motor Company)

open longer by issuing a rich command. If the exhaust is rich, the PCM will issue a lean command.

Systems float usually between rich and lean .5 to 5 times a second in a correctly functioning system. In this way the oxygen sensor keeps the air fuel ratio at **stoichometric** or the ideal air fuel ratio for the catalytic converter. Oxygen sensors are the single most important sensor when dealing with emissions. Once the sensor is heated to 600°F, its signal will be looked at by the PCM and used to modify the calculated air fuel ratio. Recent years have seen increased importance on the oxygen sensor because keeping the air fuel ratio at the ideal level is an absolute must especially if the vehicle is to pass an Enhanced Emission test. Many oxygen sensors currently in use are heated by vehicle power. This is done so that the sensor is up to operating temperature quickly. In addition heating, it prevents the sensor from cooling down at low speeds and low ambient temperatures. OBD-II has seen the addition of another oxygen sensor after the catalytic converter. We will discuss its operation within a later section of this chapter.

SWITCHES

There are numerous switches that act as input devices to the modern PCM. Generally, they will act as modifiers and are simply an on or off signal. These include a power steering pressure switch, a brake switch, and an AC clutch switch as Figure 18-29 shows. The wiring for these additional inputs is very simple and straight forward. They generally modify especially the control of idle speed based on additional loads.

For example, if the customer is parallel parking his/her vehicle and turns the steering wheel completely to one side, the pressures within the power steering system can increase in excess of 600 psi. This places a large drag or load on the engine from the power steering pump. Under these conditions, the PCM will increase the amount of idle air to prevent the

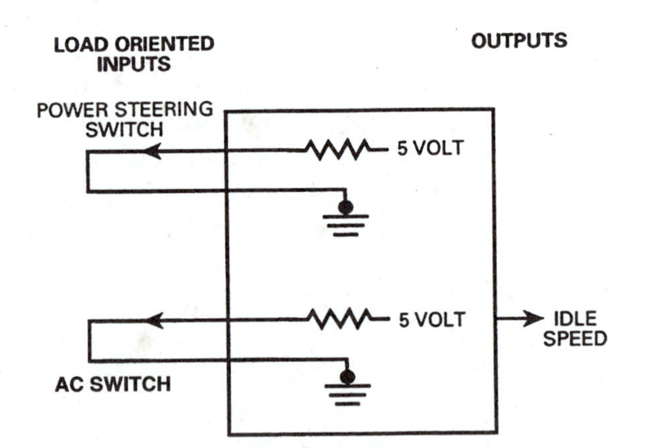

FIGURE 18-29 Typical wiring of input switches. (Courtesy of Ford Motor Company)

stalling of the engine. Generally, these switches will not have a large effect on timing or air fuel ratio, but are used as modifiers.

ENGINE OUTPUTS

The object of all of these inputs is to allow the PCM to have the information necessary to control various components called outputs. We will look at some of these outputs, starting with the all important fuel injector or injectors. Injectors are fed with pressurized fuel from a fuel pump. The pressure is controlled through the use of a pressure regulator.

Figure 18-30 shows the wiring for a set of Ford fuel injectors that are fed with battery power usually from a power relay. The ground necessary to turn the fuel injectors on will be supplied by the PCM. Internally, the injector

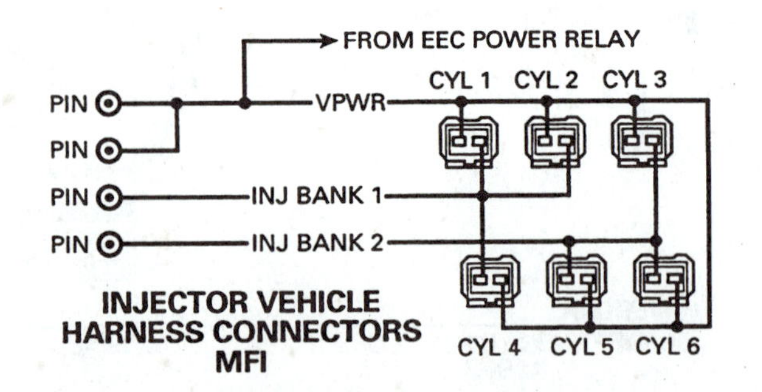

FIGURE 18-30 Wiring for a set of Ford fuel injectors fed with battery power usually from a power relay. (Courtesy of Ford Motor Company)

is a coil of wire that, when it is energized, will develop a magnetic field and open the pintle so fuel can flow. The amount of fuel injected is directly related to the amount of time that the injector is energized. This is called pulse width modulation or PWM. Increasing the pulse width increases the amount of fuel injected. This output from the PCM is directly related to engine load and temperature. For example, if the vehicle engine is cold, the amount of fuel necessary goes up. During the carburetor era, this was accomplished through the use of a choke that would restrict the amount of air that the engine could draw in. Obviously, the carburetor era is dead. Modern systems use inputs of coolant temperature and possibly intake air temperature which will be fed to the PCM. The PCM will in turn add additional fuel during cold start-up and warm up conditions. As the engine warms up, the amount of fuel injected will be reduced. Along the same lines, the way that a vehicle is driven has a lot to do with the fuel requirements. Additional fuel is necessary on hard acceleration, while virtually no fuel is necessary on a closed throttle deceleration. You have probably figured out by now that the PCM will need to have throttle information and load information to calculate the acceleration and deceleration fuel requirements. The all important information from the throttle position and manifold pressure sensors will be used to increase fuel on acceleration and reduce fuel on deceleration.

The scope pattern shown in Figure 18-31 is from a Buick 3800 fuel injector with the scope lead connected to the PCM side of the injector. Let's analyze what we are looking at. Remember that the fuel injector will be fed with B+ and that the PCM will apply a ground to turn it on. Notice the horizontal voltage line starting on the left. It is about 2-3/4 divisions from the zero line (where the black rectangle is located). It looks like our fuel injector is being fed with about 14 V, right? When the PCM turns on the ground, the voltage drops down to zero and stays down for about 1 division or 5 mS. This is the amount of time that the injector is on or open actually injecting

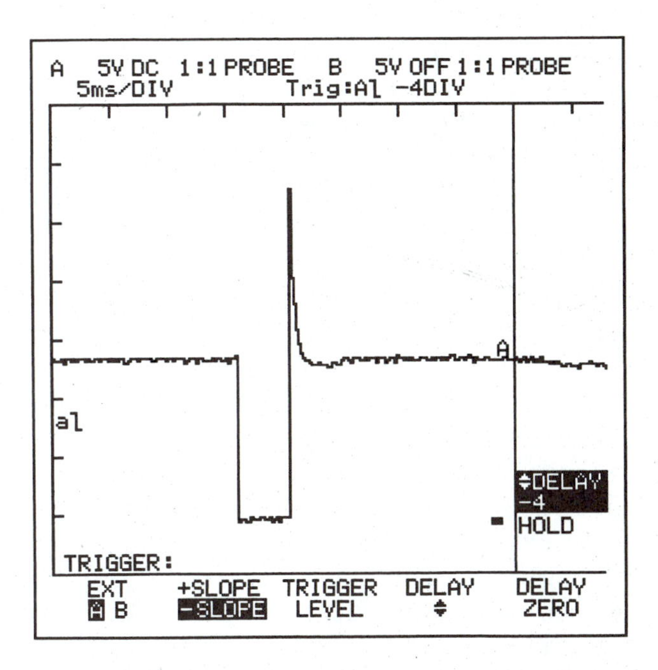

FIGURE 18-31 Scope pattern from Buick 3800 fuel injector with scope lead connected to the PCM side of the injector.

fuel and is generally referred to as the pulse width of the injector. When the PCM turns off the ground, a high voltage spike is produced by the injector which rises to about 30 volts. The amount of time that the injector remained open was not just a simple one sensor calculation, but instead a combination of applicable sensors.

Generally, the base amount of fuel is calculated off the coolant temperature sensor. This base amount is then adjusted by the PCM with the additional inputs of engine speed, load, throttle activity, and air temperature. Remember that an input from a mass airflow sensor might replace the MAP and engine speed inputs. Once the amount of fuel required has been calculated, it will be injected and looked at by the last sensor, the oxygen sensor. Hopefully you remember that this sensor will look at the remaining oxygen in the exhaust stream. This oxygen level must cycle between rich and lean so that the converter can function. This is why the signal from a functioning O_2S floats up and down as we saw in the DSO chapter. The goal is to stay in fuel control. The PCM will adjust injector pulse width based off the O_2S signal. If the signal is above 450 mV, indicating rich, the injector pulse width will be decreased. Once the O_2S voltage drops below 450 mV, indicating a lean exhaust, the injector pulse width will be increased. This cycling back and forth of the O_2S voltage is shown in Figure 18-32. The reason for the high or low voltage goes back to the fact that more fuel will use more oxygen and that is why reduced oxygen and high oxygen sensor voltage (+.450 V) usually indicates rich cylinder conditions. The opposite is also true. More than 2% oxygen indicates that the vehicle is running lean (not enough fuel). Less than the correct amount of fuel requires less oxygen to burn, leaving behind extra oxygen and a low O_2 voltage (less than .450 V). The fuel injector pulse width is obviously a major output for the PCM. Keeping the air fuel ratio within acceptable levels (usually 14.7:1) results in acceptable performance while greatly reducing tailpipe pollution. This seemingly simple calculation, how much fuel to inject, is probably the hardest and most complicated PCM function. Injector pulse width utilizes most engine sensors and is the greatest producer of pollution if the information that the PCM receives is incorrect. Keep in mind that the PCM is constantly trying to keep the vehicle at the best performance, the greatest economy and the lowest tailpipe emission, Figure 18-33.

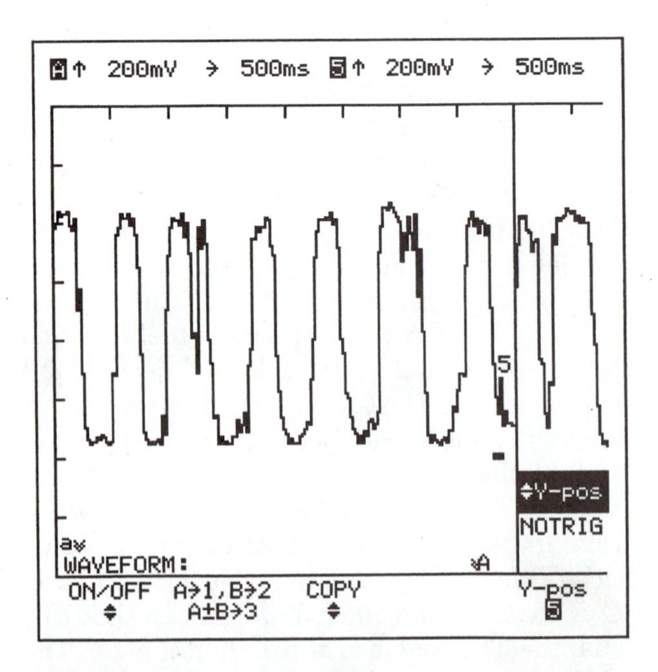

FIGURE 18-32 Scope pattern showing O_2S voltage cycling between rich and lean.

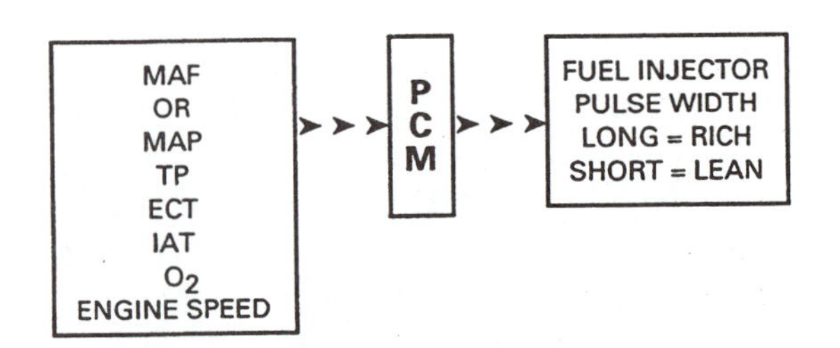

FIGURE 18-33 You can see that the fuel injector pulse width was not a simple one-input calculation, but instead, a combination of applicable sensors. (Courtesy of Ford Motor Company)

This triangle is always changing as the consumer accelerates, decelerates, cruises, starts and stops in all types of weather and driving conditions. The PCM must be constantly assessing the overall engine conditions and adjusting primarily the air fuel ratio and ignition timing.

IGNITION TIMING

Ignition timing is another very important output which must be controlled. In a past ignition chapter, we looked at the method of controlling timing with the PCM so we will not spend a lot of time here. However, let's look at some of the strategies involved starting with a cold engine, Figure 18-34.

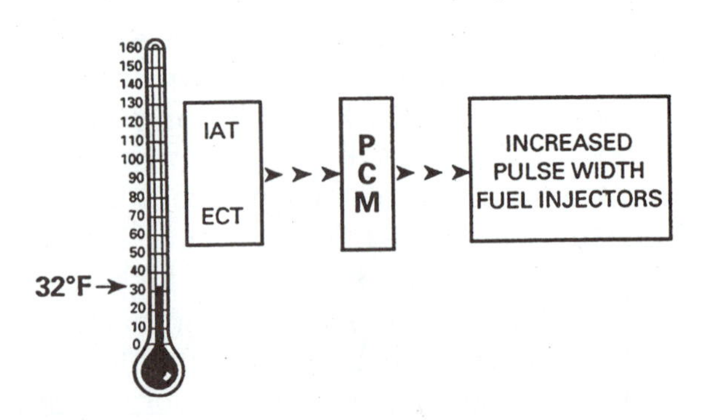

FIGURE 18-34 The PCM needs to know if the engine is cold to determine fuel and timing. (Courtesy of Ford Motor Company)

As the customer starts his/her vehicle during winter conditions, both the IAT (intake air temperature) and the ECT (engine coolant temperature) sensors send information to the PCM that the vehicle is cold. A cold engine requires that additional fuel is added to the intake via the injectors as we have seen. Keep in mind that additional fuel will require additional advance to properly burn within the cylinder. The PCM will adjust up or advance the timing to compensate for the additional fuel injected. As the engine warms up, this additional "warm-up fuel trim" will be reduced to the 14.7:1 air fuel ratio required of the catalytic converter and the timing will drop back to the less advanced curve. Remember the vacuum advance and the mechanical advance? The computerized spark output signal coming from the PCM over to the ignition module will be modified by the signal from the MAP and the engine speed sensors. The end result is the same. Additional degrees of advance during increases in either speed or engine vacuum. So the inputs of engine load and speed are used to cause timing changes. Additionally, the spark advance signal will be modified by changes in throttle and altitude. As the throttle is opened quickly from an idle position, additional fuel is usually injected to eliminate the tendency of the accelerating engine to hesitate or stumble. Remember what we stated before about a cold running engine; additional fuel can be burned more efficiently with additional degrees of timing advance. Many PCM timing programs also advance the timing during acceleration, primarily for emission reduction, but also of increased performance/driveability. The advantage of PCM controlled timing verses vacuum/mechanical advancing systems is tremendous. Older systems were only able to compensate for changing speed and load conditions. Newer systems look at temperature, rate of acceleration and deceleration, and even altitude as Figure 18-35 shows.

With all of the additional information available to the PCM, much more accurate timing adjustments can be made, with the end result of increased performance, economy, and reduced emissions.

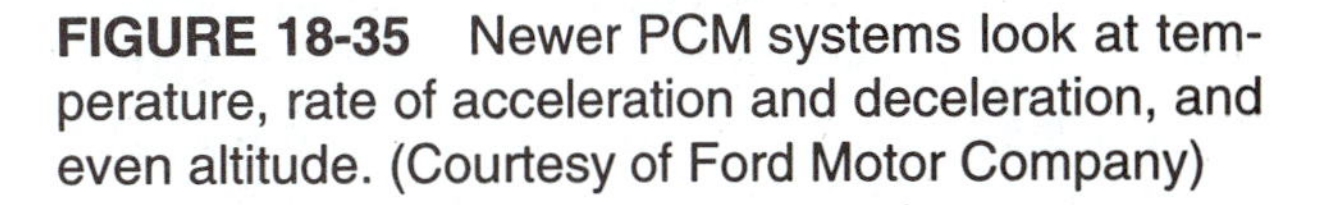

FIGURE 18-35 Newer PCM systems look at temperature, rate of acceleration and deceleration, and even altitude. (Courtesy of Ford Motor Company)

EXHAUST GAS RECIRCULATION

Another very important PCM output is the EGR or exhaust gas recirculation level. Let's look at the function of the EGR system before we discuss the PCM output. In the mid-seventies, the level of oxides of nitrogen or NO_x needed to be reduced. Up to this point, the major emphasis of the pollution control systems had been the control of the air fuel ratio and timing required to reduce the hydrocarbons and carbon monoxide emissions from the tailpipe. Generally auto manufacturers reduced the amount of fuel burned (leaned out the mixture) and HC and CO reduced proportionally. However, the leaning out of the mixture produced the undesirable effect of increasing combustion temperature, Figure 18-36. Once the temperature within the combustion chamber goes above about 2500°F, the oxygen in the air combines with the nitrogen forming NO_x.

NO_x has been shown to be a major contributor to the air pollution problem. The problem facing most vehicle manufacturers was the fact that leaning out the mixture was necessary for the reduction of HC and CO but it also raised up the NO_x production. EGR was the answer.

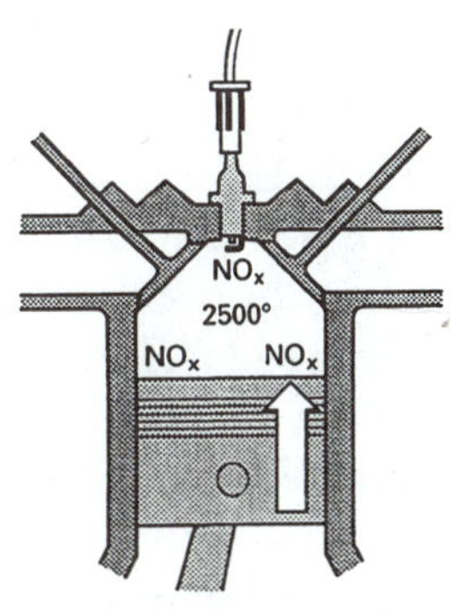

FIGURE 18-36 Increased temperature increases NO_x production. (Courtesy of Ford Motor Company)

Under most driving conditions, a small amount of exhaust gas is circulated back to the combustion chamber. Since the exhaust contains little combustible materials, and is basically inert, it will just take up space within the combustion chamber. Any exhaust gas recirculated back to the chamber will not produce any heat and will therefore reduce the internal temperature of the burn lower than the 2500°F necessary to produce NO_x. Controlling the precise amount of EGR flow was a problem until the PCM took over the metering job. Some systems utilize an EGR position sensor mounted on top of the valve. It is a type of TP sensor, which will vary the voltage based on the position of the valve. This position sensor tells the PCM the position of the EGR valve by the same voltage division system that is utilized for the throttle. The exact position of the EGR is now known by the PCM and as conditions change it can command the EGR to remain open longer or shorter as needed to maintain the less than 2500°F internal combustion chamber temperature. As Enhanced Emission testing of vehicles becomes more common, the EGR will become one of the most important emission devices. Most state's Enhanced Emission test is a dynamometer emission test that is designed to approximate the federal test procedure that vehicle manufacturers had to pass to certify new vehicles. Enhanced Emission tests test NO_x, in addition to CO and HC. Essentially, NO_x is produced if the EGR is not functioning correctly or if the timing is advanced. Advanced timing, especially with DIS systems is not really a concern of the manufacturers. The PCM is in total control of timing. It is absolutely imperative, however, that EGR be correctly calculated and measured or the vehicle will probably fail the emission test.

Just how this calculation is accomplished involves various inputs such as engine speed, load, and temperature (IAT and ECT). Once the basic information that the PCM needs has been received, an EGR flow rate will be determined. This flow rate will be sent out as a pulse width modulated signal on most applications. Now the EGR position sensor comes into operation, Figure 18-37. This sensor is essentially a TP sensor that has the EGR valve move its wiper rather than the throttle. This sensor acts like a check on the actual EGR action. For example, if the PCM analyzes all of the inputs and determines that the correct amount of flow is 50% of maximum, it will send out a 50% PWM signal to the controlling valves. The check of the system comes from analyzing the voltage signal coming back from the position sensor. It should also indicate a 50% open EGR valve.

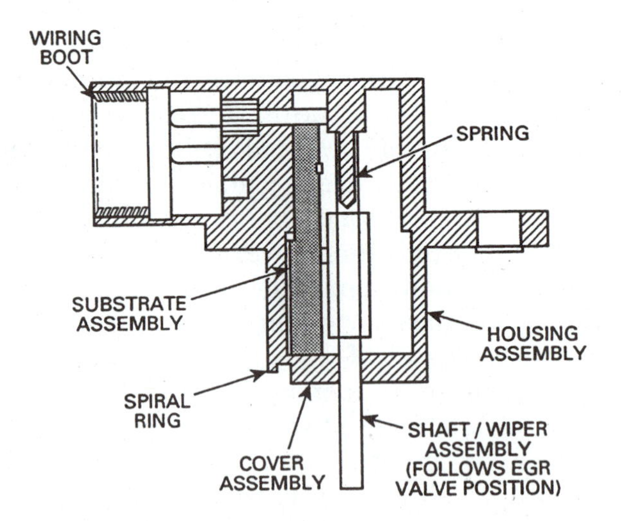

FIGURE 18-37 EGR position sensor cutaway. (Courtesy of Ford Motor Company)

If for some reason the EGR is not at a 50% position, the PCM has the option of increasing or decreasing the signal until the desired opening is achieved. The assumption here is that a 50% open EGR valve will flow 50% exhaust gas. Unfortunately, in the real world of emission control, we should realize that blocked or restricted exhaust passages are quite common and that they generally reduce the actual EGR flow to something less than the PCM feels is necessary. This greatly increases the amount of NO_x that is produced, especially under cruise conditions.

IDLE SPEED CONTROL

Another logical output of the PCM is that of controlling the idle speed. Years ago idle speed was controlled by increasing the throttle opening. Increasing the opening allowed more air to flow through the carburetor which was mixed with additional fuel and the engine ran faster. Increasing the idle speed has always been done for warm-up conditions. Computerizing the actual idle speed also allows for the setting of the correct amount of air prior to cranking the engine. In this way, the customer does not have to set the position of the throttle to get the engine started.

The PCM will automatically set the correct amount of air through either a bypass passage on most vehicles as Figure 18-38 shows, or by moving the throttle on primarily older vehicles.

Generally, this presetting of the amount of air required for engine starting is done just as the vehicle is turned off. In this way the engine is "ready" for the next start. Once the engine is actually running, the PCM will control

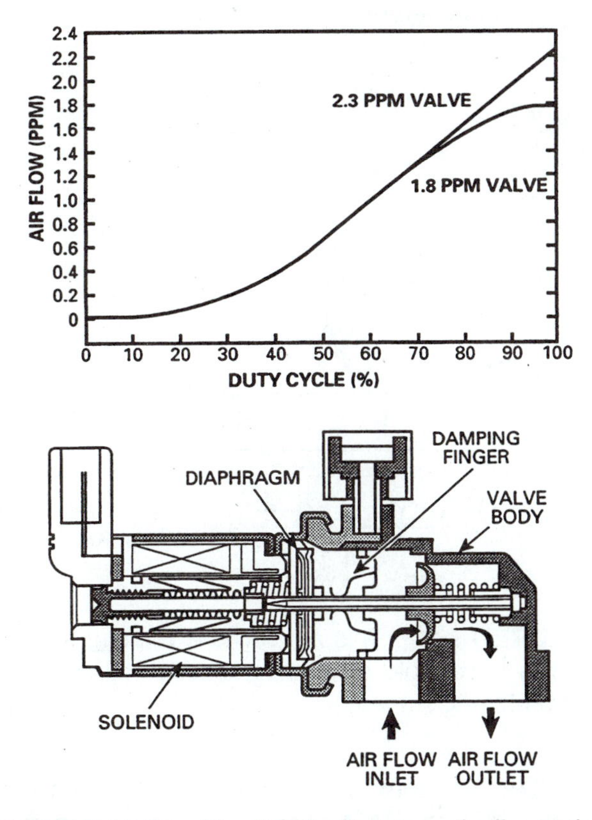

FIGURE 18-38 The PCM will automatically set the correct amount of air through either a bypass passage on most vehicles or by moving the throttle on primarily older vehicles. (Courtesy of Ford Motor Company)

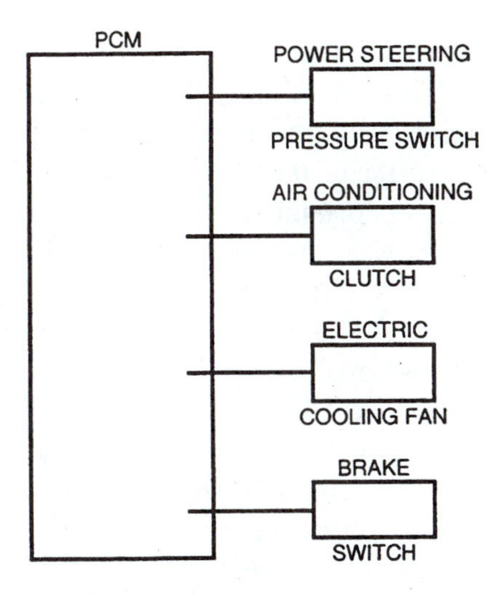

FIGURE 18-39 Typical PCM wiring diagram with additional load type sensors. (Courtesy of Ford Motor Company)

or adjust idle speed based on various sensors. Which ones? Hopefully, you realize by now that coolant temperature is the major sensor involved in determining idle speed. If the engine is cold, the idle speed is raised up to improve driveability. As the engine warms up, the idle is slowly lowered. Additionally, some systems will look at various loads that might decrease the idle speed below acceptable limits. Power steering, air conditioning, and excessive electrical loads are all examples of the need to increase idle speed. Idle speed must be correct at all times or the vehicle might produce excessive emissions in addition to producing slight driveability problems. The wiring diagrams of the PCM system frequently show additional load type sensors as inputs especially on smaller engines, where the additional load might have detrimental effects, Figure 18-39. As long as the PCM has these sensors it can control load sensitive outputs and keep the vehicle within design limits.

TRANSMISSION CONTROL

Let's look at the last major output typical of today's vehicle, the transmission. Prior to electronic control, automatic transmissions were shifted and controlled through the use of two inputs, throttle position and engine vacuum. The use of engine vacuum and throttle position allowed the transmission valve body to control mechanically the internal transmission pressures and shift points. For example, if the customer had his foot to the floor, engine vacuum would drop down to near zero. A device called the modulator would boost up the internal transmission pressure so that the clutches would not slip from the increased load. In addition, the valve body would hold the transmission in a low gear for increased torque to the wheels. Once the customer lets off the throttle, the increased vacuum to the modulator will cut back on pressure and allow a gear change to occur. Controlling the pressure and shift points within the transmission was done mechanically rather than electronically until the early 1990's when PCM control entered into the picture.

ADDITIONAL INPUTS REQUIRED FOR TRANSMISSION CONTROL

Up to this point we have looked at various inputs to the PCM which will allow it to control mostly engine functions. The PCM will require a few additional inputs if it is to adequately control the transmission. The first one of these will be transmission range (**TR**) or gear shift position. A type of throttle position sensor, mounted on the transmission, will be used that will be mechanically connected that to the gear shift lever either on the steering column or on the console, Figure 18-40.

The PCM will send out a 5-V reference signal to the TR sensor. Each shift lever position will cause a different resistance value and a different voltage delivered back to the PCM.

Another input required will be that of vehicle speed. The PCM will use the output of the vehicle speed sensor (**VSS**) to not only control shift points, but to help with fuel control, timing and speedometer. This sensor is a rotating magnet similar to variable reluctance sensors that are in use on fuel and ignition systems, Figure 18-41.

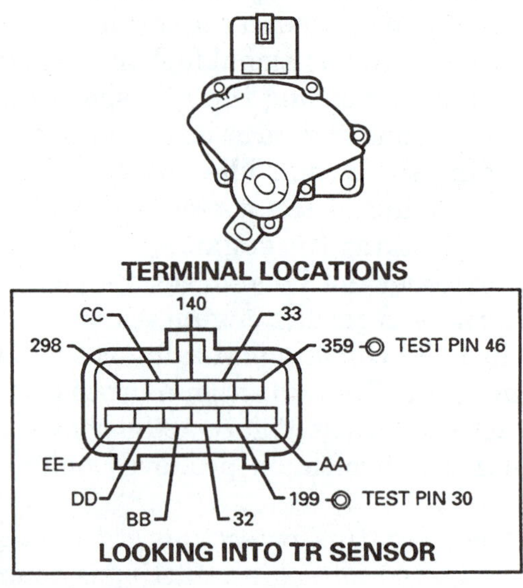

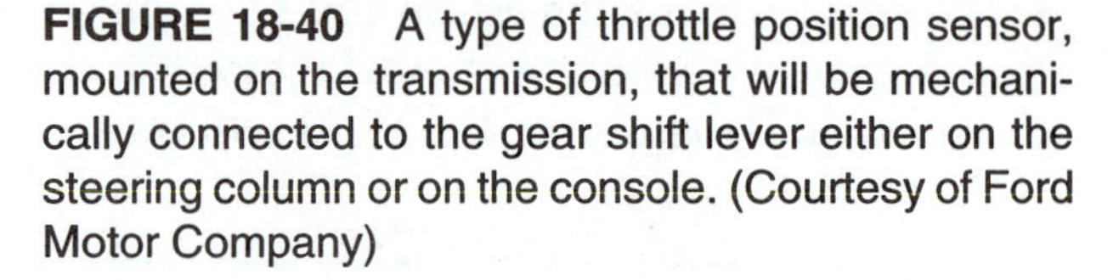

FIGURE 18-40 A type of throttle position sensor, mounted on the transmission, that will be mechanically connected to the gear shift lever either on the steering column or on the console. (Courtesy of Ford Motor Company)

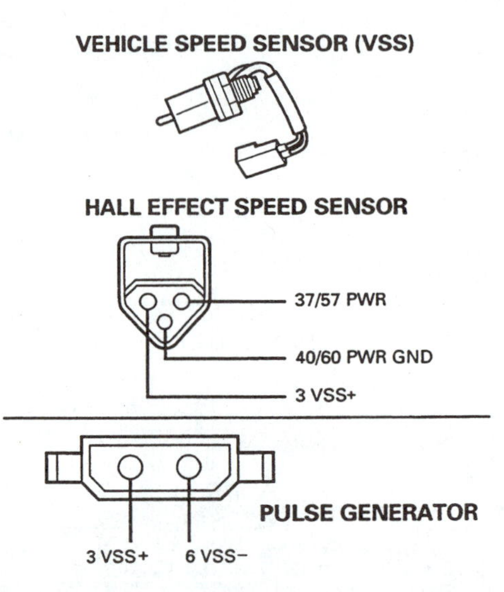

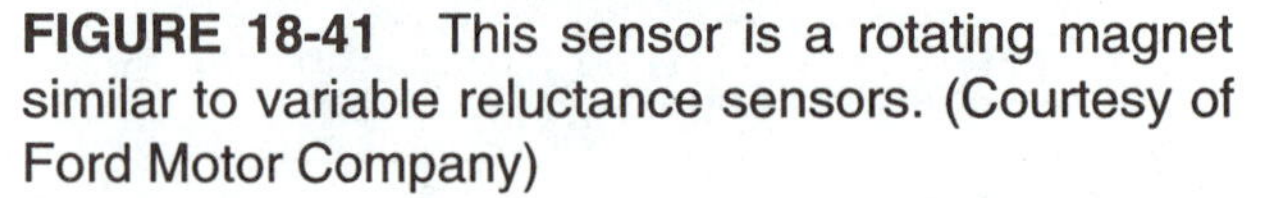

FIGURE 18-41 This sensor is a rotating magnet similar to variable reluctance sensors. (Courtesy of Ford Motor Company)

The number of pulses generated will be counted by the PCM and evaluated in terms of miles per hour. The assumption here is that the customer has not changed the final gear ratio or tire size. As the speed changes, so do the number of pulses per second (Hz) delivered to the PCM. Most applications have an additional speed type sensor called the turbine shaft speed sensor (**TSS**). Its operation and resulting frequency read by the PCM will indicate the actual transmission turbine shaft speed. This information is used by the PCM to determine the lock-up torque converter strategy. Torque converters have a slight amount of slippage even while cruising, so the vehicle manufacturers lock them up at road speed. Lock-up torque converters act much the same as a manual transmission clutch at highway, light load speeds. By eliminating the slippage, mileage increases and emissions decrease which is obviously a good addition for the customer.

The last input that we will look at is that of transmission fluid temperature (**TFT**). The TFT sensor is a negative coefficient thermistor which will operate exactly like an engine coolant (ECT) or intake air temperature sensor (IAT). As the temperature of the transmission fluid increases, the resistance of the sensor decreases resulting in a lower sensor voltage, Figure 18-42.

Our PCM controlled transmission now has all of the inputs that it will need to control not only the shift points, but the quality of shift during cold or hot weather, in addition to providing a lock-up for the torque converter. The PCM control will come in the form of output signals sent to the transmission valve body. Within the valve body are 5 solenoid type valves as Figure 18-43 shows.

Let's look at each one individually. Remember that each of the solenoid valves are

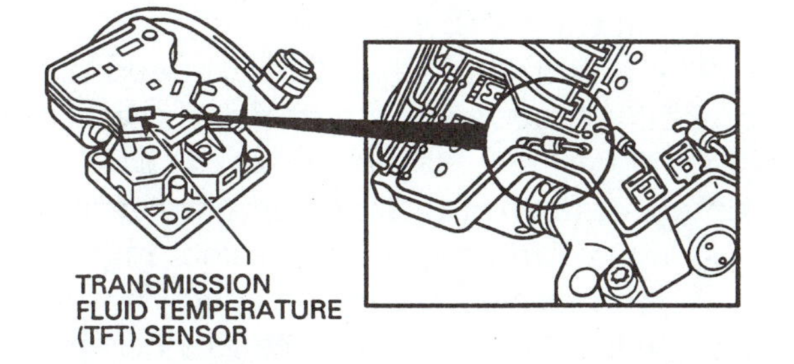

FIGURE 18-42 Transmission fluid temperature (TFT) sensor. As the temperature of the transmission fluid increases, the resistance of the sensor decreases resulting in a lower sensor voltage. (Courtesy of Ford Motor Company)

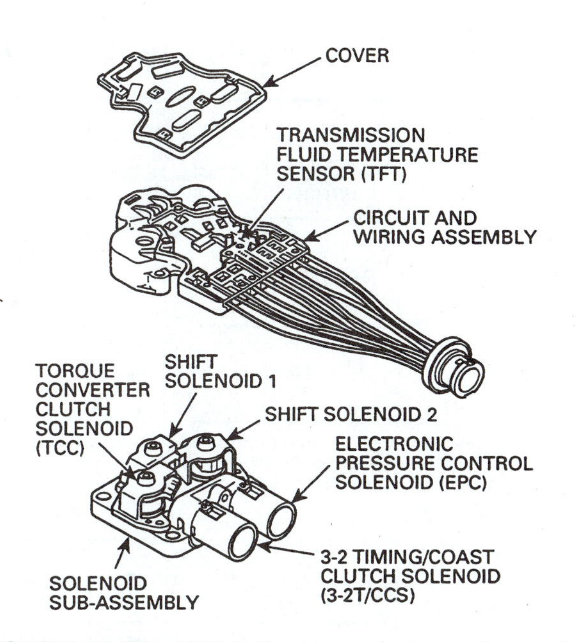

FIGURE 18-43 Within the transmission valve body are 5 solenoid type valves. (Courtesy of Ford Motor Company)

connected to and controlled by the PCM. Two of the valves are the same. These are the shift solenoids, SS1 and SS2. These solenoids are either on or off as determined by the PCM. When they are off they will block the flow of hydraulic fluid and allow fluid flow when on. Whether the SSs are on or off will determine the position of the shift valve which will determine the gear that the transmission is actually in. Notice that the two solenoid shift valves allow for 4 different combinations which will be able to control the four speeds of the transmission. Figure 18-44.

Remember that SS1 and SS2 are on or off only as commanded by the PCM. Let's follow through a simple example. The customer puts the gear select lever in drive. The TR (transmission range) sensor signals the PCM that drive has been selected. The PCM turns SS1 and SS2 both on. This places the valve body into a first gear. As the vehicle speed increases, the PCM will turn SS1 off and shift the transmission into second. SS2 remains on for second gear. Once third gear is necessary, the PCM will turn SS2 off (SS1 is still off) shifting the transmission into third. The final gear ratio is achieved by turning SS1 back on. Notice that each transmission gear change in-

SOLENOID APPLICATION CHART — VEHICLES WITH TRANSMISSION CONTROL SWITCH (TCS)

	GEAR SELECTOR POSITION	PCM COMMANDED GEAR	CD4E SOLENOIDS			
			SS1	SS2	3-2 TIMING/ COAST CLUTCH SOLENOID	TCC
D	(TCS OFF)	1	ON	ON	ON	ON
		2	OFF	ON	ON	ON
		3	OFF	OFF	ON	OFF
		4	ON	OFF	ON	OFF
	(TCS ON)	1	ON	ON	ON	OFF
		2	OFF	ON	OFF	ON or OFF*
		3	OFF	OFF	OFF	ON or OFF*

* Powertrain Control Module (PCM) assembly commanded.
□ Desired steady-state gear.
□ Fail-safe gear.

3-2 TIMING/COAST CLUTCH SOLENOID APPLICATION

PCM COMMANDED CURRENT	SOLENOID OUTPUT PRESSURE	COAST CLUTCH	3-2 TIMING
ZERO	ZERO PSI	APPLIED	—
MOD-HIGH	MOD-HIGH	RELEASED	Regulated exhaust of Direct Clutch and Servo Release circuits during shift.
MAXIMUM	MAXIMUM	RELEASED	Normal steady state: Maximum exhaust of Direct Clutch circuit during 3-2 shift, 1st, and 2nd gear operation.

FIGURE 18-44 Notice that the two solenoid shift valves allow for 4 different combinations which will be able to control the four speeds of the transmission. (Courtesy of Ford Motor Company)

volved only turning one or the other solenoid on or off.

Two additional outputs involving transmission control are the coast clutch solenoid valve (CCS) and the electronic pressure control solenoid valve (EPC). Let's look at the CCS first. The CCS is a different type of solenoid than we have looked at previously and its control is also different. Typically the PCM has supplied a pulsing signal (PWM), or an on/off signal to control an output device. Because the CCS is a variable force solenoid that will vary the pressure within the 3-2 valve circuit, the signal to it must also be variable, Figure 18-45.

By varying the current of the solenoid electrical circuit, the PCM varies the pressure in the hydraulic circuit that affects the position of the 3-2 shift timing valve. This action controls the apply of the coast clutch for engine braking during coast operation. A customer is able to manually shift into 3rd or 2nd and have the engine supply braking much the same as with a manual transmission. Normally the CCS is set so that there will be only the slightest amount of engine braking. Having engine braking all of the time would have an adverse affect on the tailpipe emissions, so it is not something that the manufacturers do automatically. The customer is indirectly in control of the CCS.

The quality of the shift and speed that it occurs is a function of the working pressure of the transmission. It is the job of the EPC (electronic pressure control) solenoid to control the pressures within the transmission. Figure 18-46. High pressures are needed for heavy loads, while lower pressures are acceptable for light loads. Reducing the pressure during light loads helps the quality of shift, match the demand. Lower pressures will allow the shift to occur early and have a softer engagement. Early shifts and soft engagement would

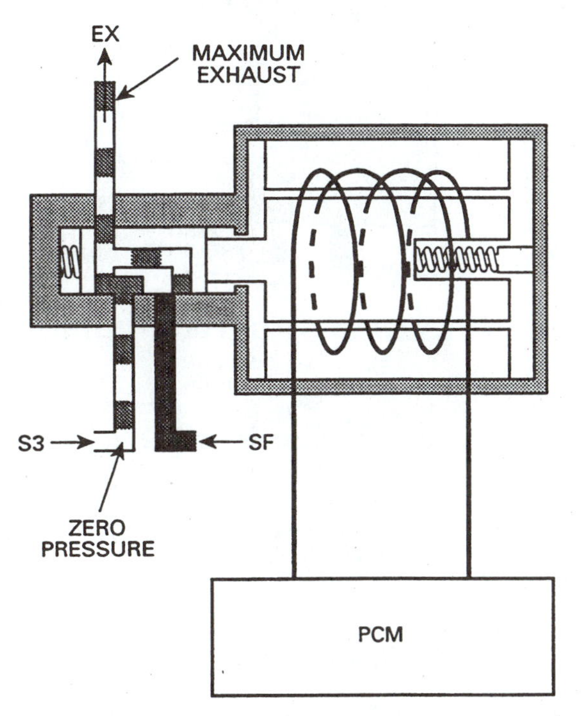

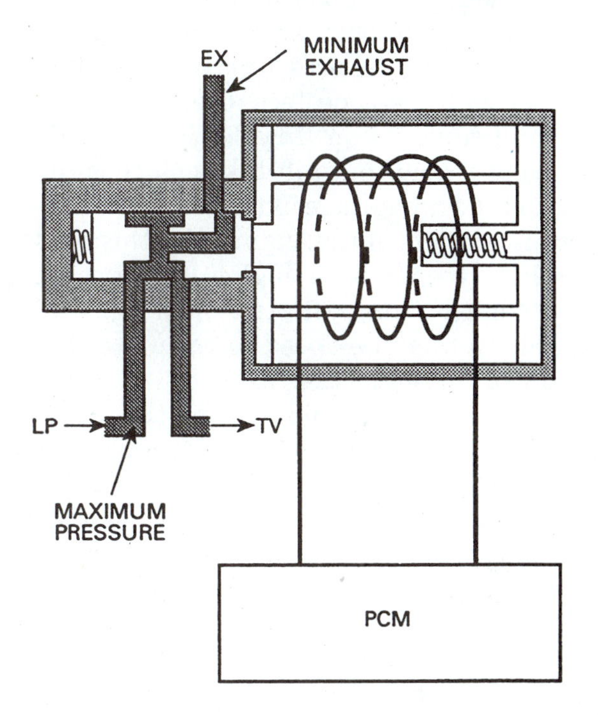

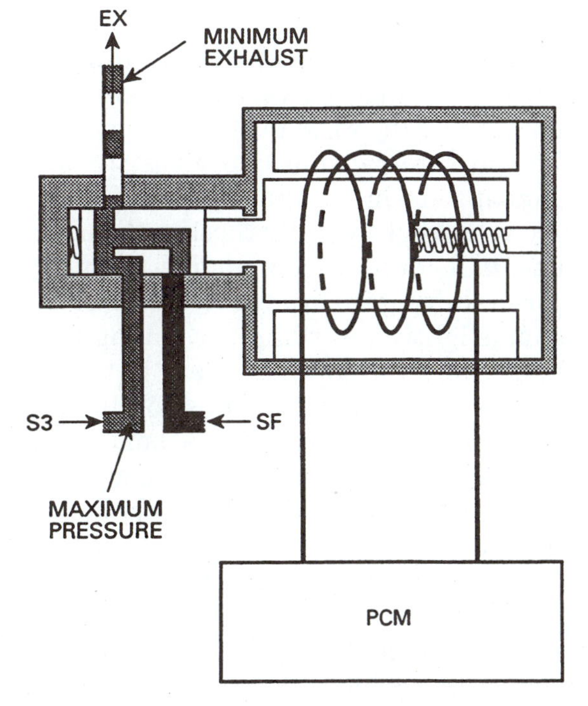

FIGURE 18-45 The CCS is a variable force solenoid that will vary the pressure within the 3-2 valve circuit. (Courtesy of Ford Motor Company)

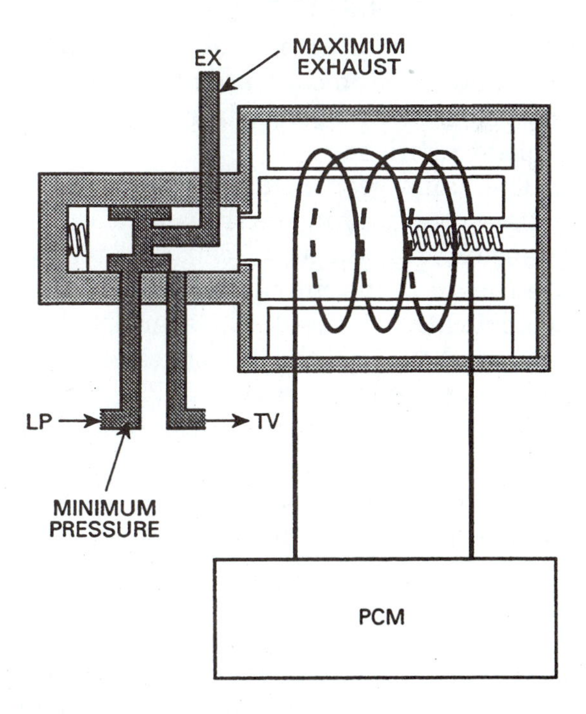

FIGURE 18-46 The EPC (electronic pressure control) solenoid controls the pressures within the transmission. (Courtesy of Ford Motor Company)

tear up an automatic, however, if the customer had his foot to the floor. Under heavy loads like wide open throttle or towing, the EPC increases the pressures which makes for harder shifts, but protects the clutches within the transmission. In addition the vehicle speed at which the shift takes place is raised up, to take advantage of the horse power and torque outputs of the engine. For this reason, the EPC responds to engine load, transmission temperature, throttle opening, vehicle speed, and engine speed to calculate the actual working pressures needed.

The final solenoid within the solenoid valve body assembly is for the torque converter clutch (TCC). Under conditions of light load, the torque converter is locked up, eliminating the slight slip that a non lock-up converter would have. By locking up the converter, a slight increase in mileage, reduction in engine speed, and corresponding reduction in tailpipe emissions takes place. TCC has been around for years in various forms. The torque converter clutch solenoid is an electro-hydraulic actuator that varies hydraulic pressure in the bypass clutch circuit, Figure 18-47. It is fed with a pulse width modulated signal (PWM) that allows for the clutch to be fully applied, fully released, or operated under a partially applied conditions.

The same sensors are involved in TCC that are used to shift the transmission. Vehicle speed, engine speed, load, and temperature are all taken into account to determine the application of the TCC. Generally, TCC is not turned on until the transmission and engine are partially warmed up. The slight increase in engine speed with TCC off is desirable during engine warm-up. The PCM will turn on the TCC in second, third, or fourth gear. TCC is not used in first or reverse and is obviously not in use in park or neutral.

Keep in mind that the PCM has interrelated many functions that were up to this point individually controlled. Look at a vehicle of 20 years ago. Timing was a function of the base timing, vacuum advance, and mechanical advance. Air fuel ratio was a function

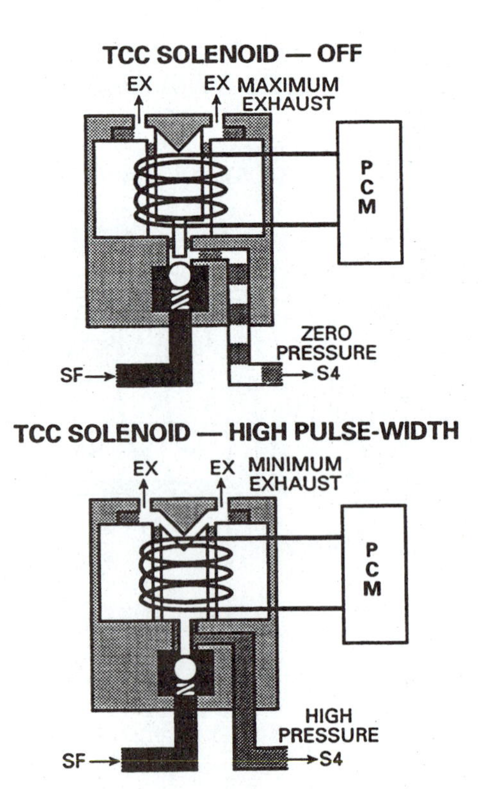

FIGURE 18-47 Typical torque converter (TCC) solenoid operation. (Courtesy of Ford Motor Company)

of the carburetor, and the transmission was controlled by manual valves and engine vacuum. Having the PCM in charge of the timing, A/F, and the transmission gives the design engineer some interesting options. Currently, some vehicles, especially those with larger engines, will "detune" the engine slightly just prior to shifting the transmission. "Detuning" means slightly retarding the timing and reducing fuel flow through the injectors. This detuning allows the transmission to shift at lower pressures, producing a smooth shift. Once the transmission is fully engaged in the next gear, full power is again applied. This system helps to save the internal components of the transmission from unnecessary wear and tear. Detuning is only possible because of the fact that the PCM is in charge and knows exactly what is happening to timing, fuel and the transmission. Split second timing is the key here, if we expect the customer not to feel anything out of the ordinary.

ON-BOARD DIAGNOSTICS—SECOND GENERATION (OBD-II)

The computer systems of the past have been extremely successful at electronic engine control and has been in use for years. They have substantially reduced emissions, increased power and decreased driveability at the same time that vehicle mileage has increased. However, additional diagnostic capabilities especially for emission control are all part of the emission law. The additional capabilities required have been added to an updated version called OBD-II. The principle difference between older systems and OBD-II is in the ability to detect misfires, and in the ability to diagnose a nonfunctioning catalytic converter.

The addition of another oxygen sensor after the catalytic converter allows the PCM to compare the output voltages and number of rich lean swings (frequency). A correctly functioning converter will have a varying voltage between 100 and 800 mV (.100 - .800 V) with wide swings between rich and lean exhaust conditions in the primary oxygen sensor (ahead of the converter) and greatly reduced voltage swings after the converter. If the PCM sees the same voltage swings before and after the converter, it will set a trouble code indicating converter failure, and turn on the MIL (malfunction indicator lamp) on the dashboard, Figure 18-48.

OBD-II also involves the addition of another crankshaft position sensor (CKP) or the programming of the PCM to look at the spacing or time between pulses which allows for the detection of misfires. Uncorrected misfires force the converter into a high heat condition that will eventually destroy the catalytic converter in addition to allowing excessive emissions to escape from the tailpipe.

The crankshaft of a correctly functioning engine without misfire will speed up on each power stroke and slow down between power strokes as Figure 18-49 shows.

By looking at the time between CKP pulses, the PCM can "see" a misfire. If a 4-cylinder engine is experiencing a misfire, three pulses will be evenly spaced with a

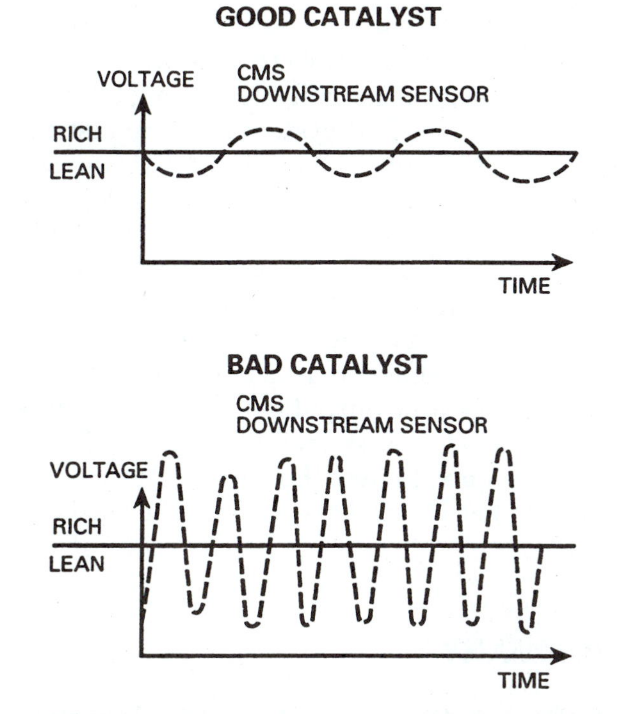

FIGURE 18-48 If the PCM sees the same voltage swings before and after the converter, it will set a trouble code indicating converter failure, and turn on the MIL (malfunction indicator lamp) on the dashboard. (Courtesy of Ford Motor Company)

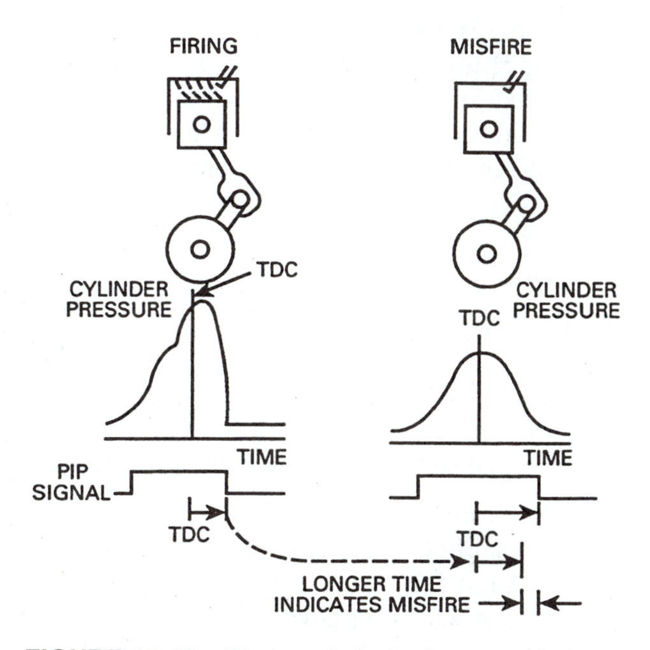

FIGURE 18-49 The crankshaft of a correctly functioning engine without misfire will speed up on each power stroke and slow down between power strokes. (Courtesy of Ford Motor Company)

longer space for the misfiring cylinder. By counting the pulses and keeping track of the spacing, the PCM is able to turn on the MIL, hopefully prior to actually destroying the converter. Most OBD-II system that are distributerless already have the CKP sensor installed. By reprogramming the PCM, it is able to get additional information from this sensor. OBD-II also incorporates various monitors of fuel and EGR functions while the vehicle is operating down the road. The future will no doubt see increased monitoring of various inputs and outputs, especially if they have emission responsibility. OBD-II has been designed to take the manufacturers into the enhanced emissions era with the least difficulty.

SUMMARY

Within this Chapter we have taken a look at the computerization of the modern vehicle. We began with a discussion of the common sensors. The major sensors of throttle position, coolant temperature, intake air temperature, and vacuum were examined. Their relationship to the actual delivery of fuel and timing changes was discussed. In addition, the mass airflow sensor was looked at. We looked at the use of an oxygen sensor to modify the air fuel ratio for best emissions based on the remaining oxygen in the exhaust stream. The major differences in speed/density and mass power control systems were analyzed. Our discussion changed from inputs to major outputs, including timing and injector fuel control. Additional outputs of EGR and idle speed, and their control was examined. The use of the PCM to control transmission shift and pressures through the use of additional inputs was looked at and discussed. The last section of the Chapter was devoted to OBD-II with the additional input of a second oxygen sensor and the reprogramming of the PCM to look at engine speed changes. These inputs were responsible for the monitoring of the catalytic converter and engine misfires, primarily for controlling the emission levels.

KEY TERMS

ECT (engine coolant temperature)
IAT (intake air temperature)
EGR (exhaust gas recirculation)
SFI (Sequential fuel injection)
OBD-II (on-board diagnostic, 2nd generation)
CMP (camshaft position)
Stoichometric
TR (transmission range)
VSS (vehicle speed sensor)
TFT (transmission fluid temperature)

CHAPTER CHECK-UP

Multiple Choice

1. The ECT is responsible for informing the PCM of
   a. exhaust levels.
   b. coolant temperature.
   c. manifold pressure.
   d. none of the above.
2. Temperature sensors are typically
   a. negative coefficient thermistors.
   b. positive coefficient thermistors.
   c. pulse width modulated devices.
   d. frequency modulated devices.

3. The PCM needs to know the temperature of
   a. transmission fluid.
   b. intake manifold air.
   c. engine coolant.
   d. all of the above.
4. A temperature of 32°F will result in a resistance of (ECT)
   a. 240,000 ohms
   b. 125,000 ohms
   c. 95,000 ohms
   d. 950 ohms
5. Position sensors are typically
   a. variable resistors.
   b. negative coefficient thermistors.
   c. pulse width modulated devices.
   d. variable force solenoids.
6. Technician A states that the MAP sensors produce a variable voltage based on the pressure in the manifold. Technician B states that the MAP sensors output is a frequency oriented signal. Who is correct?
   a. Technician A only
   b. Technician B only
   c. Both Technician A and B
   d. Neither Technician A or B
7. The oxygen sensor will generate what type of signal?
   a. pulsing or floating D.C.
   b. square wave
   c. AC
   d. frequency modulated
8. A voltage in excess of .450 V from the oxygen sensor indicates a
   a. normally running engine.
   b. rich condition.
   c. lean condition.
   d. faulty sensor.
9. The object of the coast clutch solenoid inside the transmission is to
   a. control the shift points.
   b. vary the working pressure.
   c. apply the coast clutch for engine braking.
   d. decrease the electronic pressure control.
10. Technician A states that the torque converter clutch will "lock-up" the converter under condition of light load like cruising. Technician B states that the lock-up converter will be controlled by the PCM. Who is correct?
    a. Technician A only
    b. Technician B only
    c. Both Technician A and B
    d. Neither Technician A or B

True or False

11. SS1 and SS2 are always on if the vehicle is moving.
12. IAT supplies a varying voltage to the PCM.
13. Less than .450 V from the oxygen sensor indicates a rich exhaust.
14. 2.5 V from the TP sensor indicates a 1/2 throttle position.
15. The signal from the MAP sensor may be a frequency modulated signal.
16. ECT, IAT, and TFT are all negative coefficient thermistors.
17. Fuel injectors are generally fed with 5 V from the PCM.
18. The timing is advanced when additional fuel is being injected.
19. EEC-V is capable of detecting engine misfires.
20. The additional oxygen sensor located after the converter (on EEC-V) is used to indicate converter activity.

Index

K

L

M

N

O

P

T